PRINCIPLES OF
PHYSICAL
CHEMISTRY

PRINCIPLES OF
PHYSICAL CHEMISTRY

Abhijit Mallick, Ph.D.
Associate Professor
Department of Chemistry
Academy of Technology
Hooghly

MV Learning

London • New Delhi

MV Learning
A Viva Books imprint

3, Henrietta Street
London WC2E 8LU
UK

4737/23, Ansari Road,
Daryaganj, New Delhi 110 002
India

ISBN: 978-93-87692-83-1

Printed and bound in India.

Dedicated to
my beloved parents

Acknowledgements

I am very much obliged to my family members—my father, my wife and my beloved son for their inspiration and continuous support without which it would have been impossible to complete this uphill task.

I am indebted to Viva Books Private Limited for their continuous support and inspiration to complete this book.

I must not miss this opportunity to acknowledge Prof. Jagannath Banerjee, Chairman, Academy of Technology and Prof. Anindita Banerjee, Executive Trustee of Academy of Technology, for their all-round cooperation and support.

I am also indebted to Dr Dilip Bhattacharya, Director, Academy of Technology, for his cooperation and support.

I must not miss the opportunity to acknowledge my colleagues in the Department of Chemistry and Mr Anirban Bhattacharya, Laboratory Assistant in the Department of Chemistry for their cooperation and support.

Prof. Abhijit Mallick

Preamble

This book is primarily written for B.Sc. students who are studying chemistry honours at undergraduate level in different Indian universities. However, the book is expected to be very useful for postgraduate students also. It covers by and large the physical chemistry part of the chemistry honours syllabus taught in different Indian universities. Elaborate and lucid discussion of each chapter is the strength of this book. Questions and numerical problems are also included at the end of almost every chapter. Strenuous effort has been given to derive different mathematical equations as well as to handle quantum mechanics using mathematics taught at the undergraduate level.

The book contains 20 chapters, covering major areas like thermodynamics, electrochemistry, chemical kinetics, etc.

Thermodynamics is thoroughly discussed in this book covering the 1st, 2nd and 3rd law, their applications, and thermochemistry and its applications. Applications of thermodynamics in different areas like refrigerators, compressors, power plants, IC engines, etc. are also discussed. Statistical thermodynamics is also explained in detail.

Chemical kinetics is another important part of chemistry since it covers reaction rate, order of a reaction, the theory behind the reaction rate, etc. The catalyst is also an important aspect since it has profound influence on reaction rate. Types of catalyst and mechanism of different catalyzed reactions are described elaborately. A chemical reaction reaches an equilibrium state if carried out in a closed container. However, the equilibrium is sufficiently influenced by other parameters like pressure, temperature, etc. The details of chemical equilibrium are also covered in this book.

Different physical states of matter (gaseous, liquid and solid states) are discussed thoroughly. In the solid state, behaviour of conductors and semiconductors are discussed in detail using quantum mechanics.

Detailed discussion of electrochemistry, electrochemical cell and ionic equilibria is another important aspect of this book. Application of thermodynamics in electrochemical cell and concept of buffer solutions, pH and indicators are also explained thoroughly.

Phase equilibria is another important component of physical chemistry, which is well covered in this book. The chapter includes details of phase rule, phase diagram, applications, different types of heterogeneous equilibrium system, etc.

Colligative properties of dilute solutions covering Henry's law, Raoult's law of lowering of vapour pressure, elevation of boiling point, depression of freezing point, osmotic pressure, etc. are well documented.

Surface chemistry and properties of colloidal solutions hold great importance in different chemical industries. These two sections including details of derivation of different laws, theories behind the adsorption, stability of colloidal solutions, etc. are well explained.

Nuclear reactions are different from chemical reactions and energy related to nuclear reactions is enormous, much higher than any chemical reaction. Study of different nuclear reactions including natural radioactivity, artificial radioactivity, etc. and kinetics of nuclear reactions are well described in this book. It also explores different areas of applications of nuclear reactions.

Another important aspect of chemical reactions is chemical bonding. The book covers details of covalent bonding including quantum numbers, overlapping of atomic orbitals, molecular orbitals. Besides that, ionic bonding and other types of bonding are also included.

Photochemical reactions are different from chemical reactions. Light energy is the main source of photochemical reactions. Details of it including photochemical laws, mechanism, etc. are well documented in this book.

Appendix is another asset, which provides different useful information, frequently needed for students.

I welcome all possible suggestions for improvement from the readers of this book and will be grateful if they are kind enough to point out the errors, which still might persist in the book in spite of my wholehearted effort.

Prof. Abhijit Mallick
(dramallick12@gmail.com)

Contents

9.   Chemical Kinetics   331

10. Catalysis   393

1

Ideal Gases

1.1 INTRODUCTION

Any matter may exist in any of the three states: (a) solid, (b) liquid, and (c) gas.

In solid state, molecules are closely packed and pressure has little or no effect on volume. However, volume increases only slightly with increase in temperature. Molecules in solid state are bound in 3 dimensions and hence solid has a definite shape.

In case of liquid, molecules are less compact than solid. A liquid may be considered as numerous layers, which are piled up one after another. These layers can move independently or intermixed among themselves. That is why liquid has no shape. It always takes the shape of the container where it is taken but volume of liquid remains constant whatever may be the volume of container is. Volume of liquid alters only slightly with increase in pressure and temperature. Increase in volume of liquid with temperature is more than that of solid.

In case of gas, molecules remain in discrete form and are free to move independently. Hence gas density is very low. Gas has no shape. Unlike liquid, gas always covers whole volume of the container irrespective of the amount of gas. Thus volume of container is always equal to volume of gas. Furthermore, gas pressure varies with volume as well as temperature.

1.2 GAS LAWS

1.2.1 Boyle's Law

This law was established by R. Boyle (1662).

Statement   At constant temperature volume of a definite mass of a gas is inversely proportional to its pressure. If P and V represent pressure and volume respectively of a definite mass of a gas, mathematically Boyle's law can be written as:

$$P \alpha \frac{1}{V}. \quad \text{or} \quad PV = \text{constant} \tag{1.1}$$

For a definite mass of a gas,

P_1, V_1 are the pressure and volume of the gas at state 1

P_2, V_2 are the pressure and volume of the gas at state 2

Hence, we can write that $\quad P_1 V_1 = P_2 V_2$.

Fig. 1.1 illustrates the change of pressure with volume of a gas at constant temperature.

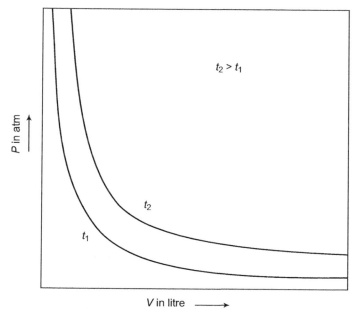

Figure 1.1 PV-isotherm according to Boyle's law

The *PV* diagram is a rectangular parabola. As temperature is constant the curve is called isothermal curve.

1.2.2 Charles's Law

This law was proposed by Jacques Charles in 1787.

Statement At constant pressure, volume of a definite mass of a gas by same relative amount for every degree rise in temperature.

Temperature is expressed in °C. It is known that a gas can maintain its identity even well below 0°C. Identity of any gas is lost at –273.15°C, which is universally taken as absolute zero. The temperature scale based on this absolute zero is known as absolute scale and absolute temperature is represented as *T*. Thus we can write that $0°K = -273.15°C \approx -273°C$.

$$t°C = (t + 273)°K \qquad (1.2)$$

The volume of a gas increases steadily from 0°K onwards and the coefficient of expansion has been assumed to be (1/273) per °C. So, volume at any temperature $t°C$ can be written as

$$V_t = V_0\left(1 + \frac{t}{273}\right) \qquad (1.3)$$

where, V_0 represents volume of the gas 0°C at a given pressure and V_t represents volume of the same gas at $t°C$ at that pressure.

At two different temperatures, t_1 and t_2, volume of a gas becomes V_1 and V_2 respectively. Then V_1 and V_2 can be expressed as

$$V_1 = V_0\left(1 + \frac{t_1}{273}\right) \quad \text{and} \quad V_2 = V_0\left(1 + \frac{t_2}{273}\right)$$

Hence, $$\frac{V_2}{V_1} = \frac{273 + t_2}{273 + t_1} = \frac{T_2}{T_1} \qquad (1.4)$$

since, $$T_1 = 273 + t_1 \text{ and } T_2 = 273 + t_2$$

So, $\dfrac{V}{T}$ = Constant at constant pressure. The above equation is the mathematical form of Charles's law.

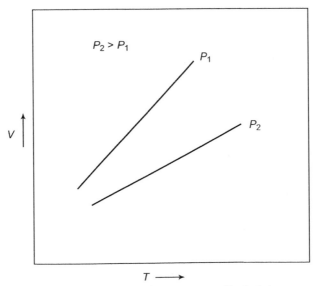

Figure 1.2   VT-diagram according to Charles's law

The straight lines are called isobars.

1.2.2.1 *P–V–T* Relation for an Ideal Gas

From Boyle's law, we have $V \propto 1/P$ at constant temperature and mass.
From Charles's law, we have $V \propto T$ at constant pressure and mass.
Combining the above two laws, we have

$$V \propto \frac{T}{P} \text{ at constant mass of the gas.}$$

Or,
$$\frac{PV}{T} = \text{Constant at constant mass of the gas} \tag{1.5}$$

1.2.3 Avogadro's Law

Statement Under constant pressure and temperature same volume of different gases must possess equal number of molecules.

Here, mass of gas is expressed in mole. It is the ratio of weight of gas to its molecular weight. If w and M represent weight of gas and molecular weight respectively, number of moles (n) can be written as
$n = w/M$. When weight and molecular weight are expressed in gm, mole is called gm – mole. When weight and molecular weight are expressed in kg, mole is called kg – mole.
According to Faraday's law of electrolysis, we have

$$e = \frac{F}{N_A}$$

e = Electronic charge. Millikan's oil-drop experiment shows that $e = 1.602 \times 10^{-12}$ Coulomb.
F = Faraday constant = 96500 Coulomb/gm. eq.
N_A = Avogadro Number = Number of molecule/gm. eq. = 6.023×10^{23}/gm. mole
So, 1 gm – mole of any gas, liquid or solid contains 6.023×10^{23} number of molecules. This number is called Avogadro Number. It's value is 6.023×10^{23}/gm – mole or 6.023×10^{26}/kg – mole.
Mathematically, Avogadro's law can be expressed as

$$V \propto n \tag{1.6a}$$

Thus for 1 gm – mole of any gas the value of PV/T is constant, which is represented as R. It is often called molar gas constant. Thus we can write

$$\frac{PV}{T} = R \text{ for 1 gm mole of any gas.}$$

Or,
$$\frac{PV}{T} = nR \text{ for } n \text{ gm moles of any gas} \tag{1.6b}$$

The above equation is a combination of Boyle's law, Charles's law and Avogadro's law. The gases which follow the above equation are known as ***ideal gases***.

Unit of R

$$P = \frac{\text{force}}{\text{area}} = \frac{\text{force}}{\text{length}^2} \quad \text{and} \quad V = \text{length}^3 \quad PV = (\text{force})\,(\text{length}) = \text{energy}$$

$$R = \frac{PV}{nT} = \frac{\text{energy}}{°K - \text{mole}} = \frac{\text{lit} - \text{atm or Joule or Cal}}{°K - \text{mole}}$$

It is universally accepted that reference value of pressure is 1 atm, which is the pressure at sea coast area.

The reference value of temperature is $0°K = -273.15°C$.

So volume is to be calculated at 1 atm pressure and 0°C. To evaluate the volume, PV of different gases are plotted against P.

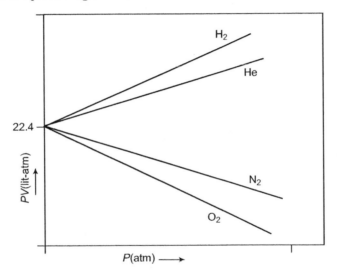

Figure 1.3   PV versus P plot of different gases

Plots are different for different gases but all the plots converge to a point at $\lim P \to 0$. The results show that volume occupied by 1 mole of any ideal gas at STP/NTP (0°C and 1 atm) is 22.4 lit. So value of R is given by

$$R = \frac{(1\,\text{atm})(22.4\,\text{lit})}{273K} = 0.082 \text{ lit} - \text{atm}/\,°K - \text{mole}$$

Again, $1 \text{ atm} = 1.0132 \times 10^5 \text{Pa}$ and $1 \text{ lit} = 1000 \text{ cm}^3 = 10^{-3}\text{m}^3$.

$$R = \frac{(1.0132 \times 10^5 \text{ Pa})(22.4 \times 10^{-3} \text{ m}^3)}{273K} = 8.314 \text{ J/mole-}°K$$

Again, $1 \text{ cal} = 4.18\text{J}$

$$R = \frac{8.314}{4.18} \text{ cal/mole} - °K = 1.987 \text{ cal/mole} - °K$$

1.2.4   Mixture of Gases   Dalton's Law of Partial Pressures

Total pressure and partial pressure   If a container of volume V is occupied by different types of gases at constant temperature, the pressure exerted by the mixture of gases is called total pressure (P). If the same container is occupied by only one type of gas, the pressure exerted by the gas is called partial pressure (p).

p_1 and n_1 are the partial pressure and number of moles of gas-1. V represents volume of the gas.

p_2 and n_2 are the partial pressure and number of moles of gas-2. V represents volume of the gas.

p_3 and n_3 are the partial pressure and number of moles of gas-3. V represents volume of the gas.

Assuming all the gases are ideal, we can write that

$p_1 V = n_1 RT$, $p_2 V = n_2 RT$, $p_3 V = n_3 RT$,...etc. In each case volume remains constant, V.

Summing up, we get

$$(p_1 + p_2 + p_3 + \cdots)V = (n_1 + n_2 + n_3 + \cdots)RT = nRT$$

Or, $(p_1 + p_2 + p_3 + \ldots)V = nRT$

Again, $PV = nRT$,   where, $n = n_1 + n_2 + n_3 + \cdots$

Hence, we have

$$P = p_1 + p_2 + p_3 + \cdots \tag{1.7}$$

The equation (1.7) is known as Dalton's law of partial pressure.

$p_1 V = n_1 RT$ and $PV = nRT$. Hence,

$$\frac{p_1}{P} = \frac{n_1}{n} = x_1 = \text{Mole fraction of gas-1} \tag{1.8a}$$

Or, $p_1 = x_1 P$. Similarly, $p_2 = x_2 P$, $p_3 = x_3 P$ ... etc., where, x_2, x_3 are known as mole fractions of gas-2, gas-3 respectively.

Hence, $x_1 + x_2 + x_3 + \cdots = 1$ (1.8b)

1.2.5   Diffusion of Gases   Graham's Law

Definition   It is the property of any substance to spread from higher concentration to lower one. However, diffusion depends on velocity of molecules. Gas molecules are free and hence can move much faster than liquid molecules, which are again moving faster than solid molecules.

In case of solids, molecules are tightly bound and diffusion occurs through jumping of atoms.

In case of liquids, diffusion depends on viscosity of liquid. Higher the viscosity, less the diffusion rate is.

In case of gases, diffusion depends on density and hence molecular weight of gas. Higher the density, lower the diffusion rate is.

Graham's law   According to T. Graham (1829), rate of diffusion (r) of any gas is inversely proportional to square root of its density (p). Mathematically, we can write that

$$r \propto \frac{1}{\sqrt{\rho}}. \quad \text{Or,} \quad r \propto \frac{1}{\sqrt{M}} \left[\text{Since,} \ \rho = \frac{PM}{RT} \right] \tag{1.9}$$

[at constant pressure and temperature]

1.2.5.1   Diffusion Coefficient

Along each axis gas molecules may move either positive direction or negative direction with equal probability. (dn/dZ) is the change in number density of gas molecules along Z-axis. It is assumed that each and every molecule undergoes collision only after travelling the mean free path l.

Number of molecules along positive Z-axis $= n(+l) = n_0 + \left(\dfrac{dn}{dZ} \right) \times l$

Number of molecules along negative Z-axis $= n(-l) = n_0 - \left(\dfrac{dn}{dZ} \right) \times l$

Molecular flux along positive Z-axis $= \dfrac{n(+l)\bar{c}}{4}$

Molecular flux along negative Z-axis $= \dfrac{n(-l)\bar{c}}{4}$

Change in flux $= J = \dfrac{n(-l)\bar{c}}{4} - \dfrac{n(+l)\bar{c}}{4} = -\dfrac{l\bar{c}}{2} \left(\dfrac{dn}{dZ} \right)$

According to Fick's law, we have

$$J = -D \left(\frac{dn}{dZ} \right)$$

Hence, we can write that

$$D = \frac{l\bar{c}}{2} \tag{1.10}$$

It is assumed that molecular collisions are perfectly elastic and there is no loss of molecules but in actual practice the situation is different. Documentations suggest that equation (1.10) can be modified into the following form.

$$D = \frac{\bar{l}\bar{c}}{3} \qquad (1.11)$$

1.2.6 Effusion

In chemistry, effusion is the process in which individual molecules are allowed to flow through a hole without collisions among themselves. This occurs if the diameter of the hole is considerably smaller than the mean free path of the molecules. According to Graham's law, gases with a lower molecular weight effuse more quickly than gases with a higher molecular weight. The root mean square molecular speed ($\sqrt{\overline{c^2}}$) of gas molecules is correlated with temperature according to the following equation

$$\frac{3}{2}kT = \frac{1}{2}m\overline{c^2} \qquad (1.12)$$

where, k is the Boltzmann's constant, m is the mass of gas molecule.

Effusion of gas molecules is assumed to follow Graham's law of diffusion. Thus, it may be concluded that higher the rate of effusion, lesser the time required for completing the effusion process. Thus, we can write that

$$t \propto \frac{1}{r}. \quad \text{Or,} \quad t \propto \sqrt{\rho}. \quad \text{Or,} \quad t \propto \sqrt{M} \qquad (1.13)$$

t = Time required for completing the effusion process.
r = Rate of effusion. ρ = Density of gas. M = Molecular weight of gas.

At constant temperature and pressure, if t_1 and t_2 are the time of effusion of two gases, we can write that

$t_1 \propto \sqrt{\rho_1}$ and $t_1 \propto \sqrt{M_1}$ for gas-1

$t_2 \propto \sqrt{\rho_2}$ and $t_2 \propto \sqrt{M_2}$ for gas-2

ρ_1 and M_1 are the density and molecular weight of gas-1 respectively.
ρ_2 and M_2 are the density and molecular weight of gas-2 respectively.
Hence, we have

$$\frac{t_2}{t_1} = \sqrt{\frac{\rho_2}{\rho_1}} = \sqrt{\frac{M_2}{M_1}} \qquad (1.14)$$

1.2.6.1 Effusion Rate and Gas Leaks

Let us consider a container, filled with an ideal gas. The container has a small hole with diameter d and the gas is slowly diffusing out through that small hole.

A = Area of the small hole = $\dfrac{\pi}{4}d^2$,

$\dfrac{dN}{dt}$ = Rate of diffusion of gas molecules = $Z_w \times A$

Z_w = Number of collision per unit area per second = $\dfrac{P}{(2\pi mkT)^{1/2}}$ [Equation (1.32a)]

As there is no wall on the hole, gas molecules, approaching to strike with the hole wall, diffuse out of the container. Thus,

Rate of diffusion of gas molecules = Collision frequency between gas molecules and area of the hole.

$$\frac{dN}{dt} = \frac{P \times A}{(2\pi mkT)^{1/2}} = \left[\frac{P}{(2\pi mkT)^{1/2}}\right]\left(\frac{\pi}{4}d^2\right) = \frac{Pd^2\sqrt{\pi}}{(32mkT)^{1/2}} \tag{1.15}$$

The above equation is called Graham's law of effusion.

As the gas molecules are diffusing out, pressure of the gas in the container decreases with time and this decrease is given by $\left(\dfrac{dP}{dt}\right)$. As the gas is an ideal gas, we can write for n moles

$$PV = nRT = nN_A kT = NkT, \text{ where, } N = nN_A.$$

$$\frac{dP}{dt} = -\frac{d}{dt}\left(\frac{NkT}{V}\right) = -\left(\frac{kT}{V}\right)\frac{dN}{dt} \tag{1.16}$$

Negative sign indicates that pressure decreases with time.

From equations (1.15) and (1.16), we get

$$\frac{dP}{dt} = -\left(\frac{kT}{V}\right)\frac{P \times A}{(2\pi mkT)^{\frac{1}{2}}}.$$

Or,

$$-\frac{dP}{P} = \left(\frac{A}{V}\right)\left(\frac{kT}{2\pi m}\right)^{\frac{1}{2}} dt = \frac{dt}{\tau}, \text{ where, } \tau = \left(\frac{A}{V}\right)\left(\frac{2\pi m}{kT}\right)^{1/2}$$

At $t = 0$, $P = P_0$ and $t = t$, $P = P$. On integration within limits, we get

$\ln \dfrac{P}{P_0} = -\dfrac{t}{\tau}$. Or, $\ln P = \ln P_0 - \dfrac{t}{\tau}$

Or, $$P = P_0 e^{-t/\tau} \tag{1.17}$$

The equation (1.17) shows how pressure of the gas decreases with time.

By plotting $\ln P$ versus t a straight line with negative slope is obtained. By measuring slope, molecular mass, m, can be calculated.

1.3 KINETIC THEORY OF GASES

1.3.1 Assumptions for Ideal Gas

Following are the assumptions made for any ideal gas:

(a) Gas molecules are considered as very small particles with non-zero mass.

(b) Gas molecules are always at random motion in every possible direction. This motion is often called Brownian motion.

(c) Gas molecules are spherical in shape and elastic in nature.

(d) Gas molecules, due to continuous motion, often collide among themselves and also with the walls of the container. Each and every collision is considered perfectly elastic.

(e) Interactions among gas molecules is negligible and hence no internal pressure is developed within gas molecules. Pressure is developed only due to collisions of gas molecules with walls of container.

(f) The total volume of all gas molecules is small and negligible compared to the volume of the container.

(g) Distance between two nearest gas molecules is much larger than thermal de Broglie wavelength and hence quantum-mechanical concept is not applicable for gas molecules. However, size of gas molecules is sufficiently large so that statistical concept is applicable.

(h) The average kinetic energy of gas molecules depends only on the temperature of the system.

(i) Time of collision between gas molecules and wall of the container is negligible compared to the collision time, i.e., the time between two successive collisions of gas molecules.

1.3.2 Maxwell's Distribution of Molecular Velocities

J.C. Maxwell (1880) established the law of distribution of molecular velocities by using theory of probability and deduced the following mathematical expression.

$$\frac{1}{n}\frac{dn}{dc} = 4\pi\left(\frac{M}{2\pi RT}\right)^{3/2} e^{-Mc^2/2RT} c^2 \tag{1.18}$$

where, M = Molecular weight of the gas, R = Molar gas constant, c = Velocity of gas molecules, n = Number of molecules per unit volume and T is the absolute temperature.

1.3.2.1 Mean Velocity ($\bar{c}$)

$$\bar{c} = \frac{1}{n}\int_0^\infty c\,dn = \int_0^\infty 4\pi\left(\frac{M}{2\pi RT}\right)^{3/2} e^{-Mc^2/2RT} c^3\,dc$$

$$= 4\pi \left(\frac{M}{2\pi RT}\right)^{3/2} \int_0^\infty e^{-Mc^2/2RT} c^3 \, dc$$

Put

$$\frac{Mc^2}{2RT} = x. \quad dx = \left(\frac{M}{RT}\right) c \, dc. \quad \text{Or,} \quad c \, dc = \left(\frac{RT}{M}\right) dx$$

Or,

$$\bar{c} = 4\pi \left(\frac{M}{2\pi RT}\right)^{3/2} \int_0^\infty e^{-x} \left(\frac{2RTx}{M}\right) \left(\frac{RT}{M}\right) dx$$

Or,

$$\bar{c} = 4\pi \left(\frac{M}{2\pi RT}\right)^{3/2} \cdot 2 \left(\frac{RT}{M}\right)^2 \int_0^\infty e^{-x} x \, dx$$

Or,

$$\bar{c} = 4\sqrt{\frac{RT}{2\pi M}} \int_0^\infty e^{-x} x \, dx = \sqrt{\frac{8RT}{\pi M}}, \quad \text{since,} \int_0^\infty e^{-x} x \, dx = 1$$

Or,

$$\bar{c} = \sqrt{\frac{8RT}{\pi M}} \tag{1.19}$$

1.3.2.2 Most Probable Velocity (c_p)

Most probable velocity is achieved at the maximum of the distribution curve.

Thus,

$$\frac{d}{dc}\left[\frac{1}{n}\frac{dn}{dc}\right] = 0$$

Or,

$$\frac{d}{dc}\left[4\pi \left(\frac{M}{2\pi RT}\right)^{3/2} e^{-Mc^2/2RT} c^2\right] = 0$$

Or,

$$e^{-Mc_p^2/2RT} \times 2c_p - c_p^2 \left(\frac{Mc_p}{RT}\right) e^{-Mc_p^2/2RT} = 0$$

Or,

$$2c_p - \frac{Mc_p^3}{RT} = 0. \quad \text{Or,} \quad c_p = \sqrt{\frac{2RT}{M}} \tag{1.20}$$

1.3.2.3 Root Mean Square Velocity ($\sqrt{\bar{c^2}}$)

$$\bar{c^2} = \frac{1}{n}\int_0^\infty c^2 \, dn = \int_0^\infty 4\pi \left(\frac{M}{2\pi RT}\right)^{3/2} e^{-Mc^2/2RT} c^4 \, dc$$

Put, $\dfrac{Mc^2}{2RT} = x.$ So, $c \, dc = \dfrac{RT}{M} dx$ and $c = \sqrt{\dfrac{2RT}{M}} x^{1/2}$

$$\overline{c^2} = \int_0^\infty \frac{4}{\sqrt{\pi}} \sqrt{\frac{M}{2RT}} \sqrt{\frac{2RT}{M}} \left(\frac{RT}{M}\right) e^{-x} \cdot x \cdot x^{1/2} \cdot dx$$

$$= \frac{4}{\sqrt{\pi}} \left(\frac{RT}{M}\right) \int_0^\infty e^{-x} \cdot x^{3/2} \cdot dx$$

Or, $\quad \overline{c^2} = \frac{4}{\sqrt{\pi}} \left(\frac{RT}{M}\right) \int_0^\infty e^{-x} \cdot x^{\frac{5}{2}-1} \cdot dx = \frac{4}{\sqrt{\pi}} \left(\frac{RT}{M}\right) \Gamma\left(\frac{5}{2}\right)$

$$\Gamma\left(\frac{5}{2}\right) = \Gamma\left(\frac{3}{2}+1\right) = \frac{3}{2}\Gamma\left(\frac{3}{2}\right) = \frac{3}{2}\Gamma\left(\frac{1}{2}+1\right) = \frac{3}{2}\times\frac{1}{2}\Gamma\left(\frac{1}{2}\right) = \frac{3}{4}\sqrt{\pi}$$

Since, $\quad \Gamma\left(\frac{1}{2}\right) = \sqrt{\pi}$

Or, $\quad \overline{c^2} = \frac{4}{\sqrt{\pi}} \left(\frac{RT}{M}\right) \Gamma\left(\frac{5}{2}\right) = \frac{3RT}{M}.$ Or, $\sqrt{\overline{c^2}} = \sqrt{\frac{3RT}{M}}$ $\quad$ (1.21)

1.4 DERIVATION OF KINETIC ENERGY

According to the assumptions, each and every collision is elastic in nature. Let us consider a molecule of mass m is moving with a velocity u along the x–axis. Its momentum is mu. It is then colliding with the wall of the container after traversing the mean free path l_x and traversing in reverse direction with the same velocity u and momentum $(-mu)$. The total distance traversed is $2l_x$ and the time taken to traverse the distance $2l_x$ is $(2l_x/u)$.

The change in momentum $= mu - (-mu) = 2mu$.

Force exerted by the molecule = Rate of change of momentum

$$= \frac{2mu}{(2l_x/u)} = \frac{mu^2}{l_x} \text{ (along } x\text{-axis)}$$

Pressure exerted by the molecule $= P = \dfrac{\text{Force}}{\text{area}} = \left(\dfrac{mu^2}{l_x}\right)\left(\dfrac{1}{l_y l_z}\right) = \dfrac{mu^2}{l_x l_y l_z}$ (along x-axis)

Or, $\quad p = \dfrac{mu^2}{V},$ where, $V = l_x l_y l_z$ $\quad$ (1.22)

Area of cross-section is $(l_y l_z)$ and V is the volume within which collision occurs.

Let us consider 1 mole of an ideal gas having Avogadro number (N_A) of gas molecules. The mass m is same for all molecules but the velocity u is different. u_1, u_2, u_3 etc. are the velocities of molecules and corresponding pressures are p_1, p_2, p_3 etc. along x-axis. According to Dalton's law of partial pressure, total pressure of all gas molecules along x-axis is given by

$$P_x = p_1 + p_2 + p_3 + \cdots = \frac{mN_A}{V}(u_1^2 + u_2^2 + u_3^2 + \cdots) = mn\bar{u}_x^2 \qquad (1.23a)$$

where $\qquad n = $ Number of molecules per unit volume $= \dfrac{N_A}{V}$

Similarly, total pressure of all gas molecules along y-axis and z-axis are given by

$$P_y = mn\bar{u}_y^2 \quad \text{and} \quad P_z = mn\bar{u}_z^2 \qquad (1.23b)$$

where, $\bar{u}_x^2, \bar{u}_y^2$ and $\bar{u}_z^2$, are the mean square velocities of gas molecules along x–, y– and z– directions respectively.

$$\bar{u}_x^2 = u_1^2 + u_2^2 + u_3^2 + \cdots, \quad \bar{u}_y^2 = v_1^2 + v_2^2 + v_3^2 + \cdots, \quad \bar{u}_z^2 = w_1^2 + w_2^2 + w_3^2 + \cdots \quad (1.23c)$$

Considering velocities along three directions, we have

$$c_{rms}^2 = \bar{u}_x^2 + \bar{u}_y^2 + \bar{u}_z^2 \qquad (1.23d)$$

As motion of gas molecule is random, we can write that

$$\bar{u}_x^2 = \bar{u}_y^2 = \bar{u}_z^2 = \frac{1}{3}c_{rms}^2 = \frac{1}{3}\bar{c^2} \qquad (1.23e)$$

Putting the value of $\bar{u}_x^2$ in the equation (1.23a), we get

$$P_x = P = \frac{1}{3}mn\bar{c^2} \text{ (Considering 1 mole of an ideal gas)} \qquad (1.24a)$$

Or, $\qquad P = \dfrac{1}{3}mn\bar{c^2} = \dfrac{1}{3}\left(\dfrac{mN_A}{V}\right)\bar{c^2} \cdot \quad$ Or, $\quad PV = \dfrac{1}{3}mN_A\bar{c^2} \qquad (1.24b)$

Density of gas (ρ) is given by $\rho = \dfrac{mN_A}{V}$. Hence the equation (1.24a) can be written as

$$P = \frac{1}{3}mn\bar{c^2} = \frac{1}{3}\left(\frac{mN_A}{V}\right)\bar{c^2} = \frac{1}{3}\rho\bar{c^2} \qquad (1.24c)$$

Again, from ideal gas equation, we have $PV = RT$ for 1 mole. Hence, we can write that

$$RT = \frac{1}{3}mN_A\bar{c^2} \cdot \quad k = \frac{R}{N_A} \cdot \quad \text{Or, } kT = \frac{1}{3}m\bar{c^2} \cdot \quad \text{Or, } \frac{3}{2}kT = \frac{1}{2}m\bar{c^2} \qquad (1.25)$$

Mean kinetic energy per gas molecule $= \dfrac{1}{2}m\overline{c^2} = \dfrac{3}{2}kT$

Mean kinetic energy per mole of gas $= \dfrac{3}{2}RT$

Mean kinetic energy for n moles of gas $= \dfrac{3}{2}nRT$

1.5   EQUIPARTITION OF ENERGY

1.5.1   The Equipartition Theorem

The equipartition theorem states that energy is shared equally amongst all energetically accessible degrees of freedom of a system and can make quantitative predictions about how much energy will appear in each degree of freedom.

One example is kinetic energy. There are three possible degrees of freedom: translation degree of freedom, rotational degree of freedom and vibrational degree of freedom.

Another example is random motion of gas molecules. Gas molecules are moving in all three directions (x, y and z) with equal probability. Thus, average kinetic energy in any one direction is $\dfrac{1}{2}m\overline{c^2}$.

According to Maxwell's distribution of molecular velocities, the probability of molecular velocity in any one direction is given by

$$f(c) = \frac{1}{n}\frac{dn}{dc} = 4\pi\left(\frac{m}{2\pi kT}\right)^{3/2} e^{-mc^2/2kT} c^2.$$

Or,
$$\frac{dn}{n} = f(c)dc = 4\pi\left(\frac{m}{2\pi kT}\right)^{3/2} e^{-mc^2/2kT} c^2 \cdot dc$$

Thus average kinetic energy of a gas molecule is given by

$$\text{K.E.} = \int_0^\infty \frac{1}{2}mc^2 \cdot f(c)dc = \frac{1}{2}m \times 4\pi\left(\frac{m}{2\pi kT}\right)^{3/2} \int_0^\infty e^{-mc^2/2kT} c^4 \cdot dc$$

Or,
$$\text{K.E.} = \left(\frac{1}{2}m\right)\left[4\pi\left(\frac{m}{2\pi kT}\right)^{3/2}\right]\left[\frac{3}{2}\left(\frac{kT}{m}\right)^2\left(\frac{2\pi kT}{m}\right)^{1/2}\right] = \frac{3}{2}kT \quad (1.26)$$

Thus total kinetic energy considering all three degrees of freedom is $\dfrac{3}{2}kT$ and hence average kinetic energy per degree of freedom is $\dfrac{1}{2}kT$. This is in agreement with the equipartition theorem. This assignment of $\frac{1}{2}kT$ amount of energy to each degree of freedom of the molecule's motion is called equipartition energy. This microscopic kinetic energy is often called thermal energy and this expression is useful in defining the term kinetic temperature.

1.5.2 Kinetic Temperature

The ideal gas law correlates PV terms with temperature. Again, PV term is also correlated with kinetic energy of gas molecules in accordance with kinetic theory of gases. For 1 mole of an ideal gas, we have

$$PV = RT \text{ and } PV = \frac{1}{3}mN\overline{c^2} = \left(\frac{2}{3}N\right)\left(\frac{1}{2}m\overline{c^2}\right).$$

Or,

$$T = \left(\frac{2}{3}\right)\left(\frac{N}{R}\right)\left(\frac{1}{2}m\overline{c^2}\right) = \left(\frac{2}{3}\right)\left(\frac{1}{k}\right)\left(\frac{1}{2}m\overline{c^2}\right)$$

Or,

$$\text{K.E.} = \frac{1}{2}m\overline{c^2} = \frac{3}{2}kT \qquad (1.27)$$

This temperature is known as **kinetic temperature**.

1.5.3 The Equipartition Theorem from Classical Physics

The internal energy of a monatomic ideal gas containing N particles is $\frac{3}{2}RT = \frac{3}{2}NkT$. This means that each particle possess, on average, $\frac{3}{2}kT$ units of energy. Monatomic particles have only three translational degrees of freedom, corresponding to their motion in three dimensions. They possess no internal rotational or vibrational degrees of freedom. Thus, the mean energy per degree of freedom in a monatomic ideal gas is $\frac{1}{2}kT$. In fact, this is a special case of a rather general result. Let us now try to prove this.

Suppose that the energy of a system is determined by some f generalized coordinates q_k and corresponding f generalized momenta p_k, so that

$$E = E(q_1, q_2, \cdots, q_f, p_1, p_2, \cdots, p_f) = E(q_1, \cdots, p_f) \qquad (1.28a)$$

Let us consider that the total energy splits additively into the form

$$E = \varepsilon_i(p_i) + E'(q_1, q_2, \cdots, p_f) \qquad (1.28b)$$

where ε_i involves only one variable p_i and the remaining part E' does not depend on p_i. The function $\varepsilon_i(p_i)$ is a quadratic function such that

$$\varepsilon_i(p_i) = bp_i^2 \tag{1.28c}$$

where, b is a constant. In one of the most common cases, p_i represents momentum. The absolute temperature is given by

$$T = \frac{1}{k\beta} \qquad \text{Or,} \qquad \beta = \frac{1}{kT} \tag{1.28d}$$

Using Boltzmann distribution function, the mean value of ε_i can be expressed as

$$\overline{\varepsilon_i} = \frac{\int_{-\infty}^{+\infty} e^{-\beta E(q_1,\cdots,p_f)} \varepsilon_i \cdot dq_1 \cdots dp_f}{\int_{-\infty}^{+\infty} e^{-\beta E(q_1,\cdots,p_f)} dq_1 \cdots dp_f} = \frac{\int_{-\infty}^{+\infty} e^{-\beta(\varepsilon_i + E')} \varepsilon_i . dq_1 \cdots dp_f}{\int_{-\infty}^{+\infty} e^{-\beta(\varepsilon_i + E')} dq_1 \cdots dp_f}$$

Or,

$$\overline{\varepsilon_i} = \frac{\int_{-\infty}^{+\infty} e^{-\beta\varepsilon_i} \varepsilon_i \cdot dp_i \int_{-\infty}^{+\infty} e^{-\beta E'} dq_1 \cdots dp_f}{\int_{-\infty}^{+\infty} e^{-\beta\varepsilon_i} dp_i \int_{-\infty}^{+\infty} e^{-\beta E'} dq_1 \cdots dp_f} = \frac{\int_{-\infty}^{+\infty} e^{-\beta\varepsilon_i} \varepsilon_i \cdot dp_i}{\int_{-\infty}^{+\infty} e^{-\beta\varepsilon_i} \cdot dp_i}$$

Again,

$$\int_{-\infty}^{+\infty} e^{-\beta\varepsilon_i} \varepsilon_i \cdot dp_i = -\frac{\partial\left[\int_{-\infty}^{+\infty} e^{-\beta\varepsilon_i} \cdot dp_i\right]}{\partial\beta}$$

Thus,

$$\overline{\varepsilon_i} = -\frac{\partial \ln\left[\int_{-\infty}^{+\infty} e^{-\beta\varepsilon_i} \cdot dp_i\right]}{\partial\beta}$$

Again

$$\varepsilon_i = bp_i^2 \qquad \text{Hence,} \qquad \int_{-\infty}^{+\infty} e^{-\beta\varepsilon_i} \cdot dp_i = \int_{-\infty}^{+\infty} e^{-\beta bp_i^2} \cdot dp_i$$

Assume,

$$x^2 = \beta p_i^2, \quad p_i = \frac{x}{\sqrt{\beta}}, \quad dp_i = \frac{dx}{\sqrt{\beta}}$$

So,

$$\int_{-\infty}^{+\infty} e^{-\beta bp_i^2} \cdot dp_i = \frac{1}{\sqrt{\beta}}\int_{-\infty}^{+\infty} e^{-bx^2} \cdot dx$$

So,

$$\ln\left[\int_{-\infty}^{+\infty} e^{-\beta\varepsilon_i} \cdot dp_i\right] = -\frac{1}{2}\ln\beta + \ln\left[\int_{-\infty}^{+\infty} e^{-bx^2} \cdot dx\right], \quad \text{where,} \quad x = \sqrt{\beta} \times p_i$$

$$\overline{\varepsilon_i} = -\frac{\partial \ln\left[\int_{-\infty}^{+\infty} e^{-\beta\varepsilon_i} \cdot dp_i\right]}{\partial\beta} = \frac{1}{2\beta} + 0 = \frac{1}{2\beta}\left[\text{since,} \int_{-\infty}^{+\infty} e^{-bx^2} \cdot dx \text{ is independent of } \beta\right]$$

Putting the value of β we get

$$\overline{\varepsilon_i} = \frac{1}{2}kT \tag{1.28e}$$

This is the famous *equipartition theorem* of classical physics. It states that the mean value of every independent quadratic term in the energy is equal to $\frac{1}{2}kT$. If all terms in the energy are quadratic then the mean energy is spread equally over all degrees of freedom (hence the name "equipartition"). The above equation holds good for one degree of freedom.

Translational energy

Linear movement of a molecule has three degrees of freedom along three axes. Along each axis energy possessed by a single molecule is $\frac{1}{2}kT$. Thus, total translational energy possessed by single molecule is

$$\overline{\varepsilon}_{\text{trans}} = 3 \times \left(\frac{1}{2}kT\right) = \frac{3}{2}kT \tag{1.29a}$$

Rotational energy

Rotation of a molecule also has three degrees of freedom and along each axis rotational energy possessed by single molecule is $\frac{1}{2}kT$. Thus, total rotational energy possessed by single molecule is

$$\overline{\varepsilon}_{\text{rot}} = 3 \times \left(\frac{1}{2}kT\right) = \frac{3}{2}kT \tag{1.29b}$$

Vibrational energy

A molecule can vibrate either in longitudinal direction or in transverse direction. So there are two degrees of freedom. Thus, total vibrational energy of single molecule is

$$\overline{\varepsilon}_{vib} = 2 \times \left(\frac{1}{2}kT\right) = kT$$

However, in case of a nonlinear polyatomic molecule vibrational energy is

$$\overline{\varepsilon}_{vib} = (3n - 6)kT \tag{1.29c}$$

where, n is the number of atoms in single molecule.

Thus for a nonlinear polyatomic molecule the average energy is

$$\overline{\varepsilon} = \overline{\varepsilon}_{\text{trans}} + \overline{\varepsilon}_{\text{rot}} + \overline{\varepsilon}_{vib} = \frac{3}{2}kT + \frac{3}{2}kT + (3n - 6)kT$$

$$= 3kT + (3n - 6)kT = 3(n - 1)kT$$

Or, $$\bar{\varepsilon} = 3(n-1)kT \qquad (1.30)$$

In case of H_2O (vapour) $n = 3$ and hence the average energy is $\bar{\varepsilon} = 6kT$.

In case of monatomic gases, gas molecules possess only translational energy. Thus total kinetic energy per mole is given by

$$E = \frac{3}{2}RT$$

1.6 COLLISION FREQUENCY (Z)

1.6.1 Collisions Among the Molecules

The distance between centre of two molecules, just touching each other, is called collision diameter, often represented by σ. This is also known as molecular diameter. All molecules are always in motion and hence they are always colliding with each other. For a single molecule, the distance between two successive collisions is called free path. However, the magnitude of this free path is not the same for all molecules, even for a single molecule it differs collision to collision. Thus the average distance between any two successive collisions of two molecules is called mean free path, represented by l. The average time between two successive collisions of two molecules is called average collision time, represented by τ. It is well illustrated in the Fig.1.4.

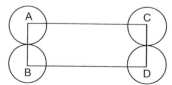

AB = Collision diameter   AC = Mean free path

Figure 1.4 Collision diameter and mean free path

Volume of the cylinder represents the collision volume, which is the total number of collisions made by a single molecule in time τ. The collision volume is given by

Collision volume (V_{col}) = Volume of cylinder = $\pi\sigma^2 l = \pi\sigma^2 c_{rel}\tau$, $\qquad (1.31a)$

[since, $l = c_{rel}\tau$] c_{rel} = relative velocity of a molecule

Thus, number of collisions made by a single molecule per unit time is $\pi\sigma^2 c_{rel}$. Again a molecule can move in any direction. However, relative velocity is calculated on the basis of horizontal and vertical motion of a molecule as shown below

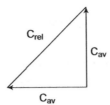

Figure 1.5   Calculation of relative velocity

Here, it is assumed that only one molecule is moving while others are at rest. Hence, $c_{rel} = \sqrt{2}\bar{c}$. So, the collision volume (V_{col}) is given by

$$V_{col} = \sqrt{2}\pi\sigma^2\bar{c}\tau \tag{1.31b}$$

Case I   The cylinder contains only one type of gas.

Consider that there are N number of molecules in the volume V_{col}.

Total number of molecules per unit volume is $n = \dfrac{N}{V_{col}}$.

Thus, the collision frequency (Z) is given by

$$Z = \sqrt{2}\pi n\sigma^2\bar{c} \tag{1.31c}$$

Two molecules are involved in collision. If only one molecule is moving the above relation is true. If both the colliding molecules are moving, the average collision frequency (Z_1) is given by

$$Z_1 = \frac{1}{2}Z = \frac{\sqrt{2}\pi n\sigma^2\bar{c}}{2} = \frac{\pi n\sigma^2\bar{c}}{\sqrt{2}} \ s^{-1} \tag{1.31d}$$

The above equation (1.31c) is deduced by considering one molecule is in motion while others are at rest. Considering all n molecules are in motion, the collision density (Z_d) is given by

$$Z_d = nZ_1 = \frac{\pi n^2\sigma^2\bar{c}}{\sqrt{2}} \ m^{-3}s^{-1} \tag{1.31e}$$

Z_d is known as collision density.

Case II   The cylinder contains two different types of gases, A and B.

In this case collision occurs in three different forms

  (a) Collision among molecules of gas A. The corresponding collision frequency, Z_A, is given by

$$Z_A = n_A Z_A = \frac{\pi n_A^2\sigma_A^2\bar{c_A}}{\sqrt{2}}$$

  (b) Collision among molecules of gas B. The corresponding collision frequency, Z_B, is given by

$$Z_B = n_B Z_B = \frac{\pi n_B^2 \sigma_B^2 \overline{c_B}}{\sqrt{2}}$$

(c) If two different molecules A and B of radii, r_A and r_B are colliding, c_{rms} should be considered instead of $\overline{c}$ and σ^2 should be replaced by $(r_A + r_B)^2$. Then c_{rms} is given by

$$c_{rms} = \sqrt{\frac{8kT}{\pi\mu}}$$

where, μ is called reduced mass and given by

$$\frac{1}{\mu} = \frac{1}{m_A} + \frac{1}{m_B} \tag{1.31f}$$

where, m_A and m_B represent masses of the molecules of A and B respectively. Again, n^2 should be replaced by $n_A n_B$. Thus the expression of collision frequency is given by

$$Z_{AB} = n_A n_B (r_A + r_B)^2 \sqrt{\frac{8\pi kT}{\mu}} \tag{1.31g}$$

Thus, the total collision frequency Z is given by

$$Z = Z_A + Z_B + Z_{AB} = \frac{\pi n_A^2 \sigma_A^2 \overline{c_A}}{\sqrt{2}} + \frac{\pi n_B^2 \sigma_B^2 \overline{c_B}}{\sqrt{2}} + n_A n_B (r_A + r_B)^2 \sqrt{\frac{8\pi kT}{\mu}} \tag{1.31h}$$

1.6.2 Collisions Between Molecules and Container Walls

v_x = Velocity of a molecule along x-axis. Δt = Time required for a gas molecule to collide with the walls.

A = Area of cross section of the container. $\overline{V}$ = Volume of the container under consideration.

$$\overline{V} = A \times (v_x \times \Delta t)$$

Considering all the gas molecules along x-axis, total volume of the container

$$V = \int_0^\infty A \times \Delta t \times v_x \times p(v_x) dv_x$$

$p(v_x)$ = Probability distribution of gas molecules along X–axis = $\sqrt{\dfrac{m}{2\pi kT}} e^{-\frac{mv_x^2}{2kT}}$

$$V = \int_0^\infty A \times \Delta t \times v_x \times \sqrt{\frac{m}{2\pi kT}} e^{-\frac{mv_x^2}{2kT}} dv_x = (A \times \Delta t)\left(\sqrt{\frac{m}{2\pi kT}}\right)\left(\frac{kT}{m}\right)$$

Thus the total volume per unit area per second $= V_{total}$

$$V_{total} = \frac{V}{A \times \Delta t} = \sqrt{\frac{kT}{2\pi m}}$$

Number density of gas molecules $= n$.   Ideal gas equation $PV = RT = N_A kT$

$$n = \frac{N_A}{V} = \frac{P}{kT}$$

Z_w = Number of collisions with walls per unit area per second

$$Z_w = n \times V_{total} = n \times \sqrt{\frac{kT}{2\pi m}}$$

$$= \left(\frac{P}{kT}\right)\left(\frac{kT}{2\pi m}\right)^{1/2} = \frac{P}{(2\pi m kT)^{1/2}} \, m^{-2}s^{-1} \tag{1.32a}$$

Average velocity $= \bar{c} = \sqrt{\dfrac{8kT}{\pi m}}$.   Hence, $Z_w = \dfrac{n\bar{c}}{4} \, m^{-2}s^{-1}$ \hfill (1.32b)

1.7 MEAN FREE PATH

According to kinetic theory, a gas molecule collides with another one only after traversing the resultant mean free path (l). If we consider a cylinder of radius equivalent to collision diameter (σ) and length equivalent to mean free path (l'), only one molecule shall be available within the volume of that cylinder. It is assumed that only one molecule is in motion while others are at rest.

l is the mean free path considering all molecules are in motion. If $\bar{c}$ is the average velocity of one molecule considering others are at rest, relative velocity is given by $c_{rel} = \sqrt{2}\bar{c}$. As relative velocity is greater than average velocity, randomness increases, collision frequency increases and hence resultant mean free path (l) decreases. So the relation between l and l' may be written as

$$l = \frac{l'}{\sqrt{2}} \tag{1.33a}$$

Volume of cylinder $= \pi\sigma^2 l'$ and only one molecule is present in that cylinder. It is shown in the Fig.1.6.

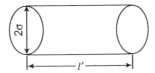

Figure 1.6   Calculation of mean free path

So, number of molecules per unit volume $= \dfrac{1}{\pi \sigma^2 l'}$

As n represents number of molecules per unit volume, we can write that

$$n = \dfrac{1}{\pi \sigma^2 l'} \qquad \text{Or,} \qquad l' = \dfrac{1}{\pi n \sigma^2} \tag{1.33b}$$

Actually all the gas molecules have erratic movement in possible directions. So relative average velocity should be considered instead of average velocity. Thus the expression of l is given by

$$l = \dfrac{l'}{\sqrt{2}} = \dfrac{1}{\sqrt{2} \pi n \sigma^2}$$

Or

$$l = \dfrac{1}{\sqrt{2} \pi n \sigma^2} \tag{1.33c}$$

1.8   VISCOSITY OF GASES

In any fluid, viscosity is the resistive force offered by lower layer of that fluid to the immediate upper layer due to which the velocity of fluid in upper layer decreases. According to the scientist Maxwell in case of gases viscosity often arises due to mixing of gas molecules among the layers. Thus the gas velocity changes as we move perpendicular to the direction of flow and hence a velocity gradient is developed perpendicular to the direction of flow. This velocity gradient is the origin of viscosity of gases.

To evaluate the gas viscosity, the scientist Maxwell assumed that
(a) Gas viscosity is independent of pressure.
(b) Gas viscosity arises due to mixing of molecules of different layers and mixing occurs only after traversing mean free path.

$l =$ Mean free path of gas molecules. $n =$ Number of molecules/volume.

$$\dfrac{\text{Number of molecules moving upward}}{\text{volume}} = \dfrac{1}{3} n$$

then, $\dfrac{\text{Number of molecules moving upward}}{\text{area} \times \text{sec}} = \dfrac{1}{3} n \bar{c}$

Since, $l = \bar{c} \times$ time and volume $=$ area $\times l$

where, $\bar{c}$ is the average velocity of gas molecules.

m = Mass of each molecule in kg.

Thus average momentum of gas molecules $= \dfrac{1}{3}mn\bar{c} \ \dfrac{\text{kg}}{m^2 \cdot s}$

Gas molecules at the point A are moving with velocity u in x-direction. These gas molecules are moving upwards and mixed with other gas molecules, having velocity $u + \dfrac{du}{dz}l$, at the point B, since mixing of gas molecules is possible only after traversing mean free path.

After mixing change of velocity of gas molecules along x-direction is given by

$$= \left(u + \frac{du}{dz}l\right) - u = l\frac{du}{dz}$$

Rate of change of momentum/area = Frictional force/area = τ with unit N/m^2.

So, $$\tau = \left(\frac{1}{3}mn\bar{c}\right) \times l\frac{du}{dz} \tag{1.34a}$$

Again τ is related with coefficient of viscosity (η) as follows

$$\tau = \eta \times \frac{du}{dz} \tag{1.34b}$$

Comparing the equations (1.34a) and (1.34b), we get

$$\eta \times \frac{du}{dz} = \left(\frac{1}{3}mn\bar{c}\right) \times l\frac{du}{dz}$$

Or, $$\eta = \frac{1}{3}mn\bar{c}l \tag{1.34c}$$

Again, $$l = \frac{1}{\sqrt{2}\pi n\sigma^2}, \qquad \text{where,} \quad \sigma = \text{collision diameter}$$

Putting the value of l in the equation (1.34c), we get

$$\eta = \frac{1}{3}mn\bar{c} \times \frac{1}{\sqrt{2}\pi n\sigma^2} = \frac{m\bar{c}}{3\sqrt{2}\pi\sigma^2} \tag{1.34d}$$

Putting the value of $\bar{c}$ in the equation (1.34d), we get

$$\eta = \frac{m}{3\sqrt{2}\pi\sigma^2} \times \sqrt{\frac{8kT}{\pi m}} = \frac{2}{3\pi\sigma^2} \times \sqrt{\frac{km}{\pi}} \times \sqrt{T} = \frac{2\sqrt{RTM}}{3\pi^{3/2}N_A\sigma^2} \tag{1.34e}$$

Or, $$\eta \propto \sqrt{T} \tag{1.34f}$$

The equation (1.34e) is based on Maxwell's distribution law. For a given gas, η can be measured at a given temperature (T), Molecular weight, M, is known. N_A is the Avogadro Number. Hence, collision diameter or molecular diameter, σ can be calculated using the equation (1.34e).

The above equation shows that coefficient of gas viscosity (η) depends on temperature. The expression of coefficient of viscosity predicts that viscosity of a gas is independent of pressure but increases with temperature since increase in temperature favours intermixing of layers. However, in actual practice increase in gas viscosity with temperature is higher than that obtained from the above equation. This deviation occurs because the relation is derived by ignoring intermolecular attractions. Considering the intermolecular attractions the scientist Sutherland formulated the following empirical relation

$$\eta = \frac{k\sqrt{T}}{1+\dfrac{C}{T}}, \quad \text{where, } k = \text{Constant and } C = \text{Sutherland constant} \quad (1.35)$$

Viscosity of gases depends on both pressure and temperature. However, this dependency on pressure and temperature varies from one gas to another. Some examples are given below

For **oxygen**, the equation is

$$\eta = 0.0190395 + 6.50043 \times 10^{-5} \times \theta - 8.97542 \times 10^{-8}\theta^2 + 8.97542 \times 10^{-7}\,P + 6.13118 \times 10^{-10}\,P^2$$

At $\theta = 20°C$ and $P = 14.7\,psi$, $\eta = 0.020317\,cP$.

For **nitrogen**, the equation is

$$\eta = 0.0167214 + 3.92728 \times 10^{-5} \times \theta + 1.22474 \times 10^{-7}\theta^2 + 8.56087 \times 10^{-7}P + 4.69295 + 10^{-10}P^2$$

At $\theta = 20°C$ and $P = 14.7\,psi$, $\eta = 0.017569$ centipoise (cP).

For **carbon dioxide**, the equation is

$$\eta = 0.0137339 + 4.41133 \times 10^{-5} \times \theta + 1.12987 + 10^{-7}\theta^2 + 4.33063 \times 10^{-8}P + 2.67625 + 10^{-9}P^2$$

At $\theta = 20°C$ and $P = 14.7\,psi$, $\eta = 0.014663\,cP$ $\eta = 0.014663\,cP$.

For **helium**, the equation is

$$\eta = 0.0185975 + 5.30773 \times 10^{-5} \times \theta - 1.04983 + 10^{-7}\theta^2 + 4.82504 \times 10^{-8}P$$

At $\theta = 20°C$ and $P = 14.7\,psi$, $\eta = 0.019616\,cP$.

For **argon**, the equation is

$$\eta = 0.0208762 + 7.02190 \times 10^{-5} \times \theta - 3.30712 + 10^{-8}\theta^2 + 1.19960 \times 10^{-6}P + 7.24294 + 10^{-10}P^2$$

At $\theta = 20°C$ and $P = 14.7\,psi$, $\eta = 0.0222185\,cP$.

1.8.1 Diameter of Gas Molecule

According to equation (1.34e), $\eta \propto \dfrac{1}{\sigma^2}$, where, σ is the collision diameter, which is assumed to be diameter of gas molecule. According to Van der Waals expression of real gas we know

$$b = 4 \times \frac{4}{3}\pi r^3 \times N_A \text{ and } \sigma = 2r. \text{ Hence, } b = \frac{2}{3}\pi\sigma^3 \times N_A.$$

Or,

$$\sigma^3 = \frac{3b}{2\pi N_A}$$

Thus equation (1.34e) becomes

$$\eta = \frac{2\sigma\sqrt{RTM}}{3\pi^{3/2} N_A \sigma^3} = \sigma \times \frac{4}{9b}\sqrt{\frac{RTM}{\pi}}. \text{ Or, } \sigma = \frac{9}{4} \times (\eta b) \times \sqrt{\frac{\pi}{RTM}}$$

$$(1.36)$$

For a given gas molecule, M is known. b and η can be determined experimentally. Thus diameter of gas molecule can be determined using equation (1.36).

1.9 THE BAROMETRIC DISTRIBUTION LAW

For any given system pressure is assumed to be constant at every point. However, in actual practice, pressure decreases with altitude due to decrease in gravity field with altitude. This decrease in pressure represents decrease in potential energy per unit volume. So, we can write that $-dP = \rho g dh$.

Again for an ideal gas, we know that

$$PV = nRT = \frac{w}{M}RT. \text{ Or, } \rho = \frac{w}{V} = \frac{PM}{RT} \tag{1.37a}$$

n = Number of moles. M = Molecular weight of the ideal gas.
Substituting this expression of ρ in the earlier equation, we get

$$-dP = \frac{PM}{RT} \cdot g \cdot dh. \text{ Or, } -\int_{P_0}^{P} \frac{dP}{P} = \frac{Mg}{RT}\int_{0}^{h} dh$$

On integration, we get

$$P = P_0 e^{-\frac{Mgh}{RT}} \tag{1.37b}$$

Ideal gas equation shows that $P \propto n$. Hence, we can write that

$$n = n_0 e^{-\frac{Mgh}{RT}} \tag{1.37c}$$

1.10 HEAT CAPACITIES OF GASES

In case of monatomic gases, gas molecules possess only translational energy. Thus total kinetic energy per mole is given by

$$U = \frac{3}{2} RT$$

According to 1st law of thermodynamics $q = \Delta U + w$

At constant volume specific heat depends only on internal energy, U, as there is no work done at constant volume.

$$C_V = \left(\frac{\partial U}{\partial T} \right)_V = \frac{3}{2} R = 3 \text{ cal} \quad (\text{since, } R = 2 \text{ cal/mole.°K})$$

At constant pressure specific heat depends on kinetic energy as well as work done by the gas. So total energy is $H = U + PV = U + RT$ (For ideal gases)

Hence, C_p is given by

$$C_P = \left(\frac{\partial H}{\partial T} \right)_P = \frac{3}{2} R + R = 5 \text{ cal} \quad (\text{since, } R = 2 \text{ cal/mole.°K})$$

Hence the ratio of C_p and C_V is given by

$$\gamma = \frac{C_P}{C_V} = \frac{5}{3} = 1.66 \tag{1.38a}$$

In case of polyatomic molecules rotational and vibrational energies are also to be considered and x represents the contribution of both rotational and vibrational energies to specific heat. Then

$$C_V = 3 + x \quad \text{and} \quad C_P = 5 + x$$

$$\gamma = \frac{C_P}{C_V} = \frac{5 + x}{3 + x} \tag{1.38b}$$

For diatomic molecules $\gamma \approx 1.4$ and for triatomic molecules $\gamma \approx 1.31$.

SOLVED PROBLEMS

Problem 1.1   Calculate the number of binary collisions per sec per ml in nitrogen gas at 1 atm and 25°C. The diameter of nitrogen molecule is 3.74 Å.

Solution

$$V = 22400 \times \left(\frac{298}{273} \right) = 24451.3 \text{ ml.} \quad n = \frac{6.023 \times 10^{23}}{24451.3} = 2.463 \times 10^{19} \text{ cm}^{-3}$$

$$\overline{c} = \sqrt{\frac{8RT}{\pi M}} = 4.8 \times 10^4 \text{ cm/sec}, \quad \sigma = 3.74 \times 10^{-8} \text{ cm}, \quad Z = \frac{\overline{c}}{l} s^{-1}$$

$$Z_d = \text{Collision density} = \frac{1}{2}\left(\frac{\overline{c}}{l}\right) \times n = \frac{1}{2} \times Z \times n = \frac{\pi n^2 \sigma^2 \overline{c}}{\sqrt{2}} = 8.98 \times 10^{28} \text{ cm}^{-3} \text{ s}^{-1}$$

Problem 1.2   Determine the mean free path of *Ar* molecules at NTP. Given diameter of *Ar* molecule is 0.4 *nm*.

Solution

$$T = 273°\text{K}, \quad P = 1 \text{ atm} = 101.325 \text{ kPa} \quad k_B = 1.38 \times 10^{-23} \text{ J/°K}.$$

$$l = \frac{1}{\sqrt{2}\pi n \sigma^2}, \; n = \frac{N_A}{V} = \frac{P}{k_B T}, \text{ since } PV = RT = N_A k_B T, \; \sigma = 0.4 \text{ nm} = 4 \times 10^{-10} \text{ m}$$

$$l = \left(\frac{k_B T}{P}\right)\left(\frac{1}{\sqrt{2}\pi \sigma^2}\right) = 52 \text{ nm}$$

Problem 1.3   Estimate the average number of collisions per second that each N_2 molecule undergoes in air at room temperature and pressure. Given diameter of N_2 molecule is 0.3 nm. and M of $N_2 = 28$.

Solution

$l = \overline{c} \times \tau$, where, $\overline{c}$ is the average velocity of gas molecule and τ is the average time of collision.

$$\overline{c} = \sqrt{\frac{8RT}{\pi M}}, \; l = \frac{1}{\sqrt{2}\pi n \sigma^2} = \left(\frac{k_B T}{P}\right)\left(\frac{1}{\sqrt{2}\sigma^2}\right)$$

$$\tau = \frac{l}{\overline{c}} = 0.2 \text{ ns}, \quad \text{Collision frequency} = Z = \frac{1}{\tau} = 5 \times 10^9 \text{ s}^{-1}$$

Problem 1.4   For molecular oxygen at 25°C. (a) Calculate the collision frequency Z and (b) the collision density Z_d at a pressure of 1atm. Given: $d_{O_2} = 0.361$ nm.

Solution

$$R = 8.314 \text{ J/gm. mole-°K}, \quad M = 32 \text{ kg/kg. mole} = 32 \times 10^{-3} \text{ kg/gm. mole}$$

$$l = \frac{1}{\sqrt{2}\pi n \sigma^2}, \; \sigma = 3.61 \times 10^{-10} \text{ m}$$

$$n = \frac{N_A}{V} = \frac{P}{k_B T} = \frac{101325}{1.38 \times 10^{-23} \times 298} = 2.46 \times 10^{25} \text{ m}^{-3},$$

Hence, $l = 70.243$ nm

$$\bar{c} = \sqrt{\frac{8RT}{\pi M}} = \sqrt{\frac{8 \times 8.314 \times 298}{3.14 \times 32 \times 10^{-3}}} = 444 \text{ m/s}$$

$$\tau = \frac{l}{\bar{c}} = \frac{70.243 \times 10^{-9}}{444} s = 0.1582 \text{ ns}, \ \ Z = \frac{1}{\tau} = 6.32 \times 10^9 \text{ s}^{-1}$$

$$Z_d = \text{Collision density} = \frac{1}{2} \times Z \times n = 7.77 \times 10^{34} \text{ m}^{-3} \text{ s}^{-1}$$

$$= \frac{7.77 \times 10^{34}}{N_A} \text{ mole m}^{-3} \text{ s}^{-1}$$

Or, $Z_d = \dfrac{7.77 \times 10^{34}}{N_A} \times 10^{-3} \text{ mole L}^{-1} \text{s}^{-1} = 1.29 \times 10^{28} \text{ mole L}^{-1} \text{s}^{-1}$

| **Problem 1.5** | Large vacuum chambers have been built for testing space vehicles at $10^{-7} Pa$. Calculate the number of molecular impacts per square meter of wall per second for molecular CO_2 with $d = 40$ nm at 35°C. |

Solution

$$J = \text{Number of collisions with wall} = \frac{n\bar{c}}{4}, \ \ \bar{c} = \sqrt{\frac{8RT}{\pi M}}, \ \ n = \frac{N_A}{V}$$

$$J = 2.25 \times 10^{15} \text{ m}^{-2}\text{s}^{-1}$$

| **Problem 1.6** | A 10 mL container with a hole of diameter $1 \mu m$ is filled with hydrogen gas. This container is placed in an evacuated chamber at 10°C. How long will it take for 80% of the hydrogen to effuse out? |

Solution

$$J_N = -\frac{1}{A} \frac{dN}{dt} = \frac{n\bar{c}}{4} = \left(\frac{N}{V}\right)\left(\frac{\bar{c}}{4}\right).$$

Or, $$-\int_{N_0}^{N} \frac{dN}{N} = A\left(\frac{\bar{c}}{4V}\right) \int_0^t dt.$$

Or,
$$\ln\frac{N_0}{N} = A\left(\frac{\bar{c}}{4V}\right)t$$

$$\frac{N_0}{N} = \frac{100}{20}, \bar{c} = \sqrt{\frac{8RT}{\pi M}}, V = 10\,ml = 10\times10^{-6}\,m^3, A = \frac{\pi}{4}(1\times10^{-6})^2\,m^2$$

Hence, $t = 47355$ sec $= 789.25$ min $= 13.15$ hr

Problem 1.7   An effusion cell has a circular hole of diameter 2.50 mm. If the molar mass of the solid in the cell is 260 gm/mole and its vapour pressure is 0.835 *Pa* at 400°K, by how much will the mass of the solid decrease in a period of 2 hr?   [0.0104 gm]

Solution

$$J_N = -\frac{1}{A}\frac{dN}{dt} = \frac{n\bar{c}}{4} = \left(\frac{N}{V}\right)\left(\frac{\bar{c}}{4}\right). \quad \text{Or,}\; \frac{\Delta w}{w} = \left(\frac{A\bar{c}}{4V}\right)\Delta t$$

Δw = Decrease in weight of the solid. w = Original weight of the solid.

$$PV = nRT = \left(\frac{w}{M}\right)RT. \quad \text{Or,}\; \frac{w}{V} = \frac{PM}{RT}$$

EXERCISES

1. Intergalactic space is nearly vacuum. Number density of H-atom is $1/cm^3$. Diameter of H-atm is 0.1 nm.
   (a) Estimate the mean free path of H – atom.   [*Ans.:* 2×10^{10} km]
   (b) Estimate $\bar{c}$ of *H* – atom at 2.7°K.   [*Ans.:* 260 m/s]
   (c) Estimate the average time of collision (in years)   [*Ans.:* $\tau = 2000$ years]
2. The mean free path of N_2 gas is 6.8×10^{-8}m at 25°C. Determine the collision frequency.
   [*Ans.:* $Z = 8.56 \times 10^{34}$ collisions/sec]
3. A gas has an average speed of 10 m/s and a collision frequency of 10 s^{-1}. What is its mean free path?   [*Ans.:* 1 m]
4. A gas has an average speed of 10 m/s and an average time of 0.1 s between collisions. What is its mean free path?   [*Ans.:* 1 m]
5. A gas has a density of 10 particles per m^3 and a molecular diameter of 0.1 m. What is its mean free path?   [*Ans.:* 2.25 m]
6. A gas in a 1 m container has a molecular diameter of 0.1 m. There are 10 molecules. What is its mean free path?   [*Ans.:* 2.25 m]
7. A gas has a molecular diameter of 0.1 m. It also has a mean free path of 2.25 m. What is its density?

8. Calculate the average speed and kinetic energy of O_2 and CCl_4 at 25°C.

   Hint:

$$\bar{c} = \sqrt{\frac{8RT}{\pi M}} \quad and \quad \overline{E_k} = \frac{1}{2}m\overline{c^2} = \frac{3}{2}RT$$

9. Calculate the mean free path for chlorine gas ($d_{Cl_2} = 0.544$ nm) at 0.1 *Pa* and 25°C.
   [*Ans.:* 0.0312 m]

10. How many molecules of oxygen gas strike the wall per unit area per unit time at 600 mm Hg at 25°C? Given $d_{O_2} = 3.60$ Å. [*Ans.:* 2.1×10^{27} m^{-2}s^{-1}]

11. A box contains H_2 (molecule *b*) and He (molecule *c*) at a total pressure of 1.4 atm. If the mixture contains 18%H_2 by weight, calculate the ratio of Z_b/Z_c. Given $d_{H_2} = 0.247$ nm and $d_{He} = 2.20$ nm. [*Ans.:* 6.97×10^{-3}]

12. Exactly 1 dm^3 of nitrogen, under a pressure of 1 bar, will take 5.80 minutes to effuse through an orifice. How long will it take for Helium to effuse under the same conditions? [*Ans.:* 2.19 min]

13. Dry air had a CO_2 mole fraction of 0.004. Calculate the total mass of CO_2 that strikes 0.5 cm^2 of one side of a green leaf in 10 sec in dry air at 39°C and 1 atm. [*Ans.:* 0.332 g]

14. A gas mixture contains H_2 at $\frac{2}{3}$ atm and O_2 at $\frac{1}{3}$ atm at 27°C. Calculate the number of collisions per second (i) among H_2 molecules, (ii) among O_2 molecules, (iii) among H_2 and O_2 molecules.

    Given $d_{H_2} = 0.272$ nm, $d_{O_2} = 0.361$ nm.
    [*Ans.:* (i) 6.5×10^9 s^{-1}, (ii) 2.1×10^9 s^{-1}, (iii) 4.7×10^9 s^{-1}]

15. A certain sample of a pure oxygen ($d = 3.61$ Å) has mean speed, $\bar{c} = 450$ m/s and the average time between two successive collisions of a given molecule with other molecules is 4.0×10^{-10} s. Find the mean free path and molecular density of this gas.
    [*Ans.:* 1.8×10^{-7} m, 9.6×10^{24} m^{-3}]

2 Real Gases

2.1 DETERMINATION OF DENSITY OF VAPOUR

2.1.1 Victor Meyer's Method

The apparatus is shown in the Fig. 2.1.

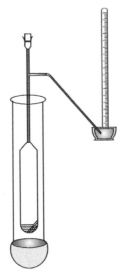

Figure 2.1 Victor Meyer's apparatus

The apparatus consists of a long-necked elongated bulb, also known as Victor Meyer's tube, surrounded by a glass or copper jacket. The top end of the tube is closed by a rubber cork and the side end is attached to a capillary tube, which is immersed in a trough of water. A graduated burette, filled with water, is inverted and held on the capillary tube under water. The surrounding jacket contains water, having boiling point much higher than that of experimental liquid. The water is heated to its boiling point. The vapourisation tube or

Victor Meyer's tube is sufficiently hot. Then the sample, vapour density of which is to be determined, is inserted into the tube.

Preparation of sample and putting the sample into the vapourisation tube is a major problem. Two methods are generally used:

(i)   The sample is sealed in a thin walled glass bulb and dropped to the bottom of the vaporization chamber. As soon as the bulb touches the bottom, it breaks and the sample is vapourised.

(ii) The sample is placed in a small glass stoppered weighing bottle and dropped onto an asbestos cushion. The weighing bottle must be removed after each determination.

In both the methods it is assumed that the volume occupied by the liquid is negligible compared to the volume of the vapourisation tube.

As the sample is vapourised, the vapour moves upwards, passes through capillary tube and ultimately collected at the top of the graduated burette by displacing water. The open end of the burette is closed by thumb under water and take it away from the trough of water. It is then immersed in a very large jar, filled with water. The open end is released and maintain the water level of the burette same as the water level of jar so that pressure is maintained at 1 atm. Under this condition, volume occupied by the vapour in the burette is measured. Suppose it is V ml and the experiment is carried out at room temperature, $t = 25°C$.

Calculation

Weight of the liquid taken = w gms.   Temperature = TK.
Aqueous tension at $TK = f\,mm$.
Thus from ideal gas equation, we have

$$\frac{P_0 V_0}{T_0} = \frac{P_1 V_1}{T_1}$$

$P_0 = 760$ mm, $T_0 = 273$ K, $P_1 = (760 - f)$ mm, $V_1 = V$ ml, $T_2 = 298$ K

Hence,
$$V_0 = \left(\frac{P_1 V}{T_1}\right)\left(\frac{T_0}{P_0}\right)$$

Thus vapour density of the sample, ρ_v is given by

$$\rho_v = \frac{w}{V_0}\ gm/\,ml \tag{2.1}$$

2.1.2 Density and Molecular Weight

Density is correlated with molecular weight according to the following equation

$$PV = nRT = \frac{w}{M}RT. \quad \rho = \frac{w}{V} = \frac{PM}{RT}. \quad M = \left(\frac{\rho}{P}\right)RT$$

This equation is valid for ideal gases only. In case of real gases, $\left(\dfrac{\rho}{P}\right)$ varies with P almost linearly as shown in the Fig. 2.2.

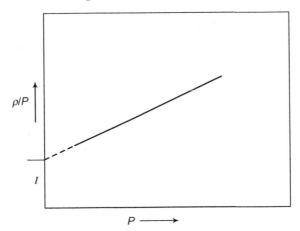

Figure 2.2 $\left(\dfrac{\rho}{P}\right)$ versus P curve of a real gas

The straight line is extrapolated to zero pressure so that the intercept, I, is given by

$$I = \lim_{P \to 0} \left(\frac{\rho}{P}\right)$$

At this low pressure gases are assumed to behave ideally and the ideal gas equation holds good. Thus molecular weight, M, can be written as

$$M = \left[\lim_{P \to 0} \left(\frac{\rho}{P}\right)\right] \tag{2.2}$$

2.1.3 Abnormal Density

There are certain substances for which density is found to be less than that calculated from their molecular formulae. This happens as these molecules undergo thermal dissociation upon heating. Following are some examples:

$$NH_4Cl \rightleftharpoons NH_3 + HCl$$

$$I_2 \rightleftharpoons 2I$$

$$N_2O_4 \rightleftharpoons 2NO_2$$

$$PCl_5 \rightleftharpoons PCl_3 + Cl_2$$

In all of the above cases, as the number of molecules increases on dissociation, volume increases though pressure remains constant. Consequently, density decreases. Consider the following dissociation reaction:

$$A(g) \rightleftharpoons nB(g)$$
$$(1 - \alpha) \qquad n\alpha$$

where, α = Degree of dissociation.

ρ_0 = Density of undissociated compound A.

ρ_0 = Density of the mixture after partial dissociation of the compound A.

V = Volume occupied by the one mole of undissociated compound A.

$[(1 - \alpha) + n\alpha] = [1 + (n - 1)\alpha]$ = Total number of moles after dissociation of the compound A, whereas, initial number of moles of undissociated A is 1.

So, after dissociation of the compound A, volume of the mixture is $V[1 + (n - 1)\alpha]$.

As the total weight remains same before and after dissociation, we have

$$w = \rho_0 V = \rho V[1 + (n - 1)a)].$$

Hence, $\qquad \rho_0 = \rho[1 + (n - 1)\alpha]$

Or, $\qquad (n - 1)\alpha = \dfrac{(\rho_0 - \rho)}{\rho}$. Or, $\alpha = \dfrac{(\rho_0 - \rho)}{\rho(n - 1)}$

If pressure and temperature remains constant, density can be replaced by molecular weight. So, the above equation can be written as

$$\alpha = \frac{(\rho_0 - \rho)}{\rho(n - 1)} = \frac{(M_0 - M)}{(n - 1)M} \qquad \text{(at constant } P \text{ and } T) \qquad (2.3)$$

where, M_0 = Molecular weight of the undissociated compound A.

M = Average molecular weight of all gases present in the mixture after dissociation of the compound A.

2.2 *PV–ISOTHERM OF IDEAL GAS ISOTHERM*

The concept of ideal gas is based on some assumptions. Out of them following are the most important:

(i)   Molecules are assumed to be point masses and hence total volume of all molecules is negligibly small compared to that of the container.

(ii) Molecules are moving independently. Intermolecular forces of attractions and repulsions are neglected.

(iii) Molecules of any ideal gas are said to have only kinetic energy.

The corresponding PV-isotherm of an ideal gas is shown in the Fig. 2.3.

In a real gas, each molecule occupies a definite volume and hence total volume of any real gas cannot be neglected with respect to volume of the container.

Each and every molecule has a positive nucleus, surrounded by electrons. Thus, strong intermolecular attractions and repulsions exist among the molecules.

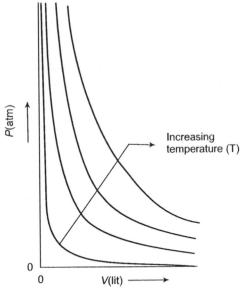

Figure 2.3 *PV*–isotherm of an ideal gas

In a real gas, molecules are said to have both potential energy and kinetic energy. The corresponding *PV*–isotherm of a real gas (CO_2) is shown in the Fig. 2.4.

The deviation of any real gas behaviour from ideal one is expressed by compressibility factor (*Z*), which is given by

$$Z = \frac{PV}{RT}$$

In case of ideal gases, molecular attractive and repulsive forces are neglected. Also volume of individual gas molecule is neglected with respect to the volume of container. According to ideal gas equation, $Z = 1$.

However, in case of real gases value of *Z* depends on both pressure and temperature.

Case I At very low pressure, gas molecules are widely separated and hence molecular attractive and repulsive forces can be neglected. So, the value of $Z \approx 1$. Thus, any real gas behaves ideally at very low pressure.

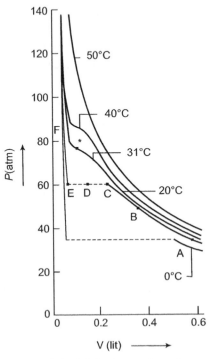

Figure 2.4 *PV*–isotherm of a real gas

Case II   At moderately high pressure attractive forces are dominating. Gas molecules are sufficiently close enough such that gas volume is less than expected from Boyle's law. So, $Z < 1$.

Case III   At very high pressure repulsive forces are dominating. Gas molecules occupy large volume, larger than that expected from Boyle's law. So, $Z > 1$.

Case IV   As temperature increases, randomness of molecules increases and gas molecules are assumed to follow kinetic theory of gases. Consequently, deviation from ideal gas behaviour also decreases as shown in the Fig. 2.5.

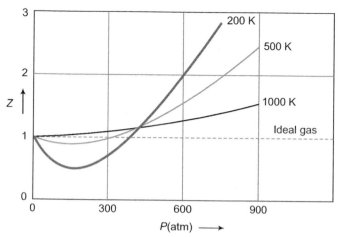

Figure 2.5   Compressibility factor *versus P* diagram at different temperatures

For every real gas there is a temperature, at which the gas behaves ideally and $Z=1$. That particular temperature is called ***Boyle temperature***, T_B. It is illustrated in the Fig. 2.6.

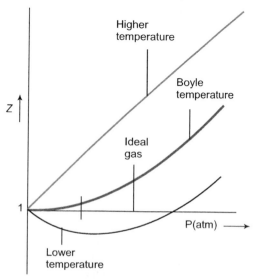

Figure 2.6   Boyle temperature in Z *versus P* diagram

Considering the above fact, Kamerlingh Onnes in 1901 proposed the following equation for any real gas

$$PV = RT\left(1 + \frac{B}{V} + \frac{C}{V^2} + \cdots\right) = RT(1 + B'P + C'P^2 + \cdots)$$

B, B' and C, C' etc. are all function of temperature and called virial coefficients

$$\frac{B}{V} \gg \frac{C}{V^2} \text{ and } B'P \gg C'P^2$$

At Boyle temperature, all the virial coefficients vanish and $PV = RT$.

2.4 CRITICAL POINT

Consider PV–isotherm of a real gas, CO_2.

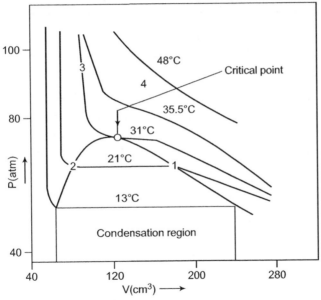

Figure 2.7 *PV*–diagram of CO_2 gas

There are two distinct phases, gas on the left hand side of the isotherm and liquid on the right hand side of the isotherm. These two phases are connected by a horizontal line, e.g., at 21°C the horizontal line is 2 – 1. At the point 2, the gas phase exists and at the point 1, the liquid phase exists. Along this horizontal line two phases coexist. As the temperature increases, the length of horizontal line decreases and ultimately merges to a point, known as critical point, above which liquid phase no longer exists. The critical point for CO_2 is 31°C. So, for any real gas critical point is the maximum point or temperature at which the gas can be liquefied.

The critical point is also known as point of inflection and the following condition holds good at the critical point.

$$\left|\left(\frac{\partial P}{\partial V}\right)_T\right|_{\text{at } C} = 0 \quad \text{and} \quad \left|\left(\frac{\partial^2 P}{\partial V^2}\right)_T\right|_{\text{at } C} = 0 \tag{2.4}$$

P_c, V_c and T_c are the critical pressure, critical volume and critical temperature respectively. Then, the term reduced parameter is defined as below

$$\text{Reduced pressure} = \frac{P}{P_c} = \pi, \quad \text{Reduced volume} = \frac{V}{V_c} = \phi$$

$$\text{Reduced temperature} = \frac{T}{T_c} = \theta$$

2.5 LAW OF RECTILINEAR DIAMETER

The critical volume of any real gas can be calculated by using the law of rectilinear diameter, developed by L. Cailletet and E. Mathias in 1886. Later on the law was verified by S. Young et al. in 1900.

Statement In a closed system, any liquid remains in equilibrium with its saturated vapour. At any temperature the densities of liquid and its saturated vapour are known as orthobaric densities and the arithmetic mean of orthobaric densities at that temperature is represented as ρ_i, which is a linear function of temperature. Mathematically, it can be written as:

$$\rho_i = \rho_0 + \alpha\theta \tag{2.5}$$

where, θ is the temperature and ρ_0 is the orthobaric density at 0°C.

The above mathematical expression, equation (2.5), is known as the law of rectilinear diameter.

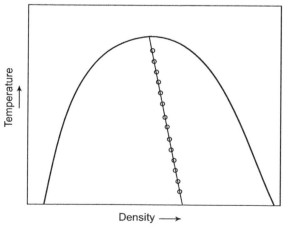

Figure 2.8 Temperature *versus* density diagram of a Liquid $\rightleftharpoons$ Vapour system

In case of *n*–pentane the values of ρ_0 and α have been found +0.3231 and –0.00046 respectively.

2.5.1 Determination of Orthobaric Densities

A known mass of a liquid (m) is taken in a sealed graduated tube at a particular temperature. At equilibrium, volume of liquid and its saturated vapour, i.e., V_l and V_g respectively are read from graduation. If ρ_l and ρ_g are the densities of liquid and its saturated vapour respectively, we can write that

$$m = \rho_l V_l + \rho_g V_g$$

In the next step, same procedure is to be followed at different mass but at that same temperature. By measuring corresponding volumes from graduation and using above equation ρ_l and ρ_g can be easily computed and hence ρ_i can also be calculated.

In the similar way ρ_i values at other temperatures are determined and by plotting ρ_i versus θ one can easily evaluate ρ_0 and α. By extrapolating the straight line to the critical point, ρ_i at the critical point can be evaluated.

2.6 COMPRESSIBILITY FACTOR

P, *V*, and *T* are the controlling parameters to define any gas. Thus, the equation of state of any gas is given by:

$$f(P, V, T) = 0 \tag{2.6a}$$

One of the solution of the above equation is $V = f(P, T)$
Differentiating, we get

$$dV = \left(\frac{\partial V}{\partial T}\right)_P dT + \left(\frac{\partial V}{\partial P}\right)_T dP$$

Or,

$$\frac{dV}{V} = \frac{1}{V}\left(\frac{\partial V}{\partial T}\right)_P dT + \frac{1}{V}\left(\frac{\partial V}{\partial P}\right)_T dP$$

Or,

$$\frac{dV}{V} = \beta dT - k dP \tag{2.6b}$$

where, $\beta = \text{Expansivity} = \dfrac{1}{V}\left(\dfrac{\partial V}{\partial T}\right)_P$

and $k = \text{Isothermal compressibility} = -\dfrac{1}{V}\left(\dfrac{\partial V}{\partial P}\right)_T$

In case of liquid, isotherms are very steep and hence the values of β and k are approximated to zero. On the basis of this assumption, liquid is often called incompressible fluid.

$P-V-T$ behaviour of different pure substances shows that the following power series expression holds good for all gases.

$$PV = a + bP + cP^2 + \cdots = a(1 + B'P + C'^{P^2} + \cdots)$$

where,

$$B' = \frac{b}{a}, \quad C' = \frac{c}{a} \text{ etc.}$$

Or

$$PV = a(1 + B'P + C'P^2 + \cdots) \tag{2.7}$$

When pressure of a real gas tends to zero, the above equation becomes

$$\lim_{P \to 0}(PV) = a = RT \tag{2.8}$$

The gas is known as ideal gas. Thus at very low pressure any real gas behaves ideally. Again for one mole of any ideal gas, we have

$$PV = RT \quad \text{and} \quad P = \frac{RT}{V}$$

Putting this value of P in the R.H.S. of equation (2.7), we get

$$PV = a\left[1 + B'\frac{RT}{V} + C'\left(\frac{RT}{V}\right)^2 + \cdots\right]$$

Or,

$$PV = a\left[1 + \frac{B}{V} + \frac{C}{V^2} + \cdots\right] \tag{2.9}$$

where, $B = B'RT$ and $C = C'(RT)^2$

Putting the value of a from the equation (2.8) in the equations (2.7) and (2.9), we get

$$PV = RT(1 + B'P + C'P^2 + \cdots) \tag{2.10a}$$

$$PV = RT\left[1 + \frac{B}{V} + \frac{C}{V^2} + \cdots\right] \tag{2.10b}$$

$$\frac{PV}{RT} = Z = \text{Compressibility factor}$$

Thus equations (2.10a) and (2.10b) become

$$Z = (1 + B'P + C'P^2 + \cdots) \tag{2.11a}$$

$$Z = \left[1 + \frac{B}{V} + \frac{C}{V^2} + \cdots\right] \tag{2.11b}$$

These above expansions are known as virial expansions and the coefficients are known as ***virial coefficients***. For an ideal gas, value of all virial coefficients are zero and hence

$$Z = \frac{PV}{RT} = 1$$

2.7 VAN DER WAALS EQUATION

Scientist Johannes Diderik van der Waals first tried to establish the equation of a real gas. According to his suggestion:

(i) Molecular attractive forces reduces the total pressure and pressure due to molecular attractive forces is called internal pressure, denoted by p. The internal pressure depends on two factors: (a) wall collision frequency, where, $p \propto (n/V)$, and (b) change in momentum at each collision, where, $p \propto (n/V)$. Here, (n/V) is the concentration of the gas. Thus,

$$p \propto \left(\frac{n}{V}\right)^2. \text{ Or, } p = a\left(\frac{n}{V}\right)^2 = \frac{an^2}{V^2}$$

Considering the internal pressure, the total pressure of any real gas can be expressed as

$$(P + p) = \left(P + \frac{an^2}{V^2}\right)$$

(ii) Volume of gas molecules cannot be neglected. So, total volume occupied by a real gas is given by $(V - b)$ for 1 mole and $(V - nb)$ for n moles. b is a constant and varies from gas to gas. It is assumed that,

$$b \propto \frac{4}{3}\pi r^3, \quad r = \text{radius of gas molecule}$$

Considering the above two propositions, van der Waals equation of state is given by

$$\left(P + \frac{an^2}{V^2}\right)(V - nb) = nRT$$

2.7.1 Reduced Equation of State

For 1 mole of a real gas van der Waals equation of state is given by

$$\left(P + \frac{a}{V^2}\right)(V - b) = RT \quad \text{(for 1 mole)} \tag{2.12}$$

Or,
$$P = \frac{RT}{V - b} - \frac{a}{V^2}$$

Differentiating with respect to V at constant T, we get

$$\left(\frac{\partial P}{\partial V}\right)_T = -\frac{RT}{(V-b)^2} + \frac{2a}{V^3} \quad \text{and} \quad \left(\frac{\partial^2 P}{\partial V^2}\right)_T = \frac{2RT}{(V-b)^3} - \frac{6a}{V^4}$$

At the critical point of $PV-$ isotherm of any real gas, we know

$$\left(\frac{\partial P}{\partial V}\right)_T = 0 \quad \text{and} \quad \left(\frac{\partial^2 P}{\partial V^2}\right)_T = 0, \quad P = P_c, \quad V = V_c, \quad T = T_c$$

$$\frac{RT_c}{(V_c-b)^2} = \frac{2a}{V_c^3} \quad \text{and} \quad \frac{2RT_c}{(V_c-b)^3} = \frac{6a}{V_c^4}.$$

Or,

$$T_c = \frac{2a}{R}\frac{(V_c-b)^2}{V_c^3} \quad \text{and} \quad T_c = \frac{3a}{R}\frac{(V_c-b)^3}{V_c^4}$$

Or,

$$\frac{2a}{R}\frac{(V_c-b)^2}{V_c^3} = \frac{3a}{R}\frac{(V_c-b)^3}{V_c^4}. \quad \text{Or,} \quad \frac{V_c-b}{V_c} = \frac{2}{3}$$

Or,

$$V_c = 3b \tag{2.13}$$

Using this value of V_c we get,

$$T_c = \left(\frac{2a}{R}\right)\left(\frac{4b^2}{27b^3}\right) = \frac{8a}{27Rb} \tag{2.14}$$

$$P_c = \frac{RT_c}{V_c-b} - \frac{a}{V_c^2} = \frac{4a}{27b^2} - \frac{a}{9b^2} = \frac{a}{27b^2} \tag{2.15}$$

$$\frac{RT_c}{P_cV_c} = \left(\frac{8a}{27b}\right)\left(\frac{1}{3b}\right)\left(\frac{27b^2}{a}\right) = \frac{8}{3} = 2.667$$

From the concept of reduced parameter, we have $P = \pi P_c$, $V = \phi V_c$ and $T = \theta T_c$. Putting these values of P, V, and T in the equation (2.12) we get,

$$\left(\pi P_c + \frac{a}{(\phi V_c)^2}\right)(\phi V_c - b) = R\theta T_c \tag{2.16}$$

$$\frac{RT_c}{P_cV_c} = \frac{8}{3} \quad \text{and} \quad V_c = 3b \quad \text{Or,} \quad RT_c = 8bP_c$$

$$P_c = \frac{a}{27b^2} \quad \text{and} \quad b = \frac{V_c}{3}. \quad \text{Or,} \quad V_c^2 = \frac{a}{3P_c}$$

Putting the values of V_c^2 and RT_c in the equation (2.16) we have

$$\left(\pi + \frac{3}{\phi^2}\right)(3\phi - 1) = 8\theta \tag{2.17}$$

Equation (2.17) is known as **reduced equation of state** for a gas following van der Waals equation.

Inference If π and θ of any two gases are same, ϕ must be same for those two gases.

Law of corresponding states All gases, when compared at same reduced temperature and reduced pressure, have same reduced volume, hence have same compressibility factor so that all of them deviate from the ideal behaviour to about the same degree.

2.7.2 Boyle Temperature

At Boyle temperature the following are true:

$$\left[\frac{\partial(PV)}{\partial P}\right]_T = 0 \text{ and } \left(\frac{\partial V}{\partial P}\right)_T \neq 0$$

Van der Waals equation for 1 mole of a real gas is

$$\left(P + \frac{a}{V^2}\right)(V - b) = RT.$$

Or, $$P = \frac{RT}{V - b} - \frac{a}{V^2}. \text{ Or, } PV = \frac{RTV}{V - b} - \frac{a}{V}$$

$$\left[\frac{\partial(PV)}{\partial P}\right]_T = \left[RT\left\{\frac{1}{V - b} - \frac{V}{(V - b)^2}\right\} + \frac{a}{V^2}\right]\left(\frac{\partial V}{\partial P}\right)_T$$

$$= \left[\frac{a}{V^2} - \frac{RTb}{(V - b)^2}\right]\left(\frac{\partial V}{\partial P}\right)_T$$

Applying the condition of Boyle temperature, we have

$$\frac{a}{V^2} - \frac{RT_B b}{(V - b)^2} = 0. \text{ Or, } T_B = \frac{a}{Rb}\left(\frac{V - b}{V}\right)^2$$

At Boyle temperature $(V - b) \approx V$. Hence, the above equation becomes

$$T_B = \frac{a}{Rb} \text{ and } \frac{T_B}{T_c} = \left(\frac{a}{Rb}\right)\left(\frac{27Rb}{8a}\right) = \frac{27}{8} = 3.375 \tag{2.18}$$

2.8   DIETERICI EQUATION OF STATE

One of the important reasons for deviation of real gas from ideal behaviour is the concept of volume. In ideal gas intermolecular forces of attraction is neglected and hence density of gas is considered constant throughout the container. However, this concept is not correct for real gases. The scientist Jeans proposed that intermolecular forces of attraction are very strong in the bulk and hence each molecule needs extra potential energy to overcome these intermolecular forces of attraction and moves towards the wall. Thus, gas density is less near the wall than that in the bulk.

n = Number of gas molecules per cm^3 near the wall.

n_0 = Number of gas molecules per cm^3 in the bulk.

Both n and n_0 are correlated by Boltzmann distribution function, i.e., $e^{-A/RT}$, where, A is the excess potential energy per molecule required to move towards the wall of the container.

Again, $P \propto n$ and $P \propto \rho$, where, P and ρ are the gas pressure and gas density respectively. So, we can write

$$\frac{n}{n_0} = \frac{\rho}{\rho_0} = \frac{P}{P_0} = e^{-A/RT}. \quad \text{Or,} \quad P = P_0 e^{-a/RT} \qquad (2.19)$$

ρ_0 = Gas density in the bulk

P_0 = Gas pressure in the bulk where all cohesive forces cancel each other and hence may be considered as ideal gas pressure.

$(V - b)$ = Volume occupied by gas molecules and b volume of gas molecules.

Using this volume correction, ideal gas law can be written as

$$P_0(V - b) = RT.$$

Or, $P e^{A/RT}(V - b) = RT.$ Or, $$P = \frac{RT}{(V - b)} e^{-\frac{A}{RT}} \qquad (2.20)$$

According to the scientist Dieterici, exponential term depends on volume and hence the equation (2.20) becomes

$$P = \frac{RT}{(V - b)} e^{-\frac{a}{RTV}} \qquad (2.21)$$

The equation (2.21) is known as Dieterici equation for 1 mole of real gases. For n moles the above equation becomes

$$P = \frac{nRT}{(V - nb)} e^{-\frac{an}{RTV}} \qquad (2.22)$$

After expansion the equation (2.21) becomes

$$P = \frac{RT}{(V - b)} \left[1 - \frac{a}{RTV} + \frac{1}{2!}\left(\frac{a}{RTV}\right)^2 - \frac{1}{3!}\left(\frac{a}{RTV}\right)^2 + \cdots \right]$$

Neglecting the higher terms, we have

$$P = \frac{RT}{(V-b)}\left[1 - \frac{a}{RTV}\right] = \frac{RT}{(V-b)} - \frac{a}{V(V-b)}$$

At low pressure, $(V-b) \approx V$ and hence the above equation becomes

$$P = \frac{RT}{(V-b)} - \frac{a}{V^2} \qquad \text{[same as van der Waals equation]}$$

Thus at low pressure, Dieterici equation of state is same as van der Waals equation of state.

2.8.1 Reduced Equation of State

At the critical point of PV–isotherm of any real gas, we know

$$\left(\frac{\partial P}{\partial V}\right)_T = 0 \quad \text{and} \quad \left(\frac{\partial^2 P}{\partial V^2}\right)_T = 0$$

Expressions of $\left(\dfrac{\partial P}{\partial V}\right)_T$ and $\left(\dfrac{\partial^2 P}{\partial V^2}\right)_T$ for the Dieterici equation for 1 mole of real gas are given as

$$\left(\frac{\partial P}{\partial V}\right)_T = -\frac{RT}{(V-b)^2}e^{-\frac{a}{RTV}} + \left(\frac{RT}{V-b}\right)\left(\frac{a}{RTV^2}\right)e^{-\frac{a}{RTV}}$$

$$= -\frac{P}{(V-b)} + \frac{aP}{RTV^2}$$

$$\left(\frac{\partial^2 P}{\partial V^2}\right)_T = -\left(\frac{1}{V-b}\right)\left(\frac{\partial P}{\partial V}\right)_T + \frac{P}{(V-b)^2} + \frac{a}{RTV^2}\left(\frac{\partial P}{\partial V}\right)_T - \frac{2aP}{RTV^3}$$

$$= \left[\frac{a}{RTV^2} - \frac{1}{(V-b)}\right]\left(\frac{\partial P}{\partial V}\right)_T + P\left[\frac{1}{(V-b)^2} - \frac{2a}{RTV^3}\right]$$

Using the critical point conditions in the Dieterici equation for 1 mole of real gas, we have

$$\left(\frac{\partial P}{\partial V}\right)_T = 0$$

Or, $\quad -\dfrac{P_c}{(V_c-b)} + \dfrac{aP_c}{RT_cV_c^2} = 0.$ Or, $\quad \dfrac{1}{(V_c-b)} = \dfrac{a}{RT_cV_c^2}$ \hfill (2.23)

Again at $T = T_c$

we have
$$\left(\frac{\partial^2 P}{\partial V^2} \right)_T = 0$$

Or, $\quad \dfrac{1}{(V_c - b)^2} - \dfrac{2a}{RT_c V_c^3} = 0.$ Or, $\dfrac{1}{(V_c - b)^2} = \dfrac{2a}{RT_c V_c^3}$ $\qquad$ (2.24)

Combining equations (2.23) and (2.24), we have
$$\frac{1}{(V_c - b)} = \frac{2}{V_c}. \text{ Or, } V_c = 2b \qquad (2.25a)$$

Putting the value of V_c in the equation (2.23), we have
$$T_c = \frac{a}{4Rb} \qquad (2.25b)$$

Again, $\qquad P_c = \dfrac{RT_c}{V_c - b} e^{-\frac{a}{RT_c V_c}}$

Putting the values of V_c and T_c in the above equation, we get
$$P_c = \frac{a}{4b^2 e^2} \qquad (2.25c)$$

Critical coefficient $= \dfrac{RT_c}{P_c V_c} = \dfrac{1}{2} e^2 = 3.695$ $\qquad$ (2.26)

2.8.2 Boyle Temperature

At Boyle temperature the following are true
$$\left[\frac{\partial (PV)}{\partial P} \right]_T = 0 \text{ and } \left(\frac{\partial V}{\partial P} \right)_T \neq 0$$

$$P = \frac{RT}{(V-b)} e^{-\frac{a}{RTV}}. \text{ Or, } PV = \frac{RTV}{(V-b)} e^{-\frac{a}{RTV}}$$

$$\left[\frac{\partial (PV)}{\partial P} \right]_T = RT \left[\left(\frac{1}{V-b} - \frac{V}{(V-b)^2} \right) e^{-\frac{a}{RTV}} + \left(\frac{V}{V-b} \right) \left(\frac{a}{RTV^2} \right) e^{-\frac{a}{RTV}} \right] \left(\frac{\partial V}{\partial P} \right)_T$$

$$= RTe^{-\frac{a}{RTV}} \left[\frac{a}{RTV(V-b)} - \frac{b}{(V-b)^2} \right] \left(\frac{\partial V}{\partial P} \right)_T$$

At Boyle temperature $T = T_B$, $\left[\dfrac{\partial (PV)}{\partial P}\right]_T = 0$

Hence, $\dfrac{a}{RT_B V(V-b)} - \dfrac{b}{(V-b)^2} = 0$. Or, $RT_B = \dfrac{a(V-b)}{bV}$

At Boyle temperature, pressure is very low and volume is very large. Thus $(V-b) \approx V$.

Hence, $$T_B = \dfrac{a}{Rb} \tag{2.27}$$

Comparing equations (2.25b) and (2.27), we get

$$\dfrac{T_B}{T_c} = 4 \tag{2.28}$$

2.9 BERTHELOT EQUATION OF STATE

The scientist Berthelot modified van der Waals equation using T term in calculating internal pressure. The modified equation is given below

$$\left(P + \dfrac{a}{TV^2}\right)(V-b) = RT \text{ for 1 mole of a real gas. Or, } P = \dfrac{RT}{(V-b)} - \dfrac{a}{TV^2}$$

$$\left(P + \dfrac{an^2}{TV^2}\right)(V - nb) = nRT \text{ for } n \text{ moles of a real gas}$$

2.9.1 Critical Coefficient

At the critical point of PV–isotherm of any real gas, we know

$$\left(\dfrac{\partial P}{\partial V}\right)_T = 0 \text{ and } \left(\dfrac{\partial^2 P}{\partial V^2}\right)_T = 0$$

Expressions of $\left(\dfrac{\partial P}{\partial V}\right)_T$ and $\left(\dfrac{\partial^2 P}{\partial V^2}\right)_T$ for the Berthelot equation for 1 mole of real gas are given as

$$\left(\dfrac{\partial P}{\partial V}\right)_T = -\dfrac{RT}{(V-b)^2} + \dfrac{2a}{TV^3} \text{ and } \left(\dfrac{\partial^2 P}{\partial V^2}\right)_T = \dfrac{2RT}{(V-b)^3} - \dfrac{6a}{TV^4}$$

Using the critical point conditions in the Berthelot equation for 1 mole of real gas, we have

$$\frac{2a}{T_c V_c^3} = \frac{RT_c}{(V_c - b)^2} \quad \text{and} \quad \frac{3a}{T_c V_c^4} = \frac{RT_c}{(V_c - b)^3} \tag{2.29a}$$

Combining above equations we get

$$\frac{2}{3} V_c = (V_c - b). \text{ Or, } V_c = 3b \tag{2.29b}$$

Putting the value of V_c in the equation (2.29a), we get

$$T_c^2 = \frac{8a}{27 Rb}. \text{ Or, } RT_c^2 = \frac{8a}{27b} \tag{2.29c}$$

At the critical point, Berthelot equation can be written as

$$P_c = \frac{RT_c}{(V_c - b)} - \frac{a}{T_c V_c^2}$$

Using equations (2.29b) and (2.29c), we get

$$\frac{P_c}{RT_c} = \frac{1}{(V_c - b)} - \frac{a}{RT_c^2 V_c^2} = \frac{1}{2b} - \frac{3}{8b} = \frac{1}{8b} \tag{2.29d}$$

$$\left(\frac{P_c}{RT_c}\right) \times (RT_c^2) = \left(\frac{1}{8b}\right) \times \left(\frac{8a}{27b}\right).$$

Or,

$$P_c T_c = \frac{a}{27b^2} = \frac{a}{3V_c^2}. \text{ Or, } a = 3P_c V_c^2 T_c$$

$$b = \frac{V_c}{3}, \quad a = 3P_c V_c^2 T_c, \quad \frac{a}{R} = \frac{27}{8} bT_c^2 = \frac{9}{8} V_c T_c^2, \quad \frac{b}{R} = \frac{T_c}{8P_c} \tag{2.29e}$$

$$\text{Critical coefficient} = \frac{RT_c}{P_c V_c} = \frac{8}{3} = 2.667 \tag{2.29f}$$

According to Berthelot, the term b represents volume unavailable for molecular motion and is equal to volume occupied by 1 mole of liquid super-cooled to $0K$. Thus, $b = V_0$. Mean density of the liquid and vapour mixture may be calculated using Cailletet-Mathias curve. Density of vapour is ρ_v and that of liquid is ρ_l. Hence mean density is given by:

$$\rho_m = \frac{\rho_l + \rho_v}{2}$$

At $0K$, $\rho_v = 0$ and $\rho_l = \rho_l^0 = 2\rho_m$. The mean density ($\rho_m$) can be calculated from the curve through extrapolation to $0K$ and hence ρ_l^0 is known and hence also V_0. It has been found that $V_c = 4V_0$.

Thus, $$b = V_0 = \frac{V_c}{4}$$ (2.30a)

So, put $V_c = \frac{4}{3} V_c$ in the equation (2.29e).

$$b = \frac{V_c}{4}, \quad \frac{a}{R} = \frac{9}{8} V_c T_c^2 = \frac{9}{8} \left(\frac{4}{3} V_c \right) T_c^2$$

Or, $$\frac{a}{R} = \left(\frac{9}{8} \right)\left(\frac{4}{3} \right)(4b)T_c^2 = 6bT_c^2. \text{ Or, } T_c^2 = \frac{a}{6Rb}$$

$$\frac{RT_c}{P_c V_c} = \frac{8}{3}. \quad \text{Put } V_c = \frac{4}{3} V_c.$$

Hence, $$\frac{RT_c}{P_c \left(\frac{4}{3} V_c \right)} = \frac{8}{3}. \quad \text{Or, } \frac{RT_c}{P_c V_c} = \frac{32}{9} = 3.556$$

$$R = \left(\frac{32}{9} \right)\left(\frac{P_c V_c}{T_c} \right). \quad \text{So, } \frac{b}{R} = \frac{9}{128} \left(\frac{T_c}{P_c} \right)$$

$$b = \frac{V_c}{4}, \quad \frac{a}{R} = 6bT_c^2, \quad \frac{b}{R} = \frac{9}{128} \left(\frac{T_c}{P_c} \right) \text{ and } \frac{RT_c}{P_c V_c} = 3.556 \quad (2.30b)$$

2.9.2 Reduced Equation of State

For 1 mole of Berthelot gas we have

$$\left(P + \frac{a}{TV^2} \right)(V - b) = RT. \text{ Or, } PV - Pb + \frac{a}{TV} - \frac{ab}{TV^2} = RT$$

Or, $$PV = RT + Pb - \frac{a}{TV} + \frac{ab}{TV^2} = RT + Pb - \frac{a}{TV}$$

(neglecting the higher term)

Or, $$PV = RT\left[1 + \frac{Pb}{RT} - \frac{a}{(TV)(RT)} \right] = RT\left[1 + \frac{Pb}{RT} - \left(\frac{1}{VT^2} \right)\left(\frac{a}{R} \right) \right]$$

Or, $$PV = RT\left[1 + \frac{Pb}{RT} - \left(\frac{1}{VT^2} \right)(6bT_c^2) \right]$$

$$= RT\left[1 + \frac{Pb}{RT} - \left(\frac{P}{RT^3} \right)(6bT_c^2) \right]\left(\text{Putting } V = \frac{RT}{P} \right)$$

Or

$$PV = RT\left[1 + \frac{Pb}{RT}\left\{1 - 6\left(\frac{T_c}{T}\right)^2\right\}\right]$$

Put $P = \pi P_c$, $V = \phi V_c$ and $T = \theta T_c$ in the above equation we get

$$\pi \phi P_c V_c = R\theta T_c\left[1 + \frac{b\pi P_c}{R\theta T_c}\left(1 - \frac{6}{\theta^2}\right)\right].$$

Or,

$$\pi\phi = \theta\left(\frac{RT_c}{P_c V_c}\right)\left[1 + \left(\frac{\pi}{\theta}\right)\left(\frac{P_c}{T_c}\right)\left(\frac{9}{128}\frac{T_c}{P_c}\right)\left(1 - \frac{6}{\theta^2}\right)\right]$$

Or

$$\pi\phi = \frac{32}{9}\theta\left[1 + \frac{9}{128}\left(\frac{\pi}{\theta}\right)\left(1 - \frac{6}{\theta^2}\right)\right]$$

Or,

$$\pi\phi = \frac{32}{9}\theta + \frac{\pi}{4}\left(1 - \frac{6}{\theta^2}\right) \tag{2.31}$$

Equation (2.31) is the reduced equation of state for a real gas following Berthelot equation.

2.9.3 Boyle Temperature

For 1 mole of real gas Berthelot equation is

$$\left(P + \frac{a}{TV^2}\right)(V - b) = RT. \quad \text{Or,} \quad P = \frac{RT}{V - b} - \frac{a}{TV^2}$$

Or,

$$PV = \frac{RTV}{V - b} - \frac{a}{TV}$$

Differentiating the above equation with respect to P at constant T, we get

$$\left[\frac{\partial(PV)}{\partial P}\right]_T = RT\left[\frac{1}{V - b} - \frac{V}{(V - b)^2} + \frac{a}{TV^2}\right]\left(\frac{\partial V}{\partial P}\right)_T$$

$$= \left[\frac{a}{TV^2} - \frac{b}{(V - b)^2}\right]\left(\frac{\partial V}{\partial P}\right)_T$$

At Boyle temperature, $T = T_B$, $\left(\dfrac{\partial V}{\partial P}\right)_T \neq 0$ and $\left[\dfrac{\partial(PV)}{\partial P}\right]_T = 0$.

Thus, the above equation becomes

$$\frac{a}{T_B V^2} = \frac{b}{(V - b)^2}$$

Again, at $T = T_B$, $(V - b) \approx V$. Hence, the above equation becomes

$$T_B = \frac{a}{b} \qquad (2.32a)$$

From equation (2.30b), we have

$$T_c^2 = \frac{a}{6Rb}. \text{ Or, } T_c = \sqrt{\frac{a}{6Rb}}$$

Hence, $\qquad \dfrac{T_B}{T_c} = \sqrt{\dfrac{6aR}{b}} \qquad (2.32b)$

2.10 KAMERLINGH-ONNES EQUATION (THE VIRIAL EQUATION)

Kamerlingh-Onnes (1901) proposed the following equation for any real gas

$$PV = RT\left(1 + \frac{B}{V} + \frac{C}{V^2} + \cdots\right) = RT(1 + B'P + C'P^2 + \cdots)$$

B, B' and C, C', etc. are all function of temperature and called virial coefficients.

$$\frac{B}{V} \gg \frac{C}{V^2} \text{ and } B'P \gg C'P^2$$

Thus, neglecting higher term we get

$$PV = RT(1 + B'P) \text{ and } PV = RT\left(1 + \frac{B}{V}\right) \qquad (2.33a)$$

Hence, $\qquad B'P = \dfrac{B}{V}. \text{ Or, } B = B'PV = B'RT$

Or, $\qquad\qquad B = B'RT \qquad (2.33b)$

At Boyle temperature, all the virial coefficients vanish and $PV = RT$

2.10.1 Determination of *B*

Consider the following virial equation neglecting the higher terms

$$PV = RT(1 + B'P) = RT + B'RTP$$

Again, $\qquad\qquad V = \dfrac{M}{\rho}, M = \text{Molar mass and } \rho = \text{Density}$

Putting the value of V in the equation (2.33a), we get

$$\frac{PM}{\rho} = RT + B'RTP.$$

Or,
$$\frac{P}{\rho} = \frac{RT}{M} + \left(\frac{B'RT}{M}\right)P = \frac{RT}{M} + \left(\frac{B}{M}\right)P \quad [\text{As,} \quad B = B'RT]$$

Thus, the plot of $\left(\dfrac{P}{\rho}\right)$ versus P results a straight line with slope $\left(\dfrac{B}{M}\right)$. Hence, both B and B' can be calculated.

2.10.2 Van der Waals Equation in the form of Virial Equation

$$\left(P + \frac{a}{V^2}\right)(V - b) = RT.$$

Or,
$$P = \frac{RT}{(V - b)} - \frac{a}{V^2}. \text{ Or, } PV = RT\left(\frac{V}{V - b}\right) - \frac{a}{V}$$

Or,
$$PV = RT\left(1 - \frac{b}{V}\right)^{-1} - \frac{a}{V} = RT\left[1 + \frac{b}{V} + \left(\frac{b}{V}\right)^2 + \cdots\right] - \frac{a}{V}$$

Or,
$$PV = RT\left[1 + \left(b - \frac{a}{RT}\right)\frac{1}{V} + \left(\frac{b}{V}\right)^2 + \cdots\right]$$

Neglecting the higher term, we have

$$PV = RT\left[1 + \left(b - \frac{a}{RT}\right)\frac{1}{V}\right] \tag{2.34a}$$

Comparing the equation (2.34a) with the virial equation (2.33a), we get

$$B = \left(b - \frac{a}{RT}\right) \tag{2.34b}$$

At the Boyle temperature $B = 0$. Hence,

$$b - \frac{a}{RT_B} = 0. \text{ Or, } T_B = \frac{a}{Rb} \tag{2.34c}$$

2.10.3 Dieterici Equation in the form of Virial Equation

$$P = \left(\frac{RT}{V - b}\right)e^{-\frac{a}{RTV}}$$

Or,
$$PV = RT\left(\frac{V}{V-b}\right)e^{-\frac{a}{RTV}} = RT\left(1-\frac{b}{V}\right)^{-1} e^{-\frac{a}{RTV}}$$

Or
$$PV = RT\left[1+\frac{b}{V}+\left(\frac{b}{V}\right)^2 +\cdots\right]$$

$$\left[1-\frac{a}{RTV}+\frac{1}{2!}\left(\frac{a}{RTV}\right)^2 -\frac{1}{3!}\left(\frac{a}{RTV}\right)^3 +\cdots\right]$$

Or,
$$PV = RT\left[1+\left(b-\frac{a}{RT}\right)\frac{1}{V}+\left(b^2 -\frac{ab}{RT}+\frac{a^2}{2!R^2T^2}\right)\frac{1}{V^2}+\cdots\right]$$

Neglecting the higher terms, we get

$$PV = RT\left[1+\left(b-\frac{a}{RT}\right)\frac{1}{V}\right] \tag{2.35a}$$

Comparing the equation (2.35a) with the virial equation (2.33a), we get

$$B = \left(b-\frac{a}{RT}\right) \tag{2.35b}$$

At the Boyle temperature $B = 0$. Hence,

$$b-\frac{a}{RT_B}=0. \quad \text{Or,} \quad T_B = \frac{a}{Rb} \tag{2.35c}$$

2.10.4 Berthelot Equation in the form of Virial Equation

$$PV = RT\left[1+\frac{Pb}{RT}\left\{1-6\left(\frac{T_c}{T}\right)^2\right\}\right]$$

where, $\dfrac{b}{R} = \dfrac{9}{128}\left(\dfrac{T_c}{P_c}\right)$ [From equation (2.30b)]

Hence,
$$PV = RT\left[1+\frac{9T_c}{128P_cT}\left\{1-6\left(\frac{T_c}{T}\right)^2\right\}P\right] \tag{2.36a}$$

Comparing the equation (2.36a) with the virial equation, we get

$$B' = \frac{9T_c}{128P_cT}\left\{1-6\left(\frac{T_c}{T}\right)^2\right\}$$

Or,
$$B = B'RT = \frac{9RT_c}{128P_c}\left\{1 - 6\left(\frac{T_c}{T}\right)^2\right\} \qquad (2.36b)$$

At the Boyle temperature $B = 0$. Again, from equation (2.30b), we have

$$T_c^2 = \frac{a}{6Rb}$$

Thus, equation (2.36b) becomes

$$1 - 6\left(\frac{T_c}{T_B}\right)^2 = 0. \quad \text{Or, } T_B^2 = 6T_c^2 = 6\left(\frac{a}{6Rb}\right) = \frac{a}{Rb} \qquad (2.36c)$$

SOLVED PROBLEMS

Problem 2.1   1 mole carbon dioxide gas at $373K$ occupies 536 ml at 50 atm pressure. Calculate % deviation of pressure using: (i) Ideal gas equation, and (ii) Van der Waals equation.

Given: Van der Waals constants for carbon dioxide $a = 3.61\ l^2.\text{atm/mole}^2$,

$$b = 0.0428\ l/\text{mole}$$

Solution

$$R = 0.082\ \text{lit. atm/mole. } K$$

(i) Using the Ideal gas equation $V = 0.536\ l,\ n = 1$ mole, $T = 373\ K$

$$P = \frac{nRT}{V} = 57.1\ \text{atm}$$

Actual pressure $= 50$ atm. % deviation $= \left(\frac{7.1}{50}\right) \times 100 = 14.2\%$

(ii) Using van der Waals equation

$$P = \frac{RT}{(V - nb)} - \frac{a}{V^2} = 49.6\ \text{atm. } \% \text{deviation} = \left(\frac{0.4}{50}\right) \times 100 = 0.8\%$$

Problem 2.2   Predict which of the substances has: (i) the smallest van der Waals "a" constant, and (ii) the largest "b" constant.

$$NH_3, N_2, CH_2Cl_2, Cl_2 \text{ and } CCl_4.$$

Solution

(i) In case of N_2 and Cl_2, covalent bonds are non-polar and hence attraction forces are minimum compared to the other molecules. Again, van der Waals attraction forces increases with number of electrons. The total number of electrons is less in N_2 than Cl_2. The van der Waals constant a depends on the intensity of van der Waals forces. Thus, the value of a is the smallest for N_2.

(ii) The van der Waals constant b is related to the molar volume of the molecule. For larger size molecule molar volume is high and consequently the value of b is also large. Among the molecules given above, CCl_4 is largest in size and hence the value of b is largest for CCl_4.

Problem 2.3

(i) Using van der Waals equation, calculate the temperature of 20 moles of helium in a 10 litre cylinder at 120 atm pressure.

Given: Van der Waals constants for helium: $a = 0.0341\ l^2.\text{atm/mole}^2$, $b = 0.0237\ l/\text{mole}$

(ii) Compare this value with the temperature calculated from the ideal gas equation.

Solution

(i) The van der Waals equation is given by

$$\left(P + \frac{an^2}{V^2}\right)(V - nb) = nRT$$

$P = 120$ atm, $n = 20$, $V = 100$ lit. $R = 0.0821$ (lit.atm)/(mole.K), $T = 696$K
[Note that the correction to P is not significant as T is well above the temperature at which helium gas will liquefy while the volume correction is significant due to the high pressure.]

(ii) The Ideal gas equation is given by
$PV = nRT$. By putting the values of P, V, n and R we get, $T = 731K$.

Problem 2.4

Derive expressions for the expansion coefficient α and the isothermal compressibility k for a real gas that obeys an equation of state given by $P(V - b) = RT$ for one mole of the gas.

Solution

$$P(V - b) = RT. \quad \text{Or,} \quad V = \frac{RT}{P} + b \quad \text{for 1 mole}$$

$$V = \frac{nRT}{P} + nb = \frac{nRT + nbP}{P} \quad \text{for } n \text{ moles}$$

$$\alpha = \frac{1}{V}\left(\frac{\partial V}{\partial T}\right)_P \quad \text{and} \quad k = -\frac{1}{V}\left(\frac{\partial V}{\partial P}\right)_T$$

$$\left(\frac{\partial V}{\partial T}\right)_P = \frac{nR}{P} \,, \quad \left(\frac{\partial V}{\partial P}\right)_T = -\frac{nRT}{P^2}$$

$$\alpha = \left(\frac{P}{nRT + nbP}\right)\left(\frac{nR}{P}\right) = \frac{1}{T}\left[\frac{1}{1 + \dfrac{bP}{RT}}\right]$$

$$k = \left(\frac{P}{nRT + nbP}\right)\left(\frac{nRT}{P^2}\right) = \frac{1}{P}\left(\frac{RT}{RT + bP}\right) = \frac{1}{P}\left[\frac{1}{1 + \dfrac{bP}{RT}}\right]$$

Problem 2.5   What is a cryogenic liquid? Describe three uses of cryogenic liquids.

Solution

Cryogenic liquids are also known as cryogens. They are gases at normal temperature and pressure. However, these gases can be converted to liquid by compression at low temperatures. These liquids are extremely cold and have boiling points less than $-100°C$. The pressure and temperature at which a gas is converted to liquid, are called cryogenic parameters. The cryogenic parameters are different for different gases. The vapours and gases released from cryogenic liquids are very cold. They often condense the moisture in air, creating a highly visible fog. The cryogenic liquid is extremely cold and small amount of this liquid can expand into very large volume of gas.

Types of cryogenic fluids

- **Inert Gases:** Inert gases do not react chemically to any great extent. They do not burn or support combustion. Examples of this group are: nitrogen, helium, neon, argon, and krypton.
- **Flammable Gases:** Some cryogenic liquids produce a gas that can burn in air. The most common examples are: hydrogen, methane, carbon monoxide, and liquefied natural gas.
- **Oxygen:** Many materials considered as non-combustible can burn in the presence of liquid oxygen. Organic materials can react explosively with liquid oxygen. The hazards and handling precautions of liquid oxygen must therefore be considered separately from other cryogenic liquids.

Applications

Cryogenic systems are the systems which are at a very low temperature region. Generally, a temperature below 123 K is considered as cryogenic systems. There are many applications for cryogenics today. Some of them are:

- Rocket propulsion
- Electronics
- High energy physics
- Space simulation
- Manufacturing process
- Recycling of materials

EXERCISES

1. A 1 litre of a real gas is compressed from 300 atm and 473K to 600 atm and 273K. The compressibility factors are 1.072 and 1.375 respectively for initial and final states. Calculate the final volume of the gas. [*Ans.:* 370.1 c.c]

2. The value of *b* of a gas is 442 lit/mole. Calculate the minimum distance between the centres of two molecules, approaching each other.

3. The normal density of oxygen is 0.01429 gm/c.c. and its critical pressure and temperature are 55 atm and 155K respectively. Calculate the molecular weight of oxygen using Berthelot equation. [*Ans.:* 31.97]

4. For an ideal gas, the product of pressure and volume should be constant, regardless of the pressure. Experimental data for methane, however, show that the value of PV decreases significantly over the pressure range 0 to 120 atm at 0°C. The decrease in PV over the same pressure range is much smaller at 100°C. Explain why PV decreases with increasing temperature. Why is the decrease less significant at higher temperatures?

5. What is the effect of intermolecular forces on the liquefaction of a gas? At constant pressure and volume, does it become easier or harder to liquefy a gas as its temperature increases? Explain your reasoning. What is the effect of increasing the pressure on the liquefaction temperature?

6. Describe qualitatively what *a* and *b*, the two empirical constants in the van der Waals equation, represent.

7. In the van der Waals equation, why is the term that corrects for volume negative and the term that corrects for pressure positive? Why is the term n/V squared in van Der Waals equation?

8. Liquefaction of a gas depends strongly on two factors. What are they? As temperature is decreased, which gas will liquefy first—ammonia, methane, or carbon monoxide? Why?

9. How can gas liquefaction facilitate the storage and transport of fossil fuels? What are potential drawbacks to these methods?

10. The van der Waals constants for xenon are: $a = 4.19$ lit^2. atm/mole^2 and $b = 0.0510$ lit/mole. If a 0.250 mole sample of xenon in a container with a volume of 3.65 lit. is cooled to –90°C, what is the pressure of the sample assuming ideal gas behaviour? What would be the *actual* pressure under these conditions?

11. The van der Waals constants for water vapour are: $a = 5.46$ lit^2. atm/mole^2 and $b = 0.0305$ lit/mole. If a 20 gm sample water in a container with a volume of 5 lit. is heated to 120°C, what is the pressure of the water vapour? Assume that water vapour behaves ideally. What would be the actual pressure under these conditions?

12. Determine the pressure in atm exerted by 1 mole of methane placed into a bulb with a volume of 244.6 ml at 25°C. Carry out two calculations: in the first calculation, assume that methane behaves as an ideal gas; in the second calculation, assume that methane behaves as a real gas and obeys the van der Waals equation.

    [*Ans.:* 100 atm; 82.9 atm]

    Given $a = 2.303$ lit^2. atm/mole^2, $b = 0.0431$ lit/mole.

13. Determine the molar volume of propane at 600K and 91 atm using: (a) the ideal gas equation, and (b) the Van der Waals equation.

    Given $a = 9.39$ lit^2. atm/mole^2, $b = 0.0905$ lit/mole

14. Use the virial equation to determine the pressure in atm of 1 mole of carbon dioxide gas contained in a volume of 5 lit. at 300K. Compare your result to the pressure that would have been obtained from the ideal gas equation.

    Given second virial coefficient B is -126 cm^3/mole   [*Ans.:* 4.80 atm; 4.92 atm]

15. By definition, the compression factor, Z, for an ideal gas equals 1. At room temperature and 1 atm pressure, by approximately what percentage does this change for diatomic nitrogen upon inclusion of the second virial coefficient term? Recalculate this result for water vapour.

    Given at $300K$, the second virial coefficient B is -3.91 cm^3/mole for diatomic nitrogen and -1126 cm^3/mole for water vapour.   [*Ans.:* 0.016%, 4.57%]

16. 2 moles of ammonia is enclosed in a 5 litre flask at 27°C. Calculate the pressure of ammonia using van der Waals equation. Given $a = 4.17$ lit^2. atm/mole^2, $b = 0.037$ lit/mole.

17. Using the Van der Waals equation calculate the pressure of 88 gms of gaseous carbon dioxide occupying a volume of 8 litres at 27°C. Compare the result with the pressure evaluated using ideal gas equation. Given $a = 3.6$ lit^2. atm/mole^2, $b = 0.043$ lit/mole.

    [*Ans.:* 5.993 atm and 6.15 atm]

18. 10 moles of ethylene are compressed to 4.86 litres at 3000K.
    (a) Find the pressure of the gas using van der Waals equation.
    (b) Compare it with the pressure obtained using the ideal gas equation.
    Given $a = 4.49$ lit^2. atm/mole^2, $b = 0.0573$ lit/mole.

3 The Liquid State

3.1.1 Concept

Let us consider a homogeneous liquid. Each and every molecule in the interior of the liquid (A in the Fig. 3.1) experiences intermolecular attraction forces from all directions so that the resultant force is zero. However, each and every molecule present on the liquid surface (B in the Fig. 3.1) is attracted inward direction and along the surface only. As there is no liquid molecules beyond the surface there is only air-liquid adhesive force existing in the upward direction. Forces along the surface are acting along two opposite directions and cancelled out. So there is liquid-liquid adhesive force, existing inward direction. As liquid-liquid adhesive force is much stronger than air-liquid adhesive force, there is a resultant force, existing along inward direction, as a result of which, liquid molecules on the surface are attracted towards the centre, creating a packed layer of molecules, acting as thin film. Thus, the liquid surface remains in tension, known as surface tension, which is same at every point and in all directions along the surface of the liquid considered. It is well illustrated in the Fig. 3.1.

Figure 3.1 Development of surface tension in liquid

Surface tension of a liquid is defined as the force, acting perpendicularly to unit length of that liquid surface and is represented by the symbol γ. The unit of γ is N/m.

In order to extend the surface area of a liquid, some work is to be done against the surface tension. The amount of work required to extend the surface area by 1 m^2 is called surface energy with unit J/m^2. Surface energy is considered as the fundamental property of a liquid. When surface energy is supplied by an external source, the liquid is said to be bubbling.

In case of two immiscible liquids, surface tension existing between two heterogeneous surfaces is called interfacial tension.

Surface energy   For a given liquid it is the amount of work required to increase the surface area by 1 m^2. Its unit is J/m^2. In case of water, the value of surface energy is $72.8 \times 10^{-3}\ J/m^2$. Considering the unit, surface energy has same meaning as surface tension.

Surface tension values of some common liquids are given below:

Table 3.1   Surface tension values of some common liquids

Liquid sample	γ $(10^{-3}\ N/m)$	Liquid sample	γ $(10^{-3}\ N/m)$
Acetic acid	27.4	Mercury	476
Acetone	23.3	Methyl alcohol	22.6
Aniline	42.9	n-Butyl alcohol	24.5
Benzaldehyde	40	n-Hexane	18.4
Benzene	28.9	Nitrobenzene	43.4
Carbon disulfide	32.3	n-Propyl alcohol	23.8
Carbon tetrachloride	26.8	o-Nitrotoluene	41.5
Chloroform	27.2	o-Xylene	30
Cyclohexane	25.5	p-Xylene	28.3
Ethyl acetate	23.8	Toluene	28.5
Ethyl alcohol	22.3	Water	72.8
Ether	16.96	n-Heptane	20.40
n-Pentane	16		

3.1.1.1 Spherical Shape of Liquid Drop

Liquid drop gains surface energy due to surface tension and hence it has a tendency to minimise this surface energy. As sphere has a minimum surface area, liquid drop always attains the spherical shape.

3.1.1.2 Capillary Action

When a capillary tube is immersed in a liquid, a contact angle is formed, due to which a component of surface tension is developed along the surface of capillary tube. This component of surface tension is responsible for rise in liquid along the capillary tube. The phenomenon is known as capillary action.

3.1.2   Liquid Droplet

Pressure inside the droplet is always greater than the pressure outside the droplet and this excess pressure is balanced by surface tension of the droplet.

Surface tension is acting parallel to the surface and hence the force due to surface tension is $\gamma \times 2\pi r$, where, r is the radius of the droplet.

Δp is the pressure difference between inside and outside the droplet and the force due to this excess pressure is $\Delta p \times \pi r^2$.

At equilibrium these two opposing forces are balanced. So, we can write

$$\gamma \times 2\pi r = \Delta p \times \pi r^2. \quad \text{Or,} \quad \Delta p = \frac{2\gamma}{r}$$

3.1.3   Soap Bubble

In this case air exists in both inside and outside the bubble. So, there are two surfaces. Hence, the force due to surface tension is $\gamma \times (2 \times 2\pi r)$, where, r is the radius of the soap bubble.

Δp is the pressure difference between inside and outside the soap bubble and the force due to this excess pressure is $\Delta p \times \pi r^2$.

At equilibrium these two opposing forces are balanced. So, we can write

$$\gamma \times 4\pi r = \Delta p \times \pi r^2. \quad \text{Or,} \quad \Delta p = \frac{4\gamma}{r} \tag{3.1}$$

3.1.4   Determination of Surface Tension

3.1.4.1   Drop Volume Method or Stalagmometric Method

In this method, stalagmometer is used. The glass capillary of the equipment is filled with a given liquid. Liquid drops are leaking out of the glass capillary of the stalagmometer. Number of drops are counted, collected and weighed. So, weight of each drop of the liquid is known.

The drops are formed slowly at the tip of the glass capillary, which is placed in a vertical direction. The pendant drop at the tip starts detaching when its weight (or volume) is balanced by the surface tension of the liquid. The weight (or volume) is dependent on the characteristics of the liquid. This method was first described in 1864 by Tate who formed an equation, which is now called the Tate's law

$$w = 2\pi r\gamma \tag{3.2a}$$

w = weight of the drop, r = radius of capillary, and γ = surface tension of the liquid.

Two cases are possible. In one case liquid wets the stalagmometer tip and in that case r is the outer radius of the capillary. In the other case, liquid does not wet the stalagmometer tip and in that case r is the inner radius of the capillary.

In actual practice a drop is formed at the stalagmometer tip and due to gravity shape of the drop changes. Ultimately the drop gets detached from the stalagmometer tip, leaving behind a fraction of volume at the stalagmometer tip. Fraction of drop volume, left on the tip may be as high as 40%. Thus a correction factor, f, has been introduced and the modified form of Tate's law is

$$w = 2\pi r f \gamma, \text{ where, } w = mg \text{ and } f = \frac{w'}{w} \tag{3.2b}$$

where, w and w' are the actual weight of the drop and weight of the falling drop respectively, m is the mass of the liquid drop.

According to Harkins and Brown, the factor, f, is a function of drop radius, r, and drop volume, V. The expression of f is given in the equation (3.3)

$$f = g\left(\frac{r}{V^{\frac{1}{3}}}\right) \tag{3.3}$$

Water and benzene are taken as reference liquids. Using several capillary tubes with different radii γ can be determined by capillary rise method. So, corresponding f values are calculated using above equation. The result is shown in the Table 3.2.

Table 3.2 f values at different capillary radii

$r/V^{\frac{1}{3}}$	f	$r/V^{\frac{1}{3}}$	f
0.00	1.000	0.95	0.6034
0.30	0.7256	1.00	0.6098
0.35	0.7011	1.05	0.6179
0.40	0.6828	1.10	0.6280
0.45	0.6669	1.15	0.6407
0.50	0.6515	1.20	0.6535
0.55	0.6362	1.25	0.6520
0.60	0.6250	1.30	0.6400
0.65	0.6171	1.35	0.6230
0.70	0.6093	1.40	0.6030
0.75	0.6032	1.45	0.5830
0.80	0.6000	1.50	0.5670
0.85	0.5992	1.55	0.5510
0.90	0.5998		

It may be observed from the Table 3.2 that the value of f changes only slightly when the value of $\left(r/V^{\frac{1}{3}}\right)$ lies within 0.6–1.2.

Thus, surface tension, γ, can be calculated using equation (3.4)

$$\gamma = \frac{mg}{2\pi r f} = \frac{V \cdot \rho \cdot g}{k} \tag{3.4}$$

For a particular stalagmometer, $k = 2\pi r f = $ known ρ is the density of the liquid and known. Thus by measuring drop volume, V, surface tension, γ, can easily be calculated using above equation.

In another approach γ of an unknown liquid can be calculated with respect to a reference liquid of known γ using the following formula assuming the factor, f, is constant.

$$\frac{\gamma_2}{\gamma_1} = \frac{V_2 \rho_2}{V_1 \rho_1} \tag{3.5}$$

However, f is not a constant. It varies from liquid to liquid. In that case the above formula is modified as below:

$$\frac{\gamma_2}{\gamma_1} = \left(\frac{m_2}{m_1}\right)^{2/3} \left(\frac{d_2}{d_1}\right)^{1/3} \tag{3.6}$$

where, m_1 and m_2 are the masses of drops of two different liquids and d_1 and d_2 are the diameters of drops of two different liquids. The equation (3.6) can be used with 99.9% accuracy.

3.1.4.2 Capillary Rise Method

If a thin capillary tube is immersed in a liquid, two cases may arise:

(i) The adhesive force between capillary wall and liquid is much stronger than cohesive force among liquid molecules. In that case, the liquid is moving up along the capillary, forming a concave meniscus as shown in the Fig. 3.2(a).

(ii) The cohesive force among liquid molecules is much stronger than adhesive force between capillary wall and liquid molecules. In that case the liquid is moving down the capillary, forming a convex meniscus as shown in the Fig. 3.2(b).

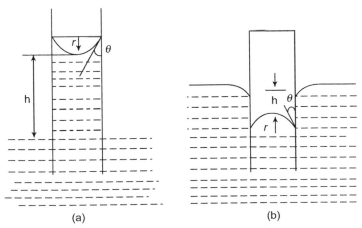

(a)   (b)

Figure 3.2 (a) Capillary rise with concave meniscus, (b) Capillary fall with convex meniscus

$h = $ Rise or fall of the capillary, $\theta = $ Contact angle and $r = $ Radius of the capillary. Cross-sectional area of the capillary is circular with a very small radius r. Then the meniscus is hemi-sphere with same radius r. If the liquid does not wet, the r value is the inner radius of the capillary.

Periphery of the semi-circular meniscus is πr and the surface tension acting along the wall is $\gamma \cos \theta$. As there are two walls, net force due to surface tension is $f_1 = 2\pi r \gamma \cos \theta$

Excess pressure due to surface tension is Δp. Hence, gravitational force is given by

$$\text{force} = f_2 = \pi r^2 \times \Delta p$$

At equilibrium, $f_1 = f_2$ and hence $2\pi r \gamma \cos \theta = \pi r^2 \times \Delta p = \pi r^2 \times (h\rho g)$

where, $\Delta p = h\rho g$ and $\rho =$ density of the liquid.

Or
$$\gamma = \frac{rh\rho g}{2\cos \theta} = \frac{rh\rho g}{2} \tag{3.7}$$

since $\theta \approx 0°$ and $\cos \theta \approx 1$

3.1.5 Applications of Surface Tension

1. Determination of solid surface energy

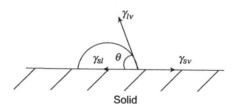

Figure 3.3   Contact angle and surface energy

When a liquid drop takes a spherical shape on solid surface, three types of interfacial tension are acting with a contact angle θ such that net force at the contact point is zero.
$\gamma_{sv} =$ Force due to interfacial tension between solid and air.
$\gamma_{lv} =$ Force due to interfacial tension between liquid and air.
$\gamma_{sl} =$ Force due to interfacial tension between solid and liquid.
At equilibrium, using force balance we get

$$\gamma_{sv} = \gamma_{sl} + \gamma_{lv} \cos \theta \tag{3.8}$$

The above equation was first established by Thomas Young in 1850.

Spreading of liquid on a solid surface depends on work of spreading, W_{sp}, which is given by

$$W_{sp} = \gamma_{sv} - (\gamma_{sl} + \gamma_{lv}) \tag{3.9}$$

The liquid drop will spread if $W_{sp} > 0$ or $\gamma_{sv} > (\gamma_{sl} + \gamma_{lv})$. If $W_{sp} < 0$ or $\gamma_{sv} < (\gamma_{sl} + \gamma_{lv})$, the liquid drop will not spread but maintains spherical shape at a definite contact angle.

2. Capillary action, e.g., rising of oil in the wick of a lamp.
3. Use of soaps and detergent for cleaning clothes.
4. Use of oily substances to set hairs.

5. Flow in capillaries and porous media is sufficiently influenced by dynamic surface tension or DST. This property is capitalised in oil recovery unit.
6. The surface tension property of liquid is used in metal and textile processing, pulp and paper production, and pharmaceutical formulations.
7. An important application of dynamic surface tension is observed in human body, where lung surfactants, where the dynamic surface tension under constant or pulsating area conditions, controls the health and stability of the alveoli.
8. In pesticide industries, the surface tension property is very important because DST plays an important role in dispersion of pesticide particles in aqueous layer. Surfactants are frequently used to spread the pesticide particles in the aqueous layer.

3.1.6 Effect of Temperature

With increase in temperature kinetic energy of molecules increases, resulting decrease in intermolecular forces of attractions. Hence, surface tension, which is based on intermolecular forces of attractions, also decreases. There are several empirical equations, describing the relation between surface tension and temperature. One of such empirical equation was given by the scientist Eötvös.

$$\gamma V^{2/3} = k(T_c - T) \tag{3.10}$$

where, V is the molar volume of the liquid, γ is the surface tension of the liquid, T_c is the critical temperature, T is the temperature of the experiment and k is a constant of value $2.1 \times 10^{-7} J. °K^{-1}.mole^{3/2}$. The scientist Eötvös assumed that surface tension of any liquid is zero at its critical point and has a linear relationship with temperature.

Ramsay and Shields modified the above equation into the following form:

$$\gamma V^{2/3} = k(T_c - T - 6) \tag{3.11}$$

where, the factor 6 is introduced for better fitness of the curve particularly at lower temperature.

Table 3.3   Change of surface tension of water with temperature

Temperature °C	γ_w $(10^{-3} N/m)$	Temperature (°C)	γ_w $(10^{-3} N/m)$	Temperature (°C)	γ_w $(10^{-3} N/m)$
10	74.22	18	73.05	26	71.82
11	74.07	19	72.90	27	71.66
12	73.93	20	72.25	28	71.50
13	73.78	21	72.59	29	71.35
14	73.64	22	72.44	30	71.18
15	73.49	23	72.28	40	69.56
16	73.34	24	72.13		
17	73.19	25	71.97		

3.1.7 Effect of Pressure

With increase in pressure entropy decreases and intermolecular forces of attractions increases, resulting increase in surface tension. However, the effect is small.

3.2 VISCOSITY OF LIQUID

3.2.1 Concept

A given volume of liquid is assumed to consisting of several layers. Strong cohesive force exists among the layers. On application of an external shear stress the layers of liquid start moving but the lowest layer cannot move due to strong frictional force offered by the surface. This frictional force is spread over the liquid layers, generating an inherent resistance to flow. This inherent resistance against liquid movement is known as viscosity of liquid. This viscous force or frictional force (f) between two successive liquid layers is proportional to the area of contact (A) between two layers as well as velocity gradient (dv/dy). Thus mathematically we can write that:

$$f \propto A\left(\frac{dv}{dy}\right). \text{ Or, } f = \eta A\left(\frac{dv}{dy}\right) \tag{3.12}$$

η is known as coefficient of viscosity, also known as shear viscosity, with unit kg/m.s or Pa.s. In cgs unit, η is expressed as g/cm.s, which is also known as poise (p).

$$1 \text{ poise} = 1\frac{\text{dyne.s}}{\text{cm}^2} = 1\frac{\text{gm}}{\text{cm.s}} = 10^{-1} \text{ Pa.s. } 1p = 100 \, cp$$

η is also known as absolute viscosity. Values of η of some fluids are listed below:

Table 3.4 η values of different liquids

Fluid	*η(Pa.s)*
Air	1.983×10^{-5}
Water	10^{-3}
Olive oil	10^{-1}
Glycerol	10^{0}
Liquid honey	10^{1}

3.2.2 Poiseuille Equation

Let us consider the liquid column as a cylinder of length l and radius r.

Area of the cylinder is $2\pi rl$. Value of $r = 0$ at the centre and $r = R =$ Radius of the pipe.

Frictional force $= f_d = \eta(2\pi rl)\left(\dfrac{dv}{dr}\right)$

Driving force $= f = \Delta p(\pi r^2)$

where, $\Delta p = (p_2 - p_1)$ is the pressure difference, where, $p_1 > p_2$

Under steady-state condition driving force is balanced by frictional force. So,

$$\Delta p(\pi r^2) = \eta(2\pi rl)\left(\frac{dv}{dr}\right). \ \text{Or,} \ \left(\frac{dv}{dr}\right) = \frac{\Delta p(\pi r^2)}{\eta(2\pi rl)} = \frac{\Delta p.r}{2\eta l}$$

As Δp is negative, (dv/dr) is also negative, indicating that velocity decreases as we move from the centre to the wall of the pipe. At the wall of the pipe velocity of the liquid is zero or $v = 0$.

$$\int_{v}^{0} dv = \frac{\Delta p}{2\eta l}\int_{r}^{R} rdr$$

On integration, we get

$$v = \frac{\Delta p}{4\eta l}(r^2 - R^2) \tag{3.13}$$

At the wall of the cylinder $\quad r = R, \quad v = 0$ and at $r = r, \quad v = v$

3.2.2.1 Liquid Flow Rate

Let us consider a thin circular element of radius r within the cross-sectional area of the cylinder and dr is the thickness of the element.

Thus, the area of the thin element is given by $2\pi rdr$.

Volumetric flow rate through this thin area $v(2\pi rdr)$, where, v is the velocity of the liquid.

Total volumetric flow rate across the cross-sectional area $= \dot{q} = \int_{0}^{R} v \cdot 2\pi rdr$

Using equation (3.13), we get

Or, $\quad \dot{q} = \int_{0}^{R}\frac{\Delta p}{4\eta l}(r^2 - R^2)2\pi rdr = \frac{\pi\Delta p}{2\eta l}\left[\int_{0}^{R} r^3 dr - \int_{0}^{R} R^2 \cdot rdr\right] = -\frac{\pi\Delta pR^4}{8\eta l}$

Or, $\quad |\dot{q}| = \dfrac{\pi|\Delta p|R^4}{8\eta l} \tag{3.14}$

The equation (3.14) is known as Poiseuille equation.

3.2.3 Viscosity Coefficients

Coefficient of viscosity can be expressed as follows:

Dynamic viscosity   It is also known as absolute viscosity. The usual unit is Pa–s. The other unit is Poise or p.

Kinematic viscosity   It is the ratio of dynamic viscosity and density. Usual unit is m^2/s, also known as Stokes or St.

Saybolt Universal Second (SUS) is another unit of kinematic viscosity and is measured by a Saybolt Universal Viscometer. Saybolt Universal Second is also known as Second Saybolt Universal or SSU.

3.2.4   Determination of Viscosity of a Liquid Sample

3.2.4.1   Determination of Liquid Viscosity by Ostwald Viscometer

For practical purpose, Ostwald viscometer (Fig. 3.4) is used to determine viscosity of a given liquid. The apparatus has a very fine capillary, where two marks are clearly visible.

V = Volume of liquid allowed to flow through the fine capillary of Ostwald viscometer in time t.

$$|\dot{q}| = \frac{V}{t}$$

L = Length of the liquid, i.e., distance between upper and lower marks.
t = Time of flow to cover the length L.

$|\Delta p| = h\rho g$, where, h is the height of capillary from the base, ρ is the density of the liquid and g is the acceleration due to gravity.

Thus, the equation (3.14) can be written as

$$\eta = \frac{\pi(h\rho g)r^4 t}{8VL}, r = \text{Radius of the capillary.}$$

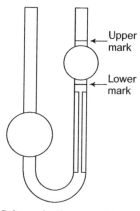

Figure 3.4   Schematic diagram of Ostwald viscometer

For a given viscometer, all the parameters on the RHS of the above equation are constant except, ρ and t. Hence the above equation can be written as $\eta \propto \rho.t$

Viscosity of a given liquid is measured against a reference liquid, preferably water. Thus, for reference liquid, we have $\quad \eta_0 \propto \rho_0.t_0$ $\qquad$ (3.15)

where, η_0 is known. ρ_0 and t_0 are measurable quantities.

For unknown liquid, we have $\quad \eta_s \propto \rho_s.t_s$ $\qquad$ (3.16)

Combining the above two equations, we have

$$\frac{\eta_s}{\eta_0} = \frac{\rho_s.t_s}{\rho_0.t_0} \qquad (3.17)$$

Only η_s is unknown and can be calculated from the equation (3.17).

3.2.4.2 Gas Viscosity

In case of gases volume depends on gas pressure and hence average pressure P_0 is used and the expression of P_0 is given by

$$P_0 = \frac{P_1 + P_2}{2}. \quad \text{Or,} \quad P_1 + P_2 = 2P_0$$

Again it is assumed that gases behave ideally. Hence, we can write that

$$P_0 V = nRT. \quad \text{Or,} \quad \frac{dn}{dt} = \left(\frac{P_0}{RT} \right) \frac{dV}{dt} = \left(\frac{P_0}{RT} \right) |\dot{q}|$$

Putting the value of $|\dot{q}|$ and P_0 in the above equation, we have

$$\frac{dn}{dt} = \left(\frac{P_0}{RT} \right) \left(\frac{\pi |\Delta p| R^4}{8\eta l} \right) = \left[\frac{(P_1 + P_2)}{2RT} \right] \left[\frac{(P_1 - P_2)R^4}{8\eta l} \right] \times \pi$$

$$= \pi \frac{(P_1^2 - P_2^2)R^4}{16RT\eta l} \quad \left[|\Delta p| = (P_1 - P_2) \right]$$

However, $P_1 \gg P_2$. Hence, $(P_1^2 - P_2^2) \approx P_1^2$.

Thus, mass flow rate of gas through capillary is given by

$$\frac{dn}{dt} = \frac{\pi P_1^2 R^4}{16RT\eta l} \qquad (3.18)$$

The equation (3.18) is the expression of viscosity of an ideal gas.

3.2.4.3 Determination of Viscosity by Falling Sphere Method

In this method, a closed container having tube A of small radius r is taken. Radius of the container is R. The tube A is filled with sample liquid. The other part of the container is filled with water. The container is equipped with a thermometer and an agitator to maintain a constant temperature. A solid sphere, preferably a steel ball is slowly inserted into the tube *A* through another small tube *B*. The ball is moving downwards, with an acceleration, *a*. The apparatus is shown in the Fig. 3.5.

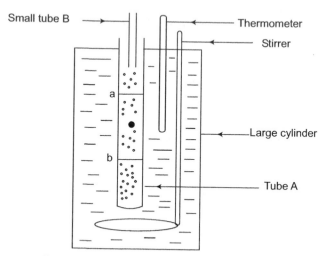

Figure 3.5   Apparatus for falling sphere method

The tube A has two marks *a* and *b*. The length *ab* is known. The time required by steel ball to cover the length *ab* is *t* sec and it is measured. Thus, settling velocity $\left(u = \dfrac{ab}{t} \right)$ can be calculated. Radius of the tube *A* is *R* and it is known.

$$\text{Forward force} = mg = \frac{4}{3}\pi r^3 \rho g$$

where, *r* is the radius of the steel ball, *m* is the mass of the steel ball and ρ is the density of the steel ball.

$$\text{Buoyant force} = \frac{4}{3}\pi r^3 \rho_l g$$

where, ρ_l is the density of the liquid and is known.

Drag force due to viscosity = $6\pi\eta r u$ (according to Stokes equation)
where, *u* is the velocity of the steel ball.

According to Newton's 2nd law of motion, we can write that

$$\frac{4}{3}\pi r^3 \rho g - \frac{4}{3}\pi r^3 \rho_l g - 6\pi\eta r u = ma$$

where, *a* is the acceleration of steel ball in downward direction.

Under steady state condition the forward and backward forces are exactly balanced and *a* = 0. Then, $u = u_s$ which is called settling velocity. Hence, we have

$$\eta = \frac{2}{9u_s} r^2 (\rho - \rho_l) g \tag{3.19}$$

Density of the steel ball (ρ) can be determined easily. Hence, viscosity of liquid, η can easily be calculated using the equation (3.19). However, the above equation is valid for laminar flow of liquid surrounding the steel ball, Reynolds Number, $R_e < 500$. R_e is calculated using the following formula:

$$R_e = \frac{2\rho_l u_s r}{\eta} \tag{3.20}$$

3.2.5 Viscosity of Liquid Blends

In case of a blend of two liquids, viscosity blending number (VBN) or viscosity blending index for each component is to be calculated first according to the following equation

$$VBN_1 = 14.534 \ln[\ln(v_1 + 0.8)] + 10.975 \tag{3.21a}$$

where, VBN_1 and v_1 are the VBN and kinematic viscosity of component 1 respectively.
Similarly for component 2, we have

$$VBN_2 = 14.534 \ln[\ln(v_2 + 0.8)] + 10.975 \tag{3.21b}$$

VBN of the blend can be calculated using the following formula

$$VBN_{blend} = [x_1 \times VBN_1] + [x_2 \times VBN_2] \tag{3.22}$$

Finally, viscosity of the blend, v_{blend} can be calculated using the following formula

$$VBN_{blend} = 14.534 \ln[\ln(v_{blend} + 0.8)] + 10.975 \tag{3.23}$$

3.2.6 Effect of Temperature on Liquid Viscosity

Intermolecular attractions among molecules are believed to be the main source of the drag force, viscosity. With increase in concentration intermolecular attraction forces increase and hence viscosity also increases. It is well illustrated in the Fig. 3.6.

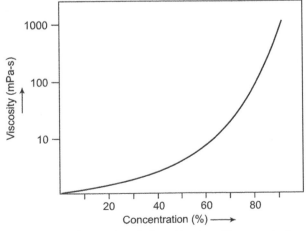

Figure 3.6 Change in solution viscosity with concentration

With increase in temperature thermal energy increases and hence viscosity decreases. At the same time density of liquid, i.e., ρ_1 also decreases as volume increases. In case of glycerin the change in density with temperature is shown in the Fig. 3.7.

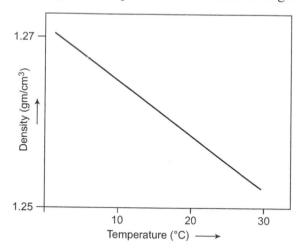

Figure 3.7   Change in density of glycerine with temperature

The dynamic viscosity (η) of a given liquid is related with temperature as follows

$$\eta = \eta_0 e^{-E/RT}, \quad \text{where, } E = \text{Activation energy}.$$

Taking ln on both sides, we get

$$\ln\eta = \ln\eta_0 - \frac{E}{RT} \tag{3.24}$$

Thus, plotting $\ln(\eta)$ versus $1/T$ results in a straight line with negative slope as shown in the Fig. 3.8.

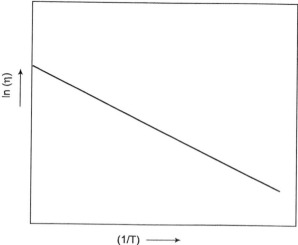

Figure 3.8   Change in viscosity with temperature

Viscosity of water decreases from 1.79 *cP* to 0.28 *cP* as temperature increases from 0 to 100°C.

3.2.7 Effect of Temperature on Gas Viscosity

In case of gases, mixing of layers increases in transverse direction with increase in temperature and hence gas viscosity increases. The increase in gas viscosity with temperature can be expressed using Sutherland's formula as given below

$$\eta = \eta_0 \left(\frac{T_0 + C}{T + C} \right) \left(\frac{T}{T_0} \right)^{3/2} = \lambda \left(\frac{T^{3/2}}{T + C} \right) \tag{3.25}$$

where, η = Viscosity of a gas at temperature T, η_0 = Viscosity of the gas at a reference temperature, T_0.

C = Sutherland's constant for that given gas.

$$\lambda = \eta_0 \left(\frac{T_0 + C}{T_0^{3/2}} \right) = \text{Constant}$$

3.2.8 Effect of Pressure on Fluid Viscosity

Liquid viscosity is almost independent of pressure since liquid is considered incompressible fluid. If the pressure is increased from 0.1 to 30 MPa, the corresponding change in liquid viscosity is same as that found by increasing temperature of about 1°C.

Viscosity of any ideal gas is independent of pressure since gas viscosity arises due to transfer and exchange of momentum from one molecule to another and pressure has no effect on momentum transfer. In case of real gases also viscosity is almost independent of pressure.

EXERCISES

Surface Tension

1. Why is hot soup more tastier than cold soup?
2. A drop of mercury of radius 5 mm is broken into 64 equal-sized droplets. What is the work done? [Surface tension of mercury is 540 dyne/cm]   [*Ans.:* 0.0005089 J]
3. Why does surface tension decreases when temperature is increased?
4. What are the exceptions to the rule that surface tension decreases as temperature rises?
5. A liquid rises to a height of 6 cm in a glass tube of radius 0.03 cm. What will be the height of liquid column in a glass capillary tube of radius 0.02 cm?   [*Ans.:* 9 cm]
6. A liquid rises to a height of 5 cm in a glass tube of radius 0.01 cm. What will be the height of liquid column in a glass capillary tube of radius 0.04 cm? [*Ans.:* 1.25 cm]
7. What is capillarity?
8. What are the differences between surface tension and viscosity?

9. Define surface tension.
10. What are the units of surface tension in CGS and SI (MKS) system?
11. What are cohesion and adhesion forces?
12. What are the factors affecting the surface tension?
13. What is the effect of temperature on the surface tension?
14. Define critical temperature.
15. Why the free surface of water is concave but that of mercury is convex?
16. What is the shape of free surface at critical temperature?
17. Why the surface of the slide should not be oily?
18. Define angle of contact.
19. Give some practical applications of surface tension.

Solutions

Ans. 1. The surface tension of hot soup is less than that of cold soup. So, hot soup spread over a larger area of tongue and therefore, is more tastier.

Ans. 3. When temperature is increased, the distance between molecules increases; cohesive force decreases; surface energy decreases and therefore surface tension decreases.

Ans. 4. There are some exceptions. Two of them are molten copper and molten cadmium.

Ans. 7. When capillary tube is dipped in water, water rises in the tube. When dipped in mercury, level falls. This rise or fall of level in tube is called capillarity.

Ans. 8. Viscosity is due to motion of molecules, whereas surface tension is due to the unbalanced intermolecular forces. Second difference is that viscosity is present only in moving fluids, whereas surface tension is observed in both moving and stationary fluids.

Ans. 9. The tangential cohesive force acting along the unit length of the surface of a liquid is called surface tension of that liquid. Mathematically, it can be written as $\gamma = \dfrac{F}{L}$, where F = total force along a line on the liquid surface and L = length of the line.

Ans. 11. Cohesion force is the attractive force between like molecules, whereas, the adhesion force is the attractive force between unlike molecules, e.g. attraction between glass slide and the liquid.

Ans. 12. (a) Nature of liquid, (b) Nature of the surface in contact, and (c) Temperature.

Ans. 14. The temperature at which the surface tension is zero.

Ans. 15. If cohesive force among the liquid molecules is less than the adhesive force between liquid and gas molecules, the free surface of liquid is concave in nature. This is true for water and hence free water surface is concave.

If cohesive force among the liquid molecules is greater than the adhesive force between liquid and gas molecules, the free surface of liquid is convex in nature. This is true for mercury and hence free mercury surface is convex.

Ans. 16. At critical temperature the surface tension becomes zero, hence the free surface is flat.

Ans. 17. In presence of oil, surface tension of water decreases. Since surface tension of oil is less than that of water. That is why, surface tension of slide surface should not be oily.

Ans. 18. Angle of contact, for a pair of solid and liquid, is defined as "the angle between tangent to the liquid surface drawn at the point of contact and the solid surface inside the liquid."

Viscosity

1. What is viscosity? What is coefficient of viscosity? How is it related to mobility of a liquid?
2. What is the direction of viscous force? How does the coefficient of viscosity of a liquid change with increase in temperature?
3. State and explain Stokes' law.
4. What is terminal velocity? On what factor terminal velocity depends?
5. Explain: (a) How viscosity of a liquid changes with increase in temperature? (b) Why viscosity of gases increases with increase in temperature?
6. Explain why Poiseuille's equation is applicable in Ostwald viscometer?
7. Discuss the factors, influencing the viscosity of a liquid.
8. Why is viscometer not rinsed with the given liquid or water?
9. Why is viscous force dissipative?
10. The intermolecular forces in oil are less than that of water but still the viscosity of oil is more than water. Justify.
11. A horizontal tube of radius 1 mm and length 40 cm is connected to the bottom of a cubical tank of side 100 cm containing water of viscosity 0.01 poise. From the full tank, water is allowed to flow through the tube. Determine the time in which the tank will be half full.

    Hint: Use Poiseuille's equation which is given by:

    $$V = \frac{\pi \Delta P r^4 t}{8 \eta L}$$

12. A fluid is flowing between two layers. Calculate the shearing stress if the shear velocity is 0.25 m/s and has fluid length 2 m and dynamic viscosity is 2 Pa-s. [*Ans.:*0.5 Pa–s]
13. A fluid moves along length 0.75 m with velocity 2m/s and has shearing stress of 2 N/m². Calculate its dynamic viscosity.   [*Ans.:*0.75 Pa–s]

Solution

Ans. 8. If the viscometer is rinsed with the given liquid or water before measuring the flow time, volume taken will be more than a definite known volume.

4 Solid State Chemistry

4.1 CRYSTALLOGRAPHY

Study of crystal structures and properties of crystalline solids is known as *crystallography*.

Crystal A solid is said to be a crystal when constituent atoms are orderly arranged in a definite sequence. To define a three-dimensional sequence three axes are required. These three axes are known as crystallographic axes.

Space lattice It is an arrangement of indefinite number of points in three dimensions such that environment of any atom or point is the reflection of that of any other point in the array. Studies on several crystal structures show that *fourteen* different kinds of arrangements or fourteen space lattices are available to describe all crystal structures. These fourteen space lattices are known as *Bravais lattices*. These fourteen Bravais lattices are divided into *seven* crystal systems. To describe any crystal lattice six parameters are essential: three crystallographic axes, *a*, *b*, *c* and angles between them, α, β, γ as shown in the Fig. 4.1.

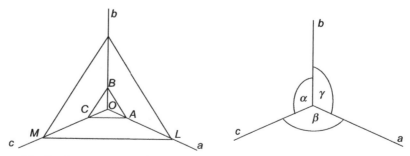

Figure 4.1 Lattice parameters of a crystal. Crystallographic axes and angles. Intercepts cut by unit cell and a crystal plane

Seven crystal systems are:

(i) **Cubic system** It has **three** space lattices, namely, Simple Cubic (SC), Body Centred Cubic (BCC) and Face Centred Cubic (FCC). In all the cases following are true: $a = b = c$ and $\alpha = \beta = \gamma = 90°$. Example: NaCl etc., where, a, b and c are edge lengths of the unit cell along three crystallographic axes.

(ii) **Trigonal or Rhombohedral system** It has **one** space lattice, i.e., Simple Trigonal, where given relations among lattice parameters are true: $a = b = c$ and $\alpha = \beta = \gamma \neq 90°$. Examples are As, Sb and Bi crystals.

(iii) **Tetragonal system** It has **two** space lattices, namely, Simple Tetragonal and Body Centred Tetragonal, where following relations among lattice parameters are true: $a = b \neq c$ and $\alpha = \beta = \gamma = 90°$. One example is SnO_2 crystal.

(iv) **Hexagonal system** It has **one** space lattice, i.e., Simple Hexagonal, where following relations among lattice parameters are true: $a = b \neq c$ and $\alpha = \beta = 90°$ and $\gamma = 120°$. One example is SiO_2 crystal.

(v) **Orthorhombic system** It has **four** space lattices: Simple Orthorhombic, End Centred Orthorhombic, Body Centred Orthorhombic and Face Centred Orthorhombic. In each case following relations among lattice parameters are true: $a \neq b \neq c$ and $\alpha = \beta = \gamma = 90°$. One example is $BaSO_4$ crystal.

(vi) **Monoclinic system** It has **two** space lattices, namely, Simple Monoclinic and End Centred Monoclinic. In each case following relations among lattice parameters are true: $a \neq b \neq c$ and $\alpha = \beta = 90° \neq \gamma$ [Ref: S. Glasstone and D. Lewis, Elements of Physical Chemistry, 2nd edition (The Macmillan Press Ltd.)] One example is $CaSO_4.2H_2O$ (Gypsum).

(vii) **Triclinic system** It has **one** space lattice, Simple Triclinic, In each case following relations among lattice parameters are true: $a \neq b \neq c$ and $\alpha \neq \beta \neq \gamma \neq 90°$. One example is $K_2Cr_2O_7$.

4.2 UNIT CELL

It is the smallest unit of a crystal, expansion of which gives rise to a 3-D real crystal. Geometry of unit cell represents the geometry of the real crystal. Lattice structure of different unit cells are described below:

(a) **Simple Cubic (SC)** eight corners of the cube constitute eight space sites.

(b) **Body Centred Cube (BCC)** (i) eight corners of the cube constitute eight space sites, (ii) center of the cube constitutes one space site.

(c) **Face Centred Cube (FCC)** (i) eight corners of the cube constitute eight space sites, (ii) each center of six faces of the cube constitutes one space site.

(d) **End Centred Orthorhombic/Monoclinic** (i) eight corners of the unit cell constitute eight space sites, (ii) centers of two opposite faces constitute two space sites.

(e) **Hexagonal** (i) twelve corners of the two hexagonal faces constitute twelve space sites, (ii) centers of two hexagonal faces constitute two space sites, (iii) centre of the unit cell constitutes three space sites. Cubic and hexagonal crystals are shown in Fig. 4.2.

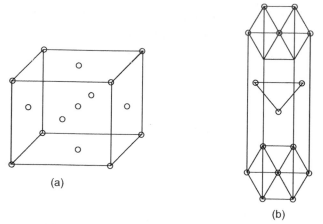

Figure 4.2 (a) Cubic crystal, (b) HCP crystal

4.2.1 Atomic Radius

In a unit cell, half of the distance between centers of two nearest neighbouring atoms is called atomic radius, represented by r.

Lattice constant For cubic system, edge length of a cube is called lattice constant, represented by a.

Relation between 'a' and 'r' This is well illustrated in Fig. 4.3.

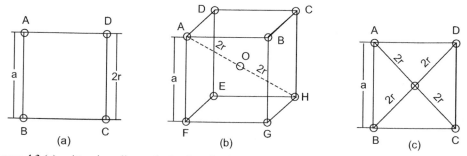

Figure 4.3 (a) Atomic radius and edge length of a SC crystal, (b) Atomic radius and edge length of a BCC crystal, (c) Atomic radius and edge length of a FCC crystal

(i) ***SC crystal*** One of the six faces of a simple cube is shown in the Fig. 4.3(a). According to Fig. 4.3(a) we have

$$AB = BC = a = 2r. \quad \text{Or, } r = \frac{a}{2} \qquad (4.1a)$$

(ii) **FCC crystal**   One of the six faces of a FCC crystal is shown in the Fig. 4.3(c), according to which we can write that

$$(\text{Face diagonal})^2 = AC^2 = AB^2 + BC^2 = AD^2 + DC^2$$

Or, $(2r + 2r)^2 = a^2 + a^2.$ Or, $r = \dfrac{a}{2\sqrt{2}}$ (4.1b)

(iii) **BCC crystal**   In the Fig. 4.3(b) *AOH* represents a cube diagonal, connecting a corner of one face to another corner of opposite face through the centre of the cube. Thus, we can write that:

$$AH^2 = AF^2 + FG^2 + GH^2$$

Or, $(2r + 2r)^2 = a^2 + a^2 + a^2.$ Or, $r = \dfrac{a\sqrt{3}}{4}$ (4.1c)

(iv) **HCP crystal**   According to Fig. 4.2(b), all triangles are equilateral triangles with arm '*a*', which is also the shortest distance between two lattice atoms. Hence, we can write

$$a = 2r. \quad \text{Or,} \ \ r = \dfrac{a}{2}$$ (4.1d)

Coordination Number   The total number of nearest neighbouring atoms, surrounding a lattice atom, is known as coordination number of that lattice atom. It is 6 for SC crystal, 8 for BCC crystal and 12 for both FCC and HCP crystals.

4.2.2 Effective Number of Atoms

Consider the lattice site *C* in the Fig. 4.4.

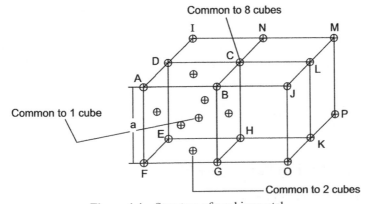

Figure 4.4   Structure of a cubic crystal

The point C is common to four cubes of faces, $DABC$, $BJLC$, $NMLC$ and $DINC$. Again each face is common to two cubes, one remains above the face and the other remains below the face. Thus, the lattice site C is common to eight cubes and hence its contribution to each cube is $\frac{1}{8}$. So, the contribution of each corner to a cube is $\frac{1}{8}$.

Each face has four corners and one centre. Each corner is common to eight cubes (as discussed above) and the centre is obviously common to two cubes. Thus, the contribution of the centre of each face to a cube is $\frac{1}{2}$.

There is a centre inside each cube and the centre is not shared by any other cube. Thus, the contribution of the cube centre to a cube is 1.

Simple Cubic (SC) Crystal   In case of SC crystal, only eight corners of the unit cell possess eight lattice points or atoms.

Thus, the total number of effective atoms $= \left(\frac{1}{8}\right) \times 8 = 1$.

Body Centred Cubic (BCC) Crystal   In case of BCC crystal, eight corners of the unit cell possess eight atoms and the centre of the unit cell possess one atom.

Thus, the total number of effective atoms $= \left(\frac{1}{8}\right) \times 8 + 1 \times 1 = 2$

Face Centred Cubic (FCC) Crystal   Eight corners of the unit cell possess eight atoms and the centre of each face possess one atom. Thus, the total number of effective atoms are:

$$\left(\frac{1}{8}\right) \times 8 + \left(\frac{1}{2}\right) \times 6 \text{ (since there are six faces)} = 1 + 3 = 4$$

Hexagonal Close Packed (HCP) Crystal   The unit cell of a HCP crystal has two hexagonal faces. Each hexagonal face contains six lattice points at six corners, each of which is common to six unit cells. The centre of each hexagonal face possesses one lattice point and the centre of each face is common to two unit cells. Furthermore, the unit cell has three interior centre atoms and each centre atom is not shared by any other unit cell [see Fig. 4.2(b)]. Thus, the total number of effective atoms are:

$$(6+6) \times \left(\frac{1}{6}\right) + (1+1) \times \frac{1}{2} + (1+1+1) \times 1 = 2 + 1 + 3 = 6$$

4.2.3   Atomic Packing Factor (APF)

It is obvious that the total volume of the unit cell cannot be occupied by lattice points or atoms. Effective number of atoms occupies only a fraction of the total volume of the unit cell. That fraction of occupied volume is known as atomic packing factor (APF). In the case of cubic system the volume of the unit cell is a^3, where a is the lattice constant or edge length of the cube. Mathematically, APF can be expressed as

$$\% APF = \frac{\text{Volume occupied by effective number of atoms}}{\text{Volume of unit cell}} \times 100 = \frac{V \times Z}{a^3} \times 100$$

where, $V =$ Volume of one lattice atom, $Z =$ Number of effective atoms

(a) **SC crystal** The effective number of atoms (Z) in the unit cell of a SC crystal is 1. Atomic radius is r and hence volume of lattice atom $V = \frac{4}{3}\pi r^3$. Thus % APF is given by

$$\% APF = \frac{\text{Volume occupied by effective number of atoms}}{\text{Volume of unit cell}} \times 100$$

$$= \frac{V \times Z}{a^3} \times 100 = \frac{V \times 1}{a^3} \times 100$$

$$= \frac{1 \times \frac{4}{3}\pi r^3}{a^3} \times 100 = \frac{\frac{4}{3}\pi \left(\frac{a}{2}\right)^3}{a^3} \times 100 \left[\text{since}, r = \frac{a}{2}\right] = \frac{\pi}{6} \times 100 = 52\%$$

(b) **BCC crystal** The effective number of atoms in the unit cell of BCC crystal is 2. Volume occupied by 2 atoms is $2 \times \frac{4}{3}\pi r^3$, where r is the atomic radius. Thus, % APF is given by

$$\% APF = \frac{\text{Volume occupied by effective number of atoms}}{\text{Volume of unit cell}} \times 100$$

$$= \frac{V \times Z}{a^3} \times 100 = \frac{V \times 2}{a^3} \times 100$$

$$= \frac{2 \times \frac{4}{3}\pi r^3}{a^3} \times 100 = \frac{\frac{8}{3}\pi \left(\frac{a\sqrt{3}}{4}\right)^3}{a^3} \times 100 \left[\text{since}, r = \frac{a\sqrt{3}}{4}\right] = \frac{\pi\sqrt{3}}{8} \times 100 = 68\%$$

(c) **FCC crystal** The effective number of atoms in the unit cell of a FCC crystal is 4. Volume occupied by 4 effective atoms is $4 \times \frac{4}{3}\pi r^3$, where r is the atomic radius. Thus, %APF is given by

$$\% APF = \frac{\text{Volume occupied by effective number of atoms}}{\text{Volume of unit cell}} \times 100$$

$$= \frac{V \times Z}{a^3} \times 100 = \frac{V \times 4}{a^3} \times 100$$

$$= \frac{4 \times \frac{4}{3}\pi r^3}{a^3} \times 100 = \frac{\frac{16}{3}\pi\left(\frac{a}{2\sqrt{2}}\right)^3}{a^3} \times 100 \left[\text{since, } r = \frac{a}{2\sqrt{2}} \right] = \frac{\pi}{3\sqrt{2}} \times 100 = 74\%$$

(d) HCP crystal   The effective number of atoms in the unit cell of a HCP crystal is 6. Each hexagonal face has six equilateral triangles of angle 60°. Height of the unit cell = Distance between two faces = c

$$\text{Total area of each hexagonal face} = 6 \times \frac{1}{2} \times a \times a \sin 60° = 6 \times \frac{1}{2} \times a \times a \frac{\sqrt{3}}{2} = \frac{3\sqrt{3}}{2}a^2$$

$$\text{Volume of the unit cell} = \frac{3\sqrt{3}}{2}a^2 \times c = \frac{3\sqrt{3}}{2}a^2 \times \sqrt{\frac{8}{3}}a = 3\sqrt{2}a^3 \left[\text{since, } c = \sqrt{\frac{8}{3}}a \right]$$

$$\text{Volume occupied by six effective atoms} = 6 \times \frac{4}{3}\pi r^3 = 6 \times \frac{4}{3}\pi\left(\frac{a}{2}\right)^3 \left[\text{since, } r = \frac{a}{2} \right]$$

Thus, %APF is given by

$$\%\text{APF} = \frac{\text{Volume occupied by effective number of atoms}}{\text{Volume of unit cell}} \times 100$$

$$= \frac{V \times Z}{3\sqrt{2}a^3} \times 100 = \frac{6 \times \frac{4}{3}\pi r^3}{3\sqrt{2}\,a^3} \times 100$$

$$= \frac{6 \times \frac{4}{3}\pi\left(\frac{a}{2}\right)^3}{3\sqrt{2}a^3} \times 100 = \frac{\pi}{3\sqrt{2}} \times 100 = 74\%$$

***c/a* ratio in HCP crystal**   Consider Fig. 4.5. The top face of the *HCP* crystal is divided into six equilateral triangles. One such equilateral triangle *AOB* is drawn separately. *GN* is the half of the vertical distance, i.e., ($c/2$). Thus, the angle $\angle AGN$ is 90°. So, we can write that

$$AN^2 = NG^2 + GA^2. \quad AN = a, \quad NG = \frac{c}{2},$$

$$BC = CO = \frac{a}{2}, \quad GA = \frac{2}{3}CA = \frac{2}{3}OA\cos 30° = \frac{a}{\sqrt{3}}$$

Putting these values of *AN*, *NG* and *GA*, we get

$$a^2 = \left(\frac{c}{2}\right)^2 + \frac{a^2}{3}. \quad \text{Or,} \quad \frac{c}{a} = \sqrt{\frac{8}{3}} = 1.633$$

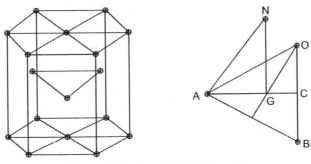

Figure 4.5 *c/a* ratio of HCP crystal

Table 4.1 Characteristics of different crystal structures

Crystal Structure	Effective number of atoms	Atomic radius (r)	Coordination number	APF(%)
Simple cubic	1	$\dfrac{a}{2}$	6	52
Body centred cubic	2	$\dfrac{a\sqrt{3}}{4}$	8	68
Face centred cubic	4	$\dfrac{a}{2\sqrt{2}}$	12	74
Hexagonal close packed crystal	6	$\dfrac{a}{2}$	12	74

4.2.4 Density of a Cubic Crystal

From the definition of density, we have

$$\text{Density}(\rho) = \frac{\text{Mass}}{\text{Volume}}$$

In case of cubic crystal, volume of the unit cell $= a^3$ and Mass = mass of effective number of atoms present in the unit cell. If Z be the effective number of atoms present in the unit cell, total mass of Z number of atoms is

$$\text{Mass} = \frac{Z \times M}{N}$$

where, M is the molar mass, i.e., mass of Avogadro number of atoms and N is the Avogadro number.

$Z = 1$ for *SC* crystal, $Z = 2$ for BCC crystal, and $Z = 4$ for FCC crystal

a = lattice parameter, i.e, edge length of the cube.

Thus, the density of the crystal = $\rho = \dfrac{Z \times M}{N \times a^3}$ gms/c.c.

4.3 FUNDAMENTAL LAWS OF CRYSTALLOGRAPHY

Each and every crystal possesses a definite geometrical shape and follows the following laws of crystallography:

 (a) Law of constancy of interfacial angle

 (b) Law of rationality of indices

 (c) Law of symmetry.

(a) **Law of constancy of interfacial angle**   Each and every crystal is surrounded by a definite number of plane surfaces, which are often termed faces. These faces intersect among themselves at a definite angle, known as interfacial angle. The interfacial angle is constant for a particular type of crystal, e.g., it is 90° for *NaCl* type of crystal structure.

Statement   A crystal may appear in different shape or size depending on the condition of crystallization but the interfacial angle between the corresponding faces remains always constant.

(b) **Law of rationality of indices**   Different lattice planes in a crystal intersect the crystallographic axes to form intercepts but in all cases ratio of intercepts, made by a lattice plane, to fundamental vectors bears a simple ratio of rational numbers. In the Fig. 4.1, *OA*, *OB*, and *OC* are the three fundamental vectors, represented by a, b and c respectively. A lattice plane makes the three intercepts *OL*, *OM*, and *ON* on the three axes *X*, *Y*, and *Z* respectively. Then $\dfrac{OL}{a}$, $\dfrac{OM}{b}$ and $\dfrac{ON}{c}$ are all integers and known as *Weiss indices*. If the lattice plane is parallel to *Z-axis*, *ON* becomes ∞ and in that case Weiss indices are difficult to handle. The scientist Miller proposed that the inverse of the ratio should be taken as indices, i.e., $\dfrac{a}{OL}$, $\dfrac{b}{OM}$ and $\dfrac{c}{ON}$. These are known as *Miller indices*, represented by (*hkl*). Miller indices are easy to handle. If *OL* = 2*a*, *OM* = 3*b* and *ON* = 4*c*. So, (234) is the Weiss indices of the plane and the corresponding Miller indices are $\left(\dfrac{1}{2}, \dfrac{1}{3}, \dfrac{1}{4} \right)$. LCM of 2, 3, 4 is 12. Multiplying with LCM the above figure becomes (643). So, the plane is designated as (643).

Characteristics of Miller Indices

 (i) If a plane meets a crystallographic axis at infinity, the corresponding Miller index is zero (0).

 (ii) Two equidistant parallel planes from the origin have same Miller indices.

(iii) If Miller indices of two planes bear the same ratio, the planes are parallel, e.g., Miller indices of (844) and (422) planes bear the same ratio (2:1:1). So (422) plane is parallel to (844) plane.

(c) **Law of symmetry**   Each and every crystal exhibits some sort of symmetry about an axis or a plane or the centre. For a particular substance all crystals show the same element of symmetry irrespective of their size and shape.

(i) **Centre of symmetry**   If any two opposite faces or planes are equidistant from a point, the latter is called centre of symmetry. A crystal may possess one centre of symmetry (Fig. 4.6a).

(ii) **Plane of symmetry**   If one part of a crystal appears as a mirror image of the other part with respect to a plane, the latter is called plane of symmetry (Fig. 4.6b). Two types of plane of symmetry are possible in a cubic crystal as shown in the Fig. 4.7.

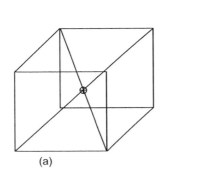
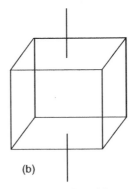

(a)   (b)

Figure 4.6   (a) Centre of symmetry, and (b) Plane of symmetry of a cubic crystal

One is called *straight plane symmetry* and the other is called *diagonal plane symmetry*. A cubic crystal has *three* straight plane symmetry and *six* diagonal plane symmetry, a total of *nine* plane of symmetry.

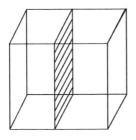
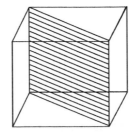

Straight plane symmetry   Diagonal plane symmetry

Figure 4.7   Straight plane symmetry and diagonal plane symmetry of a cubic crystal

(iii) **Axis of symmetry**   If a crystal is rotated by 360° about an axis such that the original configuration of the crystal returns more than once, the axis is called axis of symmetry. If the original configuration returns twice, the axis is called 2-fold axis of symmetry or *diad axis*. If the original configuration returns thrice, the axis is called 3-fold axis of symmetry or *triad axis*. If the original configuration returns four times, the axis is called 4-fold axis of symmetry or *tetrad axis*. If the original configuration returns six times, the axis is called 6-fold axis of symmetry or *hexad axis*.

Example 4.3.1 "A crystal cannot have 5–fold axis of symmetry or more than 6–fold axis of symmetry." – Explain.

Proof In the Fig. 4.8, *M, A, B, N* are the lattice points in the same plane. The crystal is rotated to an angle ϕ such that the point *M* takes the position *P*, the point *N* takes the position *Q*. If the original configuration returns the line *PQ* must be parallel to the line *MN* and the length *PQ* is an integral multiple of *AB*. Thus, $PQ = ma$ and $AB = a$, where *a* is the lattice parameter and *m* is an integer. Again from the Fig. 4.8, we have

$PQ = PR + RS + SQ$ and $PR = AP \cos\phi = a \cos\phi$, $SQ = BQ \cos\phi = a \cos\phi$, $RS = a$.

Thus, $PQ = 2a\cos\phi + a$. Again, $PQ = ma$. Hence, $2a\cos\phi + a = ma$.

Or, $2\cos\phi = (m - 1) = N$. Value of $\cos\phi$ lies between −1 and +1.

If $\cos\phi > 0$, $m > 1$ and $N > 0$.

If $\cos\phi < 0$, $m < 1$ and $N < 0$.

If $\cos\phi = 0$, $m = 1$ and $N = 0$.

$$n - \text{fold axis of symmetry} = n = \frac{360}{\varphi}$$

Table 4.2 Angle of rotation (ϕ) and corresponding value of *n*

m	*(N)*	$\cos\phi$	ϕ	*(n)*
2	1	$\frac{1}{2}$	60°	6
1	0	0	90°	4
0	−1	$-\frac{1}{2}$	120°	3
−1	−2	−1	180°	2
3	2	1	360°	1

Thus, it is evident that a crystal cannot have 5-fold axis of symmetry or more than 6-fold axis of symmetry. A cubic crystal possesses 6 numbers diad axis of symmetry, 4 numbers triad axis of symmetry and 3 numbers tetrad axis of symmetry, a total of 13 numbers axis of symmetry. Thus, a cubic crystal possesses 1 centre of symmetry, 9 planes of symmetry and 13 axis of symmetry, a total of 23 elements of symmetry.

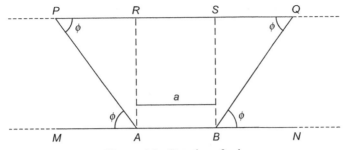

Figure 4.8   Rotation of axis

4.4 MILLER INDICES AND INTERPLANAR SPACING

Miller indices of lattice plane is usually designated by ($h\,k\,l$). $h = 0$ signifies that the plane meets the corresponding crystallographic axis at infinity. $\{h\,k\,l\}$ represents the family of all possible planes. $\bar{h}$ signifies that the plane makes the intercept at negative corresponding crystallographic axis. Thus $\{210\}$ contains the following planes: (210), (120), (201), (102), (012), (021), $(\bar{2}\,1\,0)$, $(\bar{2}\,1\,0)$, $(1\,\bar{2}\,0)$, $(1\,0\,\bar{2})$, $(01\,\bar{2})$, $(0\,\bar{2}1)$, $(2\,\bar{1}0)$, $(2\,0\,\bar{1})$, $(\bar{1}2\,0)$, $(\bar{1}\,0\,2)$, $(0\,\bar{1}\,2)$, $(0\,2\,\bar{1})$, $(\bar{2}\,\bar{1}\,0)$, $(\bar{2}\,0\,\bar{1})$, $(\bar{1}\,\bar{2}\,0)$, $(\bar{1}\,0\,\bar{2})$, $(0\,\bar{1}\,\bar{2})$, $(0\,\bar{2}\,\bar{1})$. Total number of planes are 24.

4.4.1 Interplanar Spacing

Consider two parallel planes such that one plane is passing through the origin and the other plane ABC has the Miller indices ($h\,k\,l$). A normal ON is drawn from the origin to the plane ABC, which makes the intercepts OA, OB and OC on the crystallographic axes respectively. Let the normal ON makes angles α, β and γ with the corresponding crystallographic axes as shown in the Fig. 4.9. Thus, we can write that

$$\cos\alpha = \frac{d}{OA}, \quad \cos\beta = \frac{d}{OB}, \quad \cos\gamma = \frac{d}{OC}$$

[Since any straight line in the plane ABC is perpendicular to the normal ON and d is the interplanar distance between the two parallel planes $= ON$].

Again from the Fig. 4.9 it is evident that

$$ON^2 = d^2 = x^2 + y^2 + z^2$$
$$\text{and } x = d\cos\alpha, \ y = d\cos\beta \text{ and } z = d\cos\gamma.$$

Putting these values of x, y and z in the above equation, we have

$$\cos^2\alpha + \cos^2\beta + \cos^2\gamma = 1.$$

Again, we know from the definition of Miller indices that

$$OA = \frac{a}{h}, \quad OB = \frac{b}{k} \text{ and } OC = \frac{c}{l}$$

ON is perpendicular to the plane ABC.

So, $$\angle NOA = \alpha, \ \angle NOB = \beta, \ \angle NOC = \gamma.$$
$$\angle ONA = \angle ONB = \angle ONC = 90°.$$

Thus for a cubic crystal we can write that

$$\cos\alpha = \frac{d}{OA} = \frac{d \times h}{a}, \ \cos\beta = \frac{d}{OB} = \frac{d \times k}{a} \ \text{and} \ \cos\gamma = \frac{d}{OC} = \frac{d \times l}{a}$$

[since, $a = b = c$ for cubic crystal]

Hence, $$\cos^2\alpha + \cos^2\beta + \cos^2\gamma = \frac{d^2}{a^2}[h^2 + k^2 + l^2]$$

Or, $$\frac{d^2}{a^2}[h^2 + k^2 + l^2] = 1. \ \text{Or, } d = \frac{a}{\sqrt{h^2 + k^2 + l^2}}$$

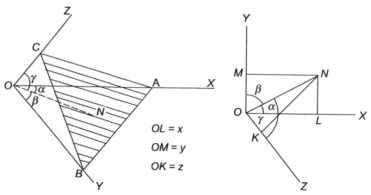

$$OL = x$$
$$OM = y$$
$$OK = z$$

Figure 4.9 Schematic diagram of the distance between two immediate parallel planes

4.4.2 Miller Indices for HCP Crystals

In the case of hexagonal crystal, a lattice plane is usually designated by four coordinate axes. The unit cell of a *HCP* crystal is shown in the Fig. 4.10. The hexagonal base is defined by three coordinate axes a_1, a_2 and a_3, making an angle 120° with each other. The height of the crystal is defined by fourth coordinate axis, represented by c-axis, making an angle 90° with the hexagonal base. The three axes a_1, a_2 and a_3 form a vector triangle such that

$$\vec{a_1} + \vec{a_2} + \vec{a_3} = 0. \ \text{Hence, } \vec{a_1} = -(\vec{a_2} + \vec{a_3}), \ \vec{a_2} = -(\vec{a_1} + \vec{a_3}), \ \vec{a_3} = -(\vec{a_1} + \vec{a_2}),$$

Thus, if two Miller indices are known the third Miller index is the negative sum of the first two. Thus, the Miller indices of any plane in a hexagonal crystal can be represented as $\vec{a_1}$, $\vec{a_2}$, c and $\vec{a_3}$, where, $\vec{a_3} = -(\vec{a_1} + \vec{a_2})$. If a plane has directions lying only in the basal plane, $c = 0$ and Miller indices are $(a_1 \ a_2 \ a_3 \ 0)$.

Type-I basal plane The basal plane has direction parallel to a_1–axis as shown in the Fig. 4.10. The Miller indices of the plane is $(h\bar{k}10)$, where $l = -(h + \bar{k})$.

Type-II basal plane   The basal plane is perpendicular to the a_2–axis. Thus the basal plane is represented by $(h0\overline{h}0)$.

Determination of indices of a diagonal axis of type-I and type-II basal planes are shown in the Fig. 4.11.

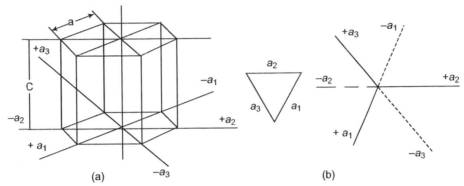

(a)   (b)

Figure 4.10   (a) Unit cell of a HCP crystal, (b) Miller indices of a HCP crystal

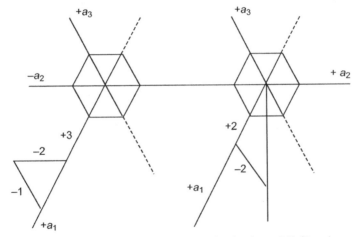

Figure 4.11   Determination of indices of a diagonal axis of type-I (left) and type-II (right)

4.5 BRAGG'S LAW

In the case of diffraction of light the diffraction grating is to be taken close to the wavelength of light source. This is the primary condition to observe diffraction of light. The scientist Laue suggested that the parallel lattice planes can be used as diffraction grating. The interplanar distance between the lattice planes is of the order of 2–3 Å. Thus X-rays can be used as light source to observe diffraction in the crystal since the wavelength of X-ray is of the order of 1 Å. WL Bragg and WH Bragg showed that if a crystal is irradiated with X-rays, electrons within the crystal are excited and start vibrating with the frequency same as that of the incident X-rays. As a result of these vibrations, X-rays are reradiated with the same frequency. This interaction between X-rays and electrons can be treated as reflection of X-rays by atomic planes. According to Bragg's proposition, any two reflected X-rays will reinforce with each other if and only if the path difference between two X-rays is an integral multiple of wavelength or $n\lambda$, where λ is the wavelength of X-ray and n is an integer, 1,2,3, etc. As it is a case of reflection angle of incidence equals the angle of reflection. Let us consider two consecutive parallel lattice planes AB and CD in a crystal and a parallel beam of X-rays are striking on them (Fig. 4.12). NOO' is normal to the planes AB and CD. RO and SO' are the two incident X-rays striking on AB and CD respectively. OR' and $O'S'$ are the corresponding reflected X-rays. Angle of incidence is θ. From the point O two perpendiculars OP and OQ are drawn on SO' and $O'S'$ respectively.

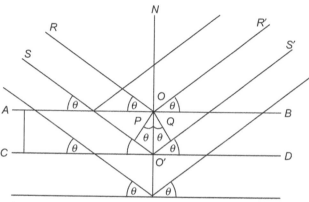

Figure 4.12 Reflection of X-rays from parallel crystal planes

Thus, $\angle ROA = \angle SO'C = \angle PO'C = \theta$ and $\angle R'OB = \angle S'O'D = \angle QO'D = \theta$.

Again, $\angle PO'C + \angle PO'O = 90°$ and $\angle PO'O + \angle POO' = 90°$.

Hence, $\angle PO'C = \angle POO' = \theta$.

Similarly, $\angle QO'D = \angle QOO' = \theta$, where $OO' = d =$ Distance between two parallel planes, AB and CD.

Hence, $PO' = QO' = OO'\sin\theta = d\sin\theta$.

Path difference between two sets of X-rays ROR' and $SO'S'$

$= PO' + QO' = d\sin\theta + d\sin\theta = 2d\sin\theta$.

WL Bragg and WH Bragg proposed that path difference is an integral multiple of wavelength, i.e.,

$$2d\sin\theta = n\lambda \qquad (4.2)$$

where n is the order of reflection.

The value of n is positive integer only. For given values of d and λ, the value of n increases with the increase in the value of θ. When $n = 1$, θ has the smallest value and the reflection is called 1st order reflection. Similarly, when $n = 2$, it is called 2nd order reflection etc. This is well illustrated in the Fig. 4.13.

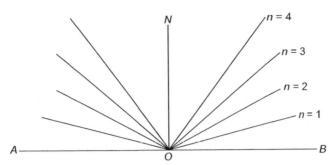

Figure 4.13 Reflection of X-rays from a crystal plane AB at different angles

Inference : $n\lambda = 2d\sin\theta$. Or, $d = \dfrac{n\lambda}{2\sin\theta}$. For $n = 1$, $d = \dfrac{\lambda}{2\sin\theta}$

Theoretically, $\theta_{max} = 90°$ and $\sin\theta = 1$. So, $\lambda = 2d$, which is practically impossible and hence wavelength of X-rays is always less than $2d$ or $\lambda < 2d$.

4.6 POINT IMPERFECTIONS

4.6.1 Imperfection

In an ideal crystal all the lattice sites are perfectly occupied by atoms or ions such that no strain is developed within the crystal. In the case of a real crystal, part of lattice sites are perfectly occupied by atoms or ions, whereas atoms or ions, occupied in other part of the lattice, leave their regular sites, creating vacancies, as a result of which, some irregularities are found on the surface as well as in the bulk of the crystal. Thus, any real crystal appears defective with respect to an ideal crystal. The defect created due to shifting of atoms/ions from their regular sites, is known as *point imperfection*. Mainly two types of defects are responsible for point imperfections:

 (a) Schottky defect
 (b) Frenkel defect.

4.6.2 Schottky Defect

In case of an ionic crystal, a pair of ions may be missing from their regular lattice sites. This type of defect is known as *Schottky defect*. In the case of a metallic crystal, Schottky defect arises due to jumping of atoms from their regular lattice sites to the vacancies. The surface of any real crystal is irregular due to which bulk atoms have a tendency to jump on to the surface to create vacancies as shown in the Fig. 4.14.

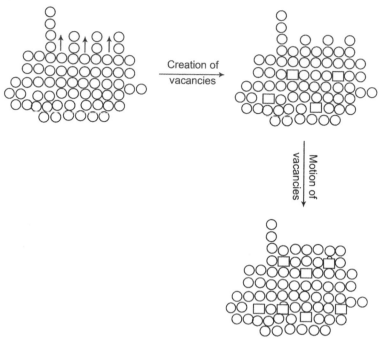

Figure 4.14   Creation and movement of vacancies

Thus a real crystal consists of lattice sites and vacancies. According to classical thermodynamics, we know that, mixing of two gases always increases entropy and the increase in entropy (ΔS^M) is given by

$$\Delta S^M = - R(n_1 \ln x_1 + n_2 \ln x_2) \qquad (4.3)$$

R = Molar gas constant. n_1 and x_1 are the number of moles and mole fraction of component (1). n_2 and x_2 are the number of moles and mole fraction of component (2). Equation (4.3) represents the change in entropy due to mixing of two ideal gases. A real crystal contains both lattice atoms and vacancies. Mixing of lattice atoms and vacancies is assumed to be equivalent to mixing of two ideal gases, since in both the cases mixing occurs at random. Suppose, n_o and n_v are the number of occupied lattice sites and vacancies respectively. Thus equation (4.3) becomes

$$\Delta S^M = - R\left(n_o \ln \frac{n_o}{n_o + n_v} + n_v \ln \frac{n_v}{n_o + n_v} \right) \qquad (4.4)$$

From thermodynamic relationship, we have

$\Delta G^M = \Delta H^M - T\Delta S^M$ and $\Delta H^M = n_v(w \times N)$, where, w = Amount of energy required to create one vacancy and N is the Avogadro number. Putting the values of ΔH^M and ΔS^M in the above equation we get

$$\Delta G^M = n_v \cdot w \cdot N + RT\left(n_o \ln \frac{n_o}{n_o + n_v} + n_v \ln \frac{n_v}{n_o + n_v} \right)$$

Or, $\quad \Delta G^M = n_v \cdot w \cdot N + RT[n_o \ln n_o + n_v \ln n_v - (n_o + n_v)\ln(n_o + n_v)]$

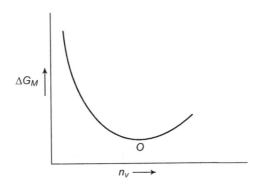

ΔG_M

O

$n_v \longrightarrow$

Figure 4.15 Variation of ΔG_M with the number of vacancies

It is evident from the Fig. 4.15 that ΔG^M reaches a minimum at the point O and hence the atoms rearrange themselves to stay at the point O.

At O, $\dfrac{\partial(\Delta G^M)}{\partial n_v} = 0$. Differentiating the above equation with respect to n_v, we get

$$\left. \frac{\partial(\Delta G^M)}{\partial n_v} \right|_{n_v,T} = Nw + RT\left[[(\ln n_v + 1) - \{\ln(n_o + n_v) + 1\}]\right]$$

$$= Nw + RT \ln \frac{n_v}{n_o + n_v}$$

At the point O, $Nw + RT \ln \dfrac{n_v}{n_o + n_v} = 0$. Hence, $\ln \dfrac{n_v}{n_o + n_v} = -\dfrac{Nw}{RT}$

Or, $\quad \dfrac{n_v}{n_o + n_v} = e^{-\frac{Nw}{RT}}$. As, $n_o \gg n_v$, $(n_o + n_v) \approx n_o$

Thus, $\quad \dfrac{n_v}{n_o} = e^{-\frac{Nw}{RT}} = e^{-\frac{E_f}{RT}}$ \hfill (4.5)

$E_f = N_w$ = Activation energy required to create 1 mole of vacancies.

Vacancy motion Due to the presence of vacancies atoms have a tendency to jump into the vacancies. The number of jumps/sec or the rate of jump (r) depends on the energy barrier of jump (E_a) as well as on (n_v/n_o). The relation among them is given by

$$r \propto \frac{n_v}{n_o} \text{ and } r \propto e^{-\frac{E_a}{RT}}. \text{ Or, } r = A\left(\frac{n_v}{n_o}\right)e^{-\frac{E_a}{RT}} \qquad (4.6)$$

where, A = frequency factor. Combining equations (4.5) and (4.6), we get

$$r = Ae^{-\frac{E_f + E_a}{RT}}$$

(4.7)

In case of ionic crystal, the equation (4.5) becomes

$$\frac{n_v}{n_o} = e^{-\frac{Nw}{2RT}} = e^{-\frac{E_f}{2RT}}$$

(4.8)

The factor 2 is introduced as a pair of ions is missing.

4.6.2.1   Characteristic Features of Schottky Defect

(a) It arises due to jumping of atoms or missing of pair of ions, leading to the formation of vacancies within the crystal lattice.

(b) As fraction of lattice sites is unoccupied, density of the crystal decreases. Thus, the observed density (ρ_{obs}) is always less than the ideal density (ρ_{id}), and this density difference helps us evaluate percent unoccupied sites in a crystal lattice.

$$\% \text{ Unoccupied sites} = \% \text{ Vacancy} = \left(\frac{\rho_{id} - \rho_{obs}}{\rho_{id}}\right) \times 100$$

$$\rho_{id} = \frac{MZ}{N_A a^3} = \text{Theoretical density}$$

M = Molecular mass of the crystal, Z = Number of effective atoms in a unit cell = 1 for SC crystal, 2 for BCC crystal and 4 for FCC crystal, N_A = Avogadro number = 6.023×10^{23}/gm. mole, a = Lattice parameter of the crystal.

(c) Ionic crystals, having cations and anions of comparable sizes, show this type of defect. Examples are NaCl, KBr, etc.

(d) This type of defect is observed in both ionic and metallic crystals.

4.6.3   Frenkel Defect

In many ionic crystals, cation is much smaller in size than corresponding anion. In that case, cations have a tendency to shift or jump to nearby interstitial voids from their regular positions. The defect, thus produced, is known as Frenkel defect.

In case of metallic crystals it may so happen that interstitial voids are filled up by foreign atoms, the imperfection is called interstitial impurity.

Sometimes, some foreign atoms replace or substitute some parent atoms within the specified crystal lattice. This type of defect is known as substitutional impurity.

Frenkel defect and interstitial impurity are well illustrated in the Fig. 4.16.

Number of Frenkel vacancies created due to jumping of cations (n_F) is given by

$$n_F = \sqrt{NN'} e^{-\frac{E_f}{2RT}}$$

(4.9)

E_f = Activation energy required to create 1 mole of vacancies.

N = Total number of available lattice sites in a given crystal per m^3.

N' = Total number of available vacancies in a given crystal per m^3.

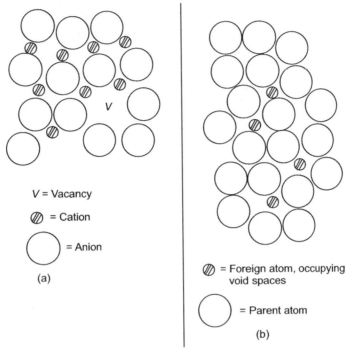

V = Vacancy

= Cation

= Anion

(a)

= Foreign atom, occupying void spaces

= Parent atom

(b)

Figure 4.16 (a) Frenkel defect: Cations jump to the nearest vacancies, (b) Interstitial impurity: Interstitial voids are filled up by foreign atoms

In the case of interstitial impurity foreign atoms occupy the interstitial voids if and only if the diameter of foreign atom is smaller than the void size.

4.6.3.1 Characteristic Features of Frenkel Defect

(a) This type of defect arises due to jumping of positive ions from their regular sites to the interstitial voids.

(b) As there is no pair of missing ions within the crystal, density of the crystal does not decreases, due to the presence of Frenkel defect.

(c) This defect is observed mainly in the ionic crystal.

(d) Ionic crystals, having cationic size much smaller than anionic size, show this type of defect. Examples are AgCl, ZnS, etc.

Schematic diagrams of Schottky and Frenkel defects are shown in the Fig. 4.17.

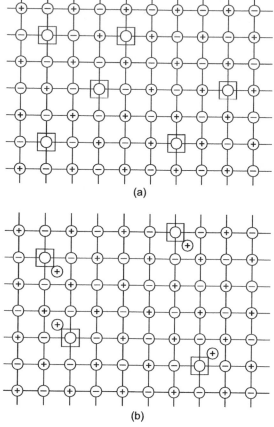

Figure 4.17   Schematic representation of (a) Schottky defect, and (b) Frenkel defect

4.6.4 Consequences of Lattice Defects

(a) Lattice imperfections in a crystal create colour centres which absorb certain wavelength of light when light is allowed to pass through the crystal. Thus the crystals appear complementary colours.

(b) At $0°K$ resistivity of a metallic conductor should be zero but actually the metal shows some resistivity even at $0°K$. This is due to the presence of imperfections, such as point defects, dislocations, grain boundaries, etc.

(c) Presence of defects increases the potential energy (or the enthalpy) of the crystal.

4.7 NON-STOICHIOMETRIC DEFECT

If crystal imperfections are so created in an ionic crystal that the composition of the ionic crystal does not follow the exact stoichiometric formula. Such defects are called non-stoichiometric defects, as a result of which, cation or anion excess always occurs. These types of defects are called non-stoichiometric defects, which can be divided into three types:

(i) Metal excess defects

(ii) Metal deficiency defects

(iii) Impurity defects.

4.7.1 Metal Excess Defect

This may occur in either of the following two ways:

4.7.1.1 Anion Vacancies in Ionic Crystals

In many ionic solid crystals, it is observed that some anions are missing from their regular sites, generating vacancies or positive holes. To maintain electrical neutrality, electrons from outside diffuse into the crystal lattice and are trapped by the positive holes. In presence of sunlight or heat energy, these trapped electrons absorb energy and jump to the nearby vacancies. Due to this absorption of energy the crystal appears coloured. Hence, the trapped electrons are often called **F-centres** or **colour centres** as they are responsible for imparting colour to the crystal.

When NaCl crystal is heated in presence of Na vapour, Cl^- diffuses out of the crystal and electrons absorb energy due to which the crystal appears yellow. Due to the same reason, LiCl appears pink in colour when it is heated in Li vapour and KCl appears violet in presence of K vapour. As the ions are missing from their regular sites, the defect is similar to *Schottky defect*.

The ionic crystal, bearing anion vacancy defect, acts as semiconductor. It is illustrated in the Fig. 4.18(a).

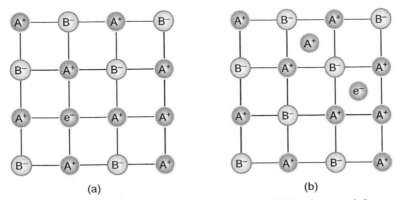

(a)   (b)

Figure 4.18   (a) Metal excess defect due to anion vacancy, (b) Metal excess defect caused by extra cation in interstitial position

4.7.1.2 Cation Excess in Ionic Crystals

When ZnO is heated, the following reaction takes place, producing Zn^{2+} ions and free electrons as shown below:

$$ZnO \rightleftharpoons Zn^{2+} + \frac{1}{2}O_2 + 2e^-$$

These Zn^{2+} ions and electrons diffuse into the crystal lattice and occupy the interstitial sites. The trapped electrons absorb colour when the ZnO crystal is heated and hence the crystal appears yellow. Again the crystal becomes white when it is cooled to room temperature. As Zn^{2+} ions occupy the interstitial voids, the defect is similar to *Frenkel defects*. Due to the presence of excess Zn^{2+} ions the formula of the crystal should be Zn_xO with $x > 1$.

The ionic crystal, bearing cation excess defect, acts as semiconductor. This type of defect is well illustrated in the Fig. 4.18(b).

4.7.2 Metal Deficiency Defect

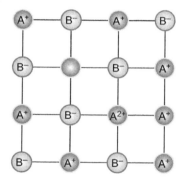

Figure 4.19 Metal deficiency defect due to the presence of multivalent cation

This type of defect is observed when metal shows variable valency. Examples are FeO, FeS, NiO. When FeO crystal is heated, some Fe^{2+} ions are oxidized to Fe^{3+} ions and hence one vacancy is created for every two Fe^{3+} ions in order to maintain the electrical neutrality. Free electrons, generated during oxidation, are trapped within these vacancies and give colour to the crystal. This type of defect is illustrated in the Fig. 4.19.

4.7.3 Characteristics of Metal Excess/Metal Deficiency Defects

(a) The free electrons are often trapped in the interstitial sites. These electrons are responsible for characteristic colours of different ionic crystals.

(b) The characteristic colours, generated due to these defects, are practically used to identify the elements, e.g., golden yellow colour indicates the presence of Na in the unknown sample or purple colour indicates the presence of K, etc.

(c) The free electrons, trapped in the interstitial sites, impart conductivity to the crystal, behaving the latter as semiconductor.

(d) As the trapped electrons reflect light, the crystals appear metallic lustrous or shining.

4.7.4 Impurity Defect in Covalent Solids

The element Si has valency 4. If Si crystal is treated with Group 13 elements like Ga, Al, etc. some Si atoms are replaced by Group 13 elements. The phenomenon is called doping. As valency of Group 13 element is 3, doped Si crystal becomes electron deficient, generating some positive holes within the crystal. These positive holes are responsible for conductivity of doped Si crystal. These are called *p*-type semiconductors.

In another case if the Si crystal is treated with Group 15 elements like P, As, etc. some Si atoms are replaced by Group 15 elements. The phenomenon is also called doping. As valency of Group 15 element is 5, doped Si crystal possesses excess electrons, generating some excess valence electrons within the crystal. These excess valence electrons are responsible for conductivity of doped Si crystal. These are called *n*-type semiconductors.

The above type of defect is called impurity defect.

4.8 SCHRÖDINGER WAVE EQUATION, ENERGY STATES AND FERMI ENERGY

4.8.1 Schrödinger Wave Equation

According to Heisenberg's uncertainty principle, position and velocity of any small particle, like electron, cannot be determined at the same time. Thus a small particle is believed to have both particulate and wavy characteristics. Schrödinger introduced probability density function (ψ^2) to describe electron density in a given energy state. ψ depends on both time and distance. It is defined as:

$$\psi = \psi_0 \sin(\omega t - kx) \tag{4.10}$$

where, $K = \dfrac{2\pi}{\lambda}$ and λ is the wavelength. ω is the angular frequency, ψ is known as probability of finding electron in a given energy state and ψ^2 is the probability density function in a given volume ($dx.\ dy.\ dz$) such that

$$\int\limits_{-\infty}^{+\infty}\int\limits_{-\infty}^{+\infty}\int\limits_{-\infty}^{+\infty} dxdydz = 1$$

In one dimension the above equation becomes

$$\int\limits_{-\infty}^{+\infty} dx = 1$$

Differentiating the equation (4.10) with respect to x we get

$$\frac{\partial \psi}{\partial x} = -k\psi_0 \cos(\omega t - kx) \text{ and } \frac{\partial^2 \psi}{\partial x^2} = -k^2\psi_0 \sin(\omega t - kx) = -k^2\psi$$

Or,
$$\frac{\partial^2 \psi}{\partial x^2} + k^2 \psi = 0 \tag{4.11}$$

where,
$$k = \frac{2\pi}{\lambda}$$

λ is the wavelength. In three dimensions the equation (4.11) takes the form

$$\frac{\partial^2 \psi}{\partial x^2} + \frac{\partial^2 \psi}{\partial y^2} + \frac{\partial^2 \psi}{\partial z^2} + k^2 \psi = 0. \text{ Or, } \nabla^2 \psi + k^2 \psi = 0 \tag{4.12}$$

According to De Broglie's concept, matter waves possess both particulate and wavy characteristics and hence we have: Total energy $= T = E + V$ and also $T = h\nu$, where, E and V are the kinetic energy and potential energy respectively. $h =$ Planck's constant $= 6.625 \times 10^{-34} J - s$.

Let us consider an electron particle in a potential well of length L such that $\psi = 0$ at both the boundaries as shown in the Fig. 4.20.

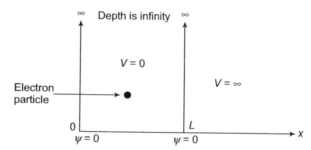

Figure 4.20   Electron particle in a potential well

According to the theory of relativity, energy associated with a moving particle with velocity v_p, is given by
$E = mc^2$ where, c is the velocity of light and m is the mass of the particle.

Momentum of the particle is $mv_p = p$. Again, $E = pv$, where, v is the velocity of the corresponding matter-wave.

So,
$$v = \frac{E}{p} = \frac{mc^2}{mv_p} = \frac{c^2}{v_p} \tag{4.13}$$

Again, according to de Broglie's concept of wave-particle duality we have

$$\lambda = \frac{h}{p} = \frac{h}{mv_p}. \text{ Or, } mv_p = \frac{h}{\lambda}$$

Total energy $= E =$ Kinetic energy as potential energy inside the potential well is zero.

Or,
$$E = \frac{1}{2}mv_p^2 = \frac{1}{2m}(mv_p)^2 = \frac{1}{2m}\left(\frac{h}{\lambda}\right)^2 = \frac{h^2}{2m\lambda^2}$$

Again wave number vector, k, is given by

$$k = \frac{2\pi}{\lambda}. \text{ Or, } \lambda = \frac{2\pi}{k}$$

Hence the above equation becomes

$$E = \frac{h^2 k^2}{2m(2\pi)^2} = \frac{h^2 k^2}{8\pi^2 m} \tag{4.14}$$

Combining equations (4.12) and (4.14), we get

$$\nabla^2 \psi + \frac{8\pi^2 m}{h^2} E\psi = 0. \text{ Or, } \nabla^2 \psi + \frac{8\pi^2 m}{h^2}(T - V)\psi = 0 \tag{4.15}$$

where, T and V represents total energy and potential energy of the particle of mass m and $T = E + V$.

The equation (4.15) is known as time independent Schrödinger wave equation. According to equation (4.14) E–k curve is a parabola as shown in the Fig. 4.21.

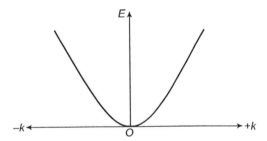

Figure 4.21 E–k curve of an electron wave

4.9 SOMMERFELD THEORY

4.9.1 Quantum Concept of Free Electron Theory

In 1900, Drude and Lorentz first proposed free electron concept of metallic bonding. Later on Sommerfeld modified the free electron theory by introducing quantum concept in 1928.

Postulates of Sommerfeld theory

(a) Free electrons in a metal are considered as waves, and energy of waves is given by: $E = Nh\nu$ according to Max Planck (1900), where, $N = 1, 2, 3, \dots$ etc.

(b) Electron waves follow wave-particle duality as proposed by de Broglie in 1924. Thus, momentum (p) of an electron is given by

$$p = \frac{h}{\lambda}. \text{ Or, } \lambda_B = \frac{h}{p} \tag{4.16}$$

where, λ_B is popularly known as de Broglie wavelength.

(c) Electron wave has only mathematical entity, governed by Schrödinger wave equation given below

$$\frac{h^2}{8\pi^2 m}\nabla^2\psi + (T - V)\psi = \frac{h}{2\pi i}\left(\frac{\partial\psi}{\partial t}\right) \qquad (4.17a)$$

[Time-dependent 3-D Schrödinger wave equation]

Time-independent 3-D, Schrödinger wave equation is given by

$$\frac{h^2}{8\pi^2 m}\nabla^2\psi + (T - V)\psi = 0 \qquad (4.17b)$$

where, ψ is an electronic wave function and given by

$$\psi(x, t) = A.e^{-2\pi i\left(ft - \frac{x}{\lambda}\right)} \qquad (4.17c)$$

f = Frequency of the wave and $\omega = 2\pi f$

T = Total energy of the electron particle or wave.

V = Potential energy of the electron particle or wave.

E = Kinetic energy of the electron. h = Planck's constant = 6.625×10^{-34} J-s.

m = Mass of an electron particle = 9.11×10^{-31} kg.

$E–k$ relation of an electron wave is governed by the equation (4.14) and the corresponding $E–k$ curve is shown in the Fig. 4.21.

According to equation (4.12), we have

$$\nabla^2\psi + k^2\psi = 0$$

Considering one dimension the above equation becomes

$$\frac{\partial^2\psi_x}{\partial x^2} + k^2\psi = 0 \qquad (4.18)$$

One of the solution of the above equation may be written as $\psi_x = A \sin kx + B \cos kx$

Applying the boundary conditions: (i) $\psi_x = 0$ at $x = 0$ and (ii) $\psi_x = L$ at $x = 0$, we get

$$B = 0, \quad kL = n_x\pi, \quad n_x = 1, 2, 3\ldots$$

Or,
$$k = \frac{n_x\pi}{L} \text{ and } \psi_x = A\sin\frac{n_x\pi x}{L} \qquad (4.19)$$

Putting the value of K from equation (4.19) into equation (4.14), we get

$$E = \frac{n_x^2 h^2}{8mL^2} \qquad (4.20a)$$

In three dimensions, the above equation becomes

$$E = \frac{(n_x^2 + n_y^2 + n_z^2)h^2}{8mL^2} \qquad (4.20b)$$

n_x, n_y and n_z are known as quantum numbers, describing different energy states in the potential well.

$$n_x = \frac{kL}{\pi} = \frac{2L}{\lambda}. \quad \text{Or,} \quad k = \frac{n_x\pi}{L}, \quad \text{since,} \quad k = \frac{2\pi}{\lambda}$$

As λ cannot be ∞, n_x cannot be zero.

As the wave function ψ represents probability density of finding electron in a given energy state, we can write that

$$\int_0^L \psi_x^2 dx = 1. \quad \text{Or,} \quad \int_0^L A^2 \sin^2 \frac{n_x\pi x}{L} dx = 1 \quad \left[\text{Since,} \ \psi_x = A \sin \frac{n_x\pi x}{L} \right]$$

Hence $$A = \sqrt{\frac{2}{L}}. \quad \text{Or,} \quad \psi_x = \sqrt{\frac{2}{L}} \sin kx = \sqrt{\frac{2}{L}} \sin \frac{n_x\pi x}{L} \qquad (4.21a)$$

and in three dimensions

$$\psi = \psi_x \cdot \psi_y \cdot \psi_z = \left(\frac{2}{L}\right)^{3/2} \sin\frac{n_x\pi x}{L} \sin\frac{n_y\pi y}{L} \sin\frac{n_z\pi z}{L} \qquad (4.21b)$$

Thus for different values of n_x, n_y and n_z, different values of E and ψ are obtained according to equations (4.20b) and (4.21b) respectively. These different values of ψ are known as **Eigenfunctions** and corresponding values of E are known as **Eigenvalues**.

(d) At $0°\,K$ electrons will not jump from one energy state to another. Thus, electrons are confined in their own energy states and the highest energy level possessing electrons is known as *Fermi level* and corresponding energy is represented as Fermi energy, E_F. However, at higher temperature jumping of electrons from one energy state to another is a common phenomenon and hence probability of finding electrons at higher energy states increases but the same decreases at lower energy states.

(e) Probability of finding electrons in any energy state is given by Fermi-Dirac formula

$$F(E) = \frac{1}{1 + e^{(E - E_F)/kT}} \qquad (4.22)$$

where, $F(E)$ = Probability of finding electrons at any given energy state (E). Free electrons, obeying the Fermi-Dirac statistics are often called **Fermions**. The following conclusions can be drawn from the Fermi-Dirac formula:

(i) At $T = 0°\,K$, $E = 0$ and $F(E) = 1$.

(ii) At $T > 0°\,K$ and $E < E_F$, the value of $e^{(E-E_F)/kT}$ increases with increase in temperature and hence $F(E)$ decreases.

(iii) At $T > 0°\,K$ and $E > E_F$, the value of $e^{(E-E_F)/kT}$ decreases with increase in temperature and hence $F(E)$ increases.

(iv) At $T > 0°\,K$ and $E = E_F$, $F(E) = \dfrac{1}{2}$ at all temperatures.

These are well illustrated in the following Fig. 4.22.

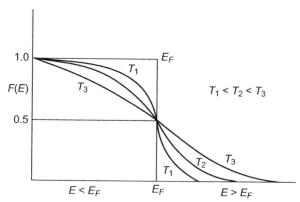

Figure 4.22 $F(E)$ versus E curves at different temperatures

(f) Filling up of energy states by electrons follows Pauli's exclusion principle, i.e., each energy state may possess a maximum of two electrons. If $Z(E)$ is the number of energy states, having energy E, the number of electrons per unit volume in the energy states lying between E and $(E + dE)$ is given by

$$dn_c = 2N(E)F(E)dE = 2\frac{Z(E)}{L^3}F(E)dE.$$

Thus, the total electron density in a given energy band is given by

$$n_c = \int_C 2N(E)F(E)dE = \int_C 2\frac{Z(E)}{L^3}F(E)dE \qquad (4.23)$$

where, $N(E)$ is known as density of energy states and $N(E) = \dfrac{Z(E)}{L^3}$

L^3 is the volume of the cube considered.

n_c is the number of electrons per unit volume, also known as *carrier concentration*.

The total number of energy states with energy E is equal to $z(E)\cdot E$.

The number of energy states may be computed from the knowledge of n_x, n_y and n_z values. All the energy states are confined in a sphere of radius r, where, $n_x^2 + n_y^2 + n_z^2 = r^2$.

As the values of n_x, n_y and n_z are only positive integers, it is evident that all the energy states are confined in one octant of total volume of sphere, i.e., within $\dfrac{1}{8}\left(\dfrac{4}{3}\pi r^3\right)$, which must be equal to the total number of energy states at energy level E.

$$Z(E).E = \frac{1}{8}\left(\frac{4}{3}\pi r^3\right) = \frac{\pi}{6}r^3 \text{ and } E = \frac{(n_x^2 + n_y^2 + n_z^2)h^2}{8mL^2} = \frac{r^2 h^2}{8mL^2}$$

Or,
$$r^2 = \frac{8mL^2}{h^2}E$$

and
$$r = \left(\frac{8mL^2}{h^2}\right)^{1/2}.E^{1/2} \qquad (4.24a)$$

$$dE = 2rdr\left(\frac{h^2}{8mL^2}\right) \text{ and } dr = \frac{1}{2r}\left(\frac{8mL^2}{h^2}\right)dE \qquad (4.24b)$$

The number of energy states at energy level $(E + dE)$ is: $Z(E)(E + dE)$. Thus, the number of energy states lying between E and $(E + dE)$ is $Z(E)(E + dE) - Z(E).E = Z(E)dE$.

Again, $Z(E).E = \frac{\pi}{6}r^3$. Differentiating both sides we get $Z(E).dE = \frac{\pi}{6}3r^2 dr = \frac{\pi}{2}r^2 dr$

Putting the values of r^2 and dr from the equations (4.24a) and (4.24b), we get

$$Z(E).dE = \frac{\pi}{2}\left(\frac{8mL^2}{h^2}E\right) \times \frac{1}{2r}\left(\frac{8mL^2}{h^2}\right)dE = \frac{\pi}{4}\left(\frac{8mL^2}{h^2}\right)^2 \times \frac{1}{r}EdE$$

Putting the value of r from the equation (4.24a) in the above equation, we get

$$Z(E).dE = \frac{\pi}{4}\left(\frac{8mL^2}{h^2}\right)^2 \left(\frac{h^2}{8mL^2}\right)^{1/2}\frac{E}{E^{1/2}}.dE = \frac{\pi}{4}\left(\frac{8mL^2}{h^2}\right)^{3/2}E^{1/2}.dE$$

Or,
$$Z(E).dE = \frac{\pi}{4}\left(\frac{8mL^2}{h^2}\right)^{3/2}E^{1/2}.dE = \frac{2\pi}{h^3}(2m)^{3/2}L^3 E^{1/2}.dE \qquad (4.25)$$

Combining equations (4.23) and (4.25), carrier concentration (n_c), is given by

$$n_c = 2 \times \left(\frac{2\pi}{h^3}\right)(2m)^{3/2}\int_C E^{\frac{1}{2}}F(E)dE = \left(\frac{4\pi}{h^3}\right)(2m)^{3/2}\int_C E^{\frac{1}{2}}F(E)dE \qquad (4.26)$$

4.9.2 Fermi Energy

As we know, Fermi energy level is the highest energy level, possessing electrons at absolute zero. Thus, $F(E) = 1$ at Fermi energy level (E_F) and $E < E_F$ but $F(E) = 0$ at $E > E_F$. Thus, the carrier concentration n_c is

$$n_c = \left(\frac{4\pi}{h^3}\right)(2m)^{3/2}\int_0^{E_F} E^{\frac{1}{2}}dE = \left(\frac{8\pi}{3h^3}\right)(2m)^{3/2}(E_F)^{3/2} \qquad (4.27a)$$

Or, $$(E_F)^{3/2} = \left(\frac{3h^3 n_c}{8\pi}\right) \times \frac{1}{(2m)^{3/2}}$$

Or, $$E_F = \left(\frac{3}{8\pi}\right)^{2/3} \left(\frac{h^2}{2m}\right)(n_c)^{2/3} \qquad (4.27b)$$

The above equation represents the expression of Fermi energy of electrons in a metal. Putting the values of h and m ($h = 6.625 \times 10^{-34}$ J–s, $m = 9.11 \times 10^{-31}$ kg) in the above equation we get

$$E_F = 0.584 \times 10^{-37} n_c^{2/3} \ J. \ \text{Again,} \ 1\,eV = 1.602 \times 10^{-19} \ J$$

So, $$E_F = 3.65 \times 10^{-19} n_c^{2/3} \ eV, \ \text{where,} \ n_c = \frac{x N_A}{\bar{V}}$$

x = Number of free electrons, N_A = Avogadro number, $\bar{V}$ = Molar volume.

4.9.2.1 Calculation of Fermi Energy

For Copper (Cu) $x = 1$, $\bar{V} = \dfrac{M}{\rho} = \dfrac{63.5 \ \text{Kg/ kg.mole}}{8930 \ \text{kg/m}^3}$, $N_A = 6.023 \times 10^{26}/\text{kg.mole}$

So, $$n_c = \frac{6.023 \times 10^{26} \times 8930}{63.5} \text{numbers/m}^3 = 8.47 \times 10^{28} \text{ numbers/m}^3$$

So, $$E_F = 3.65 \times 10^{-19} \times (8.47 \times 10^{28})^{2/3} \ eV = 7.0396 \, eV$$

For Silver (Ag) $x = 1$, $\bar{V} = \dfrac{M}{\rho} = \dfrac{108 \ \text{Kg/ kg.mole}}{10500 \ \text{kg/m}^3}$, $N_A = 6.023 \times 10^{26}/\text{kg.mole}$

So, $$n_c = \frac{6.023 \times 10^{26} \times 10500}{108} \text{numbers/m}^3 = 5.8557 \times 10^{28} \text{ numbers/m}^3$$

So, $$E_F = 3.65 \times 10^{-19} \times (5.8557 \times 10^{28})^{2/3} \ eV = 5.504 \, eV$$

For Aluminium (Al) $x = 1, \bar{V} = \dfrac{M}{\rho} = \dfrac{27 \ \text{Kg/kg.mole}}{2700 \ \text{kg/m}^3}$, $N_A = 6.023 \times 10^{26}/\text{kg.mole}$

So, $$n_c = \frac{6.023 \times 10^{26} \times 2700}{27} \text{numbers/m}^3 = 6.023 \times 10^{28} \text{ numbers/m}^3$$

So, $$E_F = 3.65 \times 10^{-19} \times (6.023 \times 10^{28})^{2/3} \ eV = 5.61 \, eV$$

4.9.3 Merits and Limitations of Sommerfeld Theory

(a) The quantum concept of Sommerfeld theory can explain the metallic properties, e.g., metallic lustre, malleability, etc.

(b) Compton Effect and Black Body Radiation can be explained with the help of this theory.

(c) The concept successfully explains the specific heat of a metal. The insignificant contribution of electrons to the specific heat can be explained with the help of Sommerfeld theory.

(d) Density of energy states $N(E)$ is given by: $N(E)dE = \dfrac{Z(E).dE}{L^3}$

(e) Putting the value of $Z(E)$ from the equation (4.25), we get

$$N(E) = \left(\frac{2\pi}{h^3}\right)(2m)^{3/2}E^{1/2}. \text{ Or, } E \propto [N(E)]^2$$

The above equation is an equation of parabola, shown in the Fig. 4.23, where, it is evident that, all energy states in a metal are continuous throughout the metallic crystal.

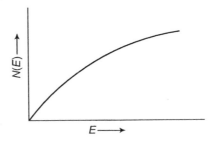

Figure 4.23 Density of energy state [$N(E)$] versus energy (E) curve of a conductor

So, it may be concluded that all the substances, having free electrons, behave as conductors. But in actual practice there are many substances, having free electrons, cannot conduct electricity. These are known as insulators. Thus, Sommerfeld theory fails to explain the behaviour of insulators.

(f) Electrical conductivity of semiconductors cannot be explained properly by Sommerfeld theory.

(g) It does not consider electron-lattice scattering. Influence of Kernels or positive ions on electrons is assumed to be negligible.

(h) According to Sommerfeld theory, potential in a metallic crystal is zero and hence electrons are free to move throughout the crystal but Bloch proposed that a periodic potential exists in a metallic crystal.

4.10 BAND THEORY OF SOLIDS

According to free electron, and Sommerfeld theory, electrons are moving throughout the crystal lattice uniformly. So electrical conductivity of semiconductors and insulators cannot be explained either by free electron theory or by Sommerfeld theory. In 1928, the scientist Bloch proposed that in any metallic crystal electrons are confined in some definite regions, known as bands and there is a definite energy gap between any two successive bands. Electrons are free to move in some bands or zones only. These are called *allowed zones*. Also electrons are not allowed to enter to some other zones. These are called *forbidden zones*. In presence of strong electric field a fraction of electrons in one *allowed zone* is able to cross the energy barrier and increases the population in the next *allowed zone*.

4.10.1   Brillouin Zones

The concept of band theory may be explained with the help of Brillouin zones. In the case of crystals atomic layers are acting as diffraction grating and *X*-rays are diffracted while travelling through the crystal. The *X*-rays are diffracted in the form of reflection and follow *Bragg's equation.*

$$n\lambda = 2d \sin\theta \qquad \text{[see equation (4.2)]}$$

where, n = order of reflection, d = interplanar distance, θ = angle of incidence.

In the case of metallic crystals also, numerous layers of Kernels are formed and electron waves are travelling through the crystals.

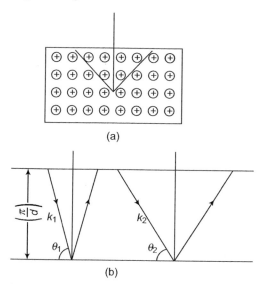

(a)

(b)

Figure 4.24   (a) Reflection of electron waves by layers of Kernels, (b) Reflection of electron waves according to Bragg's equation

We have: $k = \dfrac{2\pi}{\lambda}$ and $\lambda = \dfrac{2\pi}{k}$. Putting this value of λ in the equation (4.2) we have

$$n \times \frac{2\pi}{k} = 2d \sin\theta. \quad \text{For } n = 1, \ k \sin\theta = \frac{\pi}{d} \qquad (4.28)$$

The significance of the equation (4.28) is that whenever an electron vector satisfies the above equation, it will be reflected [illustrated in the Fig. 4.24(b)] . So we may consider a boundary wall surrounding an electron generator such that each boundary wall is separated from the electron generator by $\dfrac{\pi}{d}$ as shown in the Fig. 4.25. The electron generator is at the origin. The electron vectors k_1, k_2, k_3 touch the boundary wall and satisfy the equation (4.28) and hence are reflected. However, the electron vector k_4 does not satisfy the equation (4.28) and move unhindered. Thus, it may be concluded that electrons confined in the region *ABCD* can move freely within that region but are not allowed to leave the region.

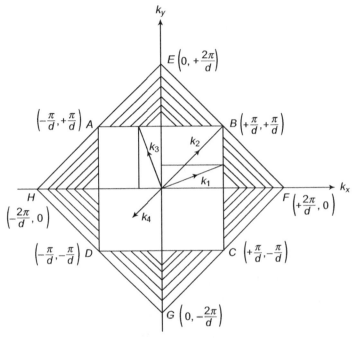

Figure 4.25   1st and 2nd Brillouin zones

The region *ABCD* is known as 1st Brillouin zone. For $n = 2$, the equation (4.28) becomes $k \sin \theta = \dfrac{2\pi}{d}$ and hence the 2nd Brillouin zone can be constructed. In two dimensions the equation (4.28) is given by

$$n_x k_x + n_y k_y = (n_x^2 + n_y^2)\frac{\pi}{d}$$

For the 1st Brillouin zone

(i) $n_x = +1, n_y = 0, \; k_x = +\dfrac{\pi}{d}, \; k_y = 0$ (ii) $n_x = -1, \; n_y = 0, \; k_x = -\dfrac{\pi}{d}, \; k_y = 0$

(i) $n_x = 0, \ n_y = +1, \ k_x = 0, \ k_y = +\dfrac{\pi}{d}$ (ii) $n_x = 0, \ n_y = -1, \ k_x = 0, \ k_y = -\dfrac{\pi}{d}$

For the 2nd Brillouin zone

(i) $n_x = +2, \ n_y = 0, \ k_x = +\dfrac{2\pi}{d}, \ k_y = 0$ (ii) $n_x = -2, \ n_y = 0, \ k_x = -\dfrac{2\pi}{d}, \ k_y = 0$

(iii) $n_x = 0, \ n_y = +2, \ k_x = 0, \ k_y = +\dfrac{2\pi}{d}$ (iv) $n_x = 0, \ n_y = -2, \ k_x = 0, \ k_y = -\dfrac{2\pi}{d}$

(v) $n_x = +1, \ n_y = +1, \ k_x = +\dfrac{\pi}{d}, \ k_y = +\dfrac{\pi}{d}$

(vi) $n_x = +1, \ n_y = -1, \ k_x = +\dfrac{\pi}{d}, \ k_y = -\dfrac{\pi}{d}$

(vii) $n_x = -1, \ n_y = +1, \ k_x = -\dfrac{\pi}{d}, \ k_y = +\dfrac{\pi}{d}$

(viii) $n_x = -1, \ n_y = -1, \ k_x = -\dfrac{\pi}{d}, \ k_y = -\dfrac{\pi}{d}$

The hatched area in the Fig. 4.25 represents the 2nd Brillouin zone.

4.10.2 Conclusions

(a) Electrons residing in the 1st Brillouin zone are confined in that zone and not allowed to leave the zone. Similarly electrons in the 2nd Brillouin zone are confined in that zone and not allowed to leave that zone. Thus it is believed that a definite energy gap exists between the two zones. Similarly, the 3rd Brillouin zone is also separated from the 2nd one by a definite energy gap (Fig. 4.26).

(b) According to Sommerfeld theory E-k curve is continuous and governed by the following equation $E = \dfrac{h^2 k^2}{8\pi^2 m}$ but according to zone theory E-k curve is discontinuous as shown in Fig. 4.26.

Zone-1, zone-2, zone-3 are the so called *allowed bands* as proposed by Bloch. All valence electrons are confined in zone-1. Usually in all metals zone-2 remains empty. Again each zone contains numerous energy levels and appears as band structure. The zone-1 is often called *valence band* and zone-2 is called *conduction band*. The energy gap between these two bands is known as *forbidden energy gap* or *band gap*. This band gap governs the electrical properties of metals and on the basis of band gap metals and solids are divided into three categories: (i) conductors, (ii) insulators, and (iii) semiconductors.

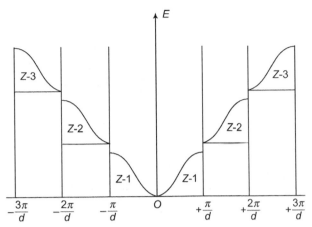

Figure 4.26 Different Brillouin zones or allowed bands in *E-k* curve of a conductor

(c) Electrons, having higher values of $\vec{k}$ vector, face the boundary wall at longer distance Fig. 4.27a. As the magnitude of $\vec{k}$ vector increases, corresponding energy E increases sharply since $E \propto k^2$ [equation (4.14)]. Obviously appearance of energy gap or break point energy, E_b depends on $\vec{k}$ value of electron wave. Higher the magnitude of $\vec{k}$ vector higher the value of E_b as is evident from the Fig. 4.27b.

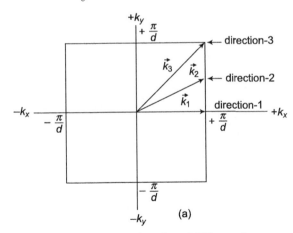

Figure 4.27 (a) Direction of motion of different electron waves

So, electrons, having wave number $\vec{k}_1$, face the energy gap, E_b, at the lowest energy level whereas electrons, travelling along the direction-3, face the energy gap, E_b at the highest energy level. From the Fig. 4.27b it is evident that there is a definite energy gap between zone 1 and zone 2 and the energy gap is popularly known as **forbidden energy gap**, represented by E_g.

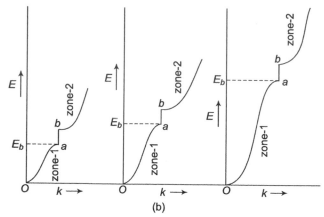

(b)

Figure 4.27   (b) Break-point energy, E_b, varies with the direction of motion of electron waves. a and b are the two points infinitesimally close to the boundary, indicating the energy gap between zone 1 and zone 2

4.11 CONDUCTORS, SEMICONDUCTORS AND INSULATORS

Considering all k vectors the partial E–k curve is shown in the Fig. 4.28.

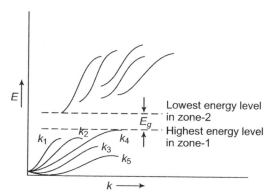

Figure 4.28   Energy levels of zones 1 and 2

In any metal, forbidden energy gap is actually the energy difference between the **highest** energy level of zone 1 and the **lowest** energy level of zone 2 (Fig. 4.28).

Conductors   There are two types of metallic conductors:

Type-I   Each metal atom has single electron in its valence shell. As each energy state may possess two electrons, only half of the total energy states of zone-1 are filled up with electrons and the energy states of other half of zone-1 remain empty (see Fig. 4.29).

In the case of metallic conductor, E_g value is very low. On the application of electric field, electrons in zone-1 easily move to the empty half of the same zone and soon reach to the end of zone-1 and easily cross the energy barrier to reach empty zone-2. The process is going on until electrons reach to the other end of the metal. This is the mechanism of conduction process in type-I metallic conductors.

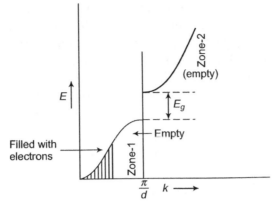

Figure 4.29 *E–k* curve of type-I metallic conductor. Shaded area indicates filled energy states

Type-II Each metal atom has two electrons in its valence shell. Hence, zone-1 is totally filled up with electrons. The corresponding *E–k* curve shows that there is a definite overlapping between lower and higher energy states, eliminating energy barrier between zones [Fig. 4.30].

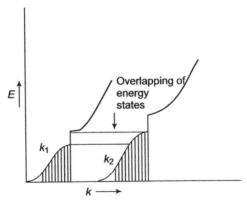

Figure 4.30 *E–k* curve of type-II metallic conductor. Overlapping of zones

Thus, on the application of electric field, electrons of a filled zone readily move up to the empty zone-2 of another plane but of equal energy. The process continues and electrons soon reach to another end of metallic conductor. Thus in type-II metallic conductor, overlapping of energy states plays the vital role in conduction process.

Insulators In the case of an insulator, zone-1 is totally filled up with electrons and forbidden energy gap is sufficiently high so that on the application of electric field electrons cannot move in zone-1, nor will they be able to jump to zone-2 due to high value of E_g. As a result of this, electrons are confined in zone-1 only and the material behaves as an insulator. One example of insulator is diamond, E_g value of which is around 6 eV.

Semiconductors In the case of semiconductors, zone-1 is filled up with electrons but E_g value lies within 3 eV. Thus, on the application of an electric field some electrons of zone-1 gain sufficient kinetic energy through collision and cross the energy barrier to move into the empty zone-2. Furthermore, fraction of those successful electrons, moving into zone-2, is able to cross the energy barrier to move into zone-3, etc. So, ultimately only few electrons are able to reach the other end of semiconducting material. Hence, semiconductors are weak conductor of electricity. Silicon and germanium are two widely used semiconductors.

E_g value of silicon and germanium are 1.1 eV and 0.74 eV respectively. Filled zone-1 is called **valence band** and empty zone-2 is called **conduction band**. Semiconductors are of two types:
  (a) Intrinsic type, and (b) Extrinsic type.

4.12 INTRINSIC SEMICONDUCTORS

As discussed earlier, intrinsic semiconductor possesses a filled Zone-1 with a narrow energy gap between valence band and conduction band (Fig. 4.31).

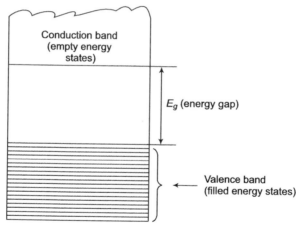

Figure 4.31 Valence band and conduction band in an intrinsic semiconductor

On application of an electric field electrons in the valence band undergo **random collision**. Some of them gain sufficient energy through collision and **jump** to the empty conduction band. At this stage the material becomes a weak conductor of electricity. An electron while jumping to conduction band leaves behind a '**positive hole**' in the valence band. So, as electrons move forward by jumping, holes are moving in backward direction. This is why, a hole is considered as a carrier of positive charge.

4.12.1 Characteristic Features of Intrinsic Semiconductors

  (i) Fraction of electrons excited and jumped to the second Brillouin zone is nothing but the probability of finding an electron in the second Brillouin zone. Thus, we can write that:

$F(E) = \dfrac{n}{N}$, where, n = Number of electrons jumped to the conduction band/m^3

N = Number of electrons available in the valence band/m^3

Second Brillouin zone is often called conduction band, and 1st Brillouin zone is known as valence band. $F(E)$ is calculated from Fermi-Dirac probability distribution, given below

$$F(E) = \frac{1}{1 + e^{(E - E_F)/kT}} \tag{4.29}$$

where, E_F is Fermi energy, k is Boltzmann's constant $= 8.614 \times 10^{-5}$ eV/K and E is any energy state. For conduction band, $E = E_C$ and for valence band, $E = E_V$

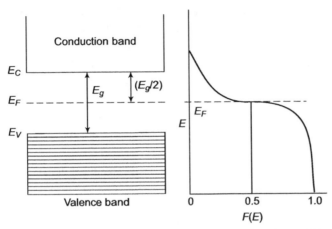

Figure 4.32 Position of Fermi level and energy gap in a semiconductor, $(E_C - E_F) = E_g/2$

Fig. 4.32 shows that probability of finding an electron in Fermi level is 50%. So, actual forbidden energy $(E_C - E_F) = E_g/2$, as shown in the Fig. 4.32. Usually, E_g value of semiconductors lies within 3 eV, e.g., for silicon, $E_g = 1.1$ eV and $E_g/2 = 0.55$ eV. At room temperature, $T = 298$K and the value of $kT = 8.614 \times 10^{-5} \times 298$ eV $= 0.0257$ eV.

Thus, $\qquad \dfrac{(E_g/2)}{kT} = \dfrac{0.55}{0.0257} = 21.4$ and $e^{E_g/2kT} = e^{10.7} \gg 1$

So, equation (4.29) becomes

$$F(E) = \frac{1}{1 + e^{(E_C - E_F)/kT}} = \frac{1}{1 + e^{E_g/2kT}} \approx \frac{1}{e^{E_g/2kT}} = e^{-E_g/2kT}.$$

Again, $\qquad F(E) = \dfrac{n}{N}$

n = Number of electrons per unit volume excited to the empty second zone, also known as **carrier concentration** with unit Number/m^3.

N = Total number of electrons per unit volume available for excitation from the top of the valence band with unit Number/m³.

Thus, above equation becomes

$$\frac{n}{N} = e^{-E_g/2kT} \tag{4.30}$$

(a) If the magnitude of ($E_g/2kt$) at 298K lies below 35, the material is said to be semiconductor.

(b) If the magnitude of ($E_g/2kt$) at 298K is very high the material is said to be an insulator, e.g., diamond ($E_g = 7.0$ eV) and ($E_g/2kT$) = 136.2.

(ii) Equation 4.30 shows that with the increase in temperature, magnitude of kT increases and hence (n/N) also increases, which means that, number of electrons in the conduction band increases with temperature. Thus, it may be concluded that with the increase in temperature conductivity of a semiconductor increases. However, in case of conductors, conductivity decreases with increase in temperature. At low temperature, electrons in a conductor are moving orderly on application of an electric field. There is a little or no intermixing of electrons during their motion. However, at higher temperature thermal vibration of lattice atoms increases, which ultimately destroys the orderly motion of electrons, facilitating intermixing of electrons, thereby, increasing in resistance to the flow of current. That is why, conductivity of a conductor decreases with increase in temperature.

(iii) As discussed earlier, conductivity of a semiconductor increases with temperature and ultimately a temperature, T_s is reached when (n/N) = ($1/e$). Or, $E_g = 2kT_s$. At this temperature the number of electrons in the conduction band is so high that the semiconductor is assumed to behave as a **conductor**.

$$E_g = 1.0 \text{ eV. Hence, } T_s = \frac{1.0 \times (1.602 \times 10^{-19})\text{J}}{2 \times 1.38 \times 10^{-23} \text{ J/K}} = 5805\text{K}$$

Thus a **semiconductor,** having $E_g = 1.0$ eV, can behave as a **conductor** at around 5805K, the temperature at which all known materials melt and vaporise. Thus, it may be concluded that a semiconductor **cannot behave as a conductor**.

(iv) According to equation (4.30) the value of (n/N) = 0 at T = 0K, which means that there is no electron in the conduction band at 0K. Thus, it may be inferred that a semiconductor behave as an **insulator** at 0K.

4.12.2 Drift Velocity and Mobility

Drift velocity   The velocity, acquired by an electron on application of an external electric field, is known as drift velocity, represented by v_d. The unit of v_d is *m/s*.

Mobility   Drift velocity, acquired by an electron under unit potential gradient, is known as mobility, represented by μ.

So,
$$\mu = \frac{v_d}{E_x} \cdot \quad \text{Unit}: \frac{m/s}{V/m} = \frac{m^2}{V\text{-}s}$$

The E_g-value of a semiconductor is low and electrons in the valence band, when excited by an external electric field or by thermal means, may be able to cross the energy barrier and occupy the vacant energy levels of conduction band. The conductivity of semiconductor depends only on the number of electrons in the conduction band. Larger the number of electrons in the conduction band higher will be the conductivity. A successful electron jump from valence band to conduction band creates one vacancy in the valence band. The vacancy in the valence band is known as *positive hole* or simply *hole*. On the other hand, due to this electron jump one vacant position in the conduction band is filled up by a negative electron. So, jumping of electron from valence band to conduction band is accompanied by jumping of a hole from conduction band to valence band. Thus, flow of electrons always occurs in opposite to the flow of holes. Thus conductivity of a semiconductor is the sum of conductivity of electrons and conductivity of positive holes. Mathematically, we can write that

$$\sigma = \sigma_e + \sigma_h \tag{4.31}$$

where, σ = conductivity of a semiconductor.

σ_e = conductivity due to electrons.

σ_h = conductivity due to positive holes.

Again, we know that $\quad \sigma = ne\mu$

where, n = number of charge carriers/m^3, also known as carrier concentration.

e = electronic charge = $1.602 \times 10^{-19} C$

μ = mobility of the charge carrier.

Thus for electrons as charge carriers, we can write that: $\quad \sigma_e = n_e e \mu_e$

And for holes as charge carriers, we can write that: $\quad \sigma_h = n_h e \mu_h$

Putting these values of σ_e and σ_h in the equation (4.31), we have

$$\sigma = n_e e \mu_e + n_h e \mu_h \tag{4.32}$$

For a pure semiconductor, $n_e = n_h = n_i$. Any pure semiconductor is called *intrinsic semiconductor*. Thus, for an intrinsic semiconductor the equation (4.32) becomes:

$$\sigma = n_i e (\mu_e + \mu_h) \tag{4.33}$$

μ_e and μ_h are electron and hole mobilities respectively and their values vary from one semiconductor to another. For example,

μ_e for Ge is 0.39 m^2/V-s and μ_e for Si is 0.14 m^2/V-s

μ_h for Ge is 0.19 m^2/V-s and μ_h for Si is 0.05 m^2/V-s

4.12.3   Law of Mass Action

Statement   In a semiconductor, the net carrier concentration always remains constant irrespective of the type of semiconductor, intrinsic or extrinsic.

If N be the total electron density in the valence band, $\left(\dfrac{n_g}{N}\right)$ and $\left(\dfrac{n_h}{N}\right)$ are the probability of finding electron in the conduction band and the probability of finding positive hole in the valence band respectively.

For an intrinsic semiconductor, $\left(\dfrac{n_e}{N}\right) = \left(\dfrac{n_h}{N}\right) = \left(\dfrac{n_i}{N}\right)$. As the probability has multiplication property, we can write that

$$\left(\frac{n_i}{N}\right)\left(\frac{n_i}{N}\right) = \left(\frac{n_e}{N}\right)\left(\frac{n_h}{N}\right). \quad \text{Or,} \quad n_i^2 = n_e \cdot n_h \tag{4.34}$$

The equation (4.34) is known as mathematical form of law of mass action. Combining equations (4.32) and (4.34), we have

$$\sigma = n_e e \mu_e + \left(\frac{n_i^2}{n_e}\right) . e . \mu_h \tag{4.35}$$

The plot of σ against n_e shows a minimum as shown in the Fig. 4.33.

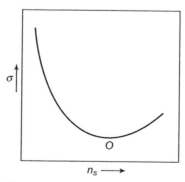

Figure 4.33   Plot of conductivity of an extrinsic semiconductor versus its carrier concentration

To find out the conductivity at the minimum, the 1st derivative of the equation (4.35) is set to zero. Thus,

$$\frac{d\sigma}{dn_e} = e\mu_e - \left(\frac{n_i^2}{n_e}\right)\left(\frac{1}{n_e}\right) . e . \mu_h = 0.$$

Or,
$$e . \mu_e = \left(\frac{n_i^2}{n_e}\right)\left(\frac{1}{n_e}\right) . e . \mu_h. \quad \text{Or,} \quad n_e . e . \mu_e = n_h . e . \mu_h$$

Or, $\sigma_e = \sigma_h$, i.e., current carried by the electrons is equal to the current carried by the positive holes. Thus at the minimum, $\sigma_e = \sigma_h$ and σ_{min} is given by

$$\sigma_{min} = \sigma_e = \sigma_h = 2\sigma_e = 2n_e.e.\mu_e \qquad (4.36)$$

4.12.4 Effect of Temperature on Conductivity of an Intrinsic Semiconductor

For an intrinsic semiconductor, we know that: $\sigma = n_i.e(\mu_e + \mu_h)$. Again, we have

$$n_i = 2\left(\frac{2\pi mkT}{h^2}\right)^{3/2} e^{-E_g/2kT} \quad \text{or,} \quad n_i = n_0 e^{-E_g/2kT},$$

where,

$$n_0 = 2\left(\frac{2\pi mkT}{h^2}\right)^{3/2}$$

So, $n_0 \propto T^{3/2}$. Furthermore, it has been documented that $\mu_e \propto \dfrac{1}{T^{3/2}}$ and $\mu_h \propto \dfrac{1}{T^{3/2}}$

Thus, $n_0 e(\mu_e + \mu_h)$ is believed to be independent of temperature. Or, $n_0 e(\mu_e + \mu_h) = A =$ Constant.

Thus equation (4.33) becomes,

$$\sigma = n_i.e(\mu_e + \mu_h) = n_0.e(\mu_e + \mu_h)e^{-E_g/2kT} \quad \text{Or,} \quad \sigma = A.e^{-E_g/2kT}$$

Taking ln on both sides we, get

$$\ln \sigma = \ln A - \frac{E_g}{2kT} \qquad (4.37a)$$

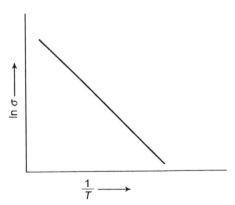

Figure 4.34 Change in conductivity of an intrinsic semiconductor with temperature

So, conductivity increases with increase in temperature as shown in the Fig. 4.34. Again, we know that

$$\sigma = \frac{1}{\rho} \quad \text{and hence } \ln \frac{1}{\rho} = \ln A - \frac{E_g}{2kT} \cdot \quad \text{Or, } \ln \rho = -\ln A + \frac{E_g}{2kT} \qquad (4.37b)$$

So, resistivity decreases with increase in temperature. Again resistivity (ρ) is related with the resistance (R) according to Ohm's law as follows

$$\rho = R\left(\frac{a}{l}\right)$$

where, l = length of the semiconductor and a = area of cross-section of the same. Putting this value of ρ in the above equation, we get

$$\ln\frac{R.a}{l} = -\ln A + \frac{E_g}{2kT}. \quad \text{Or, } \ln R = \ln C + \frac{E_g}{2kT} \tag{4.37c}$$

where, $$\ln C = \ln\frac{l}{a.A}$$

The resistance R is measured at different temperatures and these values of R are plotted against $\frac{1}{T}$ as shown in the Fig. 4.35. Measurement of slope gives rise to the value of E_g or band gap of the intrinsic semiconductor concerned.

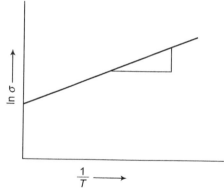

Figure 4.35 Change in resistivity of an intrinsic semiconductor with temperature

4.13 CONDUCTION MECHANISM IN AN EXTRINSIC SEMICONDUCTOR

In the case of an intrinsic semiconductor crystal lattice consists of only one type of atom. Valence band is filled up by electrons. Fermi level (E_F) exists in the midway between conduction band and valence band. If few atoms of crystal lattice are replaced by foreign atoms electron distribution in valence band is changed.

For example, in the case of *Si*-intrinsic semiconductor, the total number of valence electrons per *Si*-atom is 4. If few *Si*-atoms are replaced by *P*-atoms, number of valence electrons increases since each *P*-atom has 5 valence electrons. These extra electrons, which do not get room in the valence band, form a separate energy level just below the conduction band. This newly formed energy level is called *donor level* (E_d). Due to small energy gap between E_c and E_d heavy flow of electrons from donor level to the conduction band is realized on application of an external electric field. In that case, E_F lies between E_c and E_d as shown in the Fig. 4.36(a).

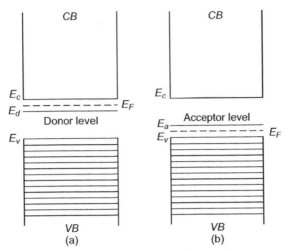

Figure 4.36 (a) Positions of donor level and Fermi level in a *n*-type semiconductor
(b) Positions of acceptor level and Fermi level in a *p*-type semiconductor

$$E_F = \frac{E_c - E_d}{2}$$

This type of semiconductor is known as *n-type semiconductor* [Fig.4.36(a)]. If N_d is the electron density in the donor level the conductivity due to donor level electrons is given by

$$\sigma_d = N_d \times e \times \mu_e \tag{4.38a}$$

As $(E_c - E_d) \ll (E_c - E_v)$, conductivity due to valence band electrons is negligible compared to that of donor level electrons and hence conductivity of an *n-type semiconductor* (σ_n) is given by

$$\sigma_n = \sigma_d = N_d \times e \times \mu_e \tag{4.38b}$$

In another case if some of *Si*-atoms of an intrinsic semiconductor are replaced by *Al*- atoms, number of valence electrons decreases since each *Al*- atom has 3 valence electrons whereas each *Si*-atom has 4 valence electrons. As a result of this positive holes are created in the valence band and these positive holes form an energy level, known as *acceptor level*, just above the valence band. If E_a represents acceptor level, the Fermi level (E_F) lies in between E_a and E_v as shown in the Fig. 4.36(b). E_F is expressed as:

$$E_F = \frac{E_a - E_v}{2}$$

This type of semiconductor is known as *p-type semiconductor* [Fig. 4.36(b)]. If N_a is the electron density in the acceptor level the conductivity due to acceptor level (σ_a) electrons is given by

$$\sigma_a = N_a \times e \times \mu_h \tag{4.39a}$$

As $(E_a-E_v) \ll (E_c-E_v)$, on application of an external electric field valence band electrons easily get transformed to acceptor level but hardly jump to the conduction band. Or in other words, positive holes can easily jump from acceptor level to valence band but jumping of positive holes from conduction band to valence band is almost negligible. So, conductivity of a *p-type semiconductor* (σ_p) is mainly governed by the positive holes of the acceptor level and is given by

$$\sigma_p = N_a \times e \times \mu_h \tag{4.39a}$$

4.13.1   Effect of Temperature on Conductivity of an Extrinsic Semiconductor

In an extrinsic semiconductor dopant atoms play major role in conductivity. In *n*-type semiconductor, dopant atoms form donor level just below the conduction band and in *p*-type semiconductor, dopant atoms form acceptor level just above the valence band. When temperature is increased electrons in donor level of *n*-type semiconductors get excited and jump to the conduction band, whereas in case of *p*-type semiconductors electrons in the valence band get excited and jump to the acceptor level. As the temperature is increased continuously, more and more electrons participate in jumping and hence conductivity in both the cases (*n*-type and *p*-type semiconductors) increases. This increase in conductivity with temperature is mainly due to donor level electrons in *n*-type semiconductor and acceptor level positive holes in *p*-type semiconductor. So this region of conductivity versus temperature plot (Fig. 4.37) is known as *extrinsic region*. This extrinsic region exists up to a certain temperature, at which all the donor electrons in an *n*-type semiconductor are excited to the conduction band or in a *p*-type semiconductor all the positive holes in acceptor level are filled up by electrons. Thus, further increase in temperature has no effect on conductivity and hence a flat region is observed in the plot (Fig. 4.37). This flat region

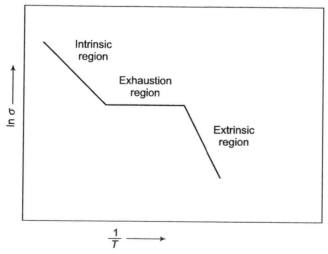

Figure 4.37   Change in conductivity of an extrinsic semiconductor with temperature

is known as *exhaustion region*. It may be mentioned that the length of the extrinsic region depends on dopant atom concentration. Higher the concentration of dopant atom, large number of charge carriers are available in the donor level or in acceptor level and hence longer will be the extrinsic region. The exhaustion region also extends up to a certain temperature above which valence electrons are excited from the valence band to the conduction band like intrinsic semiconductor. This region of the plot (Fig. 4.37) is known as *intrinsic region*.

4.14 COMPOUND SEMICONDUCTOR

Silicon and germanium form the 1st group of semiconductor where atoms of only one type of element forms the semiconductor. In the 2nd group of semiconductor more than one type of element is used to manufacture semiconductor. Usually elements of *Groups III and V* are combined to form semiconductor. This type of semiconductor is called *compound semiconductor*. Examples are: *GaP, GaAs, InP, InAs, GaSb*, etc. Compound semiconductor can also be formed by combination of *Groups II and VI* elements. Band gaps of a few compound semiconductors are given below:

Material	$E_g(eV)$	*Material*	$E_g(eV)$
GaP	2.24	CdS	2.42
GaAS	1.38	CdSe	1.74
SiC	2.86	ZnS	3.70
PbTe	0.32		

SOLVED PROBLEMS

Problem 4.1 The radius of a calcium ion is 94 pm and that of an oxide ion is 146 pm. Predict the crystal structure of calcium oxide.

Solution

The ratio $r_+/r_- = 0.64$. The prediction is an octahedral arrangement of the oxide ions around the calcium. Because the ions have equal but opposite charges, there must also been an octahedral arrangement of calcium ions around oxide ions. Thus, we can expect a rock salt (NaCl) structure.

Problem 4.2 The unit cell of silver iodide (AgI) has 4 iodine atoms in it. How many silver atoms must be there in the unit cell?

Solution

The formula, AgI, tells us that the ratio of silver atoms to iodine atoms is 1:1. Hence, if there are four iodine atoms in the unit cell, there must be four silver atoms.

Problem 4.3   The coordination number of the barium ions, Ba^{2+}, in barium fluoride (BaF_2) is 8. What must be the coordination number of the fluoride ions?

Solution

The coordination number of barium ions tells us that it is surrounded by eight fluoride ions [charge $= 8 \times (-1) = -8$]. In order to balance out the eight negative charges, we need four barium ions [charge $= 4 \times (+2) = +8$]. Hence, the coordination number of the fluoride ions must be 4. So, coordination number of F^- ions is 4.

Problem 4.4   A solid made of A and B has the following arrangement of atoms:
(i) Atoms A are arranged in CCP array, (ii) Atoms B occupy all the octahedral voids and half the tetrahedral voids. What is the formula of the compound?

Solution

In a close packing, the number of octahedral voids is equal to the number of atoms and the number of tetrahedral voids is twice the number of atoms. Since all the octahedral voids and half the tetrahedral voids are filled, there will be one atom of B in tetrahedral void and one atom in octahedral void corresponding to each A. Thus, there will be two atoms of B corresponding to each A. Hence, formula of the solid is AB_2.

Problem 4.5   In corundum, oxide ions are arranged in HCP array and the Al^{3+} ions occupy two thirds of octahedral voids. What is the formula of corundum?

Solution

In CCP or HCP packing there is one octahedral void corresponding to each atom constituting the close packing. In corundum only 2/3rd of the octahedral voids are occupied. It means corresponding to each oxide there are 2/3rd Al^{3+} ions. The whole number ratio of oxide and aluminium ion in corundum is therefore 3:2. Hence, formula of corundum is Al_2O_3.

Problem 4.6   Calculate the ratio of cation to that of anion of the alkali metal bromides on the basis of the data given below and predict the form of the crystal structure in each case. Ionic radii (in pm) are given: $Li^+ = 74$, $Na^+ = 102$, $K^+ = 138$, $Rb^+ = 148$, $Cs^+ = 170$, $Br^- = 195$.

Solution

The ratio of cation to that of anion, i.e. r_+/r_- gives the clue for crystal structure.

$$\frac{Li^+}{Br^-} = 0.379 \text{ (Tetrahedral)}, \quad \frac{Na^+}{Br^-} = 0.523 \text{ (Octahedral)}, \quad \frac{K^+}{Br^-} = 0.708 \text{ (Octahedral)}$$

$$\frac{Rb^+}{Br^-} = 0.759 \text{ (Body centered)}, \quad \frac{Cs^+}{Br^-} = 0.872 \text{ (Body centered)}.$$

Problem 4.7 In the close packed cation in an *AB* type solid have a radius of 75 pm, what would be the maximum and minimum sizes of the anions filling the voids?

Solution

For close packed *AB* type solid $r_+/r_- = 0.414 - 0.732$

Thus, maximum value of $r_+/r_- = 0.732$. Hence, $r_- = r_+/0.732 = 102.5$ pm.

Thus, minimum value of $r_+/r_- = 0.414$. Hence, $r_- = r_+/0.417 = 181.2$ pm.

Problem 4.8 NH$_4$Cl crystallizes in a body centered cubic lattice, with a unit cell distance of 387 pm. Calculate: (a) the distance between the oppositely charged ions in the lattice, and (b) the radius of the NH_4^+ ion if the radius of the Cl$^-$ ion is 181 pm.

Solution

(a) In a body centered cubic lattice oppositely charged ions touch each other along the cross-diagonal of the cube. Hence, we can write,

$$2r_+ + 2r_- = \frac{a\sqrt{3}}{2} = 335.15 \text{ pm}.$$

(b) Now, since $r_- = 181$ pm, we have $r_+ = (335.15-181)$ pm $= 154.15$ pm.

Problem 4.9 Copper has the FCC crystal structure. Assuming an atomic radius of 130 pm for copper atom (Cu – 63.54), then
(a) What is the length of unit cell of Cu? (b) What is the volume of the unit cell?
(c) How many atoms belong to the unit cell? (d) Find the density of Cu.

Solution

We know that

$$\rho = \frac{Z \times M}{N_A \times a^3} . \quad \text{Here, } Z = 4$$

(a) For FCC structure, $r = a/2\sqrt{2}$. Hence, $a = 367.64$ pm.

(b) Volume of unit cell $= a^3 = 4.968 \times 10^{-23}$ cm^{-3}.

(c) $Z = 4$. (d) $\rho = 8.54$ gm/cm^3.

EXERCISES

Multiple Choice Questions

1. A metal crystallizes with a face-centered cubic lattice. The edge of the unit cell is 408 pm. The diameter of the metal atom is:

   (i) 144 pm   (ii) 204 pm   (iii) 288 pm   (iv) 408 pm

2. AB crystallizes in a body centred cubic lattice with edge length 'a' equal to 387 pm. The distance between two oppositely charged ions in the lattice is

   (i) 300 pm   (ii) 335 pm   (iii) 250 pm   (iv) 200 pm

3. Percentage of free space in a body centred cubic unit cell is

   (i) 32%   (ii) 34%,   (iii) 28%   (iv) 20%

4. The appearance of colour in solid alkali metal halides is generally due to:

   (i) Schottky defect   (ii) Frenkel defect   (iii) Interstitial position

   (iv) F-centres

5. In a face-centered cubic lattice, centre of a face is shared equally by how many unit cells?

   (i) 2   (ii) 4   (iii) 6   (iv) 8

Short and Long Answer Type Questions

1. A compound is formed by two elements M and N. The element N forms Cubical Closest Packed (CCP) and atoms of M occupy 1/3rd of tetrahedral voids. What is the formula of the compound?

2. An element with molar mass 2.7×10^{-2} kg/mole forms a cubic unit cell with edge length 405 pm. If its density is 2.7×10^3 kg/m^3, what is the nature of the cubic unit cell?

3. The density of CaO is 3.35 gm/cm^3. The oxide crystallises in one of the cubic systems with an edge length of 4.80 Å. How many Ca^{2+} ions and O^{2-} ions belong to each unit cell, and which type of cubic system is present?   [*Ans.*: $Ca^{2+} = 4$, $O^{2-} = 4$, FCC]

4. A metal crystallizes into two cubic system-face centred cubic (FCC) and body centred cubic (BCC) whose unit cell lengths are 3.5 Å and 3.0 Å respectively. Calculate the ratio of densities of FCC and BCC.   [*Ans.*: 1.259]

5. Copper crystal has a face centred cubic structure. Atomic radius of copper atom is 128 pm. What is the density of copper metal? Atomic mass of copper is 63.5.   [*Ans.*: 8.9 gm/cm^3]

6. The first order reflections of a beam of X-rays of wavelength of 1.54 Å from the (100) face of a crystal of the simple cubic type occurs at an angle 11.29°. Calculate the length of the unit cell.   [*Ans.*: 3.93 Å]

7. X-rays of wavelength equal to 0.134 nm give a first order diffraction from the surface of a crystal when the value of θ is 10.5°. Calculate the distance between the planes in the crystal parallel to the surface examined.   [*Ans.*: 3.68 Å]

8. What is the difference in the semiconductors obtained by doping silicon with Al or with P?

9. Non-stoichiometric cuprous oxide, Cu_2O can be prepared in the laboratory. In this oxide, copper to oxygen ratio is slightly less than 2:1. Can you account of the fact that this substance is a *p*-type semiconductor?

10. Classify each of the following as being either a *p*-type or *n*-type semiconductor:
    (i) Ge doped with In, (ii) B doped with Si.

11. If NaCl is doped with 10^{-3} mole% $SrCl_2$, calculate the concentration of cation vacancies.
    [*Ans.:* 6.023×10^{-18} mole^{-1}]

12. In LiI crystal, I$^-$ ions form a cubical closest packed arrangement and Li$^+$ ions occupy octahedral holes. What is the relationship between the edge-length of the unit cells and the radii of the I$^-$ ions? Calculate the limiting ionic radii of Li$^+$ and I$^-$ if $a = 600$ pm.

Chemical Bonding

5.1 ATOM AND MOLECULE

The smallest part of an element is called atom and the smallest part of a compound is called molecule.

5.1.1 Bohr Theory of Atom

(i) Electrons are moving surrounding the positive nucleus, containing proton and neutron, in different orbits, such that centrifugal force is exactly balanced by the attractive force. Mathematically, we can write that

$$\frac{mv^2}{r} = \frac{Ze^2}{r^2} \cdot \text{ Hence, } r = \frac{Ze^2}{mv^2} \text{ and } v^2 = \frac{Ze^2}{rm} \tag{5.1a}$$

m = Mass of an electron = 9.11×10^{-31} kg, e = Electronic charge = 1.602×10^{-19} Coulomb
Z = Atomic number and r = Atomic radius.

(ii) Electrons, while revolving around the nucleus, do not radiate energy.

(iii) Angular momentum of any electron (mvr) is given by

$$mvr = \frac{nh}{2\pi} \cdot \text{ Or, } r = \frac{nh}{2\pi mv} \text{ and } v = \frac{nh}{2\pi mr} \tag{5.1b}$$

where, n is called principal quantum number, indicating the orbit, e.g., $n = 1$ indicates 1st orbit, $n = 2$ indicates the 2nd orbit, etc.

h = Planck's constant = 6.625×10^{-34} J-s.

Comparing equations (5.1a) and (5.1b) we get

$$\frac{Ze^2}{rm} = \left(\frac{nh}{2\pi mr}\right)^2 \cdot \text{ Or, } r = \frac{n^2 h^2}{4\pi^2 Ze^2 m} \tag{5.1c}$$

In case of hydrogen atom, $n = 1$ and $Z = 1$. Thus, Bohr radius r is given by

$$r = \frac{h^2}{4\pi^2 e^2 m} = 5.29 \times 10^{-11} \, m$$

Putting the value of r in equation (5.1b), we get

$$v = \frac{2\pi Z e^2}{nh} \tag{5.1d}$$

Kinetic energy for n^{th} orbit is given by

$$E_k = \frac{1}{2}mv^2 = \frac{1}{2}m\left(\frac{Ze^2}{rm}\right) = \frac{Ze^2}{2r}$$

Potential energy for n^{th} orbit is given by

$$E_p = \int_{\infty}^{r} \frac{Ze^2}{r^2} dr = -\frac{Ze^2}{r}$$

So, total energy for n^{th} orbit is given by

$$E_n = E_k + E_p = \frac{Ze^2}{2r} - \frac{Ze^2}{r} = -\frac{Ze^2}{2r}$$

Putting the value of r from equation (5.1c) in the above equation, we get

$$E_n = -\frac{2\pi^2 Z^2 e^4 m}{n^2 h^2} = -\left(\frac{2\pi^2 e^4 m}{h^2}\right)\left(\frac{Z^2}{n^2}\right) = -13.6\left(\frac{Z^2}{n^2}\right)eV \tag{5.1e}$$

(iv) An electron absorbs energy (hv) from solar radiations and jumps from lower energy orbit to higher one. In another process an electron may jump from higher energy orbit to lower one with release of energy (hv). Let us consider an electron jumps from the lower energy orbit n_1 to higher energy orbit n_2. Thus the change in energy according to equation (5.1e) is given by

$$\Delta E = hv = -\left(\frac{2\pi^2 Z^2 e^4 m}{h^2}\right)\left(\frac{1}{n_2^2} - \frac{1}{n_1^2}\right).$$

Or,

$$\frac{hc}{\lambda} = -\left(\frac{2\pi^2 Z^2 e^4 m}{h^2}\right)\left(\frac{1}{n_2^2} - \frac{1}{n_1^2}\right)$$

Or,

$$\bar{v} = \frac{1}{\lambda} = -\left(\frac{2\pi^2 Z^2 e^4 m}{ch^3}\right)\left(\frac{1}{n_2^2} - \frac{1}{n_1^2}\right) = -RZ^2\left(\frac{1}{n_2^2} - \frac{1}{n_1^2}\right) \tag{5.1f}$$

R is known as Rydberg constant and value of R is 109700 cm^{-1} and $n_2 > n_1$

If the electron jumps from higher energy orbit n_i to lower energy n_f, the equation (5.1f) can be written as

$$\Delta E = -hv = -\left(\frac{2\pi^2 Z^2 e^4 m}{h^2}\right)\left(\frac{1}{n_f^2} - \frac{1}{n_i^2}\right) \text{ and } n_f < n_i$$

Or,
$$\bar{v} = \frac{1}{\lambda} = RZ^2\left(\frac{1}{n_f^2} - \frac{1}{n_i^2}\right) \qquad (5.1g)$$

For hydrogen atom the above equation becomes

$$\bar{v} = \frac{1}{\lambda} = R\left(\frac{1}{n_f^2} - \frac{1}{n_i^2}\right) \qquad (5.1h)$$

The equation (5.1h) is known as Rydberg formula.

A series of spectra is obtained for different values of n_f.

Lyman series	$n_f = 1$	and	$n_i = 2,3,4, ...$
Balmer series	$n_f = 2$	and	$n_i = 3,4,5, ...$
Paschen series	$n_f = 3$	and	$n_i = 4,5,6, ...$
Brackett series	$n_f = 4$	and	$n_i = 5,6,7, ...$
Pfund series	$n_f = 5$	and	$n_i = 6,7,8, ...$

5.1.2   Limitations of Bohr's Theory

(i) It is established that whenever a charge particle is moving it must radiate energy and hence, centrifugal force (mv^2/r) is less than electrostatic force of attraction (Ze^2/r^2) and hence electrons are moving towards the nucleus, rendering the atom unstable. The stability of Bohr model is only 10^{-8} sec.

(ii) According to the Bohr model, position of an electron can be evaluated by calculating the radius of the orbit and velocity can be also calculated by using the momentum equation: $mvr = nh/2\pi$. This is in contrary to the Heisenberg's uncertainty principle.

5.2   ATOMIC SPECTRA

According to Bohr theory the atomic spectra of a molecule should show only single spectral line for a given energy level since it considers only primary quantum number, n, but in actual practice for a given energy level several spectral lines of different energy levels are obtained. As electron is a charged particle, it continuously exerts strong electrostatic force field surrounding itself. This strong electrostatic force field is believed to be the reason for splitting the spectral lines. To quantify these subsidiary energy levels Orbital quantum number or Azimuthal quantum number, represented by l, has been introduced. Studies of atomic spectra showed that for a given value of n the values of l ranges from 0 to $(n-1)$.

In 1896, scientist Zeeman showed that if the atomic spectra of a molecule is taken under a strong magnetic field, each subsidiary energy level further splits into sub-energy levels of same energy. To quantify these sub-energy levels Magnetic quantum number, m, has been introduced. For a given value of l, the values of m ranges from $-l$ to $+l$.

As each electron is considered as a tiny magnet, spinning of electrons may be either clockwise or anticlockwise. To quantify the direction of spinning of electron Spin quantum number, represented by the symbol s, is introduced. The value of s may be either $+\dfrac{1}{2}$ or $-\dfrac{1}{2}$.

5.3 FILLING UP OF ATOMIC ORBITALS BY ELECTRONS

Pauli's Exclusion Principle

According to this principle, no two electrons can have same four quantum numbers. Thus for a given orbital, values of n, l and m are fixed but the quantum number s has two options: $+\dfrac{1}{2}$ or $-\dfrac{1}{2}$, which signifies that one orbital may contain a maximum of two electrons.

Hund's Rule of Maximum Multiplicity

According to this rule, electrons have tendency to remain in unpaired form as long as orbitals of same energy levels (i.e., same values of n and l) are available. Thus, if more than one orbital of same energy level is available two electrons will remain in unpaired form.

Aufbau Principle

According to this principle, electrons have tendency to remain in lowest energy level and hence lowest energy level is filled up first followed by the next higher energy level and so on. As we know out of the four quantum numbers: n, l, m, s, only n and l represent the energy level, the lowest energy level is that level which has the lowest value of $(n + l)$. However, it may be possible that two energy levels have equal value of $(n + l)$. In that case energy level, having lower value of n, constitutes the lower energy level and is filled up first. $(n + l)$ values of different energy levels are given in the Table 5.1.

Table 5.1 $(n + l)$ values of different energy levels

Energy level	$(n + l)$	Energy level	$(n + l)$	Energy level	$(n + l)$	Energy level	$(n + l)$
1s	1 + 0 = 1	4s	4 + 0 = 0	5d	5 + 2 = 7	6f	6 + 3 = 9
2s	2 + 0 = 2	4p	4 + 1 = 5	5f	5 + 3 = 8	6g	6 + 4 = 10
2p	2 + 1 = 3	4d	4 + 2 = 6	5g	5 + 4 = 9	6h	6 + 5 = 11
3s	3 + 0 = 3	4f	4 + 3 = 7	6s	6 + 0 = 6	7s	7 + 0 = 7
3p	3 + 1 = 4	5s	5 + 0 = 5	6p	6 + 1 = 7	7p	7 + 1 = 8
3d	3 + 2 = 5	5p	5 + 1 = 6	6d	6 + 2 = 8		

Thus the energy levels can be arranged as follows:

$1s < 2s < 2p < 3s < 3p < 4s < 3d < 4p < 5s < 4d < 5p < 6s < 4f < 5d < 6p < 7s < 5f < 6d < 7p$....etc.

However, there are some exceptions of this rule, e.g., according to this rule, the electronic structure of Cu: $[Ar]4s^2 3d^9$, but actual electronic structure of Cu is: $[Ar]4s^1 3d^{10}$. This happens because filled $3d^{10}$-orbital is more stable than $3d^9$-orbital. Similar explanation can be proposed for the electronic structure of Cr: $[Ar]4s^1 3d^5$ instead of $[Ar]4s^2 3d^4$. In this case half-filled $3d^5$-orbital is more stable than $3d^4$-orbital.

5.4   MATTER WAVES AND SCHRÖDINGER WAVE EQUATION

5.4.1   Heisenberg's Uncertainty Principle

Each and every tiny particle, like neutron, should be considered as a wave-packet. If the wave-packet is small, position of the particle can be determined very easily but velocity (or momentum) is indeterminate. On the other hand if the size of the wave-packet is large position of the particle is indeterminate but velocity (or momentum) can be determined accurately. Thus it is impossible to determine position and velocity (or momentum) of a particle simultaneously and precisely. If Δp and Δx are the uncertainties of momentum and position of a particle respectively, we can write, according to Heisenberg, that:

$$\Delta p \cdot \Delta x \geq \frac{h}{4\pi}$$

5.4.2   de Broglie Concept

According to this concept, each and every tiny particle, like neutron, possesses both particulate characteristics and wavy characteristics. Following the wavy characteristics we can write that:

Total energy $(T) = h\nu$, where ν is the frequency of the wave and following the particulate characteristics we can write that

Total energy $(T) = mv^2$, where v is the velocity of the particle. Combining the above two concepts, we have: $mv^2 = hv = \dfrac{hv}{\lambda}$, where, λ is the wavelength of the particle and $v = v \cdot \lambda$. Hence,

$$\lambda = \frac{h}{mv} = \frac{h}{p} = \lambda_B, \text{ where } \lambda_B \text{ is often called } Broglie \ wavelength.$$

5.4.3 Time-Dependent Schrödinger Wave Equation

According to electromagnetic wave theory we have,

$$\frac{\partial^2 \psi}{\partial t^2} = c^2 \frac{\partial^2 \psi}{\partial x^2}, \text{ where, } c \text{ is the velocity of light.}$$

Schrödinger tried to explain electron-wave in the similar way as follows:

$$\frac{\partial^2 \psi}{\partial t^2} = v^2 \frac{\partial^2 \psi}{\partial x^2}, \text{ where, } \psi \text{ is the wave function of electron}$$
$$\text{and } v \text{ is the velocity of electron.}$$

At this stage, scientist Schrödinger made the following assumptions:

  (i) The atomic orbitals are governed by Heisenberg's uncertainty principle and probability distribution function is used to describe the motion of electron.
 (ii) The wavelength of electron is calculated by using de Broglie's wave-particle duality concept.
(iii) The electron waves are stationary waves like vibration of a string between two fixed points.

Considering these assumptions he described the following characteristics of ψ:

  (i) It indicates the probability of finding an electron within a specified volume under consideration. Its value may be positive, negative, zero or complex.
 (ii) The value of ψ is zero at infinity.
(iii) At any position within the specified volume under consideration ψ has a definite value, since $\psi = \infty$ has no meaning.
(iv) At any position within the specified volume under consideration ψ has only one value because more than one value of ψ at a particular point has no meaning.
 (v) Both ψ^2 and $\dfrac{\partial \psi}{\partial x}$ are continuous within the specified volume under consideration such that $\dfrac{\partial^2 \psi}{\partial x^2}$ is finite everywhere.
(vi) The point where $\psi = 0$ is called node or nodal point and hence ψ can be described as wave amplitude. As we know in the case of light, wave intensity of light is directly proportional to the square of amplitude, the term ψ^2 can be defined as intensity of electron or probability density of electron.

(vii) The function ψ^2 is considered a normalized function and from the knowledge of probability density function we can write that:

$$\int_{-\infty}^{+\infty} \psi^2 d\tau = 1 \quad \text{(where, } d\psi \text{ is the volume under consideration} = dx.dy.dz)$$

For one dimension the above equation becomes

$$\int_{-\infty}^{+\infty} \psi^2 dx = 1$$

The solution of the above equation is given by

$$\psi = C.e^{-i\omega\left(t-\frac{x}{v}\right)} \tag{5.2a}$$

where, $\omega = 2\pi v$ and v is the frequency of the wave. v is the velocity of the wave and C is a constant. Again, $v = v\lambda$. Thus, we can write that

$$\omega = \frac{2\pi v}{\lambda} \quad \text{Or,} \quad \frac{\omega}{v} = \frac{2\pi}{\lambda} = k. \quad \text{Or,} \quad \lambda = \frac{2\pi}{k} \tag{5.2b}$$

Differentiating equation (5.2a) with respect to t, we have

$$\frac{\partial \psi}{\partial t} = -i\omega C.e^{-i\omega\left(t-\frac{x}{v}\right)} = -i\omega\psi.$$

Or,

$$\psi = -\frac{1}{i\omega}\frac{\partial \psi}{\partial t} \tag{5.3}$$

Again, we know that:

$$T = E + V \tag{5.4}$$

where, T, E and V are the total energy, kinetic energy and potential energy respectively. Multiplying both sides of the equation (5.4) by ψ, we get

$$T\psi = E\psi + V\psi \tag{5.5}$$

Differentiating the equation (5.2a) with respect to x we get

$$\frac{\partial \psi}{\partial x} = \frac{i\omega}{v}C.e^{-i\omega\left(t-\frac{x}{v}\right)} = \frac{i\omega}{v}\psi \quad \text{and} \quad \frac{\partial^2 \psi}{\partial x^2} = \frac{i\omega}{v}\frac{\partial \omega}{\partial x} = -\frac{\omega^2}{v^2}\psi = -k^2\psi$$

Or,

$$\frac{\partial^2 \psi}{\partial x^2} = -k^2\psi \tag{5.6}$$

Thus, $$\psi = -\frac{1}{k^2}\frac{\partial^2\psi}{\partial x^2} \qquad (5.7)$$

Equation (5.3) is the time dependent equation of ψ, equation (5.6) is the one dimensional time independent equation of ψ and equation (5.5) contains both time-dependent and time-independent functions of ψ. Thus combining equations (5.3), (5.5) and (5.7), we get

$$-\frac{T}{i\omega}\frac{\partial\psi}{\partial t} = -\frac{E}{k^2}\frac{\partial^2\psi}{\partial x^2} + V\psi \qquad (5.8)$$

Again, kinetic energy of a particle, moving with velocity v, is given by

$$E = \frac{1}{2}mv^2 = \frac{1}{2m}(mv)^2 = \frac{p^2}{2m} \qquad (5.9)$$

$$p = mv(\text{momentum}) = \frac{h}{\lambda}(\text{according to de Broglie's concept}).$$

Thus, $$E = \frac{p^2}{2m} = \frac{h^2}{2m\lambda^2} = \frac{h^2k^2}{8\pi^2 m}, \quad \text{since, } \lambda = \frac{2\pi}{k} \quad (\text{Ch. Eqn. 5.2 b})$$

$$E = \frac{h^2k^2}{8\pi^2 m}. \quad \text{Or, } k^2 = \frac{8\pi^2 mE}{h^2} \qquad (5.10)$$

Again, we know that total energy $(T) = hv$.

$$T = hv = \frac{h\omega}{2\pi}. \quad \text{Or, } \frac{T}{\omega} = \frac{h}{2\pi} \qquad (5.11)$$

Combining equations (5.8), (5.10) and (5.11), we have

$$\frac{h}{2i\pi}\frac{\partial\psi}{\partial t} = \frac{h^2}{8\pi^2 m}\frac{\partial^2\psi}{\partial x^2} - V\psi \qquad (5.12)$$

The equation (5.12) is the time-dependent one dimensional Schrödinger wave equation. In three dimensions the equation (5.12) becomes

$$\frac{h}{2i\pi}\frac{\partial\psi}{\partial t} = \frac{h^2}{8\pi^2 m}\left[\frac{\partial^2\psi}{\partial x^2} + \frac{\partial^2\psi}{\partial y^2} + \frac{\partial^2\psi}{\partial z^2}\right] - V\psi$$

Or, $$\frac{h}{2i\pi}\frac{\partial\psi}{\partial t} = \frac{h^2}{8\pi^2 m}\nabla^2\psi - V\psi \qquad (5.13)$$

The equation (5.13) is the time-dependent three dimensional Schrödinger wave equation.

5.4.4 Time-Independent Schrödinger Wave Equation

From equation (5.2a), we have

$$\psi = C.e^{-i\omega\left(t - \frac{x}{v}\right)} \tag{5.1}$$

$$= C.e^{-i\omega t} \cdot e^{\frac{i\omega x}{v}}$$

Or,

$$\psi = C'e^{\frac{i\omega x}{v}} = C'e^{ikx} \tag{5.14}$$

where, $C' = Ce^{-i\omega t}$ = Constant, since the wave function is assumed to be independent of time and $\dfrac{\omega}{v} = k$ [equation (5.2b)].

Differentiating the equation (5.14) with respect to x we get

$$\frac{\partial \psi}{\partial x} = ikC'e^{ikx} = ik\psi \tag{5.15}$$

Differentiating equation (5.15) with respect to x we get

$$\frac{\partial^2 \psi}{\partial x^2} = ik\frac{\partial \psi}{\partial x} = -k^2\psi. \quad \text{Or,} \quad \frac{\partial^2 \psi}{\partial x^2} + k^2\psi = 0$$

Replacing k^2 from equation (5.10), we get

$$\frac{\partial^2 \psi}{\partial x^2} + \frac{8\pi^2 m}{h^2}E_x\psi = 0$$

Or,

$$\frac{h^2}{8\pi^2 m}\frac{\partial^2 \psi}{\partial x^2} + E_x\psi = 0 \tag{5.16}$$

The equation (5.16) is the one-dimensional time-independent Schrödinger wave equation.

Along x-direction: $\dfrac{h^2}{8\pi^2 m}\dfrac{\partial^2 \psi}{\partial x^2} + E_x\psi = 0$

Along y-direction: $\dfrac{h^2}{8\pi^2 m}\dfrac{\partial^2 \psi}{\partial y^2} + E_y\psi = 0$

Along z-direction: $\dfrac{h^2}{8\pi^2 m}\dfrac{\partial^2 \psi}{\partial z^2} + E_z\psi = 0$

Thus, in three dimensions, the overall equation becomes

$$\frac{h^2}{8\pi^2 m}\left[\frac{\partial^2 \psi}{\partial x^2} + \frac{\partial^2 \psi}{\partial y^2} + \frac{\partial^2 \psi}{\partial z^2}\right] + (E_x + E_y + E_z)\psi = 0$$

Or, $\dfrac{h^2}{8\pi^2 m}\left[\dfrac{\partial^2\psi}{\partial x^2}+\dfrac{\partial^2\psi}{\partial y^2}+\dfrac{\partial^2\psi}{\partial z^2}\right]+E\psi=0$, since, $E=E_x+E_y+E_z$

Or, $\dfrac{h^2}{8\pi^2 m}\left[\dfrac{\partial^2\psi}{\partial x^2}+\dfrac{\partial^2\psi}{\partial y^2}+\dfrac{\partial^2\psi}{\partial z^2}\right]+(T-V)\psi=0$, since, $E=T-V$ (Ch. Eqn. 5.4)

Or, $\nabla^2\psi+\dfrac{8\pi^2 m}{h^2}(T-V)\psi=0$ (5.17a)

where $\nabla^2\psi$ in Cartesian coordinate is given by

$$\nabla^2\psi=\frac{\partial^2\psi}{\partial x^2}+\frac{\partial^2\psi}{\partial y^2}+\frac{\partial^2\psi}{\partial z^2}$$ (5.17b)

The equation (5.17) is the three dimensional time-independent Schrödinger wave equation. The solution of time-independent Schrödinger wave equation is given by equation (5.14)

$$\psi=C'e^{\frac{i\omega x}{v}}=C'e^{ikx}$$ (5.14)

Hence, ψ can be written as

$$\psi=A\sin kx+B\cos kx=A\sin\frac{2\pi}{\lambda}x+B\cos\frac{2\pi}{\lambda}x$$ (5.18)

Consider a one dimensional potential well of length L where streams of electrons are continuously vibrating like stationary waves [assumption (iii)] in different energy levels as shown in the Fig. 5.1.

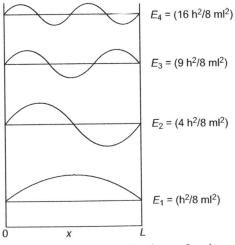

$E_4=(16\,h^2/8\,ml^2)$

$E_3=(9\,h^2/8\,ml^2)$

$E_2=(4\,h^2/8\,ml^2)$

$E_1=(h^2/8\,ml^2)$

0   x L

Figure 5.1 Potential well and wave function ψ

For any energy level, two ends of an electron wave are fixed and hence probability of finding electron at both the ends is zero. Thus we can write that at $x = 0$, $\psi = 0$, and also at $x = L$, $\psi = 0$.

Applying the first boundary condition in equation (5.18), we have, $0 = 0 + B$. Or, $B = 0$.

Thus equation (5.18) becomes

$$\psi = A \sin \frac{2\pi}{\lambda} x \qquad (5.19)$$

Applying the second boundary condition in the equation (5.19), we have

$$\frac{2\pi}{\lambda} L = n\pi. \quad \text{Or,} \quad \lambda = \frac{2L}{n} \qquad (5.20)$$

This is well illustrated in the Fig. 5.1. As λ has a definite positive value and cannot be ∞, n is always a positive integer and cannot be zero.

5.4.4.1 Significance of *n*

(i) In the potential well n indicates energy level where an electron resides.

(ii) It shows that wavelength of an electron decreases when it resides at higher energy level.

(iii) The pictorial-concept of electrons, revolving round the positive nucleus in an atom (according to Bohr theory), has been changed to well-concept of electron waves, vibrating at different energy levels.

From the equations (5.19) and (5.20) we have

$$\psi = A \sin \frac{n\pi}{L} x \qquad (5.21)$$

According to the assumption (vii), made by Schrödinger, we have

$$\int_{-\infty}^{+\infty} \psi^2 dx = 1 \quad \text{(for one dimension)}$$

In the one dimensional potential well the value of x ranges from 0 to L as shown in Fig. 5.1. Thus, the above integration becomes

$$\int_0^L \psi^2 dx = 1 \qquad (5.22)$$

From the equations (5.21) and (5.22), we have

$$\int_0^L A^2 \sin^2 \frac{n\pi}{L} x \, dx = 1$$

On integration we get $\quad A = \sqrt{\dfrac{2}{L}}$

Putting the value of A in the equation (5.21), we get

$$\psi = \sqrt{\frac{2}{L}}\,\sin\frac{n\pi}{L}x$$

Thus, Along x-direction: $\quad \psi_x = \sqrt{\dfrac{2}{L}}\,\mathrm{Sin}\dfrac{n_x\pi}{L}x$

Along y-direction: $\quad \psi_y = \sqrt{\dfrac{2}{L}}\,\mathrm{Sin}\dfrac{n_y\pi}{L}y$

Along z-direction: $\quad \psi_z = \sqrt{\dfrac{2}{L}}\,\mathrm{Sin}\dfrac{n_z\pi}{L}z$

As probability has multiplication property, we can write that

$$\psi = \psi_x\psi_y\psi_z = \sqrt{\frac{8}{L^3}}\,\sin\frac{n_x\pi}{L}x.\ \sin\frac{n_y\pi}{L}y.\ \sin\frac{n_z\pi}{L}z \qquad (5.23)$$

The equation (5.23) is the Schrödinger wave equation in Cartesian coordinate. In spherical coordinate (r, θ, ϕ) equation 5.17b becomes

$$x = r\sin\theta\cos\phi,\ \ y = r\sin\theta\sin\phi,\ \ z = r\cos\theta$$

$$\nabla^2\psi = \frac{1}{r^2}\frac{\partial}{\partial r}\left(r^2\frac{\partial\psi}{\partial r}\right) + \frac{1}{r^2\sin\theta}\frac{\partial}{\partial\theta}\left(\sin\theta\frac{\partial\psi}{\partial\theta}\right) + \frac{1}{r^2\sin\theta}\frac{\partial^2\psi}{\partial\phi^2} \qquad (5.24)$$

The electron wave function is consisted of radial and polar parts.

$$\psi(r, \theta, \phi) = \psi(r)\psi(\theta)\psi(\phi) \qquad (5.25)$$

Considering the radial part the wave functions of electron in different s-orbitals are shown in the Figs. 5.2 (a) and (b).

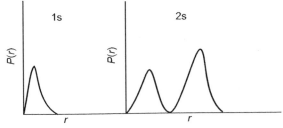

Figure 5.2 (a) The radial part of electron wave function $[\psi(r)]$ in $1s$ and $2s$ orbitals

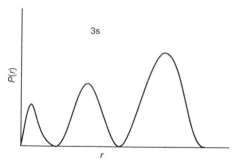

Figure 5.2   (b) The radial part of electron wave function [$\psi(r)$] in 3s orbital

Node   The points, where $\psi = 0$, are called nodes. The number of nodes for a given orbital depends on the primary quantum number (n) as shown in the following table.

Table 5.2   Number of nodes of different orbitals

Orbital	Number of nodes
s	$(n-1)$
p	$(n-2)$
d	$(n-3)$
f	$(n-4)$

Considering the angular part the Schrödinger wave functions of electron, shapes of different orbitals are shown in the Figs. 5.3 (a) and (b).

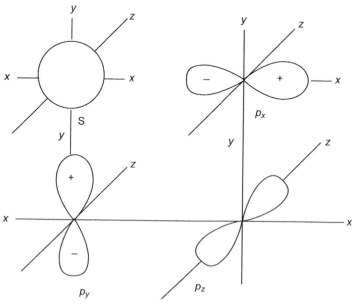

Figure 5.3   (a) The angular part of the wave function [$\psi(\theta, \phi)$] for the 2s, 2p orbitals

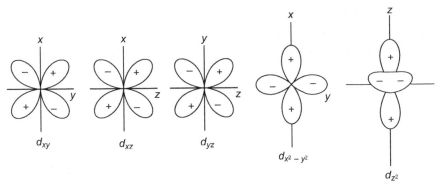

Figure 5.3   (b) The angular part of the wave function [$\psi(\theta, \phi)$ for the $3d$ orbitals

From the equation (5.10), we have

$$E = \frac{h^2 k^2}{8\pi^2 m}, \quad \text{where,} \quad k = \frac{2\pi}{\lambda} \quad \text{and} \quad \lambda = \frac{2L}{n} \quad \text{(Ch. Eqn. 5.20)}$$

Putting the values of k and λ we have

$$E = \frac{n^2 h^2}{8mL^2}$$

Along x-direction: $$E_x = \frac{n_x^2 h^2}{8mL^2}$$

Along y-direction: $$E_y = \frac{n_y^2 h^2}{8mL^2}$$

Along z-direction: $$E_z = \frac{n_z^2 h^2}{8mL^2}$$

Thus, the total energy (E) is given by: $E = E_x + E_y + E_z$

Or, $$E = \frac{(n_x^2 + n_y^2 + n_z^2)h^2}{8mL^2}$$

Or, $$E_n = \frac{n^2 h^2}{8mL^2} \qquad (5.26)$$

where, $$n^2 = n_x^2 + n_y^2 + n_z^2$$

5.4.4.2   Significance of *E*

(i) E_n indicates the energy of electron at a given energy state, *n*.

(ii) Energy difference between two successive energy states (*n*) and (*n* + 1) is given by

$$E_{n+1} - E_n = \Delta E = [(n+1)^2 - n^2]\frac{h^2}{8mL^2} = (2n+1)\frac{h^2}{8mL^2}$$

(iii) $E = \dfrac{(n_x^2 + n_y^2 + n_z^2)h^2}{8mL^2}$

Thus for different values of n_x, n_y and n_z, *E* may have same value.

For example, for an energy state ($n_x = 1$, $n_y = 2$, $n_z = 3$) the value of *E* is same as for the energy state ($n_x = 2$, $n_y = 1$, $n_z = 3$) or ($n_x = 3$, $n_y = 1$, $n_z = 2$) etc. The corresponding energy states possess same energy and they are called degenerate states.

5.5   UNCERTAINTY PRINCIPLE FROM SCHRÖDINGER WAVE EQUATION

From Heisenberg's uncertainty principle, we have $\Delta p \cdot \Delta x \geq \dfrac{h}{4\pi}$

We know from de Broglie's concept that,

$$\text{Momentum}(p_n) = \frac{h}{\lambda} = \frac{nh}{2L}, \quad \text{since} \quad \lambda = \frac{2L}{n} \quad \text{(Ch. Eqn. 5.20)}$$

Thus, uncertainty in momentum is given by

$$\Delta p = p_{n+1} - p_n = \frac{(n+1)h}{2L} - \frac{nh}{2L} = \frac{h}{2L}$$

From Fig. 5.1 it is evident that, uncertainty in position of an electron is $\Delta x = L$. Thus, the uncertainty in momentum is given by:

$$\Delta p = \frac{h}{2\Delta x}. \quad \text{Or,} \quad \Delta p \cdot \Delta x = \frac{h}{2} > \frac{h}{4\pi}$$

Thus, Schrödinger wave equation agrees with the uncertainty principle, proposed by the scientist Heisenberg.

5.6   VALIDITY OF NEIL BOHR'S THEORY

According to Bohr's theory, the radius of an atomic orbit can be calculated from the following equations:

$$\frac{Ze^2}{r^2} = \frac{mv^2}{r} \quad \text{and} \quad mvr = \frac{nh}{2}$$

where, r is the radius of nth Bohr-orbit. For hydrogen atom, atomic number, $Z = 1$.

According to Schrödinger wave equation, we have

$$E = \frac{1}{2}mv^2 = \frac{(mv)^2}{2m} = \frac{p^2}{2m}$$

Or,

$$\Delta E = \frac{(\Delta p)^2}{2m} \tag{5.27}$$

From Heisenberg's uncertainty principle, we have

$$\Delta p \cdot \Delta x \geq \frac{h}{4\pi}$$

However, in most theoretical calculations it has been accepted that

$$\Delta p \cdot \Delta x = \frac{h}{2\pi}. \quad \text{Or, } \Delta p = \frac{h}{2\pi \Delta x}$$

Putting the value of Δp in the equation (5.27), we get

$$\Delta E = \frac{h^2}{8\pi^2 m (\Delta x)^2} \tag{5.28}$$

Again, potential energy (ΔV) is given by

$$\Delta V = \int_{\infty}^{r} \frac{Ze^2}{4\pi\varepsilon_0 r^2}\,dr = -\frac{Ze^2}{4\pi\varepsilon_0 r} = -\frac{Ze^2}{4\pi\varepsilon_0 (\Delta x)}$$

Since r represents uncertainty in position (Δx). Thus,

$$\Delta V = -\frac{Ze^2}{4\pi\varepsilon_0 (\Delta x)} \tag{5.29}$$

So, the total energy (T) is given by: $T = E + V$. Or, $\Delta T = \Delta E + \Delta V$.

$$\Delta T = \frac{h^2}{8\pi^2 m (\Delta x)^2} - \frac{Ze^2}{4\pi\varepsilon_0 (\Delta x)} \tag{5.30}$$

As the electron prefers to remain in the lowest energy level corresponding to the minimum value of ΔT. To find out the minimum value of ΔT differentiate the equation (5.30) and set it to zero. Thus, we have

$$\frac{\partial \Delta T}{\partial \Delta x} = -\frac{h^2}{4\pi^2 m (\Delta x)^3} + \frac{Ze^2}{4\pi\varepsilon_0 (\Delta x)^2} = 0$$

Or,

$$\Delta x = \frac{\varepsilon_0 h^2}{\pi e^2 Zm} \tag{5.31}$$

For hydrogen atom, $Z = 1$ and the above equation reduces to

$$\Delta x = \frac{\varepsilon_0 h^2}{\pi e^2 m}$$

(5.32)

The value of Δx, calculated from the equation (5.32), is almost same as the Bohr's radius r, calculated from the equation (5.1c), although Bohr's theory of atomic structure has been discarded. This has been explained as follows: hydrogen atom has the electronic configuration $1s^1$, i.e., $n = 1$ and $s = 0$. Thus, the lone electron of hydrogen atom occupies the 1s-orbital, which is circular according to Schrödinger wave equation. As the structure of 1s-orbital is very much similar to Bohr's pictorial model, the radius of Bohr's orbit is very close to the value of Δx. Thus, it may be concluded that Bohr's theory of atomic structure is valid for hydrogen-like atoms or ions.

5.7 CONCEPT OF MOLECULAR ORBITAL

5.7.1 Bonding and Anti-bonding MO

According to Schrödinger wave equation electrons are considered as waves and these waves are overlapping in two different ways as shown in Figs. 5.4a and b.

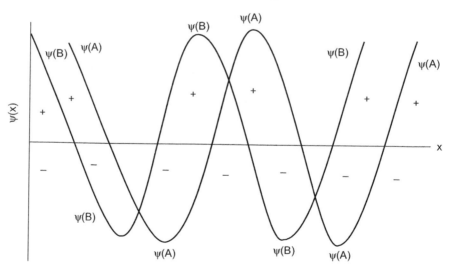

Figure 5.4   (a) Overlapping of electron waves and formation of bonding MO

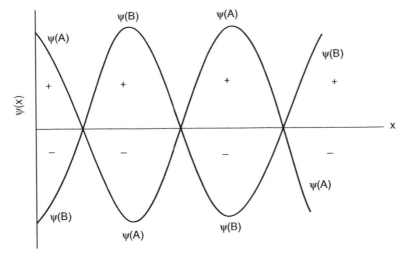

Figure 5.4   (b) Overlapping of electron waves and formation of anti-bonding MO

It is evident from the Figs. 5.4a and b that when two similar wave functions overlap with each other, constructive interference occurs, giving rise to a new resultant wave function. The situation can be considered as overlapping of two atomic orbitals, possessing similar wave function, resulting in the formation of a new orbital, known as bonding molecular orbital or bonding MO. Bonding MO is always situated at lower energy level compared to that of atomic orbitals. On the other hand, when opposite wave functions overlap with each other, destructive interference occurs and the resultant value of ψ is almost zero. Thus two atomic orbitals, having opposite wave function, always repel each other, producing another molecular orbital, known as anti-bonding molecular orbital or anti-bonding MO, which is always situated at higher energy level compared to that of atomic orbitals. Also it is evident from the Figs. 5.4a and b that the resultant wave function is totally different from either of the overlapping wave functions. Thus it may be concluded that during formation of a molecular orbital the identities of both the overlapping atomic orbitals are lost. The formation of bonding and anti-bonding molecular orbitals is well illustrated in the Fig. 5.4c.

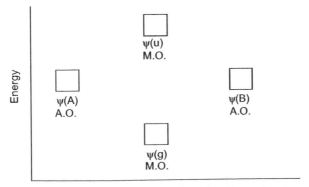

Figure 5.4   (c) Formation of bonding and anti-bonding molecular orbitals

According to Schrödinger wave equation, during overlapping of two atomic orbitals the superimposing wave functions produce constructive interference and destructive interference with equal probability and hence each overlapping of two atomic orbitals always produces two molecular orbitals, one is bonding MO and the other is anti-bonding MO.

5.7.2   Overlapping of Atomic Orbitals

Overlapping of orbitals may occur in different ways:

(i) Two atomic orbitals approach each other along an axis and overlap through head-on collision, as a result of which, electrons are distributed along the axis. This type of overlapping is called σ-overlapping and the corresponding bond is called σ-bond. The bonding MO is denoted by σ and the anti-bonding MO is denoted by σ^*.

(ii) Two atomic orbitals approach each other along an axis and overlap laterally or sidewise, as a result of which, electrons are distributed above and below the axis. This type of overlapping is called π-overlapping and the corresponding bond is called π-bond. Formation of σ- and π-bonds is well illustrated in the Figs. 5.5a, b and c.

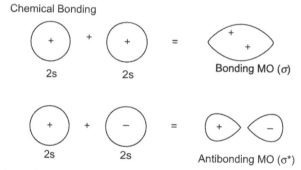

Figure 5.5   (a) Schematic representation of formation of σ-bond through overlapping of *s*-orbitals

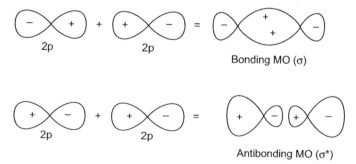

(b) Schematic representation of formation of σ-bond through overlapping of *p*-orbitals

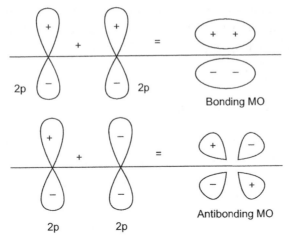

Bonding MO

Antibonding MO

Figure 5.5 (c) Schematic representation of formation of π-bond through overlapping of p-orbitals

The filling up of molecular orbitals obeys Pauli's exclusion principle, Hund's rule of maximum multiplicity and Aufbau principle. The different molecular orbitals may be arranged according to their energy levels as shown below

$$\sigma 1s < \sigma^* 1s < \sigma 2s < \sigma^* 2s < \sigma 2p_x < \pi 2p_y = \pi 2p_z < \pi^* 2p_y = \pi^* 2p_z < \sigma^* 2p_x \ldots$$

For low molecular mass, e.g., H_2, Li_2, Be_2, B_2, etc. the above arrangement of molecular orbitals becomes

$$\sigma 1s < \sigma^* 1s < \sigma 2s < \sigma^* 2s < \pi 2p_y = \pi 2p_z < \pi 2p_x < \pi^* 2p_y = \pi^* 2p_z < \pi^* 2p_x \ldots$$

Bond order   Bond order of a molecule can be evaluated by using the following formula

$$\text{Bond order} = \frac{(\text{No. of bonding orbitals} - \text{No. of antibonding orbitals})}{2}$$

Example 5.7.1 N_2 molecule

$$\sigma 1s^2 < \sigma^* 1s^2 < \sigma 2s^2 < \sigma^* 2s^2 < \sigma 2p_x^2 < \pi 2p_y^2 = \pi 2p_z^2$$

$$\text{Bond order} = \frac{(10-4)}{2} = 3.$$

Example 5.7.2 O_2 molecule

$$\sigma 1s^2 < \sigma^* 1s^2 < \sigma 2s^2 < \sigma^* 2s^2 < \sigma 2p_x^2 < \pi 2p_y^2 = \pi 2p_z^2 < \pi^* 2p_y^1 = \pi^* 2p_z^1$$

$$\text{Bond order} = \frac{(10-6)}{2} = 2.$$

As two unpaired electrons exist in the electronic structure of oxygen molecule, the latter shows paramagnetic characteristics. Thus, paramagnetic behaviour of oxygen molecule can well be explained with the help of molecular orbital theory.

Example 5.7.3 B_2 molecule

$$\sigma 1s^2 < \sigma^* 1s^2 < \sigma 2s^2 < \sigma^* 2s^2 < \sigma 2p_y^1 = \sigma 2p_z^1$$

As two degenerate π-orbitals are available and can accommodate two electrons in unpaired form, $\sigma 2p_x$ is placed at higher energy level with respect to $\pi 2p_y$ and $\pi 2p_z$.

$$\text{Bond order} = \frac{(6-4)}{2} = 1.$$

5.8 CHARACTERISTIC FEATURES OF MOLECULAR ORBITAL (MO) THEORY

(i) According to this theory, two wave functions of two atomic orbitals while interact with each other, two different orbitals are formed according to their mode of interaction. If the interaction is constructive a strong MO is formed, known as bonding MO, lying always at lower energy level than the corresponding atomic orbitals (AO). On the other hand if the interaction is destructive, a weak MO is formed, known as anti-bonding MO, lying always at higher energy level than the corresponding AOs. The formation of bonding and anti-bonding MOs are schematically shown in the Figs. 5.5a, b and c.

(ii) According to MO theory two separate molecular orbitals are formed and both the atomic orbitals lose their identity after the formation of MO.

(iii) It can explain the formation of odd electron molecules like H_2^+, NO, O_3, etc.

(iv) It can explain the paramagnetic behaviour of oxygen molecule.

(v) MO are filled up by electrons according to Aufbau principle, Hund's rule of maximum multiplicity and Pauli's exclusion principle.

(vi) Total number of bond or bond order can be evaluated by using MO theory.

(vii) It can successfully explain the metallic bond formation. On the basis of MO theory metals and materials can be classified into three categories: (a) conductors, (b) semiconductors, and (c) insulators. In the case of conductors the energy gap between conduction band and valence band is very small or does not exist due to overlapping of the two bands. So easy flow of electrons occur from valence band to conduction band. In the case of semiconductors the energy gap is small but only a few electrons are able to jump into the conduction band. This jumping of electrons from valence band to conduction band is favoured by thermal means. In the case of insulators electrons are unable to overcome the energy gap and hence they are unable to carry current.

(viii) The theory cannot explain the phenomenon "Resonance".

(ix) It cannot explain the geometry of a covalent compound.

5.9 CHARACTERISTIC FEATURES OF VALENCE BOND (VB) THEORY

(i) According to this theory, atomic orbitals of two elements overlap to form the covalent bond. The theory also considers that the electron behaves both as wave and particulate matter.

(ii) Atomic orbitals (AO) of two elements retain their identity even after overlapping.

(iii) If two atomic orbitals, lobes of which spread along the same axis, approach each other along that axis, head-on overlap occurs to form σ- bond. If two atomic orbitals, lobes of which spread along one axis, approach each other in different axis, perpendicular to that of the former one, lateral overlap occurs to form a weak bond, known as π- bond. The mechanism of formation of σ- and π-bonds are shown in the Figs. 5.5b and c.

(iv) Electrons in π- orbital are loosely bound and easily get delocalized under favourable conditions. Thus, VB theory well explains the phenomenon "Resonance".

(v) It explains the directional nature of covalent bond and hence geometry of covalent compounds.

(vi) It cannot explain the formation of odd electron molecules like H_2^+, NO, O_3, etc.

(vii) It cannot explain the paramagnetic behaviour of oxygen molecule.

(viii) It cannot explain the metallic bond formation.

5.10 PERIODIC TABLE

Atoms combine to form molecules but combination of atoms depends on their inherent characteristics. On the basis of inherent characteristics elements, found in the earth's crust, were arranged in a tabular form. Mendeleev was the first person who formed this table on the basis of atomic weight. The table is known as periodic table. Later, Henry Moseley established that atomic number (not the atomic weight) is the fundamental property of an atom. Then the periodic table was rearranged on the basis of atomic number. As atomic weight increases steadily with atomic number, only a minor alteration of Mendeleev's periodic table was enough to develop the modern periodic table.

Periodicity of Properties

All the properties, exhibited by an element, are governed by electronic configuration of that element. Any property changes continuously along the period starting from Group IA. After zero group the next element again belongs to Group IA and shows similar property with the former element of Group IA. The phenomenon is called periodicity of properties. Some of the properties are described below:

(i) **Ionisation potential (IP)** Amount of energy required to dislodge an electron from an outermost shell or valence shell of an atom is called 1st ionisation potential (IP) of that atom. IP value of an atom depends on electrostatic force of attraction between nucleus and electron, which is inversely proportional to the square of the distance between nucleus and electron (r). Down the group the distance 'r' increases and hence IP value decreases down the group. Along the period distance r decreases due to increase in electrostatic force of attraction and hence the IP value increases along the period.

Table 5.3   Ionisation energies of atoms

Atomic No. (Z)	Element	1st IP (eV)		Atomic No. (Z)	Element	1st IP (eV)
1	H	13.6		14	Si	8.15
3	Li	5.39		15	P	10.49
11	Na	5.14		16	S	10.36
19	K	4.34		17	Cl	12.97
4	Be	9.32		21	Sc	6.56
12	Mg	7.65		22	Ti	6.82
20	Ca	6.11		23	V	6.74
5	B	8.30		24	Cr	6.77
13	Al	5.99		25	Mn	7.44
31	Ga	6.09		26	Fe	7.89
6	C	11.26		27	Co	7.87
7	N	14.53		28	Ni	7.64
8	O	13.62		29	Cu	7.73
9	F	17.42		30	Zn	9.39

(ii) **Electron affinity**   It is the amount of energy released during the formation of a uninegative ion from its corresponding atom.

$$X(g) + e^- \longrightarrow X^-(g) + E$$

E is the amount of energy released, i.e., electron affinity of the atom X. Usually E is expressed as kJ/mole or eV/atom.

The above expression can be written as:

$$X^-(g) \rightarrow X(g) + e - E$$

So, numerically electron affinity is equal to the ionisation potential (but with opposite sign) of an anion.

Electron affinity increases along the period since electrostatic force of attraction increases. The Group-17 element fluorine (F) is just one electron short of stable Ne-structure and hence possesses very high electron affinity. Furthermore, electron affinity decreases down the group since metallic character increases.

Table 5.4   Electron affinity values of some selected atoms

Element	Electron affinity (eV)	Element	Electron affinity (eV)
H	0.754	F	3.620
Li	0.590	Cl	3.790
Be	−0.190	Br	3.540
K	0.520	I	3.290
Na	0.340	O	1.467
		N	−0.21

(iii) **Electronegativity**   Electronegativity of an atom is a measure of its ability to attract bond pair of electrons in a compound. Electronegativity depends on the electrostatic force of attraction, experienced by the nucleus. Down the group atomic radius (r) increases and electrostatic force of attraction decreases and hence electronegativity decreases while along the period atomic radius decreases and hence electronegativity increases.

Resonance energy
Consider a molecule $A - B$. As the molecule is heterogeneous, it is polar in nature. E_{A-B} is the bond energy, i.e., amount of energy required to break the bond $A - B$. The bonds $A - A$ and $B - B$ are 100% non-polar and E_{A-A} and E_{B-B} are the corresponding bond energies of $A - A$ and $B - B$ respectively. If the bond $A - B$ is 100% non-polar, the bond energy would be $\sqrt{E_{A-A} \cdot E_{B-B}}$. Difference between these two energies is known as resonance energy, represented by Δ with unit kcal/mole.

$$\Delta = \left[E_{A-B} - \sqrt{E_{A-A} \cdot E_{B-B}} \right] \text{kcal/ mole}$$

According to Pauling, the electronegativity difference $= (\chi_A - \chi_B) = 0.208\sqrt{\Delta}$, where, χ_A and χ_B are the electronegative values of A and B respectively.
Furthermore, polarity increases with the increase in electronegativity difference. For example % ionic character of HF is 45% but it decreases from HF to HI, as electronegativity decreases from F to I. % ionic character of HCl, HBr and HI are 19, 11, and 4% respectively.

Table 5.5   Pauling's electronegativity values of some selected elements

Element	Electro-negativity	Element	Electro-negativity	Element	Electro-negativity
Li	1.0	Ga	1.6	Bi	1.9
Na	0.9	In	1.7	O	3.5
K	0.8	Ti	1.8	S	2.5
Rb	0.8	C	2.5	Se	2.4
Cs	0.7	Si	1.8	Te	2.1
Be	1.5	Ge	1.8	Po	2.0
Mg	1.2	Sn	1.8	F	4.0
Ca	1.0	Pb	1.8	Cl	2.8
Sr	1.0	N	3.0	Br	2.6
Ba	0.9	P	2.1	I	2.5
B	2.0	As	2.0	At	2.2
Al	1.5	Sb	1.9		

(iv) Atomic radius   In true sense, atom has no definite boundary and hence it is impossible to determine its absolute dimension. So radius of an atom is a theoretically calculated value, considering Bohr's atomic model. According to this model, atom is considered as a circle with nucleus at its centre and the centre is surrounded by different orbits. The distance between the nucleus and the outermost orbit is called Bohr atomic radius. Along the period position of outermost shell is fixed. Only electrons are added into the shell. Thus, electrostatic force of attraction increases and atomic radius decreases but down the group new shell is added. So, distance between outermost shell and nucleus increases and hence atomic radius increases.

Table 5.6   Atomic radii of some selected atoms

Element	Atomic radius (in Å)	Element	Atomic radius (in Å)	Element	Atomic radius (in Å)
H	0.46	Be	1.12	He	0.31
Li	1.52	Mg	1.59	Ne	1.58
Na	1.86	Ca	1.97	Ar	1.88
K	2.31	Sr	2.15	Kr	2.00
Rb	2.44	Ba	2.17	Xe	2.17
Cs	2.62	Ra	2.17	Rn	2.17
Fr	2.50				

5.11 CHEMICAL BOND

According to octet theory, atoms of elements do not possess stable electronic configuration and hence they interact among themselves to achieve stable electronic configuration. This interaction between any two atoms leads to the formation of a bond. Any chemical bond arises mostly due to electronic interaction. The nature of electronic interaction determines the type of bond. So on the basis of electronic interaction, bonds may be classified into six different types:

(i) Ionic bond

(ii) Covalent bond

(iii) Coordinate or dative bond

(iv) Hydrogen bond

(v) Van der Waals type of bond

(vi) Metallic bond.

5.11.1 Ionic Bond

When a bond is formed between two atoms by complete transfer of electron(s) from one atom to another, the bond is called ionic bond. Usually strong electropositive element (low IP value) and strong electronegative element (high electron affinity) forms this type of bond, e.g., NaCl. Na-atom has one extra electron in the 3rd shell $[1s^2 2s^2 2p^6 3s^1]$ and Cl-atom has one electron short from its octet $[1s^2 2s^2 2p^6 3s^2 3p^5]$. Complete transfer of electron takes place from Na-atom to Cl-atom to form Na^+ and Cl^-. Then stable Na^+ ion attracts stable Cl^- ions from all directions and vice versa. This electrostatic force of attraction binds both Na^+ and Cl^- ions to form an ionic compound. Hence, the bond is called electrovalent or ionic bond.

Formation of an ionic bond depends on the following two factors:

(a) **Cationic charge and radius**   If the cationic charge is high or cationic radius is small, the cation can diffuse the electron cloud of anion, resulting in the formation of covalent bond. Thus, large cationic radius and low cationic charge favour the formation of an ionic bond.

(b) **Anionic charge and anionic radius**   If anionic radius is large or charge is high, electron cloud of the anion is easily diffused to form covalent bond. Thus, small anionic radius as well as low anionic charge favours the formation of an ionic bond. Strong repulsion exists in high anionic charge, favouring diffusion of anion.

The above two factors is collectively known as *Fajan's rule*. It is illustrated in Fig. 5.6.

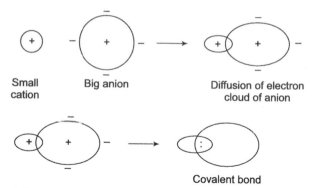

Figure 5.6 Schematic diagram of diffusion of electron cloud, explaining Fajan's rule

Characteristic Features of Ionic Bonding

(a) Ionic bond is formed by complete transfer of electron(s) between the constituent atoms. The ionic species, thus formed, are bound by electrostatic force of attraction.

(b) Ions can be regarded as charged spheres, continuously experience electrostatic forces, which are distributed uniformly in all directions in space. So any ion can attract oppositely charged ion from any direction. In other words, ionic bonds are non-directional.

(c) Each cation is surrounded by a definite number of anions and vice versa, as a result of which, a solid 3-dimensional regular crystal lattice is formed. Thus ionic compounds are usually hard, brittle, and high melting solids.

(d) In presence of a polar solvent ionic bond breaks into constituent ions. Thus ionic compounds are usually soluble in polar solvents.

(e) In solution or molten state ions are free and hence any ionic compound appears as a good conductor of electricity in solution or molten state, although it cannot conduct electricity in the solid state.

(f) Ionic compounds undergo ionic reactions in solution. Ionic reactions are fast and instantaneous.

(g) **Molar lattice energy**   Amount of energy required to dissociate one mole of an ionic crystal into its constituent ions is called molar lattice energy or in other words amount of energy evolved during formation of one mole of an ionic crystal from its constituent ions is called molar lattice energy. Also the net potential energy, contained in one mole of an ionic crystal, is called molar lattice energy of that ionic crystal. Any ionic crystal contains both positive and negative ions, bound by electrostatic force of attraction. Furthermore, each positive ion is surrounded by a definite number of negative ions and vice versa. This is called coordination number, which is defined as the total number of neighbouring anions, surrounding a cation, is called coordination number of that cation. Similarly, the total number of cations, surrounding an anion, is called coordination number of that anion. Thus, two types of forces exist in an ionic crystal:

   (i)   electrostatic forces of attraction between two oppositely charged ions, and

   (ii) repulsive forces, existing between similar ions.

If the cation and anion have the charges Z_+ and Z_- respectively, the electrostatic forces of attraction for one mole of an ionic crystal is given by

$$F = \frac{N_A \cdot A \cdot Z_+ \cdot Z_- \cdot e^2}{r^2}, \text{ where } N_A \text{ is the Avogadro number}$$

The corresponding potential energy is given by

$$W_{\text{Attraction}} = \int_r^{\infty} F \cdot dr = -\frac{N_A \cdot A \cdot Z_+ \cdot Z_- \cdot e^2}{r} \tag{5.33}$$

The factor A is introduced since attractive force is not confined within one positive ion and one negative ion but exists between one positive ion and several negative ions, surrounding it. The factor A is called Madelung constant. The equation (5.33) shows that magnitude of energy due to attractive force decreases with the increase in interionic distance.

The potential energy due to repulsive forces is given by

$$W_{\text{Repulsive}} = \frac{B}{r^n} \tag{5.34}$$

where, B is a constant and the exponent n varies from 8 to 10 depending on the crystal structure. Obviously, the repulsive potential energy decreases with the increase in the interionic distance. Combining the equations (5.33) and (5.34) the net potential energy of an ionic crystal is given by

$$V = W_{\text{Attraction}} + W_{\text{Repulsive}} = -\frac{N_A \cdot A \cdot Z_+ \cdot Z_- \cdot e^2}{r} + \frac{B}{r^n} \tag{5.35}$$

The corresponding potential energy diagram is shown in the Fig. 5.7.

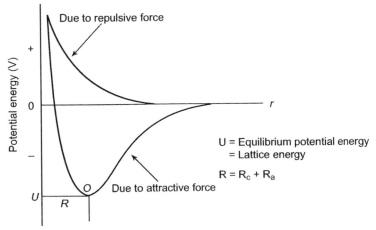

Figure 5.7   Potential energy curve of an ionic compound

It is evident from Fig. 5.7 that, there is a minimum where attractive forces are exactly balanced by the repulsive forces. The interionic distance, R, at which the minimum occurs, is called equilibrium distance between the cation and anion. Mathematically, R is given by

$$R = R_c + R_a \qquad (5.36)$$

where, R_c and R_a are the cationic radius and anionic radius respectively. At the minimum, we have

$$\left(\frac{\partial V}{\partial r}\right)_{r=R} = 0. \quad \text{Hence, we have}: \quad B = \left(\frac{N_A \cdot A \cdot Z_+ \cdot Z_- \cdot e^2 R^n}{n \cdot R}\right) \times R^n$$

Thus at equilibrium, the net potential energy is given by

$$U = -\frac{N_A \cdot A \cdot Z_+ \cdot Z_- \cdot e^2}{R} + \frac{B}{R^n} \qquad (5.37)$$

Putting the value of B in the equation (5.37), we get

$$U = -\frac{N_A \cdot A \cdot Z_+ \cdot Z_- \cdot e^2}{R}\left(1 - \frac{1}{n}\right) \qquad (5.38)$$

The value of Madelung constant 'A', varies from 1.5 to > 5, depending on the crystal structure. U, by definition, is called theoretical molar lattice energy, represented as U_{th}.

Actual Molar Lattice Energy or U_{Obs}

The actual or observed value of molar lattice energy can be evaluated experimentally and theoretical value of molar lattice energy can be determined by using Born-Haber cycle, which is based on a cyclic process of formation of an ionic crystal as shown below. Let us consider a general ionic salt, M_aX_b, where, M is usually metal or an electropositive element like Na, etc. and X is usually a diatomic electronegative element, like Cl_2, etc. ΔH_f is the molar heat of formation of the ionic solid M_aX_b,

ΔH_s is the molar heat of sublimation of the metal M,

ΔH_d is the molar heat of dissociation of X_2,

I is the molar ionization potential of the metal M,

E is the molar electron affinity of the radical $X^\bullet$,

U is the molar lattice energy of the ionic crystal M_aX_b.

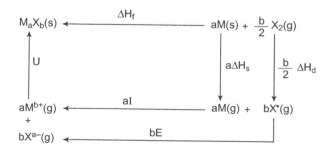

As we know that enthalpy is a state function, we can write that:

$$\Delta H_f = a\Delta H_s + \frac{b}{2}\Delta H_d + aI + bE + U \tag{5.39}$$

where, I = amount of energy required to form 1 mole of M^{b+} from 1 mole of $M(g)$.

E = amount of energy released to form 1 mole of X^{a-} from 1 mole of $X^{\bullet}(g)$.

From the equation (5.39) the lattice energy (U) can easily be evaluated if other quantities are known.

In the given cycle, the compound $M_aX_b(s)$ is considered as a perfect ionic compound. Thus the lattice energy is the theoretical lattice energy. The given cycle is known as *Born-Haber cycle*.

Significance of lattice energy

(i) It is evident from the equation (5.38) that $|U| \propto Z_+ \cdot Z_-$. So, magnitude of lattice energy increases with increase in ionic charge.

(ii) Again from equation (5.38), we have $|U| \propto \left(\dfrac{1}{R}\right)$. Thus, magnitude of lattice energy decreases with increase in interionic distance.

(iii) Percent ionic character of an ionic compound can be determined by evaluating U_{obs} (Observed lattice energy) and U_{Th} (Theoretical lattice energy). U_{obs} can be determined experimentally and U_{Th} can be determined using Born-Haber cycle.

$$\text{\% Ionic character} = \left(\frac{U_{obs}}{U_{Th}}\right) \times 100$$

5.11.2 Covalent Bond

When a bond is formed between two atoms by sharing of electrons, the bond is called covalent bond. Covalent bond is usually formed through overlapping of two atomic orbitals. Shapes of s-, p- and d-orbitals are shown in the Figs. 5.3a and b.

Electron is believed to have both wavy and particulate characteristics. The wavy characteristics of an electron is assumed to follow sinusoidal curve. If two opposite electronic waves overlap [Fig. 5.4(b)] the resultant electron density is zero in the overlapping region and the situation is known as antibonding situation. Bonding will occur if and only if two similar electronic waves overlap each other [Fig. 5.4(a)]. Furthermore, bonding overlap may be of two types:

(a) **Head-on overlap** When both the orbitals are approaching along the axis, head-on overlap takes place and the corresponding bond formed is called σ-bond [Fig. 5.8(a)].

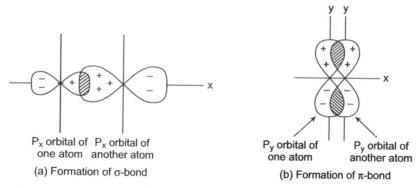

Px orbital of   Px orbital of
one atom   another atom

(a) Formation of σ-bond

Py orbital of   Py orbital of
one atom   another atom

(b) Formation of π-bond

Figure 5.8   (a) Head-on overlap: formation of σ-bond; (b) Lateral overlap: formation of π-bond

(b) **Lateral overlap**   When two orbitals overlap perpendicular to the axis of approach, a partial overlap takes place above and below the axis of approach. This type of overlap is known as lateral overlap and the corresponding bond is called π-bond [Fig. 5.8(b)].

Characteristic Features of Covalent Bonding

(a) Covalent bond is formed by sharing of electrons between constituent atoms.

(b) Covalent bond is formed when two orbitals are approaching along the same axis. Thus covalent bond is highly directional.

(c) Usually molecules of covalent compounds remain in discrete phases and hence covalent compounds exist in gaseous state. There are some covalent compounds where molecules get clustered by secondary forces, as a result of which, they exist either in liquid (e.g. water) or in low-melting solid (e.g. iodine) state.

(d) When a covalent bond is formed between two heteroatoms, bond pair of electrons is shifted to more electronegative atom, inducing polarity in the compound. Hence, compound achieves some ionic character and the % ionic character induced depends on electronegativity difference between two hetero atoms, as shown in the following chart.

Electronegativity difference	0.1	0.4	0.9	1.4	1.9	3.0	3.3
% ionic-character	0.5	4.0	19	39	60	90	100

(e) Non-polar or slightly polar compounds are usually soluble in non-polar solvents like carbon tetrachloride, benzene, ether, etc. Highly polar compounds are usually soluble in polar solvents like water, chloroform, etc.

(f) Usually covalent compounds are non-conductor of electricity. However, highly polar compounds can conduct electricity but they are not as good conductor as molten ionic compounds or aqueous solution of ionic compounds.

(g) In solution, covalent compounds undergo molecular reactions, which are slow.

5.11.3 Coordinate or Dative Bond

Sometimes it may so happen that lone pair of one atom is equally shared by two atoms. The resultant bond, thus formed, is called coordinate bond. In a coordinate bond one species (atom/molecule/anion) shares its own lone pair of electrons and is called *donor species*, while the other species accepts the lone pair of electrons of donor to its vacant orbital, and is called *acceptor species*.

Coordinate bond is of two types:

    (a) intermolecular coordinate bond, and

    (b) intramolecular coordinate bond

In intermolecular coordinate bond, one molecule of a compound shares its lone pair with another molecule of different compound.

Donor molecule is commonly known as Lewis's base and acceptor molecule is known as Lewis's acid. Examples are:

(i)
$$
\underset{\substack{\text{Lewis}\\\text{base}}}{H-\overset{\displaystyle H}{\underset{\displaystyle H}{N:}}} + \underset{\substack{\text{Lewis}\\\text{acid}}}{B-\overset{\displaystyle Cl}{\underset{\displaystyle Cl}{Cl}}} \longrightarrow \underset{\substack{\text{Lewis}\\\text{salt}}}{H-\overset{\displaystyle H}{\underset{\displaystyle H}{N:}}\longrightarrow B-\overset{\displaystyle Cl}{\underset{\displaystyle Cl}{Cl}}}
$$

(ii)
$$
\underset{\substack{\text{Lewis}\\\text{base}}}{Cl^-} + \underset{\substack{\text{Lewis}\\\text{acid}}}{FeCl_3} \longrightarrow \underset{\substack{\text{Lewis}\\\text{salt}}}{\left[Cl\longrightarrow FeCl_3\right]^-}
$$

In intramolecular coordinate bond, coordinate bond is formed within a molecule. Examples are:

$$
\underset{\text{(Nitric acid)}}{O=N\rightarrow O \atop \overset{|}{OH}} \qquad \underset{\text{(Sulphur trioxide)}}{O=S\rightarrow O \atop \overset{\downarrow}{O}} \qquad \underset{\text{(Alkyl isocyanate)}}{R-N\overset{\longrightarrow}{=}C}
$$

5.11.4 Hydrogen Bond

When hydrogen atom is covalently bonded with a strongly electronegative atom such as F, O, N, Cl, etc. polarity or partial charge separation of the molecule is sufficient to bring positively charged H-atom of one molecule close to an electronegative atom of different molecule of same compound, resulting in the formation of a unique linkage through hydrogen atom. This linkage is called hydrogen bond, e.g., hydrogen bond in HF.

$$
\cdots\cdots\overset{\delta+}{H}-\overset{\delta-}{F}\cdots\cdots\overset{\delta+}{H}-\overset{\delta-}{F}\cdots\cdots\overset{\delta+}{H}-\overset{\delta-}{F}\cdots\cdots
$$

Hydrogen bond

Characteristic Features of Hydrogen Bond

(a) Hydrogen bond is very weak and bond energy usually lies within 3–10 kcal/mole, whereas covalent bond energy is of the order of 100 kcal/mole.

(b) Hydrogen bond is directional and electrostatic in nature.

(c) There are three types of hydrogen bonds usually observed in different compounds, e.g.,

    (i) intermolecular,

    (ii) intramolecular,

    (iii) interatomic.

    (i) **Intermolecular hydrogen bond**   When hydrogen bond is formed between two different molecules, it is called intermolecular hydrogen bond. One example is *p*-Nitrophenol.

    (ii) **Intramolecular hydrogen bond**   When hydrogen bond is formed within a single molecule it is called intramolecular hydrogen bond, e.g., *o*-Nitrophenol.

    (iii) **Interatomic hydrogen bond**   In some cases hydrogen bond is strong enough to abstract the electronegative atom from another molecule and a stable anion is formed. One example is hydrogen fluoride, having formula: $(HF)_n$.

$[HF_2]^-$ is very much stable, $K[HF_2]$ crystal is known. In $[HF_2]^-$ three atoms are attached through hydrogen bond. This type of hydrogen bond is called interatomic hydrogen bond. Hydrogen bond is strongest in the case of interatomic hydrogen bond and the corresponding bond energy varies from 20 kcal/mole to 36 kcal/mole.

(d) **Association**   There are some compounds where hydrogen bond is confined within two or three molecules. Those compounds exist as dimeric or trimeric form, e.g., benzoic acid exists in dimeric form in benzene.

(e) If hydrogen bond is strong enough to associate numerous molecules at a time, boiling point or melting point of that compound increases abnormally. Among hydrogen halides, hydrogen bond is strongest in hydrogen fluoride (molecular formula HF) . Numerous molecules of HF are attached through hydrogen bond, as a result of which, hydrogen fluoride appears as liquid of BP 19°C. Strength of hydrogen bond is weaker in other hydrogen halides (HCl, HBr, HI) which are all gases. Due to the same reason, water (molecular formula H_2O) is a liquid of BP 100°C but hydrogen sulphide (H_2S) is a gas. In case of water, hydrogen bond, under certain conditions, may extend tetrahedrally in four directions, resulting in the formation of solid ice crystals of MP 0°C.

As the hydrogen bonds are longer the ice structure is quite loose and contains sufficient empty space. This is the primary reason for its low density and capacity to form clathrate compounds.

(f) New compound may be formed through hydrogen bond. One example is ammonia hydrate. Aqueous ammonia, when cooled, forms the crystals of $H_3N \cdot H_2O$, where ammonia and water are bound by hydrogen bond.

(g) Formation of hydrogen bond and its breakage takes place easily at normal temperature and this process is believed to be essential for biological process.

(h) **Solubility** If molecules of a polar solvent are able to form hydrogen bond with solute molecules, the latter partly or totally dissolves into the solvent. Phenol is sufficiently soluble in water but thiophenol is only sparingly soluble in water because water forms strong hydrogen bond with phenol.

5.11.5 Bond due to Intermolecular Forces (Van der Waals Type of Bond)

In case of a heterogeneous molecule, like H–F, partial charge separation always occurs due to electronegativity differences, resulting in the formation of polar molecule $H^{q+}-F^{q-}$. If r represents bond length, $r \times q$ represents the dipole moment and the molecule is called dipole. Apparently dipole moment of a non-polar molecule (e.g. Br_2) is zero. A molecule is a dynamic system where electrons are continuously revolving around nuclei which are in constant oscillation. Hence distribution of charges surrounding the nuclei may not be uniform, as a result of which, electron cloud shifts towards either atom with equal probability ($Br_2 \equiv Br^{\delta+}-Br^{\delta-}$ and $Br^{\delta-}-Br^{\delta+}$), rendering the formation of instantaneous micro-dipoles. As the probability of shifting electron cloud towards either atom is equal, the resultant dipole moment is zero. The positive charge of one instantaneous micro-dipole attracts the electron cloud of another molecule, rendering the formation of new instantaneous micro-dipole. These fluctuating dipoles are responsible for developing mutual attraction among the molecules.

This is known as intermolecular forces of attraction and the hypothetical bond developed among the molecules is known as van der Waals type of bond. It is mostly observed in non-polar or slightly polar compounds.

Characteristic Features of Van der Waals Type of Bond

(a) Van der Waals type of bond is very weak and energy associated with it is only about 1 kcal/mole whereas 100 kcal/mole of energy is required to disrupt covalent forces.

(b) Van der Waals type of forces are operative when intermolecular distance is about 3.5 to 5 Å.

(c) The magnitude of van der Waals force of attraction increases with increase in number of electrons per molecule and hence also with increase in molecular mass. In the cases of heavy molecules, magnitude of intermolecular forces of attraction is high enough to condense discrete molecules into solids, e.g. iodine and sulphur.

5.11.6 Metallic Bond

In the case of electronegative elements, covalent bonds are formed by sharing of valence electrons but in the case of metals this sharing of electrons is not possible due to their electropositive character. Thus to fulfil their octet, numerous metal atoms get attached in a regular manner to form a three dimensional crystal lattice. The lattice energy, thus evolved, is enough to dislodge the valence shell electrons from all metal atoms. These mobile electrons are not free enough to leave the metallic crystal but can move throughout the crystal lattice as gas molecules. This cloud of valence electrons is often called electron gas and positive metal ions, thus created, are called kernels [Fig. 5.9].

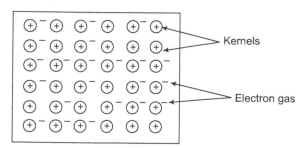

Figure 5.9 Schematic diagram of metal lattice

All the kernels are linked through electron gas and the bond is called metallic bond. The theory, explaining the metallic bond, is called free electron theory.

Merits and Demerits of Free Electron Theory

(a) **Metallic lustre** Loosely bound electrons or electron gas readily reflect sunlight, striking on the metal surface. Metals are appearing lustrous due to this reflection of light.

(b) Mobile electrons in a metal are indistinguishable and hence each and every kernel makes an attempt to share all the electrons but not at a time. In order to fulfil this, layers of kernels slip past one another, as a result of which, metals become malleable and ductile. However, presence of a foreign atom restricts the layer slip and hence brittleness appears. Thus, copper is ductile but brass is brittle.

(c) The free moving electron gas is responsible for high electrical and thermal conductivity of metals. Furthermore, thermionic and photoelectric emissions are well explained with the help of electron gas theory.

(d) Electron gas theory fails to explain the specific heat of a metal. The specific heat of an ideal gas, according to kinetic theory, is 3 cal/mole. Specific heat of a metal is expected to be very high due to the presence of free electron gas but in actual practice, specific heat of any metal is observed around the same as that an ideal gas.

SOLVED PROBLEMS

Problem 5.1 Show that Bohr's condition for the quantization of angular momentum in the hydrogen atom is given by:

$$mvr = \frac{nh}{2\pi}$$

Solution

Let us consider the nth orbit of radius r. So circumference of the orbit is $2\pi r$. Now wavelength of electron is λ. Thus to cover the $2\pi r$ distance equivalent wavelength is $n\lambda$, where, $n = 1$, 2, 3, ...

Hence, we can write $\quad 2\pi r = n\lambda$

Again, using de Broglie concept, we have

$$\lambda = \frac{h}{mv}$$

Putting this value of λ in the above equation, we get

$$2\pi r = \frac{nh}{mv}. \quad \text{Or, } mvr = \frac{nh}{2\pi}$$

Problem 5.2 Calculate the de Broglie wavelength of an electron in the first Bohr orbit in the hydrogen atom. Does quantum or classical mechanics apply to this electron?

Solution

$$mvr = \frac{nh}{2\pi} \quad \text{and } \lambda_B = \frac{h}{mv}. \quad \text{Or, } \lambda_B = \frac{2\pi r}{n}$$

For hydrogen atom, $n = 1$.

$$r = \frac{h^2}{4\pi^2 e^2 m} = 5.29 \times 10^{-11}\, m \text{ and } \lambda_B = 2\pi r = 2\pi \times 5.29 \times 10^{-11}\, m = 3.32\, \mathring{A}$$

Since the de Broglie wavelength is the same order of magnitude as the path that the electron travels (in this case, it equals the circumference of the orbit, $2\pi r$), quantum mechanics should be applied to the electron.

Problem 5.3 Assuming that the ionic character in H–Br bond is 20%. Calculate the fraction of the contribution of the ionic character to the valence bond wave function.

Solution

The combined wave function of valence bond is the sum of wave function of covalent bond and that of ionic bond. Mathematically, it can be written as

$$\psi_{total} = \psi_{covalent} + \lambda\psi_{ionic}$$

where, λ indicates fraction of contribution of ionic character. It is related to % ionic character as follows

$$\% \text{ ionic character} = \left(\frac{\lambda^2}{1+\lambda^2}\right) \times 100. \quad \text{So, } 20 = \left(\frac{\lambda^2}{1+\lambda^2}\right) \times 100$$

$$\left(\frac{1+\lambda^2}{\lambda^2}\right) = 5. \quad \text{Or, } 4\lambda^2 = 1. \quad \text{Or, } \lambda = \frac{1}{2}$$

Thus, the fraction of contribution of ionic character to ψ_{total} is 0.5.

Problem 5.4 State the rules for constructing the wave functions for the hybrid orbitals.

Solution

Rule 1: Wave function of any i^{th} hybrid orbital is formed by linear combination of corresponding atomic orbitals. For example, if hybrid orbital is formed by s- and p-orbitals, the corresponding hybrid wave function is given by

$$\psi_i = a_i\psi_s + b_i\psi_{px} + c_i\psi_{py} + d_i\psi_{pz}$$

where, ψ_s, ψ_{px}, ψ_{py} and ψ_{pz} are the atomic orbital wave functions.

Rule 2: Each wave function is normalized. So, summation of squares of coefficients of components of wave function is unity. Hence, we can write

$$a_i^2 + b_i^2 + c_i^2 + d_i^2 = 1$$

Rule 3: In a set of hybrid orbitals one hybrid orbital is orthogonal to other hybrid orbitals. Thus,

$$a_i a_j + b_i b_j + c_i c_j + d_i d_j = 0$$

Rule 4: As s-orbital is spherical, each equivalent hybrid orbital of a set contains $1/\sqrt{n}$ of the s-orbital, which is distributed among n orbitals.

EXERCISES

1. Isostructural species are those which have the same shape and hybridization. Among the given species identify the isostructural pairs with reasons. (i) NF_3 and BF_3 (ii) BF_4^- and NH_4^+ (iii) BCl_3 and $BrCl_3$ (iv) NH_3 and NO_3^-.

2. Polarity in a molecule and hence the dipole moment depends primarily on electronegativity of the constituent atoms and shape of a molecule. Which of the given has the highest dipole moment? (i) CO_2 (ii) HI (iii) H_2O (iv) SO_2.

3. In PO_4^{3-} ion the formal charge on the oxygen atom of P–O bond is: (i) +1 (ii) –1 (iii) –0.75 (iv) +0.75.

4. In NO_3^- ion, the number of bond pairs and lone pairs of electrons on nitrogen atom are:
   (i) 2, 2 (ii) 3, 1 (iii) 1, 3 (iv) 4, 0.

5. Which of the following order of energies of molecular orbitals of N_2 is correct?
   (i) $(\pi 2py) < (\sigma 2pz) < (\pi^* 2px) \approx (\pi^* 2py)$
   (ii) $(\pi 2py) > (\sigma 2pz) > (\pi^* 2px) \approx (\pi^* 2py)$
   (iii) $(\pi 2py) < (\sigma 2pz) > (\pi^* 2px) \approx (\pi^* 2py)$
   (iv) $(\pi 2py) > (\sigma 2pz) < (\pi^* 2px) \approx (\pi^* 2py)$

6. Explain the following:
   (i) He_2 is not stable but He_2^+ is expected to exist
   (ii) Be_2 is not a stable molecule
   (iii) Bond strength of N_2 is maximum amongst the homonuclear diatomic molecules belonging to the second period.

7. Amongst the following elements whose electronic configurations are given below, the one having the highest ionisation enthalpy is:
   (i) $[Ne]3s^2 3p^1$ (ii) $[Ne]3s^2 3p^2$ (iii) $[Ne]3s^2 3p^2$ (iv) $[Ar]3d^{10} 4s^2 4p^3$.

8. Calculate the bond order in the following species:
   (i) CN^- (ii) NO^+ (iii) O^{2-} (iv) O_2^{2-} (v) N_2 (vi) N_2^- (vii) F_2^+ (viii) O_2^-.

9. Identify the geometry of the following compounds:
   (i) CO_2 (ii) CCl_4 (iii) O_3 (iv) NO_2.

10. Identify magnetic properties (paramagnetic or diamagnetic) of the following species:
    (i) N_2 (ii) N_2^{2-} (iii) O_2 (iv) O_2^{2-}.

11. Explain the nonlinear shape of H_2S and non-planar shape of PCl_3 using valence shell electron pair repulsion theory.
12. Explain the shape of BrF_5.
13. Explain why PCl_5 is trigonal bipyramidal whereas IF_5 is square pyramidal.
14. "In both water and dimethyl ether, oxygen atom is central atom, and has the same hybridization, yet they have different bond angles." – explain.
15. Write Lewis structure of the following compounds and show formal charge on each atom. HNO_3, NO_2 and H_2SO_4.
16. Write the complete sequence of energy levels in the increasing order of energy in the molecule. Compare the relative stability and the magnetic behaviour of the following species. N_2, N_2^+, N_2^- and N_2^{2+}.
17. What is the effect of the following processes on the bond order in N_2 and O_2?
    (i) $N_2 \rightarrow N_2^+ + e^-$,　(ii) $O_2 \rightarrow O_2^+ + e^-$
18. Give reasons for the following:
    (i) Covalent bonds are directional bonds while ionic bonds are non-directional.
    (ii) Water molecule has bent structure whereas carbon dioxide molecule is linear.
    (iii) Ethyne molecule is linear.
19. What is an ionic bond? With two suitable examples explain the differences between an ionic and a covalent bond.
20. Arrange the following bonds in order of increasing ionic character giving reasons.
    N–H, F–H, C–H and O–H.
21. Explain why CO_3^{2-} ion cannot be represented by a single Lewis structure. How can it be best represented?
22. Elements *X*, *Y* and *Z* have 4, 5 and 7 valence electrons respectively.
    (i) Write the molecular formula of the compounds formed by these elements individually with hydrogen.
    (ii) Which of these compounds will have the highest dipole moment?
23. Draw the resonating structure of: (i) Ozone molecule, (ii) Nitrate ion.
24. Predict the shapes of the following molecules on the basis of hybridization:
    BCl_3, CH_4, CO_2 and NH_3.
25. What is meant by the term average bond enthalpy? Why does bond enthalpy of O–H bond in ethanol and water differ?
26. The first 4 lines in the visible region of atomic line spectrum of hydrogen atom occur at wavelengths of 656.2, 486.1, 434.0 and 410.2 nm (this is known as the Balmer series). Using Bohr's theory, show that these correspond to transitions in which the smaller integer in the equation is always $n = 2$.
27. The ionization potential of an atom is defined as the energy required to completely remove an electron. Using Bohr's theory, calculate the ionization potential of the hydrogen atom.
    　　　　　　　　　　　　　　　　　　　　　　　　　　　　[*Ans.:* 13.6 eV]

28. Determine the absorption frequency in wave numbers (cm^{-1}) for an electron in the hydrogen atom undergoing a transition from the $n = 2$ level to the $n = 5$ level.

[*Ans.:* $\overline{v} = 23045\,cm^{-1}$]

29. The threshold wavelength for the ejection of photoelectrons from sodium metal is 5420 Å. Calculate the velocity of photoelectrons ejected by light of wavelength 4000 Å.

[*Ans.:* 5.35×10^5 m/s]

6 Chemical Thermodynamics

INTRODUCTION

6.1.1 Thermodynamics

Study of heat energy, available from any thermal source, is called thermodynamics. Flow of heat energy between any two states depends on temperature difference ΔT and hence the latter is called the driving force for heat flow. To study the thermodynamics the following terms are to be defined:

Universe   The total material world is called universe.

System   It is that part of the universe, under consideration and separated from the rest by a definite boundary. The boundary or wall of a system can be divided into two broad categories.

Diathermal wall   It refers to that boundary which permits heat transfer. One example is copper wall.

Adiabatic wall   It refers to that boundary through which no heat transfer is allowed. One example is thermo flask.

Surroundings   Besides system the rest of the universe is called surroundings.
   Thermodynamic system can be divided into five broad categories.

Thermodynamic system	Mass transfer	Heat transfer	Work transfer
Open system	Allowed	Allowed	Allowed
Closed system	Not allowed	Allowed	Allowed
Mechanically Isolated system	Not allowed	Allowed	Not allowed
Thermally Isolated system	Not allowed	Not allowed	Allowed
Isolated system	Not allowed	Not allowed	Not allowed

Isolated system   A system is said to be isolated if both mass and energy transfer with the surroundings are not allowed. In true sense no such isolated system exists. However, the universe can be called an isolated system since it has no surroundings. Hot tea in a thermo flask is an example of an isolated system.

Closed system   A system is said to be closed if only energy but not mass transfer with the surroundings is allowed. Hot tea in a metallic container is an example of closed system.

Open system   A system is said to be open if both mass and energy transfer with the surroundings are allowed. Pond, river, etc. are the examples of open system. Human body is also an example of open system.

Thermodynamic Parameters   Thermodynamic properties of a system are studied with the help of some parameters called thermodynamic parameters, such as pressure (P), volume (V), temperature (T), etc.

Intensive property   Consider a container, full of gas of volume V and pressure P as shown in the Fig. 6.1.

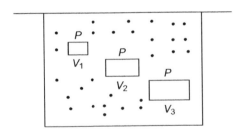

Figure 6.1   Gas pressure $= P$ and volume $= V = \sum V_i$

Measurement of gas pressure at any point within the container indicates the pressure of the whole system which means that measurement of pressure remains same at any point, irrespective of mass. Those properties, which are independent of mass, are called intensive properties. Examples are, pressure, temperature, density, etc.

Extensive property   In the previous example measurement of volume at any point within the container indicates only the partial volume and does not indicate the total volume of the system, which means that measurement of volume differs from point to point and hence depends on mass. Those properties, which depend on mass, are called extensive properties.

Examples are, volume, any form of energy, etc. For example, kinetic energy $= \dfrac{1}{2}mv^2$. So it depends on mass and so do other forms of energy.

Ratio of two extensive properties becomes an intensive property, one such example is specific volume,

$$\bar{v} = \left(\frac{V}{m}\right),$$ where, V is the total volume and m is the mass of the gas. Another example

is density, $\rho = \dfrac{m}{V}$

6.1.2 Classification of Thermodynamic Processes

Thermodynamic process depends on the conditions, employed during transfer of a thermodynamic system from one state to another.

   (a) Thermodynamic process may be carried out at constant pressure. Then it is called isobaric process.
   (b) When volume is maintained constant, the process is called isochoric process.
   (c) In some thermodynamic processes temperature is kept constant. Then it is called isothermal process.
   (d) In other thermodynamic processes heat transfer between system and surroundings is not allowed. It is called adiabatic process.

Over and above two other thermodynamic processes are also considered:
   (i) Reversible process, and (ii) Irreversible process.

(i) Reversible process   It is that process which proceeds through numerous steps and equilibrium is established in each and every step. Thus infinite time is required to complete a reversible process. So it is a non-flow process, also known as **Quasi-Static** process. Furthermore, at any point the physical state of a system remains same when the reaction proceeds in the forward direction as well as in the backward direction. It is well illustrated in the Fig. 6.2.

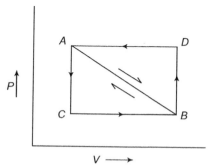

Figure 6.2   *PV–diagram of reversible and irreversible processes*

A and B are the two states of a system and AB indicates the reversible path. The double arrow indicates that forward process $A \rightarrow B$ and backward process $B \rightarrow A$ follow the same path.

Quasi-Static process   A quasi-static process is one in which: (i) the deviation from thermodynamic equilibrium is infinitesimal, (ii) all states, concerned with the process, are equilibrium states.

(ii) Irreversible process   Irreversible process is the one which proceeds spontaneously without establishment of any equilibrium. Furthermore at any point the physical state of a system differs for backward direction from that for forward direction as is evident from Fig. 6.2, where ACB shows the forward process $A \to B$ and BDA shows the backward process $B \to A$.

Cyclic process

Any process, during which a system changes its state through some changes of properties and finally returns to its original state, is called a cyclic process.

6.2   ZEROTH LAW OF THERMODYNAMICS

Statement

If two systems are in thermal equilibrium with the 3rd system separately, the above two systems will also be in thermal equilibrium. Consider two systems A and B, separated by an adiabatic wall as shown in the Fig. 6.3. Each of the two systems A and B are separately in equilibrium with the third system C through a diathermal wall. After some time thermal equilibrium (equality of temperature) is established between the systems A and B, even though they are separated by an adiabatic wall.

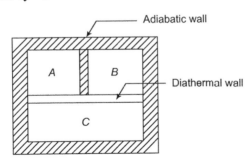

Figure 6.3   Illustration of zeroth law of thermodynamics

6.3   STATE FUNCTION

The function which is independent of path but depends only on state of a system is called state function. For any cyclic process, initial and final states are same and hence equation (6.1a) holds good for each state function.

$$\oint dM = 0 \qquad\qquad (6.1a)$$

where M is a state function. Let us assume that $M = f(x, y)$

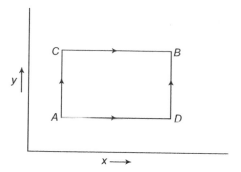

Figure 6.4   *xy*–diagram of change of state function *M* from *A → B*

Consider a system, changing its state from point *A* to point *B* (Fig. 6.4). The change of *M* may proceed along *ACB* or along *ADB*.

The path *ACB* consists of two steps *AC* and *CB*. Along *AC* the change of *M* is given by $\left(\dfrac{\partial M}{\partial y}\right)_x$ and along *CB* the change of $\left(\dfrac{\partial M}{\partial y}\right)_x$ is given by $\dfrac{\partial^2 M}{\partial x \cdot \partial y}$.

On the other hand, the path *ADB* consists of two steps *AD* and *DB*. Along *AD* the change of *M* is given by $\left(\dfrac{\partial M}{\partial x}\right)_y$ and along *DB* the change of $\left(\dfrac{\partial M}{\partial x}\right)_y$ is given by $\dfrac{\partial^2 M}{\partial y \cdot \partial x}$.

As *M* is a state function, the net change should be same, irrespective of path followed. Thus, we can write

$$\frac{\partial^2 M}{\partial x \cdot \partial y} = \frac{\partial^2 M}{\partial y \cdot \partial x} \tag{6.1b}$$

Thus the equations (6.1a) and (6.1b) hold good for any state function. Furthermore, it can be inferred that any property, *M*, for which equation (6.1b) holds good, *dM* is called perfect differential.

Example 6.3.1   Show that *P* is a state function for a real gas obeying the following equation

$$\left(P + \frac{a}{V^2}\right)V = RT$$

Answer

The equation shows that $P = f(V, T)$. The above equation can be written as

$$P = \frac{RT}{V} - \frac{a}{V^2}$$

Hence,
$$\left(\frac{\partial P}{\partial V}\right)_T = -\frac{RT}{V^2} + \frac{2a}{V^3}. \text{ Or, } \frac{\partial^2 P}{\partial T \cdot \partial V} = -\frac{R}{V^2}$$

$$P = \frac{RT}{V} - \frac{a}{V^2}$$

Again,
$$\left(\frac{\partial P}{\partial T}\right)_V = \frac{R}{V}. \text{ Or, } \frac{\partial^2 P}{\partial V \cdot \partial T} = -\frac{R}{V^2}. \text{ Thus, } \frac{\partial^2 P}{\partial V \cdot \partial T} = \frac{\partial^2 P}{\partial T \cdot \partial V}$$

So, P is a state function.

Example 6.3.2   Show that temperature is a state function.

Answer

For any cyclic process, initial and final temperatures are same. So, $\oint dT = 0$. Thus temperature is a state function.

Example 6.3.3   Show that volume is a state function for a real gas obeying the following equation. $P(V - b) = RT$ for 1 mole of a real gas.

Answer

The equation shows that $V = f(P, T)$.
The above equation can be written as

$$V = \frac{RT}{P} + b$$

Hence,
$$\left(\frac{\partial V}{\partial P}\right)_T = -\frac{RT}{P^2} \text{ and } \frac{\partial^2 V}{\partial T \cdot \partial P} = -\frac{R}{P^2}$$

Again
$$\left(\frac{\partial V}{\partial T}\right)_V = \frac{R}{P}. \text{ Or, } \frac{\partial^2 V}{\partial P \cdot \partial T} = -\frac{R}{P^2}. \text{ Thus, } \frac{\partial^2 V}{\partial P \cdot \partial T} = \frac{\partial^2 V}{\partial T \cdot \partial P}$$

So, V is a state function.

Example 6.3.4   Show that for an ideal gas work done (dW) is not a perfect differential.

Assumptions

    1. Only expansion work is considered.
    2. Gas is considered an ideal gas.

We know $V = f(P, T)$. On differentiation, we get

$$dV = \left(\frac{\partial V}{\partial T}\right)_P dT + \left(\frac{\partial V}{\partial P}\right)_T dP$$

Again for 1 mole of an ideal gas, we have

$$V = \frac{RT}{P}. \text{ Hence, } \left(\frac{\partial V}{\partial T}\right)_P = \frac{R}{P} \text{ and } \left(\frac{\partial V}{\partial P}\right)_T = -\frac{RT}{P^2}$$

Hence,

$$dV = \left(\frac{R}{P}\right)dT - \left(\frac{RT}{P^2}\right)dP$$

For expansion work, we have

$$dW = PdV = RdT - \left(\frac{RT}{P}\right)dP = RdT - VdP$$

Thus,

$$\left(\frac{\partial W}{\partial T}\right)_P = R \text{ and } \frac{\partial^2 W}{\partial P \cdot \partial T} = 0$$

Again,

$$\left(\frac{\partial W}{\partial P}\right)_T = -V \text{ and } \frac{\partial^2 W}{\partial T \cdot \partial P} = -\left(\frac{\partial V}{\partial T}\right)_P$$

Thus,

$$\frac{\partial^2 W}{\partial P \cdot \partial T} \neq \frac{\partial^2 W}{\partial T \cdot \partial P}$$

Hence, dW is not a perfect differential.

Example 6.3.5 Show that volume is a state function for a real gas obeying the following equation.

$$\left(P + \frac{a}{V^2}\right)V = RT$$

Answer

The above equation can be written as

$$PV + \frac{a}{V} = RT$$

Or,

$$PV = RT - \frac{a}{V} \tag{A}$$

Differentiating w.r.t. temperature at constant P, we get

$$P\left(\frac{\partial V}{\partial T}\right)_P = R - \frac{a}{V^2}\left(\frac{\partial V}{\partial T}\right)_P. \text{ Or, } \left(P + \frac{a}{V^2}\right)\left(\frac{\partial V}{\partial T}\right)_P = R$$

Or,

$$\left(\frac{RT}{V}\right)\left(\frac{\partial V}{\partial T}\right)_P = R. \text{ Or, } \left(\frac{\partial V}{\partial T}\right)_P = \frac{V}{T} \tag{B}$$

Hence, $$\frac{\partial^2 V}{\partial P \cdot \partial T} = \frac{1}{T}\left(\frac{\partial V}{\partial P}\right)_T \qquad (C)$$

Again, differentiating equation A, w.r.t. pressure at constant T, we get

$$V + P\left(\frac{\partial V}{\partial P}\right)_T = -\frac{a}{V^2}\left(\frac{\partial V}{\partial P}\right)_T$$

$$\left(P + \frac{a}{V^2}\right)\left(\frac{\partial V}{\partial P}\right)_T = -V. \quad \text{Or,} \quad \left(\frac{RT}{V}\right)\left(\frac{\partial V}{\partial P}\right)_T = -V$$

Or, $$\left(\frac{\partial V}{\partial P}\right)_T = -\frac{V^2}{RT} \qquad (D)$$

Hence, $$\frac{\partial^2 V}{\partial T \cdot \partial P} = \frac{V^2}{RT^2} - \frac{2V}{RT}\left(\frac{\partial V}{\partial T}\right)_P$$

Using equation (B), we get

$$\frac{\partial^2 V}{\partial T \cdot \partial P} = \frac{V^2}{RT^2} - \left(\frac{2V}{RT}\right)\left(\frac{V}{T}\right) = -\frac{V^2}{RT^2} \qquad (E)$$

Combining equations (C) and (D), we get

$$\frac{\partial^2 V}{\partial P \cdot \partial T} = -\frac{V^2}{RT^2} \qquad (F)$$

Hence, from equations (E) and (F), we get

$$\frac{\partial^2 V}{\partial T \cdot \partial P} = \frac{\partial^2 V}{\partial P \cdot \partial T}$$

Thus, V is a state function.

6.4 FIRST LAW OF THERMODYNAMICS

Statement

Sum total energy of the universe is constant. Energy cannot be created nor be destroyed. If some amount of energy disappears from one part of the universe, equivalent amount of energy reappears in different form in another part of the universe. If E represents the total energy of the universe, we have

$E = $ Constant.   Or, $dE = 0$

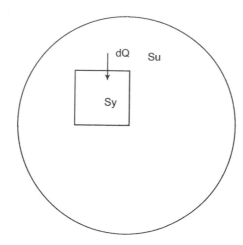

In the above figure, a system (Sy) absorbs dQ amount of heat energy from its surroundings (Su). Then dQ is called energy input to the system. Part of this dQ is converted to work, dW and the other part is used to increase the internal energy of the system, dU. Thus the total change in energy of the universe, dE is given by

dE = Energy lost by surroundings (dQ) – Energy gained by the system

Energy gained = work delivered by the system (dW) and energy absorbed by the system (dU) = $dW + dU$.

Thus, dE is given by

$$dE = dQ - (dU + dW). \quad \text{Or,} \quad dQ - (dU + dW) = 0$$

Or, $$dQ = dU + dW \tag{6.2}$$

Equation (6.2) is the mathematical form of 1st law of thermodynamics.

6.5 INTERNAL ENERGY (*U*)

Definition   Each and every system possesses some inherent energy, in the form of kinetic energy, potential energy, rotational energy, vibrational energy, etc. related to structure and chemical nature of the components. The total inherent energy content of a given system is called internal energy of that system. It is represented by the symbol U.

From equation (6.2), we have $dU = dQ - dW$.

Considering only expansion work, we have $dU = dQ - PdV$ $\tag{6.2a}$

Differentiating the equation with respect to T at constant V, we get

$$\left(\frac{\partial U}{\partial T}\right)_V = \left(\frac{\partial Q}{\partial T}\right)_V = C_V$$

Thus at constant V, we can write that $\quad dU = C_V dT \qquad\qquad$ (6.3)

The given equation holds good for 1 mole of an ideal gas only.

Hence for an isothermal process $\quad dU = 0$.

6.5.1 Physical Significance of Internal Energy

Internal energy of a system is the amount of energy possessed inherently by that system. Any molecule of a system may possess different types of energy, e.g. translational, rotational, vibrational, electronic energy and also energy due to intermolecular forces of attractions and repulsions.

Translational energy arises due to translational motion of molecules, represented by E_{tr}.

If a body changes its spatial orientation or conformation through rotation the energy associated with it is called **rotational energy,** represented by E_{rot}.

In the case of **vibrational energy,** atoms of a molecule are fixed at their positions but oscillate about their equilibrium positions. The energy associated with it is called vibrational energy, represented by E_{vib}.

Electronic energy In each molecule electrons usually remain in the ground state. In a chemical reaction electronic configuration of the product molecules differ from those of the reactant molecules and hence a change in electronic energy is observed. Energy associated with this change in electronic configuration is called electronic energy represented by E_{el}.

Intermolecular forces of attraction A molecule when comes in contact with another molecule experiences an intermolecular attraction and repulsion. Intermolecular force of attraction is maximum in a solid phase where molecules remain in most orderly state and is minimum in gaseous phase. The energy associated with it is represented by E_{interm}.

Bond energy In a molecule, chemical bond exists and energy associated with this chemical bond is called bond energy.

Thus, for a gas or liquid the total internal energy, U, is given by

$$U = E_{tr} + E_{rot} + E_{vib} + E_{el} + E_{interm} + \cdots$$

6.5.2 Absolute Value of Internal Energy

For any given system, internal energy includes all type of inherent energies, e.g., kinetic energy, potential energy, rotational energy, vibrational energy, electronic energy, etc. Absolute values of these individual energies cannot be determined and hence *absolute value of internal energy cannot be determined*. So, only *change in internal energy*, ΔU, can be determined.

6.6 ENTHALPY (*H*)

For a given thermodynamic system the total energy content of the system is called enthalpy of that system. It includes internal energy of the system as well as energy due to PV. It is represented by the symbol H.

6.6.1 Relation Between *H* and *Q* for an Ideal Gas

We know,
$$dQ = dU + dW = dU + PdV \quad \text{[considering expansion work only]}$$

$$U = f(P, T)$$

Hence,
$$dU = \left(\frac{\partial U}{\partial P}\right)_T dP + \left(\frac{\partial U}{\partial T}\right)_P dT = \left(\frac{\partial U}{\partial T}\right)_P dT,$$

$$\text{since, } \left(\frac{\partial U}{\partial P}\right)_T = 0 \text{ for ideal gases}$$

$$V = f(P, T)$$

Hence,
$$dV = \left(\frac{\partial V}{\partial P}\right)_T dP + \left(\frac{\partial V}{\partial T}\right)_P dT$$

Putting the values of dU and dV in the above equation, we get

$$dQ = \left(\frac{\partial U}{\partial T}\right)_P dT + P\left[\left(\frac{\partial V}{\partial P}\right)_T dP + \left(\frac{\partial V}{\partial T}\right)_P dT\right]$$

$$= P\left(\frac{\partial V}{\partial P}\right)_T dP + \left[\left(\frac{\partial U}{\partial T}\right)_P + P\left(\frac{\partial V}{\partial T}\right)_P\right] dT$$

Again,
$$\left(\frac{\partial Q}{\partial T}\right)_P = \left(\frac{\partial U}{\partial T}\right)_P + P\left(\frac{\partial V}{\partial T}\right)_P$$

So,
$$dQ = P\left(\frac{\partial V}{\partial P}\right)_T dP + \left(\frac{\partial Q}{\partial T}\right)_P dT = P\left(\frac{\partial V}{\partial P}\right)_T dP + C_P dT$$

$$\left[\text{Since, } \left(\frac{\partial Q}{\partial T}\right)_P = C_P \text{ (See sec. 6.10)}\right]$$

For 1 mole of an ideal gas, we have $PV = RT$

Hence,
$$P\left(\frac{\partial V}{\partial P}\right)_T = -V$$

Thus,
$$dQ = -VdP + C_P dT. \quad \text{Or, } \quad C_P dT = dQ + VdP.$$

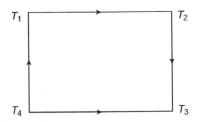

As T is a state function, $\oint dT = 0$. Hence, $\oint C_p dT = 0$. Thus, $C_p dT$ is a state function represented as dH.

So,
$$C_p dT = dH = dQ + VdP. \quad \text{Or,} \quad dH = dQ + VdP \tag{6.4}$$

Again, $dH = C_p dT$. Or, $C_P = \left(\dfrac{\partial H}{\partial T}\right)_P$. Hence, $dQ = C_p dT - VdP$ (6.5)

Equation (6.4) is also another mathematical expression of 1st law of thermodynamics. So there are three mathematical expressions of 1st law of thermodynamics, given below

$$dQ = dU + dW \tag{6.2},$$

$$dQ = dU + PdV \tag{6.2a}$$

$$dQ = dH - VdP \tag{6.4}$$

6.6.2 Relation Between *H* and *U* for an Ideal Gas

From equations (6.2a) and (6.4), we have

$$dH = dQ + VdP = (dU + PdV) + VdP = dU + (PdV + VdP) = dU + d(PV) = d(U + PV)$$

Hence, we have $\qquad H = U + PV$ and $\Delta H = \Delta U + \Delta(PV)$ (6.6)

Equation (6.6) represents the relation between two state functions U and H.

Inference

(i) As absolute value of internal energy cannot be determined, it is impossible to determine absolute value of enthalpy also. Only a change in enthalpy can be determined.

(ii) For an isothermal process, $dH = 0$, according to equation (6.5).

(iii) ΔH can be used to calculate heat of a reaction of a process.

(iv) ΔH can be used to determine whether the process is exothermic or endothermic.

Example 6.6.1 Prove that U and H are state functions.

For any cyclic process $\oint dT = 0$. Hence, $\oint C_V dT = 0$. As we know, $dU = C_V dT$, so, $\oint dU = 0$. Thus, U is a state function.

For any cyclic process, $\oint dT = 0$. Hence, $\oint C_P dT = 0$. As we know, dH $= C_P dT$, so, $\oint dH = 0$. Thus, H is a state function.

Example 6.6.2 State the physical meaning of ΔU and ΔH.

Answer "In a closed system of constant mass, heat absorbed at constant volume is equal to the increase in internal energy or ΔU considering expansion work only." Mathematically, we can write $q_V = \Delta U$.

"In a closed system of constant mass, heat absorbed at constant pressure is equal to the increase in enthalpy or ΔH considering expansion work only." Mathematically, we can write $q_P = \Delta H$.

6.7 CONCLUSIONS OF 1ST LAW OF THERMODYNAMICS

6.7.1 Corollaries of 1st Law of Thermodynamics

(a) $dQ = dU + dW$. For isothermal process, $dU = C_V dT = 0$. So, $dQ = dW$. Or, $q_T = w$. Thus, total heat absorbed by a system can be converted to equivalent amount of work in an isothermal process.

(b) $dQ = dU + dW$. For adiabatic process, $dQ = 0$. So, $dW = -dU$. Or $w_a = -\Delta U$. Thus in adiabatic process, change in internal energy gives rise to equivalent amount of work. As U is a state function, adiabatic work (w_a) is also a state function.

(c) $dQ = dU + PdV$. For isochoric process, $dV = 0$ and $dQ = dU$. Or, $q_V = \Delta U$. As U is a state function, q_V is also a state function.

(d) Again, $dQ = dH - VdP$. For isobaric process, $dP = 0$ and $dQ = dH$. Or, $q_P = \Delta H$. As H is a state function, q_P is also a state function. Thus, q_P and q_V are state functions but q is not a state function.

6.7.2 Advantages of 1st Law of Thermodynamics

(a) Total energy of the universe remains constant.
(b) Different forms of energy are interconvertible.
(c) When one form of energy disappears, an equivalent amount of energy in another form appears and hence heat and work can be correlated.

6.7.3 Limitations of 1st Law of Thermodynamics

(a) The law cannot predict the characteristics of a chemical reaction, for example, spontaneity or reversibility of the reaction.
(b) It cannot predict whether a reaction occurs or not.
(c) It cannot predict the direction of heat flow, i.e., heat flow from hotter body to colder one or vice versa.

(d) It cannot predict whether energy transformation occurs or not. If energy transformation occurs, it is not possible to evaluate how much energy is transformed from one form to the other in a given chemical reaction.

(e) According to 1st law of thermodynamics, total heat absorbed can be completely transferred into equivalent amount of work but in actual practice, total heat cannot be transformed into equivalent amount of work without permanent change in the system or surrounding which is not explained by this law.

6.8 EQUATION OF STATE

6.8.1 Chain Rule

Any physical state of a thermodynamic system can be expressed by thermodynamic parameters P, V and T. The general form of equation of state, relating to P, V and T is given by

$$f(P, V, T) = 0 \qquad (6.7)$$

One of the solution of the above equation is given by $V = \phi(P, T)$

The total differentiation is given by

$$dV = \left(\frac{\partial V}{\partial T}\right)_P dT + \left(\frac{\partial V}{\partial P}\right)_T dP$$

Again differentiating with respect to T at constant volume we have

$$0 = \left(\frac{\partial V}{\partial T}\right)_P + \left(\frac{\partial V}{\partial P}\right)_T \left(\frac{\partial P}{\partial T}\right)_V$$

Or, $\left(\frac{\partial V}{\partial P}\right)_T \left(\frac{\partial P}{\partial T}\right)_V = -\left(\frac{\partial V}{\partial T}\right)_P$. Or, $\left(\frac{\partial V}{\partial P}\right)_T \left(\frac{\partial P}{\partial T}\right)_V \left(\frac{\partial T}{\partial V}\right)_P = -1$

Or, $\left(\frac{\partial P}{\partial V}\right)_T \left(\frac{\partial V}{\partial T}\right)_P \left(\frac{\partial T}{\partial P}\right)_V = -1 \qquad (6.8)$

The above equation is also known as chain rule or cyclic rule.

6.8.2 Coefficients of Expansion

Coefficient of expansion at constant pressure (α_P)

If V_0 is the volume at temperature T_0 and pressure P_0, $(\partial V/\partial T)_P$ indicates the increase in volume with increase in temperature at constant pressure. Coefficient of expansion at constant pressure (α_P) is defined as

$$\alpha_P = \frac{1}{V_0}\left(\frac{\partial V}{\partial T}\right)_P \cdot \text{ Or, } \left(\frac{\partial V}{\partial T}\right)_P = \alpha_P \cdot V_0$$

Coefficient of expansion at constant volume (α_V)

If P_0 is the pressure at temperature T_0 and volume V_0, $(\partial V/\partial T)_V$ indicates the increase in pressure with temperature at constant volume. Coefficient of expansion at constant volume (α_V) is defined as

$$\alpha_V = \frac{1}{P_0}\left(\frac{\partial P}{\partial T}\right)_V. \text{ Or, } \left(\frac{\partial P}{\partial T}\right)_V = \alpha_V . P_0. \text{ Or, } \left(\frac{\partial T}{\partial P}\right)_V = \frac{1}{\alpha_V \cdot P_0}$$

Isothermal compressibility (k)

It is the decrease in volume with increase in pressure at constant temperature. It is expressed as

$$k = -\frac{1}{V_0}\left(\frac{\partial V}{\partial P}\right)_T. \text{ Or, } \left(\frac{\partial P}{\partial V}\right)_T = -\frac{1}{k \cdot V_0}$$

Putting the values of $\left(\dfrac{\partial V}{\partial T}\right)_P$, $\left(\dfrac{\partial T}{\partial P}\right)_V$ and $\left(\dfrac{\partial P}{\partial V}\right)_T$ in equation (6.8), we get

$$\frac{(\alpha_P V_0)}{(\alpha_V P_0)(-kV_0)} = -1. \text{ Or, } \frac{\alpha_P}{\alpha_V kP_0} = 1. \text{ Or, } \frac{\alpha_V kP_0}{\alpha_P} = 1 \tag{6.9}$$

Equation (6.9) represents the relation among α_P, α_V and k.

6.8.3 The Joule's Experiment

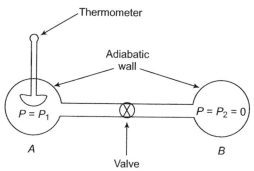

Figure 6.5 Chamber A: Filled with gas, Chamber B: Vacuum

In 1843, Joule performed an experiment to determine $\left(\dfrac{\partial U}{\partial V}\right)_T$ for a gas, assumed to behave ideally. The experiment set-up consists of two chambers A and B separated by a valve and each chamber is surrounded by an adiabatic wall. Initially chamber A is filled with a gas and chamber B is evacuated. Then valve between the chambers is opened and the gas of chamber A readily occupies the chamber B. The gas flow occurs till equilibrium is reached. The process is irreversible. Temperature change in the system is measured by a thermometer.

There is no heat transfer between the system and the surroundings. So, $\Delta Q = 0$. The work done due to this irreversible process $= w = P_2(V_2 - V_1)$. As $P_2 = 0$, there is no work done. Hence from 1st law of thermodynamics [equation (6.3)], we have $\Delta U = 0$.

Or, $U =$ Constant.

Hence, $$\left(\frac{\partial U}{\partial V}\right)_T = 0 \qquad\qquad (6.10a)$$

In this experiment Joule coefficient (μ_J) is defined as $\mu_J = \left(\frac{\partial T}{\partial V}\right)_U$

Since U is a state function using cyclic rule we can write that

$$\left(\frac{\partial T}{\partial V}\right)_U \left(\frac{\partial V}{\partial U}\right)_T \left(\frac{\partial U}{\partial T}\right)_V = -1. \ \ \text{Or,} \ \left(\frac{\partial U}{\partial V}\right)_T = -\left(\frac{\partial U}{\partial T}\right)_V \left(\frac{\partial T}{\partial V}\right)_U$$

Or, $$\left(\frac{\partial U}{\partial V}\right)_T = -C_V \mu_J \quad \left[\text{since, } C_V = \left(\frac{\partial U}{\partial T}\right)_V\right] \qquad (6.10b)$$

Using equation (6.10a), we can write $\mu_J = 0$.

Or, $$\left(\frac{\partial T}{\partial V}\right)_U = 0 \qquad\qquad (6.10c)$$

Conclusion

The term, $\left(\frac{\partial U}{\partial V}\right)_T$ arises due to internal pressure of gas. According to kinetic theory of ideal gases, mutual attraction among the gas molecules is neglected and hence there is no internal pressure due to gas molecules. Thus, in an isothermal process change in internal energy of an ideal gas does not depend on its volume or pressure. Thus for ideal gases we can write that

$$PV = nRT \ \ \text{and} \ \left(\frac{\partial U}{\partial V}\right)_T = 0 \qquad\qquad (6.11)$$

However, in case of real gases, mutual attraction among the gas molecules is considered and hence, there is a definite internal pressure due to gas molecules. Thus, $\left(\frac{\partial U}{\partial V}\right)_T$ has a definite positive value, which represents the internal pressure of real gas.

Inference

For any ideal gas the following are true

$$\left(\frac{\partial U}{\partial V}\right)_T = 0, \ \left(\frac{\partial U}{\partial P}\right)_T = 0, \ \left(\frac{\partial H}{\partial V}\right)_T = 0 \ \text{and} \ \left(\frac{\partial H}{\partial P}\right)_T = 0$$

6.9   WORK

The total work delivered by the system can be divided into two parts:
(a) Expansion work.
(b) Elementary useful work.

6.9.1   Expansion Work

When work is obtained due to expansion of volume of the system the work is called expansion work. It is well illustrated in the Fig. 6.6.

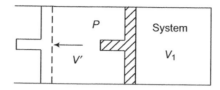

V_1 = Initial volume of the system. P = Piston.

$V_2 = V_1 + V'$ = Final volume of the system after expansion.

Figure 6.6   Volume of the system increases due to movement of the piston, P

Thus, mathematically the expansion work is represented as

$$dW = PdV \tag{6.12}$$

6.9.2   Elementary Useful Work

The useful work may be divided into different categories:
(i) Work due to change in potential energy, represented by $(-Mgdh)$, where, Mg = weight and dh = change in height.
(ii) Work due to increase in surface area, represented by $(-\sigma dA)$, where, σ = surface tension and dA = change in area.
(iii) Work due to increase in charge, represented by $(-\phi dq)$, where ϕ = potential and dq = increment of charge.
(iv) Mechanical work, represented by $(F.dx)$, where, F = Force and dx = displacement.

6.10   SPECIFIC HEAT

Amount of heat required to increase the temperature of unit mass of any substance by 1°K is known as specific heat of that substance. As it is independent of mass, it is an *intensive property*.

Amount of heat required to increase the temperature of one mole of any substance by 1°K is known as molar heat capacity of that substance. Value of 1 mole is different for different substances. Thus it is dependent of mass and hence an ***extensive property***.

In case of gases, molar heat capacity can be measured either at constant volume or at constant pressure and corresponding molar heat capacities are C_V and C_P respectively. If dQ amount of heat is required to increase the temperature of a gas by dT, C_V and C_P are represented as

$$C_V = \left(\frac{\partial Q}{\partial T}\right)_V \text{ and } C_P = \left(\frac{\partial Q}{\partial T}\right)_P$$

6.10.1 Relation Between C_P and C_V for Ideal Gases

From the 1st law of thermodynamics, we have $\quad dQ = dW + dU$

Considering only expansion work we can write that $\quad dQ = dU + PdV$

Differentiating the above equation with respect to (w.r.t.) temperature at constant volume (V), we get

$$C_V = \left(\frac{\partial Q}{\partial T}\right)_V = \left(\frac{\partial U}{\partial T}\right)_V \tag{6.13}$$

Since U is a state function, we can write that $\quad U = f(V, T)$

Differentiating, we have

$$dU = \left(\frac{\partial U}{\partial V}\right)_T dV + \left(\frac{\partial U}{\partial T}\right)_V dT \tag{6.14}$$

Again, we can write that $\quad U = f(P, T)$

Differentiating, we have

$$dU = \left(\frac{\partial U}{\partial P}\right)_T dP + \left(\frac{\partial U}{\partial T}\right)_P dT \tag{6.15a}$$

From equations (6.14) and (6.15a) we have

$$\left(\frac{\partial U}{\partial P}\right)_T dP + \left(\frac{\partial U}{\partial T}\right)_P dT = \left(\frac{\partial U}{\partial V}\right)_T dV + \left(\frac{\partial U}{\partial T}\right)_V dT \tag{6.15b}$$

Again, $\quad \left(\frac{\partial U}{\partial T}\right)_P = \left(\frac{\partial Q}{\partial T}\right)_P - P\left(\frac{\partial V}{\partial T}\right)_P \quad$ [since, $dU = dQ - PdV$] and $C_V = \left(\frac{\partial U}{\partial T}\right)_V$

Putting the value of $\left(\frac{\partial U}{\partial T}\right)_P$ and $\left(\frac{\partial U}{\partial T}\right)_V$ in the equation (6.15b) and knowing $C_P = \left(\frac{\partial Q}{\partial T}\right)_P$, we get

$$\left(\frac{\partial U}{\partial P}\right)_T dP + \left[C_P - P\left(\frac{\partial V}{\partial T}\right)_P\right] dT = \left(\frac{\partial U}{\partial V}\right)_T dV + C_V dT$$

Differentiating w.r.t. T at constant P, we get

$$C_P - P\left(\frac{\partial V}{\partial T}\right)_P = \left(\frac{\partial U}{\partial V}\right)_T \left(\frac{\partial V}{\partial T}\right)_P + C_V$$

Or, $$C_P - C_V = \left[\left(\frac{\partial U}{\partial V}\right)_T + P\right]\left(\frac{\partial V}{\partial T}\right)_P \qquad (6.16)$$

Gay-Lussac-Joule's law states that in an isothermal process change in internal energy of an ideal gas is independent of its volume. Thus for an ideal gas $\left(\frac{\partial U}{\partial V}\right)_T = 0$ and equation (6.16) becomes

$$C_P - C_V = P\left(\frac{\partial V}{\partial T}\right)_P \qquad (6.17)$$

Again for 1 mole of an ideal gas, we know that $PV=RT$. Differentiating w.r.t. T at constant P, we get

$$P\left(\frac{\partial V}{\partial T}\right)_P = R \qquad (6.18)$$

Thus equation (6.17) becomes

$$C_P - C_V = R \qquad (6.19)$$

6.11 P–V–T RELATIONS IN ADIABATIC PROCESS

6.11.1 Ideal Gas

From 1st law of thermodynamics we know that

$$dQ = dU + PdV. \quad \text{Or,} \quad dQ = C_V dT + PdV$$

[since, $dU = C_V dT$ for 1 mole and $dU = nC_V dT$ for n *moles* of an ideal gas]

For any adiabatic process the net heat transfer is zero or $dQ = 0$. Hence the above equation becomes

$$C_V dT + PdV = 0 \quad \text{[For 1 mole of a gas] and}$$

$$n\,C_V dT + PdV = 0 \quad \text{[For } n \text{ moles of a gas]}$$

For n *moles* of an ideal gas we know that

$$PV = nRT. \quad \text{Or,} \quad P = \frac{nRT}{V}$$

Putting the value of P in the above equation, we get

$$nC_V dT + \frac{nRT}{V} dV = 0. \quad \text{Or,} \quad d\ln T = -\frac{R}{C_V} d\ln V \qquad (6.20)$$

Again from 1st law of thermodynamics, we know that $dQ = dH - VdP$

Again applying the adiabatic condition, i.e. $dQ = 0$, we get

$dH = VdP.$ Or, $C_p dT = VdP$ [For 1 mole of a gas] and $nC_p dT = VdP$ [For n *moles* of a gas]

[since, $dH = C_p dT$ for 1 mole and $dH = nC_p dT$ for n *moles* of an ideal gas]

For 1 mole of an ideal gas, we have, $V = \frac{RT}{P}$. Putting the value of V in the above equation we get

$$C_p dT = \frac{RT}{P} dP \text{ and for } n \text{ moles of an ideal gas, } nC_p dT = \frac{nRT}{P} dP$$

Or, $$d\ln T = \frac{R}{C_p} d\ln P \qquad (6.21)$$

From equations (6.20) and (6.21) we get

$$\frac{C_V d\ln P}{C_p d\ln V} = -1. \quad \text{Or,} \quad d\ln P + \frac{C_p}{C_V} d\ln V = 0$$

On integration we get

$$\ln P + \gamma \ln V = \text{Constant.} \quad \text{Or,} \quad \ln PV^\gamma = \text{Constant.}$$

Or, $$PV^\gamma = \text{Constant} \qquad (6.22a)$$

where, $\gamma = \dfrac{C_p}{C_V}$ = Ratio of specific heats. Again, $C_p - C_V = R$ for 1 mole of an ideal gas.

So, $$\frac{R}{C_V} = \gamma - 1 \, [\text{For 1 mole of an ideal gas}] \text{ and}$$

$$\frac{nR}{nC_V} = \gamma - 1 \, [\text{For } n \text{ moles of an ideal gas}]$$

Thus equation (6.20) becomes $d\ln T = -(\gamma - 1)d\ln V$. On integration we get

$$\ln TV^{\gamma-1} = \text{Constant.} \quad \text{Or,} \quad TV^{\gamma-1} = \text{Constant} \qquad (6.22b)$$

Again $$\frac{R}{C_p} = 1 - \frac{C_V}{C_p} = 1 - \frac{1}{\gamma} = \left(\frac{\gamma - 1}{\gamma}\right)$$

Putting this value of $\dfrac{R}{C_P}$ in equation (6.21) we get

$$d \ln T = \left(\frac{\gamma - 1}{\gamma} \right) d \ln P$$

On integration, we get

$$\ln TP^{\frac{1-\gamma}{\gamma}} = \text{Constant}. \text{ Or, } TP^{\frac{1-\gamma}{\gamma}} = \text{Constant} \tag{6.22c}$$

So, we can write

$$PV^\gamma = \text{Constant}, \ TV^{\gamma-1} = \text{Constant} \quad \text{and} \quad TP^{\frac{1-\gamma}{\gamma}} = \text{Constant} \tag{6.22}$$

So the equation (6.22) is valid equally for 1 mole as well as *n moles* of any ideal gas.

6.11.2 Real Gas

For a real gas obeying van der Waals equation we know that

$$\left(P + \frac{an^2}{V^2} \right)(V - nb) = nRT, \text{ where, Internal pressure} = \frac{an^2}{V^2}$$

Again, Internal pressure $= \left(\dfrac{\partial U}{\partial V} \right)_T$. Or, $\left(\dfrac{\partial U}{\partial V} \right)_T = \dfrac{an^2}{V^2}$

Again, $U = f(V, T)$ and hence dU can be written as

$$dU = \left(\frac{\partial U}{\partial T} \right)_V dT + \left(\frac{\partial U}{\partial V} \right)_T dV = nC_V dT + \left(\frac{\partial U}{\partial V} \right)_T dV$$

$$\left[\text{since, } \left(\frac{\partial U}{\partial T} \right)_V = C_V \text{ for 1 mole of gas} \right]$$

Thus dU for *n moles* of a van der Waals gas is given by

$$dU = nC_V dT + \frac{an^2}{V^2} dV \tag{6.23}$$

From 1st law of thermodynamics, we have $dQ = dU + PdV$ [considering only expansion work]
Adiabatic condition $dQ = 0$. Hence, we have $PdV = -dU$.

Putting the value of dU from equation (6.23) we have

$$PdV = -nC_V dT - \frac{an^2}{V^2} dV. \text{ Or, } \left(P + \frac{an^2}{V^2}\right) dV = -nC_V dT \quad (6.24)$$

Using van der Waals equation, equation (6.24) can be written as

$$\frac{nRT}{(V-nb)} dV = -nC_V dT. \text{ Or, } \frac{dV}{(V-nb)} = -\left(\frac{C_V}{R}\right)\left(\frac{dT}{T}\right).$$

Or, $\qquad d\ln T + (\gamma - 1)d\ln(V - nb) = 0$

Or, $\qquad T(V - nb)^{(\gamma - 1)} = \text{Constant}$ $\hspace{5cm}$ (6.25a)

Replacing T from van der Waals equation in equation (6.25a), we get

$$\left(P + \frac{an^2}{V^2}\right)(V - nb)^\gamma = \text{Constant} \hspace{4cm} (6.25b)$$

Replacing $(V - nb)$ from van der Waals equation in equation (6.25a), we get

$$T\left(P + \frac{an^2}{V^2}\right)^{\left(\frac{1-\gamma}{\gamma}\right)} = \text{Constant} \hspace{4cm} (6.25c)$$

The equations (6.25a), (6.25b) and (6.25c) are the adiabatic relations for van der Waals equation

$$T(V - nb)^{(\gamma - 1)} = \text{Constant}, \left(P + \frac{an^2}{V^2}\right)(V - nb)^\gamma = \text{Constant}$$

$$T\left(P + \frac{an^2}{V^2}\right)^{\left(\frac{1-\gamma}{\gamma}\right)} = \text{Constant}$$

6.12 POLYTROPIC PROCESS

Any thermodynamic process is called polytropic process, which can be divided into four categories:
  (i) Adiabatic process: condition, $dQ = 0$
 (ii) Isothermal process: condition, $dT = 0$
(iii) Isochoric process: condition, $dV = 0$
 (iv) Isobaric process: condition, $dP = 0$

In general for an ideal gas the mathematical expression of any polytropic process is given by

$$PV^n = \text{Constant.} \quad \text{Or,} \quad VP^{\frac{1}{n}} = \text{Constant} \tag{6.26}$$

Different polytropic processes are summarized in the Fig. 6.7.

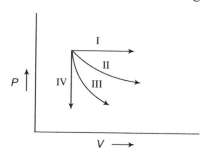

Figure 6.7 *PV*–diagram for different polytropic processes

Curve-I   (a) It shows an isobaric process, i.e., $P = $ Constant. Hence, $n = 0$ in equation (6.26).

(b) Molar specific heat, $C_P = \left(\dfrac{\partial Q}{\partial T}\right)_P$.

(c) Area under the curve is maximum. So maximum work is obtained in isobaric process.

Curve-II   (a) It shows an isothermal process, i.e., $T = $ Constant and $PV = $ Constant. Hence, $n = 1$ in equation (6.26).

(b) Molar specific heat, $C = \dfrac{dQ}{dT} = \infty$, since, $dT = 0$.

(c) Area under the curve is lower than that of isobaric process. So work obtained in isothermal process is less than that of isobaric process.

Curve-III   (a) It shows an adiabatic process, i.e., $Q = $ Constant, and $PV^\gamma = $ Constant. Hence, $n = \gamma$ in equation (6.26).

(b) Molar specific heat, $C = \dfrac{dQ}{dT} = 0$, since, $dQ = 0$

(c) Area under the curve is lower than that of isothermal process. So work obtained in adiabatic process is less than that of isothermal process.

Curve-IV   (a) It shows an isochoric process, i.e., $V = $ Constant and hence $n = \infty$ in equation (6.26).

(b) Molar specific heat, $C_V = \left(\dfrac{\partial Q}{\partial T}\right)_V$

(c) Area under the curve is zero. So, no work is obtained in isochoric process.

6.13 CALCULATION OF WORK

Work due to expansion of gases is given by $dW = PdV$. For different polytropic processes work will be different.

(a) **Isochoric process** The condition of isochoric process is $dV = 0$. Hence, $dW = PdV = 0$. Thus no net work is obtained in an isochoric process.

(b) **Isobaric process** Pressure is constant in an isobaric process. So, $dP = 0$. Thus during expansion of a gas from volume V_1 to volume V_2, work done is

$$w_P = \int_{V_1}^{V_2} PdV = P(V_2 - V_1). \quad \text{Or,} \quad W = P(V_2 - V_1) \tag{6.27}$$

(c) **Isothermal reversible process** In isothermal process temperature is kept constant or $dT = 0$.

(i) For 1 mole of an ideal gas, we know that $PV = RT$. Or $P = \dfrac{RT}{V}$.
So, work done is given by

$$|w_r|_T = \int_{V_1}^{V_2} PdV = \int_{V_1}^{V_2} \frac{RT}{V} dV = RT \ln\frac{V_2}{V_1} = RT \ln\frac{P_1}{P_2}, \text{ since, } \frac{V_2}{V_1} = \frac{P_1}{P_2}$$

For *n moles* of an ideal gas

$$|w_r|_T = nRT \ln\frac{V_2}{V_1} = nRT \ln\frac{P_1}{P_2} \tag{6.28a}$$

(ii) For *n moles* of a gas, obeying van der Waals equation, we know that

$$\left(P + \frac{an^2}{V^2}\right)(V - nb) = nRT. \quad \text{Or,} \quad P = \frac{nRT}{(V - nb)} - \frac{an^2}{V^2}$$

where a and b are constants. Substituting this value of P in the expression of work we get

$$|w_r|_T = \int_{V_1}^{V_2} PdV = \int_{V_1}^{V_2} \frac{nRT}{(V - nb)} dV - \int_{V_1}^{V_2} \frac{an^2}{V^2} dV = nRT \ln\frac{(V_2 - nb)}{(V_1 - nb)} + an^2\left[\frac{1}{V_2} - \frac{1}{V_1}\right]$$

Or, $$|w_r|_T = nRT \ln\frac{(V_2 - nb)}{(V_1 - nb)} + an^2\left[\frac{1}{V_2} - \frac{1}{V_1}\right] \tag{6.28b}$$

(d) **Isothermal irreversible process** Let us consider a thermodynamic system, which is transferred from state A to state B irreversibly as shown in the Fig. 6.8a. The system first follows an isochoric process to move from state A to state C. Corresponding work done is $w_{AC} = 0$. Then the system follows an isobaric expansion process and corresponding work done is $w_{CB} = P_2(V_2 - V_1)$.

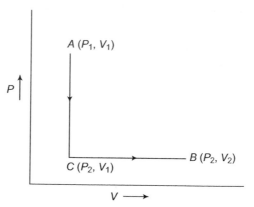

Figure 6.8a   *PV*–diagram of an irreversible process

Total work done, $w_{ir} = w_{AC} + w_{CB} = 0 + P_2(V_2 - V_1) = P_2(V_2 - V_1)$. Thus for an isothermal irreversible expansion or compression process, work done is always given by

$$\left|w_{ir}\right|_T = P_2(V_2 - V_1) \tag{6.29}$$

(e) **Adiabatic process**   From 1st law of thermodynamics we know that $dQ = dU + dW$. As the process is adiabatic, $dQ = 0$ and hence, $dW = -dU$.

Again, $U = U(V, T)$ [since U is a state function]. Differentiating, we have

$$dU = \left(\frac{\partial U}{\partial V}\right)_T dV + \left(\frac{\partial U}{\partial T}\right)_V dT$$

1.   Ideal Gas

Method 1

For 1 mole of an ideal gas, we have

$$\left(\frac{\partial U}{\partial V}\right)_T = 0 \ [Cf.\,\text{equation (6.10a)}] \ \text{and hence,} \ dU = \left(\frac{\partial U}{\partial T}\right)_V dT = C_V dT$$

For *n moles* of an ideal gas $dU = nC_V dT$. Thus work is given by

$$dW = -dU = -nC_V dT. \ \text{Or,} \ \int dW = -nC_V \int_{T_1}^{T_2} dT. \ \text{Or,} \ w_a = -nC_V (T_2 - T_1)$$

$$w_a = -nC_V(T_2 - T_1) \tag{6.30a}$$

Method 2

Again, we know that $dW = PdV$ (considering only expansion work).

For adiabatic expansion process, we know that $PV^\gamma = \text{Constant} = K$

Or, $\qquad\qquad P = \dfrac{K}{V^\gamma}$. Hence, $dW = PdV = \dfrac{K}{V^\gamma}dV$

On integration we get

$$\int dW = \int_{V_1}^{V_2}\frac{K}{V^\gamma}dV. \text{ Or, } w_a = \frac{K}{1-\gamma}\left[\frac{1}{V_2^{\gamma-1}} - \frac{1}{V_1^{\gamma-1}}\right]$$

$$= \frac{1}{1-\gamma}\left[\frac{KV_2}{V_2^\gamma} - \frac{KV_1}{V_1^\gamma}\right] = \frac{P_2V_2 - P_1V_1}{1-\gamma}$$

$$\left[\text{Since, } \frac{K}{V_2^\gamma} = P_2 \text{ and } \frac{K}{V_1^\gamma} = P_1\right]$$

$$w_a = \frac{P_2V_2 - P_1V_1}{1-\gamma} \qquad\qquad (6.30b)$$

Again, for *n moles* of an ideal gas, we know $P_1V_1 = nRT_1$ and $P_2V_2 = nRT_2$

Hence, $\qquad\qquad w_a = \dfrac{nRT_2 - nRT_1}{1-\gamma} = \dfrac{nR(T_2 - T_1)}{1-\gamma} \qquad\qquad (6.30c)$

All the equations (6.30a), (6.30b) and (6.30c) represent the adiabatic work for *n moles* of an ideal gas. Furthermore, the equation (6.30b) represents the adiabatic work for 1 mole as well as *n moles* of an ideal gas.

2. **Real gas**

For a real gas obeying van der Waals equation we know that

$$dU = nC_V dT + \frac{an^2}{V^2}dV \qquad\qquad (6.23)$$

Thus on integration we get

$$\Delta U = \int dU = \int_{T_1}^{T_2} nC_V dT + \int_{V_1}^{V_2}\frac{an^2}{V^2}dV = nC_V(T_2 - T_1) - an^2\left(\frac{1}{V_2} - \frac{1}{V_1}\right)$$

As the process is adiabatic, work is given by

$$w_a = -\Delta U = -nC_V(T_2 - T_1) + an^2\left(\frac{1}{V_2} - \frac{1}{V_1}\right) \qquad\qquad (6.31)$$

The equation (6.31) represents the adiabatic work for *n moles* of a real gas obeying van der Waals equation.

6.14 REVERSIBLE AND IRREVERSIBLE PROCESSES

In another angle of classification all thermodynamic processes can be divided into two broad categories:

(i) Reversible process, and (ii) Irreversible process.

Reversible process

A process is said to be reversible if the system undergoes a change of state through numerous steps such that equilibrium holds good in each and every step. That means in a reversible process both forward and backward processes proceed through same physical states. To perform a process reversibly only infinitesimal change in pressure or temperature is sufficient to execute any chemical change slowly and reversibly. This is well illustrated in the Fig. 6.8b.

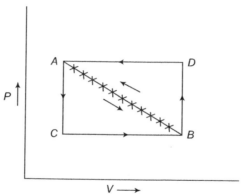

Figure 6.8b *PV–diagram for reversible and irreversible process*

The path *AB* indicates the reversible path since it passes through numerous steps. Furthermore both the forward process $A \to B$ and the backward process $B \to A$ follow the same path. Thus infinite time is required to complete a reversible process. So it is a non-flow process, also known as **Quasi-Static** process.

Example 6.14.1 What is quasi-static process?

Answer A quasi-static process is that process which proceeds through numerous states with the following conditions.

(i) the deviation from thermodynamic equilibrium is infinitesimal, and

(ii) all states are equilibrium states.

Irreversible Process

A process is said to be irreversible if the system undergoes a change of state to a definite direction spontaneously and never returns to its original state following the same path. Usually irreversible process is a single-step or two-step process. In the Fig. 6.8b, the system changes its state $A \to B$ through one step process *ACB* but returns to its original state through another single-step process *BDA*.

Example 6.14.2   State the differences between reversible and irreversible processes.

Answer   The differences between reversible and irreversible processes are tabulated in the following table.

Differences between reversible and irreversible processes

Reversible process	*Irreversible process*
The process is very slow, non-spontaneous and occurs only when one of the state parameters is only slightly changed.	The process is very fast and spontaneous.
It proceeds through maximum number of steps and equilibrium exists in each step.	It proceeds through only one or two steps and equilibrium does not exist in either step.
Work obtained is maximum.	Work obtained is always less than that of reversible process.
Forward and backward reactions proceed through same physical states.	Forward and backward reactions proceed through different physical states.
Net entropy change is zero for a reversible cyclic process.	Net entropy change is increasing for an irreversible cyclic process.

Example 6.14.3   Show that maximum work is available in a reversible process.

Answer

Method-1   Graphical Approach

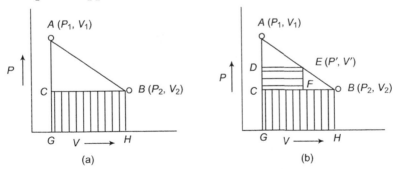

Figure 6.9   *PV*–diagram for (a) one-step process, and (b) two-step process

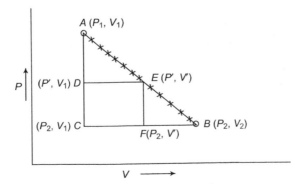

Figure 6.10   *PV*–diagram for one step and two step processes

Fig. 6.9(a) shows a single step irreversible process ACB, where AB represents reversible path. The work done is the area under the curve ACB, represented by vertical hatching lines. Fig. 6.9(b) shows the two step irreversible process ($ADEFB$) where the work done is the area under the curve $ADEFB$. Comparing the two figures, the extra work obtained in two step process is the area of the rectangle $CDEF$, represented by horizontal hatching lines. Thus, work done in two step process is more than that in one step process.

The corresponding reversible process, represented by AB in the Fig. 6.9b, consists of numerous steps and hence maximum work is available in the reversible process.

Method-2 Numerical Approach

In the Fig. 6.10, ACB and $ADEFB$ represent single step and two step processes, respectively.

Work done along $ACB = W_1 =$ Work done along $AC +$ Work done along CB.
$$= \text{Work done along } CB = P_2(V_2 - V_1).$$
[Since AC shows an isochoric process and work in an isochoric process is zero]
Work along $ADEFB = W_2$

$= $ Work done along $AD +$ Work done along $DE +$ Work done along $EF +$ Work done FB

$= $ Work done $DE +$ work done FB [since AD and EF represent isochoric processes]

$$= P'(V' - V_1) + P_2(V_2 - V').$$

Difference between two-step-work and one-step-work $= W_2 - W_1$.

$$W_2 - W_1 = [P'(V' - V_1) + P_2(V_2 - V')] - [P_2(V_2 - V_1)]$$

$$= P'V' - P'V_1 + P_2V_2 - P_2V' - P_2V_2 + P_2V_1$$

$$= (P'V' - P_2V') - (P'V_1 - P_2V_1)$$

$$= V'(P' - P_2) - V_1(P' - P_2) = (V' - V_1)(P' - P_2)$$

Or,
$$W_2 - W_1 = (V' - V_1)(P' - P_2)$$

From the graph it is evident that $P' > P_2$ and $V' > V_1$. Hence, $(W_2 - W_1) > 0$.
Or, $W_2 > W_1$.

So, work obtained in a two-step process is more than that obtained in a one-step process. As the reversible path (AB in the Fig. 6.10) consists of numerous steps, maximum work is available in the reversible process.

Example 6.14.4 Show that for an ideal gas PV–adiabat is steeper than PV–isotherm.

Answer

Steepness of PV-diagram is given by $\left| \dfrac{dP}{dV} \right|$.

In case of isothermal process the ideal gas relation is $PV = $ constant $= C$. Hence,

$$V\left(\frac{dP}{dV}\right)_{\text{isothermal}} + P = 0. \text{ Or, } \left(\frac{dP}{dV}\right)_{\text{isothermal}} = -\frac{P}{V}. \text{ Or, } \left|\left(\frac{dP}{dV}\right)_{\text{isothermal}}\right| = \frac{P}{V}$$

In case of adiabatic process the ideal gas relation is $PV^{\gamma} = $ Constant $= K$. Hence,

$$V^{\gamma}\left(\frac{dP}{dV}\right)_{\text{adiabatic}} + P\gamma V^{\gamma-1} = 0. \text{ Or, } \left(\frac{dP}{dV}\right)_{\text{adiabatic}} = -\frac{\gamma P}{V}. \text{ Or, } \left|\left(\frac{dP}{dV}\right)_{\text{adiabatic}}\right| = \frac{\gamma P}{V}$$

As $\gamma > 1$, $\left(\frac{dP}{dV}\right)_{\text{adiabatic}} > \left(\frac{dP}{dV}\right)_{\text{isothermal}}$. Thus, PV–adiabat is steeper than PV–isotherm.

6.15 HEAT ENGINE AND ENTROPY

Heat engine   An engine, which undergoes a cyclic process to convert heat energy to mechanical work, is called heat engine. Fig. 6.11a shows the schematic representation of a heat engine as proposed by the scientist Carnot. Fig. 6.11b shows the schematic representation of working principle of a Carnot heat engine.

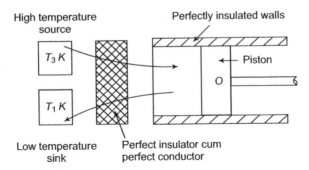

Figure 6.11a   Schematic diagram of Carnot heat engine

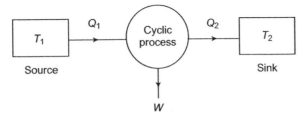

Figure 6.11b   Schematic diagram of working principle of a Carnot heat engine

A heat engine absorbs Q_1 amount of heat from a heat source at temperature T_1 and delivers W amount of work. At the end Q_2 amount of heat is rejected to the heat sink at temperature T_2 and $T_2 < T_1$. So the efficiency of the heat engine, η is the fraction of energy input that appears as mechanical work. Thus, η is given by

$$\eta = \frac{W}{Q_1}, \text{ where } Q_1 \text{ is the amount of heat energy input.}$$

The heat engine was first invented by the scientist **Nicolas Léonard Sadi Carnot** in 1824. The scientist Carnot was born in 1796 and died of cholera in 1832. He imagined a cyclic process with four reversible steps through which an engine delivers work. Hence, the cyclic process is also known as reversible Carnot cycle. The engine is popularly known as **Carnot engine**. A **Carnot cycle** is defined as a reversible cycle, consisting of two isothermal steps operating at two different temperatures and two adiabatic steps. It is well illustrated in the Figs. 6.12a and 6.12b.

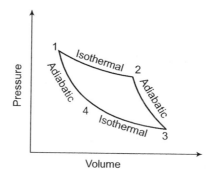

Figure 6.12a Four steps of Carnot cycle

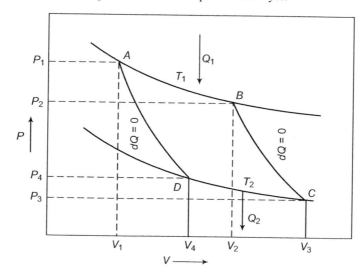

Figure 6.12b *PV*–diagram of the Carnot cycle ($T_1 > T_2$)

Assumptions of Carnot cycle

(a) All the steps in the Carnot cycle are reversible.

(b) The medium used to execute the cyclic process is an ideal gas.

6.15.1   Process Description

There are two isothermal steps *AB* and *CD*. Along *AB*, Q_1 amount of heat is absorbed at temperature T_1 and along *CD*, Q_2 amount of heat is rejected at temperature T_2.

There are two adiabatic steps, *BC* and *DA*. Applying adiabatic *TV*–relation [equation (6.22)] along *BC* and along *DA* we get,

Along *BC* $\qquad T_1 V_2^{\gamma-1} = T_2 V_3^{\gamma-1}.$ Or, $\dfrac{T_1}{T_2} = \left(\dfrac{V_3}{V_2}\right)^{\gamma-1}$

Along *DA* $\qquad\qquad T_2 V_4^{\gamma-1} = T_1 V_1^{\gamma-1}.$ Or, $\dfrac{T_1}{T_2} = \left(\dfrac{V_4}{V_1}\right)^{\gamma-1}$

Hence, $\qquad\qquad \left(\dfrac{V_3}{V_2}\right) = \left(\dfrac{V_4}{V_1}\right).$ Or, $\left(\dfrac{V_3}{V_4}\right) = \left(\dfrac{V_2}{V_1}\right)$ $\qquad\qquad$ (6.32)

Calculation of total work and efficiency

(i) Along the path *AB*, $dT = 0$. Or, temperature is kept constant at T_1. Work done is W_1 and heat absorbed is Q_1.

$$W_1 = \int_{V_1}^{V_2} PdV = nRT_1 \ln\frac{V_2}{V_1} \quad \text{for } n \text{ moles of an ideal gas.}$$

From 1st law of thermodynamics, we know that $\quad dQ = dU + dW$.
On integration, we get $\quad Q_1 = \Delta U + W_1$.
Again we know that $\quad dU = nC_V \, dT$. Along *AB*, $dT = 0$. Hence, $\Delta U = 0$ along *AB*.

Hence, $\qquad\qquad Q_1 = W_1 = nRT_1 \ln\frac{V_2}{V_1}$

As $V_2 > V_1$, $W_1 > 0$, which indicates that work is done by the system.

(ii) The curve *BC* shows an adiabatic path. Hence work done along *BC* is given by

$$W_2 = -\Delta U = -nC_V \Delta T = -nC_V (T_2 - T_1) \quad \text{since,} \quad dQ = 0$$

(iii) Along the curve *CD* temperature is kept constant at T_2. According to 1st law of thermodynamics, we have

$$Q_2 = W_3 = nRT_2 \ln\frac{V_4}{V_3}, \quad [\text{since, } dU = 0]$$

Or, $\qquad\qquad Q_2 = W_3 = nRT_2 \ln\frac{V_1}{V_2} \ [\text{Using equation(6.32)}]$

As, $V_4 < V_3$, $W_3 < 0$, which means that work is done on the system.

(iv) The curve DA shows an adiabatic path. Hence, work done along DA is given by

$$W_4 = -\Delta U = -nC_V \Delta T = -nC_V(T_1 - T_2) \quad \text{since,} \quad dQ = 0$$

Or,

$$W_4 = -nC_V(T_1 - T_2) = nC_V(T_2 - T_1) = -W_2. \quad \text{Or,} \quad W_2 + W_4 = 0$$

Thus the total work, W is given by

$$W = W_1 + W_2 + W_3 + W_4 = W_1 + W_3 \quad [\text{since,} \quad W_2 + W_4 = 0]$$

Or,

$$W = nRT_1 \ln \frac{V_2}{V_1} + nRT_2 \ln \frac{V_4}{V_3} = nRT_1 \ln \frac{V_2}{V_1} - nRT_2 \ln \frac{V_2}{V_1}$$

$$\left[\text{since,} \quad \left(\frac{V_3}{V_4} \right) = \left(\frac{V_2}{V_1} \right) \text{equation (6.32)} \right]$$

Or,

$$W = nR(T_1 - T_2) \ln \frac{V_2}{V_1} \tag{6.33}$$

Again,

$$Q_1 = nRT_1 \ln \frac{V_2}{V_1} \quad \text{and} \quad |Q_1| = nRT_1 \ln \frac{V_2}{V_1},$$

$$Q_2 = -nRT_2 \ln \frac{V_2}{V_1} \quad \text{and} \quad |Q_2| = nRT_2 \ln \frac{V_2}{V_1}$$

The negative sign indicates that heat is rejected to the surroundings, i.e. to the heat sink.

Thus the work done by the engine can be written as

$$W = |Q_1| - |Q_2| \tag{6.34}$$

Thus the efficiency, η_r, is given by

$$\eta_r = \frac{W}{|Q_1|} = \frac{|Q_1| - |Q_2|}{|Q_1|} = \frac{T_1 - T_2}{T_1} = 1 - \frac{T_2}{T_1}, \quad \text{Or,} \quad \eta_r = 1 - \frac{T_2}{T_1} \tag{6.35}$$

Using adiabatic relations along BC, we have

$$\frac{T_1}{T_2} = \left(\frac{V_3}{V_2} \right)^{\gamma - 1} = r^{\gamma - 1}, \quad \frac{T_1}{T_2} = \left(\frac{P_3}{P_2} \right)^{\frac{1 - \gamma}{\gamma}}$$

Or,

$$\frac{T_1}{T_2} = \left(\frac{P_2}{P_3} \right)^{\frac{\gamma - 1}{\gamma}} = r_P^{\frac{\gamma - 1}{\gamma}}, \quad \text{since,} \quad \frac{V_3}{V_2} = r \quad \text{and} \quad \frac{P_2}{P_3} = r_P$$

where, r and r_p are known as compression ratio with respect to volume and pressure respectively.

Thus,

$$\eta_r = 1 - \frac{T_2}{T_1} = 1 - \frac{1}{\dfrac{T_1}{T_2}} = 1 - \frac{1}{r^{\gamma - 1}} = 1 - \frac{1}{r_P^{\frac{\gamma - 1}{\gamma}}} \tag{6.36}$$

Inferences

(i) As all the steps are reversible, maximum work is obtained in Carnot heat engine. In any practical heat engine all the steps cannot be maintained reversible and hence efficiency of Carnot heat engine is always greater than that of any heat engine.

(ii) Efficiency of Carnot heat engine is less than unity.

(iii) Efficiency of Carnot engine is independent of type of fluid used as medium.

(iv) Efficiency of Carnot engine can be improved either by increasing source temperature (T_1) or by decreasing sink temperature (T_2). [Cf. equation (6.35)].

(v) As $(\gamma - 1) > 0$, efficiency of Carnot cycle increases as compression ratio, r, increases, which signifies that efficiency of large engine volume is greater than that of small engine volume.

(vi) As $\left(\dfrac{\gamma - 1}{\gamma}\right) > 0$, efficiency of Carnot cycle increases as pressure ratio, r_p, increases.

Thus, Carnot cycle should be operated at high peak pressure to avail high efficiency.

6.15.2 Carnot's Theorem

Statement   There is no such engine, operating between two heat reservoirs, which is more efficient than Carnot engine. In other words, any reversible heat engine, operating between two heat reservoirs, has same efficiency as that of Carnot engine.

Proof   Let us consider two heat engines and one of them is Carnot engine, which is assumed to be less efficient than the other engine. The two engines are connected with two heat reservoirs such that more efficient engine (marked as X in the Fig. 6.13) acts as heat engine while less efficient Carnot engine (marked as Carnot in the same figure) acts as heat pump as shown.

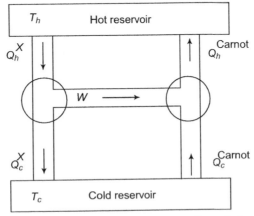

Figure 6.13   Schematic diagram to prove Carnot's theorem

Q_h^X = Amount of heat absorbed by the engine X from the hot reservoir.
Q_c^X = Amount of heat rejected by the engine X to the cold reservoir.
W = Work output by the engine; X = Work absorbed by the Carnot engine.

$$\eta_X = \text{Efficiency of the engine } X = \frac{W}{Q_h^X}$$

Q_c^{Carnot} = Amount of heat absorbed by the Carnot engine from the cold reservoir.
Q_h^{Carnot} = Amount of heat rejected by the Carnot engine to the hot reservoir.

$$\eta_{\text{Carnot}} = \text{Efficiency of the Carnot engine} = \frac{W}{Q_c^{\text{Carnot}}}$$

According to the assumption, $\eta_X > \eta_{\text{Carnot}}$. Thus, we can write

$$\frac{W}{Q_h^X} > \frac{W}{Q_c^{\text{Carnot}}} \cdot \text{Or, } \frac{1}{Q_h^X} > \frac{1}{Q_c^{\text{Carnot}}} \cdot \text{ Or, } Q_c^{\text{Carnot}} > Q_h^X. \text{ Or, } (Q_c^{\text{Carnot}} - Q_h^X) > 0$$

So, heat is transferred from cold reservoir to hot reservoir without any external work. So this is a violation of 2nd law of thermodynamics. Thus, it may be concluded that $\eta_X \leq \eta_{\text{Carnot}}$.

Example 6.15.1   Show that increase in efficiency by decreasing sink temperature is greater than increase in efficiency by increasing source temperature.

Method-1

η is the efficiency of Carnot engine, operating between the temperatures T_1 and T_2.
η_1 is the efficiency of Carnot engine due to change in source temperature from T_1 to $(T_1 + \Delta T)$.
η_2 is the efficiency of Carnot engine due to change in sink temperature from T_2 to $(T_2 + \Delta T)$.

$$\eta = \frac{T_1 - T_2}{T_1}, \ \eta_1 = \frac{(T_1 + \Delta T) - T_2}{(T_1 + \Delta T)} = \frac{(T_1 - T_2) + \Delta T}{(T_1 + \Delta T)},$$

$$\eta_2 = \frac{T_1 - (T_2 - \Delta T)}{T_1} = \frac{(T_1 - T_2) + \Delta T}{T_1}$$

Thus,

$$\frac{\eta_2}{\eta_1} = \frac{(T_1 + \Delta T)}{T_1} > 1$$

Method-2

$$\eta = \frac{T_1 - T_2}{T_1} = 1 - \frac{T_2}{T_1}$$

So, η increases with increase in T_1. Thus, differentiating η w.r.t. T_1 at constant T_2, we have

$$\left(\frac{\partial \eta}{\partial T_1}\right)_{T_2} = \frac{T_2}{T_1^2}, \quad \left|\left(\frac{\partial \varsigma}{\partial T_1}\right)_{T_2}\right| = \frac{T_2}{T_1^2}$$

Again, η increases with decrease in T_2. Thus, differentiating η w.r.t. T_2 at constant T_1, we have

$$\left(\frac{\partial \eta}{\partial T_2}\right)_{T_1} = -\frac{1}{T_1}, \quad \left|\left(\frac{\partial \eta}{\partial T_2}\right)_{T_1}\right| = \frac{1}{T_1}$$

$$\left|\left(\frac{\partial \eta}{\partial T_2}\right)_{T_1}\right| - \left|\left(\frac{\partial \eta}{\partial T_1}\right)_{T_2}\right| = \frac{1}{T_1} - \frac{T_2}{T_1^2} = \frac{T_1 - T_2}{T_1^2} > 0, \quad \text{since, } T_1 > T_2$$

Thus, increase in efficiency by decreasing sink temperature is greater than increase in efficiency by increasing source temperature.

6.16 ENTROPY—A STATE FUNCTION

From the expression of efficiency of Carnot engine, we have

$$\eta = \frac{W}{|Q_1|} = \frac{|Q_1| - |Q_2|}{|Q_1|} = \frac{T_1 - T_2}{T_1} \cdot \text{ Or, } 1 - \frac{|Q_2|}{|Q_1|} = 1 - \frac{T_2}{T_1} \cdot \text{ Or, } \frac{|Q_2|}{|Q_1|} = \frac{T_2}{T_1}$$

Hence,

$$\frac{|Q_1|}{T_1} = \frac{|Q_2|}{T_2} \cdot \text{ Or, } \frac{|Q_1|}{T_1} - \frac{|Q_2|}{T_2} = 0$$

Thus, for a complete reversible cycle we have

$$\oint \left(\frac{dQ}{T}\right)_{rev} = 0$$

So, $\left(\dfrac{dQ}{T}\right)_{rev}$ is a state function and represented by dS, where S is called entropy of the system. Thus, for any reversible process, we can write that

$$dS = \left(\frac{dQ}{T}\right)_{rev} \quad \text{and } \oint dS = 0 \tag{6.37}$$

Thus, for any reversible process $\quad dQ = TdS \tag{6.38}$

Example 6.16.1   Show that $dS > \left(\dfrac{dQ}{T}\right)_{ir}$ for any irreversible process.

From the definition of efficiency (η) of any heat engine we have

$$\eta = \frac{W}{|Q_1|}$$

If one of the steps in a cyclic process is irreversible the whole cyclic process is considered as an irreversible cyclic process. For an irreversible cycle, efficiency is given by

$$\eta_{ir} = \frac{W_{ir}}{|Q_1|} = \frac{|Q_1| - |Q_2|}{|Q_1|}$$

For a reversible heat engine the efficiency can be written as

$$\eta_r = \frac{W_{rev}}{|Q_1|} = \frac{T_1 - T_2}{T_1}$$

As discussed in the sec.6.14, we have $W_{ir} < W_{rev}$. Hence, $\eta_{ir} < \eta_r$. So, we can write that

$$\frac{|Q_1| - |Q_2|}{|Q_1|} < \frac{T_1 - T_2}{T_1} \cdot \text{ Or, } 1 - \frac{|Q_2|}{|Q_1|} < 1 - \frac{T_2}{T_1} \cdot \text{ Or, } -\frac{|Q_2|}{|Q_1|} < -\frac{T_2}{T_1} \cdot \text{ Or, } \frac{|Q_2|}{|Q_1|} > \frac{T_2}{T_1}$$

Or, $\left[\dfrac{|Q_1|}{T_1} - \dfrac{|Q_2|}{T_2}\right] < 0.$ Thus for a complete cycle, $\oint\left(\dfrac{dQ}{T}\right)_{ir} < 0$

For a reversible cycle, we have $\oint dS = 0$

Combining the above two equations, we get

$$\oint\left(\frac{dQ}{T}\right)_{ir} < \oint dS. \text{ Or, } dS > \left(\frac{dQ}{T}\right)_{ir}. \text{ Or, } dQ_{ir} < TdS \qquad (6.39)$$

Inferences

(a) For a reversible adiabatic process, $dQ = 0$ and hence $dS = 0$ according to equation (6.38). The process is also called *isentropic process*, where both S and Q are constant.

(b) For an irreversible adiabatic process, $dQ = 0$ but $dS > 0$ according to equation (6.39).

(c) Combining the above two conditions we can write that $dS_a \geq 0$, where dS_a is the change in entropy in an adiabatic process. This is known as **Clausius inequality**.

6.16.1 Physical Significance of Entropy (S)

From the definition of entropy we have $dS = \left(\dfrac{dQ}{T}\right)_{rev}$. The term $\dfrac{dQ}{T}$ is often called **reduced heat**. In adiabatic process, $dQ = 0$ but in non-adiabatic process absorption of heat energy increases the temperature of the system. Only a part of absorbed heat energy is utilized to increase the temperature of the system. Thus entropy increases, since variation of temperature is small compared to amount of heat absorbed.

In an isothermal process heat is absorbed by the system at constant temperature, e.g. sublimation process, in which a solid acquires latent heat and sublimes to gas. The process is irreversible. So according to 2nd law of thermodynamics entropy increases. In solids, molecules are fixed at their positions and remain in orderly state or compact form but after sublimation molecules are in gaseous state and execute ceaseless random motion, called Brownian motion. Thus, in a sublimation process orderly state of molecules changes to most disorderly state or random state. Thus, increase in entropy signifies the disorder or randomness of a system. In other words entropy is a measure of disorder or randomness of a system.

Example 6.16.2   Show that entropy is a state function.

$$\eta_r = \frac{|Q_1| - |Q_2|}{|Q_1|} = \frac{T_1 - T_2}{T_1}, \quad \text{Or,} \quad \frac{|Q_2|}{|Q_1|} = \frac{T_2}{T_1}, \quad \text{Or,} \quad \frac{|Q_1|}{T_1} = \frac{|Q_2|}{T_2}.$$

$$\frac{|Q_1|}{T_1} - \frac{|Q_2|}{T_2} = 0, \quad \text{Or,} \quad \oint \left(\frac{dQ}{T}\right)_r = 0, \quad \text{Or,} \quad \oint dS = 0$$

Thus, entropy is a state function.

6.17 SECOND LAW OF THERMODYNAMICS

Statement 1

According to equation (6.38), the efficiency of a Carnot engine is given by

$$\eta_r = \frac{W}{|Q_1|} = \frac{|Q_1| - |Q_2|}{|Q_1|} = \quad \text{So, } \eta_r < 1$$

Thus, $\eta_r \neq 1$ even for Carnot engine, which is considered the most efficient heat engine.

Kelvin-Planck Statement

There is no such device, operating in a cyclic process, which absorbs heat from a hot reservoir and converts the same to equivalent amount of work. It is illustrated in the Fig. 6.14.

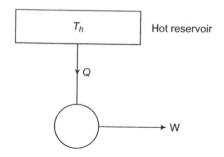

Figure 6.14 Illustration of Kelvin-Planck statement

Statement 2

Clausius Statement

There is no such device, operating in a cyclic process, which absorbs heat from a cold reservoir and transfers the same to a hot reservoir without any external work. It is illustrated in the Fig. 6.15.

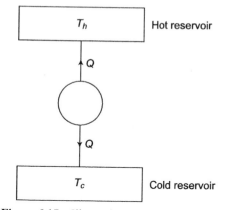

Figure 6.15 Illustration of Clausius statement

Statement 3

According to Clausius inequality, we have $dS_a > 0$ for spontaneous process.

"All natural processes are non-equilibrium and spontaneous and hence entropy of the universe is increasing. So entropy can be created but cannot be destroyed".

6.17.1 Zeroth Law of Thermodynamics

If any two systems, maintaining thermal equilibrium with each other, come in contact with the 3rd system, the latter also holds thermal equilibrium with each other.

After establishment of 1st, 2nd and 3rd laws of thermodynamics the above statement is required for the development of thermodynamics but the statement logically precedes the other three. According to Zeroth law of thermodynamics temperature is a state function. The law is discussed in detail in the sec.6.2.

6.17.2 Thermodynamic Temperature Scale

For a reversible Carnot engine the efficiency, η_r is given by

$$\eta_r = \frac{W}{|Q_1|} = \frac{|Q_1| - |Q_2|}{|Q_1|} = \frac{T_1 - T_2}{T_1}$$

Thus according to 2nd law of thermodynamics, $\eta_r < 1$ for all kinds of heat engines. T represents absolute temperature but there was no reference value of T with respect to which, temperature of any system can be determined. Pressure P has a reference value. The pressure of surrounding atmosphere above the sea level has been universally accepted as 1 atm. Volume V is an extensive property and can be determined by using the formula $V = M/\rho$, where M = Mass of a substance and ρ = density of that substance.

At this situation the scientist Lord Kelvin attempted to find out the reference value of absolute temperature. In the above equation $|Q_2|$ is the amount of heat rejected to the sink and T_2 is the absolute temperature of the sink. Lord Kelvin imagined a reversible cycle such that no heat is rejected to the sink and hence $|Q_2| = 0$.

Thus, $\eta_r = \dfrac{W}{|Q_1|} = \dfrac{|Q_1| - |Q_2|}{|Q_1|} = \dfrac{|Q_1|}{|Q_1|} = 1$. Again, $\eta_r = \dfrac{T_1 - T_2}{T_1} = 1$. Or, $T_2 = 0$

According to 2nd law of thermodynamics, the above situation is impossible to occur. According to Lord Kelvin the temperature, at which the 2nd law of thermodynamics is just violated, is called absolute zero, represented by $0°K$ in the name of Lord Kelvin. Later, it was found that, $T°K = t°C + 273.15$, where $t°C$ is the temperature expressed in °C. Thus, the thermodynamic temperature scale had been set.

6.18 THIRD LAW OF THERMODYNAMICS

(a) **Nernst-Simon statement (1907)**   For any isothermal process that involves only pure substances, each in internal equilibrium, the entropy change goes to zero as absolute temperature tends to zero.

Mathematically, it can be expressed as $\lim_{T \to 0} \Delta S = 0$

(b) **Max Planck statement (1912)**   At absolute zero entropy of a perfectly crystalline solid or an ideal solid is zero, where an ideal solid is a crystal possessing regularly oriented atoms or molecules without any crystal defect.

From equation (6.5) we have $dQ = C_p dT - VdP$.
At constant pressure, we can write that $dQ = C_p dT$.
For a reversible process we know that, $dQ = TdS$.
Thus, we can write that $TdS = C_p dT$.

Or, $$dS = \frac{C_P}{T}dT \qquad (6.40)$$

According to 3rd law of thermodynamics, $\Delta S \to 0$ as $T \to 0$ but according to equation (6.40), $\Delta S \to \infty$ as $T \to 0$, provided C_P is constant. Thus, equation (6.40) holds good if and only if $C_P \to 0$ as $T \to 0$.

Corollary-1 Heat capacity of an ideal solid or any pure substance approaches zero if the absolute temperature approaches zero. Thus, C_P must be a function of temperature and the relation is assumed to be: $C_P \propto T^a$

(i) At $a = 1$, $(C_P/T) = $ Constant. Thus according to equation (6.40) we have

$$\frac{dS}{dT} = \frac{C_P}{T} = \text{Constant}$$

which means that entropy does not change with temperature and has a definite value even as $T \to 0$ and hence the 3rd law of thermodynamics is violated.

(ii) At $a > 1$, $\frac{C_P}{T} = f(T)$. Or, $\lim_{T \to 0} \frac{C_P}{T} = \lim_{T \to 0} f(T) = 0$.

Hence from equation (6.40) we have

$$\lim_{T \to 0} \Delta S = \int \left(\lim_{T \to 0} \frac{C_P}{T} \right) dT = 0$$

Thus the 3rd law of thermodynamics holds good.

Corollary-2 Heat capacity of any crystalline substance or pure substance is directly proportional to T^a, where $a > 1$.

6.19 CALCULATION OF ENTROPY

From the 1st law of thermodynamics we have

$$dQ = dU + PdV \qquad (6.2a)$$

[Considering only expansion work]
Again from equation (6.4) we have

$$dQ = dH - VdP \qquad (6.4)$$

From the definition of entropy, we can write that

$$dS = \left(\frac{dQ}{T} \right)_{rev}$$

6.19.1 Ideal Gas

(a) Isochoric process $dV = 0$ for isochoric process and the equation (6.2a) becomes
$dQ_V = dU = nC_V dT$ [for n *moles* of an ideal gas]

$$dS_V = \left(\frac{dQ_V}{T}\right)_{rev} = \frac{dU}{T} = nC_V \frac{dT}{T}$$

On integration, within the limits T_1 and T_2 we get

$$\Delta S_V = nC_V \ln\frac{T_2}{T_1} \qquad (6.41a)\ [\text{assuming } C_V \text{ is independent of temperature}]$$

As U is a state function, dQ also behaves as a state function in isochoric process, since in an isochoric process, $dQ_V = dU$.

(b) Isobaric process $dP = 0$ for isobaric process and equation (6.4) becomes
$dQ_P = dH = nC_P dT$

$$dS_P = \left(\frac{dQ_P}{T}\right)_{rev} = \frac{dH}{T} = nC_P \frac{dT}{T}$$

Assuming that C_P is independent of temperature and integrating within the limits T_1 and T_2 we get

$$\Delta S_P = nC_P \ln\frac{T_2}{T_1} \qquad (6.41b)$$

(c) Isothermal process $dT = 0$ for isothermal process and hence $dU = 0$, $dH = 0$.

$$dS_T = \left(\frac{dQ}{T}\right)_{rev} = \frac{PdV}{T} \qquad \text{From equation (6.2a)}$$

Again, $$dS_T = \left(\frac{dQ}{T}\right)_{rev} = -\frac{VdP}{T} \qquad \text{From equation (6.4)}$$

For n *moles* of an ideal gas we know that

$$PV = nRT. \text{ Hence, } V = \frac{nRT}{P} \text{ and } P = \frac{nRT}{V}$$

Putting the values of P and V in the above equations (6.2a) and (6.4) respectively, we get

$$dS_T = \left(\frac{nRT}{V}\right)\left(\frac{dV}{T}\right) \text{ and } dS_T = -\left(\frac{nRT}{P}\right)\left(\frac{dP}{T}\right)$$

On integration, we get

$$\Delta S_T = nR\ln\frac{V_2}{V_1} = -nR\ln\frac{P_2}{P_1} = nR\ln\frac{P_1}{P_2} \qquad (6.41c)$$

(d) Total entropy   In general the total entropy of *n moles* of an ideal gas is given by

$$\Delta S = \Delta S_T + \Delta S_V = nR \ln \frac{V_2}{V_1} + nC_V \ln \frac{T_2}{T_1} \qquad \text{[From equations (6.41a) and (6.41c)]}$$

$$\Delta S = \Delta S_T + \Delta S_P = nR \ln \frac{P_1}{P_2} + nC_P \ln \frac{T_2}{T_1} \qquad \text{[From equations (6.41b) and (6.41c)]}$$

Or, $S = -nR \ln P + nC_P \ln T + S_0$

Molar entropy $= \bar{S} = \dfrac{S}{n} = -R \ln P + C_P \ln T + \bar{S}_0$ 　　　　　　　(6.41d)

$\bar{S}_0$ is known as molar entropy constant.

6.19.2   Real Gas

(a) Isothermal process

For *n moles* of a van der Waals gas, we have

$$\left(P + \frac{an^2}{V^2} \right)(V - nb) = nRT. \quad \text{Or,} \quad P = \frac{nRT}{(V - nb)} - \frac{an^2}{V^2}$$

Thus isothermal entropy change dS_T is given by

$$dS_T = \left(\frac{dQ}{T} \right)_{rev} = \frac{PdV}{T} \qquad \text{From equation (6.2a)}$$

Or, $$dS_T = \frac{PdV}{T} = \left[\frac{nRT}{(V - nb)} - \frac{an^2}{V^2} \right]\left(\frac{dV}{T} \right) = nR\left(\frac{dV}{(V - nb)} \right) - \left(\frac{an^2}{T} \right)\left(\frac{dV}{V^2} \right)$$

On integration within the limits V_1 and V_2, we get

$$\Delta S_T = nR \ln \frac{V_2 - nb}{V_1 - nb} + \left(\frac{an^2}{T} \right)\left(\frac{1}{V_2} - \frac{1}{V_1} \right) \qquad (6.42a)$$

Neglecting the internal pressure term, i.e., $\dfrac{an^2}{V^2}$, the van der Waals equation becomes

$$P(V - nb) = nRT. \quad \text{Or,} \quad V = \frac{nRT}{P} + nb$$

Thus isothermal entropy change dS_T is given by

$$dS_T = \left(\frac{dQ}{T} \right)_{rev} = -\frac{VdP}{T} \qquad \text{From equation (6.4)}$$

Or, $$dS_T = -\frac{VdP}{T} = -\left(\frac{nRT}{P} + nb\right)\frac{dP}{T} = -nR\frac{dP}{P} - \frac{nb}{T}dP$$

On integration within the limits P_1 and P_2, we get

$$\Delta S_T = nR\ln\frac{P_1}{P_2} + \frac{nb}{T}(P_1 - P_2) \qquad\qquad (6.42a)$$

(b) Isochoric process   We know that

$$dS_V = \frac{dQ_V}{T} = \frac{dU}{T}$$

For *n moles* of a van der Waals gas, we have

$$dU = nC_V dT + \frac{an^2}{V^2}dV \quad \text{[From equation (6.23)]}.$$

Again, $U = f(V, T)$, since U is a state function.

Hence, $$dU = \left(\frac{\partial U}{\partial T}\right)_V dT + \left(\frac{\partial U}{\partial V}\right)_T dV. \text{ At constant volume, } dU = \left(\frac{\partial U}{\partial T}\right)_V dT$$

$$dU = C_V dT \left[\text{Since, } C_V = \left(\frac{\partial U}{\partial T}\right)_V \text{ for 1 mole of a gas}\right].$$

$$dU = nC_V dT \text{ [for } n \text{ moles of a gas]}$$

Thus at constant volume, $dU = nC_V dT$

Hence, $$dS_V = nC_V \frac{dT}{T}$$

On integration within the limits T_1 and T_2, and assuming that C_V is independent of temperature, we get

$$\Delta S_V = nC_V \ln\frac{T_2}{T_1} \qquad\qquad \text{[Identical with equation(6.41a)]}$$

As there is no P or V term in the above equation, the equation (6.41a) is valid both for *n moles* of an ideal gas as well as *n moles* of a real gas.

So, for *n moles* of a van der Waals gas, the total entropy is given by

$$\Delta S = \Delta S_V + \Delta S_T = nC_V \ln\frac{T_2}{T_1} + nR\ln\frac{V_2 - nb}{V_1 - nb} + \left(\frac{an^2}{T}\right)\left(\frac{1}{V_2} - \frac{1}{V_1}\right) \qquad (6.42c)$$

(c) Isobaric process   We know that,

$$dS_P = \frac{dQ_P}{T} = \frac{dH}{T} \qquad\qquad \text{[From equation (6.23)]}.$$

Again, $H = f(P, T)$, since H is a state function.

Hence, $dH = \left(\dfrac{\partial H}{\partial T}\right)_P dT + \left(\dfrac{\partial H}{\partial P}\right)_T dP$. At constant pressure, $dH = \left(\dfrac{\partial H}{\partial T}\right)_P dT$

$$dH = C_P dT \left[\text{Since, } C_P = \left(\dfrac{\partial H}{\partial T}\right)_P \text{ for 1 mole of a gas} \right].$$

$$dH = nC_P dT \quad [\text{for } n \text{ moles of a gas}]$$

Thus, $$dS_P = nC_P \dfrac{dT}{T}$$

On integration within the limits T_1 and T_2, and assuming that C_P is independent of temperature, we get

$$\Delta S_P = nC_P \ln\dfrac{T_2}{T_1} \quad [\text{Identical with equation(6.41b)}]$$

As there is no P or V term in the above equation the equation (6.41b) is valid both for *n moles* of an ideal gas as well as *n moles* of a real gas.

Thus, the total entropy for *n moles* of a van der Waals gas

$$\Delta S = \Delta S_P + \Delta S_T = nC_P \ln\dfrac{T_2}{T_1} + nR\ln\dfrac{P_1}{P_2} + \dfrac{nb}{T}(P_1 - P_2) \quad (6.42d)$$

6.19.3   Entropy Change Due to Phase Transition

Phase transition is an isothermal as well as isobaric process, so, $dT = 0$ and $dP = 0$. One example is: evaporation of water $H_2O(l) \rightleftharpoons H_2O(g)$.

So, $$dS_{T,P} = \dfrac{dQ_{T,P}}{T}. \quad \text{Or, } \Delta S = \dfrac{\Delta H}{T} = \dfrac{L}{T} \quad (6.43)$$

where L is the molar latent heat. The equation (6.43) is applicable for evaporation or boiling of liquid, melting of solid and sublimation of solid.

6.19.4   Transition of One Mole of a Super-Cooled Liquid Water at $(-t°C)$ into Ice at $(-t°C)$

Step-I   Heating of one mole of the super cooled liquid water from $(-t°C)$ to $(0°C)$. Process is isobaric, i.e., $dP = 0$. As the process is isobaric, the entropy change for the liquid water may be calculated using equation (6.41b). Thus, we can write that

$$\Delta S_1 = C_{P,l} \ln\dfrac{T_2}{T_1}, \quad \text{where, } T_1 = (-t + 273)°K \text{ and } T_2 = 273°K$$

$C_{P,l}$ = Molar specific heat of the super-cooled liquid.

Step-II Fusion of super-cooled liquid to ice at 0°C. It is an isothermal and isobaric process. So, $dT = 0$ and $dP = 0$.

$$\Delta S_2 = -\frac{L}{T_2} \text{ [From equation (6.43)], where, } T_2 = 273°K$$

The negative sign indicates that latent heat is rejected to the surroundings.

Step-III Cooling of ice from 0°C to $(-t°C)$, i.e., from T_2 to T_1. It is an isobaric process, i.e., $dP = 0$. So, entropy change is given by

$$\Delta S_3 = C_{P,s} \ln \frac{T_1}{T_2} \text{ [From equation (6.41}b)],$$

$$\text{where, } T_1 = (-t + 273)°K \text{ and } T_2 = 273°K$$

where, $C_{P,s}$ = Molar specific heat of ice.

Thus, the total entropy change for the whole process is given by

$$\Delta S = \Delta S_1 + \Delta S_2 + \Delta S_3 = C_{P,l} \ln \frac{T_2}{T_1} - \frac{L}{T_2} + C_{P,s} \ln \frac{T_1}{T_2} = (C_{P,l} - C_{P,s}) \ln \frac{T_2}{T_1} - \frac{L}{T_2} \quad (6.44)$$

6.19.5 Entropy Change Due to Mixing of Ideal Gases

At first, different ideal gases are stored in different chambers maintaining constant pressure, P, and temperature, T, in each chamber.

For ideal gas 1, n_1 = Number of moles, $C_{P(1)}$ = Molar specific heat, $\overline{S}_1^0$ = Standard molar entropy.

For ideal gas 2, n_2 = Number of moles, $C_{P(2)}$ = Molar specific heat, $\overline{S}_2^0$ = Standard molar entropy.

For ideal gas i, n_i = Number of moles, $C_{P(i)}$ = Molar specific heat, $\overline{S}_i^0$ = Standard molar entropy.

We know that for 1 mole of an ideal gas, $dS = C_P d \ln T - R d \ln P$, Or, $\overline{S} = C_P \ln T - R \ln P + \overline{S}^0$

For ideal gas 1, $n_1 \overline{S}_1 = n_1 C_{P(1)} \ln T - n_1 R \ln P + n_1 \overline{S}_1^0$

For ideal gas 2, $n_2 \overline{S}_2 = n_2 C_{P(2)} \ln T - n_2 R \ln P + n_2 \overline{S}_2^0$

For ideal gas i, $n_i \overline{S}_i = n_i C_{P(i)} \ln T - n_i R \ln P + n_i \overline{S}_i^0$

So, the total entropy before mixing, S_A, is given by

$$S_A = \ln T \sum n_i C_{P(i)} - R \sum n_i \ln P + \sum n_i \overline{S}_i^0$$

Now all these ideal gases are mixed in a single compartment at constant pressure, P and constant temperature, T. Now entropy calculation is based on partial pressure, p, since, $\Sigma p_i = P$. So,

For ideal gas 1, $\qquad n_1 S_1' = n_1 C_{P(1)} \ln T - n_1 R \ln p_1 + n_1 \bar{S}_1^0$

For ideal gas 2, $\qquad n_2 S_2' = n_2 C_{P(2)} \ln T - n_2 R \ln p_2 + n_2 \bar{S}_2^0$

For ideal gas i, $\qquad n_i S_i' = n_i C_{P(i)} \ln T - n_i R \ln p_i + n_i \bar{S}_i^0$

where, p_i is the partial pressure of ith ideal gas. So the total entropy after mixing, S_B, is given by

$$S_B = \ln T \, \Sigma n_i C_{P(i)} - R \Sigma n_i \ln p_i + \Sigma n_i \bar{S}_i^0$$

Thus, entropy change due to mixing of ideal gases is

$$\Delta S^M = S_B - S_A$$

$$\Delta S^M = \left[\ln T \, \Sigma n_i C_{P(i)} - R \Sigma n_i \ln p_i + \Sigma n_i \bar{S}_i^0 \right]$$
$$- \left[\ln T \, \Sigma n_i C_{P(i)} - R \Sigma n_i \ln P + \Sigma n_i \bar{S}_i^0 \right]$$

Or, $\qquad \Delta S^M = - R \Sigma n_i \ln p_i + R \Sigma n_i \ln P = - R \Sigma n_i \ln \dfrac{p_i}{P} = - nR \Sigma \dfrac{n_i}{n} \ln \dfrac{p_i}{P}$

As, $\qquad x_i = \dfrac{n_i}{n} = \dfrac{p_i}{P} = $ Mole fraction of ith ideal gas and $x_i < 1$, $\Sigma x_i = 1$, $\ln x_i < 0$

Or, $\qquad \Delta S^M = - R \Sigma n_i \ln x_i = - nR \Sigma x_i \ln x_i \qquad\qquad (6.45)$

Thus, $\Delta S^M > 0$, which signifies that entropy is increasing on mixing, of gases, which is a spontaneous non-equilibrium process since after mixing gases cannot be separated. Thus, 2nd law of thermodynamics holds good in mixing of gases.

Conclusion Mixing of gases is an irreversible process.

Example 6.19.1 Draw the *TS*-diagram of Carnot cycle.

Answer The *TS*-diagram of Carnot cycle is shown in the Fig. 6.16.

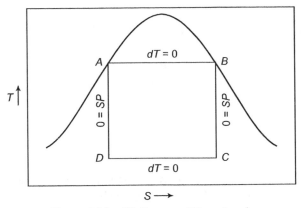

Figure 6.16 *TS*-diagram of Carnot cycle

Example 6.19.2 Prove that entropy of the universe is increasing.

Answer Let us consider an isolated system, having two compartments with same volume, separated by a diathermal wall. In compartment-I, the temperature is T_A, which is greater than T_B, the temperature of the compartment-II. dQ amount of heat flows from the compartment-I to the compartment-II to maintain the thermal equilibrium. Thus, we can write

Heat change in compartment-I $= -dQ$ and entropy change $= dS = -\dfrac{dQ}{T_A}$

Heat change in compartment-II $= +dQ$ and entropy change $= dS = +\dfrac{dQ}{T_B}$

Thus the overall change in entropy is given by

$$dS = -\frac{dQ}{T_A} + \frac{dQ}{T_B} = dQ\left(\frac{1}{T_B} - \frac{1}{T_A}\right)$$

$dQ > 0$ and $T_B < T_A$. So, $dS > 0$, indicating that entropy of an isolated system is increasing. As the universe is considered as an isolated system, entropy of the universe is also increasing.

Example 6.19.3 If q cal of heat is absorbed by the surrounding, show that,

$$\Delta S_{surr} = -\frac{\Delta H_{sys}}{T}$$

Answer If q cal of heat is absorbed by the surrounding, entropy of the surrounding can be written as $\Delta S_{surr} = \dfrac{q}{T}$. Surrounding is considered to have a large volume with constant pressure and temperature. Thus, we can write

$$q = q_P = \Delta H_{surr} = -\Delta H_{sys}. \quad \text{Thus,} \quad \Delta S_{surr} = \frac{q}{T_{sur}} = -\frac{\Delta H_{sys}}{T_{sur}}$$

As pressure and temperature of surrounding are constant, q_P is a state function and hence ΔS_{surr} depends only on ΔH_{sys}. The value of ΔS_{surr} is same in both reversible and irreversible processes.

6.20 OTHER STATE FUNCTIONS

Besides U, H and S there are two more state functions. One is called Helmholtz energy or work function, represented by A and the other is Gibbs free energy, represented by G.

6.20.1 Helmholtz Free Energy or Work Function (A)

The Helmholtz free energy or work function is represented by A and expressed by

$$A = U - TS. \quad \text{Or,} \quad U = A + TS \tag{6.46}$$

As U, S and T are all state functions, A is also a state function.

Example 6.20.1 Show that $W_{max} = -A$, where, W_{max} is the maximum work available from a given thermodynamic system.

Answer From the definition of entropy we can write that

$$dS \geq \frac{dQ}{T}. \quad \text{Or,} \quad dQ \leq TdS \quad \text{[Clausius inequality]}$$

Again, from the 1st law of thermodynamics, we have $dW = dQ - dU$. Putting the value of dQ in the above equation, we get

$$dW + dU \leq TdS. \quad \text{Or,} \quad dW \leq TdS - dU$$

For a reversible process or equilibrium process, $dQ = TdS$ and $dW = dW_{max}$. Or, $dW_{max} = TdS - dU$. On integration we get

$$W_{max} = T(S_2 - S_1) - (U_2 - U_1) = (U_1 - TS_1) - (U_2 - TS_2)$$

Or, $$W_{max} = A_1 - A_2 = -(A_2 - A_1) = -\Delta A$$

Or, $$W_{max} = -\Delta A \tag{6.47}$$

where, W_{max} = Expansion work + useful work and ΔA represents the ***maximum work*** available from a system. Hence it is called ***work function***.

A is the amount of energy available which can be partly or totally converted into work. Hence A is also known as Helmholtz energy. TS is called bound energy or unavailable energy since it cannot be converted into work but transformed only into heat. Entropy S of the term TS is known as ***capacity factor of bound energy***.

Example 6.20.2 Show that $(\Delta A)_{V,T} \leq 0$.

Solution Differentiating the equation (6.46) we get

$$dA = dU - TdS - SdT = (dQ - dW) - TdS - SdT, \text{ since, } dQ = dU + dW$$

Or, $\qquad dA = (dQ - TdS) - dW - SdT = (dQ - TdS) - PdV - SdT$

where, $dW = PdV$, considering only expansion work.

Thus at constant V and T, we have $\quad (dA)_{V,T} = dQ - TdS$.

(i) $(dA)_{V,T} = 0$ for reversible process, since, $dQ = TdS$ for reversible process.

(ii) $(dA)_{V,T} < 0$ for irreversible process, since, $dQ < TdS$ for irreversible process.

In general $\quad (dA)_{V,T} \leq 0, \quad$ or, $\quad (\Delta A)_{V,T} \leq 0$ $\qquad\qquad$ (6.48)

Conclusion

For a non-equilibrium or spontaneous process, Helmholtz energy decreases but does not change for equilibrium process.

Example 6.20.3 Deduce the expression of A for an ideal gas undergoing reversible isothermal expansion process.

Solution Differentiating the equation (6.46) we get

$$dA = dU - TdS - SdT = (dQ - dW) - TdS - SdT, \text{ since, } dQ = dU + dW$$

Or, $\qquad\qquad\qquad dA = (dQ - TdS) - dW - SdT = (dQ - TdS) - PdV - SdT$

$\qquad\qquad\qquad$ [since $dW = PdV$ for expansion work]

For a reversible process $\quad dQ = TdS$

Hence, $\qquad\qquad\qquad dA = -PdV - SdT$ $\qquad\qquad\qquad\qquad$ (6.49a)

At constant temperature, i.e., for isothermal process $\quad dA_T = -PdV$. For n *moles* of an ideal gas, we have

$$PV = nRT. \text{ Or, } P = \frac{nRT}{V}$$

Putting the value of P in the above equation we get

$$dA_T = -nRT\frac{dV}{V}$$

On integration within the limits V_1 and V_2 we get

$$A_T = nRT \ln\frac{V_1}{V_2}$$ $\qquad\qquad\qquad$ (6.49b)

6.20.2 Gibbs Free Energy (*G*)

The Gibbs free energy is represented by *G* and expressed by

$$G = H - TS. \quad \text{Or,} \quad H = G + TS \tag{6.50}$$

As *H*, *S* and *T* are all state functions, *G* is also a state function.

Example 6.20.4 Show that $W'_{max} = -\Delta G$, where, W'_{max} is the maximum useful work available from a given thermodynamic system.

Solution From the definition of entropy we can write that

$$dS \geq \frac{dQ}{T}. \quad \text{Or,} \quad dQ \leq TdS \quad \text{[Clausius inequality]}$$

Again, from the 1st law of thermodynamics, we have $dW = dQ - dU$. Putting the value of dQ in the above equation, we get

$$dW + dU \leq TdS. \quad \text{Or,} \quad dW \leq TdS - dU$$

For a reversible process or equilibrium process, $dQ = TdS$ and $dW = dW_{max}$.

Or, $$dW_{max} = TdS - dU \tag{6.51a}$$

Again, $dW =$ Expansion work + Useful work $= PdV + dW'$.

For reversible process, $dW = dW_{max}$ and $dW' = dW'_{max}$.

Thus, $$dW_{max} = PdV + dW'_{max} \tag{6.51b}$$

Combining equations (6.51a) and (6.51b), we get

$$PdV + dW'_{max} = TdS - dU. \quad \text{Or,} \quad dW'_{max} = TdS - dU - PdV$$

On integration at constant *P* and *T*, we get

$$W'_{max} = T(S_2 - S_1) - (U_2 - U_1) - P(V_2 - V_1)$$

$$= T(S_2 - S_1) - [(U_2 + PV_2) - (U_1 + PV_1)]$$

Or, $$W'_{max} = T(S_2 - S_1) - (H_2 - H_1) = -[(H_2 - TS_2) - (H_1 - TS_1)]$$

$$= -(G_2 - G_1)$$

Or, $$W'_{max} = -\Delta G \tag{6.52}$$

where, $G = H - TS$ and *G* is known as Gibbs free energy. ΔG represents the ***maximum useful work*** available from the system.

Example 6.20.5 Show that $(\Delta G)_{P,T} \leq 0$.

Solution We know $G = H - TS$. Differentiating the equation we get

$$dG = dH - TdS - SdT = (dQ + VdP) - TdS - SdT, \text{ since, } dH = dQ + VdP$$

Or, $$dG = (dQ - TdS) + VdP - SdT = (dQ - TdS) + VdP - SdT$$

Thus at constant P and T, we have $(dG)_{P,T} = dQ - TdS$.

(i) $(dG)_{P,T} = 0$ for reversible process, since, $dQ = TdS$ for reversible process.

(ii) $(dG)_{P,T} < 0$ for irreversible process, since, $dQ < Tds$ for irreversible process.

In general $(dG)_{P,T} \leq 0$, or, $(\Delta G)_{P,T} \leq 0$ (6.53)

Example 6.20.6 State the physical significance of $(\Delta G)_{P,T} < 0$.

Answer For non-equilibrium or spontaneous process $(\Delta G)_{P,T} < 0$, i.e., free energy change decreases. The physical significance of the above equation is that energy always flows from higher to lower energy state but the reverse is not true. So any process, where free energy change decreases, is a spontaneous non-equilibrium process and hence the thermodynamic condition for a spontaneous process is $\Delta G < 0$.

Example 6.20.7 State the conditions of equilibrium and spontaneous processes.

Answer The following are true in case of any irreversible or non-equilibrium process:

(i) $dS > \left(\dfrac{dQ}{T}\right)_{ir}$ Or, $\Delta S_{ir} > 0$ [From 2nd law of thermodynamics]

(ii) $(\Delta A)_{V,T} < 0$ [From equation (6.48)]

(iii) $(\Delta G)_{P,T} < 0$ [From equation (6.53)]

For any reversible or equilibrium process the following are true

(i) $dS = \left(\dfrac{dQ}{T}\right)_{r}$. Or, $\Delta S_r = 0$ [From equation (6.37)]

(ii) $(\Delta A)_{V,T} = 0$ [From equation (6.48)]

(iii) $(\Delta G)_{P,T} = 0$ [From equation (6.53)]

$\Delta G = 0$ signifies the thermodynamic condition of equilibrium since at this condition free energy of either side is the same.

Example 6.20.8 Deduce the expression of **G** for an ideal gas undergoing reversible isothermal expansion process.

Solution We know $G = H - TS$. Differentiating the equation we get,

$$dG = dH - TdS - SdT = (dQ + VdP) - TdS - SdT, \text{ since, } dH = dQ + VdP$$

Or, $$dG = (dQ - TdS) + VdP - SdT = (dQ - TdS) + VdP - SdT$$

For a reversible process $dQ = TdS$

Hence, $$dG = VdP - SdT$$ (6.54a)

At constant temperature, i.e., for isothermal process $dG = VdP$. For *n moles* of an ideal gas, we have

$$PV = nRT. \text{ Or, } V = \frac{nRT}{P}$$

Putting the value of P in the above equation we get

$$dG = nRT\frac{dP}{P}$$

On integration we get

$$G = G^0 + nRT \ln P \tag{6.54b}$$

where, G^0 is termed **standard free energy**, which is defined as the free energy of a given thermodynamic system at a constant pressure of 1 atm.

In case of aqueous solution, the above equation becomes

$$G = G^0 + nRT \ln a \tag{6.54c}$$

where, a is the activity of the solute and G^0 is the standard free energy of the aqueous solution at unit activity of the solute.

On integration within the limits P_1 and P_2 we get

$$\Delta G = nRT \ln\frac{P_2}{P_1} \tag{6.54d}$$

6.21 GIBBS HELMHOLTZ EQUATIONS

6.21.1 Gibbs Helmholtz Equation in Terms of Work Function, *A*

From equation (6.46), we have $\quad A = U - TS \tag{6.46}$

We know from equation (6.49a) that $\quad dA = -PdV - SdT \tag{6.49a}$

If we consider two thermodynamic states, having work functions, internal energies and entropies are A_1 & A_2, U_1 & U_2, and S_1 & S_2 respectively, equation (6.46) can be written as

$$A_1 = U_1 - TS_1 \quad \text{and} \quad A_2 = U_2 - TS_2.$$

Hence, $\qquad (A_2 - A_1) = (U_2 - U_1) - T(S_2 - S_1)$

Or, $\qquad\qquad \Delta A = \Delta U - T\Delta S \tag{6.55}$

Differentiating the equation (6.49a) w.r.t. T at constant V we get

$$\left(\frac{\partial A}{\partial T}\right)_V = -S. \text{ Or, } S = -\left(\frac{\partial A}{\partial T}\right)_V, \ S_1 = -\left(\frac{\partial A_1}{\partial T}\right)_V, \ S_2 = -\left(\frac{\partial A_2}{\partial T}\right)_V$$

Hence, $\quad (S_2 - S_1) = \Delta S = -\left[\frac{\partial(A_2 - A_1)}{\partial T}\right]_V = -\left[\frac{\partial(\Delta A)}{\partial T}\right]_V. \text{ Or, } \Delta S = -\left[\frac{\partial(\Delta A)}{\partial T}\right]_V$

Putting the value of S and ΔS in the equations (6.46) and (6.55) respectively we get

$$A = U + T\left(\frac{\partial A}{\partial T}\right)_V \quad \text{and} \quad \Delta A = \Delta U + T\left[\frac{\partial(\Delta A)}{\partial T}\right]_V \qquad (6.56a)$$

Another form in terms of A.

Differentiating the term (A/T) w.r.t. T at constant V, we get

$$\left[\frac{\partial(A/T)}{\partial T}\right]_V = -\frac{A}{T^2} + \frac{1}{T}\left(\frac{\partial A}{\partial T}\right)_V = -\frac{1}{T^2}\left[A - T\left(\frac{\partial A}{\partial T}\right)_V\right]$$

Comparing the above equation with (6.56a) we get

$$\left[\frac{\partial(A/T)}{\partial T}\right]_V = -\frac{U}{T^2} \qquad (6.56b)$$

Similarly, differentiating the term $(\Delta A/T)$ w.r.t. T at constant V, we get

$$\left[\frac{\partial(\Delta A/T)}{\partial T}\right]_V = -\frac{\Delta A}{T^2} + \frac{1}{T}\left[\frac{\partial(\Delta A)}{\partial T}\right]_V = -\frac{1}{T^2}\left[\Delta A - T\left[\frac{\partial(\Delta A)}{\partial T}\right]_V\right]$$

Comparing the above equation with (6.56a) we get

$$\left[\frac{\partial(\Delta A/T)}{\partial T}\right]_V = -\frac{\Delta U}{T^2} \qquad (6.56c)$$

Equations (6.56a), (6.56b) and (6.56c) are known as Gibbs Helmholtz equations in terms of A.

6.21.2 Gibbs Helmholtz Equation in Terms of Gibbs Free Energy, G

We know that $\quad G = H - TS$ $\qquad\qquad\qquad\qquad\qquad\qquad$ (6.50)

If we consider two thermodynamic states having free energies G_1 & G_2, enthalpies H_1 & H_2 and entropies S_1 & S_2 respectively, the Gibbs free energies of these two states can be written as

$G_1 = H_1 - TS_1$ $\quad$ and $\quad$ $G_2 = H_2 - TS_2$. $\quad$ Hence, $\quad$ $(G_2 - G_1) = (H_2 - H_1) - T(S_2 - S_1)$

Or, $\qquad\qquad\qquad\qquad\qquad \Delta G = \Delta H - T\Delta S$ $\qquad\qquad\qquad\qquad$ (6.57)

The above equation is popularly known as Helmholtz equation.

From equation (6.54a) we have

$$dG = VdP - SdT \qquad (6.54a)$$

Differentiating the above equation w.r.t. T at constant P, we have

$$\left(\frac{\partial G}{\partial T}\right)_P = -S. \text{ Or, } S = -\left(\frac{\partial G}{\partial T}\right)_P, \ S_1 = -\left(\frac{\partial G_1}{\partial T}\right)_P, \ S_2 = -\left(\frac{\partial G_2}{\partial T}\right)_P$$

Hence,
$$(S_2 - S_1) = \Delta S = -\left[\frac{\partial (G_2 - G_1)}{\partial T}\right]_P = -\left[\frac{\partial(\Delta G)}{\partial T}\right]_P . \text{ Or, } \Delta S = -\left[\frac{\partial(\Delta G)}{\partial T}\right]_P$$

Putting the values of S and ΔS in the equations (6.50) and (6.57) respectively, we get

$$G = H + T\left(\frac{\partial G}{\partial T}\right)_P \text{ and } \Delta G = \Delta H + T\left[\frac{\partial(\Delta G)}{\partial T}\right]_P \quad (6.58a)$$

Another form in terms of G.

Differentiating the term (G/T) w.r.t. T at constant P, we get

$$\left[\frac{\partial(G/T)}{\partial T}\right]_P = -\frac{G}{T^2} + \frac{1}{T}\left(\frac{\partial G}{\partial T}\right)_P = -\frac{1}{T^2}\left[G - T\left(\frac{\partial G}{\partial T}\right)_P\right]$$

Comparing the above equation with (6.58a) we get

$$\left[\frac{\partial(G/T)}{\partial T}\right]_P = -\frac{H}{T^2} \quad (6.58b)$$

Similarly, differentiating the term $(\Delta G/T)$ w.r.t. T at constant P, we get

$$\left[\frac{\partial(\Delta G/T)}{\partial T}\right]_V = -\frac{\Delta G}{T^2} + \frac{1}{T}\left[\frac{\partial(\Delta G)}{\partial T}\right]_P = -\frac{1}{T^2}\left[\Delta G - T\left[\frac{\partial(\Delta G)}{\partial T}\right]_P\right]$$

Comparing the above equation with (6.58a) we get

$$\left[\frac{\partial(\Delta G/T)}{\partial T}\right]_P = -\frac{\Delta H}{T^2} \quad (6.58c)$$

Equations (6.58a), (6.58b) and (6.58c) are also known as Gibbs Helmholtz equations in terms of G.

6.22 CONDITION OF SPONTANEITY FOR DIFFERENT CHEMICAL REACTIONS

The Helmholtz equation is given by:

$$\Delta G = \Delta H - T\Delta S \quad (6.59)$$

It has been discussed in the earlier section that the thermodynamic condition for a spontaneous process is $\Delta G < 0$.

(i) Consider a chemical reaction where, $\Delta H > 0$ and $\Delta S > 0$. Thus, to maintain $\Delta G < 0$ according to equation (6.59), the required thermodynamic condition is $|T\Delta S| > |\Delta H|$.

(ii) Consider a chemical reaction where, $\Delta H > 0$ and $\Delta S < 0$. According to equation (6.59), the chemical reaction ***never occurs*** since in this case, ΔG is always positive at any temperature.

(iii) Consider a chemical reaction where, $\Delta H < 0$ and $\Delta S > 0$. The chemical reaction ***always occurs*** spontaneously, since ΔG is always negative irrespective of temperature.

(iv) Consider a chemical reaction where, $\Delta H < 0$ and $\Delta S < 0$. Thus, to maintain $\Delta G < 0$ according to equation (6.59), the required thermodynamic condition is $|\Delta H| > |T\Delta S|$.

6.23 PHASE TRANSITION CLAUSIUS-CLAPEYRON EQUATION

If a pure substance exists in two or more phases in equilibrium there is a possibility of transition from one phase to another. Such transition takes place at constant temperature and pressure but entropy changes due to change in molecular disorderness. If phase transition takes place in a closed system, equilibrium will be established between the phases. In that case we can write $dQ = TdS$. From the 1st law of thermodynamics, we have

$$dU = dQ - dW = dQ - PdV \quad \text{[Considering only expansion work]}$$

Assuming the process is reversible we can write that $dU = TdS - PdV$.
On integration of the above equation, we get

$$U_2 - U_1 = T(S_2 - S_1) - P(V_2 - V_1). \quad \text{Or,} \quad (U_2 + PV_2) - TS_2 = (U_1 + PV_1) - TS_1$$

Or, $(H_2 - TS_2) = (H_1 - TS_1)$. Or, $G_2 = G_1$. Or, $G_1 = G_2$ (6.60a)

Now the pressure and temperature are changed slightly from P to $(P + dP)$ and T to $(T + dT)$ respectively, G_1 changes to $(G_1 + dG_1)$ and G_2 changes to $(G_2 + dG_2)$. According to Le Chatelier's principle, the system rearranges within itself to restore the equilibrium. Thus, we can write that

$$(G_1 + dG_1) = (G_2 + dG_2)$$

Comparing the above equation with the equation (6.60a) we get

$$dG_2 = dG_2 \quad (6.60b)$$

From equation (6.56a), we know that $dG = VdP - SdT$. Hence, the equation (6.60b) can be written as

$$V_1 dP - S_1 dT = V_2 dP - S_2 dT. \quad \text{Or,} \quad \frac{dP}{dT} = \frac{(S_2 - S_1)}{(V_2 - V_1)} \quad (6.60c)$$

$(S_2 - S_1) = \Delta S$ = Entropy change due to phase transition. According to equation (6.43), we have

$$\Delta S = (S_2 - S_1) = \frac{L}{T}$$

where L is the molar latent heat of phase transition at temperature T.

Thus the equation (6.60c) becomes

$$\frac{dP}{dT} = \frac{L}{T(V_2 - V_1)} \tag{6.60d}$$

The equation (6.60d) is known as Clapeyron equation, which holds good for any pure substance existing in two phases. The phases may be different type of solid phases or different type of liquid phases or gas phase.

The scientist Clausius modified the above equation by considering equilibrium system comprising of one gas phase and one liquid or solid phase.

Consider the following phase equilibrium   Liquid $\rightleftharpoons$ Vapour

The equation (6.60d) can be written as

$$\frac{dP}{dT} = \frac{L_v}{T(V_g - V_l)}, \text{ where, } L_v = \text{Molar latent heat of vapourisation}.$$

where, V_l = Volume of liquid and V_g = Volume of vapour. Since, $V_g \gg V_l$, it may be approximated that $(V_g - V_l) \approx V_g$. So, we can write that

$$\frac{dP}{dT} = \frac{L_v}{TV_g}$$

Assuming that the vapour behaves ideally, we can write that $PV_g = RT$ for one mole of vapour.

Or,
$$V_g = \frac{RT}{P}$$

Putting the value of V_g in the above equation we get

$$\frac{dP}{dT} = \frac{LP}{RT^2}. \text{ Or, } \frac{dP}{P} = \frac{L}{RT^2}dT$$

On integration, we get

$$\int_{P_1}^{P_2} d\ln P = \int_{T_1}^{T_2} \frac{L}{RT^2} dT. \text{ Or, } \ln \frac{P_2}{P_1} = -\frac{L}{R}\left[\frac{1}{T_2} - \frac{1}{T_1}\right] \tag{6.60e}$$

The equation (6.60e) is known as **Clausius-Clapeyron** equation.

Application   One of the most important applications of the above equation is to determine the latent heat of vaporization of a solvent or liquid by measuring vapour pressures at two different temperatures.

6.24 MAXWELL RELATIONS

Let us consider a state function M such that $M = f(x, y)$. Then according to equation (6.1b), we can write that

$$\frac{\partial^2 M}{\partial x \cdot \partial y} = \frac{\partial^2 M}{\partial y \cdot \partial x}$$

U, H, A and G are the thermodynamic state functions.

(i) In case of free energy, we have

$$dG = VdP - SdT \quad \text{and} \quad G = f(P, T)$$

Hence,
$$\left(\frac{\partial G}{\partial P}\right)_T = V \quad \text{and} \quad \left(\frac{\partial G}{\partial T}\right)_P = -S$$

Again differentiating, we have

$$\frac{\partial^2 G}{\partial T \cdot \partial P} = \left(\frac{\partial V}{\partial T}\right)_P \quad \text{and} \quad \frac{\partial^2 G}{\partial P \cdot \partial T} = -\left(\frac{\partial S}{\partial P}\right)_T$$

As G is a state function, we have

$$\frac{\partial^2 G}{\partial P \cdot \partial T} = \frac{\partial^2 G}{\partial T \cdot \partial P}$$

Hence
$$\left(\frac{\partial S}{\partial P}\right)_T = -\left(\frac{\partial V}{\partial T}\right)_P \tag{6.61a}$$

(ii) In case of work function, we have

$$dA = PdV - SdT \quad \text{and} \quad A = f(V, T)$$

Hence,
$$\left(\frac{\partial A}{\partial V}\right)_T = -P \quad \text{and} \quad \left(\frac{\partial A}{\partial T}\right)_V = -S$$

Again differentiating, we have

$$\frac{\partial^2 A}{\partial T \cdot \partial V} = -\left(\frac{\partial P}{\partial T}\right)_V \quad \text{and} \quad \frac{\partial^2 A}{\partial V \cdot \partial T} = -\left(\frac{\partial S}{\partial V}\right)_T$$

As A is a state function, we have

$$\frac{\partial^2 A}{\partial V \cdot \partial T} = \frac{\partial^2 A}{\partial T \cdot \partial V}$$

Hence,
$$\left(\frac{\partial S}{\partial V}\right)_T = \left(\frac{\partial P}{\partial T}\right)_V \tag{6.61b}$$

(iii) In case of enthalpy, we have

$$dH = dQ + VdP. \quad \text{Or}, \quad dH = TdS + VdP. \text{ Hence, } H = f(S, P)$$

Hence, $\qquad \left(\dfrac{\partial H}{\partial S}\right)_P = T \text{ and } \left(\dfrac{\partial H}{\partial P}\right)_S = V$

Again differentiating, we have

$$\frac{\partial^2 H}{\partial P \cdot \partial S} = \left(\frac{\partial T}{\partial P}\right)_S \quad \text{and} \quad \frac{\partial^2 H}{\partial S \cdot \partial P} = \left(\frac{\partial V}{\partial S}\right)_P$$

As H is a state function, we have

$$\frac{\partial^2 H}{\partial S \cdot \partial P} = \frac{\partial^2 H}{\partial P \cdot \partial S}$$

Hence, $\qquad\qquad \left(\dfrac{\partial V}{\partial S}\right)_P = \left(\dfrac{\partial T}{\partial P}\right)_S \qquad\qquad$ (6.61c)

(iv) In case of internal energy, we have

$$dU = dQ - PdV. \quad \text{Or}, \quad dU = TdS - PdV. \text{ Hence, } U = f(S, V)$$

Hence, $\qquad \left(\dfrac{\partial U}{\partial S}\right)_V = T \text{ and } \left(\dfrac{\partial U}{\partial V}\right)_S = -P$

Again differentiating, we have

$$\frac{\partial^2 U}{\partial V \cdot \partial S} = \left(\frac{\partial T}{\partial V}\right)_S \quad \text{and} \quad \frac{\partial^2 U}{\partial S \cdot \partial V} = -\left(\frac{\partial P}{\partial S}\right)_V$$

As U is a state function, we have

$$\frac{\partial^2 U}{\partial S \cdot \partial V} = \frac{\partial^2 U}{\partial V \cdot \partial S}$$

Hence, $\qquad\qquad \left(\dfrac{\partial P}{\partial S}\right)_V = -\left(\dfrac{\partial T}{\partial V}\right)_S \qquad\qquad$ (6.61d)

So, following are the Maxwell's relations

$$\left(\frac{\partial S}{\partial P}\right)_T = -\left(\frac{\partial V}{\partial T}\right)_P \qquad\qquad (6.61a)$$

$$\left(\frac{\partial S}{\partial V}\right)_T = \left(\frac{\partial P}{\partial T}\right)_V \qquad\qquad (6.61b)$$

$$\left(\frac{\partial V}{\partial S}\right)_P = \left(\frac{\partial T}{\partial P}\right)_S \qquad (6.61c)$$

$$\left(\frac{\partial P}{\partial S}\right)_V = -\left(\frac{\partial T}{\partial V}\right)_S \qquad (6.61d)$$

The above equations are known as Maxwell's equations or Maxwell's relations. It is interesting to note that the Maxwell's relations are the relations among four parameters P, V, T, and S. Further, the slopes of **P–S curve** in both isothermal and isochoric processes are **negative**, i.e., $\left(\frac{\partial S}{\partial P}\right)_T$ and $\left(\frac{\partial P}{\partial S}\right)_V$ terms are **negative** in Maxwell's relations and the slopes of **V–S curve** in both isothermal and isobaric processes are **positive**, i.e., $\left(\frac{\partial S}{\partial V}\right)_T$ and $\left(\frac{\partial V}{\partial S}\right)_P$ terms are **positive** in Maxwell's relations.

6.24.1 Applications of Maxwell's Relation

6.24.1.1 Calculation of Internal Pressure, i.e., $\left(\frac{\partial U}{\partial V}\right)_T$ and ΔU

As U is a state function we can write that $U = f(V, T)$. Thus on differentiation, we get

$$dU = \left(\frac{\partial U}{\partial T}\right)_V dT + \left(\frac{\partial U}{\partial V}\right)_T dV \qquad (6.62)$$

From 1st law of thermodynamics, we have $dU = dQ - PdV$ [Considering only expansion work]

Or, $\qquad dU = TdS - PdV$ [since, $dQ = TdS$ for reversible process]

Differentiating the above equation w.r.t. V at constant T, we have

$$\left(\frac{\partial U}{\partial V}\right)_T = T\left(\frac{\partial S}{\partial V}\right)_T - P$$

From Maxwell's relations we have

$$\left(\frac{\partial S}{\partial V}\right)_T = \left(\frac{\partial P}{\partial T}\right)_V$$

So, $\qquad \left(\frac{\partial U}{\partial V}\right)_T = T\left(\frac{\partial P}{\partial T}\right)_V - P$, where, $\left(\frac{\partial U}{\partial V}\right)_T$ is termed internal pressure.

(a) For n moles of an ideal gas, we know that $PV = nRT$. Differentiating w.r.t. T at constant V, we get

$$V\left(\frac{\partial P}{\partial T}\right)_V = nR. \quad \text{Or,} \quad T\left(\frac{\partial P}{\partial T}\right)_V = \frac{nRT}{V} = P.$$

Hence,
$$\left(\frac{\partial U}{\partial V}\right)_T = T\left(\frac{\partial P}{\partial T}\right)_V - P = P - P = 0$$

Or,
$$\left(\frac{\partial U}{\partial V}\right)_T = 0 \text{ for ideal gases}$$

Scientist Joule found the same result in his experiment [discussed in the sec. 6.8.3]. Thus in the case of an ideal gas, equation (6.62) becomes

$$dU = \left(\frac{\partial U}{\partial T}\right)_V dT = nC_V dT$$

Assuming C_V is independent of temperature, integration within temperature limits, T_1 and T_2, we get

$$\Delta U = nC_V(T_2 - T_1) \tag{6.63}$$

(b) For n moles of a van der Waals gas, we have

$$\left(P + \frac{an^2}{V^2}\right)(V - nb) = nRT. \text{ Or, } P = \frac{nRT}{(V - nb)} - \frac{an^2}{V^2}$$

Differentiating w.r.t. T at constant V, we get

$$\left(\frac{\partial P}{\partial T}\right)_V = \frac{nR}{(V - nb)} \cdot \text{ Or, } T\left(\frac{\partial P}{\partial T}\right)_V = \frac{nRT}{(V - nb)} = P + \frac{an^2}{V^2}$$

Thus,
$$\left(\frac{\partial U}{\partial V}\right)_T = T\left(\frac{\partial P}{\partial T}\right)_V - P = \left[P + \frac{an^2}{V^2}\right] - P = \frac{an^2}{V^2}$$

Or,
$$\left(\frac{\partial U}{\partial V}\right)_T = \frac{an^2}{V^2} \tag{6.64a}$$

Putting this value of $\left(\frac{\partial U}{\partial V}\right)_T$ in the equation (6.62), we get

$$dU = \left(\frac{\partial U}{\partial T}\right)_V dT + \frac{an^2}{V^2} dV = nC_V dT + \frac{an^2}{V^2} dV$$

On integration of the above equation assuming C_V is independent of temperature, we get

$$\Delta U = nC_V(T_2 - T_1) - an^2\left[\frac{1}{V_2} - \frac{1}{V_1}\right] \tag{6.64b}$$

6.24.1.2 Calculation of $\left(\dfrac{\partial H}{\partial P}\right)_T$ and ΔH

As H is a state function we can write that, $H = f(P, T)$. Thus on differentiation, we get

$$dH = \left(\frac{\partial H}{\partial T}\right)_P dT + \left(\frac{\partial H}{\partial P}\right)_T dP = nC_P dT + \left(\frac{\partial H}{\partial P}\right)_T dP \quad (6.65)$$

Again, from 1st law of thermodynamics, we have $\quad dH = dQ + VdP$

Again, $\quad dQ = TdS$ [Considering reversible process].

Thus, $\quad dH = TdS + VdP$. Differentiating this w.r.t. P at constant T we have

$$\left(\frac{\partial H}{\partial P}\right)_T = T\left(\frac{\partial S}{\partial P}\right)_T + V$$

From Maxwell's relation we have

$$\left(\frac{\partial S}{\partial P}\right)_T = -\left(\frac{\partial V}{\partial T}\right)_P$$

Thus, $\qquad \left(\dfrac{\partial H}{\partial P}\right)_T = -T\left(\dfrac{\partial V}{\partial T}\right)_P + V$

(a) For n moles of an ideal gas, we have $PV = nRT$. Differentiating w.r.t. T at constant P, we get

$$P\left(\frac{\partial V}{\partial T}\right)_P = nR. \ \text{ Or, } \ T\left(\frac{\partial V}{\partial T}\right)_P = \frac{nRT}{P} = V,$$

Hence, $\qquad \left(\dfrac{\partial H}{\partial P}\right)_T = -T\left(\dfrac{\partial V}{\partial T}\right)_P + V = -V + V = 0$

$$\left(\frac{\partial H}{\partial P}\right)_T = 0 \ \text{ for ideal gases.}$$

Thus in the case of an ideal gas, equation (6.65) becomes

$$dH = \left(\frac{\partial H}{\partial T}\right)_P dT = nC_P dT$$

On integration of the above equation assuming C_P is independent of temperature, we get

$$\Delta H = nC_P(T_2 - T_1) \qquad (6.66a)$$

(b) For n moles of a van der Waals gas, we have

$$\left(P + \frac{an^2}{V^2}\right)(V - nb) = nRT$$

Neglecting the internal pressure terms, the above equation becomes

$$P(V - nb) = nRT. \quad \text{Or, } V = \frac{nRT}{P} + nb$$

Differentiating w.r.t. T at constant P, we have

$$\left(\frac{\partial V}{\partial T}\right)_P = \frac{nR}{P}. \quad \text{Or, } T\left(\frac{\partial V}{\partial T}\right)_P = \frac{nRT}{P} = (V - nb)$$

Hence,

$$\left(\frac{\partial H}{\partial P}\right)_T = -T\left(\frac{\partial V}{\partial T}\right)_P + V = -(V - nb) + V = nb$$

Or,

$$\left(\frac{\partial H}{\partial P}\right)_T = nb \text{ for van der Waals gas}$$

Putting this value of $\left(\dfrac{\partial H}{\partial P}\right)_T$ in the equation (6.65), we get $dH = nC_p dT + nb dP$

On integrating the above equation assuming C_p is independent of temperature, we get

$$\Delta H = nC_p(T_2 - T_1) + nb(P_2 - P_1) \tag{6.66b}$$

6.24.1.3 Calculation of $\left(\dfrac{\partial S}{\partial V}\right)_T$

From Maxwell's relation [equation (6.61b)], we know that

$$\left(\frac{\partial S}{\partial V}\right)_T = \left(\frac{\partial P}{\partial T}\right)_V$$

(a) For n *moles* of an ideal gas, we know that $PV = nRT$. Differentiating w.r.t. T at constant V we get

$$V\left(\frac{\partial P}{\partial T}\right)_V = nR. \quad \text{Or, } \left(\frac{\partial P}{\partial T}\right)_V = \frac{nR}{V}. \quad \text{Or, } \left(\frac{\partial S}{\partial V}\right)_T = \left(\frac{\partial P}{\partial T}\right)_V = \frac{nR}{V} \tag{6.67a}$$

(b) For n *moles* of a van der Waals gas, we have

$$\left(P + \frac{an^2}{V^2}\right)(V - nb) = nRT. \quad \text{Or, } P = \frac{nRT}{(V - nb)} - \frac{an^2}{V^2}$$

Differentiating w.r.t. T at constant V, we get

$$\left(\frac{\partial P}{\partial T}\right)_V = \frac{nR}{(V - nb)} \cdot \text{ Or, } \left(\frac{\partial S}{\partial V}\right)_T = \left(\frac{\partial P}{\partial T}\right)_V = \frac{nR}{(V - nb)} \quad\quad (6.67b)$$

Example 6.24.1 Show that C_P is independent of pressure.

Solution

$$C_P = \left(\frac{\partial H}{\partial T}\right)_P \cdot \text{ Or, } \left(\frac{\partial C_P}{\partial P}\right)_T = \left[\frac{\partial}{\partial P}\left(\frac{\partial H}{\partial T}\right)_P\right]_T = \left[\frac{\partial}{\partial T}\left(\frac{\partial H}{\partial P}\right)_T\right]_P$$

As H is a state function, we can write that

$$\left[\frac{\partial}{\partial P}\left(\frac{\partial H}{\partial T}\right)_P\right]_T = \left[\frac{\partial}{\partial T}\left(\frac{\partial H}{\partial P}\right)_T\right]_P$$

$$\left(\frac{\partial C_P}{\partial P}\right)_T = \left[\frac{\partial}{\partial T}\left(\frac{\partial H}{\partial P}\right)_T\right]_P$$

Again, $dH = dQ + VdP.$ Or, $dH = TdS + VdP$

Hence,

$$\left(\frac{\partial H}{\partial P}\right)_T = T\left(\frac{\partial S}{\partial P}\right)_T + V$$

$$\left(\frac{\partial C_P}{\partial P}\right)_T = \left[\frac{\partial}{\partial T}\left(\frac{\partial H}{\partial P}\right)_T\right]_P = \left(\frac{\partial}{\partial T}\left[T\left(\frac{\partial S}{\partial P}\right)_T\right]\right)_P + \left(\frac{\partial V}{\partial T}\right)_P$$

$$= \left(\frac{\partial S}{\partial P}\right)_T + \left(\frac{\partial V}{\partial T}\right)_P$$

$$\left[T\frac{\partial}{\partial T}\left(\frac{\partial S}{\partial P}\right)_T = 0, \text{ since}\left(\frac{\partial S}{\partial P}\right)_T \text{ is constant}\right]$$

According to Maxwell's relation, we have

$$\left(\frac{\partial S}{\partial P}\right)_T = -\left(\frac{\partial V}{\partial T}\right)_P$$

Thus,

$$\left(\frac{\partial C_P}{\partial P}\right)_T = \left(\frac{\partial S}{\partial P}\right)_T + \left(\frac{\partial V}{\partial T}\right)_P = 0$$

So, C_P is independent of pressure.

6.24.1.4 Evaluation of $(C_P - C_V)$

For reversible process, we have $dQ = TdS$

From the definition of specific heat we know that

$$C_V = \left(\frac{\partial Q}{\partial T}\right)_V = \left[\frac{\partial(TdS)}{\partial T}\right]_V \cdot \text{ So, } C_V = T\left(\frac{\partial S}{\partial T}\right)_V$$

$$C_P = \left(\frac{\partial Q}{\partial T}\right)_P = \left[\frac{\partial(TdS)}{\partial T}\right]_P \cdot \text{ So, } C_P = T\left(\frac{\partial S}{\partial T}\right)_P$$

$$C_P - C_V = T\left(\frac{\partial S}{\partial T}\right)_P - T\left(\frac{\partial S}{\partial T}\right)_V = T\left[\left(\frac{\partial S}{\partial T}\right)_P - \left(\frac{\partial S}{\partial T}\right)_V\right]$$

As S is a state function, we can write that $S = S(V, T)$

Or,
$$dS = \left(\frac{\partial S}{\partial T}\right)_V dT + \left(\frac{\partial S}{\partial V}\right)_T dV$$

Differentiating w.r.t. T at constant P, we get

$$\left(\frac{\partial S}{\partial T}\right)_P = \left(\frac{\partial S}{\partial T}\right)_V + \left(\frac{\partial S}{\partial V}\right)_T \left(\frac{\partial V}{\partial T}\right)_P$$

Or,
$$\left(\frac{\partial S}{\partial T}\right)_P - \left(\frac{\partial S}{\partial T}\right)_V = \left(\frac{\partial S}{\partial V}\right)_T \left(\frac{\partial V}{\partial T}\right)_P$$

From Maxwell's relation, we have

$$\left(\frac{\partial S}{\partial V}\right)_T = \left(\frac{\partial P}{\partial T}\right)_V \quad \text{[Equation (6.61b)]}$$

Thus,
$$C_P - C_V = T\left[\left(\frac{\partial S}{\partial T}\right)_P - \left(\frac{\partial S}{\partial T}\right)_V\right] = T\left(\frac{\partial S}{\partial V}\right)_T \left(\frac{\partial V}{\partial T}\right)_P$$

$$= T\left(\frac{\partial P}{\partial T}\right)_V \left(\frac{\partial V}{\partial T}\right)_P$$

Or,
$$C_P - C_V = T\left(\frac{\partial P}{\partial T}\right)_V \left(\frac{\partial V}{\partial T}\right)_P \quad\quad (6.68a)$$

For *n moles* of an ideal gas, we know that $PV = nRT$. Differentiating w.r.t. T at constant V, we get

$$V\left(\frac{\partial P}{\partial T}\right)_V = nR. \text{ Or, } \left(\frac{\partial P}{\partial T}\right)_V = \frac{nR}{V} \cdot \text{ Or, } T\left(\frac{\partial P}{\partial T}\right)_V = \frac{nRT}{V} = P$$

Again differentiating $PV = nRT$ w.r.t. T at constant P, we get

$$P\left(\frac{\partial V}{\partial T}\right)_P = nR. \text{ Or, } \left(\frac{\partial V}{\partial T}\right)_P = \frac{nR}{P}$$

Putting these values of $T\left(\dfrac{\partial P}{\partial T}\right)_V$ and $\left(\dfrac{\partial V}{\partial T}\right)_P$ in the equation (6.68a), we get

$$C_P - C_V = P\left(\frac{nR}{P}\right) = nR. \text{ For 1 mole of an ideal gas, } C_P - C_V = R \qquad (6.68b)$$

For *n moles* of a van der Waals gas, we know that

$$\left(P + \frac{an^2}{V^2}\right)(V - nb) = nRT. \text{ Or, } P = \frac{nRT}{(V - nb)} - \frac{an^2}{V^2}$$

Differentiating both sides w.r.t. T at constant V, we get

$$\left(\frac{\partial P}{\partial T}\right)_V = \frac{nR}{(V - nb)} \cdot \text{ Or, } T\left(\frac{\partial P}{\partial T}\right)_V = \frac{nRT}{(V - nb)} = P + \frac{an^2}{V^2}$$

Again, $\left(P + \dfrac{an^2}{V^2}\right)(V - nb) = nRT.$ Or, $PV - Pnb + \dfrac{an^2}{V} - \dfrac{abn^3}{V^2} = nRT$

Differentiating both sides w.r.t. T at constant P, we get

$$P\left(\frac{\partial V}{\partial T}\right)_P - \frac{an^2}{V^2}\left(\frac{\partial V}{\partial T}\right)_P + \frac{2abn^3}{V^3}\left(\frac{\partial V}{\partial T}\right)_P = nR$$

As V is very large, $(1/V^2)$ and $(1/V^3)$ terms can be ignored. Thus, the above equation becomes

$$P\left(\frac{\partial V}{\partial T}\right)_P = nR. \text{ Or, } \left(\frac{\partial V}{\partial T}\right)_P = \frac{nR}{P}$$

Putting these values of $T\left(\dfrac{\partial P}{\partial T}\right)_V$ and $\left(\dfrac{\partial V}{\partial T}\right)_P$ in the equation (6.68a), we get

$$C_P - C_V = \left(P + \frac{an^2}{V^2}\right)\left(\frac{nR}{P}\right) = nR + \frac{an^3 R}{PV^2}$$

For 1 mole of van der Waals gas

$$C_P - C_V = R + \frac{aR}{PV^2} \qquad (6.68c)$$

Example 6.24.2 Prove that

(i) $C_p - C_V = -T\left(\dfrac{\partial P}{\partial V}\right)_T\left[\left(\dfrac{\partial V}{\partial T}\right)_P\right]^2$ and (ii) $C_p - C_V = \left[V - \left(\dfrac{\partial H}{\partial P}\right)_T\right]\left(\dfrac{\partial P}{\partial T}\right)_V$

Solution

(i) Using cyclic rule we can write

$$\left(\frac{\partial P}{\partial T}\right)_V \left(\frac{\partial T}{\partial V}\right)_P \left(\frac{\partial V}{\partial P}\right)_T = -1. \quad \text{Or,} \quad \left(\frac{\partial P}{\partial T}\right)_V = -\left(\frac{\partial P}{\partial V}\right)_T \left(\frac{\partial V}{\partial T}\right)_P$$

Putting this value of $\left(\dfrac{\partial P}{\partial T}\right)_V$ in the equation (6.68a), we get

$$C_p - C_V = T\left(\frac{\partial P}{\partial T}\right)_V \left(\frac{\partial V}{\partial T}\right)_P = -T\left(\frac{\partial P}{\partial V}\right)_T \left[\left(\frac{\partial V}{\partial T}\right)_P\right]^2$$

(ii) From 1st law of thermodynamics, we have

$dH = dQ + VdP$. Or, $dH = TdS + VdP$. Differentiating the equation w.r.t. P at constant T, we get

$$\left(\frac{\partial H}{\partial P}\right)_T = T\left(\frac{\partial S}{\partial P}\right)_T + V$$

From Maxwell's relations, we have

$$\left(\frac{\partial S}{\partial P}\right)_T = -\left(\frac{\partial V}{\partial T}\right)_P$$

Thus the above equation becomes

$$\left(\frac{\partial H}{\partial P}\right)_T = -T\left(\frac{\partial V}{\partial T}\right)_P + V. \quad \text{Or,} \quad T\left(\frac{\partial V}{\partial T}\right)_P = V - \left(\frac{\partial H}{\partial P}\right)_T$$

Putting this value of $T\left(\dfrac{\partial V}{\partial T}\right)_P$ in the equation (6.68a), we get

$$C_p - C_V = T\left(\frac{\partial P}{\partial T}\right)_V \left(\frac{\partial V}{\partial T}\right)_P = \left[V - \left(\frac{\partial H}{\partial P}\right)_T\right]\left(\frac{\partial P}{\partial T}\right)_V$$

6.24.1.5 Derivation of Stefan-Boltzmann law, $u \propto T^4$ using Maxwell's relation

Let us consider that black body radiation is assumed to follow kinetic theory of gases. u and P are the energy of radiation per unit volume and radiation pressure respectively.

According to kinetic theory of gases, we have

$$P = \frac{1}{3}mn\overline{c^2} = \frac{1}{3}u, \quad \text{where}$$

m = Mass of each molecule

n = Number of molecule/volume

$\overline{c^2}$ = Mean square velocity of gas molecules

Thus, we can write $u = \left(\dfrac{\partial U}{\partial V}\right)_T$, $P = \dfrac{1}{3}u$. From equation (6.62), we have

$$dU = \left(\frac{\partial U}{\partial T}\right)_V dT + \left(\frac{\partial U}{\partial V}\right)_T dV$$

From 1st law of thermodynamics, we have

$$dU = dQ - PdV \qquad \text{[Considering only expansion work]}$$

Or, $\qquad dU = TdS - PdV$, since, $dQ = TdS$ for reversible process.

Differentiating the above equation w.r.t. V at constant T, we have

$$\left(\frac{\partial U}{\partial V}\right)_T = T\left(\frac{\partial S}{\partial V}\right)_T - P$$

From Maxwell's relation we have

$$\left(\frac{\partial S}{\partial V}\right)_T = \left(\frac{\partial P}{\partial T}\right)_V. \quad \text{So,} \quad \left(\frac{\partial U}{\partial V}\right)_T = T\left(\frac{\partial P}{\partial T}\right)_V - P$$

Putting the value of u and P in the above equation, we get

$$u = T\frac{d(u/3)}{dT} - \frac{u}{3} = \frac{T}{3}\frac{du}{dT} - \frac{u}{3}. \quad \text{Or,} \quad 4u = T\frac{du}{dT}. \quad \text{Or} \quad \frac{du}{u} = 4\frac{dT}{T}$$

On integration, we get

$$\ln u = 4\ln T + \ln C. \quad \text{Or,} \quad \ln u = \ln C.T^4. \quad \text{Hence,} \ u \propto T^4$$

Example 6.24.3 Show that for 1 mole of an ideal gas $\left(\dfrac{\partial V}{\partial S}\right)_P = \dfrac{RT}{PC_P}$

Solution   Using chain rule we have

$$\left(\frac{\partial T}{\partial P}\right)_S \left(\frac{\partial P}{\partial S}\right)_T \left(\frac{\partial S}{\partial T}\right)_P = -1. \quad \text{Or,} \quad \left(\frac{\partial T}{\partial P}\right)_S \left(\frac{\partial P}{\partial S}\right)_T \left[T\left(\frac{\partial S}{\partial T}\right)_P\right] = -T \qquad (A)$$

$$C_P = \left(\frac{\partial Q}{\partial T}\right)_P = T\left(\frac{\partial S}{\partial T}\right)_P, \quad \text{since,} \ dQ = TdS$$

Using Maxwell's equation, we have

$$\left(\frac{\partial T}{\partial P}\right)_S = \left(\frac{\partial V}{\partial S}\right)_P$$

Hence, equation (A) becomes

$$\left(\frac{\partial V}{\partial S}\right)_P \left(\frac{\partial P}{\partial S}\right)_T \left[T\left(\frac{\partial S}{\partial T}\right)_P\right] = -T. \text{ Or, } \left(\frac{\partial V}{\partial S}\right)_P = -\frac{T}{C_P}\left(\frac{\partial S}{\partial P}\right)_T$$

Or, $$\left(\frac{\partial V}{\partial S}\right)_P = \frac{T}{C_P}\left(\frac{\partial V}{\partial T}\right)_P \quad \left[\text{Using Maxwell's equation } \left(\frac{\partial S}{\partial P}\right)_T = -\left(\frac{\partial V}{\partial T}\right)_P\right]$$

For 1 mole of an ideal gas, we have $V = \dfrac{RT}{P}$.

Hence, $$\left(\frac{\partial V}{\partial T}\right)_P = \frac{R}{P}$$

Putting the value of $\left(\dfrac{\partial V}{\partial T}\right)_P$ in the above equation we get

$$\left(\frac{\partial V}{\partial S}\right)_P = \frac{RT}{PC_P}$$

6.25 JOULE-THOMSON (*JT*) EFFECT

6.25.1 *JT*-Expansion Process is an Isenthalpic Process

In 1853, Joule and William Thomson performed an experiment which involved a slow throttling of a gas through a rigid porous plug, illustrated in the Fig. 6.17.

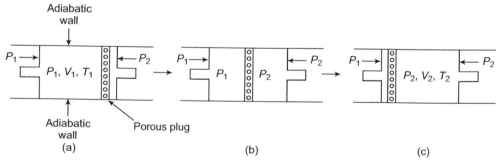

Figure 6.17 The Joule-Thomson experiment

There are two chambers having fixed pressures of P_1 and P_2 respectively such that $P_2 < P_1$. The chambers are separated by a porous plug, as shown in the Fig. 6.17. Initially, the volume of left chamber is V_1 and that of right chamber is **zero.** The gas at the left chamber is allowed to expand through the porous plug and hence gas volume of the left chamber decreases while that of the right chamber increases. At the end, the gas volume of left chamber becomes **zero** and that of the right chamber is V_2.

As the system is covered by adiabatic walls there is no heat transfer between system and surroundings. So, we can write that $dQ = 0$. From the first law of thermodynamics, we have

$$dQ = dU + dW. \quad \text{Hence,} \quad dW = -dU. \quad \text{Or,} \quad \Delta W = -(U_2 - U_1) \qquad (6.69a)$$

Net work done by the gas $= \Delta W = W_1 + W_2$

$W_1 = $ Work done on the gas in the left chamber, where pressure is fixed at P_1 and volume changes from V_1 to 0.

$$W_1 = \int_{V_1}^{0} P_1 dV = -P_1 V_1$$

$W_2 = $ Work done by the gas in the right chamber, where pressure is fixed at P_2 and volume changes from 0 to V_2.

$$W_2 = \int_{0}^{V_2} P_2 dV = P_2 V_2$$

$$\Delta W = W_1 + W_2 = P_2 V_2 - P_1 V_1$$

Putting this value of ΔW in the equation (6.69a), we get

$$P_2 V_2 - P_1 V_1 = -(U_2 - U_1). \quad \text{Or,} \quad U_2 + P_2 V_2 = U_1 + P_1 V_1. \quad \text{Or,} \quad H_2 = H_1. \quad \text{Or,} \quad dH = 0$$

Thus, Joule-Thomson expansion process or throttling process is an **isenthalpic** process. During this process temperature changes with decrease in pressure and this change is termed μ_{JT}. This μ_{JT} is called Joule-Thomson coefficient or *JT*-coefficient which is expressed as:

$$\mu_{JT} = \left(\frac{\partial T}{\partial P} \right)_H$$

During *JT*-expansion process pressure decreases and hence dP is always negative.

6.25.2 Significance of μ_{JT}

(i) When $\mu_{JT} > 0$, temperature decreases or $dT < 0$, since $dP < 0$. Thus, cooling effect is observed during *JT*-expansion process.

(ii) When $\mu_{JT} < 0$, temperature increases or $dT > 0$, since $dP < 0$. Thus, heating effect is observed during *JT*-expansion process.

(iii) When $\mu_{JT} = 0$, temperature does not change or $dT = 0$, since $dP < 0$. Thus, neither heating nor cooling effect is observed during *JT*-expansion process.

6.25.3 Mathematical Expression of μ_{JT}

As enthalpy H is state function, the equation of state is given by $f(H, P, T) = 0$.
Using cyclic rule, we can write

$$\left(\frac{\partial T}{\partial P}\right)_H \left(\frac{\partial P}{\partial H}\right)_T \left(\frac{\partial H}{\partial T}\right)_P = -1. \quad \text{Hence,} \quad \left(\frac{\partial T}{\partial P}\right)_H = -\frac{\left(\frac{\partial H}{\partial P}\right)_T}{\left(\frac{\partial H}{\partial T}\right)_P}$$

Again, $\mu_{JT} = \left(\dfrac{\partial T}{\partial P}\right)_H$ and $C_P = \left(\dfrac{\partial H}{\partial T}\right)_P$. So, $\mu_{JT} = -\dfrac{1}{C_P}\left(\dfrac{\partial H}{\partial P}\right)_T$ (6.69b)

From 1st law of thermodynamics, we have

$$dH = dQ + VdP. \quad \text{Or,} \quad dH = TdS + VdP, \quad [\text{since,} \quad dQ = TdS].$$

Differentiating the above equation w.r.t. P at constant T, we get

$$\left(\frac{\partial H}{\partial P}\right)_T = T\left(\frac{\partial S}{\partial P}\right)_T + V = -T\left(\frac{\partial V}{\partial T}\right)_P + V, \quad [\text{using Maxwell relation (6.61}a)]$$

Putting the value of $(\partial H/\partial P)_T$ in the equation (6.69b), we get

$$\mu_{JT} = -\frac{1}{C_P}\left[-T\left(\frac{\partial V}{\partial T}\right)_P + V\right] = \frac{1}{C_P}\left[T\left(\frac{\partial V}{\partial T}\right)_P - V\right] \quad (6.70)$$

This is the mathematical expression of μ_{JT}

6.25.3.1 Calculation of μ_{JT} for Ideal Gas

In the case of an ideal gas, we have $PV = nRT$. Differentiating w.r.t. T at constant P, we get

$$P\left(\frac{\partial V}{\partial T}\right)_P = nR. \quad \text{Or,} \quad T\left(\frac{\partial V}{\partial T}\right)_P = \frac{nRT}{P} = V$$

Thus, $\mu_{JT} = \dfrac{1}{C_P}\left[T\left(\dfrac{\partial V}{\partial T}\right)_P - V\right] = \dfrac{1}{C_P}(V - V) = 0$

So, $\mu_{JT} = 0$ is valid for any ideal gas at all temperatures. Thus, when an ideal gas is allowed to expand through a throttle valve, the temperature of the gas remains constant. In other words, neither cooling nor heating effect is observed on execution of a throttling process of an ideal gas.

6.25.3.2 Calculation of μ_{JT} for a van der Waals Gas

For 1 mole of a van der Waals gas, we have

$$\left(P + \frac{a}{V^2}\right)(V - b) = RT. \quad \text{Or,} \quad \left(P + \frac{a}{V^2}\right) = \frac{RT}{(V - b)} \tag{6.71}$$

Differentiating the above equation w.r.t. T at constant P, we get

$$-\frac{2a}{V^3}\left(\frac{\partial V}{\partial T}\right)_P = \frac{R}{(V - b)} - \frac{RT}{(V - b)^2}\left(\frac{\partial V}{\partial T}\right)_P$$

Or,

$$\left[\frac{RT}{(V - b)^2} - \frac{2a}{V^3}\right]\left(\frac{\partial V}{\partial T}\right)_P = \frac{R}{(V - b)}. \quad \text{Or,} \quad T\left(\frac{\partial V}{\partial T}\right)_P = \frac{\dfrac{RT}{(V - b)}}{\left[\dfrac{RT}{(V - b)^2} - \dfrac{2a}{V^3}\right]}$$

Or,

$$T\left(\frac{\partial V}{\partial T}\right)_P - V = \frac{\dfrac{RT}{(V - b)}}{\left[\dfrac{RT}{(V - b)^2} - \dfrac{2a}{V^3}\right]} - V = \frac{\dfrac{RT}{(V - b)} - \dfrac{RTV}{(V - b)^2} + \dfrac{2a}{V^2}}{\left[\dfrac{RT}{(V - b)^2} - \dfrac{2a}{V^3}\right]}$$

Or,

$$T\left(\frac{\partial V}{\partial T}\right)_P - V = \frac{\dfrac{2a}{V^2} - \dfrac{RTb}{(V - b)^2}}{\left[\dfrac{RT}{(V - b)^2} - \dfrac{2a}{V^3}\right]}$$

Or,

$$\mu_{JT} = \frac{1}{C_P}\left[T\left(\frac{\partial V}{\partial T}\right)_P - V\right] = \frac{\dfrac{2a}{V^2} - \dfrac{RTb}{(V - b)^2}}{C_P\left[\dfrac{RT}{(V - b)^2} - \dfrac{2a}{V^3}\right]} \tag{6.72}$$

6.25.3.3 Inversion Temperature

The temperature at which JT-coefficient (μ_{JT}) of a real gas is zero, is called inversion temperature, represented by T_i. Above T_i, $\mu_{JT} < 0$ and hence heating effect will be observed on execution of throttling process and below T_i, $\mu_{JT} > 0$ and hence cooling effect will be observed on execution of throttling process. Each and every real gas has a particular inversion temperature. If throttling process is executed on a real gas at its T_i or inversion temperature, neither cooling nor heating effect will be observed.

Thus, from equation (6.72), $\mu_{JT} = 0$ appears when $T = T_i$, which signifies that numerator of equation (6.72) is zero. Hence, we can write

$$\frac{2a}{V^2} - \frac{RT_ib}{(V-b)^2} = 0. \text{ Or, } T_i = \frac{2a}{Rb}\left(\frac{V-b}{V}\right)^2 . \text{ Or, } \left(\frac{V-b}{V}\right)^2 = \frac{Rb}{2a}T_i \quad (6.73a)$$

Or,
$$Y^2 = 4mX, \text{ where, } Y = \left(\frac{V-b}{V}\right), \quad 4m = \frac{Rb}{2a} \text{ and } X = T_i$$

Thus, the plot of $[(V-b)/V]$ against T is a parabola. Similar parabolic curve is also obtained by plotting T vs. P, as shown in the Fig. 6.18.

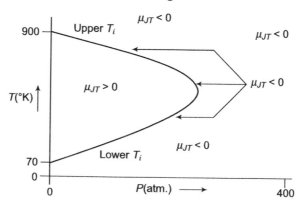

Figure 6.18 *T–P diagram of a real gas undergoing a throttling process*

From the Fig. 6.18, following observations can be made:

(a) If the throttling process is executed at any point within the area bounded by the curve and T-axis, the cooling effect will be observed, whereas if the same is executed at any point outside the curve, heating effect will be observed. On the line of the curve, $\mu_{JT} = 0$ and no effect will be observed.

(b) The upper curve is called the upper T_i curve and the lower curve is called lower T_i curve. Thus except at nose there are two T_i values at a particular pressure.

Usually throttling process is executed at around room temperature and all the real gases show cooling effect except hydrogen and helium.

In order to get cooling effect of upper T_i value should be above room temperature but in the cases of hydrogen and helium upper T_i value is well below the room temperature.

The upper T_i value of hydrogen is –73°C (200°K) at 1 atm and –92°C (181°K) at 100 atm. In case of helium, the upper T_i value is –222°C at 1 atm.

As throttling process is executed at around room temperature, which is always higher than upper T_i values of hydrogen and helium, heating effect is observed in case of hydrogen and helium according to the Fig. 6.18.

If we consider $(V-b) \approx V$, the equation (6.73a) becomes

$$1 = \frac{Rb}{2a}T_i . \text{ Or, } T_i = \frac{2a}{Rb} \quad (6.73b)$$

(c) As $(V - b) \approx V$, $[RT/(V - b)^2] > [2a/V^3]$, since volume of a gas is usually very large. Thus, denominator of equation (6.72) is always positive.

Hence, $\mu_{JT} > 0$, when $2a > RTb$. Or, $T < (2a/Rb)$. Or, $T < T_i$.

Thus, when a real gas is subjected to throttling process at a temperature below its inversion temperature, T_i, cooling effect will be observed.

(d) As $(V - b) \approx V$, $[RT/(V - b)^2] > [2a/V^3]$, since volume of a gas is usually very large. Thus, denominator of equation (6.72) is always positive.

Hence, $\mu_{JT} < 0$, when $2a < RTb$. Or, $T > (2a/Rb)$. Or, $T > T_i$.

Thus, when a real gas is subjected to throttling process at a temperature above its inversion temperature, T_i, heating effect will be observed.

6.26 FUGACITY

According to equation (6.54a), we have $dG = VdP - SdT$. For 1 mole of an ideal gas, we know that

$$PV = RT. \quad \text{Or,} \quad V = \frac{RT}{P}$$

For an isothermal process, $dT = 0$ and the expression of dG becomes

$$dG = VdP = RT\frac{dP}{P}. \quad \text{Or,} \quad dG = RTd \ln P$$

On integration, we get $G = G° + RT \ln P$.

At a particular pressure, the value of G of a real gas differs from that calculated from the above equation. Thus, for 1 mole of a real gas, the above equation can be written as

$$G = G° + RT \ln f \tag{6.74}$$

where, f is known as fugacity.

As we know at very low pressure any real gas behaves ideally, we can write that

$$\lim_{P \to 0}\left(\frac{f}{P}\right) = 1$$

Differentiating equation (6.74), we get $dG = RTd \ln f$

Again we know that $dG = VdP$ at constant temperature. Thus, the above equation becomes

$$VdP = RTd \ln f. \quad d \ln f = \frac{V}{RT}dP \tag{6.75}$$

where, V_r = real volume of gas = (Total volume) − (Volume of Avogadro No. of molecules)

Or, $V_r = (V_{id} - \alpha)$

V_{id} = Total volume = Volume of 1 mole of an ideal gas = (RT/P).

Or,
$$V_r = (V_{id} - \alpha) = \left(\frac{RT}{P} - \alpha\right)$$

Putting this value of V in the equation (6.75), we get

$$d \ln f = \left(\frac{RT}{P} - \alpha\right)\frac{dP}{RT}. \quad \text{Or, } d \ln f = d \ln P - \left(\frac{\alpha}{RT}\right)dP$$

On integration, we get

$$\ln \frac{f_2}{f_1} = \ln \frac{P_2}{P_1} - \frac{\alpha}{RT}\int_{P_1}^{P_2} dP. \quad \text{At } P_1 \to 0, \frac{f_1}{P_1} = 1 \text{ and } \ln \frac{f_1}{P_1} = 0$$

$$\ln f_2 = \ln P_2 + \ln \frac{f_1}{P_1} - \frac{\alpha}{RT}\int_{P_1}^{P_2} dP. \quad \text{Or, } \ln f_2 = \ln P_2 + \lim_{P_1 \to 0}\ln \frac{f_1}{P_1} - \frac{\alpha}{RT}\int_{P_1 \to 0}^{P_2} dP$$

Or, $\ln f_2 = \ln P_2 - \dfrac{\alpha}{RT}\displaystyle\int_{P_1 \to 0}^{P_2} dP$ [assuming that α is constant between limits P_1 and P_2]

Or,
$$\ln f_2 = \ln P_2 - \frac{\alpha P_2}{RT}$$

Eliminating the subscript, the above equation becomes

$$\ln f = \ln P - \frac{\alpha P}{RT}. \quad \text{Or, } \ln \frac{f}{P} = -\frac{\alpha P}{RT}$$

Or,
$$\frac{f}{P} = e^{-\frac{\alpha P}{RT}} = \gamma \qquad (6.76)$$

α is known as activity coefficient. Again, expansion of equation (6.76) gives rise to

$$\alpha = e^{-\frac{\alpha P}{RT}} = 1 - \frac{\alpha P}{RT} + \frac{1}{2!}\left(\frac{\alpha P}{RT}\right)^2 - \frac{1}{3!}\left(\frac{\alpha P}{RT}\right)^3 + \cdots$$

Neglecting the square and higher terms, we get

$$\gamma = 1 - \frac{\alpha P}{RT} = \frac{P}{RT}\left(\frac{RT}{P} - \alpha\right) = \frac{\left(\dfrac{RT}{P} - \alpha\right)}{\left(\dfrac{RT}{P}\right)} = \frac{V_r}{V_{id}}$$

Or,
$$\gamma = 1 - \frac{\alpha P}{RT} = \frac{V_r}{V_{id}} \qquad (6.77)$$

V_r = Real volume of a gas and can be experimentally determined.

$V_{id} = (RT/P)$. So, V_{id} can be easily calculated at a particular temperature and pressure.

Thus, γ can be calculated using equation (6.77). Knowing this value of γ, α can easily be evaluated from equation (6.77). Once γ is evaluated the fugacity, f, can easily be calculated from equation (6.76).

6.27 DEPENDENCE OF SPECIFIC HEAT ON VOLUME AND PRESSURE

From 1st law of thermodynamics, we have $dU = dQ - dW$. Considering only expansion work, we have

$dU = dQ - PdV$. Again, $dQ = TdS$. Thus, $dU = TdS - PdV$

Hence $\left(\dfrac{\partial U}{\partial V}\right)_T = T\left(\dfrac{\partial S}{\partial V}\right)_T - P = T\left(\dfrac{\partial P}{\partial T}\right)_V - P$ [using equation (6.61b)]

From equation (6.62), we have

$$dU = \left(\frac{\partial U}{\partial V}\right)_T dV + \left(\frac{\partial U}{\partial T}\right)_V dT \quad \text{[Equation (6.62)]}$$

From the above equations, we get

$$dU = \left[T\left(\frac{\partial P}{\partial T}\right)_V - P\right]dV + C_V dT, \text{ where, } C_V = \left(\frac{\partial U}{\partial T}\right)_V \quad (6.78a)$$

Differentiating w.r.t. T at constant V, we get

$$\left(\frac{\partial U}{\partial T}\right)_V = C_V. \text{ Or, } \frac{\partial^2 U}{\partial V \cdot \partial T} = \left(\frac{\partial C_V}{\partial V}\right)_T$$

Differentiating w.r.t. V at constant T, we get

Again, $\left(\dfrac{\partial U}{\partial V}\right)_T = T\left(\dfrac{\partial P}{\partial T}\right)_V - P.$ Or, $\dfrac{\partial^2 U}{\partial T \cdot \partial V} = \left[\dfrac{\partial}{\partial T}\left\{T\left(\dfrac{\partial P}{\partial T}\right)_V - P\right\}\right]_V$

As U is a state function, we have

$$\frac{\partial^2 U}{\partial T \cdot \partial V} = \frac{\partial^2 U}{\partial V \cdot \partial T}$$

Thus, we have

$$\left[\frac{\partial}{\partial T}\left\{T\left(\frac{\partial P}{\partial T}\right)_V - P\right\}\right]_V = \left(\frac{\partial C_V}{\partial V}\right)_T. \text{ Or, } T\left(\frac{\partial^2 P}{\partial T^2}\right)_V = \left(\frac{\partial C_V}{\partial V}\right)_T \quad (6.78b)$$

Similarly, in case of enthalpy, we have

$$dH = dQ + VdP \text{ and } dQ = TdS. \text{ Hence, } dH = TdS + VdP$$

Hence,

$$\left(\frac{\partial H}{\partial P}\right)_T = T\left(\frac{\partial S}{\partial P}\right)_T + V = -T\left(\frac{\partial V}{\partial T}\right)_P + V \quad \text{[using equation (6.61a)]}$$

From equation (6.65), we have

$$dH = \left(\frac{\partial H}{\partial P}\right)_T dP + \left(\frac{\partial H}{\partial T}\right)_P dT \quad \text{[Equation (6.65)]}$$

Combining the above two equations, we get

$$dH = \left[-T\left(\frac{\partial V}{\partial T}\right)_P + V\right]dP + C_P dT, \quad \text{where, } C_P = \left(\frac{\partial H}{\partial T}\right)_P \quad (6.78c)$$

Hence,

$$\left(\frac{\partial H}{\partial T}\right)_P = C_P. \quad \text{Or, } \frac{\partial^2 H}{\partial P \cdot \partial T} = \left(\frac{\partial C_P}{\partial P}\right)_T$$

Again,

$$\left(\frac{\partial H}{\partial P}\right)_T = -T\left(\frac{\partial V}{\partial T}\right)_P + V. \quad \text{Or, } \frac{\partial^2 H}{\partial T \cdot \partial P} = \left[\frac{\partial}{\partial T}\left\{-T\left(\frac{\partial V}{\partial T}\right)_P + V\right\}\right]_P$$

As *H* is a state function, we have

$$\frac{\partial^2 H}{\partial T \cdot \partial P} = \frac{\partial^2 H}{\partial P \cdot \partial T}$$

Thus, we have

$$\left[\frac{\partial}{\partial T}\left\{-T\left(\frac{\partial V}{\partial T}\right)_P + V\right\}\right]_P = \left(\frac{\partial C_P}{\partial P}\right)_T. \quad \text{Or, } -T\left(\frac{\partial^2 V}{\partial T^2}\right)_P = \left(\frac{\partial C_P}{\partial P}\right)_T \quad (6.78d)$$

(a) For 1 mole of an ideal gas, we know that $PV = RT$

   Differentiating both sides w.r.t. *T* at constant *V*, we get

$$V\left(\frac{\partial P}{\partial T}\right)_V = R. \quad \text{Or, } \left(\frac{\partial P}{\partial T}\right)_V = \frac{R}{V}$$

Again differentiating w.r.t. *T* at constant *V*, we get

$$\left(\frac{\partial^2 P}{\partial T^2}\right)_V = 0. \quad \text{Hence from equation (6.78b), we have} \left(\frac{\partial C_V}{\partial V}\right)_T = 0$$

which indicates that in an isothermal process C_V of an ideal gas is independent of volume.

Similarly, differentiating the ideal gas equation w.r.t. T at constant P, we get

$$P\left(\frac{\partial V}{\partial T}\right)_P = R. \quad \text{Or,} \quad \left(\frac{\partial V}{\partial T}\right)_P = \frac{R}{P}$$

Further differentiating w.r.t. T at constant P, we get

$$\left(\frac{\partial^2 V}{\partial T^2}\right)_P = 0. \text{ Hence from equation (6.78d), we have } \left(\frac{\partial C_P}{\partial P}\right)_T = 0$$

which signifies that in an isothermal process C_P of an ideal gas is independent of pressure.

(b) For 1 mole of a van der Waals gas, we have

$$\left(P + \frac{a}{V^2}\right)(V - b) = RT. \quad \text{Or,} \quad P = \frac{RT}{(V-b)} - \frac{a}{V^2}$$

Differentiating w.r.t. T at constant V, we get

$$\left(\frac{\partial P}{\partial T}\right)_V = \frac{R}{(V-b)}$$

Further differentiating w.r.t. T at constant V, we get

$$\left(\frac{\partial^2 P}{\partial T^2}\right)_V = 0. \text{ Hence from equation (6.78b), we have } \left(\frac{\partial C_V}{\partial V}\right)_T = 0$$

which signifies that in an isothermal process C_V of a real gas, obeying van der Waals equation, is independent of volume.

Neglecting the internal pressure term the van der Waals equation becomes

$$P(V - b) = RT. \quad \text{Or,} \quad V = \frac{RT}{P} + b \quad \text{[For 1 mole of gas]}$$

Differentiating both sides w.r.t. T at constant P, we get

$$\left(\frac{\partial V}{\partial T}\right)_P = \frac{R}{P}$$

Further differentiating w.r.t. T at constant P, we get

$$\left(\frac{\partial^2 V}{\partial T^2}\right)_P = 0. \text{ Hence from equation (6.78d), we have } \left(\frac{\partial C_P}{\partial P}\right)_T = 0$$

which signifies that in an isothermal process C_P of a real gas, obeying van der Waals equation, is independent of pressure.

6.28 PARTIAL MOLAR QUANTITIES AND GIBBS-DUHEM EQUATION

In the case of an intensive property sum of all partial quantities gives the total quantity. For example, total pressure (P) of a gas mixture of m different gases, is given by

$$P = p_1 + p_2 + p_3 + \ldots + p_m$$

where, p_1, p_2, p_3, etc. are the partial pressures of the corresponding gases.

In the case of any extensive property mass or mole must appear in the additive sequence, e.g., the total volume of the same gas mixture, as described above, is given by

$$V = n_1 \overline{V}_1 + n_2 \overline{V}_2 + n_3 \overline{V}_3 + \cdots + n_i \overline{V}_i$$

where, $\overline{V}_i$ is the molar volume of ith species in the mixture and n_i is number of moles of ith species. Differentiating both sides w.r.t. n_1 at constant n_2, n_3, etc. we have

$$\left(\frac{\partial V}{\partial n_1} \right)_{n_2} = \overline{V}_1$$

As pressure (P) and temperature (T) remain constant during mixing, we can write that

$$\left(\frac{\partial V}{\partial n_1} \right)_{P,T,n_2} = \overline{V}_1. \text{ In general, } \left(\frac{\partial V}{\partial n_i} \right)_{P,T,n_j} = \overline{V}_i \qquad (6.79a)$$

The subscript n_j indicates that number of moles of all species excepting the ith species are kept constant. $\overline{V}_i$ is known as partial molar volume. Similarly, we have

Partial molar internal energy $\qquad \overline{U}_i = \left(\dfrac{\partial U}{\partial n_i} \right)_{P,T,n_j}$

Partial molar enthalpy $\qquad \overline{H}_i = \left(\dfrac{\partial H}{\partial n_i} \right)_{P,T,n_j}$

Partial molar entropy $\qquad \overline{S}_i = \left(\dfrac{\partial S}{\partial n_i} \right)_{P,T,n_j}$

Partial molar work function $\qquad \overline{A}_i = \left(\dfrac{\partial A}{\partial n_i} \right)_{P,T,n_j}$

Partial molar Gibb's free energy $\qquad \overline{G}_i = \left(\dfrac{\partial G}{\partial n_i} \right)_{P,T,n_j}$

$\overline{G}_i$ is specially termed **chemical potential** of ith species, represented by μ_i. Thus, we may write that

$$\mu_i = \left(\frac{\partial G}{\partial n_i} \right)_{P,T,n_j} \tag{6.79b}$$

Euler's theorem   For any homogeneous function of degree m, e.g., $\phi = f(x^m y^m z^m)$, we can write according to Euler's theorem that

$$x\frac{\partial \phi}{\partial x} + y\frac{\partial \phi}{\partial y} + z\frac{\partial \phi}{\partial z} = m\phi$$

In the case of a mixture of several components Gibb's free energy, G, is a homogeneous function of degree 1 as given below

$$G = G(P, T, n_1, n_2 \ldots)$$

At constant pressure and temperature, we have

$$G = G(n_1, n_2 \ldots)$$

Hence, by Euler's theorem we can write that

$$G = n_1 \left(\frac{\partial G}{\partial n_1} \right)_{P,T,\,n_2} + n_2 \left(\frac{\partial G}{\partial n_2} \right)_{P,T,\,n_3} + \cdots + n_i \left(\frac{\partial G}{\partial n_i} \right)_{P,T,n_j} + \cdots$$

From the definition of chemical potential, the above equation becomes

$$G = n_1\mu_1 + n_2\mu_2 + \cdots + n_i\mu_i + \cdots \ \text{ Or, } \ G = \Sigma n_i\mu_i \tag{6.79c}$$

Similarly, we can write that

$$V = \Sigma n_i \bar{V}_i, \ \ U = \Sigma n_i \bar{U}_i, \ \ H = \Sigma n_i \bar{H}_i, \ \ S = \Sigma n_i \bar{S}_i, \ \ A = \Sigma n_i \bar{A}_i$$

Again we know that

$$G = G(n_1, n_2, \ldots, n_i) \qquad \text{[at constant temperature and pressure]}$$

Differentiating we get

$$dG = \left(\frac{\partial G}{\partial n_1} \right)_{P,T,n_2} dn_1 + \left(\frac{\partial G}{\partial n_2} \right)_{P,T,n_3} dn_2 + \cdots + \left(\frac{\partial G}{\partial n_i} \right)_{P,T,n_j} dn_i + \cdots$$

$$= \mu_1 dn_1 + \mu_2 dn_2 + \cdots + \mu_i dn_i + \cdots = \Sigma \mu_i dn_i$$

$$dG = \Sigma \mu_i dn_i \tag{6.79d}$$

Differentiating equation (6.79c) we get

$$dG = \Sigma \mu_i dn_i + \Sigma n_i d\mu_i$$

Putting the value of dG from equation (6.79d) into the above equation we get

$$\Sigma \mu_i dn_i = \Sigma \mu_i dn_i + \Sigma n_i d\mu_i. \ \text{ Or, } \ \Sigma n_i d\mu_i = 0 \tag{6.79e}$$

Thus we have the following set of equations

$$G = \sum n_i \mu_i, \quad dG = \sum \mu_i dn_i, \quad \sum n_i d\mu_i = 0 \qquad (6.79f)$$

The set of equations (6.79f) is known as Gibbs-Duhem equation.

6.29   THERMODYNAMIC PROPERTIES OF IDEAL GAS MIXTURE

Ideal gas mixture   When two or more ideal gases are mixed there will be no intermolecular interaction among the gas molecules. That type of mixture is called ideal gas mixture.

Ideal solution   A solution contains molecules of different components. In the case of an ideal solution, molecules of one component behave identically with those of other components and molecules of all components change their positions without changing the energy of intermolecular interactions in the solution.

Thus, from the definition of an ideal solution it is clear that at constant temperature and pressure, volume (V) and internal energy (U) do not change during the formation of an ideal solution from the corresponding pure components. So, mathematically we can write that: $\Delta V_{id} = 0$ and $\Delta U_{id} = 0$.

As we know, $\Delta H = \Delta U + P\Delta V$, we have for ideal solution $\Delta H_{id} = \Delta U_{id} + P\Delta V_{id} = 0$.

From Helmholtz equation, we have $\Delta G = \Delta H - T\Delta S$. In the case of an ideal solution, having a mixture of two or more components, the above equation becomes

$$\Delta G^M = \Delta H^M - T\Delta S^M = -T\Delta S^M \quad \text{[since, } \Delta H^M = 0]. \text{ Or, } \Delta G^M = -T\Delta S^M$$

From the equation (6.45), we have $\Delta S^M = -nR \sum x_i \ln x_i$.

Putting the value of ΔS^M in the above equation we get

$$\Delta G^M = nRT \sum x_i \ln x_i = RT \sum n_i \ln x_i \qquad (6.80)$$

where, $$nx_i = n\left(\frac{n_i}{n}\right) = n_i$$

$\Delta G^M =$ Free energy change during mixing of pure components to form an ideal solution.
As $x_i < 1$, $\Delta G^M < 0$. Thus, free energy decreases during mixing. Before mixing all components are pure components. The total free energy is given by the equation (6.79c).

$G^* = \sum n_i \mu_i^*$, where, $\mu_i^* =$ Chemical potential of pure liquid of ith species.

After mixing the free energy is given by $G = \sum n_i \mu_i$.

Thus, the change in free energy is given by

$$\Delta G^M = G - G^* = \sum n_i \mu_i - \sum n_i \mu_i^*. \text{ Or, } \Delta G^M = \sum n_i (\mu_i - \mu_i^*) \qquad (6.81)$$

In the case of an ideal solution, equations (6.80) and (6.81) are the same and hence we have

$$\sum n_i (\mu_i - \mu_i^*) = nRT \sum x_i \ln x_i = RT \sum n_i \ln x_i$$

So, $(\mu_i - \mu_i^*) = RT \ln x_i$. Or, $\mu_i = \mu_i^* + RT \ln x_i$

As pressure and temperature are constant, we can write that

$$\mu_i = \mu_i^*(T, P) + RT \ln x_i$$

Standard state In case of an ideal solution standard state of any component is nothing but the pure form of that component. So, when $x_i = 1$, the ith species is in the pure form.

So, $$(\mu_i)_{x_i = 1} = \mu_i^0 = \mu_i^*(T, P)$$

Thus, the above equation becomes

$$\mu_i = \mu_i^0 + RT \ln x_i \qquad (6.82a)$$

This is the general expression of chemical potential of ith species for an ideal solution. Conversely, a solution is said to be ideal if the chemical potential of each component of the solution obeys the equation (6.82a) for all solution compositions and for a range of T and P.

(i) In the case of an ideal gas mixture, we have $x_i = (p_i/P)$. Putting this value of x_i in the above equation (6.82a), we get

$$\mu_i = \mu_i^0 + RT \ln \frac{p_i}{P} = (\mu_i^0 - RT \ln P) + RT \ln p_i = \mu_i^0(P) + RT \ln p_i \qquad (6.82b)$$

where, $\mu_i^0(P) = \mu_i^0 - RT \ln P$
P is total pressure of the gas mixture and p_i is the partial pressure of the ith species in the mixture.

In the case of an ideal solution, we have $x_i = \left(\dfrac{a_i}{a}\right)$.

a_i = activity of the ith species in the solution = $m_i \gamma_i$, where, m and γ indicate the concentration (in molality) and activity coefficient respectively.

a = Total activity of all the components.

Using activity term the equation (6.82a) becomes

$$\mu_i = \mu_i^0 + RT \ln \frac{a_i}{a} = (\mu_i^0 - RT \ln a) + RT \ln a_i = \mu_i^0(a) + RT \ln a_i \qquad (6.82c)$$

where, $\mu_i^0(a) = (\mu_i^0 - RT \ln a)$
Again using concentration term the equation (6.82a) becomes

$$\mu_i = \mu_i^0 + RT \ln \frac{m_i}{m} = (\mu_i^0 - RT \ln m) + RT \ln m_i = \mu_i^0(m) + RT \ln m_i \qquad (6.82d)$$

where, $\mu_i^0(m) = (\mu_i^0 - RT \ln m)$ and m = concentration of the solution.

The change in thermodynamic properties for a two-component ideal solution is shown in the Fig. 6.19.

$\frac{T\Delta S^M}{n}$
$\frac{\Delta H^M}{n}$
$\frac{\Delta G^M}{n}$
Cal/mole
J/mole
0
0
0
0.5
1.0
X_B

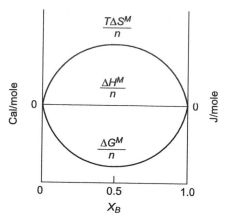

Figure 6.19 Thermodynamic properties of a two component system as a function of composition at 25°C. x_B = Mole fraction of the component B in the solution

6.29.1 Other Thermodynamic Properties of Ideal Solutions

6.29.1.1 Raoult's Law and Henry's Law

Consider a system containing a multicomponent ideal solution remaining in equilibrium with its vapours.

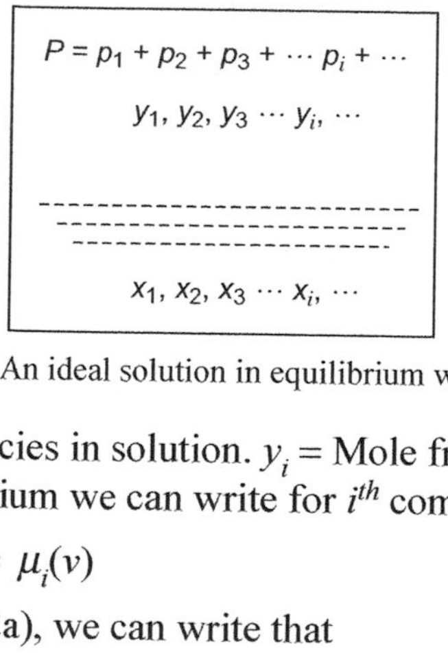

$P = p_1 + p_2 + p_3 + \cdots p_i + \cdots$
$y_1, y_2, y_3 \cdots y_i, \cdots$
$x_1, x_2, x_3 \cdots x_i, \cdots$

Figure 6.20 An ideal solution in equilibrium with its vapour

x_i = Mole fraction of ith species in solution. y_i = Mole fraction of ith species in vapour. As the system is in equilibrium we can write for i^{th} component that

$$\mu_i(l) = \mu_i(v)$$

By using the equation (6.82a), we can write that

$$\mu_{i,l}^0(T, P) + RT\ln x_i = \mu_{i,v}^0(T, P) + RT\ln y_i$$

Or,
$$\mu_{i,l}^0(T, P) + RT\ln x_i = \mu_{i,v}^0(T, P) + RT\ln \frac{p_i}{P} \tag{6.83a}$$

If the above equilibrium exists between pure liquid and vapours of ith component, we have $x_i = 1$ and $\mu_{i\,l}^0(T, P) = \mu_{i\,l}^0(T, P^0)$, where, P^0 is the vapour pressure of the pure liquid of ith component at temperature T. Thus, the equation (6.83a) becomes

$$\mu_{i,l}^0(T, P^0) = \mu_{i,v}^0(T, P) + RT \ln \frac{P^0}{P} \qquad (6.83b)$$

Subtracting the equation (6.83b) from the equation (6.83a), we get

$$\mu_{i,l}^0(T, P) - \mu_{i,l}^0(T, P^0) + RT \ln x_i = RT \ln \frac{P_i}{P^0}$$

As the pressure difference $(P - P^0)$ is very low, we can write that

$$\mu_{i,l}^0(T, P) = \mu_{i,l}^0(T, P^0)$$

Thus, the above equation becomes

$$RT \ln x_i = RT \ln \frac{P_i}{P^0}. \quad \text{Or,} \quad p_i = x_i P^0 \qquad (6.84a)$$

The equation (6.84a) is the mathematical form of Raoult's law.

Statement For a given ideal solution, partial vapour pressure of a component is directly proportional to the vapour pressure of the pure liquid of that component.

From the equation (6.83a) we have

$$\mu_{i,l}^0(T, P) - \mu_{i,v}^0(T, P) = RT \ln p_i - RT \ln P - RT \ln x_i$$

Or, $$RT \ln \frac{p_i}{x_i} = \mu_{i,l}^0(T, P) - \mu_{i,v}^0(T, P) + RT \ln P.$$

Or, $$\ln \frac{p_i}{x_i} = \frac{1}{RT} \left[\mu_{i,l}^0(T, P) - \mu_{i,v}^0(T, P) + RT \ln P \right]$$

Or, $$\ln \frac{p_i}{x_i} = \ln K, \quad \text{where,} \quad K = \text{Constant at given } P \text{ and } T.$$

Or $$p_i = K x_i \qquad (6.84b)$$

The equation (6.84b) is the mathematical form of Henry's law.

Statement For a given ideal solution, partial vapour pressure of a solute i is directly proportional to the mole fraction of that solute present in that ideal solution.

Inference In an ideal dilute solution, the solvent obeys Raoult's law and each solute obeys Henry's law.

6.29.2 Applications of Gibbs-Duhem Equation

One of the major applications of Gibbs-Duhem equation is to evaluate the thermodynamic properties of components in a solution. Consider a binary system of two components such that one component (say component 1) behaves ideally. Thus for component 1, equation (6.82a) can be written as

$$\mu_1 = \mu_1^0 + RT \ln p_1$$

As the component 1 behaves ideally, Raoult's law holds good. So from equation (6.84a) we can write that

$p_1 = x_1 P_1^0$, where P_1^0 is the vapour pressure of pure liquid of component 1.
Putting this value of p_1 into the above equation, we get

$$\mu_1 = \mu_1^0 + RT \ln x_1 P_1^0 = (\mu_1^0 + RT \ln P_1^0) + RT \ln x_1. \text{ Or, } \mu_1 = \mu_{11}^0 + RT \ln x_1$$

Differentiating both sides, we get

$$d\mu_1 = RTd \ln x_1 = RT \frac{dx_1}{x_1} \tag{6.85a}$$

Again sum of the mole fractions of the two components is unity, i.e., $x_1 + x_2 = 1$. Differentiating both sides, we get

$$dx_1 + dx_2 = 0 \quad \text{Or,} \quad dx_1 = -dx_2$$

Putting this value of dx_1 in the equation (6.85a), we get

$$d\mu_1 = - RT \frac{dx_2}{x_1} \tag{6.85b}$$

Applying Gibbs-Duhem equation in this binary system, we get

$$n_1 d\mu_1 + n_2 d\mu_2 = 0 \quad \text{[Ch. Equation (6.79f)]}$$

Or,
$$d\mu_1 = -\frac{n_2}{n_1} d\mu_2 = -\frac{n_2/n}{n_1/n} d\mu_2$$

Or,
$$d\mu_1 = -\frac{x_2}{x_1} d\mu_2 \tag{6.85c}$$

n = Total number of moles = $n_1 + n_2$, x_1 and x_2 are the mole fractions of the two components 1 and 2 respectively.

Combining the equations (6.85b) and (6.85c), we get

$$\frac{x_2}{x_1} d\mu_2 = RT \frac{dx_2}{x_1}. \quad \text{Or, } d\mu_2 = RT \frac{dx_2}{x_2}$$

On integration we get

$$\mu_2 = \mu_{22}^0 + RT \ln x_2 \tag{6.85d}$$

The equation (6.85d) signifies that the component 2 behaves ideally and hence according to Raoult's law, we can write that $P_2 = x_2 P_2^0$, where, P_2^0 is the vapour pressure of the pure liquid of component 2.

Thus, for a 2-component system if one component behaves ideally the 2nd component of the system also behaves ideally.

It can be shown that in a multicomponent system if one component behaves ideally the other components of the system also behave ideally.

Inference In any multicomponent system all the components behave in a similar manner. Thus, study of one component is enough to evaluate the thermodynamic properties of the whole system.

6.30 VAN'T HOFF EQUATION

6.30.1 Equilibrium Constant (K_a) and Standard Free Energy Change (ΔG^0)

Consider a general chemical reaction

$$bB + cC + eE + \cdots \rightleftharpoons lL + mM + nN + \cdots$$

According to Gibbs-Duhem equation [equation (6.79f), total free energy of the reactants (G_r) is given by

$$G_r = \Sigma b\mu_B = b\mu_B + c\mu_C + e\mu_E \cdots, \text{ where, } \mu_B = \mu_B^0 + RT \ln a_B \text{(equation 6.82c)}$$

Or, $$G_r = b(\mu_B^0 + RT \ln a_B) + c(\mu_C^0 + RT \ln a_C) + e(\mu_E^0 + RT \ln a_E) + \cdots$$

Or, $$G_r = (b\mu_B^0 + c\mu_C^0 + e\mu_E^0 + \cdots) + RT \ln a_B^b + RT \ln a_C^c + RT \ln a_E^e + \cdots$$

Or, $$G_r = \Sigma(b\mu_B^0) + RT \ln(a_B^b \cdot a_C^c \cdot a_E^e \cdots)$$

Similarly, the total free energy of the products (G_p) is given by

$$G_p = \Sigma l\mu_L = \Sigma(l\mu_L^0) + RT \ln(a_L^l \cdot a_M^m \cdot a_N^n \cdots)$$

The net free energy change for the forward reaction (ΔG) is given by

$$\Delta G = (G_p - G_r) = \Sigma n_p \mu_p - \Sigma n_r \mu_r = [\Sigma(l\mu_L^0) - \Sigma(b\mu_B^0)] + RT \ln \frac{a_L^l \cdot a_M^m \cdot a_N^n \cdots}{a_B^b \cdot a_C^c \cdot a_E^e \cdots}$$

$$\Delta G^0 = [\Sigma(l\mu_L^0) - \Sigma(b\mu_B^0)] = \text{Change in standard free energy.}$$

Thus, $$\Delta G = \Delta G^0 + RT \ln Q, \text{ where, } Q = \frac{a_L^l \cdot a_M^m \cdot a_N^n \cdots}{a_B^b \cdot a_C^c \cdot a_E^e \cdots} = \text{Reaction quotient} \quad (6.86a)$$

When equilibrium is established, $\Delta G = 0$ and $Q = K_a = $ Equilibrium constant.

$$K_a = \frac{a_L'^l \cdot a_M'^m \cdot a_N'^n \cdots}{a_B'^b \cdot a_C'^c \cdot a_E'^e \cdots}, \text{ where, } a_L'^l \text{ is the equilibrium activity of the component } L, \text{etc.}$$

Hence, at equilibrium, equation (6.86a) becomes

$$0 = \Delta G^0 + RT \ln K_a. \text{ Or, } \Delta G^0 = - RT \ln K_a \qquad (6.86b)$$

The equation (6.86b) shows the relation between change in standard free energy and equilibrium constant.

6.30.2 Temperature Dependence of the Equilibrium Constant (K_a)

From the equation (6.86b), we have

$$\ln K_a = - \frac{\Delta G^0}{RT}$$

Differentiating w.r.t. T at constant P, we get

$$\left(\frac{\partial \ln K_a}{\partial T} \right)_P = - \frac{1}{R} \left[\frac{\partial (\Delta G^0 / T)}{\partial T} \right]_P. \text{ Again, } \left[\frac{\partial (\Delta G / T)}{\partial T} \right]_P = - \frac{\Delta H}{T^2} \text{ [Equation (6.58c)]}$$

Thus, $\left[\dfrac{\partial (\Delta G^0 / T)}{\partial T} \right]_P = - \dfrac{\Delta H^0}{T^2}.$ So, $\left(\dfrac{\partial \ln K_a}{\partial T} \right)_P = \dfrac{\Delta H^0}{RT^2}$

At constant pressure we can write that,

$$\frac{d \ln K_a}{dT} = \frac{\Delta H^0}{RT^2} \qquad (6.87a)$$

The equation (6.87a) is known as Van't Hoff equation.

Case 1   Gas phase reaction

For gas-phase reactions ΔH^0 is assumed to remain constant within a small temperature difference. Thus, on integration of the equation (6.87a) we get

$$\int_{K_{a1}}^{K_{a2}} d \ln K_a = \frac{\Delta H^0}{R} \int_{T_1}^{T_2} \frac{dT}{T^2}. \text{ Or, } \ln \frac{K_{a2}}{K_{a1}} = - \frac{\Delta H^0}{R} \left(\frac{1}{T_2} - \frac{1}{T_1} \right) \qquad (6.87b)$$

Case 2 ΔH^0 is a function of temperature

In general ΔH^0 is a function of temperature as given below

$$\Delta H^0 = A + BT + CT^2 + DT^3 + \frac{E}{T}, \text{ where, } A, B, C, D, \text{ and } E \text{ are constants.}$$

Putting this value of ΔH^0 in the equation (6.87a) and on integration we get,

$$\ln \frac{K_{a2}}{K_{a1}} = -\frac{A}{R}\left(\frac{1}{T_2} - \frac{1}{T_1}\right) + \frac{B}{R}\ln\frac{T_2}{T_1} + \frac{C}{R}(T_2 - T_1) + \frac{D}{2R}(T_2^2 - T_1^2) - \frac{E}{2R}\left(\frac{1}{T_2^2} - \frac{1}{T_1^2}\right) \quad (6.87c)$$

6.31 THERMOCHEMISTRY

6.31.1 Introduction

Study on heat of chemical reactions is called thermochemistry. Any chemical reaction may proceed either at constant temperature or with change in temperature. At constant temperature, a chemical reaction may proceed either at constant volume or at constant pressure.

Case 1 Reaction proceeds at constant temperature and at constant volume, i.e., $dT = 0$ and $dV = 0$.

From the 1st law of thermodynamics, we have

$$dQ = dU + dW = dU + PdV \qquad \text{[Considering only expansion work]}$$

As, $\qquad dV = 0, \quad \text{Or}, \quad Q_V = \Delta U \qquad\qquad\qquad (6.88a)$

Case 2 Reaction proceeds at constant temperature and at constant pressure, i.e., $dT = 0$ and $dP = 0$.

We know that $H = U + PV$. Differentiating, we have $dH = dU + PdV + VdP$.

At constant pressure the above equation becomes $dH = dU + PdV$.

Or, $\qquad \Delta H = \Delta U + P\Delta V$.

Considering only gas-phase reaction and assuming all the components behave ideally, we can write that

$PV = nRT \quad$ and $\quad P\Delta V = \Delta nRT$, where, $\Delta n = $ (No. of moles of products) – (No. of moles of reactants).

Putting this value of $P\Delta V$ in the above equation we get

$$\Delta H = \Delta U + \Delta nRT \qquad\qquad\qquad (6.88b)$$

(a) For a chemical reaction, if $\Delta n = 0$, $\Delta H = \Delta U$. Example: $H_2 + I_2 = 2HI$

(b) For a chemical reaction, if $\Delta n > 0$, $\Delta H > \Delta U$. Example: $2SO_3 = 2SO_2 + O_2$

(c) For a chemical reaction, if $\Delta n < 0$, $\Delta H < \Delta U$. Example: $N_2 + 3H_2 = 2NH_3$

Again, according to the 1st law of thermodynamics, we have

$$dH = dQ + VdP. \quad \text{Or}, \quad dQ = dH - VdP$$

At constant pressure, $dP = 0$. Hence, $dQ_p = dH$. Or, $q_p = \Delta H \qquad (6.88c)$

$\Delta H = $ Enthalpy of the products – Enthalpy of the reactants $= H_p - H_r$.

(i) If $\Delta H > 0$, the reaction is called endothermic reaction, i.e., heat is absorbed during the chemical reaction.

(ii) If $\Delta H < 0$, the reaction is called exothermic reaction, i.e., heat is released during the chemical reaction.

6.31.2 Reactions Occurring at Constant Temperature and Pressure

Law of Lavoisier and Laplace

The net enthalpy change of a given forward reaction is equal in magnitude but opposite in sign to that of the corresponding backward reaction.

Hess's Law: The law of constant heat summation

Statement In any given reaction, products can be obtained from the corresponding reactants by following different paths but the total heat change at constant pressure (q_P) of the process following one path is equal to that following any other path, i.e., heat change of a process is independent of path.

We know from equation (6.88c) that $q_P = \Delta H$. As H is a state function, q_P is also a state function and independent of path.

Thus Hess's law, in other words, may be stated that, net enthalpy change of any chemical reaction is constant and independent of reaction path.

One example is the formation of solid ammonium chloride from ammonia gas and hydrogen chloride gas.

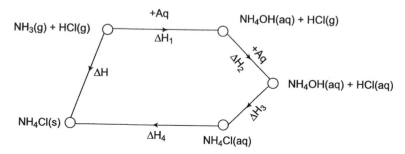

For the above cyclic process, we have $\Delta H = \Delta H_1 + \Delta H_2 + \Delta H_3 + \Delta H_4$.

The algebraic value of ΔH of any process is either positive (+ve) or negative (–ve), depending on the nature of the process, i.e., endothermic or exothermic respectively.

6.31.3 Applications of Hess's Law

(a) To determine heat of formation of a chemical compound

Definition Molar heat of formation of a chemical compound is the heat of reaction of formation of one mole of that compound from the corresponding elements. One example is the formation of liquid benzene from its elements.

$$6C(gr.) + 3H_2(g) \rightleftharpoons C_6H_6(l)$$

The cyclic process of the reaction is shown below.

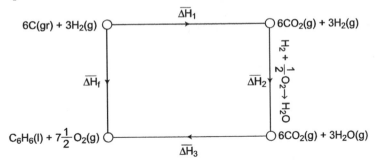

According to Hess's law we have,

$$\Delta \overline{H}_f = 6\Delta \overline{H}_1 + 3\Delta \overline{H}_2 + \Delta \overline{H}_3$$

$\Delta \overline{H}_1$ indicates the change in enthalpy due to formation of 1 mole of CO_2.

$\Delta \overline{H}_2$ is the change in enthalpy due to formation of 1 mole of H_2O.

$\Delta \overline{H}_3$ is the change in enthalpy due to formation of 1 mole of C_6H_6.

Algebraic values of $\Delta \overline{H}$ should be used.

Usually $\Delta \overline{H}_f$ is computed at a standard condition, i.e., at 25°C(298°K) and 1 atm pressure, i.e., $\left| \Delta \overline{H} \right|_{298°K,\, 1\,atm}$, which is frequently called standard heat of formation and are designated as $\Delta \overline{H}_{298}^0$.

(b) Calculation of lattice energy of an ionic crystal

Definition   The amount of energy evolved during formation of one mole of an ionic crystal from its constituent ions is known as molar lattice energy of that crystal. The cyclic process of formation of sodium chloride crystal lattice is shown below.

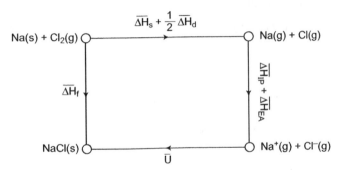

$\Delta \overline{H}_f$ = Molar heat of formation of NaCl(*s*).

$\Delta \overline{H}_s$ = Molar heat of sublimation of Na(*s*).

$\Delta \overline{H}_d$ = Molar heat of dissociation of $Cl_2(g)$.

$\Delta \overline{H}_{IP}$ = Molar ionization potential of Na(*g*).

$\Delta \overline{H}_{EA}$ = Molar electron affinity of Cl(*g*).

$\overline{U}$ = Molar lattice energy of NaCl(*s*).

According to Hess's law we can write that,

$$\Delta \overline{H}_f = \Delta \overline{H}_s + \frac{1}{2}\Delta \overline{H}_d + \Delta \overline{H}_{IP} + \Delta \overline{H}_{EA} + \overline{U}$$

Or, $$\overline{U} = \Delta \overline{H}_f - (\Delta \overline{H}_s + \frac{1}{2}\Delta \overline{H}_d + \Delta \overline{H}_{IP} + \Delta \overline{H}_{EA})$$

(c) Calculation of bond energy

Definition   For a compound, amount of energy required to dissociate a chemical bond into its constituent atoms, is called the bond energy. It may also be defined as the amount of energy released during formation of a chemical bond. To determine bond energies of *C–H* and C–C the following cycles have been considered.

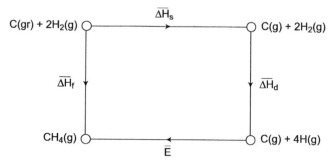

Applying Hess's law, we have

$$\Delta \overline{H}_f = \Delta \overline{H}_s + 2\Delta \overline{H}_d + \overline{E}. \quad \text{Or,} \quad \overline{E} = \Delta \overline{H}_f - (\Delta \overline{H}_s + 2\Delta \overline{H}_d)$$

$\Delta \overline{H}_f$ = Molar heat of formation of methane (CH_4).

$\Delta \overline{H}_s$ = Molar heat of sublimation of solid carbon, graphite.

$\Delta \overline{H}_d$ = Molar heat of dissociation of $H_2(g)$.

$\overline{E} = 4\overline{E}_{C-H}$ (Since methane molecule has four C–H bonds).

$\overline{E}_{C-H} =$ Molar bond energy of C–H bond.

Thus $\overline{E}_{C-H}$ can easily be calculated from the above equation.
In the next cycle let us consider the formation of ethane.

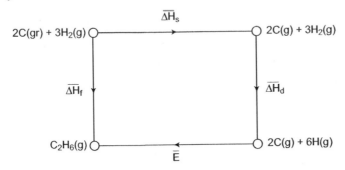

Applying Hess's law, we have

$\Delta \overline{H}_f = 2\Delta \overline{H}_s + 3\Delta \overline{H}_d + \overline{E}$. Or, $\overline{E} = \Delta \overline{H}_f - (2\Delta \overline{H}_s + 3\Delta \overline{H}_d)$

where, $\overline{E} = \overline{E}_{C-C} + 6\overline{E}_{C-H}$ (Since one ethane molecule has one C–C bond and six C–H bonds).

$\overline{E}_{C-C} =$ Bond energy of 1 mole of C–C bond.

$\overline{E}_{C-H} =$ Bond energy of 1 mole of C–H bond and is known from the earlier cyclic process.

Thus, $\overline{E}_{C-C}$ can easily be evaluated from the above equation.

6.31.4 Reactions Occurring with the Change in Temperature at Constant Pressure

If a reaction proceeds with a change in temperature the heat of reaction changes accordingly. A chemical reaction, proceeding with the change in temperature may follow any one of the following paths as shown below:

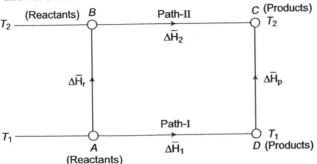

In path-I, reactants are allowed to react to form products at temperature T_1 and then the products are heated from temperature T_1 to temperature T_2.

Along Path-I, the net enthalpy change $= \Delta H_1 + \Delta H_p$, where, $\Delta H_p = C_P(p)\,(T_2 - T_1)$ $C_P(p) = $ Total heat capacity of all the products. $\Delta H_1 = $ Heat of reaction at temperature T_1.

In Path-II, reactants are heated along AB from temperature T_1 to temperature T_2 and then they are allowed to react to form products along BC, at constant temperature T_2.

Along path-II the net enthalpy change $= \Delta H_r + \Delta H_2$, where, $\Delta H_r = C_P(r)\,(T_2 - T_1)$ $C_P(r) = $ Total heat capacities of all the reactants. $\Delta H_2 = $ Heat of reaction at temperature T_2. As enthalpy is a state function, we can write that

$$\Delta H_2 + \Delta H_r = \Delta H_1 + \Delta H_p. \quad \text{Or,} \quad \Delta H_2 + C_{P(r)}\,(T_2 - T_1) = \Delta H_1 + C_{P(p)}\,(T_2 - T_1)$$

Or,

$$\Delta H_2 - \Delta H_1 = [C_{P(p)} - C_{P(r)}]\,(T_2 - T_1)$$

Or,

$$\left[\frac{\partial(\Delta H)}{\partial T} \right]_P = \Delta C_P, \quad \text{where,} \quad \Delta C_P = [C_{P(p)} - C_{P(r)}]$$

At constant pressure we can write that

$$\frac{d(\Delta H)}{dT} = \Delta C_P. \quad \text{Or,} \quad d(\Delta H) = \Delta C_P dT \tag{6.89a}$$

The equation (6.89a) is known as differential form of **Kirchhoff's equation**. Let us consider the following equation:

$$lL + mM + \cdots \longrightarrow xX + yY + \cdots$$

The temperature dependence of C_P is commonly expressed by a power series of the following form

$$C_P = A + BT + CT^2 \quad \text{(for 1 mole of a component)}$$

Hence,

$$C_{P(L)} = A_L + B_L T + C_L T^2, \quad C_{P(M)} = A_M + B_M T + C_M T^2, \text{ etc.}$$

Thus, $C_{P(r)}$ is given by

$$C_{P(r)} = (lA_L + mA_M + \cdots) + (lB_L + mB_M + \cdots)T + (lC_L + mC_M + \cdots)T^2$$

$$= \Sigma lA_L + (\Sigma lB_L)T + (\Sigma lC_L)T^2$$

Again,

$$C_{P(X)} = A_X + B_X T + C_X T^2, \quad C_{P(Y)} = A_Y + B_Y T + C_Y T^2, \text{ etc.}$$

Thus, $C_{P(p)}$ is given by

$$C_{P(p)} = (xA_X + yA_Y + \cdots) + (xB_X + yB_Y + \cdots)T + (xC_X + yC_Y + \cdots)T^2$$

$$= \sum xA_X + (\sum xB_X)T + (\sum xC_X)T^2$$

$$\Delta C_P = C_{P(p)} - C_{P(r)} = a + bT + cT^2$$

where, $a = \sum xA_X - \sum lA_L$, $b = \sum xB_X - \sum lB_L$, $c = \sum xC_X - \sum lC_L$.

Putting this value of ΔC_P in the equation (6.89a), we get

$$d(\Delta H) = (a + bT + cT^2)dT$$

On integration within the temperature limits 298°K and T°K, we get

$$\Delta H_T - \Delta H_{298} = a(T - 298) + \frac{b}{2}(T^2 - 298^2) + \frac{c}{3}(T^3 - 298^3) \qquad (6.89b)$$

The above equation is known as integrated form of **Kirchhoff's equation**.

SOLVED PROBLEMS

Problem 6.1   A Carnot heat engine operates between a source at 1000°K and a sink at 300°K. If the heat engine is supplied with heat at a rate of 800 kJ/min, determine (a) the thermal efficiency and (b) power output of the heat engine.

Solution

$$\eta = \frac{T_1 - T_2}{T_1} = \frac{1000 - 300}{1000} = 0.7, \quad \eta = \frac{W}{|Q_1|}. \text{ Or, } W = \eta \times |Q_1| = 560 \,\text{kJ/min}$$

Problem 6.2   A Carnot heat engine receives 500 kJ of heat from a source of unknown temperature and rejects 200 kJ of heat to a sink at 17°C. Determine (a) the temperature of the source and (b) the thermal efficiency of the heat engine.

Solution

$$\eta = \frac{|Q_1| - |Q_2|}{|Q_1|} = \frac{500 - 200}{500} = 0.6, \quad \eta = \frac{T_1 - T_2}{T_1} = 1 - \frac{T_2}{T_1}, \text{ Or, } T_1 = 725°K$$

Problem 6.3   Three Carnot heat engines, having same thermal efficiency, are connected in series. The 1st engine absorbs 2400 kJ of heat from a thermal reservoir at 1250°K and the 3rd engine rejects 300 kJ of heat to a sink at 150°K. Determine the work output in each engine.

Solution

$$\eta_1 = \frac{|Q_1| - |Q_2|}{|Q_1|} = 1 - \frac{|Q_2|}{|Q_1|}. \text{ Or, } \frac{|Q_2|}{|Q_1|} = 1 - \eta_1$$

Similarly,

$$\frac{|Q_3|}{|Q_2|} = 1 - \eta_2 \text{ and } \frac{|Q_4|}{|Q_3|} = 1 - \eta_3. \text{ Again, } \eta_1 = \eta_2 = \eta_3 = \eta$$

So,

$$(1 - \eta)^3 = \left(\frac{|Q_2|}{|Q_1|}\right) \times \left(\frac{|Q_3|}{|Q_2|}\right) \times \left(\frac{|Q_4|}{|Q_3|}\right) = \frac{|Q_4|}{|Q_1|} = \frac{300}{2400} = \frac{1}{8} = \left(\frac{1}{2}\right)^3.$$

Or,

$$\eta = \frac{1}{2} = 0.5$$

Problem 6.4 $H_2O(l) + 40.8\,J \longrightarrow H_2O(g)$ is energetically unfavourable but still it is spontaneous – explain.

Solution

As heat is required, the process is energetically unfavourable but randomness increases and hence entropy also increases. Thus, the process is thermodynamically favourable and thermodynamic condition overshadows the effect of kinetic condition, making the process spontaneous.

Problem 6.5   Explain why ice melts at 0°C but water does not freeze at 0°C.

Solution

Entropy is increasing on melting of ice and hence the process is spontaneous. On the other hand, entropy is decreasing on freezing of water and hence the process does not occur.

Problem 6.6   A real gas, obeying the following relation: $P(V - nb) = nRT$.
Show that $\Delta H = nC_p dT + nb(P_2 - P_1)$.

Solution

$$H = f(P, T)$$

$$dH = \left(\frac{\partial H}{\partial T}\right)_P dT + \left(\frac{\partial H}{\partial P}\right)_T dP \cdots (1) \quad dH = dQ + VdP = TdS + VdP$$

$$\left(\frac{\partial H}{\partial P}\right)_T = T\left(\frac{\partial S}{\partial P}\right)_T + V. \text{ Or, } \left(\frac{\partial H}{\partial P}\right)_T = -T\left(\frac{\partial V}{\partial T}\right)_P + V$$

Since,
$$\left(\frac{\partial S}{\partial P}\right)_T = \left(\frac{\partial V}{\partial T}\right)_P$$

Applying $T\left(\frac{\partial V}{\partial T}\right)_P$ in the equation $P(V - nb) = nRT$, we get

$$T\left(\frac{\partial V}{\partial T}\right)_P = V - nb. \text{ Thus, } \left(\frac{\partial H}{\partial P}\right)_T = nb$$

So, equation (1) becomes, $dH = nC_P dT + nbdP$. On integration within limits, we get
$$\Delta H = nC_P(T_2 - T_1) + nb(P_2 - P_1).$$

Problem 6.7 The normal freezing point of Hg is –38.9°C. $\Delta H_{\text{fusion}} = 2.29$ kJ/mole. Calculate the entropy change of the system when 50 gm of Hg(l) freezes at normal freezing point. Molecular weight of Hg is 200.59.

Solution

$$q = \Delta \bar{H} = -2.29 \times 1000 \text{ J/mole. Or, } \Delta H = -\frac{2.29 \times 1000 \times 50}{200.59} \text{ J}$$

Or, $\Delta H = -571$ J. $\Delta S = \dfrac{\Delta H}{T} = -\dfrac{571}{234.1} = -2.44 \text{ J/°K}$

Problem 6.8 Calculate the entropy change for the transition of 1 mole of super-cooled liquid water at –5°C into ice at –5°C. Given: $C_{P(s)} = 2.05$ J/g.°K, $C_{P(l)} = 4.184$ J/g.°K , $L = 6.06$ kJ/mole.

Solution

$\mathrm{H_2O}(l)$ at –5°C $\rightarrow$ $\mathrm{H_2O}(l)$ at 0°C $\qquad \Delta S_1 = C_{P(l)} \ln \dfrac{273}{268} = 0.0773 \text{ J/°K}$

$\mathrm{H_2O}(l)$ at 0°C $\rightarrow$ Ice at 0°C $\qquad \Delta S_2 = \dfrac{L}{273} = -22.198 \text{ J/°K}$

Ice at 0°C $\rightarrow$ Ice at –5°C $\qquad \Delta S_3 = C_{P(s)} \ln \dfrac{268}{273} = -0.0379 \text{ J/°K}$

$\Delta S = \Delta S_1 + \Delta S_2 + \Delta S_3 = -22.158 \text{ J/°K}$

Problem 6.9　At NTP 3.4l of oxygen is mixed with 12.4l of hydrogen. Calculate the increase in entropy.

Solution

At NTP 3.4l of oxygen contains $\dfrac{3.4}{22.4} = 0.152$ mole of oxygen. So, $n_1 = 0.152$.

At NTP 12.4l of oxygen contains $\dfrac{12.4}{22.4} = 0.554$ mole of oxygen. So, $n_2 = 0.554$.

$$x_1 = \frac{n_1}{n_1 + n_2} = 0.215, \quad x_2 = \frac{n_2}{n_1 + n_2} = 0.785.$$

$$\Delta S^M = -R(n_1 \ln x_1 + n_2 \ln x_2) = 3.06 \text{ J/°K}.$$

Problem 6.10　Calculate the entropy change of the surroundings for the following reaction at 25°C.

$$N_2(g) + 3H_2(g) \longrightarrow 2NH_3(g). \text{ Given}: \Delta \overline{H} = -46.11 \text{ kJ/ mole}$$

Solution

$$\Delta H = 2(-46.11)\frac{\text{kJ}}{\text{mole}} = -92.22 \text{ kJ/ mole}. \, \Delta S_{surr} = -\frac{\Delta H_{sys}}{T_{surr}} = \frac{92220}{298} = 309.46 \text{ J/°K}$$

EXERCISES

1st law of thermodynamics

1. A gas during expansion from 10 litres to 20 litres at 2 atm pressure absorbs heat energy of amount 2256 J. Find the change in internal energy.　　　　　　　[*Ans.*: 228.2 J]

   Hint: $W = P(V_2 - V_1), \quad \Delta E = Q - W, \quad 1 \text{ lit–atm} = 101.39 \text{ J}.$

2. Calculate the maximum work done when 0.5 mole of a gas expands isothermally and reversibly from a volume of 2l to 5l at 27°C. What is the change in internal energy if 400 cal of heat is absorbed?　　　　　　　[*Ans.*: 125 cal]

   Hint: $\Delta W = nRT \ln \dfrac{V_2}{V_1} \quad \Delta U = \Delta Q - \Delta W = (400 - 275) \text{ cal} = 125 \text{ cal}$

3. A real gas obeys the relation

   $$PV = RT + \frac{a}{V}, \quad \text{where,} \quad a = 3.6\frac{\text{atm} \cdot \text{lit}^2}{\text{mole}^2}$$

Calculate the work done when one mole of the gas expands from 0.224 lit to 22.4 lit at 127°C. Given: $R = 0.082$ lit–atm/(mole-°K) $= 8.314$ J/(mole-°K).

[*Ans.:* 16.93 kJ]

4. Determine the values ΔU and ΔH for the reversible isothermal evaporation of 90 g of water at 100°C. Assume that water vapour behaves as an ideal gas and the heat of evaporation of water is 540 cal/g. [*Ans.:* $\Delta H = 48600$ cal, $\Delta U = 44870$ cal]

Hint: $\Delta H = 90 \times 540$ cal $= 48600$ cal. $\Delta U = \Delta H - P\Delta V$. $\Delta V = V_g - V_l \approx V_g$

$$\Delta U = \Delta H - P\Delta V \text{ Or, } \Delta U = \Delta H - PV_g = \Delta H - nRT$$

5. For an isothermal compression of an ideal gas which one of the following is true?
   (a) $\Delta U = 0$ (b) $\Delta H = 0$ (c) $\Delta U = \Delta H = 0$ (d) $q = \Delta U$

6. Extensive properties of a thermodynamic system depend on:
   (a) specific volume   (b) temperature   (c) mass   (d) pressure.

7. 5 ml concentrated HCl solution is added into 5 ml concentrated NaOH solution in a beaker. Temperature rise is found 5°C. Now, 10 ml concentrated HCl solution is added into 10 ml concentrated NaOH solution in another beaker. What will be the rise in temperature?
   (a) 5°C   (b) 10°C   (c) 20°C   (d) Cannot be determined.

   Hint: As temperature is an intensive property, rise in temperature does not depend on volume or mass of the material. So, temperature rise will be 5°C.

8. Show that work done in reversible isothermal process is higher than that of reversible adiabatic process.

9. The amount of heat released when 20 ml 0.5 M NaOH mixed with 100 ml 0.1 M HCl is 10 kJ. The heat of neutralization is:
   (a) +1000 kJ/mole   (b) –1000 kJ/mole   (c) –500 kJ/mole   (d) +500 kJ/mole.

   Hint: 100 ml 0.1 M HCl $\equiv$ 0.01 mole of HCl $\equiv$ 0.01 mole of NaCl. ΔH for 0.01 mole of NaCl is 10 kJ. So, for 1 mole of NaCl $\Delta H = -10,000$ kJ. As the reaction is exothermic ΔH is negative.

10. (a) How much heat is needed to warm 250 g of water from 22°C to near its boiling point, 98°C? The specific heat of water is 4.18 J/g.°K. (b) Also calculate the molar heat capacity of water. [*Ans.:* $\Delta H = 79$ kJ, $C_P(H_2O) = 75.2$ J mol^{-1}, °K^{-1}]

11. How much work is done when 0.020 mole of Ar, initially at 25°C expands adiabatically from 0.50 dm^3 to 1.00 dm^3? The molar heat capacity of argon is $C_V = 12.48$ J/(mole.°K) and $\gamma = 1.67$.

    Hint: $T_2 = T_1 \left(\dfrac{V_1}{V_2}\right)^{\gamma-1} = 298\left(\dfrac{1}{2}\right)^{0.67} = 187.29°K$,

    $$\Delta T = (298 - 187.29)°K = -110.7°K$$

    $$W = -nC_V\Delta T = 27.63 \text{ J.}$$

12. Which of the following processes must violate the 1st law of thermodynamics?
    (a) $W < 0$, $q < 0$ and $\Delta U = 0$ (b) $W < 0$, $q < 0$ and $\Delta U > 0$
    (c) $W > 0$, $q > 0$ and $\Delta U < 0$ (d) $W < 0$, $q > 0$ and $\Delta U < 0$

13. An ideal gas undergoes expansion from an initial pressure, P_i atm and initial volume of V_i lit. to final volume of V_f lit. in two different processes: (i) isothermal process, where, final pressure is P_T and (ii) adiabatic process, where, final pressure is P_q. Which one of the following is true?
    (a) $P_T > P_q$ (b) $P_T < P_q$ (c) $P_T = P_q$
    (d) No relationship. Explain your answer.

    Hint: For isothermal process $\dfrac{P_i}{P_T} = \dfrac{V_f}{V_i}$. For adiabatic process $\dfrac{P_i}{P_q} = \left(\dfrac{V_f}{V_i}\right)^{\gamma}$.
    As $\gamma > 1$, $P_T > P_q$.

2nd law of thermodynamics

14. Calculate the change in entropy when 5 moles of an ideal gas expands from a volume of $4l$ to $40l$ at 27°C. Given: $R = 2$ cal/(mole-°K).   [23.03 cal-°K]

    Hint: $\Delta S = nC_V \ln\dfrac{T_2}{T_1} + nR\ln\dfrac{V_2}{V_1}$

15. (a) Calculate the entropy of fusion of ice if its heat of fusion is 6.01 kJ/mole at 0°C.

    Hint: $\Delta G = \Delta H - T\Delta S$. $\Delta G = 0$ at equilibrium. $\Delta S = \dfrac{\Delta H}{T} = \dfrac{6.01}{273}$ kJ/(mole-°K)
    $$= 22.01 \text{ J/(mole-°K)}$$

    (b) Calculate the increase in entropy when 1 mole of water evaporates at 100°C.
    Given: latent heat of fusion is 540 cal/g at 100°C.

    Hint: Latent Heat $= 540 \times 18$ cal/ mole. $\Delta S = \dfrac{L}{T} = \dfrac{9720}{373} = 26.06$ cal/(mole-°K)

16. For ammonia $C_p = 6.2 + 7.9$ T Cal/mole. Assuming ideal behaviour estimate the entropy change in heating 34 gm of the gas from a volume of 100 lit at 300 K to a volume of 60 lit at 800 K.   [*Ans.:* 7906 e.u]

17. Calculate the change in entropy when 100 kg of water at 27°C is converted to 100 kg of steam at 200°C under constant atmospheric pressure.   [7.5×10^5 J/°K]

    Given: $C_{P(l)} = 4184$ J/kg. °K, $C_{P(v)} = (1670 + 0.49T)$ J/kg. °K, $L = 23 \times 10^5$ J/kg.

18. Explain how entropy of a system decreases during a spontaneous process or increases during a non-spontaneous process.

19. Calculate the change in entropy of a large iron block at 20°C when 1.5 kJ of energy is released as heat.

20. Which of the following has higher entropy in each pair?
    (a) 1 gm pure solid NaCl   (b) 1 gm NaCl dissolved in 100 ml distilled water
    (c) 1 gm water at 25°C   (d) 1 gm water at 50°C.

21. An ideal gas is expanded from one state to another state.
    State 1: $V_1 = 0.5$ lit $P_1 = 1$ atm, $t_1 = 25°C$
    State 2: $V_2 = 1$ lit $P_2 = 1$ atm, $t_2 = 100°C$
    Calculate the change in entropy for the expansion process.

    Hint: $n = \dfrac{PV}{RT} = 0.204,\ \Delta S = nC_V \ln\dfrac{T_2}{T_1} + nR\ln\dfrac{V_2}{V_1}$

22. 1 mole of water vapour at 17°C and 1.2 atm pressure undergoes a cyclic process for which $w = 123\ J$. Calculate q and ΔS.   [*Ans.: q = 123 J, ΔS = 0*]

23. $TdS = dU + PdV$ is applicable to
    (a) Open system   (b) Closed system
    (c) Single phase system   (d) Both open and closed systems.

24. If 1 mole of $NH_3(g)$ and 1 mole of $HCl(g)$ are mixed in a closed container to form 1 mole of $NH_4Cl(g)$, which one of the following is true?
    (a) $\Delta H > \Delta U$ (b) $\Delta H < \Delta U$
    (c) $\Delta H = \Delta U$ (d) Cannot be correlated.

25. Enthalpy and entropy changes of a reaction are 49.57 kJ/mole and 123.2 J/°K. Calculate the free energy change of the reaction at 27°C.   [*Ans.: 12.61 kJ*]

26. For the reaction $A(g) + 3B(g) \rightarrow 2C(g)$, the enthalpy change is –90.2 kJ/mole and ΔS is –0.1584 kJ/°K.mole. Predict whether the reaction is feasible or not at 298°K.

27. Ethanoic acid and hydrochloric acid react with sodium hydroxide solution. The enthalpy of neutralisation of ethanoic acid is –55.8 kJ/mole while that of hydrochloric acid is –57.3 kJ/mole. Can you think of the difference?

28. ΔG^0 and ΔH^0 for a chemical reaction at 300°K is – 66.9 kJ/mole and –41.8 kJ/mole respectively. Calculate ΔG^0 for the same reaction at 330°K.   [*Ans.: –69.41 kJ/mole*]

29. For a chemical reaction, ΔH and ΔS have been found to be 9.08 kJ/mole and 35.7 J/mole.°K respectively. Which of the following statements is true at 25°C?
    (a) Reversible and isothermal, (b) Reversible and exothermic, (c) Spontaneous and endothermic, (d) Spontaneous and exothermic.

30. Determine whether the following reaction is spontaneous.

    $2KClO_3(s) \rightarrow 2KCl(s) + 3O_2(g)$. Given: ΔG_f^0 of $KCl(s) = -408.3$ kJ/mole.

    [*Ans.: ΔG^0 = –236.8 kJ/mole*]

31. In a closed system the following reaction occurs:
    $$CaCO_3(s) \rightarrow CaO(s) + CO_2(g)$$
    Given: $\Delta H^0 = 177.8$ kJ/mole, $\Delta S^0 = 160.5$ J/mole.°K.
    Calculate the equilibrium temperature, T_m. Predict the spontaneity of the reaction at 25°C.
    [*Ans.: 1108°K*]

32. $H_2(g) + Cl_2(g) \rightarrow 2HCl(g)$. At 25°C the following data are given:

$p_{H_2} = 0.25$ atm, $p_{Cl_2} = 0.45$ atm, $p_{HCl} = 0.30$ atm. ΔG_f^0 of HCl $= -95.27$ kJ/mole.

Calculate ΔG at 25°C and equilibrium constant.   [*Ans.: -191.1 kJ/mole, 2.476×10^{33}*]

33. At 1 atm liquid water is heated above 100°C. ΔS_{surr} for the process

(a) greater than zero   (b) less than zero   (c) equal to zero   (d) cannot be answered.

Thermochemistry

34. What is the enthalpy of combustion of methane at 25°C?

Given: enthalpy of formation of methane is -74.81 kJ/mole, that of water is -187.78 kJ/mole and the enthalpy of formation of carbon dioxide is -393.51 kJ/mole.

[*Ans.: -694 kJ/mole*]

35. Calculate the standard enthalpy change, ΔH^0, for the formation of 1 mole of strontium carbonate from its elements as shown below: [*Ans.: -1220 kJ*]

$$Sr(s) + C(graphite) + \frac{3}{2}O_2(g) \longrightarrow SrCO_3(s)$$

The available information is

$$Sr(s) + \frac{3}{2}O_2(g) \longrightarrow SrO(s) \qquad\qquad\qquad [\Delta H^0 = -592 \text{ kJ}]$$

$$SrO(s) + CO_2(g) \longrightarrow SrCO_3(s) \qquad\qquad\qquad [\Delta H^0 = -234 \text{ kJ}]$$

$$C(graphite) + O_2(g) \longrightarrow CO_2(g) \qquad\qquad\qquad [\Delta H^0 = -394 \text{ kJ}]$$

36. One reaction involved in the conversion of iron ore to the metal is

$$FeO(s) + CO(g) \longrightarrow Fe(s) + CO_2(g)$$

Calculate the standard enthalpy change for the above reaction using standard enthalpies of the following reactions: [*Ans.: -11 kJ*]

$$3Fe_2O_3(s) + CO(g) \longrightarrow 2Fe_3O_4(s) + CO_2(g) \qquad\qquad [\Delta H^0 = -47 \text{ kJ}]$$

$$Fe_2O_3(s) + 3CO(g) \longrightarrow 2Fe(s) + 3CO_2(g) \qquad\qquad [\Delta H^0 = -25 \text{ kJ}]$$

$$Fe_3O_4(s) + CO(g) \longrightarrow 3FeO(s) + CO_2(g) \qquad\qquad [\Delta H^0 = +19 \text{ kJ}]$$

37. The equation for the production of coal gas is

$$2C(s) + 2H_2O(g) \longrightarrow CH_4(g) + CO_2(g)$$

Determine the standard enthalpy change for this reaction using standard enthalpies of the following reactions: [*Ans.: -20.7 kJ*]

$$C(s) + H_2O(g) \longrightarrow CO(g) + H_2(g) \qquad\qquad\qquad [\Delta H^0 = +113.3 \text{ kJ}]$$

$$CO(g) + H_2O(g) \longrightarrow CO_2(g) + H_2(g) \qquad\qquad\qquad [\Delta H^0 = -41.2 \text{ kJ}]$$

$$CH_4(g) + H_2O(g) \longrightarrow 3H_2(g) + CO(g) \qquad\qquad\qquad [\Delta H^0 = +206.1 \text{ kJ}]$$

38. Determine ΔH^0 at 298 K for the reaction. [*Ans.:* –74.83 kJ/mole]

$$C(\text{graphite}) + 2H_2(g) \longrightarrow CH_4(g)$$

The following data is given

$$C(\text{graphite}) + O_2(g) \longrightarrow CO_2(g) \qquad [\Delta H^0 = -393.53 \text{ kJ/mole}]$$

$$H_2(g) + \frac{1}{2}O_2(g) \longrightarrow H_2O(l) \qquad [\Delta H^0 = -285.8 \text{ kJ/mole}]$$

$$CO_2(g) + 2H_2O(l) \longrightarrow CH_4(g) + 2O_2(g) \qquad [\Delta H^0 = +890.3 \text{ kJ/mole}]$$

39. Calculate lattice energy of NaCl crystal using the following data:

[*Ans.:* –183.3 cal/mole]

$$\Delta\overline{H}_f = -98.2 \text{ kcal/mole}, \ \ \Delta\overline{H}_s = 26 \text{ kcal/mole}, \ \ \Delta\overline{H}_d = 57.2 \text{ kcal/mole}$$

$$\Delta\overline{H}_{IP} = 117.9 \text{ kcal/mole}, \ \ \Delta\overline{H}_{EA} = -87.4 \text{ kcal/mole}$$

40. For the following reaction, ΔH has been found to be –67850 cal at 25°C.

$$CO + \frac{1}{2}O_2 \rightarrow CO_2$$

Determine ΔH of the above reaction at 100°C. Given [*Ans.:* –67962.5 cal]

$$C_p(CO) = 6.97 \text{ cal/mole.°K},$$

$$C_p(CO_2) = 8.97 \text{ cal/mole.°K},$$

$$C_p(O_2) = 7.00 \text{ cal/mole.°K}$$

41. Heat of formation of $H_2O(g)$ is –58000 cal at 500°K. What will be the heat of formation at 1000°K? [*Ans.:* –59052.5 cal]

Given: $C_p(H_2) = 6.94 - 0.2 \times 10^{-3}T,$

$$C_p(O_2) = 6.150 + 3.2 \times 10^{-3}T,$$

$$C_p(H_2O) = 7.25 + 2.28 \times 10^{-3}T$$

Miscellaneous

1. 2 moles of an ideal monatomic gas initially at 100°C and 5 atm pressure expands adiabatically and reversibly to 2 atm pressure. Calculate: (i) work done by the gas, (ii) final molar volume and (iii) change in enthalpy. Given: $\gamma = 1.67$ and $R = 8.314$ J/mole.°K. [$W = 2854$ J, $\overline{V} = 10.585l$, $\Delta H = -4766.3$ J]

2. Calculate ΔU and ΔH for the transformation of one mole of an ideal gas from 27°C, 1 atm to 327°C, 17 atm. Given: $\overline{C}_V = 12.6 + 0.042T$ J/mole. [$\Delta U = 9450$J/mole, $\Delta H = 11944$ J/mole]

3. $\Delta H = \Delta U$ for condensed phases—explain.

4. C_p and C_V of liquid water are same at 4°C—explain.

5. The cycle in the PV–diagram is for an ideal gas. Find the work done during each of the four stages of the process and the net work done in the cycle. 1 atm = 10^5 Pa.

$$[W_{AB} = 4.5 \ MJ, \ W_{BC} = -1.2 \ MJ, \ W_{CD} = -0.6 \ MJ, \ W_{DA} = -0.6 \ MJ, \ W_{Total} = 2.1 \ MJ]$$

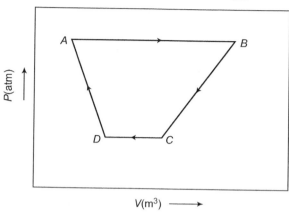

Clausius-Clapeyron equation

6. Determine the slope of the melting curve of water at the melting point (0°C, 1 atm). $\overline{\Delta H}_{fusion}$ is 6010 J/mole and $\overline{\Delta V}_{fusion} = -1.63 \ cm^3$ / mole. Also calculate the fusion temperature at 1000 atm.

   Hint: $\dfrac{dP}{dT} = \dfrac{\Delta H}{T \Delta V}, \quad \Delta V = -1.63 \times 10^{-6} \ m^3$ / mole

   $$\frac{dT}{dP} = -\frac{273 \times 1.63 \times 10^{-6}}{6010} \ °K/Pa = -\frac{273 \times 1.63 \times 10^{-6}}{6010 \times 10^{-5}} \ °K/atm = -0.0074 \ °K/atm$$

   $$\int_{273}^{T} dT = -0.0074 \int_{1}^{1000} dP, \ \text{or} \ T = 265.6°K.$$

7. A substance has a vapour pressure of 0.2020 atm at 261.5°K. If its heat of vapourisation is 20.94 kJ/mole, calculate its vapour pressure at 240.9°K. [*Ans.:* 0.08843 atm]

8. A substance has a vapour pressure of 77.86 mm Hg at 318.3°K and a vapour pressure of 161.3 mm Hg at 340.7°K. Calculate its heat of vapourisation in kJ/mole.

   [*Ans.:* 29.38 kJ/mole]

Thermodynamics Part-II

7.1 PROPERTIES OF FLUIDS AND EQUATION OF STATE

7.1.1 Triple Point and Critical Point

Any pure material can exist in any of the three phases: solid or liquid or vapour. From the knowledge of phase rule we have:

$$F = C - P + 2 \tag{7.1}$$

where, C is number of components, P is number of phases, and F is degrees of freedom.

For a pure material, $C = 1$. If it exists in a single phase, $P = 1$ and hence according to the equation (7.1), we have, $F = 2$. Then, the system is called *bivariant*, i.e., the system can be investigated by controlling any two parameters, e.g., pressure (P) and volume (V), pressure and temperature (T) or volume and temperature.

If the pure material exists in two phases in equilibrium, $C = 2$ and hence $F = 1$. The system is said to be *univariant*, i.e., the system may be investigated by controlling only one of the three parameters, P, V or T and a two dimensional curve is obtained, as shown in Fig. 7.1. When vapour and liquid phases are in equilibrium, the curve is called *vaporisation curve*. When solid and liquid phases are in equilibrium, the curve is called *fusion curve*. When solid and vapour phases coexist the curve is called *sublimation curve*.

If three phases coexist at a point, $C = 3$ and hence $F = 0$. The system is said to be *invariant*, i.e., all the parameters are fixed. The invariant point, O, is called *triple point* (see Fig. 7.1).

The vaporisation curve starts from the triple point, O and ends at the point C, known as *critical point*, beyond which no liquid phase exists. The corresponding pressure and temperature at the critical point C are known as critical pressure (P_c) and critical temperature (T_c) respectively. Above the point C there is no demarcation between liquid and gas. So, P_c and T_c are the highest pressure and temperature at which the corresponding gas can be liquefied. Volume of gas at the critical point is called critical volume, V_c.

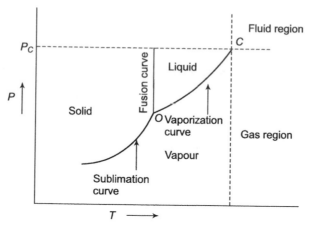

Figure 7.1 *PT*-diagram of a pure substance

Beyond the critical point *C* a wide region is left, demarcated by dashed lines. The material, existing in this region, cannot be liquefied. This region is called *fluid region*. In Fig. 7.1, a vertical dashed line, which extends beyond the critical point *C*, is drawn from T_c. Below the critical point and left to the dashed line, there exists liquid region, which can be vaporized by reduction of pressure at constant temperature. Below the curve *OC* there exists vapour region, which can be liquefied by reduction of temperature at constant pressure.

The critical point is also known as point of inflection and characterized by the following condition.

$$\left|\left(\frac{\partial P}{\partial V}\right)_T\right|_{at\,C} = 0 \text{ and } \left|\left(\frac{\partial^2 P}{\partial V^2}\right)_T\right|_{at\,C} = 0 \tag{7.2}$$

P_c, V_c and T_c are the critical pressure, critical volume, and critical temperature respectively. Then the term reduced parameter is defined as below:

$$\text{Reduced pressure} = \frac{P}{P_c} = \pi, \text{ Reduced volume} = \frac{V}{V_c} = \phi$$

$$\text{Reduced temperature} = \frac{T}{T_c} = \theta$$

7.1.2 Equation of State

P, *V*, and *T* are the controlling parameters to define any gas. Thus, the equation of state of any gas is given by

$$f = (P,V,T) = 0 \tag{7.3}$$

One of the solution of the above equation is $V = f(P, T)$

Differentiating, we get

$$dV = \left(\frac{\partial V}{\partial T}\right)_P dT + \left(\frac{\partial V}{\partial P}\right)_T dP$$

Or, $$\frac{dV}{V} = \frac{1}{V}\left(\frac{\partial V}{\partial T}\right)_P dT + \frac{1}{V}\left(\frac{\partial V}{\partial P}\right)_T dP$$

Or, $$\frac{dV}{V} = \beta dT - k dP \qquad (7.4)$$

where, β = Expansivity, also known as volume expansion coefficient $= \frac{1}{V}\left(\frac{\partial V}{\partial T}\right)_P$

and k = Isothermal compressibility $= -\frac{1}{V}\left(\frac{\partial V}{\partial P}\right)_T$

In case of liquid, isotherms are very steep and hence the values of β and k are approximated to zero. On the basis of this assumption liquid is often called incompressible fluid.

P–V–T behaviour of different pure substances shows that the following power series expression holds good for all gases.

$$PV = a + bP + cP^2 + \cdots\cdots = a(1 + B'P + C'P^2 + \cdots\cdots)$$

$$\text{where, } B' = \frac{b}{a}, \quad C' = \frac{c}{a}, \quad \text{etc.}$$

Or, $$PV = a(1 + B'P + C'P^2 + \ldots\ldots) \qquad (7.5)$$

When pressure of a real gas tends to zero the above equation becomes

$$\lim_{P \to 0}(PV) = a = RT \qquad (7.6)$$

The gas is known as ideal gas. Thus, at very low pressure any real gas behaves ideally. Again for one mole of any ideal gas we have

$$PV = RT \text{ and } P = \frac{RT}{V}$$

Putting this value of P in the R.H.S. of equation (7.5), we get

$$PV = a\left[1 + B'\frac{RT}{V} + C'\left(\frac{RT}{V}\right)^2 + \cdots\cdots\right]$$

Or, $$= a\left[1 + \frac{B}{V} + \frac{C}{V^2} + \cdots\cdots\right] \qquad (7.7a)$$

where, $B = B'RT$, $C = C'(RT)^2$, etc.
Putting this value of a from the equation (7.6) in the equations (7.5) and (7.7a), we get

$$PV = RT(1 + B'P + C'P^2 + \ldots\ldots) \qquad (7.7b)$$

$$PV = RT\left[1 + \frac{B}{V} + \frac{C}{V^2} + \cdots\cdots\right] \tag{7.7c}$$

$$\frac{PV}{RT} = Z = \text{Compressibility factor}$$

$$Z = (1 + B'P + C'P^2 + \cdots\cdots) \tag{7.8a}$$

$$Z = \left[1 + \frac{B}{V} + \frac{C}{V^2} + \cdots\cdots\right] \tag{7.8b}$$

The above expansions are known as virial expansions and the coefficients are known as *virial coefficients*. For an ideal gas values of all virial coefficients are zero and hence,

$$Z = \frac{PV}{RT} = 1$$

In most of the cases, real gas is expressed by van der Waals equation of state, shown below

$$\left(P + \frac{a}{V^2}\right)(V - b) = RT \ (for \ 1 \ mole) \tag{7.9}$$

At the critical point the above equation becomes

$$\left(P_c + \frac{a}{V_c^2}\right)(V_c - b) = RT_c \ (for \ 1 \ mole) \tag{7.10}$$

From the equation (7.9), we have

$$PV - Pb + \frac{a}{V} - \frac{ab}{V^2} = RT$$

Differentiating with respect to V at constant T, we get

$$P + V\left(\frac{\partial P}{\partial V}\right)_T - b\left(\frac{\partial P}{\partial V}\right)_T - \frac{a}{V^2} + \frac{2ab}{V^3} = 0 \tag{7.11}$$

At $T = T_c$, we know $P = P_c$ and $(\partial P/\partial V)_T = 0$. Thus, the above equation becomes

$$P_c - \frac{a}{V_c^2} + \frac{2ab}{V_c^3} = 0 \tag{7.12}$$

Further differentiating equation (7.11) with respect to V at constant T, we get

$$\left(\frac{\partial P}{\partial V}\right)_T + V\left(\frac{\partial^2 P}{\partial V^2}\right)_T + \left(\frac{\partial P}{\partial V}\right)_T - b\left(\frac{\partial^2 P}{\partial V^2}\right)_T + \frac{2a}{V^3} - \frac{6ab}{V^4} = 0$$

At critical state, $T = T_c$, $P = P_c$ and $V = V_c$. Thus, from equation (7.2) we have

$$\left(\frac{\partial P}{\partial V}\right)_T = 0 \ \ and \ \ \left(\frac{\partial^2 P}{\partial V^2}\right)_T = 0$$

Using the critical criteria in the above equation, we get

$$V_c = 3b. \ \ Or, \ \ b = \frac{V_c}{3}$$

Putting this value of b in the equation (7.12), we get

$$P_c - \frac{a}{V_c^2} + \frac{2aV_c}{3V_c^3} = 0. \ \ Or, \ \left(a - \frac{2a}{3}\right) = P_c V_c^2. \ \ Or, \ a = 3P_c V_c^2. \ Or, \ V_c^2 = \frac{a}{3P_c}$$

Thus, $\quad V_c = 3b$ and $V_c^2 = \dfrac{a}{3P_c}$ $\hfill$ (7.13a)

Putting the values of V_c and V_c^2 in the equation (7.10), we get

$$(P_c + 3P_c)(3b - b) = RT_c. \ \ Or, \ \ b = \frac{RT_c}{8P_c}. \ \ a = 3P_c V_c^2 = 3P_c(3b)^2 = 27P_c b^2$$

$$a = 27P_c b^2 \ \ and \ \ b = \frac{RT_c}{8P_c} \hfill (7.13b)$$

From the concept of reduced parameter, we have $P = \pi P_c$, $V = \phi V_c$ and $T = \theta T_c$.
Putting these values of P, V, and T in the equation (7.9) we get

$$\left(\pi P_c + \frac{a}{(\phi V_c)^2}\right)(\phi V_c - b) = R\theta T_c, \quad Again, \ V_c^2 = \frac{a}{3P_c} \ \ and \ V_c = 3b$$

Thus, the above equation becomes

$$P_c\left(\pi + \frac{3}{\phi^2}\right)(3\phi - 1) \, b = \theta(RT_c). \ \ Again, \ RT_c = 8bP_c \ [\text{Ch. equation}(7.13b)]$$

Or, $\qquad \left(\pi + \dfrac{3}{\phi^2}\right)(3\phi - 1) = 8\theta$ $\hfill$ (7.14)

Equation (7.14) is known as **reduced equation of state**.
Inference If π and θ of any two gases are same, ϕ must be same for those two gases.

Law of corresponding states

All gases, when compared at same reduced temperature and reduced pressure, have same reduced volume and hence have same compressibility factor so that all of them deviate from the ideal behaviour to about the same degree.

7.2 THERMODYNAMIC PROPERTIES

7.2.1 Residual Property

Free energy, G, is related with enthalpy, H and entropy, S, as follows

$$G = H - TS \tag{7.15}$$

The differential form is given by

$$dG = VdP - Sdt \tag{7.16}$$

Again, $d\left(\dfrac{G}{RT}\right) = \dfrac{dG}{RT} - \dfrac{GdT}{RT^2} = \dfrac{(VdP - SdT)}{RT} - \dfrac{GdT}{RT^2} = \dfrac{VdP}{RT} - \dfrac{(G+TS)}{RT^2} = \dfrac{VdP}{RT} - \dfrac{HdT}{RT^2}$

Or, $d\left(\dfrac{G}{RT}\right) = \dfrac{VdP}{RT} - \dfrac{HdT}{RT^2}$ \hfill (7.17)

Hence, $\dfrac{V}{RT} = \left[\dfrac{\partial(G/RT)}{\partial P}\right]_T$, $\dfrac{H}{RT^2} = -\left[\dfrac{\partial(G/RT)}{\partial T}\right]_P$ \hfill (7.18a)

Again, $\dfrac{G}{RT} = \dfrac{H}{RT} - \dfrac{S}{R}$.   Or, $\dfrac{S}{R} = \dfrac{H}{RT} - \dfrac{G}{RT}$ \hfill (7.18b)

Suppose M is a parameter of any real system and M^{id} is the value of same parameter of that system assuming the latter be an ideal system. Then the difference $(M - M^{id})$ is known as the **residual property** of that system and is represented as M^r. So we can write that $M^r = M - M^{id}$.

Thus, $V^r = V - V^{id}$, $G^r = G - G^{id}$, etc. Hence, equations (7.18a) and (7.18b) can be rewritten as

$$\frac{V^r}{RT} = \left[\frac{\partial(G^r/RT)}{\partial P}\right]_T ,\quad \frac{H^r}{RT^2} = -\left[\frac{\partial(G^r/RT)}{\partial T}\right]_P ,\quad \frac{S^r}{R} = \frac{H^r}{RT} - \frac{G^r}{RT} \tag{7.19}$$

V^r, G^r, H^r and S^r are the residual volume, residual free energy, residual enthalpy, and residual entropy respectively of that system. Again for 1 mole of an ideal gas, we have

$$V^{id} = V^{ig} = \frac{RT}{P},\ V^r = V - V^{ig} = V - \frac{RT}{P}.\ \text{Again,}\ Z = \frac{PV}{RT}.\ \text{Or,}\ V = \frac{RTZ}{P}$$

Or, $V^r = \dfrac{RTZ}{P} - \dfrac{RT}{P} = \dfrac{RT}{P}(Z-1)$ \hfill (7.20a)

Comparing equations (7.19) and (7.20a), we get

$$\left[\frac{\partial(G^r/RT)}{\partial P}\right]_T = \frac{1}{P}(Z-1)$$

Thus on integration at constant temperature, T we have

$$\frac{G^r}{RT} = \int_0^P \frac{(Z-1)}{P} dP \qquad (7.20b)$$

Differentiating equation (7.20b) with respect to T at constant P, we get

$$\left[\frac{\partial(G^r/RT)}{\partial T}\right]_P = \int_0^P \left(\frac{\partial Z}{\partial T}\right)_P \frac{dP}{P}$$

Comparing this with equation (7.19), we get

$$\frac{H^r}{RT^2} = -\left[\frac{\partial(G^r/RT)}{\partial T}\right]_P = -\int_0^P \left(\frac{\partial Z}{\partial T}\right)_P \frac{dP}{P}$$

Or, $$\frac{H^r}{RT} = -T\int_0^P \left(\frac{\partial Z}{\partial T}\right)_P \frac{dP}{P} \qquad (7.20c)$$

Again from equation (7.19), we have

$$\frac{S^r}{R} = \frac{H^r}{RT} - \frac{G^r}{RT} = -T\int_0^P \left(\frac{\partial Z}{\partial T}\right)_P \frac{dP}{P} - \int_0^P \frac{(Z-1)}{P} dP$$

Or, $$\frac{S^r}{R} = -\int_0^P \left[T\left(\frac{\partial Z}{\partial T}\right)_P + (Z-1)\right] d\ln P \qquad (7.20d)$$

Equations (7.20a to d) are known as fundamental residual property relations. V^r, G^r, H^r and S can be evaluated by using appropriate conditions. Thus, the actual property M can be determined by using the equation

$M = M^{ig} + M^r$. For example $S = S^{ig} + S^r$

S is the actual molar entropy of the pure substance, S^{ig} is the molar entropy of the same pure substance which is assumed to be an ideal one.

$$S^{ig} = S_0^{ig} + \int_0^T C_P^{ig} d\ln T - R\int_0^P d\ln P$$

If C_P^{ig} and S_0^{ig} are known, S^{ig} can easily be evaluated by using the above equation.

Thus,
$$S = \left[S_0^{ig} + \int_0^T C_P^{ig} d\ln T - R \int_0^P d\ln P \right] + S^r \tag{7.21}$$

Similarly, the other actual thermodynamic properties of any pure substance can easily be evaluated.

7.2.2 Two-phase System

Consider a system where saturated liquid remains in equilibrium with its saturated vapour. If x and y are the mole fractions of the liquid and vapour phases respectively, we have $x + y = 1$.

Any extensive property, say volume, V, can be represented as

$$V = xV^l + yV^g = (1-y)V^l + yV^g = V^l + (V^g - V^l)y \tag{7.22}$$

For two phase system of a pure substance we have

$$\frac{dP^{sat}}{dT} = \frac{L}{T\Delta V} = \frac{L}{T(V^g - V^l)}$$

Again, $V^g \gg V^l$ and assuming that the saturated vapour behaves ideally, we can write that

$$(V^g - V^l) \approx V^g \quad \text{and} \quad V^g = \frac{RT}{P^{sat}}$$

Hence, the above equation becomes

$$\frac{dP^{sat}}{dT} = \frac{LP^{sat}}{RT^2} \cdot \quad \text{Or,} \quad d\ln P^{sat} = \frac{L}{R} d\left(\frac{1}{T^2}\right)$$

where, L is the molar latent heat of vaporization. On integration we get,

$$\ln P^{sat} = -\frac{L}{RT} + C \tag{7.23a}$$

When $T = T_b =$ The boiling point of the pure substance, the above equation takes the following form in accordance with **Riedel**

$$\frac{L/T_b}{R} = \frac{1.092\,(\ln P_c - 1.013)}{0.930 - \theta_b} \tag{7.23b}$$

$P_c =$ Critical pressure in *bar*, $\theta_b =$ Reduced temperature at T_b, $R =$ Molar gas constant.

Two important thermodynamic properties S and H of any pure substance can be plotted against pressure (P) and temperature (T). There are different types of plots, namely, (i) TS-plot, (ii) $\ln P$ vs. H plot, and (iii) HS-plot. The HS-plot is known as **Mollier diagram** as shown in the Fig. 7.4.

The tie line *AB*, which is parallel to *S*-axis, indicates the phase change from liquid to saturated vapour at constant pressure, *P* and temperature, *T*. The line *DE* represents triple point line. All three phases coexist along this line. The curve *XY* represents *TS*-plot of vapour at constant *H*. The total phase transition of a pure substance is represented by the curve $A_2A_1ABB_1$. The pure substance remains in solid state at A_2 and slowly transforms to liquid state along A_2A_1. The phase change takes place at constant temperature, *T* and pressure, *P*. So, the tie line A_2A_1 is parallel to *S*-axis. With the increase in temperature, *S* of the pure liquid increases and reaches the point *A*, from which another phase transition (liquid to vapour) begins. Phase transition is completed at the point *B*. The line BB_1 indicates superheated vapour.

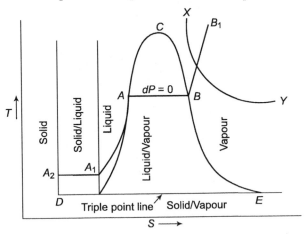

Figure 7.2   *TS*-diagram of a pure substance. The point *C* represents the critical point

Fig. 7.3 represents ln *P vs. H* plot of a pure substance.

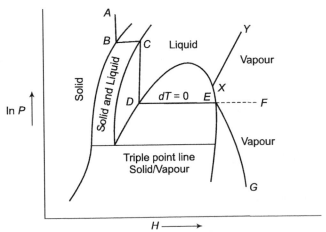

Figure 7.3   ln *P vs. H* diagram of a pure substance. The point *C* represents the critical point

The line *XY* in ln *P vs. H* plot represents vapour at constant entropy, *S*. The total phase change of a pure substance is represented by ABCDEG. The dotted line *EF* represents superheated vapour. Temperature, T and pressure, P remain constant during phase change, i.e., along BC and DE.

Fig. 7.4 represents HS-plot of a pure substance. The line AB represents a smooth phase change from liquid phase to vapour phase at constant temperature, T and pressure, P. The line BB' represents superheating of vapour at constant pressure, P, whereas, BD represents increase in entropy at constant temperature, T. At constant temperature, H of an ideal gas or ideal vapour remains constant. So the line BD runs almost parallel to S-axis. The slight deviation of BD in upward direction indicates that enthalpy increases slightly with pressure at constant temperature, indicating that the vapour behaves as a non-ideal gas.

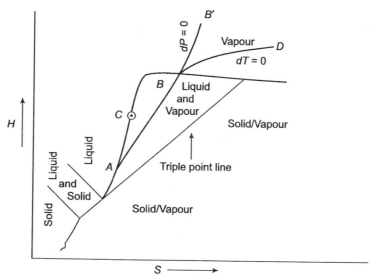

Figure 7.4   *HS*-diagram (Mollier diagram) of a pure substance. The point *C* represents the critical point

7.3   ENERGY IN FLOW PROCESS

7.3.1   Conservation of Energy

For a system like flowing fluid through a conduit, the net energy balance of the system can be written as

$$\frac{d(mU)}{dt} = \dot{m}(E_1 - E_2) + \dot{m}(Q - W) \tag{7.24}$$

$\dfrac{d(mU)}{dt}$ = Rate of change in internal energy within the system; m = Mass of the fluid within the system.

U = Specific internal energy; E_1 and E_2 are the specific energies of the fluid at the entrance and exit respectively; Q = Net heat flow per unit mass of the fluid within the system.

W = Net work output per unit mass of the fluid within the system; $\dot{m}$ = Mass flow rate of the fluid.

The specific energy, E, consists of internal energy per unit mass, i.e., U, kinetic energy per unit mass, i.e., $\frac{1}{2}u^2$ and potential energy per unit mass, Zg, where, Z is height of the fluid from a reference point, u is velocity of the fluid and g is gravitational constant. So, we can write that

$$E = U + \frac{1}{2}u^2 + Zg \tag{7.25}$$

At the steady state process, we can write that $\frac{d(mU)}{dt} = 0$. Hence, equation (7.24) becomes

$$\dot{m}(E_1 - E_2) + \dot{m}(Q - W) = 0. \ \text{Or,} \ E_2 - E_1 = Q - W. \ \text{Or,} \ dE = dQ - dW$$

$$W = \text{Net work output} = \text{shaft work} + \text{work due to } PdV = W_s + \int PdV$$

Thus, $dE = dQ - (dW_s + PdV)$. Or, $dE + PdV = dQ - dW_s$

Using equation (7.25) we get

$$d\left(U + \frac{1}{2}u^2 + Zg\right) + PdV = dQ - dW_s. \ \text{Or,} \ (dU + PdV) + d\left(\frac{1}{2}u^2 + Zg\right) = dQ - dW_s$$

Again, $dH = dU + PdV$. Thus, $dH + d\left(\frac{1}{2}u^2 + Zg\right) = dQ - dW_s$

Or, $$d\left(H + \frac{1}{2}u^2 + Zg\right) = dQ - dW_s$$

Or, $$\Delta\left(H + \frac{1}{2}u^2 + Zg\right) = \Delta Q - \Delta W_s \tag{7.26a}$$

Equation (7.26a) represents the net energy balance of a control system for a steady state flow process. Again we know from 1st law of thermodynamics, $dH = dQ + VdP$

$$\int dH = \int dQ + \int_{P_1}^{P_2} VdP. \ \text{Or,} \ \Delta H = \Delta Q + \int_{P_1}^{P_2} VdP$$

Putting the value of ΔH in the equation (7.26a), we get

$$-\Delta W_s = \int_{P_1}^{P_2} VdP + \frac{1}{2}\Delta u^2 + g\Delta Z$$

Considering the frictional energy F the above equation takes the form

$$-\Delta W_s = \int_{P_1}^{P_2} VdP + \frac{1}{2}\Delta u^2 + g\Delta Z + F \tag{7.26b}$$

The equation (7.26b) is the general equation of energy balance for any steady state flow process.

Consider a fluid flowing through a conduit such that $\Delta Z = 0$ and $\Delta W_s = 0$. If the flow process is assumed to be adiabatic, (i.e., $\Delta Q = 0$), equation (7.26a) becomes

$$\Delta H + \frac{1}{2}\Delta u^2 = 0. \quad \text{Or,} \quad dH = -udu \qquad (7.27a)$$

ρ_1, u_1, A_1 are the density of the fluid, velocity of the fluid, and area of cross-section at the entrance and ρ_2, u_2, A_2 are the density of the fluid, velocity of the fluid, and area of cross-section at the exit. Again, $\rho = 1/V$ where V is the specific volume. From the law of conservation of mass, we have

$$\dot{m} = \text{Constant.} \quad \text{Hence,} \quad \rho_1 u_1 A_1 = \rho_2 u_2 A_2. \quad \text{Or,} \quad \frac{u_1 A_1}{V_1} = \frac{u_2 A_2}{V_2}. \quad \text{Or,} \quad d\left(\frac{uA}{V}\right) = 0$$

Assuming that area of cross-section is constant, the above equation becomes

$$Ad\left(\frac{u}{V}\right) = 0. \quad \text{Or,} \quad d\left(\frac{u}{V}\right) = 0. \quad \text{Or,} \quad \frac{du}{V} - \frac{udV}{V^2} = 0. \quad \text{Or,} \quad du - \frac{udV}{V} = 0$$

Or, $$udu = \frac{u^2 dV}{V} \qquad (7.27b)$$

Again, from the knowledge of thermodynamics, the expression of enthalpy is given by

$$dH = TdS + VdP \qquad (7.27c)$$

For reversible process, $dQ = TdS$ and for irreversible adiabatic process, $dQ = 0$ but $dS > 0$. Combining the equations (7.27a), (7.27b), and (7.27c), we get

$$TdS = dH - VdP = -udu - VdP = -\frac{u^2 dV}{V} - VdP \qquad (7.27d)$$

ΔP is the driving force for flowing a fluid through a conduit and hence it is a negative quantity. Thus 2nd term of right hand side of equation (7.27d), $(-VdP)$, is always positive.

Again with the decrease in pressure, volume V increases since for an ideal gas $PV^\gamma = $ Constant for adiabatic process.

Thus, as the pressure of fluid at exit decreases, both volume, V and the velocity, u increase, although increase in volume, V, is much larger than that of velocity, u. So there is a possibility to achieve a condition, such that $dS = 0$ and at that condition velocity of the fluid, u becomes maximum, represented as u_{max}. The corresponding fluid flow is known as isentropic flow.

Thus, at $u = u_{max}$ the equation (7.27d) becomes

$$\frac{u_{max}^2 dV}{V} = -VdP. \quad \text{Or,} \quad u_{max}^2 = u_{sonic}^2 = -V^2 \left(\frac{\partial P}{\partial V}\right)_S \qquad (7.27e)$$

The above expression of u_{max} is same as the expression of speed of sound and hence u_{max} is conventionally written as u_{sonic}. The ratio of fluid velocity, u to sonic velocity, u_{sonic}, is known as **Mach number**, represented as M.

So, $$M = \frac{u}{u_{sonic}}$$

When $M > 1$, the fluid flows with **supersonic velocity** and when $M < 1$, the fluid flows with **subsonic velocity**.

7.3.2 Euler's Equation

Euler's equation is based on Newton's 2nd law of motion. It is applicable for steady flow of an ideal fluid along a streamline.

Assumptions:
  (a) The fluid is non-viscous, i.e., there is no frictional loss.
  (b) The fluid is homogeneous and incompressible.
  (c) The flow is steady, streamline, and continuous.
  (d) Only gravity and pressure difference are the driving forces for fluid flow.

Let us consider a fluid element of length dl as shown in the Fig. 7.5. If dt is the time taken by the fluid to cover the length dl, velocity of the fluid is given by $v = \dfrac{dl}{dt}$.

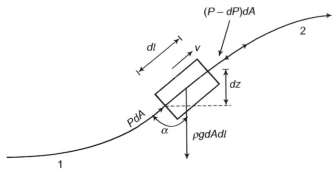

Figure 7.5 Flow of a fluid element from high pressure to low pressure

The fluid is flowing from point 1 to the point 2.

dA is the cross-sectional area of the fluid element, dl is the length of the fluid element and dz is the differential change in height along the streamline.

α = Angle between streamline flow direction and gravitational force direction as shown in the Fig. 7.5.

ρ = Density of the fluid.

Mass of the fluid element $= dm = \rho dA \cdot dl$

$$v = \text{Velocity of the fluid} = \frac{dl}{dt} \quad \text{and} \quad \frac{dv}{dt} = \text{Acceleration of the fluid}$$

$$\frac{dv}{dt} = \left(\frac{dv}{dl}\right)\left(\frac{dl}{dt}\right) = v\left(\frac{dv}{dl}\right)$$

Thus the motion of the fluid element is given by

$$dm\left(\frac{dv}{dt}\right) = (\rho dA \cdot dl)v\left(\frac{dv}{dl}\right) = \rho dA v dv \qquad (7.28a)$$

Gravitational force is $-\rho g dA \cdot dl$. The negative sign indicates that gravitational force acts as retarding force.

So, the retarding force acting along streamline is given by $-\rho g dA \cdot dl \cos a$

Again, $\quad \cos\alpha = \dfrac{dz}{dl}$

Hence, $\quad$ retarding force $= -\rho g dA \cdot dl \cos\alpha = -\rho g dA \cdot dl\left(\dfrac{dz}{dl}\right) = -\rho g dA dz$

$P_1 = $ Pressure of the fluid at the point 1 $= P$

$P_2 = $ Pressure at the point 2 $= [P - dP]$

As the fluid is flowing from the point 1 to the point 2, the pressure difference is $\Delta P = -dP$. This pressure difference is the driving force for the fluid flow.

Hence, forwarding force $= -dP \cdot dA$

Thus, the net force acting along the streamline is given by

$$F = \text{forwarding force} + \text{retarding force} = -dP \cdot dA - \rho g dA dz \qquad (7.28b)$$

According to Newton's 2nd law of motion, equations (7.28a) and (7.28b) can be combined as follows

$$-dP \cdot dA - \rho g dA dz = \rho dA v dv$$

Dividing both sides by $(-\rho dA)$ and rearranging the above equation, we get

$$\frac{dP}{\rho} + v dv + g dz = 0 \qquad (7.28c)$$

The equation (7.28c) is known as Euler's equation.

7.3.3 Bernoulli's Equation

Integration of the equation (7.28c) gives the following result

$$\frac{1}{\rho}\int dP + \int v dv + g\int dz = \text{Constant} \quad [\text{since, } \rho \text{ of an incompressible fluid is constant}]$$

Or, $\dfrac{P}{\rho} + \dfrac{v^2}{2} + gz = \text{Constant}$ (7.29a)

Integration within two limits we get

$$\frac{P_1}{\rho} + \frac{v_1^2}{2} + gz_1 = \frac{P_2}{\rho} + \frac{v_2^2}{2} + gz_2$$ (7.29b)

The equations (7.29a) and (7.29b) are known as Bernoulli's equations which are valid for ideal inviscid fluid.

7.3.4 Flow Through a Nozzle

Nozzle has a converging section, diverging section and a throat region. Thus, cross-sectional area varies both in converging and diverging sections but is assumed to be constant at throat region.

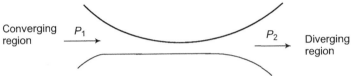

Figure 7.6 Converging-diverging nozzle

Fluid in the converging region is in the subsonic state and achieves sonic velocity in the throat region and energies with supersonic velocity in the diverging region. It is assumed that area of cross-section remains constant in the throat region. Thus for an isentropic flow, i.e., $dS = 0$, the maximum velocity achievable in the throat region is:

$$u_{\text{sonic}}^2 = -V^2 \left(\frac{\partial P}{\partial V} \right)_S = u_{\text{throat}}^2$$ (7.30a)

Suppose P_1 and P_2 are the entrance and exit fluid pressures in the nozzle, the fluid flow depends on the pressure ratio P_2/P_1. If $P_2/P_1 = 1$, there will be no fluid flow. As the ratio decreases the fluid flow increases with fluid velocity and ultimately the ratio P_2/P_1 reaches a critical value, when the sonic flow persists at the exit of the nozzle. Further decrease in pressure ratio has no effect on fluid flow or fluid velocity. For steam, this critical value $(P_2/P_1)_{cr}$ has been found 0.55 at moderate temperature and pressure. As the adiabatic condition is maintained in the nozzle, the following relation can be conveniently used for an ideal gas flow.

$$PV^\gamma = \text{Constant} = K$$

Differentiating with respect to V at constant S, we have

$$\left(\frac{\partial P}{\partial V} \right)_S V^\gamma + \gamma V^{\gamma-1} P = 0. \quad \text{Or,} \quad \left(\frac{\partial P}{\partial V} \right)_S = -\frac{\gamma P}{V}$$ (7.30b)

Combining equations (7.30a) and (7.30b), we get

$$V^2 \left(\frac{\partial P}{\partial V} \right)_S = -\gamma P V$$

Hence, $\quad u_{throat}^2 = -V^2 \left(\frac{\partial P}{\partial V} \right)_S = \gamma P_2 V_2$ (7.30c)

[where, V is the specific volume with unit *lit/kg*]

[Since u_{throat} is achieved only at the exit of the nozzle where pressure and volume of fluid are P_2 and V_2 respectively.]

If the fluid is gas, the frictional energy can be neglected. Further it is assumed that fluid does not transfer any shaft work with the surroundings. Again entrance and exit of a nozzle remain at the same height.

So, $\Delta W_S = 0$, $F = 0$, and $\Delta Z = 0$. Hence, the equation (7.26b) reduces to

$$\frac{1}{2} \Delta u^2 = -\int_{P_1}^{P_2} V dP$$

Or, $\quad u_2^2 - u_1^2 = -2 \int_{P_1}^{P_2} \frac{K}{P^{1/\gamma}} dP,$

$$\left[\text{using adiabatic condition } PV^\gamma = K'. \text{ Or, } V = \frac{K}{P^{1/\gamma}} \text{ and } K = (K')^{\frac{1}{\gamma}} \right]$$

Or, $\quad u_2^2 - u_1^2 = -2K \left[\frac{P^{-\frac{1}{\gamma}+1}}{-\frac{1}{\gamma}+1} \right]_{P_1}^{P_2} = -\frac{2\gamma}{\gamma-1} \left[KP_2 P_2^{-\frac{1}{\gamma}} - KP_1 P_1^{-\frac{1}{\gamma}} \right]$

$$= -\frac{2\gamma}{\gamma-1} KP_2 P_2^{-\frac{1}{\gamma}} \left[1 - \left(\frac{P_1}{P_2} \right)^{-\frac{1}{\gamma}+1} \right]$$

Or, $\quad u_2^2 - u_1^2 = \frac{2\gamma P_2 V_2}{\gamma-1} \left[\left(\frac{P_1}{P_2} \right)^{\frac{\gamma-1}{\gamma}} - 1 \right]$ (7.31)

To achieve the sonic velocity at the throat of a nozzle assume that $u_1 = 0$.

So, $u_2^2 = u_{throat}^2 = u_{sonic}^2 = \gamma P_2 V_2$ [according to equation (7.30c)]. Putting these conditions in the equation (7.31), we get

$$u^2_{\text{throat}} = \frac{2\gamma P_2 V_2}{\gamma - 1}\left[\left(\frac{P_1}{P_2}\right)^{\frac{\gamma-1}{\gamma}} - 1\right]$$

Or, $$\gamma P_2 V_2 = \frac{2\gamma P_2 V_2}{\gamma - 1}\left[\left(\frac{P_1}{P_2}\right)^{\frac{\gamma-1}{\gamma}}_{cr} - 1\right] \text{ [Ch. equation(7.30c)]}$$

Or, $$\left(\frac{P_1}{P_2}\right)^{\frac{\gamma-1}{\gamma}}_{cr} - 1 = \frac{\gamma - 1}{2}. \text{ Or, } \left(\frac{P_2}{P_1}\right)_{cr} = \left(\frac{2}{\gamma+1}\right)^{\frac{\gamma}{\gamma-1}} \tag{7.32}$$

where, $(P_2/P_1)_{cr}$ is the critical pressure ratio at which the sonic velocity is achieved at the throat region.

Thus, for any ideal fluid critical pressure ratio can easily be evaluated just by knowing the γ-value. For varying cross-sectional area the equation of conservation of mass is given by:

$$\frac{dm}{dt} = 0. \text{ Or, } dm = 0. \text{ Or, } d(\rho u A) = 0. \text{ Or, } d\left(\frac{uA}{V}\right) = 0.$$

Or, $$\frac{udA + Adu}{V} - \frac{uA}{V^2}dV = 0. \text{ Or, } \frac{udA + Adu}{uA} = \frac{dV}{V}$$

Or, $$\frac{dA}{A} + \frac{du}{u} = \frac{dV}{V} \tag{7.33a}$$

The above equation holds good at any position of converging or diverging section of the pipe. At the throat of the nozzle $dA = 0$ and sonic velocity is achieved. Thus, at the throat the equation (7.33a) becomes

$$\frac{udu}{u^2_{sonic}} = \frac{dV}{V} \tag{7.33b}$$

Putting this value of $\dfrac{dV}{V}$ in the equation (7.33a), we get

$$\frac{dA}{A} + \frac{du}{u} = \frac{udu}{u^2_{sonic}}$$

Or, $$\frac{dA}{A} = \frac{udu}{u^2_{sonic}} - \frac{du}{u} = \left(\frac{u^2}{u^2_{sonic}} - 1\right)\frac{du}{u} = (M^2 - 1)\frac{du}{u} \tag{7.33c}$$

For the converging section of the nozzle dA is negative and hence $M < 1$. So, fluid flow is subsonic in the converging section. For the diverging section of the nozzle, dA is positive and hence $M > 1$. So, fluid flow is supersonic in the diverging section.

7.3.5 Expanders and Compressors

7.3.5.1 Expanders

If a gas is allowed to expand through a nozzle, internal energy of the gas gets converted into shaft work. The device which delivers shaft work through expansion of gas is known as expander. In different chemical or petrochemical plants, ammonia or ethylene is used as working fluid in expanders. In power plant, usually steam is used as a working fluid and then the expander is specially called turbine.

The expansion process is fully adiabatic and hence $dQ = 0$. Neglecting changes in kinetic and potential energies the equation (7.26a) becomes

$$\Delta H = -\Delta W_s \tag{7.34a}$$

Maximum shaft work, $(\Delta W_s)_{max}$, is available if and only if the expansion process is reversible adiabatic, i.e., isentropic or $dS = 0$. So, for isentropic expansion process $|(\Delta W_s)_{max}| = (\Delta H)_S$. Hence, efficiency of an expander is given by:

$$\eta = \frac{|\Delta W_s|}{|(\Delta W_s)_{max}|} = \frac{\Delta H}{(\Delta H)_S} \tag{7.34b}$$

The *HS*-diagram of an expander is shown in the Fig. 7.7.

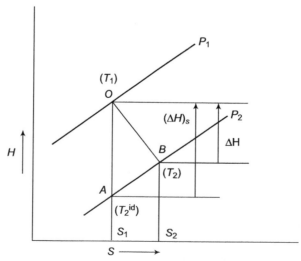

Figure 7.7   Adiabatic expansion process in an expander, where, $P_2 < P_1$

P_1 is the pressure of incoming gas of the nozzle and P_2 is the pressure of outgoing gas of the nozzle. The path *OA* represents the isentropic expansion process and the path *OB* represents the actual expansion process. $(S_2 - S_1)$ = Entropy change associated with the irreversible process. For 1 mole of an ideal gas, we have

$$\Delta S = C_P \ln \frac{T_2}{T_1} - R \ln \frac{P_2}{P_1}$$

For isentropic process, i.e., along the path OA, we have $T_2 = T_2^{id}$ and $\Delta S = 0$.

Hence,
$$T_2^{id} = T_1 \left(\frac{P_2}{P_1} \right)^{\frac{R}{C_P}} = T_1 \left(\frac{P_2}{P_1} \right)^{\frac{C_P - C_V}{C_P}} = T_1 \left(\frac{P_2}{P_1} \right)^{1 - \frac{1}{\gamma}} \qquad (7.35a)$$

Thus,
$$(\Delta W_s)_{\max} = -(\Delta H)_S = -C_P (T_2^{id} - T_1) \qquad (7.35b)$$

$$(W_s)_{\max} = C_P T_1 \left[1 - \left(\frac{P_2}{P_1} \right)^{1 - \frac{1}{\gamma}} \right] \qquad (7.35c)$$

Again, actual shaft work = $\Delta W_s = -\Delta H = -C_P (T_2 - T_1)$.

Hence,
$$\eta = \frac{\Delta W_s}{(\Delta W_s)_{\max}} = \frac{-C_P (T_2 - T_1)}{-C_P (T_2^{id} - T_1)} = \frac{(T_2 - T_1)}{(T_2^{id} - T_1)}$$

Or,
$$T_2 = T_1 + \eta (T_2^{id} - T_1) = \eta T_2^{id} + (1 - \eta) T_1 \qquad (7.36)$$

T_1 and T_2 are experimentally determined quantities. P_1 and P_2 are known. So, T_2^{id} can easily be calculated by using equation (7.35a) and hence the efficiency (η) of an expander can easily be evaluated by using equation (7.36).

7.3.5.2 Compressors

The operating system of a compressor is just reverse to that of an expander. In an expander, shaft work is realized at the cost of internal energy of working fluid. Whereas, in the case of a compressor, shaft work is introduced at one end and very high kinetic energy of working fluid is received at the other end. This kinetic energy is mostly used in rotating blades in different equipment like pumps, blowers, compressors, etc. Thus, it is obvious that volume of working fluid decreases or pressure increases during compression process but in expansion process pressure of working fluid decreases or volume increases. Like expanders, in a compressor also the process may be visualized in two ways:

   (i) **Isentropic process ($dS=0$)** This is the ideal process where minimum shaft work is required to achieve the final pressure.

   (ii) **Irreversible adiabatic process ($dQ=0$ but $dS > 0$)** This is the real process where a definite shaft work, greater than that required in the isentropic process, is required to achieve the final pressure. Thus, the efficiency of a compressor can be written as:

$|\Delta W_s|_{\text{isentropic}} = (\Delta H)_S$ and $|\Delta W_s| = \Delta H$. Hence, efficiency, η is given by

$$\eta = \frac{(\Delta W_s)_{\text{isentropic}}}{\Delta W_s} = \frac{(\Delta H)_S}{\Delta H}. \text{ Or, } \Delta W_s = -\frac{(\Delta H)_S}{\eta} \tag{7.37}$$

where, $(\Delta W_s)_{\text{isentropic}} = -(\Delta H)_S = -C_P(T_2^{id} - T_1), \Delta W_s = -\Delta H = -C_P(T_2 - T_1)$

T_1 = Initial temperature, T_2 = Final temperature achieved in a real process, i.e., in an irreversible adiabatic process. T_2^{id} = Final temperature achieved in an ideal process, i.e., in an isentropic process.

Efficiency of a compressor is given by

$$\eta = \frac{(\Delta W_s)_{\text{isotropic}}}{\Delta W_s} = \frac{C_P(T_2^{id} - T_1)}{C_P(T_2 - T_1)} = \frac{(T_2^{id} - T_1)}{(T_2 - T_1)}. \text{ Or, } (T_2 - T_1) = \frac{(T_2^{id} - T_1)}{\eta}$$

Or $\quad T_2 = \dfrac{T_2^{id}}{\eta} - \left(\dfrac{1}{\eta} - 1\right)T_1 \tag{7.38}$

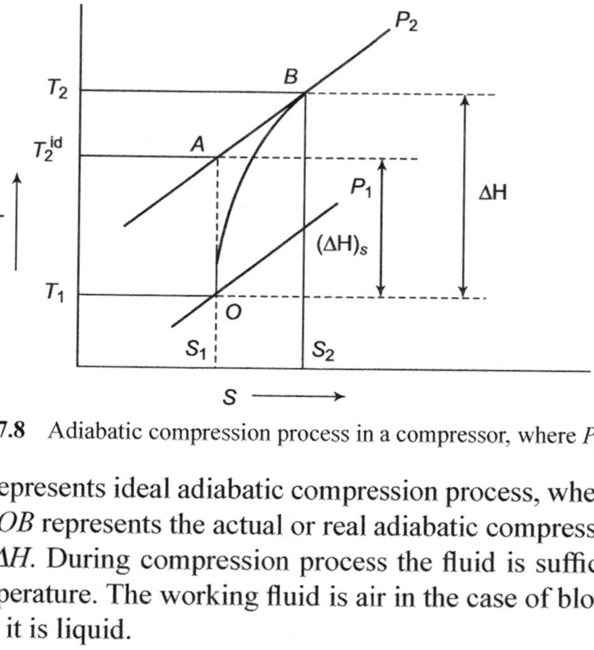

Figure 7.8 Adiabatic compression process in a compressor, where $P_2 > P_1$

In Fig. 7.8, *OA* represents ideal adiabatic compression process, where enthalpy change is $(\Delta H)_S$. The curve *OB* represents the actual or real adiabatic compression process, where enthalpy change is ΔH. During compression process the fluid is sufficiently compressed with increase in temperature. The working fluid is air in the case of blowers, fans, etc. but in the case of pumps it is liquid.

$(\Delta H)_S$ may be evaluated by using thermodynamic property relations.

$dH = dQ + VdP$. For reversible process, $dQ = TdS$. Hence, $dH = TdS + VdP$.

Or $\quad \left(\dfrac{\partial H}{\partial P}\right)_T = V + T\left(\dfrac{\partial S}{\partial P}\right)_T$

Again, from Maxwell's relations, we have

$$\left(\frac{\partial S}{\partial P}\right)_T = -\left(\frac{\partial V}{\partial T}\right)_T = -\beta V,$$

where, $\beta = \frac{1}{V}\left(\frac{\partial V}{\partial T}\right)_T$. Or, $\left(\frac{\partial H}{\partial P}\right)_T = V - \beta VT = V(1 - \beta T)$

Again we can write that $H = f(P, T)$.

Hence, $dH = \left(\frac{\partial H}{\partial T}\right)_P dT + \left(\frac{\partial H}{\partial P}\right)_T dP = C_P dT + (V - \beta TV)dP$ \hfill (7.39a)

ΔH is the actual enthalpy change and $(\Delta H)_S$ is the enthalpy change for a reversible isentropic process. So, we can write that

$$(\Delta H)_S = C_P(T_2^{id} - T_1) + (1 - \beta T)V\Delta P \hfill (7.39b)$$

Thus from equation (7.37), we can write that

Or, $(\Delta W_s)_{\text{isotropic}} = -(\Delta H)_S = -[C_P(T_2^{id} - T_1) + V(1 - \beta T)\Delta P]$ \hfill (7.39c)

Again, from the 1st law of thermodynamics, we have $dH = dQ + VdP$

We know that for a reversible process

$$dS = \frac{dQ}{T}$$

Thus using the 1st law of thermodynamics and equation (7.39a), we get

$$dS = \frac{dQ}{T} = \frac{dH - VdP}{T} = \frac{[C_P dT + (V - \beta TV)dP] - VdP}{T} = \frac{C_P dT}{T} - \beta V dP$$

On integration, we get

$$\Delta S = C_P \ln\frac{T_2}{T_1} - \beta V \Delta P \hfill (7.40)$$

Shaft work $= \Delta W_s = -\frac{(\Delta H)_S}{\eta}$, where, η = efficiency of the compressor. Using equation (7.39b), we have

$$\Delta W_s = -\frac{(\Delta H)_S}{\eta} = -\frac{1}{\eta}\left[C_P(T_2^{id} - T_1) + V(1 - \beta T)\Delta P\right] \hfill (7.41)$$

The equations (7.40) and (7.41) represent the entropy change and actual shaft work respectively of any real adiabatic compression process. However, equations (7.39b), (7.39c), (7.40) and (7.41) are mostly used in those adiabatic compression processes, where liquid is used as working fluid since in many cases enthalpy values of compressed fluids or gases are not available. Thus, the above equations are very much useful in the case of pump.

7.4 POWER CYCLES

7.4.1 Introduction

In any conventional power plant, heat energy is released either through combustion of fossil fuel or through nuclear fission or fusion process. The function of heat engine is to convert this part of heat energy into work. In all work producing devices, conversion of heat energy into work takes place through cyclic processes, which are commonly known as **power cycles.** The heat engines can be divided into two broad categories:

(a) Steam engine

(b) Internal-combustion (IC) engine

In the case of steam engine, working fluid, i.e., steam is enclosed in a cylinder and heat is supplied from outside the cylinder. So to reach the working fluid, heat has to cross the metallic boundary. The working fluid, after receiving the heat energy, executes a cyclic process, delivering a definite amount of work.

In the case of IC-engine, the source of heat energy is available within heat engine itself. Chemical energy of fuel is converted into heat energy through combustion, and combustion products serve as the working medium.

Disadvantages of Steam Engine

(a) Huge amount of heat loss is inevitable during conduction process.

(b) Since metallic wall has to withstand high temperature and pressure, temperature, at which heat absorption occurs, must not be very high.

7.4.2 Steam Engine

The most ideal and efficient heat engine is the Carnot heat engine, where water gets converted into steam in the boiler. The steam then undergoes an expansion process in a turbine and finally condensed into water, which is then pumped to the boiler. The schematic diagram of the steam power plant and the corresponding TS-diagram are shown in the Figs.7.9 and 7.10 respectively.

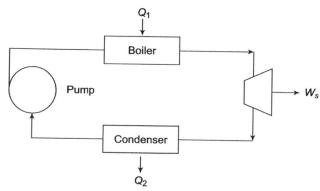

Figure 7.9 Schematic diagram of a steam power plant

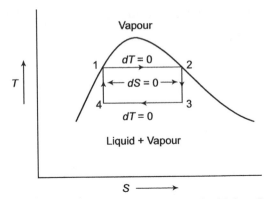

Figure 7.10 *TS*-diagram of Carnot heat engine. 1 → Saturated liquid, 2 → Saturated vapour, 3 → Liquid/vapour mixture with high percentage of vapour, 4 → Liquid/vapour mixture with high percentage of liquid

(a) At stage 2, the saturated vapour enters into the turbine and the exhaust contains high liquid content, bringing about severe damage to the turbine.

(b) In step 4 → 1 pump accepts liquid/vapour mixture and delivers saturated liquid to the boiler. It is next to impossible to design such an engine to execute the step 4 → 1 isentropically.

The above two problems can be avoided by modifying the cycle as follows:

(a) Super-heated vapour is used in the turbine, which delivers works isentropically.

(b) Saturated liquid is delivered to the pump, which absorbs work isentropically.

Thus, the corresponding *TS*-diagram is shown in Fig. 7.11.

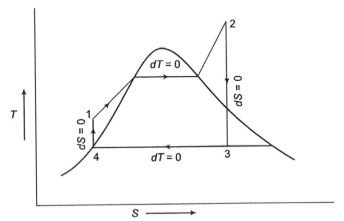

Figure 7.11 *TS*-diagram of Rankine cycle

The cycle, for which the above *TS*-diagram (Fig. 7.11) is shown, is known as **Rankine cycle** and is accepted as a standard cycle for heat engines.

In actual practice the expansion process in turbine is an irreversible process and hence the process is adiabatic but not isentropic. Furthermore, work requiring step, i.e., step 4 → 1 is also irreversible and hence the process is also not isentropic. The *TS*-diagram of actual power cycle is shown in the Fig. 7.12.

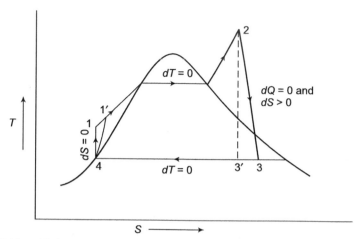

Figure 7.12   *TS*-diagram of actual power cycle (1′–2–3′–4–1′) and (1–2–3–4–1) indicates Rankine cycle

Work absorbed at step $4 \rightarrow 1' = W_1 = -\Delta H_1$

Work absorbed at step $4 \rightarrow 1 = W_1^R = -(\Delta H_1)_S$

Work delivered at step $2 \rightarrow 3' = W_2 = -\Delta H_2$

Work delivered at step $2 \rightarrow 3 = W_2^R = -(\Delta H_2)_S$

$\Delta H_{2 \rightarrow 3} = \phi(\Delta H_2)_S$ and $\phi < 1$. Hence, $W_2 = \phi W_2^R$

Net work available at Rankine cycle $= W_1^R + W_2^R = W_S^R$

Net work available at actual cycle $W_1 + W_2 = W_{net}$

$(\Delta H_1)_S = (H_1 - H_4) > 0$ and hence, $W_1^R < 0,$ $\because$ $W_1^R = -(\Delta H_1)_S$

$\Delta H_1 = (H_1' - H_4) > 0$ and hence, $W_1 < 0,$ $\because$ $W_1 = -\Delta H_1$

$(\Delta H_2)_S = (H_3 - H_2) < 0$ and hence, $W_2^R > 0,$ $\because$ $W_2^R = -(\Delta H_2)_S$

$\Delta H_2 = (H_3' - H_2) < 0$ and hence, $W_2 > 0,$ $\because$ $W_2 = -\Delta H_2$

$\Delta H_{4 \rightarrow 1'} = \Delta H_1 = (\Delta H_1)_S / \zeta$ and $\zeta < 1$. Hence, $W_1 = (W_1^R)/\zeta$ and $\Delta H_1 > (\Delta H_1)_S$

Thus, $W_{net} = W_1 + W_2 = W_1^R/\zeta + \phi W_2^R.$ [$\phi < 1$ and $\zeta < 1$]

Hence, it is obvious that

$$W_S^R = (W_1^R + W_2^R) \gg (W_1 + W_2). \text{ Or, } W_{net} \ll W_S^R$$

So, heat absorbed in the boiler in Rankine cycle $= H_2 - H_1 = Q_1^R.$

Heat absorbed in the boiler in actual cycle $= H_2 - H_1' = Q_1$

$\Delta H_1 = (H_1' - H_4)$ and $(\Delta H_1)_S = (H_1 - H_4).$

As, $\Delta H_1 > (\Delta H_1)_S$ we have $H_1' > H_1.$

Hence, $(H_2 - H_1) > (H_2 - H_1')$. Or, $Q_1^R > Q_1$.

Thermal efficiency of a Rankine cycle (η_R) is

$$\eta_R = \frac{W_S^R}{Q_1^R}$$

Thermal efficiency of an actual power cycle (η_{net}) is given by

$$\eta_{net} = \frac{W_{net}}{Q_1}, \quad Q_1 < Q_1^R \text{ and } W_{net} \ll W_S^R. \text{ Hence, } \eta < \eta_R \qquad (7.42)$$

7.4.3 Internal Combustion Engines

Two commercially important IC-engines are: (a) The Otto engine, and (b) the Diesel engine.

7.4.3.1 The Otto Engine

Working The combustion chamber consists of a cylinder with a movable piston. In the 1st step, the piston moves outward with the intake of air/fuel mixture into the cylinder.

In the 2nd step, the mixture is compressed almost adiabatically up to the pressure P_2. The mixture is then ignited and combustion occurs very rapidly so that pressure P_2 rises to P_3 at constant volume.

In the 3rd step, the combustion products expand almost adiabatically till the pressure falls to P_4 at which point valve opens and pressure falls rapidly to P_1 at constant volume.

In the 4th and final step, the piston moves inward and disposes all combustion products to the atmosphere.

The Otto cycle is shown in the Fig. 7.13.

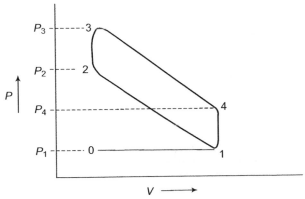

Figure 7.13 Otto cycle for internal combustion engine. $0 \rightarrow 1$ indicates step-I, $1 \rightarrow 2 \rightarrow 3$ indicates step-II, $3 \rightarrow 4 \rightarrow 1$ indicates step-III and $1 \rightarrow 0$ indicates step-IV

The working fluid is petrol.

As the behaviour of combustion products is not known, an idealized standard cycle is considered where air is used as working medium. The efficiency and other characteristics of that idealized standard cycle is equivalent to that of Otto cycle. That idealized standard cycle is known as air-standard Otto cycle. The *PV*-diagram of air-standard Otto cycle is shown in the Fig. 7.14.

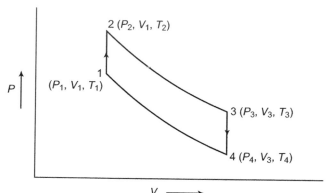

Figure 7.14   Air-standard Otto cycle

Step 1 → 2 Absorption of heat at constant volume = $|Q_1|$

Step 2 → 3 Reversible adiabatic expansion of air

Step 3 → 4 Rejection of heat or cooling at constant volume = $|Q_2|$

Step 4 → 1 Reversible adiabatic compression of air

The thermal efficiency of the above cycle is given by:

$$\eta = \frac{W_{net}}{|Q_1|} = \frac{|Q_1| - |Q_2|}{|Q_1|} = 1 - \frac{|Q_2|}{|Q_1|}$$

Again, $|Q_1| = C_V(T_2 - T_1)$ for 1 mole of an ideal gas. Assume that air, used in the above cycle, behaves as an ideal gas.

$$Q_2 = C_V(T_4 - T_3) = -C_V(T_3 - T_4). \text{ Or, } |Q_2| = C_V(T_3 - T_4), \text{ since, } T_3 > T_4$$

Or,
$$\frac{|Q_2|}{|Q_1|} = \frac{(T_3 - T_4)}{(T_2 - T_1)}$$

Again, along the adiabatic curve 2 → 3, we have

$$T_2 V_1^{\gamma-1} = T_3 V_3^{\gamma-1}, \ \frac{V_3}{V_1} = r = \text{Compression ratio}, \ T_2 = T_3\left(\frac{V_3}{V_1}\right)^{\gamma-1} = T_3 r^{\gamma-1}$$

$$r = \frac{\text{Volume of gas at the beginning of compression stroke}}{\text{Volume of gas after compression}}$$

Along the adiabatic curve 4 → 1, we have

$$T_4 V_3^{\gamma-1} = T_1 V_1^{\gamma-1} \text{ Or, } T_1 = T_4\left(\frac{V_3}{V_1}\right)^{\gamma-1} = T_4 r^{\gamma-1}$$

Hence, $$T_2 - T_1 = T_3 r^{\gamma-1} - T_4 r^{\gamma-1} = (T_3 - T_4)r^{\gamma-1}. \text{ Or, } \frac{(T_3 - T_4)}{(T_2 - T_1)} = \left(\frac{1}{r}\right)^{\gamma-1}$$

Thus, the efficiency of an air standard Otto cycle, η, is given by

$$\eta = 1 - \frac{|Q_2|}{|Q_1|} = 1 - \frac{(T_3 - T_4)}{(T_2 - T_1)} = 1 - \left(\frac{1}{r}\right)^{\gamma-1} \tag{7.43}$$

It is evident that with the increase in compression ratio (r) the efficiency of the Otto cycle increases.

7.4.3.2 The Diesel Engine

In contrast to the Otto engine high compression ratio has been set in the case of Diesel engine in order to achieve high temperature so that combustion occurs spontaneously. Furthermore, in the case of Diesel engine, fuel is injected at the end of compression step but slow enough to complete heat absorption process at constant pressure.

At the same compression ratio, efficiency of the Otto engine is higher than that of the Diesel engine but the Otto engine cannot be operated at high compression ratio due to pre-ignition problem. As the Diesel engine can be operated at very high compression ratio safely, higher efficiency can be obtained in the case of the Diesel engine compared to that of the Otto engine.

In order to evaluate efficiency of the Diesel engine an idealized standard cycle is considered where air is used as working medium. The PV–diagram of air-standard Diesel cycle is shown in the Fig. 7.15.

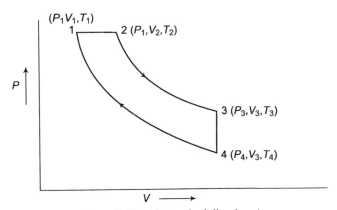

Figure 7.15 Air-standard diesel cycle

Step 1 → 2 Absorption of heat at constant pressure = $|Q_P|$
Step 2 → 3 Reversible adiabatic expansion of air
Step 3 → 4 Rejection of heat at constant volume = $|Q_V|$
Step 4 → 1 Reversible adiabatic compression of air.

The internal efficiency, η, of the above cycle is given by

$$\eta = \frac{W_{net}}{|Q_P|} = \frac{|Q_P| - |Q_V|}{|Q_P|} = 1 - \frac{|Q_V|}{|Q_P|} \tag{7.44a}$$

Again, $|Q_P| = C_P(T_2 - T_1)$ for 1 mole of an ideal gas and assume that air, used in the above cycle, behaves as an ideal gas.

Similarly, $|Q_V| = C_V(T_3 - T_4)$ for 1 mole of an ideal gas and $T_3 > T_4$. $\gamma = C_P/C_V$

So,

$$\eta = 1 - \frac{|Q_V|}{|Q_P|} = 1 - \frac{C_V(T_3 - T_4)}{C_P(T_2 - T_1)} = 1 - \frac{(T_3 - T_4)}{\gamma(T_2 - T_1)}$$

Along the adiabatic path $2 \rightarrow 3$, we have

$$P_1 V_2^\gamma = P_3 V_3^\gamma. \quad \text{Or,} \quad \frac{P_1}{P_3} = r_e^\gamma, \quad r_e = \text{Expansion ratio} = \frac{V_3}{V_2}$$

Along the adiabatic path $4 \rightarrow 1$, we have

$$P_4 V_3^\gamma = P_1 V_1^\gamma. \quad \text{Or,} \quad \frac{P_1}{P_4} = r_c^\gamma, \quad r_c = \text{Compression ratio} = \frac{V_3}{V_1}. \quad \text{Hence,} \quad \frac{r_c}{r_e} = \frac{V_2}{V_1}$$

$$\frac{P_1}{P_3} = r_e^\gamma, \quad \frac{P_1}{P_4} = r_c^\gamma, \quad \frac{P_3}{P_4} = \left(\frac{r_c}{r_e}\right)^\gamma = \left(\frac{V_2}{V_1}\right)^\gamma \qquad (7.44b)$$

For 1 mole of air, we have

At the point 3, $P_3 V_3 = RT_3$. At the point 4, $P_4 V_3 = RT_4$.

Hence,

$$\frac{T_3}{T_4} = \frac{P_3}{P_4} = \left(\frac{r_c}{r_e}\right)^\gamma. \quad \text{Or,} \quad T_3 = T_4\left(\frac{r_c}{r_e}\right)^\gamma \qquad (7.44c)$$

Again, along the adiabatic path $2 \rightarrow 3$, we have

$$T_2 V_2^{\gamma-1} = T_3 V_3^{\gamma-1}. \quad \text{Or,} \quad T_2 = T_3\left(\frac{V_3}{V_2}\right)^{\gamma-1} = T_3\, r_e^{\gamma-1}$$

Along the adiabatic path $4 \rightarrow 1$, we have

$$T_4 V_3^{\gamma-1} = T_1 V_1^{\gamma-1}. \quad \text{Or,} \quad T_1 = T_4\left(\frac{V_3}{V_1}\right)^{\gamma-1} = T_4\, r_c^{\gamma-1}$$

So, $\qquad T_2 = T_3\, r_e^{(\gamma-1)}$ and $T_1 = T_4\, r_c^{(\gamma-1)}$ $\qquad\qquad (7.44d)$

Thus, combining equations (7.44a), (7.44c) and (7.44d), we have

$$\eta = 1 - \frac{|Q_V|}{|Q_P|} = 1 - \frac{(T_3 - T_4)}{\gamma(T_2 - T_1)}$$

$$\eta = 1 - \frac{T_4 \left(\dfrac{r_c}{r_e}\right)^\gamma - T_4}{\gamma \left[T_3 \, r_e^{\gamma-1} - T_4 \, r_c^{\gamma-1}\right]} = 1 - \frac{T_4 \left[\left(\dfrac{r_c}{r_e}\right)^\gamma - 1\right]}{\gamma \left[T_3 \, r_e^{\gamma-1} - T_4 \, r_c^{\gamma-1}\right]} = 1 - \frac{T_4 \left[\left(\dfrac{r_c}{r_e}\right)^\gamma - 1\right]}{\gamma T_4 \left[\left(\dfrac{T_3}{T_4}\right) r_e^{\gamma-1} - r_c^{\gamma-1}\right]}$$

$$\eta = 1 - \frac{\left[\left(\dfrac{r_c}{r_e}\right)^\gamma - 1\right]}{\gamma \left[\left(\dfrac{r_c}{r_e}\right)^\gamma r_e^{\gamma-1} - r_c^{\gamma-1}\right]} = 1 - \frac{\left[\left(\dfrac{1}{r_e}\right)^\gamma - \left(\dfrac{1}{r_c}\right)^\gamma\right]}{\gamma \left(\dfrac{1}{r_e} - \dfrac{1}{r_c}\right)}$$

Thus, the efficiency of an air standard diesel cycle is given by

$$\eta = 1 - \frac{\left[\left(\dfrac{1}{r_e}\right)^\gamma - \left(\dfrac{1}{r_c}\right)^\gamma\right]}{\gamma \left(\dfrac{1}{r_e} - \dfrac{1}{r_c}\right)} \tag{7.45}$$

7.4.4 Gas Turbine Power Plant

The advantage of internal combustion engine is the direct use of energy at high temperature and pressure without any heat transfer. Simultaneously, the IC-engine has the disadvantage of loss of energy due to friction between reciprocating piston and cylinder but the turbine does not have this disadvantage.

The gas turbine engine enjoys both the advantages of IC-engine and turbine. The schematic diagram of a gas turbine power plant is shown in Fig. 7.16.

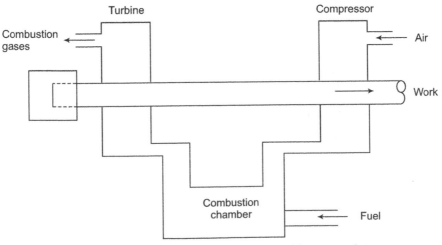

Figure 7.16 Schematic diagram of a gas-turbine power plant

Working principle

(a) Air is compressed to several atmospheres in a compression chamber and then allowed to pass into combustion chamber where fuel is injected.

(b) The fuel is burnt in the combustion chamber and the combustion gases enter into the turbine, where expansion occurs due to which mechanical work is obtained. Part of this mechanical work is used to compress air in the compressor.

(c) Efficiency of the power plant increases with the increase in inlet temperature of the combustion gases in the turbine.

The *PV*-diagram of (air-standard idealized) gas-turbine cycle, popularly known as Brayton cycle, is shown in the Fig. 7.17.

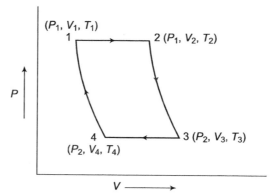

Figure 7.17   *PV*-diagram of Brayton cycle

Step 1 → 2 Addition of heat at constant pressure $= Q_1$. We know that $\Delta H_P = \Delta Q_P$. Thus, $Q_1 = H_2 - H_1 = C_P(T_2 - T_1)$ for 1 mole of an ideal gas.

Step 4 → 1 Work done in isentropic compression process (i.e. reversible and adiabatic process) $= W_{4 \to 1}$

We know from equation (7.26a) that

$$\Delta\left(H + \frac{u^2}{2} + zg\right) = \Delta Q - \Delta W_s$$

As the process is adiabatic, $\Delta Q = 0$. Neglecting potential energy and kinetic energy terms, we get $-\Delta W_s = \Delta H$. Thus, $W_{(4 \to 1)} = -(\Delta H)_{(4 \to 1)} = -C_P(T_1 - T_4)$.

Step 2 → 3 Work done in isentropic process $= W_{(2 \to 3)} = -(\Delta H)_{(2 \to 3)} = -C_P(T_3 - T_2)$.

$W = $ Total work $= W_{(2 \to 3)} + W_{(4 \to 1)} = -C_P(T_3 - T_2) - C_P(T_1 - T_4) = C_P(T_2 - T_3) - C_P(T_1 - T_4)$

Along the adiabatic curve $2 \to 3$ we have,

$$T_2 P_1^{\frac{1-\gamma}{\gamma}} = T_3 P_2^{\frac{1-\gamma}{\gamma}}. \quad \text{Or,} \quad T_3 = T_2\left(\frac{P_1}{P_2}\right)^{\frac{1-\gamma}{\gamma}}$$

Along the adiabatic curve $4 \rightarrow 1$, we have

$$T_4 P_2^{\frac{1-\gamma}{\gamma}} = T_1 P_1^{\frac{1-\gamma}{\gamma}} . \quad \text{Or,} \quad T_4 = T_1 \left(\frac{P_1}{P_2} \right)^{\frac{1-\gamma}{\gamma}}$$

Putting these values of T_3 and T_4 in the total work (W), we get

$$W = C_P \left[T_2 - T_2 \left(\frac{P_1}{P_2} \right)^{\frac{1-\gamma}{\gamma}} \right] - C_P \left[T_1 - T_1 \left(\frac{P_1}{P_2} \right)^{\frac{1-\gamma}{\gamma}} \right] = C_P (T_2 - T_1) \left[1 - \left(\frac{P_1}{P_2} \right)^{\frac{1-\gamma}{\gamma}} \right]$$

$$\text{Efficiency} = \eta = \frac{W}{Q_1} = \left[1 - \left(\frac{P_1}{P_2} \right)^{\frac{1-\gamma}{\gamma}} \right] \tag{7.46}$$

where, $|Q_1| = C_P (T_2 - T_1)$

It is evident that higher the value of P_1, i.e., higher the inlet pressure, higher is the efficiency of the Brayton cycle.

7.5 REFRIGERATION CYCLE

7.5.1 Carnot Cycle

Refrigeration means maintenance of temperature below the temperature of surroundings. This is reverse of the heat engine cycle. In heat engine cycle, heat is absorbed at higher temperature and rejected at lower temperature but in refrigeration cycle, heat is absorbed at lower temperature and rejected at higher temperature. Efficiency of a refrigeration cycle is expressed as coefficient of performance (COP).

$$COP = \frac{Amount\ of\ heat\ absorbed}{Net\ work}$$

Two different types of refrigeration cycles are widely used commercially:
(i) Vapour-compression cycle, and (ii) Absorption refrigeration cycle.

Ideal refrigeration cycle is the Carnot refrigeration cycle, where Q_2 is the amount of heat absorbed at lower temperature T_2, and Q_1 is the amount of heat rejected at higher temperature T_1 followed by two adiabatic steps. According to 2nd law of thermodynamics this cycle cannot be operated without use of any external work. So some work is required to operate the cycle. Let the work be W.

According to 1st law of thermodynamics we know that $\Delta Q = \Delta U + W$

For a cyclic process $\Delta U = 0$ and $W = \Delta Q = |Q_1| - |Q_2|$, since, $|Q_1| > |Q_2|$.

Hence,
$$COP = \frac{|Q_2|}{W} = \frac{|Q_2|}{|Q_1| - |Q_2|} = \frac{1}{\dfrac{|Q_1|}{|Q_2|} - 1}$$

Again from Carnot cycle we know that

$$\frac{|Q_1|}{|Q_2|} = \frac{T_1}{T_2}, \text{ where, } T_1 > T_2$$

Hence,
$$COP = \frac{1}{\dfrac{T_1}{T_2} - 1} = \frac{T_2}{T_1 - T_2} \tag{7.47}$$

Carnot refrigerator is independent of working medium.

7.5.2 Vapour-Compression Cycle

The most useful commercial refrigeration cycle is the vapour-compression cycle. In this cycle, liquid refrigerant absorbs heat from the surroundings and gets converted to saturated vapour, which is then compressed sufficiently to superheat the refrigerant vapour. That superheated vapour is allowed to pass through a condenser where heat is rejected at higher temperature and vapour is converted to saturated liquid, which is then passed through a throttle valve where expansion occurs in isenthalpic process, as a result of which, a sharp drop in temperature is observed. During evaporation process, liquid refrigerant absorbs heat from the surroundings, as a result of which, surrounding atmosphere gets cooled. The whole process is shown in the Fig. 7.18 and the corresponding TS-diagram is shown in the Fig. 7.19.

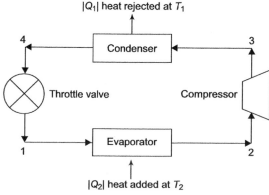

Figure 7.18 Schematic diagram of a vapour-compression cycle

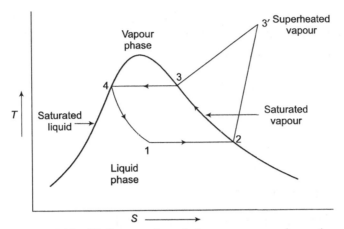

Figure 7.19 *TS*-diagram of a typical vapour-compression cycle

Step 1 → 2 Heat is absorbed from the surroundings by the refrigerant isothermally to form saturated vapour. Amount of heat added = $|Q_2|$.

Step 2 → 3′ → 3 Saturated vapour is heated to form superheated vapour (3′), which is then cooled to saturated vapour (state 3).

Step 3 → 4 The saturated vapour is allowed to condense to form saturated liquid (state 4). Amount of heat rejected isothermally = $|Q_1|$.

Step 4 → 1 The saturated liquid is then passed through a throttle valve due to which a sharp drop of temperature is observed. The process is isenthalpic but entropy increases, as shown in the Fig. 7.19.

Condenser Heat is rejected at constant pressure and temperature. H_1, H_2, H_3, and H_4 are the enthalpy values per unit mass flow rate at the states 1, 2, 3, and 4 respectively.

We know that $dH = dQ + VdP$. At constant pressure, $dH = dQ$. Or, $\Delta Q = \Delta H$.

Thus at condenser $Q_1 = (H_4 - H_3)$ and $|Q_1| = (H_3 - H_4)$, since, $H_3 > H_4$.

Evaporator Heat is absorbed from the surroundings at constant pressure and temperature. So, $Q_2 = (H_2 - H_1)$ and $|Q_2| = (H_2 - H_1)$, since, $H_2 > H_1$

Throttle valve Isenthalpic process. So, $H_1 = H_4$. Hence, net work required to operate the cycle is given by $W_{net} = |Q_1| - |Q_2| = (H_3 - H_4) - (H_2 - H_1) = (H_3 - H_2)$

Hence, $$COP = \frac{|Q_2|}{W_{net}} = \frac{H_2 - H_1}{H_3 - H_2} \tag{7.48}$$

It is evident that, lower the value of H_3, higher the value of COP is.

Again, $|Q_2| = |H_2 - H_1|$, where H_1 and H_2 are the enthalpy values in *kJ* per unit mass flow rate. If $\dot{m}$ is the mass flow rate of refrigerant in (kg/hr), we can write that

$$|Q_2| = \dot{m}(H_2 - H_1)\text{kJ/hr. Or, } \dot{m} = \frac{|Q_2|}{(H_2 - H_1)}\text{kg/hr} \tag{7.49}$$

Ton of refrigeration 1 ton of refrigeration implies that amount of heat which is to be removed to freeze 1 ton or 1000 kg of water to 1 ton of ice at 0°C in 24 hrs.

Latent heat of water = 80 cal/g = 80 kcal/kg = 336 kJ/kg.

Total heat to be removed to freeze 1 ton of water = 336 × 1000 kJ = 336,000 kJ. This amount of heat is to be removed in 24 hrs.

So, amount of heat to be removed per hour per ton of water $= \dfrac{336,000}{24}$ kJ = 14,000 kJ

In actual process amount of heat removed per hour $= |Q_2| = \dot{m}(H_2 - H_1)$ kJ

Thus,   ton of refrigeration $= \dfrac{\dot{m}(H_2 - H_1)}{14,000}$

7.5.3 Absorption Refrigerator

In the absorption refrigeration process, refrigerant is absorbed in a relatively non-volatile solvent and refrigerant is regenerated as vapour, which is then executed as vapour-compression cycle. The schematic diagram of an absorption refrigeration cycle is shown in the Fig. 7.20.

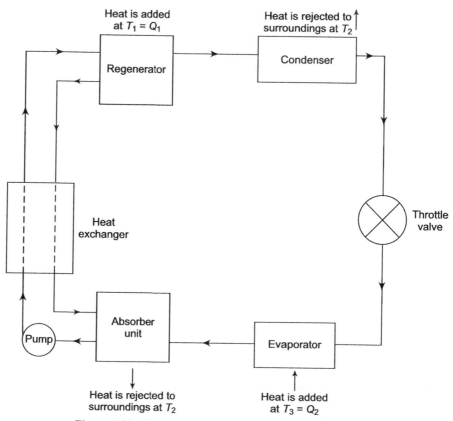

Figure 7.20   Schematic diagram of an absorption refrigerator

Evaporator At this unit, liquid refrigerant is evaporated with the absorption of heat $|Q_2|$

at temperature T_3.

Absorber Vapour of the refrigerant is absorbed by a relatively non-volatile solvent and heat of absorption is rejected to the surroundings at temperature T_2.

Pump and heat exchanger Solvent with refrigerant is compressed and pumped to the regenerator through heat exchanger. In the heat exchanger two streams are flowing. In the upstream, solvent, enriched with refrigerant, absorbs heat. In the downstream, solvent, having low concentration of refrigerant and flowing from the regenerator, rejects heat in the heat exchanger and enters into the absorber unit.

Regenerator Pressure in the regenerator is the same as that in the condenser. $|Q_1|$ amount of heat is added isothermally at temperature T_1. In this unit, solvent, enriched with refrigerant, is heated to reach the temperature T_1 such that refrigerant gets evaporated and leaves the regenerator, moving towards the condenser unit. So, the concentration of refrigerant in the solvent falls and reaches a minimum. This solvent, with low concentration of refrigerant, is sent to the absorber unit through heat exchanger. Net work available can be computed by using Carnot engine efficiency equation

$$\eta = \frac{W}{|Q_1|} = \frac{T_1 - T_2}{T_1}. \text{ Or, } W = |Q_1| \left(\frac{T_1 - T_2}{T_1} \right) \tag{7.50}$$

This work is used in the next section of the refrigeration cycle, consisting of condenser and evaporator units.

Condenser In this unit, refrigerant vapour gets condensed with the release of latent heat to the surroundings, at temperature T_2. After condensation the saturated liquid refrigerant is allowed to pass through the throttle valve, as a result of which, temperature of the liquid refrigerant drops to T_3 and the liquid is sent to the evaporator unit. Amount of work required for refrigeration is given by the equation (7.48).

$$COP = \frac{|Q_2|}{W} = \frac{T_3}{T_2 - T_3}. \text{ Or, } |Q_2| = W \left(\frac{T_3}{T_2 - T_3} \right) \tag{7.51}$$

Putting the value of W from equation (7.50) in the equation (7.51), we get

$$|Q_2| = |Q_1| \left(\frac{T_1 - T_2}{T_1} \right) \left(\frac{T_3}{T_2 - T_3} \right). \text{ Or, } \frac{|Q_1|}{|Q_2|} = \left(\frac{T_1}{T_1 - T_2} \right) \left(\frac{T_2 - T_3}{T_3} \right) \tag{7.52}$$

All the processes considered in the above cycle are reversible and hence *COP* becomes maximum for this cycle. As *COP* is maximum $|Q_2|$ is also maximum according to equation (7.51).

Hence, $\dfrac{|Q_1|}{|Q_2|}$ is minimum for Carnot absorption refrigerator.

Example 7.5.1 Carnot engine is coupled to a Carnot refrigerator so that all of the work produced by the engine is used by the refrigerator in extraction of heat from a heat reservoir at 270 K at the rate of 4 kJ/s. The source of energy for the Carnot engine is a heat reservoir at 500 K. If both devices discard heat to the surroundings at 300 K, how much heat does the engine absorb from the 500 K–reservoir?

If the actual coefficient of performance of the refrigerator is $\omega = \omega_{Carnot}/1.5$ and if the thermal efficiency of the engine is $\eta = \eta_{Carnot}/1.5$, how much heat does the engine absorb from the 500 K–reservoir?

We know that

$$\frac{|Q_1|}{|Q_2|} = \left(\frac{T_1}{T_1 - T_2}\right)\left(\frac{T_2 - T_3}{T_3}\right)$$

$|Q_2|$ = Rate of heat absorbed at T_3 is 4 kJ/s, where, $T_3 = 270$ K.
$|Q_1|$ = Rate of heat absorbed at T_1, where, $T_1 = 500$ K.
T_2 = Temperature at which heat is rejected to the surroundings = 300 K.

$$|Q_1| = 4 \times \left(\frac{500}{500 - 300}\right)\left(\frac{300 - 270}{270}\right) = 1.11 \text{kJ/s}$$

Amount of heat absorbed at 500 K = 1.11 kJ/s. The work obtained from Carnot engine is W, which is given by

$$W = |Q_1|\left(\frac{T_1 - T_2}{T_1}\right)$$

$$\text{Actual work} = W_s = \frac{W}{1.5} = \frac{|Q_1|}{1.5}\left(\frac{T_1 - T_2}{T_1}\right)$$

$$COP \text{ of Carnot refrigerator} = \frac{|Q_2|}{W} = \frac{T_3}{T_2 - T_3}$$

$$COP \text{ of actual refrigerator} = \frac{|Q_2|}{W_s} = \frac{1}{1.5}\left(\frac{T_3}{T_2 - T_3}\right). \text{ Or, } |Q_2| = \frac{W_s T_3}{1.5(T_2 - T_3)}$$

Or, $$|Q_2| = \frac{|Q_1|}{1.5}\left(\frac{T_1 - T_2}{T_1}\right)\frac{T_3}{1.5(T_2 - T_3)}. \text{ Or, } |Q_1| = 2.25|Q_2|\frac{T_1(T_2 - T_3)}{T_3(T_1 - T_2)}$$

Or, $$|Q_1| = \frac{2.25 \times 4 \times 500(300 - 270)}{270(500 - 300)} = 2.5 \text{kJ/s}$$

So, the actual amount of heat absorbed by the refrigerator = 2.5 kJ/s.

Refrigerants In most cases methane and ethane based halogenated hydrocarbons are used as refrigerants. Nomenclature of different refrigerants is based on number of hydrogen atoms and halogen atoms present in the compound. Commonly used halogenated hydrocarbon is **chlorofluorocarbon,** which is written as **CFC.**

For CFC − 12, add 90 with 12, i.e. 90 + 12 = 102

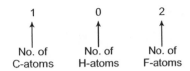

So, two fluorine atoms are attached with one carbon atom. The remaining two positions are occupied by chlorine atoms. Thus, formula of the compound is CCl_2F_2. Technically it is represented as $R12$.

Similarly for $R143$, add 90 with 143, i.e. $90 + 143 = 233$

$$\begin{array}{ccc} 2 & 3 & 3 \\ \uparrow & \uparrow & \uparrow \\ \text{No. of} & \text{No. of} & \text{No. of} \\ \text{C-atoms} & \text{H-atoms} & \text{F-atoms} \end{array}$$

Thus, formula of the compound will be $C_2H_3Cl_2F_3$.

CFCs are very much stable, readily move up to stratosphere and destroy the ozone layer. That is why the use of CFC is very much limited. Ammonia is also a good refrigerant, particularly in absorption refrigeration system but it is toxic, inflammable and very much unfriendly to environment. Sulphur dioxide (SO_2), methyl chloride (CH_3Cl), propane (C_3H_8), butane (C_4H_{10}), etc. are also used as refrigerants.

Example 7.5.2 An engine working on Otto cycle is supplied with air at 0.1 MPa and 35°C. The compression ratio is 8. Heat supplied is 2100 kJ/kg. Calculate the maximum pressure and temperature of the cycle, the cycle efficiency and the mean effective pressure (mep).

Given: For air, $C_p = 1.005$ kJ/kg.°K, $C_V = 0.718$ kJ/kg.°K, and $R = 0.287$ kJ/kg.°K. The PV–diagram of Otto-cycle is shown below:

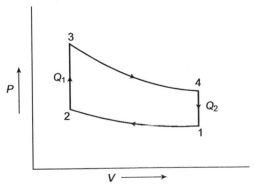

PV–diagram of Otto-cycle

At the state 1: $P_1 = 0.1$ MPa, $T_1 = (273 + 35)\text{K} = 308$ K, $Q_1 = 2100$ kJ/kg

Efficiency of the cycle (η) is given by

$$\eta = 1 - \left(\frac{1}{r}\right)^{\gamma-1}, \ r = \frac{V_1}{V_3} = \frac{V_1}{V_2} = 8, \ \gamma = \frac{C_P}{C_V} = \frac{1.005}{0.718} = 1.4. \ \text{Hence, } \eta = 1 - \left(\frac{1}{8}\right)^{0.4} = 0.565$$

Heat is absorbed at constant volume.

So, $Q_1 = C_V(T_3 - T_2)$. Or, $2100 = 0.718(T_3 - T_2)$ (1)

Process $1 \rightarrow 2$ is an adiabatic process.

So, $T_1 V_1^{\gamma-1} = T_2 V_2^{\gamma-1}$. Or, $\dfrac{T_2}{T_1} = \left(\dfrac{V_1}{V_2}\right)^{\gamma-1} = (8)^{0.4}$. Or, $T_2 = 308 \times (8)^{0.4} = 708.4$ K

Putting the value of T_2 in equation (1) we get

$$(T_3 - 708.4) = \frac{2100}{0.718} = 2925. \ \text{Or,} \ \ T_3 = 3633 \text{ K}$$

T_3 is the maximum temperature of this cycle. Again, applying adiabatic expression in the process $1 \rightarrow 2$ we have,

$$P_1 V_1^{\gamma} = P_2 V_2^{\gamma}. \ \text{Or,} \ \ P_2 = P_1 \left(\frac{V_1}{V_2}\right)^{\gamma} = 0.1 \times (8)^{0.4} \text{ MPa} = 1.837 \text{ MPa}$$

Again, along $2 \rightarrow 3$, isothermal heat transfer occurs. Thus, we can write

$$\frac{P_2 V_2}{T_2} = \frac{P_3 V_3}{T_3} \cdot V_2 = V_3, \ T_3 = 3633 \text{K}, \ T_2 = 708.4 \text{K}, P_2 = 1.837 \text{MPa}$$

Or, $P_3 = 1.837 \times \left(\dfrac{3633}{708.4}\right) = 9.421 \text{ MPa}$

P_3 is the maximum pressure of this cycle. Work $= W = \eta \times Q_1 = 0.565 \times 2100 = 1186.5$ kJ/kg

If P_m is the mean effective pressure, $W = P_m(V_1 - V_2)$.

$$P_1 = 0.1 \text{MPa} = 100 \text{kPa} = 100 \text{kJ/m}^3. \ T_1 = 308 \text{K}. \ V_1 = \frac{RT_1}{P_1} = \frac{0.287 \times 308}{100} = 0.884 \text{ m}^3$$

$\dfrac{V_1}{V_2} = 8.$ Or, $V_2 = \dfrac{V_1}{8} = \dfrac{0.884}{8} = 0.11 \text{ m}^3.$

Thus, $W = 1186.5$ kJ/kg, $V_1 = 0.884$ m^3, $V_2 = 0.11$ m^3

Again, $W = P_m(V_1 - V_2)$. Or, $1186.5 = P_m (0.884 - 0.11)$. Or, $P_m = 1533$ kPa $= 1.533$ MPa

Example 7.5.3   An air-standard dual cycle has a compression ratio of 16 and the compression begins at 1 bar and 50°C. The maximum pressure is 70 bar. The heat transferred to air at constant pressure is equal to that at constant volume. Estimate: (a) the pressures and temperatures at the cardinal points of the cycle, (b) the cycle efficiency, and (c) the mean effective pressure of the cycle.

Given: $C_V = 0.718$ kJ/(kg.°K), $C_p = 1.005$ kJ/(kg.°K)

The *PV*-diagram of dual cycle is shown below:

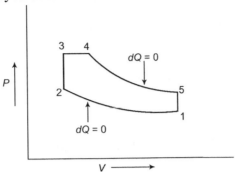

PV-diagram of dual cycle

Heat is absorbed partly at constant volume (process $2 \to 3$) and partly at constant pressure (process $3 \to 4$).

So, net heat absorbed $= Q_1 = mC_V(T_3 - T_2) + mC_P(T_4 - T_3)$.

Heat is rejected at the end of the cycle (process $5 \to 1$) at constant volume.

So, $Q_2 = mC_V(T_1 - T_5)$. $[T_5 > T_1]$. Or, $|Q_2| = mC_V(T_5 - T_1)$

$Q_1 = mC_V(T_3 - T_2) + mC_P(T_4 - T_3)$. $[T_3 > T_2$ and $T_4 > T_3]$.

Thus,   efficiency of the cycle $= \eta = 1 - \dfrac{|Q_2|}{|Q_1|} = 1 - \dfrac{mC_V(T_1 - T_5)}{mC_V(T_3 - T_2) + mC_P(T_4 - T_3)}$

Or, $\eta = 1 - \dfrac{(T_1 - T_5)}{(T_3 - T_2) + \gamma(T_4 - T_3)}$, $T_1 = (270 + 50)°K = 323°K$.

$P_1 = 1$ bar, $P_3 = P_4 = P_{max} = 70$ bar. Compression ratio $= r = \dfrac{V_1}{V_2} = \dfrac{V_1}{V_3} = 16$

$\gamma = \dfrac{C_P}{C_V} = \dfrac{1.005}{0.718} = 1.4$

Along $1 \to 2$

$$P_1 V_1^\gamma = P_2 V_2^\gamma. \text{ Or, } P_2 = P_1 \left(\frac{V_1}{V_2}\right)^\gamma = 1 \times 16^{1.4} = 48.5 \,\text{bar}$$

$$T_1 V_1^{\gamma-1} = T_2 V_2^{\gamma-1}. \text{ Or, } T_2 = T_1 \left(\frac{V_1}{V_2}\right)^{\gamma-1} = 323 \times 16^{1.4} = 979°K.$$

Along $2 \to 3$

$\dfrac{P_2 V_2}{T_2} = \dfrac{P_3 V_3}{T_3}$, $V_2 = V_3$, $P_3 = 70\,\text{bar}, T_2 = 979°K, P_2 = 48.5\,\text{bar}$. So, $T_3 = 979\left(\dfrac{70}{48.5}\right) = 1413°K$

Again, according to given condition $Q_{2\to3} = Q_{3\to4}$. Or, $mC_V(T_3 - T_2) = mC_P(T_4 - T_3)$

$$T_4 = \frac{(1+\gamma)T_3 - T_2}{\gamma} = \frac{2.4 \times 1413 - 979}{1.4} = 1723°K,$$

$$V_1 = \frac{RT_1}{P_1} = \frac{0.287 \times 323}{100} = 0.927 \text{ m}^3/\text{kg}$$

$[P_1 = 1 \text{ bar} = 10^5 \text{ Pa} = 100 \text{ kPa} = 100 \text{ kJ/m}^3]$

$$V_2 = \frac{0.927}{16} = 0.0579 \text{ m}^3/\text{kg}, V_3 = V_2 = 0.0579 \text{ m}^3/\text{kg}, P_3 = P_4 = 70 \text{ bar}$$

Along $3 \to 4$

$$\frac{P_3V_3}{T_3} = \frac{P_4V_4}{T_4}, V_4 = \frac{V_3T_4}{T_3} = \frac{0.0579 \times 1723}{1413} = 0.0706 \text{ m}^3/\text{kg}, P_1 = 1 \text{ bar} = 100 \text{ kPa}$$

$$V_5 = V_1 = \frac{RT_1}{P_1} = 0.927 \text{ m}^3/\text{kg},$$

Along $4 \to 5$

$$\eta = 1 - \frac{(T_1 - T_5)}{(T_3 - T_2) + \gamma(T_4 - T_3)} = 1 - \frac{(615 - 323)}{(1413 - 979) + 1.4(1723 - 1413)} = 0.664$$

$$Q_1 = 2C_V(T_3 - T_2) = 2 \times 0.718(1413 - 979) = 623.2 \text{ kJ/kg}$$

$$W = \eta \times Q_1 = 0.664 \times 623.2 \text{ kJ/kg} = 413.82 \text{ kJ/kg}$$

Mean effective pressure $= P_m$. So, $P_m(V_1 - V_2) = W = 413.82$.

Or, $P_m(0.927 - 0.0579) = 413.82$. Or, $P_m = 476.15 \text{ kPa} = 4.7615 \text{ bar}$.

SOLVED PROBLEMS

Problem 7.1 Air at a temperature of 15°C passes through a heat exchanger at a velocity of 30 m/s where its temperature is raised to 800°C. It then passes through a turbine with the same velocity of 30 m/s and expands until the temperature falls to 650°C. On leaving the turbine, the air is taken at a velocity of 60 m/s to a nozzle where it expands until its temperature has fallen to 500°C. If the air flow rate is 2 kg/s, find (a) rate of heat transfer from the heat exchanger (b) the power output from the turbine (c) velocity at nozzle exit assuming no heat loss. Assume for air $C_P = 1.005 \text{ kJ/kg.}°\text{K}$.

Solution

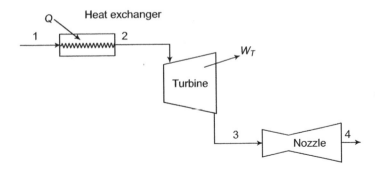

$T_1 = (273 + 15)°K = 288°K,\ T_2 = (273 + 800)°K$
$= 1073°K,\ T_3 = (273 + 650)°K = 923°K$
$T_4 = (273 + 500)°K = 773°K,\ v_1 = 30\ m/s,\ v_2 = 30\ m/s,\ v_3 = 60\ m/s.$
According to equation (7.26a), we have

$$\Delta\left(H + \frac{1}{2}u^2 + Zg\right) = q - w$$

In case of heat exchanger, $w_S = 0$, $u = 0$, and $Z = 0$. Thus, the above equation becomes

$$\dot{q} = \dot{m}\Delta H = \dot{m}(h_2 - h_1) = \dot{m}C_P(T_2 - T_1) = 2 \times 1.005 \times (1073 - 288) = 1580\ kJ/s.$$

In case of turbine, $q = 0$ and $Z = 0$. Thus, we have

$$W_T = \dot{m}\left[(h_2 - h_3) + \frac{1}{2}(v_2^2 - v_3^2)\right] = \dot{m}\left[C_P(T_2 - T_3) + \frac{1}{2}(30^2 - 60^2)\right]$$

Putting the values of $\dot{m}$, C_P, T_2, and T_3, we get

$W_T = 298.8\ kW$

At nozzle $q = 0$, $w = 0$ and $Z = 0$. Thus, we have

$$\Delta H = -\frac{\Delta u^2}{2}.\ \text{Or, } \dot{m}C_P(T_4 - T_3) = \frac{v_3^2}{2} - \frac{v_4^2}{2}$$

Only v_4 is unknown. Hence, $v_4 = 554\ m/s.$

Problem 7.2   The food compartment of a refrigerator is maintained at 4°C by removing heat from it at a rate of 360 kJ/min. If the required power input to the refrigerator is 2 kW, determine (a) the coefficient of performance of the refrigerator, and (b) the rate of heat rejection to the room that houses the refrigerator.

Solution

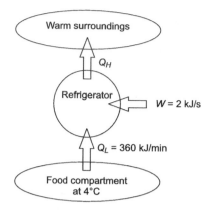

COP of the refrigerator is given by

$$COP = -\frac{\text{Amount of heat removed in kW}}{\text{Work supplied in kW}} = \frac{6\,\text{kW}}{2\,\text{kW}} = 3$$

According to the given figure, we have

$$Q_H = Q_L + W = 8\,\text{kW}.$$

Problem 7.3   A heat engine is used to drive a heat pump. The work (W) is transferred from the heat engine to the heat pump as shown in the figure. The efficiency of the heat engine is 27% and the COP of the heat pump is 4. Determine the ratio of the total heat rejection rate to the heat transfer to the heat engine.

Solution

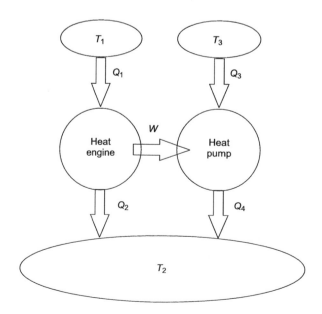

Efficiency of heat engine

$$\eta = \frac{W}{Q_1} = 0.27. \text{ Or, } W = 0.27Q_1. \text{ Again, } \eta = 1 - \frac{Q_2}{Q_1}. \text{ Or, } Q_2 = Q_1(1 - \eta) = 0.73Q_1$$

COP of heat pump

$$COP = \frac{Q_4}{W} = 4. \text{ Or, } W = 0.25Q_4$$

Or, $$\frac{Q_4}{Q_1} = 1.08. \text{ Or, } Q_4 = 1.08Q_1$$

$$Q_2 + Q_4 = (1.08 + 0.73)Q_1 = 1.81Q_1. \text{ Hence, } \frac{Q_2 + Q_4}{Q_1} = 1.81$$

EXERCISES

1. In an air standard Diesel cycle the compression ratio is 16. At the beginning of compression, the temperature and pressure are 15°C and 0.1 MPa respectively. Heat is added at constant pressure, until the temperature reaches to 1480°C. Calculate: (a) the cut-off ratio, (b) the heat supplied per kg of air, (c) the cycle efficiency, and (d) the mean effective pressure.

2. An air standard Diesel cycle absorbs 1500 J/mole of heat at constant pressure. The pressure and temperature at the beginning of the compression step are 1 bar and 20°C. The pressure at the end of the compression step is 4 bar. Assuming air to be an ideal gas for which $C_P = (7/2)R$ and $C_V = (5/2)R$, what are the compression ratio and expansion ratio of the cycle?

3. A refrigerated space is maintained at –12°C and cooling water is available at 21°C. The evaporator and condenser are of sufficient size that –12°C temperature difference can be maintained for heat transfer in each case. The refrigeration capacity is 120,000 kJ/hr.

    (a) What is the value of COP or (ω) for a Carnot refrigerator?

    (b) Calculate ω and mass flow rate ($\dot{m}$) for the vapour compression cycle where expansion occurs iscntropically.

    (c) Calculate ω and $\dot{m}$ for the vapour compression cycle where expansion occurs at constant enthalpy with compressor efficiency 80%.

4. A mass of 8 kg gas expands within a flexible container as per $PV^{1.2}$ = constant. The initial pressure is 1000 kPa and the initial volume is 1 m³. The final pressure is 5 kPa. If the specific internal energy of the gas decreases by 40 kJ/kg, find the heat transfer in magnitude and direction. [*Ans.: 2615 kJ*]

5. Air at 10°C and 80 kPa enters the diffuser of a jet engine steadily with a velocity of 200 m/s. The inlet area of the diffuser is 0.4 m². The air leaves the diffuser with a velocity that

is very small compared with the inlet velocity. Determine: (a) the mass flow rate of the air, and (b) the temperature of the air leaving the diffuser.

[*Ans.:* (a) 78.8 kg/s, (b) 303°K]

6. A refrigerator is maintained at a temperature of 2°C. Each time the door is opened, 420 kJ of heat is introduced inside the refrigerator, without changing the temperature of the refrigerator. The door is opened 20 times a day and the refrigerator operates at 15% of the ideal COP. The cost of work is Rs. 2.50 per kWh. Determine the monthly bill for this refrigerator if the atmosphere is at 30°C. [*Ans.:* Rs. 118.8]

7. An automobile engine consumes fuel at a rate of 28 L/h and delivers 60 kW of power to the wheels. If the fuel has a heating value of 44000 kJ/kg and a density of 0.8 g/cm^3, determine the efficiency of this engine. [*Ans.:* 21.9%]

8 Statistical Thermodynamics

8.1 MOLECULAR PARTITION FUNCTION

Atoms and molecules have several energy levels. Atoms have only electronic energy levels, which can be quantized into several states while molecules have electronic, translational, rotational, and vibrational energy levels. All these energy levels can be quantized into different states.

The probability of a system, having energy level, E_i, is assumed to be proportional to Boltzmann factor, i.e., $e^{-\beta E_i}$, where, $\beta = \dfrac{1}{kT}$ and k is called Boltzmann constant. The total probability of all energy levels is proportional to $\sum e^{-\beta E_i}$, which is known as partition function, q. Mathematically, q can be written as

$$q = \sum e^{-\beta E_i} = \sum e^{-E_i/kT}, \quad \text{where,} \quad \beta = \frac{1}{kT} \tag{8.1}$$

If number of possible degenerate states is $g(E_i)$, the partition function may be written as

$$q = \sum g(E_i) = e^{-\beta E_i} \tag{8.2}$$

Differentiating equation (8.1) with respect to β, we get

$$\frac{\partial q}{\partial \beta} = -\sum E_i e^{-\beta E_i} \tag{8.3}$$

However, the above expression of q is valid for individual partition function for each atom considering only one degenerate state.

Thus, the probability of individual energy level is given by

$$\text{Probability,} \ P(E_i) = \frac{e^{-\beta E_i}}{\sum e^{-\beta E_i}} = \frac{e^{-\beta E_i}}{q} = \frac{e^{-E_i/kT}}{q}$$

Thus, the average energy of the system is given by

$$\bar{\varepsilon} = \frac{\sum E_i\, e^{-\beta E_i}}{\sum e^{-\beta E_i}} = -\frac{1}{q}\left(\frac{\partial q}{\partial \beta}\right) \quad \text{[Using equations (8.1) and (8.3)]}$$

or,
$$\bar{\varepsilon} = -\left(\frac{\partial \ln q}{\partial \beta}\right) \tag{8.4}$$

In case of molecules there are four different types of partition functions, namely, electronic, translational, rotational, and vibrational. Thus, E and q can be written as

$$E = E_{el} + E_{vib} + E_{rot} + E_{trans} \tag{8.5a}$$

$$q = e^{-E/kT} = e^{-(E_{el}+E_{vib}+E_{rot}+E_{trans})/kT} = e^{-E_{vib}/kT} \times e^{-E_{rot}/kT} \times e^{-E_{trans}/kT} \times e^{-E_{el}/kT}$$

Or,
$$q = q_{el} \times q_{vib} \times q_{rot} \times q_{trans} \tag{8.5b}$$

Larger the value of q larger the number of energy states available for a given system. Furthermore,

$$E_{el} > E_{vib} > E_{rot} > E_{trans}$$

As the translational energy state is the lowest energy level, maximum number of energy states are available for translational motion and hence q_{trans} has the maximum value.

The value of q_{el} is nearly unity, while values of q_{vib} and q_{rot} lie in the range of 1 to 100. The value of q_{trans} is generally in the order of 10^{20}.

8.2   TRANSLATIONAL PARTITION FUNCTION

Translational motion has three degrees of freedom, x, y and z. Thus, we have

$$q_{trans} = q_x q_y q_z \tag{8.6a}$$

Again from quantum mechanics, we have

$$E_x = \frac{n_x^2 h^2}{8\,mL^2} \cdot \text{ Hence, } q_x = \int_0^\infty e^{-n_x^2 h^2/8\,mL^2\,kT}\cdot dn_x = \int_0^\infty e^{-an_x^2} \times dn_x = \frac{1}{2}\sqrt{\frac{\pi}{a}}$$

where,
$$a = \frac{h^2}{8\,mL^2\,kT}$$

So, $q_x = \dfrac{1}{2}\sqrt{\dfrac{\pi}{a}} = \dfrac{1}{2}\left(\dfrac{\sqrt{8\,\pi mkT}}{h}\right)\cdot L = \left(\dfrac{\sqrt{2\,\pi mkT}}{h}\right)\times L = \dfrac{L}{\Lambda}$, where, $\Lambda = \dfrac{h}{\sqrt{2\pi mkT}}$

where, Λ is called Broglie thermal wavelength, L = Length of the potential well and V = Volume of the potential box = L^3.

So, $$q_{\text{trans}} = q_x q_y q_z = \left(\frac{L}{\Lambda}\right)\left(\frac{L}{\Lambda}\right)\left(\frac{L}{\Lambda}\right) = \frac{L^3}{\Lambda^3} = \frac{V}{\Lambda^3} \qquad (8.6b)$$

8.2.1 Mean Translational Energy

Along x-direction,

$$q_{\text{trans}} = \frac{X}{\Lambda}, \quad \text{where,} \quad \Lambda = \frac{h}{\sqrt{2\pi mkT}} = \frac{h \times \sqrt{\beta}}{\sqrt{2\pi m}} \quad \text{and} \quad \frac{1}{\Lambda} = \left(\frac{\sqrt{2\pi m}}{h}\right) \times \beta^{-1/2}$$

According to equation (8.4), we can write

$$\varepsilon_{\text{trans}} = -\frac{1}{q_{\text{trans}}}\left(\frac{\partial q_{\text{trans}}}{\partial \beta}\right)_V = -\left(\frac{\Lambda}{X}\right)(X)\left(-\frac{1}{2}\right) \times \left(\frac{\sqrt{2\pi m}}{h}\right)\beta^{-3/2} = (\Lambda)\left(\frac{1}{2\beta}\right)\left(\frac{1}{\Lambda}\right) = \frac{1}{2\beta}$$

Or, $$\varepsilon_{\text{trans}} = \frac{1}{2\beta} = \frac{1}{2}kT \qquad (8.7a)$$

Considering three dimensions, we have

$$\varepsilon_{\text{trans}} = \frac{3}{2}kT \qquad (8.7b)$$

8.3 ROTATIONAL PARTITION FUNCTION

In case of diatomic molecules, rotational energy is given by

$$\varepsilon_{\text{rot}} = \bar{B}hcJ(J+1), \quad \text{where,} \quad \bar{B} = \frac{h}{8\pi^2 Ic} = \text{Rotational constant with unit cm}^{-1}$$

$I = \mu r^2$ = Moment of inertia of the molecule
μ = Reduced mass, r = Bond length
Rotational energy is often expressed in terms of rotational temperature θ_{rot} which is given by

$$\theta_{\text{rot}} = \frac{h^2}{8\pi^2 Ik} = \frac{hc\bar{B}}{k} \qquad (8.8a)$$

Each energy level has $(2J+1)$ number of degenerate states. So, q_{rot} is given by

$$q_{\text{rot}} = \int_0^\infty (2J+1)e^{-hc\bar{B}J(J+1)/kT} \times dJ$$

Putting $x = J(J+1)$ and on integration, we get

$$q_{\text{rot}} = \frac{kT}{hc\bar{B}} \qquad (8.8b)$$

In case of homonuclear diatomic molecule like H_2, rotating the molecule by $180°$ with respect to bond leads to a configuration same as original configuration. This is called symmetry effect. Considering this symmetry effect the equation (8.8b) can be modified as

$$q_{rot} = \frac{kT}{\sigma hc\bar{B}} = \frac{T}{\sigma \cdot \theta_{rot}} \quad \text{[where, } \sigma = \text{Symmetry number]} \tag{8.8c}$$

Symmetry Number   It refers to number of indistinguishable orientations of a molecule. It is represented by σ. Table 8.1 shows σ values of some common compounds.

Table 8.1   Symmetry number of some common compounds

Molecule	σ
H_2O	2
NH_3	3
CH_4	12
C_6H_6	12

8.3.1   Mean Rotational Energy

Case-I   At low temperature, $T < \theta_{rot}$ and q_{rot} is given by

$$q_{rot} = \sum_{0}^{\infty}(2J+1)e^{-hc\bar{B}J(J+1)/kT} = \sum_{0}^{\infty}(2J+1)e^{-\beta hc\bar{B}J(J+1)}$$

Or,
$$q_{rot} = 1 + 3e^{-2\beta hc\bar{B}} + 5e^{-6\beta hc\bar{B}} + 7e^{-12\beta hc\bar{B}} + \cdots$$

$$\varepsilon_{rot} = -\left(\frac{1}{q_{rot}}\right)\left(\frac{\partial q_{rot}}{\partial \beta}\right)_V = \frac{hc\bar{B}(6e^{-2\beta hc\bar{B}} + 30e^{-6\beta hc\bar{B}} + 84e^{-\beta hc\bar{B}} + \cdots)}{(1 + 3e^{-2\beta hc\bar{B}} + 5e^{-6\beta hc\bar{B}} + 7e^{-12\beta hc\bar{B}} + \cdots)} \tag{8.9a}$$

Case-II   At high temperature, $T \gg \theta_{rot}$ and q_{rot} is given by equation (8.8c)

$$q_{rot} = \frac{kT}{\sigma hc\bar{B}} = \frac{1}{\beta \sigma hc\bar{B}}$$

$$\varepsilon_{rot} = -\left(\frac{1}{q_{rot}}\right)\left(\frac{\partial q_{rot}}{\partial \beta}\right)_V = (\beta\sigma hc\bar{B})\left(\frac{1}{\beta^2}\right)\left(\frac{1}{\sigma hc\bar{B}}\right) = \frac{1}{\beta} = kT$$

Or,
$$\varepsilon_{rot} = \frac{1}{\beta} = kT \tag{8.9b}$$

8.4 VIBRATIONAL PARTITION FUNCTION FOR DIATOMIC MOLECULE

The vibrational energy level of a diatomic molecule is given by

$$E_{vib} = \left(n + \frac{1}{2}\right)hv, \quad [\text{where, } n = 0, 1, 2, \ldots\ldots] \qquad (8.10a)$$

Thus, q_{vib} is given by

$$q_{vib} = \sum_{n=0}^{n=\infty} e^{-\left(n+\frac{1}{2}\right)hv/kT} = e^{-hv/2kT}\left[1 + e^{-hv/kT} + e^{-2hv/kT} + e^{-3hv/kT} \ldots\ldots\right]$$

Put, $e^{-hv/kT} = x$

$$q_{vib} = e^{-hv/2kT}\left(1 + x + x^2 + x^3 + \cdots\right) = \frac{e^{-hv/2kT}}{(1-x)} = \frac{e^{-hv/2kT}}{1 - e^{-hv/kT}} = \frac{1}{1 - e^{-hv/kT}}$$

If the zero energy scale is at $\dfrac{hv}{2kT}$, we have,

$$\frac{hv}{2kT} = 0 \quad \text{and hence, } e^{-\frac{hv}{2kT}} = 1$$

Or,

$$q_{vib} = \frac{1}{1 - e^{-hc\bar{v}/kT}} = \frac{1}{1 - e^{-\theta_{vib}/T}} \qquad (8.10b)$$

Vibrational temperature, θ_{vib} is given by

$$\theta_{vib} = \frac{hc\bar{v}}{k}, \quad \text{where, } \bar{v} = \text{Vibrational frequency in cm}^{-1}$$

8.4.1 Mean Vibrational Energy

According to equation (8.10b), we have

$$q_{vib} = \frac{1}{1 - e^{-hc\bar{v}/kT}} = \frac{1}{1 - e^{-\beta hc\bar{v}}} \quad \left[\text{since, } \beta = \frac{1}{kT}\right]$$

$$\varepsilon_{vib} = -\frac{1}{q_{vib}}\left(\frac{\partial q_{vib}}{\partial \beta}\right)_V = (1 - e^{-\beta hc\bar{v}})\left[\frac{(hc\bar{v})(e^{-\beta hc\bar{v}})}{(1 - e^{-\beta hc\bar{v}})^2}\right] = \frac{hc\bar{v}}{e^{\beta hc\bar{v}} - 1}$$

or, $$\varepsilon_{vib} = \frac{hc\bar{v}}{e^{\beta hc\bar{v}} - 1} \tag{8.10c}$$

Zero-point energy is measured at $T = 0°$ K

$$\left(\varepsilon_{vib}\right)_0 = \frac{1}{2}\,(hc\bar{v}) \tag{8.10d}$$

At $T \gg \theta_{vib}$, we have

$$\varepsilon_{vib} = \frac{hc\bar{v}}{e^{\beta hc\bar{v}} - 1} = \frac{hc\bar{v}}{\beta hc\bar{v} + \dfrac{(\beta hc\bar{v})^2}{2!} + \cdots} \approx \frac{hc\bar{v}}{\beta hc\bar{v}} = \frac{1}{\beta} = kT$$

[As $T \gg \theta_{vib}$, higher terms of $(\beta hc\bar{v})$ can be neglected]

or, $$\bar{\varepsilon}_{vib} = \frac{1}{\beta} = kT \tag{8.10e}$$

8.5 ELECTRONIC AND OVERALL PARTITION FUNCTION

$$q_{el} = 1 \tag{8.11}$$

Thus, overall partition function is given by

$$q = q_{el} \times q_{trans} \times q_{rot} \times q_{vib} = 1 \times \left(\frac{V}{\Lambda^3}\right) \times \left(\frac{T}{\sigma \cdot \theta_{rot}}\right) \times \left(\frac{1}{1 - e^{-\theta_{vib}/T}}\right) \tag{8.12}$$

8.6 CANONICAL PARTITION FUNCTION

All the above calculations are based on only one diatomic molecule. If there are N identical indistinguishable diatomic molecules (preferably observed in gas) present in a given system, the total partition function or canonical partition function, Q, is given by

$$Q = \frac{q^N}{N!} \tag{8.13a}$$

The above equation was postulated by the scientist Gibbs. The denominator $N!$ is introduced as the molecules are indistinguishable.

In case of distinguishable molecules (preferably in case of solid) the canonical partition function, Q, is given by

$$Q = q^N \tag{8.13b}$$

8.7 PARTITION FUNCTION AND ENTROPY

According to the definition of internal energy we can write,

$$U = \text{Sum of all possible energies} = \bar{\varepsilon}$$

Using equation (8.4), we can write

$$\bar{\varepsilon} = -\left(\frac{\partial \ln Q}{\partial \beta}\right)_V \tag{8.14a}$$

Hence,

$$U = -\left(\frac{\partial \ln Q}{\partial \beta}\right)_V$$

Considering ground state energy the above equation becomes

$$U - U(0) = -\left(\frac{\partial \ln Q}{\partial \beta}\right)_V \tag{8.14b}$$

Again,

$$Q = Q(\beta, V, N)$$

Hence,

$$d \ln Q = \left(\frac{\partial \ln Q}{\partial \beta}\right)_{V,N} d\beta + \left(\frac{\partial \ln Q}{\partial V}\right)_{\beta,N} dV + \left(\frac{\partial \ln Q}{\partial N}\right)_{\beta,V} dN$$

As N is constant, $dN = 0$. Thus, we have

$$d \ln Q = \left(\frac{\partial \ln Q}{\partial \beta}\right)_{V,N} d\beta + \left(\frac{\partial \ln Q}{\partial V}\right)_{\beta,N} dV$$

Using equation (8.14b), we have

$$d \ln Q = -\left[U - U(0)\right] d\beta + \beta P dV,$$

$$\left[\text{since, } P = kT \left(\frac{\partial \ln Q}{\partial V}\right)_T \text{ according to equation (8.17)}\right]$$

Again,

$$d \{[U - U(0)]\beta\} = \beta dU + [U - U(0)] d\beta.$$

or,

$$[U - U(0)]d\beta = d \{[U - U(0)]\beta\} - \beta dU$$

$$d \ln Q = -d\{[U - U(0)]\beta\} + \beta dU + \beta P dV$$

$$d(\ln Q + [U - U(0)]\beta) = \beta dU + \beta P dV = \beta(dU + P dV) = \beta dQ = \frac{dQ}{kT}$$

[According to 1st law of thermodynamics, $dQ = dU + PdV$]

Again,

$$dQ = TdS \text{ for reversible process.}$$

Hence,
$$d\left(\ln Q + [U - U(0)]\beta\right) = \frac{dQ}{kT} = \frac{dS}{k}$$

or
$$dS = kd(\ln Q + [U - U(0)]\beta) \quad \text{and} \quad S = k(\ln Q + [U - U(0)]\beta) + S^0 \quad (8.15a)$$

As
$$T \to 0° \text{ K}, S^0 \to 0.$$

As entropy difference is measured, $S^0 = 0$.   So, $\Delta S = S - S^0 = S$

So,
$$S = k(\ln Q + [U - U(0)]\beta) = k \ln Q + \frac{[U - U(0)]}{T}$$

or,
$$S = k \ln Q + \frac{[U - U(0)]}{T} \qquad (8.15b)$$

or,
$$TS = kT \ln Q + [U - U(0)] \qquad (8.15c)$$

8.8   HELMHOLTZ ENERGY

From the definition of Helmholtz energy, we can write
$$A = U - TS = -kT \ln Q + U(0) \quad \text{[using equation (8.15b)]}$$

At
$$T = 0° \, k, A = A(0) = U(0)$$

$$A - A(0) = -kT \ln Q \qquad (8.16a)$$

Differentiating equation (8.16a) at constant T, we get
$$dA = -kTd \ln Q \qquad (8.16b)$$

Again from thermodynamics, we know $dA = -PdV - SdT$

or,
$$P = -\left(\frac{\partial A}{\partial V}\right)_T = kT\left(\frac{\partial \ln Q}{\partial V}\right)_T \qquad (8.17)$$

[Using equation (8.16b)

8.9   GIBBS FREE ENERGY

From the definition of Gibbs free energy, we can write $G = H - TS$
Again, from the definition of enthalpy, we have $H = U + PV$
Putting the value of H in the expression of G, we get

$$G = H - TS = (U + PV) - TS = (U - TS) + PV = A + PV, \text{ [since, } A = U - TS]$$

Considering ground state energy the above equation becomes

$$G - G(0) = A - A(0) + PV$$

Putting the value of $[A - A(0)]$ from the equation (8.16a) in the above equation, we get

$$G - G(0) = -kT \ln Q + PV \qquad (8.18a)$$

Combining equations (8.17) and (8.18a) we get

$$G - G(0) = -kT \ln Q + kTV \left(\frac{\partial \ln Q}{\partial V} \right)_T \qquad (8.18b)$$

8.9.1 Ideal Gas

For an ideal gas, molecules are moving independently and $PV = nRT$

Thus, the equation (8.18a) becomes

$$G - G(0) = -kT \ln Q + nRT \qquad (8.18c)$$

Putting the value of Q from the equation (8.13a) in the equation (8.18c), we get

$$G = -kT \ln \frac{q^N}{N!} + nRT = -kT (N \ln q - \ln N!) + nRT$$

Using Stirling's approach $\ln N! = N \ln N - N$, we get

$$G = -kT (N \ln q - N \ln N + N) + nRT$$

At $\qquad\qquad\qquad T = 0°\ \text{K},\ G = G(0) = 0$

Again, $\qquad\qquad\qquad N = nN_A \text{ and } R = N_A \cdot k$

$$\Delta G = G - G(0) = G = -RT(n \ln q - n \ln N + n) + nRT = -nRT \ln \frac{q}{N} \qquad (8.18d)$$

For 1 mole of a gas $q = q_M =$ Molar partition function, $n = 1$ and $N = N_A =$ Avogadro Number

$$\Delta \bar{G} = \bar{G} - \bar{G}(0) = \Delta \bar{G} = -RT \ln \left(\frac{q_M}{N_A} \right) \qquad (8.18e)$$

Again, from thermodynamics, we know that

$$dG = VdP - SdT. \quad \text{Or, } V = \left(\frac{\partial G}{\partial P} \right)_T \qquad (8.19a)$$

Differentiating equation (8.18c) at constant T, we get

$$dG = -kTd \ln Q$$

Hence, $\qquad \left(\frac{\partial G}{\partial P} \right)_T = -kT \left(\frac{\partial \ln Q}{\partial P} \right)_T \qquad (8.19b)$

Combining equations (8.19a) and (8.19b) we get

$$V = -kT \left(\frac{\partial \ln Q}{\partial P} \right)_T \qquad (8.20)$$

8.10   AVERAGE CHEMICAL POTENTIAL

From the definition of chemical potential, we have

$$\mu = \left(\frac{\partial G}{\partial N}\right)_{T,V} \tag{8.21a}$$

Again, we know that $G - G(0) = -kT \ln Q + nRT$ [Ch. equation (8.18c)]
Differentiating equation (8.18c) at constant T, we get

$$dG = -kTd \ln Q$$

Hence, $\mu = \left(\frac{\partial G}{\partial N}\right)_{V,T} = -\frac{1}{\beta}\left(\frac{\partial \ln Q}{\partial N}\right)_{V,T} = -\frac{1}{\beta Q}\left(\frac{\partial Q}{\partial N}\right)_{V,T}$ $\left[\text{Since, } \beta = \frac{1}{kT}\right]$

or, $$\mu = -\frac{1}{b}\left(\frac{\partial \ln Q}{\partial N}\right)_{V,T} \tag{8.21b}$$

8.11   IDEAL GAS LAW

From equation (8.17), we know that

$$P = kT\left(\frac{\partial \ln Q}{\partial V}\right)_T$$

We know

$$Q = \frac{q^N}{N!} \text{ [Equation (8.13a)]} \text{ and } q_{trans} = \frac{V}{\Lambda^3} \text{ [Equation (8.6b)]}$$

Hence, $\ln Q = N \ln q - \ln N! = N \ln q - N \ln N + N$

[Using Stirling's approach, $\ln N! = N \ln N - N$]
For a given mass of gas, $N = $ Constant and $dN = 0$
Thus, $d \ln Q = Nd \ln q.$

Hence, $\left(\frac{\partial \ln Q}{\partial V}\right)_T = N\left(\frac{\partial \ln q}{\partial V}\right)_T = \frac{N}{q}\left(\frac{\partial q}{\partial V}\right)_T$ $\tag{8.22a}$

$q = q_{trans} \times q_{rot} \times q_{vib} \times q_{el} \approx q_{trans}$, since $q_{trans} \gg q_{rot}$, $q_{trqans} \gg q_{vib}$ and $q_{trans} \gg q_{el}$

Thus, $$q = q_{trans} = \frac{V}{\Lambda^3}$$

Differentiating with respect to V at constant T, we get

$$\left(\frac{\partial q}{\partial V}\right)_T = \left(\frac{1}{\Lambda^3}\right)$$

Putting the values of q and $\left(\dfrac{\partial q}{\partial V}\right)_T$ in the equation (8.22a), we get

$$\left(\frac{\partial \ln Q}{\partial V}\right)_T = \frac{N}{V} \tag{8.22b}$$

Hence, $\qquad P = kT\left(\dfrac{\partial \ln Q}{\partial V}\right)_T = kT\left(\dfrac{N}{V}\right) = \dfrac{nRT}{V}, \quad$ since, $N = nN_A$ and $N_A k = R$

or, $\qquad\qquad\qquad PV = nRT \tag{8.22c}$

The equation (8.22c) is the well-known ideal gas equation for n moles.

8.12 INTERNAL ENERGY AND ENTHALPY

According to the equation (8.14b), we know that

$$U - U(0) = -\left(\frac{\partial \ln Q}{\partial \beta}\right)_V \tag{8.14b}$$

where, $U = \overline{\varepsilon}_{trans} + \overline{\varepsilon}_{rot} + \overline{\varepsilon}_{vib} + \overline{\varepsilon}_{el}$

Using equations (8.7b), (8.9b), and (8.10e), we know that

$$\overline{\varepsilon}_{trans} = \frac{3}{2}kT, \quad \overline{\varepsilon}_{rot} = kT \quad \text{and} \quad \overline{\varepsilon}_{vib} = kT, \quad \overline{\varepsilon}_{el} \approx 1$$

$$H = U + PV. \quad \text{Or,} \quad H - H(0) = [U - U(0)] + PV \tag{8.23a}$$

Putting the values of P from equation (8.17) and $[U - U(0)]$ from equation (8.14b), we have

$$H - H(0) = -\left(\frac{\partial \ln Q}{\partial \beta}\right)_V + kTV\left(\frac{\partial \ln Q}{\partial V}\right)_T \tag{8.23b}$$

where, $U(0)$ and $H(0)$ are the internal energy and enthalpy of gases at ground state respectively.

8.12.1 Ideal Monatomic Gas

According to kinetic theory of gases, ideal gas molecules are independent. Considering monatomic gas molecules, we have only translational energy. Thus,

$$U - U(0) = \frac{3}{2}nRT \quad \text{and} \quad PV = nRT$$

or, $\qquad\qquad H - H(0) = \dfrac{3}{2}nRT + nRT = \dfrac{5}{2}nRT$

Thus internal energy and enthalpy for an ideal monatomic gas are given by

$$U - U(0) = \frac{3}{2}nRT \quad \text{and} \quad H - H(0) = \frac{5}{2}nRT \tag{8.23c}$$

8.13   HEAT CAPACITIES

$$C_V = N_A \left(\frac{\partial U}{\partial T} \right)_V, \quad \beta = \frac{1}{kT} \cdot \text{ Or, } \frac{d\beta}{\partial T} = -\frac{1}{kT^2} = -k\beta^2$$

$$C_V = N_A \left(\frac{\partial U}{\partial T} \right)_V = N_A \left(\frac{\partial U}{\partial \beta} \right)_V \left(\frac{d\beta}{dT} \right)_V = -N_A k\beta^2 \left(\frac{\partial U}{\partial \beta} \right)_V$$

or,

$$C_V = -N_A k\beta^2 \left(\frac{\partial U}{\partial \beta} \right)_V$$

$$U = e_{\text{trans}} + e_{\text{rot}} + e_{\text{vib}}$$

$$C_V^{\text{trans}} = -N_A k\beta^2 \left(\frac{\partial e_{\text{trans}}}{\partial \beta} \right)_V = \frac{3}{2} R$$

$$C_V^{\text{rot}} = -N_A k\beta^2 \left(\frac{\partial e_{\text{rot}}}{\partial \beta} \right)_V = R \text{ (for linear molecules)}$$

$$C_V^{\text{rot}} = -N_A k\beta^2 \left(\frac{\partial e_{\text{rot}}}{\partial \beta} \right)_V = \frac{3}{2} R \text{ (for non-linear molecules)}$$

$$C_V^{\text{vib}} = -N_A k\beta^2 \left(\frac{\partial e_{\text{vib}}}{\partial \beta} \right)_V = R$$

$$C_V = C_V^{\text{trans}} + C_V^{\text{rot}} + C_V^{\text{vib}} = \frac{3}{2} R + R + R = \frac{7}{2} R \text{ (for linear molecules)}$$

Heat capacity values can be plotted against temperature as shown in the Fig. 8.1.

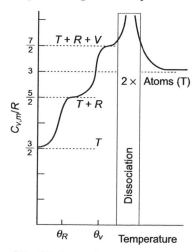

Figure 8.1   Heat capacity versus temperature plot

8.14 PARTITION FUNCTION AND EQUILIBRIUM CONSTANT

From thermodynamics, we have

$$\Delta G^0 = -RT \ln K \tag{8.24}$$

Again from equation (8.18e), we have

$$\bar{G} = \Delta \bar{G} = -RT \ln \frac{q_M}{N_A} \tag{8.18e}$$

The above equation is applicable for a particular component, M. Thus, at standard state the above equation becomes

$$\bar{G}^0 = -RT \ln \frac{q_M^0}{N_A} \tag{8.25a}$$

For n moles, the above equation becomes

$$G^0 = n\bar{G}^0 = -nRT \ln \frac{q_M^0}{N_A} = -RT \ln \left(\frac{q_M^0}{N_A} \right)^n \tag{8.25b}$$

Consider the following equation

$$aA + bB \rightleftharpoons cC + dD$$

For the components A and B, we have

$$G_A^0 = -RT \ln \left(\frac{q_{A,M}^0}{N_A} \right)^a, \quad G_B^0 = -RT \ln \left(\frac{q_{D,M}^0}{N_A} \right)^b$$

Similarly for the components C and D, we have

$$G_C^0 = -RT \ln \left(\frac{q_{A,M}^0}{N_A} \right)^c, \quad G_D^0 = -RT \ln \left(\frac{q_{D,M}^0}{N_A} \right)^d$$

Hence, $\Delta G^0 = (G_C^0 + G_D^0) - (G_A^0 + G_B^0) = -RT \ln \dfrac{(q_{C,M}^{\circ}/N_A)^c \times (q_{D,M}^{\circ}/N_A)^d}{(q_{A,M}^{\circ}/N_A)^a \times (q_{B,M}^{\circ}/N_A)^b}$ (8.26)

The ratio of numbers of product and reactant molecules, (N_p/N_r) is related to the corresponding partition function ratio as follows:

$$\frac{N_p}{N_r} = \left(\frac{q_p}{q_r} \right) e^{-\Delta U_0 / RT} \tag{8.27}$$

Combining equations (8.24), (8.26) and (8.27), we get

$$K = \frac{(q^{\circ}_{C,M}/N_A)^c \times (q^{\circ}_{D,M}/N_A)^d}{(q^{\circ}_{A,M}/N_A)^a \times (q^{\circ}_{B,M}/N_A)^b}$$

(8.28a)

where, ΔU_0 is the difference in molar energies of the ground state of the products and the reactants.

In general the above equation can be written as

$$K = \Pi \left(\frac{q^{\circ}_{C,M}}{N_A} \right)^{v} e^{-\Delta U_0/RT}$$

(8.28b)

where, v = Sum of exponents of products – Sum of exponents of reactants.

8.14.1 Dissociation Equilibrium

Let us consider dissociation of iodine molecule into atoms as given below:

$$I_2 \rightleftharpoons 2I$$

Using equation (8.28a), we get

$$K = \frac{(q^{\circ}_{I,M}/N_A)^2}{q^{\circ}_{I_2,M}/N_A} e^{-\Delta U_0/RT} = \frac{(q^{\circ}_{I,M})^2}{q^{\circ}_{I_2,M} \times N_A} e^{-\Delta U_0/RT}, \quad \text{where, } \Delta U_0 = 2U_0(I) - U_0(I_2)$$

SOLVED PROBLEMS

Problem 8.1 Evaluate the rotational partition function of HCl at 25°C, given that $\bar{B} = 10.591 \, \text{cm}^{-1}$.

Hint: At 25°C, the following are true:

Solution

$$\frac{kT}{hc} = 207.22 \, \text{cm}^{-1}, q_{\text{rot}} = \frac{kT}{hc\bar{B}} = \frac{207.22}{10.591} = 19.6$$

Problem 8.2 The wave numbers of the three normal modes of H_2O are 3656.7 cm^{-1}, 1594.8 cm^{-1} and 3755.8 cm^{-1}. Evaluate the vibrational partition function at 1500° K.

Solution

$$\text{At } 1500° \text{ K}, \frac{kT}{hc} = 1042.6 \, \text{cm}^{-1}, q_{\text{vib}} = \frac{1}{1 - e^{-hc\bar{v}/kT}}$$

Mode	1	2	3
$\bar{v}(cm^{-1})$	3656.7	1594.8	3755.8
$hc\bar{v}/kT$	3.507	1.530	3.602
q_{vib}	1.031	1.276	1.028

Thus overall vibrational partition function is

$$q_{vib} = 1.031 \times 1.276 \times 1.028 = 1.353.$$

Problem 8.3 Calculate the value of $\left[\bar{G} - \bar{G}(0)\right]$ for $H_2O(g)$ at 1500 °K using the following data:

$$\left(\frac{q_M}{N_A}\right)_{trans} = 1.706 \times 10^8, \left(\frac{q_M}{N_A}\right)_{rot} = 486.7 \text{ and } \left(\frac{q_M}{N_A}\right)_{vib} = 1.352$$

Solution

According to equation (8.18d), we can write that

$$\bar{G} - \bar{G}(0) = -RT \ln\left[\left(\frac{q_M}{N_A}\right)_{trans} \times \left(\frac{q_M}{N_A}\right)_{rot} \times \left(\frac{q_M}{N_A}\right)_{vib}\right]$$

Using the given data we have

$$\bar{G} - \bar{G}(0) = -8.314 \times 1500 \ln\left[(1.706 \times 10^8) \times 486.7 \times 1.352\right] kJ/mole$$

or

$$\bar{G} - \bar{G}(0) = -317.3 \text{ kJ/mole}.$$

Problem 8.4 The three characteristic rotational temperatures for NO_2 are 11.5°K, 0.624°K and 0.590°K. The three vibrational temperatures are 1900°K, 1980°K and 2330°K. Calculate rotational and vibrational partition functions at 300°K. Given σ for NO_2 is 2.

Solution

The rotational temperature and rotational partition function are given by

$$\theta_A = \frac{h^2}{8\pi^2 I_A k}, q_{rot} = \left(\frac{\sqrt{\pi}}{\sigma}\right)\left(\sqrt{\frac{T}{\theta_A}}\right)\left(\sqrt{\frac{T}{\theta_B}}\right)\left(\sqrt{\frac{T}{\theta_C}}\right)$$

or,

$$q_{rot} = \left(\frac{1.772}{2}\right)\left(\sqrt{\frac{300}{11.5}}\right)\left(\sqrt{\frac{300}{0.624}}\right)\left(\sqrt{\frac{300}{0.590}}\right) = 2242.4$$

The vibrational partition function is calculated by taking zero point energies as the reference point. Thus, at a particular temperature vibrational partition function is given by

$$q_{vib} = \frac{1}{1 - e^{-\theta_{vib}/T}}$$

Considering three different θ_{vib} values the overall vibrational partition function is given by

$$q_{vib} = \Pi \left(\frac{1}{1 - e^{-\theta_{vib}/T}} \right) = \left(\frac{1}{1 - e^{-1900/300}} \right) \left(\frac{1}{1 - e^{-1980/300}} \right) \left(\frac{1}{1 - e^{-2330/300}} \right) = 1.0035.$$

9 Chemical Kinetics

INTRODUCTION

What is chemical kinetics?

It basically deals with the rate at which reactant molecules are consumed to form products. There are certain conditions to execute this phenomenon. One condition is thermodynamic condition. As we know, for a spontaneous chemical reaction to occur the required thermodynamic condition is $\Delta G < 0$ but this condition is not enough to bring about a chemical reaction unless we have sufficient knowledge on the following:

(a) Rate at which the molecules interact
(b) Factors controlling the reaction rate
(c) Mechanism through which the reactants are converted into products.

On the basis of the above information, reaction conditions, popularly known as kinetic conditions, (e.g. temperature, pressure, catalyst, solvent, etc.) are set. Thus, to occur a chemical reaction the reactants must satisfy the thermodynamic condition as well as kinetic condition. Chemical kinetics or reaction kinetics deals with the study on the rates and mechanisms of chemical reactions.

Classification of chemical reactions

Chemical reactions are broadly classified into two categories:

(a) Homogeneous reaction, where all the reacting species are in same phase.
(b) Heterogeneous reaction, where the reacting species exist in two or more phases.

In another way of classification chemical reactions can be divided into two categories:

(a) One way reaction, which proceeds along one direction only, represented by a single headed arrow ($\rightarrow$), e.g., $aA + bB \longrightarrow lL + mM$.
   It is also called irreversible (non-equilibrium) reaction.
(b) Two way reaction, which proceeds both in the forward and backward directions, represented by a double headed arrow ($\rightleftharpoons$), e.g., $aA + bB \rightleftharpoons lL + mM$.
   It is also called reversible (equilibrium) reaction.

Assumptions

Following are the assumptions made in studying kinetics of chemical reactions:
   (i) Volume of the whole system (V) remains constant throughout the reaction.
   (ii) Only homogeneous gas phase reactions are considered.

9.2 FUNDAMENTAL POSTULATES

Let us consider the following irreversible homogeneous reaction:

$$aA + bB \cdots \cdots \cdots \cdots \cdots \longrightarrow lL + mM \cdots \cdots \cdots \cdots \cdots$$

(a) The rate of consumption per mole is same for each reactant and also equals to the rate of formation per mole of any of the products. If n_A, n_B, ..., etc. are the number of moles of reactants A, B, ..., etc. present at time t, we have

$$-\frac{1}{a}\frac{dn_A}{dt} = -\frac{1}{b}\frac{dn_B}{dt} = \cdots\cdots\cdots\cdots$$

The negative sign indicates that the reactants are consumed. Again, if n_L, n_M, ..., etc. are the number of moles of L, M, ..., etc. present at time t, we have

$$\frac{1}{l}\frac{dn_L}{dt} = \frac{1}{m}\frac{dn_M}{dt} = \cdots\cdots\cdots\cdots$$

Combining the above two equations, we get

$$-\frac{1}{a}\frac{dn_A}{dt} = -\frac{1}{b}\frac{dn_B}{dt} = \cdots\cdots\cdots\cdots = \frac{1}{l}\frac{dn_L}{dt} = \frac{1}{m}\frac{dn_M}{dt} = \cdots\cdots\cdots\cdots$$

Hence, we have

$$\frac{|dn_A|}{a} = \frac{|dn_B|}{b} = \cdots\cdots\cdots\cdots = \frac{|dn_L|}{l} = \frac{|dn_M|}{m} = \cdots\cdots\cdots\cdots = d\varepsilon$$

$d\varepsilon$ is known as **reaction coordinate**. It indicates per cent (%) conversion of reactants to products. However, the above relation does not hold good for a reaction occurring in more than one step. Nevertheless, the above relation holds good for multi-step reactions if the concentrations of all reaction intermediates are small so that their effect on stoichiometry can be neglected.

(b) Average reaction rate

   Let us consider the following reaction:

$$Br_2\,(aq) + HCOOH(aq) \longrightarrow 2Br^-\,(aq) + 2H^+\,(aq) + CO_2$$

Bromine appears as reddish-brown colour while the other species in the above reaction are colourless. Thus, as the reaction proceeds, Br_2 is being consumed and the reddish-brown colour becomes fade. By measuring the concentration of Br_2 at two different time intervals the average rate of the reaction can be computed as below:

$$\text{Average rate} = -\frac{\Delta[Br_2]}{\Delta t} = -\left(\frac{[Br_2]_{final} - [Br_2]_{initial}}{t_{final} - t_{initial}}\right)$$

The following table shows the change in rate and rate constant with bromine concentration.

Table 9.1 Change of rate of reaction between bromine and formic acid at 25°C

Time	$[Br_2]$ M	Rate (m/s)
0	0.0120	4.20×10^{-5}
50	0.0101	3.52×10^{-5}
100	0.00846	2.96×10^{-5}
150	0.00710	2.49×10^{-5}
200	0.00596	2.09×10^{-5}
250	0.00500	1.75×10^{-5}
300	0.00420	1.48×10^{-5}
350	0.00353	1.23×10^{-5}
400	0.00296	1.04×10^{-5}

(c) Let us consider the reaction: $aA + bB \longrightarrow lL + mM$. Rate of the reaction is given by

$$J = \frac{1}{V}\left(-\frac{1}{a}\frac{dn_A}{dt}\right) = kC_A^{n_1}C_B^{n_2} \tag{9.1}$$

Again, $\dfrac{n_A}{V} = C_A$. Hence, $J = -\dfrac{1}{a}\dfrac{dC_A}{dt} = -\dfrac{1}{b}\dfrac{dC_B}{dt} = \dfrac{1}{l}\dfrac{dC_L}{dt} = \dfrac{1}{m}\dfrac{dC_M}{dt}$

J = Velocity or rate of the chemical reaction

k = Velocity constant or rate constant or specific reaction rate. It is defined as the rate of a reaction at unit concentrations of reactants.

C_A, C_B, etc. are the concentrations of reactants left at time t. n_1 is the exponent to C_A and is termed **order of the reaction** with respect to the reactant A. Similarly, n_2 is the **order of the reaction** with respect to the reactant B. n_1, n_2 are also called partial orders. The overall order or simply order of the whole reaction, n, is the sum of the partial orders and is given by $n = n_1 + n_2$.

The equation (9.1) is called the **rate law**.

9.3 MOLECULARITY AND ORDER

Consider the reaction: $aA + bB \longrightarrow lL + mM$.

Rate of the reaction is given by $J = kC_A^{n_1} C_B^{n_2}$

Definition of Molecularity

Total number of molecules involved in a stoichiometrically balanced equation is called molecularity of that reaction. In other words, molecularity of a reaction is the total number of molecules that are taking part in that given reaction. Thus, molecularity of the given reaction is $(a + b)$. As it is a sum of stoichiometric coefficient, molecularity is a theoretical quantity.

Definition of Order

Order of a reaction is the sum of exponents of concentrations of reactants to which the rate of that chemical reaction depends. In the given example order of a reaction is given by $n = n_1 + n_2$. As n_1 and n_2 are determined experimentally, the order of a chemical reaction is an experimental quantity.

Table 9.2 Differences between molecularity and order

Molecularity	Order
It is a theoretical concept.	It is an experimentally determined quantity.
It is always a positive whole number and cannot be zero or fraction.	It may be positive whole number/zero/fractional number/negative.
Molecularity has no limit since it is obtained from stoichiometrically balanced equation, e.g., $KClO_3 + 6FeCl_2 + 6HCl \rightarrow KCl + 6FeCl_3 + 3H_2O$. Molecularity of the above reaction is 13 but order is only 3.	The order of a chemical reaction is usually limited within +3. Only very few reactions, known to us, have order greater than 3.
Rate of a reaction does not depend on molecularity.	Rate of a reaction depends on order.
It provides idea about reaction mechanism, through which product is formed.	It does not throw any light on reaction mechanism.

9.4 IRREVERSIBLE 1ST ORDER REACTIONS

9.4.1 Case-I Only One Reactant with Molecularity 1

Consider the reaction: $A \longrightarrow P + Q + R + \cdots\cdots\cdots$

Time (t)	No. of moles of reactant A	No. of moles of any product (say P)
0	a	0
t	$(a - x)$	x
$t_{1/2}$	$(a/2)$	$a/2$

x = Number of moles of reactant A consumed during time t.

V = Volume of the system = Constant.

C_A = Concentration of reactant A left at time $t = \left(\dfrac{a-x}{V}\right)$ moles/litre.

$C_P = \dfrac{x}{V}$ = Concentration of product P at time t.

Boundary conditions: (i) at $t = 0$, $(a-x) = a$ and $x = 0$.

                         (ii) at $t = t$, $(a-x) = (a-x)$ and $x = x$.

The rate equation is given by

$$J = -\frac{dC_A}{dt} = kC_A. \quad \text{Again,} \quad J = \frac{dC_P}{dt} = \frac{d(x/V)}{dt} = \frac{1}{V}\frac{dx}{dt}$$

Putting the value of C_A in the above equation, we get

$$-\frac{d[(a-x)/V]}{dt} = k\left(\frac{a-x}{V}\right). \quad \text{Or,} \quad -\frac{1}{V}\frac{d(a-x)}{dt} = k\left(\frac{a-x}{V}\right). \quad \text{Or,} \quad -\frac{d(a-x)}{dt} = k(a-x)$$

Method-1   Using the boundary conditions of $(a-x)$, we get

$$-\int_{a}^{(a-x)} \frac{d(a-x)}{(a-x)} = k\int_{0}^{t} dt. \quad \text{Hence,} \quad \ln\frac{a}{(a-x)} = kt.$$

Or, $t = \dfrac{1}{k}\ln\dfrac{a}{(a-x)}$ (9.2)

Equation (9.2) is the rate equation of 1st order of type shown above.

Method-2   Using the boundary conditions of x, we get

$$-\frac{d(a-x)}{dt} = \frac{dx}{dt} = k(a-x). \quad \text{Hence,} \quad \int_{0}^{x}\frac{dx}{(a-x)} = k\int_{0}^{t} dt. \quad \text{Or,} \quad \ln\frac{a}{(a-x)} = kt$$

Or, $t = \dfrac{1}{k}\ln\dfrac{a}{(a-x)}$ (9.2)

9.4.2   Case-II   Only One Reactant with Molecularity Greater than 1

Consider the reaction: $lA \longrightarrow Pp + qQ + rR + \cdots\cdots\cdots$, where, l, p, q, r, etc. are the stoichiometric coefficients.

Time (t)	No. of moles of reactant A	No. of moles of any product (say P)
0	a	0
t	$(a-x)$	$(p/l)x$
$t_{1/2}$	$(a/2)$	$(p/l)(a/2)$

x = No. of moles of reactant A consumed during time t.

V = Volume of the system = Constant.

C_A = Concentration of reactant A left at time $t = \left(\dfrac{a-x}{V}\right)$ moles/litre.

$C_P = \left(\dfrac{p}{l}\right)\left(\dfrac{x}{V}\right)$ = Concentration of product P at time t.

Boundary conditions: (i) at $t = 0$, $(a-x) = a$ and $x = 0$.

(ii) at $t = t$, $(a-x) = (a-x)$ and $x = x$.

The rate equation is given by

$$J = -\frac{1}{l}\frac{dC_A}{dt} = kC_A. \quad \text{Again,} \quad J = \frac{1}{p}\frac{dC_P}{dt} = \frac{1}{p}\cdot\frac{d}{dt}\left(\frac{px}{lV}\right) = \frac{1}{lV}\frac{dx}{dt}$$

Putting the value of C_A in the above equation, we get

$$-\frac{1}{l}\frac{d[(a-x)/V]}{dt} = k\left(\frac{a-x}{V}\right). \quad \text{Or,} \quad -\frac{1}{lV}\frac{d(a-x)}{dt} = k\left(\frac{a-x}{V}\right). \quad \text{Or,} \quad -\frac{d(a-x)}{dt} = kl(a-x)$$

Or, $-\dfrac{d(a-x)}{(a-x)} = k\cdot l\,dt = k_L\,dt$, where, $k_L = k\cdot l$

On integration of the above equation w.r.t. $(a-x)$ using the boundary conditions we get

$$t = \frac{1}{k_L}\ln\frac{a}{(a-x)} \tag{9.3}$$

Again, $\quad J = \dfrac{1}{p}\dfrac{dC_P}{dt} = \dfrac{1}{lV}\dfrac{dx}{dt}$. Again, $J = kC_A = k\left(\dfrac{a-x}{V}\right)$. Or, $\dfrac{1}{lV}\dfrac{dx}{dt} = k\left(\dfrac{a-x}{V}\right)$

Or, $\quad \dfrac{dx}{(a-x)} = k\cdot l\,dt = k_L\,dt$

On integration of the above equation w.r.t. x using the boundary conditions we get,

$$\int_0^x \frac{dx}{(a-x)} = \int_0^t k_L\,dt. \quad \text{Or,} \quad t = \frac{1}{k_L}\ln\frac{a}{(a-x)} \tag{9.3}$$

The equation (9.3) is the rate equation of 1st order reaction of type shown above. The only difference between equations (9.2) and (9.3) is in the value of rate constant, k and k_L, which are related as $k_L = k\cdot l$

9.4.3 Half-life Period

The time taken to consume a reactant to half of its amount is called half-life period of that reactant and it is designated by $t_{1/2}$. Thus, at $t = t_{1/2}$, $x = (a/2)$. Hence, $(a-x) = (a/2)$. Putting the value of $(a-x)$ at $t_{1/2}$ in the equation (9.3) we get,

$$t_{1/2} = \frac{1}{k_L}\ln\frac{a}{a/2} = \frac{2.303\log 2}{k_L} = \frac{0.693}{k_L} \tag{9.4}$$

Unit of specific reaction rate (k_L) From the equation (9.4), we get $k_L = 0.693/t_{1/2}$. The unit of $t_{1/2}$ is time and hence the unit of k_L time^{-1}, e.g., sec^{-1}, min^{-1}, hr^{-1}.

9.4.4 Average Life Period

Consider the total number of molecules present in a moles of reactant A is N_0 and N is the number of molecules present in $(a - x)$ moles of reactant A. Thus, the equation (9.3) becomes

$$\ln\frac{N_0}{N} = k_L . t. \quad \text{Or,} \quad \ln\frac{N}{N_0} = -k_L . t. \quad \text{Or,} \quad N = N_0 e^{-k_L t}$$

Differentiating both sides w.r.t. t, we get

$$\frac{dN}{dt} = -N_0 k_L e^{-k_L t}. \quad \text{Or,} \quad dN = -N_0 k_L e^{-k_L t} . dt \tag{9.5}$$

All N_0 molecules are not taking part in the reaction at the same time and hence all molecules do not have same lifetime. Thus, N_0 molecules are divided into numerous windows, such that first window contains dN_1 number of molecules, each of which has a lifetime t_1, which means that each of dN_1 molecules is converted into products after time t_1.

Thus, the total lifetime of dN_1 molecules is $t_1(-dN_1)$. The negative sign indicates that the reactant molecules are consumed.

Similarly, number of molecules in the 2nd window is dN_2 and lifetime of each molecule is t_2.

Total lifetime of all dN_2 molecules is $t_2(-dN_2)$.

Table 9.3 Lifetime of reactant molecules

Compartment	1	2	3			(i + 1)	
No. of molecules	dN_1	dN_2	dN_3	... ...	... ...	dN_i	... ...
Lifetime of each molecule	t_1	t_2	t_3	... ...	... ...	t_i	... ...
Total lifetime	$t_1(-dN_1)$	$t_2(-dN_2)$	$t_3(-dN_3)$	... ...	... ...	$t_i(-dN_i)$	... ...

So, $N_0 = dN_1 + dN_2 + dN_3 + \cdots\cdots\cdots\cdots + dN_{i+1} + \cdots\cdots$. Or, $N_0 = \displaystyle\sum_{i=0}^{i=\infty} dN_{i+1}$

Total lifetime of all N_0 molecules $= T = t_1(-dN_1) + t_2(-dN_2) + t_3(-dN_3) + \cdots\cdots + t_{(i+1)}$
$$(-dN_{(i+1)}) + \cdots\cdots$$

Or, $\qquad T = \displaystyle\sum_{i=0}^{i=\infty} t_{i+1}(-dN_{i+1})$

As the series is continuous, we can write that $T = -\displaystyle\int_0^{\infty} t . dN$

Putting the value of dN [from equation (9.5)] into the above equation, we get

$$T = \int_0^\infty t \cdot N_0 k_L e^{-k_L \cdot t} \cdot dt$$

Thus, the average life-period per molecule $= \tau = T/N_0$.

$$\tau = \frac{T}{N_0} = \int_0^\infty t \cdot k_L e^{-k_L \cdot t} \cdot dt = k_L \int_0^\infty t \cdot e^{-k_L \cdot t} \cdot dt = k_L \left[\frac{t \cdot e^{-k_L \cdot t}}{-k_L} \right]_0^\infty - k_L \int_0^\infty \frac{e^{-k_L \cdot t}}{-k_L} dt = \int_0^\infty e^{-k_L \cdot t} dt$$

Or, $$\tau = \left[\frac{e^{-k_L \cdot t}}{-k_L} \right]_0^\infty = \frac{1}{k_L} \tag{9.6}$$

9.4.5   Examples of 1st Order Reaction

(i) Dissociation of dimethyl ether: $CH_3OCH_3 \longrightarrow CH_4 + H_2 + CO$

(ii) Dissociation of nitrogen pentoxide: $2N_2O_5 \longrightarrow 4NO_2 + O_2$

(iii) Decomposition of hydrogen peroxide in presence of *Pt* catalyst.

$$H_2O_2 \xrightarrow{\ Pt\ } H_2O + O_2$$

(iv) Conversion of *N*-chloroacetanilide to *P*-chloroacetanilide.

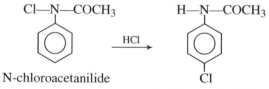

$$\text{N-chloroacetanilide} \qquad\qquad \text{P-chloroacetanilide}$$

9.5   IRREVERSIBLE 2ND ORDER REACTIONS

9.5.1   Case-I   Only One Reactant with Molecularity Greater than 1

Consider the reaction: $lL \longrightarrow Pp + qQ + rR + \cdots \cdots \cdots \cdots$, where, l, p, q, r, etc. are the stoichiometric coefficients. x moles of the reactant L gets reacted producing (px/l) moles of product P, (qx/l) moles of product Q, etc.

Time (t)	No. of moles of reactant L	No. of moles of any product (say P)
0	a	0
t	$(a - x)$	$(p/l)x$
$t_{1/2}$	$(a/2)$	$(p/l)(a/2)$

$x = $ No. of moles of reactant L consumed during time t.

$V = $ Volume of the system $= $ Constant.

$C_L = $ Concentration of reactant L left at time $t = \left(\dfrac{a - x}{V} \right)$ moles/litre.

$$C_P = \left(\frac{p}{l}\right)\left(\frac{x}{V}\right) = \text{Concentration of product } P \text{ at time } t.$$

Boundary conditions: (i) at $t = 0$, $(a-x) = a$ and $x = 0$.

(ii) at $t = t$, $(a-x) = (a-x)$ and $x = x$.

The rate equation is given by

$$J = -\frac{1}{l}\frac{dC_L}{dt} = -\frac{1}{l}\frac{d[(a-x)/V]}{dt} = -\frac{1}{lV}\frac{d(a-x)}{dt} = \frac{1}{l.V}\frac{dx}{dt}$$

Again, $$J = \frac{1}{p}\frac{dC_P}{dt} = \frac{1}{p}\frac{d(px/lV)}{dt} = \frac{1}{l.V}\frac{dx}{dt}$$

The rate equation is given by $J = kC_L^2$

Or, $$-\frac{1}{lV}\frac{d(a-x)}{dt} = k\left(\frac{a-x}{V}\right)^2. \text{ Or, } -\frac{d(a-x)}{(a-x)^2} = \frac{kl}{V}dt = \frac{k_L}{V}dt, \text{ where, } k_L = k \cdot l$$

$$-\int_a^{(a-x)}\frac{d(a-x)}{(a-x)^2} = \frac{k_L}{V}\int_0^t dt$$

On integration, we get

$$\frac{1}{a-x} - \frac{1}{a} = \frac{k_L.t}{V}. \text{ Or, } t = \frac{V}{k_L}\frac{x}{a(a-x)} \tag{9.7a}$$

Equation (9.7a) is the rate equation of 2nd order reaction of type shown above.

Inference

If $l = 1$, i.e., the stoichiometric coefficient of reactant 'L' is unity, $k_L = k$ and equation (9.7a) becomes

$$t = \frac{V}{k}\frac{x}{a(a-x)} \tag{9.7b}$$

Half-life period At $t = t_{1/2}$, $x = (a/2)$. Thus, at half-life period equation (9.7a) becomes

$$t_{1/2} = \frac{V}{k_L}\frac{(a/2)}{a(a/2)} = \frac{V}{k_L \cdot a} \tag{9.8}$$

Unit of specific reaction rate (k_L) From equation (9.8), we have $k_L = \dfrac{V}{a.t_{1/2}}$

The unit of a is moles, V is litre and $t_{1/2}$ is sec. Thus, the unit of k_L is litre/mole–sec.

9.5.2 Case-II Two Reactants with Different Molecularity

Consider the reaction: $lL + mM \longrightarrow Pp + qQ + rR + \text{............}$, where, l, m, p, q, r, etc. are the stoichiometric coefficients. x moles of the reactant L gets reacted with (mx/l) moles of M producing (px/l) moles of product P, (qx/l) moles of Q, etc.

Time (t)	No. of moles of reactant L	No. of moles of reactant M	No. of moles of any product (say P)
0	a	b	0
t	$(a-x)$	$b-(m/l)x$	$(p/l)\cdot x$
$(t_{1/2})_L$	$a/2$	$b-(m/l)(a/2)$	$(p/l)(a/2)=(pa/2l)$
$(t_{1/2})_M$	$[a-(b/2)\,(l/m)]$	$b/2$	$[(b/2)(l/m)\,(p/l)=pb/2m]$

x = Number of moles of reactant L consumed during time t.

V = Volume of the system = Constant.

$C_L = (a-x)/V$ = Concentration of the reactant L at time t.

$C_M = [b-(mx/l)]/V$ = Concentration of the reactant M at time t.

$C_P = (px/lV)]$ = Concentration of the product P at time t.

Boundary conditions: (i) at $t = 0$, $(a-x) = a$ and $x = 0$.

(ii) at $t = t$, $(a-x) = (a-x)$ and $x = x$.

The rate is given by

$$J = -\frac{1}{l}\frac{dC_L}{dt} = -\frac{1}{lV}\frac{d(a-x)}{dt} = \frac{1}{lV}\frac{dx}{dt}$$

$$J = -\frac{1}{m}\frac{dC_M}{dt} = -\frac{1}{mV}\frac{d[b-(m/l)x]}{dt} = \frac{1}{lV}\frac{dx}{dt}$$

$$J = \frac{1}{p}\frac{dC_P}{dt} = \frac{1}{pV}\frac{d(px/l)}{dt} = \frac{1}{lV}\frac{dx}{dt}$$

The rate equation is given by $J = kC_L C_M$.

This equation holds good when order of the reaction with respect to each reactant is unity. So, the rate equation becomes

$$\frac{1}{lV}\frac{dx}{dt} = \left[\frac{k(a-x)}{V}\right]\left[\frac{\{b-(mx/l)\}}{V}\right]. \text{ Or, } \frac{dx}{dt} = \left(\frac{kl}{V}\right)(a-x)\left(\frac{m}{l}\right)\left(\frac{lb}{m}-x\right)$$

Or, $\qquad \dfrac{dx}{dt} = \dfrac{km}{V}(a-x)(b'-x)$, where, $b' = \dfrac{lb}{m}$

Or, $\qquad \dfrac{dx}{dt} = \dfrac{k_M}{V}(a-x)(b'-x)$, where, $k_M = k.m$

$$\int_0^x \frac{dx}{(a-x)(b'-x)} = \frac{k_M}{V}\int_0^t dt. \text{ Or, } \frac{1}{(a-b')}\int_0^x\left[\frac{1}{(b'-x)}-\frac{1}{(a-x)}\right]dx = \frac{k_M}{V}\int_0^t dt$$

On integration, we get

$$\frac{k_M \cdot t}{V}(a-b') = -\ln\frac{(b'-x)}{b'} + \ln\frac{(a-x)}{a} = \ln\frac{b'(a-x)}{a(b'-x)}$$

Or, $$t = \frac{V}{k_M(a-b')} \ln \frac{b'(a-x)}{a(b'-x)}$$ (9.9a)

Equation (9.9a) is the rate equation of 2nd order reaction of type shown above.

Case-II(a) $l = m = 1$. So, $k_M = k$ and $b' = b$. $a \neq b$.
Hence, equation (9.9a) becomes

$$t = \frac{V}{k(a-b)} \ln \frac{b(a-x)}{a(b-x)}$$ (9.9b)

Case-II(b) $a = b, l \neq m$. So, $b' = \left(\frac{lb}{m}\right) = \left(\frac{la}{m}\right) = a' \text{(say)}$.
Thus, equation (9.9a) becomes

$$t = \frac{V}{k_M(a-a')} \ln \frac{a'(a-x)}{a(a'-x)}$$ (9.9c)

9.5.2.1 Half-life Period

Half-life period can be calculated using equation (9.9a). As initial concentrations of the reactants are different, there are two half-lives as described below:

(a) Calculation of $t_{1/2}$ with respect to the reactant L
At $t = (t_{1/2})_L$, $x = (a/2)$. Hence,

$$C_L = \frac{a}{2V}, \quad C_M = \frac{\left(b - \frac{ma}{2l}\right)}{V}$$

$$(t_{1/2})_L = \frac{V}{k_M(a-b')} \ln \frac{b'(a-a/2)}{a(b'-a/2)} = \frac{V}{k_M(a-b')} \ln \frac{b'}{(2b'-a)}$$ (9.10a)

(b) Calculation of $t_{1/2}$ with respect to the reactant M.
At $t = (t_{1/2})_M$, $(m/l)x = (b/2)$. Or, $x = (bl/2m)$. Hence,

$$C_L = \frac{\left(a - \frac{bl}{2m}\right)}{V}, \quad C_M = \frac{b}{2V}$$

$$(t_{1/2})_M = \frac{V}{k_M(a-b')} \ln \frac{b'(a-bl/2m)}{a(b'-bl/2m)}$$ (9.10b)

9.5.3 Examples of 2nd Order Reaction

(i) Formation of hydrogen iodide from its elements. $H_2 + I_2 \longrightarrow 2HI$
(ii) Saponification of an ester by an alkali.
$$CH_3COOC_2H_5 + NaOH \longrightarrow CH_3COONa + C_2H_5OH$$

9.6   IRREVERSIBLE 3RD ORDER REACTIONS

In the case of 3rd order reaction there are several possible cases. Only three of them are discussed here.

9.6.1   Case-I   Only One Reactant with Molecularity Greater than 1

Consider the reaction: $lL \longrightarrow Pp + qQ + rR + \cdots\cdots\cdots$, where, l, p, q, r, etc. are the stoichiometric coefficients. x moles of the reactant L gets reacted producing (px/l) moles of product P, (qx/l) moles of Q, etc. Rate equation is given by:

$$J = -\frac{1}{l}\frac{dC_L}{dt} = kC_L^3$$

Time (t)	No. of moles of reactant L	No. of moles of any product (say P)
0	a	0
t	$(a-x)$	$(p/l)x$
$t_{1/2}$	$(a/2)$	$(p/l)(a/2)$

$x =$ No. of moles of reactant L consumed during time t.

$V =$ Volume of the system $=$ Constant.

$C_L =$ Concentration of reactant L left at time $t = \left(\dfrac{a-x}{V}\right)$ moles/litre.

$C_P = \left(\dfrac{p}{l}\right)\left(\dfrac{x}{V}\right) =$ Concentration of product P at time t.

Boundary conditions: (i) at $t = 0$, $(a - x) = a$ and $x = 0$.

(ii) at $t = t$, $(a - x) = (a - x)$ and $x = x$.

Putting the value of C_L in the rate equation we get

$$-\frac{1}{lV}\frac{d(a-x)}{dt} = k\left(\frac{a-x}{V}\right)^3. \text{ Or, } -\frac{d(a-x)}{(a-x)^3} = \frac{kl}{V^2}dt = \frac{k_L}{V^2}dt, \text{ where, } k_L = k \cdot l$$

$$-\int_a^{(a-x)}\frac{d(a-x)}{(a-x)^3} = \frac{k_L}{V^2}\int_0^t dt$$

On integration, we get

$$\frac{1}{2}\left[\frac{1}{(a-x)^2} - \frac{1}{a^2}\right] = \frac{k_L \cdot t}{V^2}. \text{ Or, } t = \left(\frac{V^2}{2k_L}\right)\frac{x(2a-x)}{a^2(a-x)^2} \tag{9.11a}$$

Half-life period   Putting $x = \dfrac{a}{2}$ in the equation (9.11a), we get

$$t_{1/2} = \left(\frac{V^2}{2k_L}\right)\frac{3}{a^2} \tag{9.11b}$$

Unit of specific reaction rate From equation (9.11b), we have

$$k_L = \frac{3V^2}{2a^2 t_{1/2}}$$

Unit of a is moles, unit of V is litre and that of $t_{1/2}$ is sec. Thus, the unit of k_L is litre²/(mole²-sec).

9.6.2 Case-II Two Reactants with Different Partial Order

Consider the reaction: $L + M \longrightarrow P + Q + R + \cdots \cdots \cdots \cdots$ Rate equation may be written as

either $\quad J = -\dfrac{dC_L}{dt} = kC_L^2 C_M \;$ or $\; J = -\dfrac{dC_L}{dt} = kC_L C_M^2$

Time (t)	No. of moles of reactant L	No. of moles of reactant M	No. of moles of any product (say P)
0	a	b	0
t	$(a-x)$	$b-x$	x
$(t_{1/2})_L$	$a/2$	$b-(a/2)$	$a/2$
$(t_{1/2})_M$	$[a-(b/2)]$	$b/2$	$b/2$

x = No. of moles of reactant L consumed during time t.

V = Volume of the system = Constant.

C_L = Concentration of reactant L left at time $t = \left(\dfrac{a-x}{V}\right)$ moles/litre.

C_M = Concentration of reactant M left at time $t = \left(\dfrac{b-x}{V}\right)$ moles/litre.

$C_P = \left(\dfrac{x}{V}\right)$ = Concentration of product P at time t.

Boundary conditions: (i) at $t = 0$, $(a-x) = a$, $(b-x) = b$ and $x = 0$.

(ii) at $t = t$, $(a-x) = (a-x)$, $(b-x) = (b-x)$ and $x = x$.

Case-II(a) Consider the rate equation $J = -\dfrac{dC_L}{dt} = kC_L^2 C_M$

Putting the values of C_L and C_M in the rate equation $J = -\dfrac{dC_L}{dt} = kC_L^2 C_M$ we get

$$-\frac{1}{V}\frac{d(a-x)}{dt} = k\left(\frac{a-x}{V}\right)^2 \left(\frac{b-x}{V}\right). \text{ Or, } \frac{dx}{(a-x)^2(b-x)} = \left(\frac{k}{V^2}\right)dt$$

$$\int_a^x \frac{dx}{(a-x)^2(b-x)} = \left(\frac{k}{V^2}\right)\int_0^t dt. \text{ Or, } \frac{1}{(b-a)}\int_0^x \left[\frac{1}{(a-x)^2} - \frac{1}{(a-x)(b-x)}\right] dx = \left(\frac{k}{V^2}\right)\int_0^t dt$$

Or, $$\frac{k(b-a)t}{V^2} = \int_0^x \frac{dx}{(a-x)^2} - \int_0^x \frac{dx}{(a-x)(b-x)}$$

Or, $$\frac{k(b-a)t}{V^2} = -\int_a^{(a-x)} \frac{d(a-x)}{(a-x)^2} - \frac{1}{(b-a)}\left[\int_0^x \frac{dx}{(a-x)} - \int_0^x \frac{dx}{(b-x)}\right]$$

Or, $$\frac{k(b-a)t}{V} = \frac{x}{a(a-x)} + \frac{1}{(b-a)}\ln\frac{(a-x)}{(b-x)}$$

$$t = \frac{V}{k(b-a)}\left[\frac{x}{a(a-x)} + \frac{1}{(b-a)}\ln\frac{(a-x)}{(b-x)}\right] \qquad (9.12a)$$

9.6.2.1 Half-life Period for Case-II(a)

(a) Calculation of $t_{1/2}$ with respect to the reactant L.

At $t = (t_{1/2})_L$, $x = a/2$

$$(t_{1/2})_L = \frac{V}{k(b-a)}\left[\frac{1}{a} + \frac{1}{(b-a)}\ln\left(\frac{a}{2b-a}\right)\right] \qquad (9.12b)$$

(b) Calculation of $t_{1/2}$ with respect to the reactant M.

With respect to the reactant M, at $t = (t_{1/2})_M$, $x = b/2$

$$(t_{1/2})_M = \frac{V}{k(b-a)}\left[\frac{b}{a(2a-b)} + \frac{1}{(b-a)}\ln\left(\frac{2a-b}{b}\right)\right] \qquad (9.12c)$$

Case-II(b) Consider the rate equation $J = -\dfrac{dC_L}{dt} = kC_L C_M^2$

Putting the value of C_L and C_M in the rate equation $J = -\dfrac{dC_L}{dt} = kC_L C_M^2$, we get

$$-\frac{1}{V}\frac{d(a-x)}{dt} = k\left(\frac{a-x}{V}\right)\left(\frac{b-x}{V}\right)^2. \text{ Or, } \frac{dx}{(a-x)(b-x)^2} = \left(\frac{k}{V^2}\right)dt$$

$$\int_a^x \frac{dx}{(a-x)(b-x)^2} = \left(\frac{k}{V^2}\right)\int_0^t dt. \text{ Or, } \frac{1}{(a-b)}\int_0^x \left[\frac{1}{(b-x)^2} - \frac{1}{(a-x)(b-x)}\right] dx = \left(\frac{k}{V^2}\right)\int_0^t dt$$

Or, $$\frac{k(a-b)t}{V^2} = \int_0^x \frac{dx}{(b-x)^2} - \int_0^x \frac{dx}{(a-x)(b-x)}$$

Or, $\dfrac{k(a-b)t}{V^2} = -\displaystyle\int_{b}^{b-x}\dfrac{d(b-x)}{(b-x)^2} - \dfrac{1}{(b-a)}\left[\displaystyle\int_{0}^{x}\dfrac{dx}{(a-x)} - \displaystyle\int_{0}^{x}\dfrac{dx}{(b-x)}\right]$

Or, $\dfrac{k(a-b)t}{V} = \dfrac{x}{b(b-x)} + \dfrac{1}{(b-a)}\ln\dfrac{(a-x)}{(b-x)}$

Or, $t = \dfrac{V}{k(a-b)}\left[\dfrac{x}{b(b-x)} + \dfrac{1}{(b-a)}\ln\dfrac{(a-x)}{(b-x)}\right]$ 　　　　(9.13a)

9.6.2.2 Half-life Period for Case-II(b)

(a) Calculation of $t_{1/2}$ with respect to the reactant L.
At $t = (t_{1/2})_L$, $x = a/2$

$$(t_{1/2})_L = \dfrac{V}{k(a-b)}\left[\dfrac{a}{b(2b-a)} + \dfrac{1}{(b-a)}\ln\left(\dfrac{a}{2b-a}\right)\right] \qquad (9.13b)$$

(b) Calculation of $t_{1/2}$ with respect to the reactant M.
At $t = (t_{1/2})_M$, $x = b/2$

$$(t_{1/2})_M = \dfrac{V}{k(b-a)}\left[\dfrac{1}{b} + \dfrac{1}{(b-a)}\ln\left(\dfrac{2a-b}{b}\right)\right] \qquad (9.13c)$$

9.6.3　Case-III　Three Reactants with Partial Order 1 Each

Consider the reaction: $L + M + N \longrightarrow P + Q + R + \cdots\cdots\cdots$ Rate equation may be given by

$$J = -\dfrac{dC_L}{dt} = kC_L C_M C_N$$

Time (t)	No. of moles of reactant L	No. of moles of reactant M	No. of moles of reactant N	No. of moles of any product (say P)
0	a	b	c	0
t	$(a-x)$	$b-x$	$c-x$	x
$(t_{1/2})_L$	$a/2$	$b-(a/2)$	$c-(a/2)$	$a/2$
$(t_{1/2})_M$	$[a-(b/2)]$	$b/2$	$c-(b/2)$	$b/2$
$(t_{1/2})_N$	$[a-(c/2)]$	$b-(c/2)$	$c/2$	$c/2$

V = Volume of the system = Constant.

C_L = Concentration of reactant L left at time $t = \left(\dfrac{a-x}{V}\right)$ moles/litre.

C_M = Concentration of reactant M left at time $t = \left(\dfrac{b-x}{V}\right)$ moles/litre.

C_N = Concentration of reactant N left at time $t = \left(\dfrac{c-x}{V}\right)$ moles/litre.

$C_P = \left(\dfrac{x}{V}\right)$ = Concentration of product P at time t.

Boundary conditions: (i) at $t = 0$, $(a-x) = a$, $(b-x) = b$, $(c-x) = c$ and $x = 0$

(ii) at $t = t$, $(a-x) = (a-x)$, $(b-x) = (b-x)$, $(c-x) = (c-x)$ and $x = x$

Putting the values of C_L: C_M and C_N in the rate equation $J = -\dfrac{dC_L}{dt} = kC_L C_M C_N$ we get

$$-\frac{1}{V}\frac{d(a-x)}{dt} = k\left(\frac{a-x}{V}\right)\left(\frac{b-x}{V}\right)\left(\frac{c-x}{V}\right). \text{ Or, } \frac{dx}{(a-x)(b-x)(c-x)} = \left(\frac{k}{V^2}\right)dt$$

$$\int_a^x \frac{dx}{(a-x)(b-x)(c-x)} = \left(\frac{k}{V^2}\right)\int_0^t dt.$$

Or, $\qquad \dfrac{1}{(a-x)(b-x)(c-x)} = \dfrac{A}{(a-x)} + \dfrac{Bx+D}{(b-x)(c-x)}$

$A(b-x)(c-x) + Bx(a-x) + D(a-x) = 1$

$Abc + aD = 1, \quad -A(b+c) + aB - D = 0, \quad A - B = 0$

Solving we get,

$$A = B = \frac{1}{(c-a)(b-a)}, \quad D = -\frac{(b+c-a)}{(c-a)(b-a)} \text{ and } Bx+D = \frac{x-(b+c-a)}{(c-a)(b-a)}$$

$$\frac{Bx+D}{(b-x)(c-x)} = \frac{E}{(b-x)} + \frac{F}{(c-x)}. \text{ Or, } Bx+D = Ec - Ex + Fb - Fx$$

Or, $\qquad Bx+D = \dfrac{x}{(c-a)(b-a)} - \dfrac{(b+c-a)}{(c-a)(b-a)} = Ec - Ex + Fb - Fx$

Hence, $\qquad E + F = -\dfrac{1}{(c-a)(b-a)}$ and $Ec + Fb = -\dfrac{(b+c-a)}{(c-a)(b-a)}$

Solving we get,

$$E = -\frac{1}{(c-b)(b-a)} \text{ and } F = \frac{1}{(c-b)(c-a)}$$

Thus,

$$A = -\frac{1}{(c-a)(a-b)}, \quad E = -\frac{1}{(b-c)(a-b)} \text{ and } F = -\frac{1}{(b-c)(c-a)}$$

Thus, the above integration becomes

$$\int_{a}^{x} \frac{dx}{(a-x)(b-x)(c-x)} = \int_{0}^{x} \frac{A\,dx}{(a-x)} + \int_{0}^{x} \frac{E}{(b-x)} + \int_{0}^{x} \frac{F\,dx}{(c-x)}$$

$$= -\frac{1}{(c-a)(a-b)} \times \int_{0}^{x} \frac{dx}{(a-x)} - \frac{1}{(b-c)(a-b)} \times \int_{0}^{x} \frac{dx}{(b-x)} - \frac{1}{(b-c)(c-a)} \times \int_{0}^{x} \frac{dx}{(c-x)}$$

$$= -\frac{1}{(a-b)(b-c)(c-a)} \left[\int_{0}^{x} \frac{(b-c)dx}{(a-x)} + \int_{0}^{x} \frac{(c-a)dx}{(b-x)} + \int_{0}^{x} \frac{(a-b)dx}{(c-x)} \right]$$

$$= \frac{1}{(a-b)(b-c)(c-a)} \left[(b-c)\ln\left(\frac{a}{a-x}\right) + (c-a)\ln\left(\frac{b}{b-x}\right) + (a-b)\ln\left(\frac{c}{c-x}\right) \right]$$

Thus, the rate equation becomes

$$t = \frac{V^2}{k(a-b)(b-c)(c-a)} \left[(b-c)\ln\left(\frac{a}{a-x}\right) + (c-a)\ln\left(\frac{b}{b-x}\right) + (a-b)\ln\left(\frac{c}{c-x}\right) \right] \quad (9.14)$$

9.6.3.1 Half-life Period for Case-III

(a) Calculation of $t_{1/2}$ with respect to the reactant L.

At $t = (t_{1/2})_L$, $x = a/2$

$$(t_{1/2})_L = \frac{V^2}{k(a-b)(b-c)(c-a)} \left[(b-c)\ln 2 + (c-a)\ln\left(\frac{2b}{2b-a}\right) + (a-b)\ln\left(\frac{2c}{2c-a}\right) \right]$$

$$(9.15a)$$

(b) Calculation of $t_{1/2}$ with respect to the reactant M.

At $t = (t_{1/2})_M$, $x = b/2$

$$(t_{1/2})_M = \frac{V^2}{k(a-b)(b-c)(c-a)} \left[(b-c)\ln\left(\frac{2a}{2a-b}\right) + (c-a)\ln 2 + (a-b)\ln\left(\frac{2c}{2c-b}\right) \right]$$

$$(9.15b)$$

(c) Calculation of $t_{1/2}$ with respect to the reactant N.

At $t = (t_{1/2})_N$, $x = c/2$

$$(t_{1/2})_N = \frac{V^2}{k(a-b)(b-c)(c-a)} \left[(b-c)\ln\left(\frac{2a}{2a-c}\right) + (c-a)\ln\left(\frac{2b}{2b-c}\right) + (a-b)\ln 2 \right]$$

$$(9.15c)$$

9.6.4 Examples of 3rd Order Reaction

(i) Oxidation of stannous chloride by ferric chloride.

$$2FeCl_3 + SnCl_2 \longrightarrow SnCl_4 + 2FeCl_2$$

(ii) Reactions between nitric oxide with one of the elements, chlorine, bromine, oxygen or hydrogen, e.g.,

$$NO + NO \longrightarrow (NO)_2, \ (NO)_2 + O_2 \longrightarrow 2NO_2$$

(iii) Reaction between sulphur dioxide and oxygen in presence of nitric oxide.

$$2SO_2 + O_2 \xrightarrow{\ NO\ } 2SO_3$$

9.7 IRREVERSIBLE ZERO ORDER REACTION

Consider the reaction: $lL \longrightarrow Pp + qQ + rR + \cdots\cdots$, where, l, p, q, r, etc. are the stoichiometric coefficients.

Time (t)	*No. of moles of reactant L*	*No. of moles of any product (say P)*
0	a	0
t	$(a-x)$	$(p/l)x$
$t_{1/2}$	$(a/2)$	$(p/l)\,(a/2)$

x = Number of moles of reactant L consumed during time interval t.

V = Volume of the system = Constant.

C_L = Concentration of reactant L after the time $t = (a - x)/V$.

Boundary conditions: (i) at $t = 0$, $(a - x) = a$ and $x = 0$.

(ii) at $t = t$, $(a - x) = (a - x)$ and $x = x$.

The rate equation is given by

$$J = -\frac{1}{l}\frac{dC_L}{dt} = kC_L^0 = k. \ \ \text{Or,} \ \ -\frac{1}{l}\frac{d(a-x)/V}{dt} = \frac{1}{lV}\frac{dx}{dt} = k. \ \ \text{Or,} \ \ dx = klVdt$$

$$\int_0^x dx = klV \int_0^t dt = k_L V \int_0^t dt, \ \ \text{where,} \ \ k_L = lk$$

On integration, we get

$$x = k_L Vt. \ \ \text{Or,} \ \ t = \frac{x}{k_L V} \tag{9.16}$$

The equation (9.16) is the rate equation for zero order reaction.

9.7.1 Half-life Period

At $t = t_{1/2}$, $x = (a/2)$. Hence, equation (9.16) becomes

$$t_{1/2} = \frac{a}{2k_L V} \tag{9.17}$$

9.7.2 Unit of Specific Reaction Rate

From equation (9.17), we have

$$k_L = \frac{a}{2V t_{1/2}}$$

Unit of a is moles, unit of V is litre and that of $t_{1/2}$ is sec. Thus, the unit of k_L is moles/(litre-sec).

9.7.3 Examples of Zero Order Reaction

(i) Thermal decomposition of hydrogen iodide on gold surface.

(ii) Decomposition of ammonia over molybdenum surface.

(iii) Iodination of acetone in acidic medium.

$$CH_3COCH_3 + I_2 \xrightarrow{\text{H}^+} CH_3COCH_2I + HI.$$ The reaction is zero order with respect to iodine.

9.8 EXAMPLES OF NEGATIVE ORDER, FRACTIONAL ORDER REACTIONS AND CATALYST

9.8.1 Examples of Negative Order Reaction

(i) Oxidation of iodide with hypochlorite in presence of an alkali.

$$OCl^- + I^- \xrightarrow{\text{OH}^-} OI^- + Cl^-.$$ The rate is given by:

$$J = k \frac{[I^-][OCl^-]}{[OH^-]}$$

Thus, the order of the reaction with respect to OH^- is -1. Increase in concentration of OH^-, results in decrease in rate and hence OH^- is called a retarder to the above reaction.

(ii) Oxidation of mercurous ion by the thallium ion.

$$Hg_2^{2+} + Tl^{3+} \longrightarrow 2Hg^{2+} + Tl^+.$$ The rate is given by:

$$J = k \frac{[Hg_2^{2+}][Tl^{3+}]}{[Hg^{2+}]}$$

The presence of mercuric ion decreases the rate and hence it acts as a retarder. Order of the reaction with respect to Hg^{2+} is -1.

Inference Any species for which the partial order is negative is called retarder.

9.8.2 Example of Catalytic Reaction

Oxidation of sulphur dioxide in presence of nitric oxide.

$$2SO_2(g) + O_2(g) \xrightarrow{\text{NO}} 2SO_3(g). \text{ The rate is given by } J = k[O_2][NO]^2$$

The presence of NO speeds up the reaction and hence NO is called *catalyst*.

9.8.3 Example of Reaction Having Non-integer Order

(i) Dissociation of acetaldehyde.

$$CH_3CHO \longrightarrow CH_4 + CO. \text{ The rate is given by:}$$

$J = k[CH_3CHO]^{3/2}$. Order of the reaction is $\dfrac{3}{2}$.

(ii) Formation of phosgene.

$$CO + Cl_2 \longrightarrow COCl_2. \text{ The rate is given by:}$$

$J = k[CO][Cl_2]^{1/2}$. Partial order with respect to Cl_2 is 1/2 and the total order is $\left(1 + \dfrac{1}{2}\right) = \dfrac{3}{2}$.

9.9 IRREVERSIBLE n^{th} ORDER REACTION

9.9.1 Case-I Only One Reactant with Molecularity Greater than 1

Consider the reaction: $lL \longrightarrow Pp + qQ + rR + \cdots\cdots\cdots$, where, l, p, q, r, etc. are the stoichiometric coefficients.

Time (t)	No. of moles of reactant L	No. of moles of any product (say P)
0	a	0
t	$(a - x)$	$(p/l)x$
$t_{1/2}$	$(a/2)$	$(p/l)(a/2)$

$x =$ Number of moles of reactant L consumed during time interval t.

$V =$ Volume of the system = Constant.

$C_L =$ Concentration of reactant L after the time $= (a - x)/V$.

Boundary conditions: (i) at $t = 0$, $(a - x) = a$ and $x = 0$.

(ii) at $t = t$, $(a - x) = (a - x)$ and $x = x$.

The rate equation is given by:

$$J = -\frac{1}{l}\frac{dC_L}{dt} = kC_L^n. \text{ Or, } -\frac{1}{l}\frac{d(a-x)/V}{dt} = k\left(\frac{a-x}{V}\right)^n \text{ Or, } -\frac{1}{lV}\frac{d(a-x)}{dt} = k\left(\frac{a-x}{V}\right)^n$$

Or, $\qquad -\dfrac{d(a-x)}{(a-x)^n} = \dfrac{kl}{V^{n-1}}\,dt. \text{ Or, } -\int_a^{a-x}\dfrac{d(a-x)}{(a-x)^n} = \dfrac{k_L}{V^{n-1}}\int_0^t dt$

On integration, we get

$$\frac{1}{(n-1)}\left[\frac{1}{(a-x)^{n-1}} - \frac{1}{a^{n-1}}\right] = \frac{k_L t}{V^{n-1}}. \quad t = \frac{V^{n-1}}{k_L(n-1)}\left[\frac{1}{(a-x)^{n-1}} - \frac{1}{a^{n-1}}\right] \quad (9.18)$$

The equation (9.18) is the rate equation for n^{th} order reaction of type shown above.

9.9.1.1   Half-life Period

At $t = t_{1/2}$, $x = (a/2)$. Thus, equation (9.18) becomes

$$t_{1/2} = \frac{V^{n-1}}{k_L(n-1)}\left[\frac{1}{(a-a/2)^{n-1}} - \frac{1}{a^{n-1}}\right] = \frac{v^{n-1}}{k_L(n-1)}\left[\frac{2^{n-1}-1}{a^{n-1}}\right] \quad (9.19)$$

9.9.1.2   Unit of Specific Reaction Rate (k_L)

From equation (9.19) we have

$$k_L = \frac{V^{n-1}}{t_{1/2}(n-1)}\left[\frac{2^{n-1}-1}{a^{n-1}}\right]$$

The unit of a is mole, V is litre, and $t_{1/2}$ is sec. Thus, the unit of k_L is litre^{n-1}/(mol^{n-1}-sec).

9.9.2   Case-II Two Reactants with Different Stoichiometric Coefficients

Consider the reaction: $lL + mM \longrightarrow Pp + qQ + rR + \cdots$, where, l, p, q, r, etc. are the stoichiometric coefficients.

Time (t)	No. of moles of reactant L	No. of moles of reactant M	No. of moles of any product (say P)
0	a	b	0
t	$(a-x)$	$b-(m/l)x$	$(p/l)\cdot x$
$(t_{1/2})_L$	$a/2$	$b-(m/l)(a/2)$	$(p/l)(a/2) = (pa/2l)$
$(t_{1/2})_M$	$[a-(b/2)(l/m)]$	$b/2$	$[(b/2)(l/m)](p/l) = pb/2m$

x = Number of moles of reactant L consumed during time interval t.

V = Volume of the system = Constant.

C_L = Concentration of reactant L after the time $t = (a-x)/V$.

$C_M = [b - (mx/l)]/V$ = Concentration of the reactant M at time t.

$C_p = (px/lV)$ = Concentration of the product P at time t.

Boundary conditions: (i) at $t = 0$, $(a - x) = a$ and $x = 0$.

(ii) at $t = t$, $(a - x) = (a - x)$ and $x = x$.

The rate is given by

$$J = -\frac{1}{l}\frac{dC_L}{dt} = -\frac{1}{lV}\frac{d(a - x)}{dt} = \frac{1}{l \cdot V}\frac{dx}{dt}$$

$$J = -\frac{1}{m}\frac{dC_M}{dt} = -\frac{1}{mV}\frac{d[b - (m/l)x]}{dt} = \frac{1}{l \cdot V}\frac{dx}{dt}$$

$$J = \frac{1}{p}\frac{dC_P}{dt} = \frac{1}{pV}\frac{d(px/l)}{dt} = \frac{1}{l \cdot V}\frac{dx}{dt}$$

The rate equation is given by $J = kC_L^{n_1} C_M^{n_2}$, where, n_1 and n_2 are partial orders and $n = n_1 + n_2$ is the total order of the reaction shown above. Putting the values of J, C_L, and C_M in the rate equation we have

$$\frac{1}{l \cdot V}\frac{dx}{dt} = k\left[\frac{a - x}{V}\right]^{n_1}\left[\frac{b - (m/l)x}{V}\right]^{n_2} = \frac{k}{V^n}[a - x]^{n_1}[b - (m/l)x]^{n_2}$$

$$\frac{dx}{[a - x]^{n_1}[b - (m/l)x]^{n_2}} = \frac{kl}{V^{n-1}}dt = \frac{k_L}{V^{n-1}}dt \qquad (9.20)$$

where, $k_L = kl$

According to stoichiometric equation we have, a minimum l moles of L and m moles of M are required to bring about the chemical reaction, where l and m are positive integers. Reactant L, having number of moles less than l, cannot undergo reaction. Similar explanation is true for the reactant M also. Thus, the reaction takes place between

l moles of L and m moles of M or

$2l$ moles of L and $2m$ moles of M or

$3l$ moles of L and $3m$ moles of M

....etc.

Hence, x = Number of moles of reactant L consumed = sl, where, s = 1, 2, 3, ..., etc. Thus,

$$(a - x) = (a - sl) \quad \text{and} \quad \left(b - \frac{mx}{l}\right) = \left(b - \frac{m}{l} \times sl\right) = (b - sm)$$

a and b are the number of moles of reactants L and M taken to bring about the chemical reaction. To make the equation (9.20) simpler, a and b are so chosen that,

$$\frac{\text{Number of moles of reactant } L}{\text{Number of moles of reactant } M} = \frac{a}{b} = \frac{l}{m} = w. \text{ Or, } a = wl, \ b = wm$$

Thus, $(a - x) = (a - sl) = (wl - sl) = l(w - s)$

$$\left(b - \frac{mx}{l} \right) = (b - sm) = (wm - sm) = m(w - s)$$

Putting these values of $(a - x)$ and $\left(b - \frac{mx}{l} \right)$ in the equation (9.20) we get,

$$\frac{dx}{[l(w - s)]^{n_1} [m(w - s)]^{n_2}} = \frac{k_L}{V^{n-1}} dt. \text{ Or, } \frac{dx}{(w - s)^{n_1 + n_2}} = \frac{k_L}{V^{n-1}} l^{n_1} m^{n_2} dt$$

Again, $x = sl$. Or, $dx = lds$. When (i) $x = 0$, $s = 0$, since $l \neq 0$ and when (ii) $x = x$, $s = s$.

$$\frac{lds}{(w - s)^n} = \frac{k_L}{V^{n-1}} l^{n_1} m^{n_2} dt \quad [\text{since, } n_1 + n_2 = n].$$

Or, $$\frac{ds}{(w - s)^n} = \frac{k_L}{V^{n-1}} l^{n_1 - 1} m^{n_2} dt$$

$$\int_0^s \frac{ds}{(w - s)^n} = \frac{k_{LM}}{V^{n-1}} \int_0^t dt, \text{ where, } k_{LM} = k_L l^{n_1 - 1} m^{n_2}$$

$$\frac{k_{LM}}{V^{n-1}} t = - \int_w^{(w-s)} \frac{d(w - s)}{(w - s)^n} = \frac{1}{(n - 1)} \left[\frac{1}{(w - s)^{n-1}} - \frac{1}{w^{n-1}} \right]$$

Or, $$\left(\frac{k_{LM}}{V^{n-1}} \right) t = \frac{1}{(n - 1)} \left[\frac{1}{[(a - x)/l]^{n-1}} - \frac{1}{(a/l)^{n-1}} \right]$$

Or, $$\frac{k_L l^{n_1 - 1} m^{n_2}}{V^{n-1}} \times t = \frac{l^{n-1}}{(n - 1)} \left[\frac{1}{(a - x)^{n-1}} - \frac{1}{a^{n-1}} \right]$$

Or, $$k_L \left(\frac{m}{l} \right)^{n_2} \times \frac{t}{V^{n-1}} = \frac{1}{(n - 1)} \left[\frac{1}{(a - x)^{n-1}} - \frac{1}{a^{n-1}} \right]$$

Or, $$t = \frac{V^{n-1}}{k_{M/L} (n - 1)} \left[\frac{1}{(a - x)^{n-1}} - \frac{1}{a^{n-1}} \right] \tag{9.21}$$

where, $$k_{M/L} = k_L \left(\frac{m}{l} \right)^{n_2}$$

Equation (9.21) is similar to equation (9.18).

9.10 METHODS FOR DETERMINING ORDER OF A REACTION

9.10.1 Half-life Period Method

Consider the irreversible reaction: $lL + mM \longrightarrow Pp + qQ + rR + \cdots$

The rate equation is given by $J = kC_L^{n_1} C_M^{n_2}$, $n = n_1 + n_2 = $ Total order of the reaction.

To find out n_1, the reaction is carried out in large excess of the reactant M so that C_M remains almost constant throughout the reaction. Thus, the above rate equation becomes

$$J = k'C_L^{n_1}, \quad \text{where,} \quad k' = kC_M^{n_2}, \quad a = \text{Number of moles of the reactant } L \text{ at } t = 0.$$

The half-life period of the reactant L is given by equation (9.19)

$$t_{1/2} = \frac{V^{n_1-1}}{k_L'(n-1)}\left[\frac{2^{n_1-1}-1}{a^{n_1-1}}\right], \quad \text{where,} \quad k_L' = k'l$$

Thus, we have

$$t_{1/2} \propto \frac{1}{n^{n_1}-1}$$

Let us carry out the reaction in two phases, starting with the two initial concentrations of the reactant L, namely (a_1/V) mole/litre and (a_2/V) mole/litre. Thus, we have two different half-life periods $(t_{1/2})_1$ and $(t_{1/2})_2$ for the two phases. In both the cases we can write that,

$$(t_{1/2})_1 \propto \frac{1}{a_1^{n_1-1}} \quad \text{and} \quad (t_{1/2})_2 \propto \frac{1}{a_2^{n_1-1}}. \quad \text{Or,} \quad \frac{(t_{1/2})_2}{(t_{1/2})_1} = \left(\frac{a_1}{a_2}\right)^{n_1-1}$$

Taking $\log_{10}$ on both sides we get

$$\log\frac{(t_{1/2})_2}{(t_{1/2})_1} = (n_1-1)\log\left(\frac{a_1}{a_2}\right) \tag{9.22}$$

a_1 and a_2 are known, $(t_{1/2})_1$ and $(t_{1/2})_2$ are measurable quantities. Thus, the partial order n_1 can easily be determined using equation (9.22).

In the next step the reaction is carried out in large excess of the reactant L so that C_L remains almost constant throughout the reaction. Thus, the above rate equation becomes

$$J = k''C_M^{n_2}, \quad \text{where,} \quad k'' = kC_L^{n_1}, \quad b = \text{Number of moles of the reactant } M \text{ at } t = 0.$$

The half-life period of the reactant M is given by equation (9.19)

$$t_{1/2} = \frac{V^{n_2-1}}{k_M'(n-1)}\left[\frac{2^{n_2-1}-1}{b^{n_2-1}}\right], \quad \text{where,} \quad k_M' = k''m$$

Proceeding in similar way, we get

$$\log \frac{(t_{1/2})_2}{(t_{1/2})_1} = (n_2 - 1)\log\left(\frac{b_1}{b_2}\right) \tag{9.23}$$

b_1 and b_2 are known, $(t_{1/2})_1$ and $(t_{1/2})_2$ are measurable quantities. Thus, the partial order n_2 can easily be determined using equation (9.23).

Thus, the total order of the reaction

$$n = n_1 + n_2. \tag{9.24}$$

9.10.2 Graphical Method

For the same reaction mentioned in the earlier section rate expression, in large excess of the reactant M, is given by

$$t = \frac{V^{n_1-1}}{k_L'(n_1-1)}\left[\frac{1}{(a-x)^{n_1-1}} - \frac{1}{a^{n_1-1}}\right] \tag{9.25}$$

Or, $\quad t = mX - C$, where, $m = \dfrac{V^{n_1-1}}{k_L'(n_1-1)} = \text{Constant}, \quad C = \dfrac{m}{a^{n_1-1}} = \text{Constant}$

$$X = \frac{1}{(a-x)^{n_1-1}}$$

Or, we can write that

$$X = \frac{t}{m} + \frac{C}{m} \tag{9.26}$$

Thus, the plot of X versus t is a straight line with slope $(1/m)$ and intercept (C/m) as shown in the Fig. 9.1.

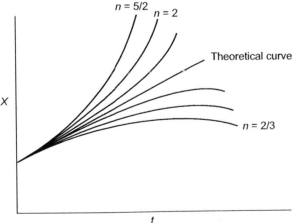

Figure 9.1 X versus t plot at different values of n. Equation (9.25) is valid for all values of n except $n = 1$

At different time intervals amount of product formed, i.e., x can be measured. The initial amount of the reactant L is known (a mole). Thus, ($a-x$) can be calculated at different time intervals, t. All these values are listed in the Table 9.4.

Table 9.4 List of parameters x, ($a-x$), etc. at different values of t

t	x	($a-x$)	X^b at $n_1 = \dfrac{1}{2}$	X^b at $n_1 = \dfrac{2}{3}$	X^b at $n_1 = \dfrac{3}{2}$	X^b at $n_1 = 2$	X^b at $n_1 = \dfrac{5}{2}$
t_1	x_1	($a-x_1$)	X_1	X_1	X_1	X_1	X_1
t_2	x_2	($a-x_2$)	X_2	X_2	X_2	X_2	X_2
t_3	x_3	($a-x_3$)	X_3	X_3	X_3	X_3	X_3
t_4	x_4	($a-x_4$)	X_4	X_4	X_4	X_4	X_4
t_5	x_5	($a-x_5$)	X_5	X_5	X_5	X_5	X_5
.	.	.	.	.	.	.	.
.	.	.	.	.	.	.	.
.	.	.	.	.	.	.	.
t_{15}	x_{15}	($a-x_{15}$)	X_{15}	X_{15}	X_{15}	X_{15}	X_{15}

$b: X = 1/(a-x)^{n_1-1}$. The values of X are calculated at different values of x by taking different values of n.

The values of X are plotted against t values and a separate curve for each value of n is obtained as shown in the Fig. 9.1. For a particular value of n_1, the curve becomes straight line and matches with the theoretical curve. That value of n_1 is called the *order of the reaction* with respect to the reactant L. Slope of the straight line $= (1/m) = [k'_L (n_1 - 1)/V^{n_1 - 1}]$. The value of $(1/m)$ can be evaluated from the curve. V is known and n_1 has been evaluated. Thus, from the above equation the rate constant k'_L of the said reaction can easily be calculated.

Proceeding in the same manner the partial order n_2, *order of the reaction* with respect to the reactant M, can also be calculated. The total order of the reaction, n, can be evaluated from the equation $n = n_1 + n_2$.

9.11 SIGNIFICANCES OF HALF-LIFE PERIOD ($t_{1/2}$)

(a) For a reactant, participating in a particular chemical reaction, $t_{1/2}$ is the time taken by that reactant to decay to half-of its amount.

(b) For any reactant, $t_{1/2}$ depends on rate constant k of the chemical reaction. A reactant may participate in different chemical reactions. As k-values for different chemical reactions are different, the $t_{1/2}$ of that reactant varies from reaction to reaction.

(c) The value of $t_{1/2}$ of any reactant depends on the initial concentration of that reactant, e.g., $t_{1/2} \propto a$ for zero order reaction, $t_{1/2}$ is independent of a for 1st order reaction and $t_{1/2} \propto (1/a)$ for 2nd order reaction, ..., etc.

(d) For 1st order reaction we have

$$t_{1/2} = \frac{0.693}{k_L} \qquad (9.4)$$

Thus, by measuring $t_{1/2}$ one can easily evaluate the rate constant of that reaction.

(e) By measuring half-life period at different initial concentrations of a reactant both partial orders and total order of any irreversible chemical reaction can be evaluated.

9.12 ELEMENTARY REACTIONS

There are many examples, where partial orders are totally different from the corresponding stoichiometric coefficients. This is because those reactions occur in multiple steps and reaction in each step is known as elementary reaction. An elementary reaction is said to be unimolecular or bimolecular depending on the number of species involved in that reaction. The elementary reaction $A \rightarrow$ Products is unimolecular since one molecule is involved in the reaction but the elementary reaction $2A \rightarrow$ Products, or $A + B \rightarrow$ Products is bimolecular since two molecules are involved in the elementary reaction.

According to collision theory, molecules are involved in random collisions through which the reactant molecules gain necessary activation energy to form products. Usually, collision comes about between two molecules of same reacting species (i.e., unimolecular) or between two molecules of different reacting species (i.e., bimolecular). Trimolecular elementary reaction of type $A + B + C \rightarrow$ Products is uncommon since simultaneous three-body collision rarely occurs. Obviously elementary reaction, involving more than three molecules, is unknown.

Let us consider a bimolecular reaction $A + B \rightarrow$ Products.

If Z_{A-B} is the rate of A–B collisions per unit volume, rate of reaction, J, is given by $J \propto Z_{A-B}$.

Again from the kinetic theory of ideal gas, we know that

$$Z_{A-B} \propto \left(\frac{n_A}{V} \right) \left(\frac{n_B}{V} \right)$$

where, n_A and n_B are the number of moles of the reactants A and B respectively and V is the volume of the system.

Hence, $J = k C_A C_B$

Thus, for any ideal elementary reaction the partial orders are nothing but the corresponding stoichiometric coefficients.

For an ideal elementary reaction of type $aA + bB \rightarrow$ Products rate is given by

$$J = k C_A^a C_B^b, \text{ where, } a + b = 1 \text{ or } 2 \text{ or } 3.$$

Let us consider a reversible elementary reaction of type $lL + mM \rightleftharpoons pP + qQ$.

At equilibrium, rate of forward reaction = rate of backward reaction.

Or, $\quad k_f C_L^l C_M^m = k_b C_P^p C_Q^q.$ Equilibrium constant $= K_c = \dfrac{k_f}{k_b} = \dfrac{C_P^p C_Q^q}{C_L^l C_M^m}.$

9.13 STUDY OF REACTION MECHANISM

There are two concepts to study a multi-step reaction mechanism:
  (i) Rate-determining step approach
  (ii) Steady-state approach.

9.13.1 Rate-determining Step Approach

According to this concept a reaction takes place in multiple steps and one of them is the slowest one. The rate of the slowest step actually determines the rate of the whole reaction. One of such example is.

Example 9.13.1 Decomposition of hydrogen peroxide by iodide ion in acidic medium.

The overall reaction is:

$$H_2O_2 + 2H^+ + 2I^- \longrightarrow I_2 + 2H_2O$$

The elementary reactions are:
1st step

$$H^+ + I^- \rightleftharpoons HI \qquad \text{[Rapid equilibrium]}$$

Forward reaction rate constant $= k_1$ and backward reaction rate constant $= k_{-1}$.
Then the reaction proceeds as follows:

$$HI + H_2O_2 \xrightarrow{k_2} H_2O + HOI \quad \text{[Slow]}$$

$$HOI + I^- \xrightarrow{k_3} I_2 + OH^- \qquad \text{[Fast]}$$

$$OH^- + H^+ \xrightarrow{k_4} H_2O \qquad \text{[Fast]}$$

The second step is the slowest step and hence rate of the whole reaction is given by

$$J = k_2 [HI] [H_2O_2] \tag{9.27}$$

As equilibrium holds in the first step we can write that

$$K_c = \frac{[HI]}{[H^+][I^-]}. \quad \text{Or,} \quad [HI] = K_c[H^+][I^-]$$

Putting the value of [HI] in the above equation we get

$$J = k_2 K_c [H^+] [I^-] [H_2O_2] = k[H^+] [H_2O_2] [I^-] \tag{9.28}$$

where, $k = k_2 K_c$

Example 9.13.2 Nitration of benzene in acidic medium.

$$HNO_3 + C_6H_6 \xrightarrow{H^+} C_6H_5NO_2 + H_2O$$

The elementary reactions are

$$H^+ + OH - NO_2 \rightleftharpoons H_2O - NO_2^+ \qquad \text{[Rapid equilibrium]}$$

Forward reaction rate constant $= k$ and backward reaction rate constant $= k_{-1}$.
Then the reaction proceeds as follows

$$H_2O - NO_2^+ \xrightarrow{k_2} H_2O + NO_2^+ \qquad \text{[Slow]}$$

$$NO_2^+ + C_6H_6 \xrightarrow{k_3} C_6H_5NO_2 + H^+ \qquad \text{[Fast]}$$

The second step is the slowest step and hence the rate of the whole reaction is given by

$$J = k_2[H_2O - NO_2^+] \tag{9.29}$$

As equilibrium exists in the 1st step we can write that

$$K_c = \frac{k_1}{k_{-1}} = \frac{[H_2O - NO_2^+]}{[H^+][HNO_3]}. \quad \text{Or,} \quad [H_2O - NO_2^+] = \frac{k_1}{k_{-1}}[H^+][HNO_3]$$

Putting the value of $[H_2O - NO_2^+]$ in the above equation (9.29) we get

$$J = \frac{k_2 k_1}{k_{-1}}[H^+][HNO_3]. \quad \text{Or,} \quad J = k[H^+][HNO_3] \tag{9.30}$$

where, $\qquad k = \dfrac{k_2 k_1}{k_{-1}}$

9.13.2 Steady-state Approach

According to this concept: (i) intermediate, formed during a chemical reaction is short-lived and has transient existence, (ii) rate of formation of any reaction intermediate, formed in elementary reactions, is equal to the rate of consumption of that intermediate species since intermediate species are very much reactive and do not appear in the final phase of the reaction.

By using this concept, rate of reactions discussed in the earlier section, can be determined.

Example 9.13.3 Decomposition of hydrogen peroxide by iodide ion in acidic medium.

The overall reaction is:

$$H_2O_2 + 2H^+ + 2I^- \longrightarrow I_2 + 2H_2O$$

Elementary reactions are:
1st step

$$H^+ + I^- \rightleftharpoons HI \qquad \text{[Rapid equilibrium]}$$

Forward reaction rate constant $= k_1$ and backward reaction rate constant $= k_{-1}$.

Then the reaction proceeds as follows:

$$HI + H_2O_2 \xrightarrow{k_2} H_2O + HOI \quad [Slow]$$

$$HOI + I^- \xrightarrow{k_3} I_2 + OH^- \quad [Fast]$$

$$OH^- + H^+ \xrightarrow{k_4} H_2O \quad [Fast]$$

Rate of formation of product

$$J = k_3[HOI][I^-] \tag{9.31}$$

According to steady-state concept, we have

$$\frac{d[HOI]}{dt} = 0 = k_2[HI][H_2O_2] - k_3[HOI][I^-]. \quad \text{Or,} \quad [HOI] = \frac{k_2[HI][H_2O_2]}{k_3[I^-]}$$

Again, $\quad \dfrac{d[HI]}{dt} = 0 = k_1[H^+][I^-] - k_{-1}[HI] - k_2[HI][H_2O_2].$

Or, $\qquad [HI] = \dfrac{k_1[H^+][I^-]}{k_{-1} + k_2[H_2O_2]}$

Hence, $\quad [HOI] = \dfrac{k_1 k_2}{k_3[I^-]} \dfrac{[H^+][I^-][H_2O_2]}{\{k_{-1} + k_2[H_2O_2]\}} = \dfrac{k_1 k_2}{k_3} \dfrac{[H^+][H_2O_2]}{\{k_{-1} + k_2[H_2O_2]\}}$

Putting this value of [HOI] in the rate equation, we get

$$J = \frac{k_1 k_2[H^+][I^-][H_2O_2]}{k_{-1} + k_2[H_2O_2]} \tag{9.32}$$

Equations (9.28) and (9.32) resembles with each other but not the same. These two equations will be the same if $k_{-1} \gg k_2[H_2O_2]$.

Example 9.13.4   Nitration of benzene in acidic medium.

$$HNO_3 + C_6H_6 \xrightarrow{H^+} C_6H_5NO_2 + H_2O$$

Elementary reactions are

$$H^+ + OH - NO_2 \rightleftharpoons H_2O - NO_2^+ \quad [Rapid\ equilibrium]$$

Forward reaction rate constant $= k_1$ and backward reaction rate constant $= k_{-1}$.

$$H_2O - NO_2^+ \xrightarrow{k_2} H_2 + NO_2^+ \quad [Slow]$$

$$NO_2^+ + C_6H_6 \xrightarrow{k_3} C_6H_5NO_2 + H^+ \quad [Fast]$$

Rate of formation of the product

$$J = k_3[NO_2^+][C_6H_6] \tag{9.33a}$$

According to steady-state concept, we have

$$\frac{d[NO_2^+]}{dt} = 0. \quad \text{Or,} \quad 0 = k_2[H_2O - NO_2^+] - k_3[NO_2^+][C_6H_6]$$

Or, $\qquad k_2[H_2O - NO_2^+] - J = 0 \qquad$ [Using equation (9.33a)]

Or, $\qquad J = k_2[H_2O - NO_2^+] \qquad\qquad\qquad\qquad\qquad$ (9.33b)

Again, $\qquad \dfrac{d[H_2O - NO_2^+]}{dt} = 0.$

Or, $\qquad k_1[H^+][HNO_3] - k_{-1}[H_2O - NO_2^+] - k_2[H_2O - NO_2^+] = 0$

Or, $\qquad [H_2O - NO_2^+] = \dfrac{k_1[H^+][HNO_3]}{k_{-1} + k_2}$

Putting this value of $[H_2O - NO_2^+]$ in the above equation (9.33b) we get

$$J = \frac{k_1 k_2[H^+][HNO_3]}{k_{-1} + k_2} = k'[H^+][HNO_3] \qquad\qquad (9.34)$$

where, $\qquad k' = \dfrac{k_1 k_2}{k_{-1} + k_2}$

Equations (9.30) and (9.34) are same but differ with the magnitude of rate constant. $k' = k$, if $k_{-1} \gg k_2$.

9.13.3 Free Radical Mechanism

Besides ionic mechanism there are many reactions which proceed through free radical mechanism. One such example is gas-phase reaction between hydrogen and bromine. The overall reaction is $H_2 + Br_2 \longrightarrow 2HBr$
The elementary reactions are

$$Br_2 \xrightarrow{hv/k_1} 2Br^\bullet \qquad\qquad \text{Rate} = J_1 = k_1[Br_2]$$

$$Br^\bullet + H_2 \xrightarrow{k_2} HBr + H^\bullet \qquad \text{Rate} = J_2 = k_2[Br^\bullet][H_2]$$

$$H^\bullet + Br_2 \xrightarrow{k_3} HBr + Br^\bullet \qquad \text{Rate} = J_3 = k_3[H^\bullet][Br_2]$$

$$Br^\bullet + Br^\bullet \xrightarrow{k_4} Br_2 \qquad\qquad \text{Rate} = J_4 = k_4[Br^\bullet]^2$$

$$H^\bullet + HBr \xrightarrow{k_5} H_2 + Br^\bullet \qquad \text{Rate} = J_5 = k_5[H^\bullet][HBr]$$

The elementary reactions show that reactive intermediates, generated from the reactants, are converted into products and regenerated. Thus, the whole process is repeating continuously. These types of reactions are called chain reactions. Usually chain reactions proceed through free radical mechanism.

Rate of formation of product is given by:

$$\frac{1}{2}\frac{d[HBr]}{dt} = J = \frac{1}{2}(J_2 + J_3 - J_5) \tag{9.35a}$$

According to the steady-state concept, we can write that

$$\frac{d[H^\bullet]}{dt} = 0 = (J_2 - J_3 - J_5) \tag{9.35b}$$

$$\frac{d[Br^\bullet]}{dt} = 0 = (J_1 + J_3 + J_5 - J_4 - J_2) \tag{9.35c}$$

Adding the above two equations, we get $J_1 = J_4$.

Or, $\quad k_1[Br_2] = k_4[Br^\bullet]^2$. Or, $[Br^\bullet] = \left(\dfrac{k_1}{k_4}\right)^{\frac{1}{2}} [Br_2]^{\frac{1}{2}}$ $\tag{9.35d}$

Again from equation (9.35b) we have $J_3 + J_5 = J_2$

Or, $\quad k_3[H^\bullet][Br] + k_5[H^\bullet][HBr] = k_2[Br^\bullet][H_2]$. Or, $[H^\bullet] = \dfrac{k_2[Br^\bullet][H_2]}{k_3[Br_2] + k_5[HBr]}$

Putting the value of [Br*] in the above equation we get

$$[H^\bullet] = \frac{k_2 \left(\dfrac{k_1}{k_4}\right)^{\frac{1}{2}} [Br_2]^{\frac{1}{2}}[H_2]}{k_3[Br_2] + k_5[HBr]}$$

Or, $\quad k_3[H^\bullet][Br_2] = \dfrac{k[H_2][Br_2]^{\frac{1}{2}}}{1 + k'\dfrac{[HBr]}{[Br_2]}}$, $k = k_2\left(\dfrac{k_1}{k_4}\right)^{\frac{1}{2}}$ and $k' = \dfrac{k_5}{k_3}$

We know from equation (9.35a)

$$J = \frac{1}{2}\frac{d[HBr]}{dt} = \frac{1}{2}(J_2 + J_3 - J_5)$$

Again from equation (9.35b) we have $J_5 = (J_2 - J_3)$. Putting the value of J_5 in the above equation we get

$$J = \frac{1}{2}(2J_3) = J_3 = k_3[H^\bullet][Br_2] = \frac{k[H_2][Br_2]^{\frac{1}{2}}}{1 + k'\dfrac{[HBr]}{[Br_2]}} \tag{9.35e}$$

Here it is interesting to note that an increase in concentration of HBr decreases the rate J and hence the product HBr acts as a **retarder**.

9.14 MECHANISM TO EXPLAIN UNIMOLECULAR REACTION (LINDEMANN HYPOTHESIS)

Most of the reactions are bimolecular, where two molecules A and B collide with each other to form products. But in the case of unimolecular reactions only one species is involved and hence apparently collision theory fails to explain the unimolecular reactions. In 1922, Lindemann proposed a scheme to explain unimolecular reaction on the basis of collision theory.

All types of isomerization reactions and decomposition reactions are usually unimolecular reactions, e.g.,

(i) *cis*-CHCl = CHCl to *trans*-CHCl = CHCl

(ii) $CH_3CH_2I \longrightarrow CH_2 = CH_2 + HI$

Consider the following unimolecular reaction $A \longrightarrow$ Products.

The corresponding elementary reactions are:

Step-I $\quad A + A \xrightarrow{k_1} A^* + A \quad$ (activation)

Step-II $\quad A^* + A \xrightarrow{k_2} A + A \quad$ (deactivation)

Step-III $\quad A^* \xrightarrow{k_3}$ Products

1 mole of the reacting species A contains Avogadro number of molecules. According to Lindemann, collisions continuously take place within the molecules of the same species and these collisions lead to the above elementary reactions. The species A^* is an activated or energized molecule that means in A^* the molecule A remains in high vibrational energy level. The rate of formation of product, J is given by:

$$J = k_3 C_A^*$$

According to steady-state approach, we can write that

$$\frac{d[C_A^*]}{dt} = 0 = k_1 C_A^2 - k_2 C_A C_A^* - k_3 C_A^*. \quad \text{Or, } C_A^* = \frac{k_1 C_A^2}{k_2 C_A + k_3}$$

Putting the value of C_A^* in the rate equation we get

$$J = \frac{k_1 k_3 C_A^2}{k_2 C_A + k_3} \tag{9.36}$$

Case-I $k_3 \gg k_2 C_A$ and hence, $k_2 C_A + k_3 \approx k_3$ which means that only a few molecules get deactivated through step-II. Thus, equation (9.36) becomes

$$J = \frac{k_1 k_3 C_A^2}{k_2 C_A + k_3} = \frac{k_1 k_3 C_A^2}{k_3} = k_1 C_A^2 \tag{9.37a}$$

So, the reaction follows 2nd order kinetics.

Again for a unimolecular reaction, the rate is given by

$$J = k_u C_A \tag{9.37b}$$

Comparing equations (9.37a) and (9.37b) we get,

$$k_u = k_1 C_A = k_1 p_A \text{ (for gas phase reaction)} \tag{9.37c}$$

p_A is the partial pressure of the reactant at time t. p_A^0 is the initial pressure of the reactant.

Figure 9.2 shows the plot of k_u versus p_A^0 for the gas phase decomposition of N_2O_5. It is evident that the equation (9.37c) holds good only at very low pressure, $p_{N_2O_5} < 0.4$ mm of Hg.

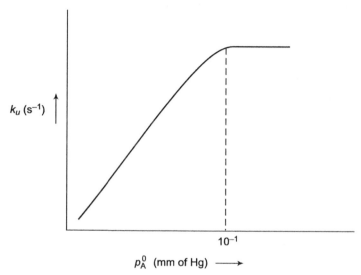

Figure 9.2 Plot of k_u versus initial pressure p_A^0 of decomposition of N_2O_5

Case-II $k_2 C_A \gg k_3$, which means that maximum number of molecules undergo deactivation through step-II. Hence, $k_2 C_A + k_3 \approx k_2 C_A$.

Thus, the equation (9.36) becomes

$$J = \frac{k_1 k_3 C_A^2}{k_2 C_A + k_3} = \frac{k_1 k_3 C_A^2}{k_2 C_A} = k C_A, \quad \text{where, } k = \frac{k_1 k_3}{k_2} \tag{9.38}$$

So the reaction follows 1st order kinetics. Comparing equations (9.37b) and (9.38), we get

$$k_u = k \tag{9.38b}$$

Thus, k_u is constant, which is reflected in the Fig. 9.2. It has been found that the value of k_u and k are independent of $p_{N_2}O_5$ in a wide range of pressure (0.6 mm of Hg to 70 mm of Hg).

Nitrogen pentoxide decomposes into nitrogen dioxide through the following elementary reactions

$$N_2O_5(g) \longrightarrow N_2O_3(g) + O_2(g)$$
$$N_2O_3(g) + N_2O_5(g) \longrightarrow 4NO_2(g)$$

The value of rate constant, k, has been investigated over a wide pressure range (0.0002 mm–70 mm of Hg).

At very low pressure of $N_2O_5(g)$, i.e., $p^0_{N_2O_5(g)} < 0.4$ mm of Hg, the above reaction follows 2nd order kinetics and at high pressure of $N_2O_5(g)$, i.e., $p^0_{N_2O_5(g)} > 0.6$ mm of Hg, the above reaction follows 1st order kinetics.

Inferences

(i) Order of a unimolecular reaction depends on initial pressure of the reactant only.

(ii) Collision theory explains unimolecular, bimolecular, and trimolecular reactions satisfactorily. Thus, the mechanism of formation of product is independent of molecularity of a reaction.

(iii) The time lag between two successive collisions is called collision time. If collision time is greater than the lifetime of activated or energized complex, the latter splits into products before colliding with another molecule. Thus, step-III is dominating over step-II. So, $k_3 \gg k_2 C_A$ and $k_2 C_A + k_3 \approx k_3$. The reaction follows 2nd order kinetics. This usually comes true at low concentration of reactant.

(iv) If the initial concentration of reactant is very high and the collision time is very short, as a result of which, molecules prefer colliding with each other to form products. So, $k_2 C_A \gg k_3$ and $k_2 C_A + k_3 \approx k_3 C_A$. The reaction follows 1st order kinetics.

9.15 KINETIC STUDY OF SOME SPECIAL REACTIONS

9.15.1 Pseudo 1st Order Reactions

There are some reactions where more than one reactant is involved but rate depends on one reactant only. In those cases, actual order of the reaction is more than 1 but the apparent order of the reaction is 1. Those reactions are called pseudo 1st order reactions. Two such cases are discussed below:

9.15.1.1 Inversion of Cane Sugar in Acidic Medium

The following reaction occurs in presence of acid catalyst:

$$C_{12}H_{22}O_{11}(\text{Sucrose}) + H_2O \xrightarrow{H^+} C_6H_{12}O_6(\text{Glucose}) + C_6H_{12}O_6(\text{Fructose})$$
$$\text{(+)dextro} \qquad\qquad\qquad \text{(+)dextro} \qquad \text{(−)levo}$$

The rate equation is given by $J = k'[C_{12}H_{22}O_{11}][H_2O][H^+]$

For a given concentration of acid, $[H^+]$ is fixed since catalyst is regenerated at the end of the reaction. As the reaction is carried out in profuse amount of water concentration of water, i.e., $[H_2O]$ is assumed to be constant. Thus, the above rate equation becomes

$$J = k[C_{12}H_{22}O_{11}] \tag{9.39}$$

where, $\qquad k = k'[H_2O][H^+]$

Thus, the apparent order of the above reaction is 1. Hence it is called pseudo first order reaction. Again the rate depends only on the concentration of one reactant though more than one reactant is involved in actual reaction. Thus, pseudo first order reaction is also called pseudo unimolecular reaction.

Determination of rate constant (k) of the above reaction

Sucrose is an optically active substance and shows dextrorotatory (+) characteristics. On hydrolysis it gives glucose (dextrorotatory) and fructose (levorotatory).

When a plane polarized light is passed through sucrose solution, the net angle of rotation of the plane polarized light decreases with the increase in fructose concentration. Thus, the concentration of sucrose is directly proportional to the net angle of rotation. As the reaction is 1st order, the rate equation is given by:

$$t = \frac{1}{k} \ln \frac{a}{a-x}$$ [Equation (9.2)]

a = Initial number of moles of sucrose.

x = Amount of sucrose reacts during the time interval t sec.

θ_0 = Angle of rotation of plane polarized light at $t = 0$, i.e., at the beginning of hydrolysis.

θ_t = Angle of rotation measured at any time t.

θ_∞ = Angle of rotation measured at $t = \infty$, i.e., at the end of hydrolysis.

As the reaction proceeds, angle of rotation decreases with increase in fructose concentration. Thus, $\theta_0 > \theta_t \gg \theta_\infty$

At infinite time, hydrolysis is complete and angle of rotation (θ) becomes minimum.

Thus, the initial concentration of sucrose is $(a/V) \propto (\theta_0 - \theta_\infty)$, where V is the volume of the total solution and assumed to be constant.

Concentration of the product formed at any time t is $(x/V) \propto (\theta_0 - \theta_t)$, where x is the number of moles of product formed after the time interval t.

So, $\left(\dfrac{a-x}{V}\right) \propto [(\theta_0 - \theta_\infty) - (\theta_0 - \theta_t)]$. Or, $\left(\dfrac{a-x}{V}\right) \propto (\theta_t - \theta_\infty)$

Thus, $\left(\dfrac{a}{a-x}\right) = \left(\dfrac{\theta_0 - \theta_\infty}{\theta_t - \theta_\infty}\right)$

Putting this value of $(a/a - x)$ in the above rate equation, we get

$$t = \frac{1}{k} \ln \left(\frac{\theta_0 - \theta_\infty}{\theta_t - \theta_\infty}\right) = \frac{2.303}{k} \log \left(\frac{\theta_0 - \theta_\infty}{\theta_t - \theta_\infty}\right) \tag{9.40}$$

θ_0 and θ_∞ can easily be measured with the help of polarimeter and the values of θ_t are measured at different time intervals. By plotting log $(\theta_0 - \theta_\infty/\theta_t - \theta_\infty)$ versus t a straight line with slope $(k/2.303)$ is obtained. The rate constant k can easily be evaluated from the slope of the straight line.

9.15.1.2 Hydrolysis of Ester in Acidic Medium

Consider the hydrolysis of ethyl acetate in acidic medium.

$$CH_3COOC_2H_5 + H_2O \xrightarrow{\ H^+\ } CH_3COOH + C_2H_5OH$$

The rate equation is given by

$$-\frac{d[CH_3COOC_2H_5]}{dt} = k'[CH_3COOC_2H_5][H_2O][H^+]$$

As the reaction is carried out in huge excess of H_2O, the concentration of H_2O, i.e., $[H_2O]$ is assumed to be constant during the course of reaction. In this reaction, H^+ ion acts as catalyst and the mechanism shows that rate constant (k) depends on catalyst concentration. However, the concentration of catalyst remains unchanged during the course of reaction. Thus, the above rate equation becomes

$$-\frac{d\left[CH_3COOC_2H_5\right]}{dt} = k\left[CH_3COOC_2H_5\right], \quad \text{where, } k = k'[H_2O][H^+]$$

Determination of rate constant (k) of the above reaction

Acetic acid (CH_3COOH) is produced during hydrolysis and the rate can be studied by neutralizing acetic acid by sodium hydroxide (NaOH). So, volume of NaOH consumed indicates the amount of ester hydrolyzed. As the reaction is 1st order the rate equation is given by

$$t = \frac{2.303}{k} \log \frac{a/V}{(a-x)/V} = \frac{2.303}{k} \log \frac{a}{a-x} \qquad \text{[Equation (9.2)]}$$

where, a = initial number of moles of ester.

x = Concentration of the product, i.e., acetic acid in moles after the time interval t.

V = Volume of the system in litre.

$[CH_3COOC_2H_5]$ = Concentration of ester left after the time t = Concentration of acetic acid formed after the time $t = (a-x)/V$ moles/litre.

Both a and $(a-x)$ may be taken in terms of equivalent volume of NaOH consumed. As concentration ratio is used, standardization of NaOH solution is not required.

V_0 = The volume of NaOH required in ml to neutralise the *acidic medium* only in presence of ester but before starting the hydrolysis, i.e., at $t = 0$. So, V_0 is the volume of NaOH consumed by the catalyst, i.e., acid only.

After addition of ester, V_∞ and V_t are the equivalent volume of NaOH consumed in ml at time $t = \infty$ and $t = t$ respectively. Both V_∞ and V_t include V_0 and hence,

$$[CH_3COOH]_{t=\infty} \propto (V_\infty - V_0) \text{ and } [CH_3COOH]_{t=t} \propto (V_t - V_0)$$

At $t = \infty$, hydrolysis is completed, i.e., all ester is converted to equivalent amount of acetic acid. Under this circumstance, $[CH_3COOH]_{t=\infty}$ = Initial concentration of ester = a/V.

At $t = t$, hydrolysis is partially completed, i.e., some amount of ester is converted to equivalent amount of acetic acid. Under this circumstance, $[CH_3COOH]_{t=t} = x/V$ and concentration of ester left $= (a - x)/V$.

Thus, $\qquad \dfrac{a}{V} \propto (V_\infty - V_0), \quad \dfrac{x}{V} \propto (V_t - V_0),$ and $\dfrac{(a-x)}{V} \propto [(V_\infty - V_0) - (V_t - V_0)],$

i.e., $\qquad \dfrac{(a-x)}{V} \propto (V_\infty - V_t)$

So, we can write

$$\frac{a}{a-x} = \frac{V_\infty - V_0}{V_\infty - V_t}$$

Putting this value of $(a/a - x)$ in the equation (9.2), we get

$$t = \frac{2.303}{k} \log \frac{V_\infty - V_0}{V_\infty - V_t} \qquad\qquad (9.41)$$

Determination of V_0 is difficult and hence equation (9.41) is modified as follows:

At $t = t_1, V_t = V_1, t = t_2, V_t = V_2, t = t_3, V_t = V_3, \ldots,$ etc. $t = t_n, V_t = V_n$

$\Delta t = (t_n - t_1)$ and V_n = Volume of NaOH consumed at $t = t_n$.

$$\Delta t = t_n - t_1 = \frac{2.303}{k} \log \frac{V_\infty - V_1}{V_\infty - V_n}. \text{ Or, } \log \frac{V_\infty - V_1}{V_\infty - V_n} = \left(\frac{k}{2.303}\right) \times \Delta t$$

A straight line is obtained by plotting log $(V_\infty - V_1/V_\infty - V_n)$ against Δt. The rate constant k can easily be obtained from the slope of the straight line.

9.15.2   Competing Reactions/Parallel Reactions/Side Reactions

There are many reactions where a reactant gives two products at different rates, e.g., dissociation of potassium chlorate as shown below:

$$6KClO_3 \begin{array}{c} \xrightarrow{k_1} 2KCl + 3O_2 \\[2em] \xrightarrow{k_2} 3KClO_4 + KCl \end{array}$$

Another example is chlorination of toluene.

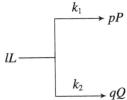

These types of reactions are known as parallel reactions.

Rate equation of a competing reaction Let us consider a parallel reaction of general form

$$lL \begin{array}{c} \xrightarrow{\quad k_1 \quad} pP \\ \\ \xrightarrow{\quad k_2 \quad} qQ \end{array}$$

k_1 and k_2 are the rate constants for the two products P and Q respectively.

Time(t)	No. of moles of reactant L	No. of moles of product P	No. of moles of product Q
0	a	0	0
t	$(a-x)$	$(p/l)x$	$(q/l)x$
$t_{1/2}$	$(a/2)$	$(p/l)(a/2)$	$(q/l)(a/2)$

x = No. of moles of reactant L consumed during time t.

V = Volume of the system = Constant.

C_L = Concentration of reactant L left at time $t = \left(\dfrac{a-x}{V}\right)$ moles/litre.

$C_P = \left(\dfrac{p}{l}\right)\left(\dfrac{x}{V}\right) = $ Concentration of product P at time t.

$C_Q = \left(\dfrac{q}{l}\right)\left(\dfrac{x}{V}\right)$

Boundary conditions: (i) at $t = 0$, $(a-x) = a$, and $x = 0$

(ii) at $t = t$, $(a-x) = (a-x)$, and $x = x$.

The rate equation is given by

$$J = -\frac{1}{l}\frac{dC_L}{dt} = -\frac{1}{l}\frac{d[(a-x)/V]}{dt} = -\frac{1}{lV}\frac{d(a-x)}{dt} = \frac{1}{lV}\frac{dx}{dt}$$

Again, $\quad J = \frac{1}{p}\frac{dC_P}{dt} = \frac{1}{p}\frac{d(px/lV)}{dt} = \frac{1}{lV}\frac{dx}{dt}$. Or, $J = \frac{1}{q}\frac{dC_Q}{dt} = \frac{1}{q}\frac{d(qx/lV)}{dt} = \frac{1}{lV}\frac{dx}{dt}$

Considering the product P, the rate equation is given by $J_1 = k_1 C_L$, assuming that the parallel reaction is 1st order.

Considering the product Q, the rate equation is given by $J_2 = k_2 C_L$, assuming that the parallel reaction is 1st order.

Thus, the total rate of decomposition of the reactant, L, is given by

$$J = J_1 + J_2 = k_1 C_L + k_2 C_L = (k_1 + k_2)C_L = kC_L, \text{ where, } k = k_1 + k_2$$

$$-\frac{1}{l}\frac{d[(a-x)/V]}{dt} = k\left(\frac{a-x}{V}\right). \text{ Or, } -\frac{1}{lV}\frac{d(a-x)}{dt} = k\left(\frac{a-x}{V}\right).$$

Or, $\quad -\frac{1}{l}\frac{d(a-x)}{dt} = k(a-x)$

Using the boundary conditions of $(a-x)$, we get

$$-\int_a^{(a-x)} \frac{d(a-x)}{(a-x)} = lk\int_0^t dt. \quad \text{Hence, } \ln\frac{a}{(a-x)} = k_L t, \text{ where, } k_L = lk$$

Or, $\quad t = \frac{1}{k_L}\ln\frac{a}{a-x} = \frac{2.303}{k_L}\log\frac{a}{a-x}$ \hfill (9.42a)

Again, we have

$$\frac{\text{Amount of product } P \text{ formed at time } t}{\text{Amount of product } Q \text{ formed at time } t} = \frac{(px/l)}{(qx/l)} = \frac{p}{q}$$

Or, $\quad \dfrac{p}{q} = \dfrac{J_1}{J_2} = \dfrac{k_1 C_L}{k_2 C_L} = \dfrac{k_1}{k_2}$ \hfill (9.42b)

p and q are the stoichiometric coefficients and known. So, (k_1/k_2) is known. Again, k_L can easily be evaluated by plotting $\log(a/a - x)$ versus t. Also, $k_L = l(k_1 + k_2)$ and l is stoichiometric coefficient and known. Thus, using equations (9.42a) and (9.42b), k_1 and k_2 can easily be determined separately.

9.15.3 Consecutive Reactions

There are many reactions which occur through multiple steps and each step is called elementary reaction. One example is conversion of *tert-butyl chloride* to *tert-butyl alcohol*.

$$H_3C-\underset{\underset{CH_3}{|}}{\overset{\overset{CH_3}{|}}{C}}-Cl \quad \xrightarrow[\text{Slow}]{k_1} \quad H_3C-\underset{\underset{CH_3}{|}}{\overset{\overset{CH_3}{|}}{C^+}} + Cl^-$$

$$H_3C-\underset{\underset{CH_3}{|}}{\overset{\overset{CH_3}{|}}{C^+}} + OH^- \quad \xrightarrow[\text{Fast}]{k_2} \quad H_3C-\underset{\underset{CH_3}{|}}{\overset{\overset{CH_3}{|}}{C}}-OH$$

In that case, the slowest step is considered as rate-determining step and the rate of the rate-determining step is the rate of the whole reaction. The detailed discussion has been made in sec. 9.13. This type of multi-step reaction is known as consecutive reaction.

Radioactive nuclear reactions occur through multi-steps, e.g.,

$$U_{92}^{239} \longrightarrow Np_{93}^{239} + e_{-1}^0$$

$$Np_{93}^{239} \longrightarrow Pu_{94}^{239} + e_{-1}^0$$

This type of nuclear reaction can also be termed consecutive reaction.

The simplest consecutive reaction is represented as $A \xrightarrow{k_1} B \xrightarrow{k_2} D$

Initially, at $t = 0$, $[A] = C_A^0$, $[B] = 0$ and $[D] = 0$.

After the time interval t, $[A] = C_A$, $[B] = C_B$ and $[D] = C_D$ such that

$$C_A + C_B + C_D = C_A^0 \tag{9.43a}$$

The rate equations are:

Step I: $-\dfrac{dC_A}{dt} = k_1.C_A$, Step II: $\dfrac{dC_B}{dt} = k_1.C_A - k_2.C_B$, Step III: $\dfrac{dC_D}{dt} = k_2.C_B$

$$\int_{C_A^0}^{C_A} \frac{dC_A}{C_A} = -k_1 \int_0^t dt. \text{ Or, } C_A = C_A^0 e^{-k_1 t} \tag{9.43b}$$

Again, $\dfrac{dC_B}{dt} + k_2.C_B = k_1.C_A = k_1.C_A^0 e^{-k_1 t}$

Solving the differential equation we get

$$C_B = \left(\frac{k_1}{k_2 - k_1}\right)[e^{-k_1 t} - e^{-k_2 t}]C_A^0 \tag{9.43c}$$

From equations (9.43a), (9.43b), and (9.43c) we get

$$C_D = C_A^0 - (C_A + C_B) = \left[1 + \left(\frac{k_1 e^{-k_2 t} - k_2 e^{-k_1 t}}{k_2 - k_1}\right)\right]C_A^0 \tag{9.43d}$$

Case-I $k_1 \gg k_2$

Thus, $(k_2 - k_1) \approx -k_1$ and $(k_1 e^{-k_2 t} - k_1 e^{-k_2 t}) \approx k_1 e^{-k_2 t}$

Hence, $C_D = (1 - e^{-k_2 t})C_A^0$ \hfill (9.44)

Thus, rate of production of D or $\dfrac{dC_D}{dt}$ is independent of k_1. As the $\dfrac{dC_D}{dt}$ depends on step I and step II, the overall rate depends on k_1. So, step II is the rate determining step. The change in concentrations of A, B, and D with time is shown in the Fig. 9.3.

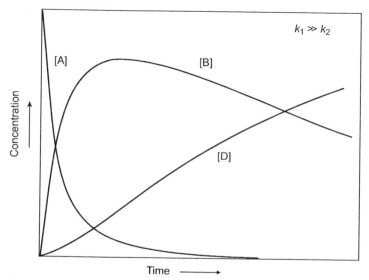

Figure 9.3 Change in concentration of reactants and products with time

Case-II $k_2 \gg k_1$

Thus, $(k_2 - k_1) \approx k_2$ and $(k_1 e^{-k_2 t} - k_2 e^{-k_1 t}) \approx -k_2 e^{-k_1 t}$

Hence, $C_D = (1 - e^{-k_1 t})C_A^0$ \hfill (9.45)

Thus, rate of production of D or $\dfrac{dC_D}{dt}$ is independent of k_2. As the $\dfrac{dC_D}{dt}$ depends on step I and step II, the overall rate is independent of k_2. So, step I is the rate determining step. The change in concentrations of A, B, and D with time is shown in the Fig. 9.4.

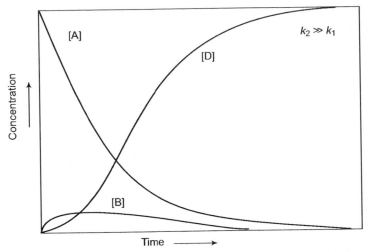

Figure 9.4 Change in concentration of reactants and products with time

9.15.4 Opposing Reactions

9.15.4.1 1st Order Reversible Reactions

Let us consider 1st order reversible reaction of type $A \rightleftharpoons B$, where, k_1 and k_2 are the forward and backward reaction rate constants respectively. Total volume of the system is V.

Time (t)	No. of moles of reactant A	No. of moles of product B
0	a	b
t	$(a-x)$	$(b+x)$

x = No. of moles of reactant A consumed during time t.

V = Volume of the system = Constant.

C_A = Concentration of reactant A left at time $t = \left(\dfrac{a-x}{V}\right)$ mole/litre.

C_B = Concentration of product B at time $t = \left(\dfrac{b+x}{V}\right)$ mole/litre.

If several reactions are occurring simultaneously in a system, each of them is independent of others and the rate of each reaction is directly proportional to the concentrations of the corresponding reactants only.

The rate of formation of B is given by

$$J_1 = -\frac{dC_A}{dt} = -\frac{d[(a-x)/V]}{dt} = -\frac{1}{V}\frac{d(a-x)}{dt} = \frac{1}{V}\frac{dx}{dt}.$$

Again $\qquad J_1 = k_1 C_A = k_1 \left(\dfrac{a-x}{V} \right)$

Or, $\qquad J_1 = \dfrac{dx}{dt} = k_1(a-x)$ $\hfill$ (9.46a)

The rate of disappearance of B is given by

$$J_2 = -\frac{dC_B}{dt} = -\frac{d[(b+x)/V]}{dt} = -\frac{1}{V}\frac{d(b+x)}{dt} = -\frac{1}{V}\frac{dx}{dt}$$

Again, $\qquad J_2 = k_2 C_B = k_2 \left(\dfrac{b+x}{V} \right)$

Or, $\qquad J_2 = -\dfrac{1}{V}\dfrac{dx}{dt} = k_2 \left(\dfrac{b+x}{V} \right)$. Or, $\dfrac{dx}{dt} = -k_2(b+x)$ $\hfill$ (9.46b)

Combining the above two equations, we get $J = J_1 + J_2 = \dfrac{1}{V}\dfrac{dx}{dt}$

Or, $\qquad \dfrac{dx}{dt} = k_1(a-x) - k_2(b+x) = (k_1 a - k_2 b) - (k_1 + k_2)x = (k_1 + k_2)\left[\dfrac{k_1 a - k_2 b}{k_1 + k_2} - x \right]$

Or, $\qquad \dfrac{dx}{dt} = (k_1 + k_2)(L - x)$ $\hfill$ (9.46c)

where, $\qquad L = \dfrac{k_1 a - k_2 b}{k_1 + k_2} = \dfrac{(k_1/k_2)a - b}{(k_1/k_2)+1} = \dfrac{Ka - b}{K+1}$, where, $K = \dfrac{k_1}{k_2}$

Or, $\qquad \dfrac{dx}{L-x} = (k_1 + k_2)dt$. Or, $-\dfrac{d(L-x)}{L-x} = (k_1 + k_2)dt$

Boundary conditions: (i) at $t = 0$, $x = 0$, and $L - x = L$

$\qquad\qquad\qquad\qquad$ (ii) at $t = t$, $x = x$, and $L - x = L - x$.

$$-\int_{L}^{(L-x)} \frac{d(L-x)}{(L-x)} = (k_1 + k_2)\int_{0}^{t} dt.$$

Hence, $\qquad \ln \dfrac{L}{(L-x)} = (k_1 + k_2)t$

Or, $\qquad 2.303 \log (L-x) = -(k_1 + k_2)t + 2.303 \log L$

Or, $\qquad \log(L-x) = -\dfrac{(k_1 + k_2)}{2.303}t + \log L$ $\hfill$ (9.46d)

By measuring x at different time t, one can plot $log(L-x)$ versus t. The plot is a straight line according to equation (9.46d) and shown in the Fig. 9.5.

At equilibrium, $x = x_e$ and $dx/dt = 0$. Thus, the equation (9.46c) becomes

$$(k_1 + k_2)(L - x_e) = 0. \quad \text{Or, } L = x_e$$

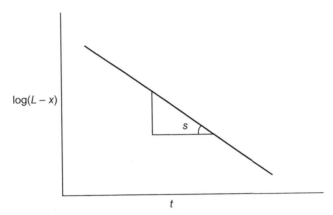

Figure 9.5 Plot of $log(L-x)$ versus t

Thus, the expression of L is given by

$$L = \frac{Ka - b}{K + 1} = x_e. \quad \text{Or, } Ka - b = Kx_e + x_e. \quad \text{Or, } K = \frac{b + x_e}{a - x_e} \quad (9.46e)$$

The value of x_e can be determined experimentally. Hence K can be calculated from equation (9.46e). Using this value of K the value of L can be evaluated. By measuring x at different time interval t, one can plot $log(L - x)$ versus t. The plot will be straight line according to equation (9.46d). The slope of the straight line (Fig. 9.5) is given by

$$\text{Slope} = s = -\frac{k_1 + k_2}{2.303}. \quad \text{Or, } k_1 + k_2 = -2.303s. \quad \text{Again, } K = \frac{k_1}{k_2}$$

Thus, from the above two equations k_1 and k_2 can easily be calculated.

9.16 THEORY OF REACTION RATE

Potential energy For each element or compound there is a potential well, which contains numerous energy levels. Each energy level contains a definite number of molecules, which are continuously vibrating. A typical potential energy (PE) diagram is shown in the Fig. 9.6.

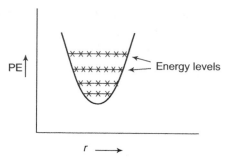

Figure 9.6   Potential energy diagram of an element or compound, r = Internuclear distance

Consider the reaction: $X - Y + Z \rightleftharpoons X + Y - Z$

As atom Z approaches to the compound $X-Y$, the bond between Y and Z becomes stronger and stronger, while the bond between X and Y becomes weaker and weaker, leading to the formation of an activated complex, also known as transition state (TS), shown below

$$X - Y + Z \longrightarrow [X\cdots Y\cdots Z] \text{ (TS)}$$

Both the bonds $X–Y$ and $Y–Z$ are weak and hence TS is placed at high energy level, as shown in the Fig. 9.7. Thus, activated complex is unstable and readily dissociates into products.

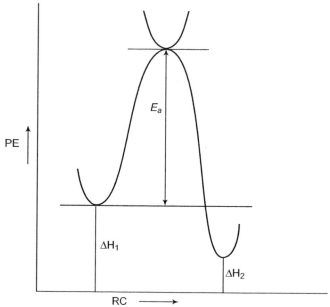

Figure 9.7   Reaction path profile for the reaction: $X - Y + Z \longrightarrow [X\cdots Y\cdots Z] \longrightarrow X + Y - Z$
RC = Reaction coordinate = $d\varepsilon \cdot E_a$ = Activation energy

E_a, shown in the Fig. 9.7, is called the activation energy.

Activation energy For a given chemical reaction, it is the minimum amount of energy, excluding the inherent potential energy of reactants, required for reactants to cross the energy barrier and take part in a chemical reaction, producing the products. E_a in the Fig. 9.7 is the activation energy of the chemical reaction.

Fig. 9.7 shows that $|\Delta H_1|$ and $|\Delta H_2|$ are the enthalpy of reactants and enthalpy of products respectively.

If $(|\Delta H_2| - |\Delta H_1|)$ is negative, some amount of energy is released in the form of heat. The reaction is then said to be an **exothermic reaction**.

If $(|\Delta H_2| - |\Delta H_1|)$ is positive, some amount of energy is absorbed during chemical reaction. The reaction is then said to be an **endothermic reaction**.

Threshold energy It is the minimum energy, including the inherent potential energy of reactants, required for reactants to cross the energy barrier and take part in a chemical reaction, producing the products. Or in other words, it is the potential energy of the activated complex. It is represented by $E_{threshold}$ and related to E_a as follows $E_{threshold} = (\Delta H_1 + E_a)$.

There are two popular theories to correlate reaction rate constant and activation energy. These are: (i) Collision theory, and (ii) Transition state theory (TST).

9.16.1 Collision Theory

Postulates

(a) According to this theory, molecules are considered as hard and rigid spheres.
(b) Reactant molecules do not interact among themselves nor do they influence the movement of other molecules.
(c) During collision of two molecules possibility of intervention of third molecule is neglected. Or in other words, only binary collisions are considered.
(d) The above assumptions suggest that the theory can be restricted to only homogeneous bimolecular gas phase reactions.
(e) According to collision theory, all the molecules are colliding with each other. Each and every molecule possesses a secluded volume (Fig. 9.8), into which no other molecule is allowed to enter.

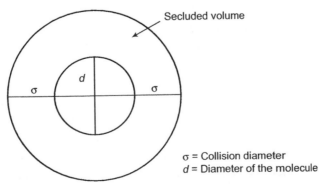

σ = Collision diameter
d = Diameter of the molecule

Figure 9.8 Secluded volume and collision diameter of a molecule

For a reaction of type $2A \rightarrow$ Products, the total number of collisions, Z_0, is given by

$$Z_0 = 2n^2\sigma_T^2\sqrt{\frac{\pi RT}{M}} \qquad\qquad (9.47a)$$

n = Number of molecules/c.c., M = Molecular mass, σ_T = Collision diameter at temperature $T°K$.

Z_0 = Total number of collisions per unit area per sec.

(f) **Observations of Svante Arrhenius**

(i) Reaction rate does not increase with temperature linearly. He found an exponential relationship between reaction rate constant and temperature as shown in the Fig. 9.9.

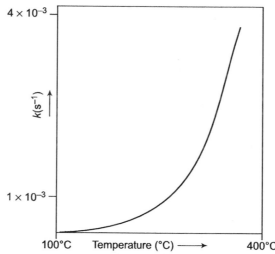

Figure 9.9 Rate constant versus temperature plot for a 1st order reaction

(ii) Reaction rate depends on fraction of molecules, able to gain potential energy higher than activation energy, E_a.

(iii) Reaction rate depends on fraction of those excited molecules, having proper orientation. $f_{\text{orientation}}$ represents orientation factor.

On the basis of above observations, **Svante Arrhenius** proposed the following mathematical relationship between k and E_a.

According to Arrhenius approach, in each collision one molecule gains energy and goes to higher energy level. But only a fraction of total number of activated molecules gains the necessary activation energy E_a. Only those activated molecules, gaining the activation energy E_a, form the activated complex and ultimately break into the products as shown below:

$$A + A \longrightarrow A^*(\text{TS}) \longrightarrow \text{Products [For unimolecular reactions]}$$
$$A + B \longrightarrow (AB)^* (\text{TS}) \longrightarrow \text{Products [For bimolecular reactions]}$$

Collisions, which are effective to help reactant molecules gain necessary activation energy, are called **effective collisions**, represented by Z_a. According to Arrhenius, Z_a can be correlated to Z_0 by Boltzmann distribution factor as given below:

$$\frac{Z_a}{Z_0} \propto e^{-E_a/RT}, \text{ where, } E_a = \text{Activation energy.}$$

(g) According to this theory, only "effective collisions" will lead to a chemical reaction. Two factors are to be considered in calculating effective collisions: (i) gaining necessary activation energy through collisions, (ii) collisions with proper orientation. Considering the above two factors, number of effective collisions, represented as Z_a, is related to total number of collisions, Z_0, as follows:

$$\frac{Z_a}{Z_0} = e^{-(E_a/RT)} \times f_{\text{orientation}}. \quad \text{Or, } Z_a = Z_0 e^{-(E_a/RT)} \times f_{\text{orientation}} \qquad (9.47b)$$

Inferences

1. Only a fraction of collisions is effective in bringing about a chemical change. These collisions are called effective collisions.
2. With the increase in number of reactant molecules the number of effective collisions increases and hence concentration of activated complex increases and so does reaction rate, J. Thus, rate of a chemical reaction is directly proportional to the concentration of the reactants.
3. Since the rate of a chemical reaction, J, depends on the number of effective collisions only, i.e., Z_a. Arrhenius proposed that rate constant, k, is directly related to Z_a as follows:

$$k = f_{\text{orientation}} \times Z_a. \quad \text{Or, } k = f_{\text{orientation}} \times Z_0 \times e^{-E_a/RT}$$

[From equation (9.47b)]

or, $\qquad k = Ae^{-E_a/RT}$ $\qquad\qquad\qquad\qquad\qquad\qquad\qquad$ (9.47c)

where, A is called Arrhenius factor, also known as Frequency factor.

$$A = Z_0 \times f_{\text{orientation}}$$

Taking ln on both sides of the equation (9.47c), we get

$$\ln k = \ln A - \frac{E_a}{RT}. \quad \text{Or, } \frac{d\ln k}{dT} = \frac{E_a}{RT^2} \qquad\qquad (9.47d)$$

Boundary conditions: (i) $T = T_1$, $k = k_1$, (ii) $T = T_2$, $k = k_2$.
Using these boundary conditions, we have

$$\ln k_1 = \ln A - \frac{E_a}{RT_1} \text{ and } \ln k_2 = \ln A - \frac{E_a}{RT_2}. \quad \text{Or, } \ln\frac{k_2}{k_1} = -\frac{E_a}{R}\left[\frac{1}{T_2} - \frac{1}{T_1}\right]$$

Or, $\qquad \log\frac{k_2}{k_1} = -\frac{E_a}{2.303R}\left[\frac{1}{T_2} - \frac{1}{T_1}\right]$ $\qquad\qquad\qquad$ (9.47e)

For a chemical reaction, if k_1 and k_2 are known, activation energy, E_a, can easily be evaluated from equation (9.47e).

4. Reactant molecules gain necessary activation energy through collisions but even after acquiring activation energy reactant molecules may not undergo chemical reaction. Suitable orientation of reactant molecules is essential to occur a chemical reaction. Thus, orientation factor or steric factor, p, is introduced in the expression of rate constant.

$$k = Z_0 \times f_{orientation}\ e^{-(E_a/RT)}$$

In case of fast reactions, the value of $f_{orientation}$ is very high of the order of 10^2 to 10^6 but in case of slow reactions, the value of $f_{orientation}$ is very low of the order of 10^{-6} to 10^{-9}.

9.16.2 Transition State Theory (TST)

(a) According to this theory, reactant molecules weaken their translational and rotational degrees of freedom and execute only vibrational degrees of freedom. Under this circumstance, reactant molecules collide at a specific direction to form an activated complex with high potential energy. The energy difference between activated complex and reactant molecules is known as ***activation energy***. It is illustrated in Fig. 9.7.

(b) Activated complex is formed through the formation of weak bonds among reactant molecules as shown in the following example. As the theory is based on the formation of activated complex, it is also known as **Activated complex theory**.

$$X - Y + Z \rightleftharpoons [X \cdots Y \cdots Z]$$

(c) Activated complex remains in equilibrium with the reactant molecules. That means activated complex may dissociate into reactant molecules. However, activated complex is unstable and short-lived. Finally, it is decomposed into products and once it is decomposed it cannot be regenerated. That means breakdown of activated complex into products is an irreversible process. However, formation of product does not affect the equilibrium. Hence, it is called **quasi-equilibrium state** and concentration of activated complex can be calculated by using statistical mechanics.

(d) Activated complex is also known as transient species, which has only theoretical entity and hence cannot be isolated. It is assumed that thermodynamic properties, degrees of freedom of activated complex are similar to those of reactant molecules.

(e) According to this theory a bimolecular reaction can be represented as:

$$A + B \rightleftharpoons AB^* \longrightarrow Products$$

$$K = \frac{C_{AB^*}}{C_A.C_B}. \text{ Or, } C_{AB^*} = K.C_A.C_B \tag{9.48a}$$

(f) Activation energy is the additional energy required for the reactant molecules to form activated complex. According to this theory, activation energy is the energy difference between the lowest vibrational energy of activated complex and that of reactant molecules. It does not correspond to experimental activation energy.

(g) Activated complex is continuously vibrating and finally splits into products. The vibrational energy is related to temperature according to the following equation:

$$hv = k_B T. \text{ Or, } v = \frac{k_B T}{h}, \text{ where, } k_B = \text{Boltzmann constant.}$$

K is the equilibrium constant and is related to thermodynamic free energy as follows:

$$\Delta G^0 = -RT \ln K. \text{ Again, } \Delta G^0 = \Delta H^0 - T\Delta S^0. \text{ So, } K = e^{-\Delta G^0/RT} = e^{-\Delta H^0/RT} \cdot e^{\Delta S^0/R}$$

(h) According to this theory the reaction rate is given by:

$$\text{Rate} = J = v.C_{AB^*} = \left(\frac{k_B T}{h}\right)(K \cdot C_A \cdot C_B) \qquad [\text{using equation 9.48(a)}]$$

Putting the value of K the rate equation becomes

$$J = \left(\frac{k_B T}{h}\right)e^{-\Delta H^0/RT} \cdot e^{\Delta S^0/R} \cdot C_A \cdot C_B \tag{9.48b}$$

Again according to reaction kinetics the reaction rate is given by:

$$J = k \cdot C_A \cdot C_B \tag{9.49}$$

Comparing equations (9.48b) and (9.49), we get

$$k = \left(\frac{k_B T}{h}\right)e^{-\Delta H^0/RT} \cdot e^{\Delta S^0/R} = \left(\frac{RT}{N_A h}\right)e^{\Delta S^0/R} \cdot e^{-\Delta H^0/RT} \tag{9.50a}$$

Taking ln on both sides we get

$$\ln k = \ln\left(\frac{k_B}{h}\right) + \ln T - \frac{\Delta H^0}{RT} + \frac{\Delta S^0}{R}$$

Differentiating the above equation with respect to temperature, we get

$$\frac{d \ln k}{dT} = \frac{\Delta H^0}{RT^2} + \frac{1}{T} = \frac{(\Delta H^0 + RT)}{RT^2} \tag{9.50b}$$

The equation (9.50b) is known as **Wynne-Jones and Eyring's** equation developed in 1935.

According to Arrhenius equation, we have

$$\frac{d \ln k}{dT} = \frac{E_a}{RT^2} \tag{9.47d}$$

Comparing the equations (9.47d) and (9.50b), we have

$$E_a = \Delta H^0 + RT \quad \text{and} \quad \text{Frequency factor} = A = \left(\frac{RT}{N_A h}\right)e^{\Delta S^0/R}$$

(i) There is no steric factor in the expression of rate constant because it does not consider how the activated complex is formed. However in collision theory, expression of rate constant includes a steric factor.

(j) Absolute value of rate constant can be determined by using this theory. Hence, it is also known as **Absolute reaction rate theory**.

9.17 FACTORS INFLUENCING THE REACTION RATE

(a) **Activation energy** As we have seen in the earlier discussion, the rate constant, k, is related to the activation energy, E_a, according to the following equation:

$$k = Ae^{-(E_a/RT)} \tag{9.47c}$$

Thus, it is evident that with the decrease in activation energy, the rate constant, k, increases and hence the reaction rate increases.

(b) **Catalyst** In presence of a catalyst, the path of a chemical reaction changes and also the activation energy. Figure 9.10 shows how the reaction path changes in presence of a catalyst.

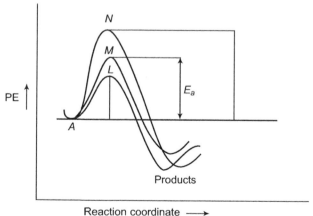

Figure 9.10 PE diagram of reaction of type $2A \rightarrow$ Products; E_a = Activation energy in absence of any catalyst

The curve M indicates the normal reaction path in absence of any catalyst. In presence of a catalyst if the reaction path changes from M to L the activation energy decreases and the catalyst is called positive catalyst. The catalyst is said to be negative catalyst if its presence changes the reaction path from M to N such that activation energy increases. Normal reaction without catalyst may be written as:

$$A + A \longrightarrow A^* + A$$
$$A^* \longrightarrow \text{Products}$$

Whereas in the catalyzed reaction, collision occurs between reactant molecules, A, and catalyst molecules, C to form activated complex, $(AC)^*$, which ultimately breaks into products with the regeneration of catalyst.

$$A + C \longrightarrow (AC)^*$$
$$(AC)^* \longrightarrow \text{Products} + C$$

If the catalyst is a positive catalyst, the potential energy of $(AC)^*$ is much lower than that of A^* and the reaction occurs at much faster rate than the reaction without catalyst, e.g., esterification reaction in acidic medium.

$$RCOOH + R'OH \xrightarrow{H^+} RCOOR' + H_2O$$

Here acid acts as a positive catalyst.

In presence of a negative catalyst, the potential energy of $(AC)^*$ is much higher than that of A^* and hence the reaction occurs at much slower rate than the reaction without catalyst. One example is oxidation of hydrochloric acid by potassium permanganate. The reaction rate is sufficiently retarded in presence of manganous sulphate.

$$KMnO_4 + HCl \xrightarrow{MnSO_4} MnCl_2 + Cl_2$$

Here, $MnSO_4$ acts as a negative catalyst.

(c) **Temperature**　From equation (9.47c) it is evident that the rate constant, k increases with the increase in temperature. It has been found that with just $10°$ rise in temperature the rate of reaction is nearly doubled. The ratio $(k_{\theta+10}/k_\theta)$ lies between 2 to 3 for several chemical reactions. The ratio $(k_{\theta+10}/k_\theta)$ is known as **temperature coefficient of reaction**.

(d) **Concentration of reactants**　Rate of a reaction is directly related to the concentration of the reactants. The reactant to which order of a reaction is positive the reaction rate increases with the increase in the reactant concentration but the same decreases with the increase in reactant concentration, to which order of the reaction is negative. That type of reactant is called inhibitor. One example is:

$$OCl^- + I^- \xrightarrow{OH^-} OI^- + Cl^-$$

The rate of reaction, J, is given by:

$$J = k\frac{[I^-][OCl^-]}{[OH^-]}$$

The order of reaction with respect to $[OH^-]$ is -1 and hence the reaction rate, J, decreases with the increase in $[OH^-]$. Thus, OH^- is called inhibitor to the above reaction.

(e) **State of subdivision**　If one of the reactants remains in solid state, the reaction rate depends on the particle size of the solid. Smaller the particle size larger the area and hence faster the reaction rate. In case of heterogeneous catalyst system, catalyst often remains in solid state. Thus, finely subdivided catalyst provides a very large surface area, accelerating the reaction rate significantly.

(f) **Radiation**　If the reactants of a reaction are irradiated with a monochromatic light the reaction proceeds through free radical mechanism. One such example is formation of hydrochloric acid from its elements.

$$Cl_2 \xrightarrow{hv} 2Cl^\bullet \qquad \text{(Initiation)}$$

$$Cl^\bullet + H_2 \longrightarrow HCl + H^\bullet \quad \text{(Propagation)}$$

$$H^{\bullet} + Cl_2 \longrightarrow HCl + Cl^{\bullet} \quad \text{(Propagation)}$$

$$Cl^{\bullet} + Cl^{\bullet} \longrightarrow Cl_2 \quad \text{(Termination)}$$

Radiation intensity has significant influence on reaction rate since mechanism is initiated through the formation of free radicals by radiation. So, reaction rate is also influenced by radiation intensity.

SOLVED PROBLEMS

Problem 9.1   Write the rate expressions for the following reactions in terms of the disappearance of the reactants and the appearance of the products.

(a) $I^- (aq) + OCl^- (aq) \longrightarrow Cl^- (aq) + OI^- (aq)$

(b) $4NH_3(g) + 5O_2(g) \longrightarrow 4NO(g) + 6H_2O(g)$

Solution

(a) Here the stoichiometric coefficients equals 1. Thus,

$$\text{Rate} = -\frac{\Delta[I^-]}{\Delta t} = -\frac{\Delta[OCl^-]}{\Delta t} = +\frac{\Delta[Cl^-]}{\Delta t} = +\frac{\Delta[OI^-]}{\Delta t}$$

(b) Here the stoichiometric coefficients are 4,5,4, and 6.

$$\text{Rate} = -\frac{1}{4}\frac{\Delta[NH_3]}{\Delta t} = -\frac{1}{5}\frac{\Delta[O_2]}{\Delta t} = +\frac{1}{4}\frac{\Delta[NO]}{\Delta t} = +\frac{1}{6}\frac{\Delta[H_2O]}{\Delta t}$$

Problem 9.2   Consider the reaction: $4NO_2(g) + O_2(g) \longrightarrow 2N_2O_5(g)$

Suppose that, at a particular moment during the reaction, molecular oxygen is reacting at the rate of 0.024 M/s. (a) At what rate is N_2O_5 being formed? (b) At what rate is NO_2 reacting?

Solution

$$\text{Rate} = -\frac{1}{4}\frac{\Delta[NO_2]}{\Delta t} = -\frac{\Delta[O_2]}{\Delta t} = +\frac{1}{2}\frac{\Delta[N_2O_5]}{\Delta t}. \quad \text{Given:} -\frac{\Delta[O_2]}{\Delta t} = 0.024 \text{ mole/s}$$

$$\text{Hence,} \frac{1}{2}\frac{\Delta[N_2O_5]}{\Delta t} = 0.024 \text{ mole/s.} \quad \text{Or,} \frac{\Delta[N_2O_5]}{\Delta t} = 0.048 \text{ mole/s}$$

$$\text{Again,} -\frac{1}{4}\frac{\Delta[NO_2]}{\Delta t} = -\frac{\Delta[O_2]}{\Delta t} = 0.024 \text{ mole/s.} \quad \text{Or,} \frac{\Delta[NO_2]}{\Delta t} = -0.096 \text{ mole/s}$$

Problem 9.3   Consider the following reaction $2A + B \longrightarrow C + D$. The reaction rate is given by $J = k\,[A][B]$. If the volume of the container is reduced to $\dfrac{1}{4}$ th of the initial, what will be the reaction rate compared to initial one?

Solution

As volume is reduced to $\dfrac{1}{4}$ th of the initial, concentration of each reactant is increased to 4 times. Thus, the reaction rate is increased to 16 times.

Problem 9.4   The reaction of nitric oxide with hydrogen is given by:

$$2NO(g) + 2H_2(g) \longrightarrow N_2(g) + 2H_2O(g)$$

From the following data collected at this temperature, determine (a) the rate law, (b) the rate constant, and (c) the rate of the reaction when $[NO] = 12 \times 10^{-3}$ M and $[H_2] = 6 \times 10^{-3}$ M.

Observation	*[NO] M*	*[H₂] M*	*Initial rate M/s*
1	5×10^{-3}	2×10^{-3}	1.3×10^{-5}
2	10×10^{-3}	2×10^{-3}	5×10^{-5}
3	10×10^{-3}	4×10^{-3}	10×10^{-5}

Solution

The rate law is given by Rate $= J = k[NO]^x[H_2]^y$

Observations 1 and 2 show that on doubling the concentration of [NO] at constant concentration of H_2, the rate increases 4 times. Taking the ratio of the rates from these two observations, we have

$$\frac{J_2}{J_1} = \frac{5 \times 10^{-5}}{1.3 \times 10^{-5}} \approx 4 = 2^2. \text{ Again, } \frac{J_2}{J_1} = \frac{[2NO]^x}{[NO]^x} = 2^x. \text{ Or, } x = 2$$

Observations 2 and 3 show that on doubling the concentration of H_2 at constant concentration of NO, the rate increases 2 times. Taking the ratio of the rates from these two observations, we have

$$\frac{J_3}{J_2} = \frac{10 \times 10^{-5}}{5 \times 10^{-5}} = 2 = 2^1. \text{ Again, } \frac{J_3}{J_2} = \frac{[2H_2]^y}{[H_2]^y} = 2^y. \text{ Or, } y = 1$$

So, the rate expression is Rate $= J = k[NO]^2[H_2]$

$$k = \frac{J}{[NO]^2[H_2]} = \frac{5 \times 10^{-5}}{[10 \times 10^{-3}]^2[2 \times 10^{-3}]} = 2.5 \times 10^2 \text{ l}^2/\text{mole}^2 \cdot \text{s}$$

Using this value of k and known concentrations of NO and H_2, we have

$$\text{Rate} = J = (2.5 \times 10^2)(12 \times 10^{-3})^2(6 \times 10^{-3}) = 2.2 \times 10^{-4} \text{ mole/s}.$$

Problem 9.5 Develop the rate equation for the following chemical reaction:

$$2NO(g) + Br_2(g) \longrightarrow 2NOBr(g)$$

Solution

Step-I $NO + Br_2 \underset{k_2}{\overset{k_1}{\rightleftharpoons}} NOBr_2$ (Fast)

Step-II $NO + NOBr_2 \xrightarrow{k_3} 2NOBr$ (Slow)

In case of Step-I, equilibrium exists and hence we can write that
Rate of forward reaction = Rate of backward reaction.

Or, $k_1[NO][Br_2] = k_2[NOBr_2].$

Hence, $[NOBr_2] = \left(\dfrac{k_1}{k_2}\right)[NO][Br_2]$

Rate of formation of product $= J = k_3[NOBr_2][NO]$. Putting the value of $[NOBr_2]$, we have

Or, $J = \left(\dfrac{k_1 k_3}{k_2}\right)[NO]^2[Br_2] = k[NO]^2[Br_2]$, where, $k = \left(\dfrac{k_1 k_3}{k_2}\right)$.

Problem 9.6 Does a catalyst change the heat of reaction?

Solution

Heat of reaction $= \Delta H = H_P - H_R$. As H_R and H_P are constants irrespective of presence of any catalyst, the latter cannot change the heat of reaction.

Problem 9.7 For the reaction $N_2 + 3H_2 \longrightarrow 2NH_3$, $\{d[NH_3]/dt\} = 2 \times 10^{-4}$ mole. $L^{-1}s^{-1}$. What will be the value of $\{-d[H_2]/dt\}$?

Solution

$$-\dfrac{d[N_2]}{dt} = -\dfrac{1}{3}\dfrac{d[H_2]}{dt} = \dfrac{1}{2}\dfrac{d[NH_3]}{dt}, \quad -\dfrac{d[H_2]}{dt} = \dfrac{3}{2}\dfrac{d[NH_3]}{dt} = 3 \times 10^{-4} \text{ mole.} L^{-1}s^{-1}$$

Problem 9.8 $BrO_3^-(aq) + 5Br^-(aq) + 6H^+ \longrightarrow 3Br_2(l) + 3H_2O(l)$. Find the relation between rate of appearance of Br_2 and rate of disappearance of Br^-.

Solution

$$\dfrac{1}{3}\dfrac{d[Br_2(l)]}{dt} = -\dfrac{1}{5}\dfrac{d[Br^-]}{dt}, \quad \dfrac{d[Br_2(l)]}{dt} = -\dfrac{3}{5}\dfrac{d[Br^-]}{dt}$$

Problem 9.9 $t_{1/2}$ for a 1st order reaction is 10 minutes. Initial concentration of reactant is 10 M. What will be the concentration after 30 minutes?

Solution

$$t_{1/2} = 10 \text{ minutes}, \quad t = 3t_{1/2}, \quad \frac{C}{C_0} = \left(\frac{1}{2}\right)^3, \quad C = \frac{C_0}{8} = \left(\frac{10}{8}\right) M = 1.25 \, M.$$

Problem 9.10 For a chemical reaction $A \longrightarrow B$, the rate of reaction is 2×10^{-3} mol dm^{-3}s^{-1}, when the initial concentration is 0.05 mol.dm^{-3}. The rate of the same reaction is 1.6×10^{-2} mol.dm^{-3}s^{-1}, when the initial concentration is 0.1 mol.dm^{-3}. What is the order of the reaction?

Solution

$$\frac{J_2}{J_1} = \left(\frac{a_2}{a_1}\right)^n, \quad \text{Or,} \quad \frac{1.6 \times 10^{-2}}{2 \times 10^{-3}} = \left(\frac{0.1}{0.05}\right)^n, \quad \text{Or,} \quad 2^3 = 2^n, \quad \text{Or,} \quad n = 3.$$

Problem 9.11 Find the order of the following reactions:

$$A_2 \rightleftharpoons 2A \qquad \text{(Fast)}$$
$$A + B_2 \rightarrow AB + B \quad \text{(Slow)}$$
$$A + B \rightarrow AB \qquad \text{(Fast)}$$

Solution

$$K = \frac{C_A^2}{C_{A_2}}, \quad C_A = K^{1/2} C_{A_2}^{1/2}, \quad J = kC_A C_{B_2} = kK^{1/2} C_{A_2}^{1/2} C_{B_2}, \quad \text{Order} = \frac{3}{2}.$$

Problem 9.12 The gas phase decomposition of dimethyl ether follows first order kinetics.

$$CH_3OCH_3(g) \longrightarrow CH_4(g) + H_2(g) + CO(g)$$

The reaction is carried out in a constant volume container at 500°C and has a half-life of 14.5 minutes. Initially only dimethyl ether is present at a pressure of 0.40 atm. What is the total pressure of the system after 12 minutes ? Assume ideal gas behaviour, what is the partial pressure of $CH_3OCH_3(g)$ after 1420 sec?   [*Ans.:* 0.749 atm, 0.942 atm]

Solution

$$CH_3OCH_3(g) \longrightarrow CH_4(g) + H_2(g) + CO(g)$$
$$\quad (a-x) \qquad\qquad x \qquad\quad x \qquad\quad x$$

Total number of moles $= (a + 2x)$ and $a \propto 0.40$ atm.

$$k = \frac{0.693}{t_{1/2}} = \frac{0.693}{14.5} \text{min}^{-1} = 0.0478 \text{min}^{-1}. \quad t = 12 = \frac{1}{0.0478} \ln \frac{a}{(a-x)}.$$

Or, $x = 0.4365a$

$(a + 2x) = a + 2 \times 0.4365a = 1.873a \propto 1.873 \times 0.40$ atm $\propto 0.749$ atm.

EXERCISES

1. The reaction of peroxydisulfate ion $(S_2O_8^{2-})$ with iodide ion (I^-) is given below:

$$S_2O_8^{2-}(aq) + 3I^-(aq) \longrightarrow 2SO_4^{2-}(aq) + I_3^-(aq)$$

Using the following data, determine the rate law and calculate the rate constant.

Observations	$S_2O_8^{2-}$ (M)	I^-(M)	Rate (mole/lit.s)
1	0.080	0.034	2.2×10^{-4}
2	0.080	0.017	1.1×10^{-4}
3	0.16	0.017	2.2×10^{-4}

2. The conversion of cyclopropane to propene in the gas phase is a 1st order reaction with a rate constant of $6.7 \times 10^{-4} s^{-1}$ at 500°C.

(a) If the initial concentration of cyclopropane was 0.25M, what is the concentration after 8.8 minutes?

(b) How long (in minutes) will it take for the concentration of cyclopropane to decrease from 0.25 M to 0.15 M?

(c) How long (in minutes) will it take to convert 74% of the starting material?

[*Ans.:* 0.18 M, 13 minutes, 33 minutes]

3. Ethyl iodide (C_2H_5I) decomposes at a certain temperature in the gas phase as follows:

$$C_2H_5I(g) \longrightarrow C_2H_4(g) + HI(g)$$

Using the following data determine the order of the reaction and the rate constant.

Time (min)	$[C_2H_5I]$ (M)
0	0.36
15	0.30
30	0.25
48	0.19
75	0.13

4. Iodine atoms combine to form molecular iodine in the gas phase:

$$I(g) + I(g) \longrightarrow I_2(g)$$

This reaction follows second-order kinetics and has the rate constant of 7×10^9 mole/lit.s at 23°C. (a) If the initial concentration of $I(g)$ was 0.086 M, calculate the concentration after 2 minutes. (b) Calculate the half-life of the reaction if the initial concentration of $I(g)$ is 0.060 M and (c) if it is 0.42 M.

[*Ans.:* (a) 1.2×10^{-12} M, (b) 2.4×10^{-10} s, (c) 3.4×10^{-10} s]

5. Half-life time for substance A in a second order reaction A → Products at 20°C is 25 seconds when $[A]_0 = 0.8$ mole/l.
   (a) Calculate the time required so that the concentration of A is reduced to one-fifth of its initial value.
   (b) If the rate of the reaction is doubled when temperature increases from 20°C to 30°C, calculate the activation energy of the process.

   [*Ans.* (a) 100 s, (b) 51.2 kJ/mole]

6. Sucrose or common sugar $(C_{12}H_{22}O_{11})$ is hydrolyzed with water to form fructose $(C_6H_{12}O_6)$ and glucose $(C_6H_{12}O_6)$. The mix of fructose and glucose is known as inverted sugar and it is very important because it does not crystallize, so soft candies can be produced. Using the given kinetic data, determine:
   (a) The order of the reaction.
   (b) Time required for 50% of sucrose to hydrolyze.
   (c) For an ideal perfect mix batch reactor, what is the reactor volume required to obtain 2000 kg of inverted sugar in every batch (for a conversion rate of 93%)?

Reaction time (minutes)	$[C_{12}H_{22}O_{11}]$ *(M)*
0	0.57
60	0.45
96.4	0.39
157.5	0.30

   [*Ans:* (a) 1st order, (b) 170.3 minutes, (c) 139.2 m³]

7. A 1st order reaction has a chemical equation: $A \longrightarrow B + C$. The rate constant for this reaction is 1.43×10^{-2} s^{-1} and the initial concentration of A is 0.567 M. (a) What is the concentration of A, 2.5 min. later? (b) How long does it take to drop the concentration from 0.5 M to 0.00463 M?

   [*Ans.:* (a) 0.0664 M, (b) 327 s]

8. Consider this reaction: $C_2H_6 \longrightarrow C_2H_4 + H_2$. Deduce the rate expression as the rate of production of C_2H_4 if this reaction involves the following mechanisms:
   Initiation: $C_2H_6 \longrightarrow 2CH_3^*$, Transfer: $CH_3^* + C_2H_6 \longrightarrow CH_4 + C_2H_5^*$
   Propagation: $C_2H_5^* \longrightarrow C_2H_4 + H^*$, $H^* + C_2H_6 \longrightarrow C_2H_5^* + H_2$
   Termination: $H^* + C_2H_5^* \longrightarrow C_2H_6$

   $$\left[\text{Rate} = k_3 \left(\frac{2k_1 + k_4 k'}{k_3 + k_5 k'} \right) [C_2H_6], \text{ where, } k' = \frac{-k_1 k_5 \pm \sqrt{k_1 k_5 (k_1 k_5 + 4k_3 k_4)}}{2k_4 k_5} \right]$$

9. The optical rotation of sucrose in 0.5 (N)-hydrochloric acid at 35°C and at various time intervals are given below. Find out the order of the reaction.

Time (min)	0	10	20	30	40	∞
Rotation (Degree)	+32.4	+28.8	+25.5	+22.4	+19.6	−11.1

10. The decomposition of a certain insecticide in water follows first-order kinetics with a rate constant of 1.45 yr^{-1} at 12°C. A quantity of this insecticide is washed into a lake

on June 1, leading to a concentration of 5×10^{-7} gm/cm^3. Assume that the average temperature of the lake is 12°C.

(a) What is the concentration of the insecticide on June 1 of the following year?

(b) How long will it take for the concentration of the insecticide to drop to 3×10^{-7} gm/cm^3?

[*Ans.:* (a) 1.2×10^{-7} gm/cm^3, (b) 0.35 yr]

11. At 100°C, the gaseous reaction $A \longrightarrow B + C$ is observed to be 1st order. On starting with pure A it is found that at the end of 10 minutes the total pressure of the system is 176 mm Hg and after a long time it is 270 mm Hg. Calculate the half-life period of the reaction.

[*Ans.:* 10.7 minutes]

12. Two substances A and B undergo a bimolecular reaction step. The following table gives the concentration of A at various times for an experiment carried out at constant temperature of 17°C.

$[A] \times 10^4$ *(mole/dm^3)*	10.00	7.94	6.31	5.01	3.98
Time (min)	0	10	20	30	40

The initial concentration of B is 2.5 mole/dm^3. Calculate the value of 2nd order rate constant.

[*Ans.:* 1.54×10^{-4} dm^3/mole·s]

13. The gas-phase reaction of decomposition of $NO_2Cl(g)$ is given below:

$$2NO_2Cl(g) \longrightarrow 2NO_2(g) + Cl_2(g).$$ Rate equation is $J = k[NO_2Cl]$

Devise two mechanisms consistent with this rate law.

14. The reaction $2Cr^{2+} + Tl^{3+} \longrightarrow 2Cr^{3+} + Tl^+$ in aqueous solution has the rate expression $J = k[Cr^{2+}][Tl^{3+}]$. Devise two mechanisms consistent with this rate law.

15. The gas-phase reaction: $2NO_2 + F_2 \longrightarrow 2NO_2F$ has the rate expression $J = k[NO_2][F_2]$. Devise two mechanisms consistent with this rate law.

16. In a 2nd order reaction the initial concentration of reactants is 0.1 mole/l. The reaction is found to be 20% complete in 40 minutes. Calculate: (i) the rate constant, (ii) half-life period, (iii) time required to complete 75% of the reaction.

[*Ans.:* (i) 0.0625 l/(mole–min), (ii) 160 minutes, (iii) 480 minutes]

17. The 1st order decomposition of N_2O_4 to NO_2 has a k value of 4.5×10^3 s^{-1} at 1°C and an activation energy of 58 kJ/mole. At what temperature its half-life would be 6.93×10^{-5} s?

[*Ans.:* 287.9° K]

18. For a 2nd order reaction values of k are 1.2×10^{-3} and 3×10^{-5} l/(mole–s) at 700 K and 629 K respectively. Estimate E_a and frequency factor.

[*Ans.:* $E_a = 45.5$ kcal/mole and $A = 1.9 \times 10^{11}$ mole-s/l]

19. Two 2nd order reactions A and B have identical frequency factors. The activation energy of A exceeds that of B by 10.46 kJ/mole. At 100 °C the reaction is 30% completed after 1 hr when the initial concentration of the reactant is 0.1 moles/l. How long will it take for the reaction B to reach 70% completion at the same temperature for an initial concentration of 0.05 moles/l?

[*Ans.:* 1340 s]

20. The specific reaction rate of a chemical reaction at $0°C$ and $33°C$ are 2.45×10^{-5} sec^{-1} and 16.2×10^{-4} sec^{-1} respectively. Calculate the energy of activation in kJ. (Given: R $= 2$ cal/(mole-k)]

   [*Ans.:* 88.66 kJ]

21. The specific rate constant of a 2nd order reaction ($2A \rightarrow$ Products) is given by:

   $$\log k = -\frac{3163}{T} + 11.899$$

   The initial concentration of the reactant is 0.005 M. Calculate activation energy and $t_{1/2}$ at $25°C$. Given: $R = 2$ cal/(mole–K). [*Ans.:* $E = 14.568$ kcal, $t_{1/2} = 10.4$ minutes]

22. The half-life period of a 1st order reaction is 15 minutes. Calculate the rate constant and time taken to complete 80% of the reaction. [*Ans.:* 34.8 minutes]

23. Show that for a 1st order reaction the time required for 75% reaction is twice the time for 50% reaction.

24. The 1st order decomposition of H_2O_2 in a suitable medium is characterized by a rate constant 3×10^{-2} min^{-1}. Find the time to complete one-third of the reaction.

   [*Ans.:* 13.5 minutes]

25. What are effective collisions?

   [*Ans.:* Effective collisions are those collisions, that gives rise to products.]

26. State the properties of catalyst influenced by promoters.

   [*Ans.:* (a) activity, (b) selectivity, and (c) stability.]

27. For a 1st order reaction, $t_{1/2}$ is 14 sec. Calculate the time period for the initial concentration to reduce to $\frac{1}{8}$th of its value.

   $$\left[Ans: \frac{N_t}{N_0} = \left(\frac{1}{2}\right)^n = \frac{1}{8}. \text{ Or, } n = 3. \ t = 3 \times t_{1/2} = 42\,\text{sec} \right]$$

28. $N_2(g) + 3H_2(g) \rightleftharpoons 2NH_3(g) + 22$ kcal. The activation energy for the forward reaction is 50 kcal. What is the activation energy for the backward reaction? [*Ans.:* 72 kcal]

29. The rate constant for a 1st order reaction becomes 6 times when the temperature is raised from $350°K$ to $400°K$. Calculate E_a in kJ/mole. [*Ans.:* 41.721 kJ/mole]

30. Temperature coefficient of a reaction has been found 2.6. The reaction is carried out at $400°K$. Calculate E_a in kcal/mole. [*Ans.:* 31.34 kcal/mole]

31. The rate constant of a 1st order reaction at $17°C$ is 10^{-3} min^{-1}. The temperature coefficient of this reaction is 2. What is the rate constant at $27°C$ for this reaction?

32. The half-life period of decomposition of a compound is 50 minutes. If initial concentration is increased 4 times, the half-life period reduces to 25 minutes. What is the order of the reaction?

   $$\left[Ans.: t_{1/2} \propto \frac{1}{a^{n-1}}. \text{ Or, } \frac{50}{25} = \left(\frac{a}{4a}\right)^{1-n}. \text{ Or, } n = \frac{3}{2} \right]$$

33. Half-life period of decomposition of a compound was found to be 50 minutes. With initial concentration a mole/l. If the initial concentration is halved, half-life period becomes 100 minutes. Calculate order of the reaction. [*Ans.:* 2nd order]

34. Consider the neutralization reaction of a strong acid by a strong base $H^+ + OH^- \longrightarrow H_2O$. Rate constant of the reaction has been found to be 1.3×10^{11} l/mole.s. Calculate the half-life period of the above reaction when initial concentration of strong acid is $10^{-4}N$, same as that of strong base. [*Ans.:* 7.7×10^{-8} s]

35. The decomposition of ammonia on tungsten at 1200°K is a zero order reaction. The reaction is carried out in a rigid vessel. The initial pressure of ammonia is 200 mmHg. After 160 minutes the total pressure has been found to be 300 mmHg. What will be the partial pressure of ammonia after 1 hour if the initial pressure of ammonia is 150 mmHg? [*Ans.:* 112.5 mmHg]

36. In gas phase, kinetics pressures instead of concentrations are used in rate laws. Consider the gas phase reaction $A \rightarrow$ Products. The rate equation is given by $-\dfrac{dp_A}{dt} = k_P p_A^n$, where, k_p is the rate constant with respect to pressure and p_A is the partial pressure of A.

    (a) Show that $k_p = k_C (RT)^{1-n}$.

    (b) Is that valid for n^{th} order reaction?

37. The half-life of ^3H is 12.4 years. (a) Calculate the activity of 20 gm HNO_3, containing 0.2 mole% of $^3HNO_3$.

    (b) Calculate the activity of $^3HNO_3$ after 6.2 years.
    [*Ans.:* (a) 6.77×10^{11} s^{-1}, (b) 4.79×10^{11} s^{-1}]

38. Consider the reaction: $2Cr^{2+} + Tl^{3+} \longrightarrow Cr^{3+} + Tl^+$. In aqueous solution the reaction rate is given by $r = k [Cr^{2+}][Tl^{3+}]$. Devise two mechanisms consistent with this rate law.

39. Consider the gas phase reaction: $2NO_2 + F_2 \longrightarrow 2NO_2F$. The rate equation is given by $r = k [NO_2][F_2]$. Devise a mechanism consistent with this rate law.

10 Catalysis

10.1.1 Definition

Studies on several chemical reactions under different circumstances reveal that rate of a chemical reaction may be significantly altered in presence of a specific foreign substance, although the latter is not at all consumed and remains unchanged chemically at the end of the reaction. The phenomenon is known as catalysis and the foreign substance is called catalyst.

10.1.2 Characteristic Features of a Catalyst

(i) It remains unchanged chemically and quantitatively at the end of reaction.

(ii) It cannot start a chemical reaction but can only change the velocity of the reaction.

(iii) Only a trace amount of catalyst is enough to produce a massive impact on reaction rate.

(iv) A catalyst takes part in chemical reaction and changes the reaction path such that activation energy changes.

(v) It speeds up both forward and backward reactions equally such that equilibrium constant (K) remains unaltered and hence also the position of equilibrium, since we know that $\Delta G^0 = -RT \ln K$, where, ΔG^0 is the standard free energy change.

(vi) A catalyst can change the nature of product since action of catalyst on chemical reactions is highly specific. For example, ethanol gives acetaldehyde in presence of Ni catalyst but gives ethylene in presence of alumina (Al_2O_3).

10.2 CLASSIFICATION

(a) On the basis of **activity** catalysts can be divided into two main categories: (i) Positive catalyst, and (ii) Negative catalyst.

 (i) **Positive catalyst** A catalyst is said to be positive catalyst when it accelerates the reaction rate appreciably with significant reduction in activation energy. It is illustrated in the Fig. 10.1.

 Example-1 During manufacturing of sulphuric acid by contact process, platinum is used as a positive catalyst in the conversion of sulphur dioxide to sulphur trioxide.

$$2SO_2(g) + O_2(g) \xrightarrow{[Pt]} 2SO_3(g)$$

 (ii) **Negative catalyst** A catalyst is said to be negative catalyst when it retards the reaction rate noticeably with significant increase in activation energy. It is illustrated in the Fig. 10.1.

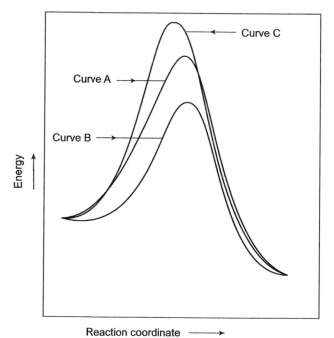

Figure 10.1 Change in activation energy in presence of positive and negative catalysts. Curve-A represents energy diagram for a reaction in absence of any catalyst, Curve-B represents energy diagram in presence of positive catalyst and Curve-C represents energy diagram in presence of negative catalyst

 Example-1 Decomposition of hydrogen peroxide is sufficiently restricted in presence of orthophosphoric acid or a trace amount of glycerin. So orthophosphoric acid or glycerin acts as negative catalyst in case of decomposition of hydrogen peroxide.

$$2H_2O_2 \xrightarrow{H_3PO_4/Glycerin} 2H_2O + O_2$$

Example-2 Sodium sulphite (Na_2SO_3) is oxidised to sodium sulphate (Na_2SO_4) spontaneously in air but the oxidation process is sufficiently suppressed in presence of ethanol. Thus, ethanol acts as negative catalyst.

$$2Na_2SO_3 + O_2(g) \xrightarrow{C_2H_5OH} 2Na_2SO_4$$

(b) On the basis of **physical state** catalysts can be broadly divided into two main categories: (i) Homogeneous catalyst, and (ii) Heterogeneous catalyst.

   (i) **Homogeneous catalyst** When the catalyst and reactants remain in the same phase, the catalyst is called homogeneous catalyst, e.g., in the chamber process of manufacturing sulphuric acid, gaseous nitric oxide (NO) acts as a catalyst, which remains in the same phase with reactants (SO_2 and O_2).

$$2SO_2(g) + O_2(g) \xrightarrow{[NO]} 2SO_3(g)$$

   (ii) **Heterogeneous catalyst** When catalyst remains in different phase with reactants, the catalyst is called heterogeneous catalyst, e.g., finely divided iron acts as a catalyst in the Haber process of manufacturing ammonia.

$$N_2(g) + 3H_2(g) \xrightarrow{[Fe]} 2NH_3(g)$$

(c) **Induced catalyst**
   There are some reactions, which usually never occur due to high activation energy. However, those reactions occur spontaneously in presence of other chemical reactions. The phenomenon is called induced catalysis and the reaction, which catalyzes the other chemical reaction, is called induced catalyst. One example is oxidation of Na_3AsO_3 in air. The oxidation reaction of pure Na_3AsO_3 to Na_3AsO_4 never occurs alone but oxidation of pure Na_2SO_3 in air occurs spontaneously according to the following equation.

$$Na_2SO_3 + \frac{1}{2}O_2 \xrightarrow{Air} Na_2SO_4$$

If the mixture of Na_2SO_3 and Na_3AsO_3 is kept open in atmosphere, both are oxidized.

$$Na_2SO_3 + \frac{1}{2}O_2 \xrightarrow{Air} Na_2SO_4 \quad \text{and} \quad Na_3AsO_3 + \frac{1}{2}O_2 \xrightarrow{Air} Na_3AsO_4$$

10.3 AUTOCATALYSIS

There are some chemical reactions where one of the reaction products acts as a catalyst. The phenomenon is known as autocatalysis.

Example-1 Hydrolysis of ester.

$$RCOOR' + H_2O \rightleftharpoons RCOOH + R'OH$$

Acid (RCOOH), produced, during the course of hydrolysis, accelerates the rate of hydrolysis appreciably.

Example-2 **Redox reaction** between oxalic acid and potassium permanganate in acidic medium.

$$MnO_4^- + (COOH)_2 + 8H^+ \rightleftharpoons Mn^{2+} + 2CO_2 + 4H_2O$$

Initially the forward reaction is very slow but as manganous ion concentration increases the reaction rate becomes faster and faster. The reaction is self-catalyzed by manganous ion (Mn^{2+}).

10.4 ACID AND BASE CATALYZED REACTIONS

There are some reactions which are catalyzed by acids as well as bases. One such example is hydrolysis of ester.

(i) **Reaction in absence of catalyst**

$$RCOOR' + H_2O \rightleftharpoons RCOOH + R'OH$$

$$\text{Rate} = -\frac{d[\text{Ester}]}{dt} = k_0'[\text{Ester}], \quad \text{where,} \quad k_0 = k_0'[H_2O]$$

[Since the reaction is carried out in huge excess of water.]

(ii) **Acid catalyzed reaction**

$$\text{Rate} = -\frac{d[\text{Ester}]}{dt} = k_{H^+}[\text{Ester}][H^+] = k[\text{ester}], \quad \text{where,} \quad k = k_{H^+}[H^+]$$

Here, k is the overall rate constant. Taking log on both sides, we get

$$\log k = \log k_{H^+} + \log C_{H^+} = +\log k_{H^+} - pH \tag{10.1}$$

k_{H^+} is constant but the overall rate constant, k, decreases linearly with the increase in pH. One can plot $\log k$ versus pH at different concentration of acid, where, $\log k_{H^+}$ is the intercept. Hence, the value of $\log k_{H^+}$ can easily be evaluated.

(iii) **Base catalyzed reaction**

$$R-\underset{\underset{O}{\|}}{C}-OR' \xrightarrow[\text{Slow}]{OH^-} R-\underset{\underset{OH}{\|}}{C}-OR' \xrightarrow{\text{Fast}} RCOOH + OR^-$$

$$OR'^- + H_2O \xrightarrow{\text{Fast}} R'OH + OH^-$$

$$\text{Rate} = -\frac{d[\text{Ester}]}{dt} = k_{OH^-}[\text{Ester}][OH^-], = k[\text{Ester}], \quad \text{where,} \quad k = k_{OH^-}[OH^-]$$

Here, k is the overall rate constant. Taking log on both sides, we get

$$\log k = \log k_{OH^-} + \log C_{OH^-}$$
$$= \log k_{OH^-} - pOH$$
$$= \log k_{OH^-} - (pK_w - pH) \quad [\text{Since,} \ pK_w = pH + pOH]$$

Or, $\quad\quad\quad \log k = \log (k_{OH^-} \cdot K_w) + pH \quad\quad [\text{Since,} \ pK_w = -\log K_w] \quad (10.2)$

k_{OH^-} and K_w are constants but the overall rate constant (k) increases linearly with increase in pH. One can plot $\log k$ versus pH at different concentrations of base, where, $\log (k_{OH^-} \cdot K_w)$ is the intercept. Hence, the value of (k_{OH^-}) can easily be evaluated.

10.5 ENZYME CATALYZED REACTION

Enzymes are complex protein substances. They usually reside in living cells in colloidal form. Enzymes are essential to execute numerous biochemical reactions, constantly occurring in living cells of human body. In most of the reactions enzymes are acting as catalyst but their action as catalyst is highly specific in nature. One enzyme can catalyze only one chemical reaction and remains inert in other reactions, e.g., enzyme urease hydrolyses only urea but cannot hydrolyze even methyl urea or protein.

10.5.1 Michaelis-Menten Equation

In an enzyme catalyzed reaction usually one substrate undergoes chemical change. At first, enzyme (E) reacts with the substrate (S) to form an intermediate unstable complex [ES], which further dissociates into products (P) with the regeneration of enzyme. Schematically, it is represented as:

$$E + S \underset{k_2}{\overset{k_1}{\rightleftharpoons}} [ES] \xrightarrow{k_3} E + P$$

Rate of formation of product

$$\frac{d[P]}{dt} = k_3 C_{[ES]} \quad\quad\quad\quad (10.3)$$

As the complex [ES] is unstable, at steady state we can write using steady-state concept that

Rate of formation of [ES] = Rate of disappearance of [ES].

$$k_1 C_E \cdot C_S = k_2 C_{[ES]} + k_3 C_{[ES]} = (k_2 + k_3) \, C_{[ES]} \qquad (10.4)$$

Again, if C_E^0 is the initial concentration of enzyme, at time t, we can write that $C_E^0 = C_E + C_{[ES]}$, where C_E is the concentration of free enzyme at time t.

Or, $$C_E = C_E^0 - C_{[ES]}$$

Putting this value of C_E in the equation (10.4), we get

$$k_1 (C_E^0 - C_{[ES]}) C_S = (k_2 + k_3) \, C_{[ES]}$$

Hence, $$C_{[ES]} = \frac{k_1 C_E^0 C_S}{k_1 C_S + k_2 + k_3} = \frac{C_E^0 C_S}{\dfrac{k_2 + k_3}{k_1} + C_S} = \frac{C_E^0 C_S}{k_M + C_S} \qquad (10.5)$$

$$\left[\text{where,} \quad k_M = \frac{k_2 + k_3}{k_1} \right]$$

Combining equations (10.3) and (10.5) we get

$$v = \frac{d[P]}{dt} = \frac{k_3 C_E^0 C_S}{k_M + C_S} \qquad (10.6)$$

where, v is called reaction velocity.

Case-I   If the concentration of the substrate, i.e., C_S, is very small, it may be approximated that

$k_M + C_S \approx k_M$. Thus, equation (10.6) becomes

$$\frac{d[P]}{dt} = \frac{k_3 C_E^0 C_S}{k_M} = k C_S \qquad (10.7)$$

$$\left[\text{where,} \quad k = \frac{k_3}{k_M} C_E^0, \ \text{since } C_E^0 \text{ is constant.} \right]$$

k is constant for a given initial concentration of enzyme.

Conclusion   The enzyme-catalyzed reaction follows first-order kinetics and is independent of the concentration of enzyme, i.e., C_E at low concentration of the substrate.

Case-II   If the concentration of the substrate, i.e., C_S, is very large, it may be approximated that

$k_M + C_S \approx C_S$. Thus, equation (10.6) becomes

$$\frac{d[P]}{dt} = k_3 C_E^0 \qquad (10.8)$$

Comparing equations (10.3) and (10.8), we have $C_{[ES]} = C_E^0$, which means that all enzyme is converted to the complex [*ES*] and there is no free enzyme.

Conclusion The enzyme-catalyzed reaction follows zero-order kinetics, with respect to the substrate at high concentration of the substrate. The reaction velocity reaches maximum under this condition and is represented as

$$v_{max} = \left| \frac{d[P]}{dt} \right|_{max} = k_3 C_E^0$$

Thus, equation (10.6) can be written as

$$v = v_{max} \left(\frac{C_S}{k_M + C_S} \right). \quad \text{At } C_S = k_M, \quad v = \frac{v_{max}}{2} \tag{10.9}$$

The equation (10.9) is known as Michaelis-Menten equation. The corresponding velocity-concentration graph is shown in the Fig. 10.2.

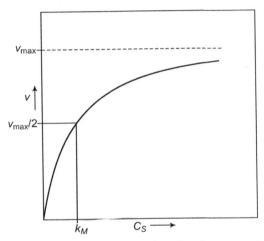

Figure 10.2 Reaction velocity versus concentration plot of enzyme catalyzed reaction

10.5.2 Characteristic Features of Enzyme Catalyzed Reactions

(i) Catalytic activity of enzyme is highly specific and selective. One enzyme can catalyze only one reaction. The enzyme, invertase breaks sucrose into glucose and fructose but unable to break another disaccharide maltose, dissociation of which needs another enzyme, known as maltase.

(ii) Catalytic activity of an enzyme is worst affected in presence of a trace amount of HCN, H_2S, etc.

(iii) Catalytic activity of an enzyme very much depends on pH of the medium, usually enzymes are most active near its **isoelectric point**, e.g., the enzyme **trypsin** is most active at pH around 8.0, which is also the isoelectric point of trypsin.

(iv) Catalytic activity of an enzyme is strongly pronounced within 30°C to 50°C. Reaction rate of an enzyme catalyzed reaction remarkably decreases below and above the temperature range.

(v) Catalytic power of a catalyst increases with increase in temperature but after a certain temperature, catalytic power begins to decrease. Each and every catalyst has a particular temperature, at which catalytic activity is maximum. This temperature is called **optimum temperature**.

10.6 PROMOTERS AND INHIBITORS

Promoters

There are substances, which promote the catalytic activity of catalysts. These substances are known as promoters. Promoter has no catalytic activity. It usually interacts with active component of catalyst, due to which, electronic configuration or crystal structure of catalyst changes. Ultimately performance of catalyst is improved.

Example-1   Haber's process of manufacturing ammonia.

$$N_2(g) + 3H_2(g) \xrightarrow[\text{[Promoter]}]{\text{[Catalyst]}} 2NH_3(g)$$

Catalyst is finely divided iron and promoter is Mo.

Example-2   Contact process of manufacturing sulphuric acid (H_2SO_4),

$$2SO_2(g) + O_2(g) \xrightarrow[\text{[Promoter]}]{\text{[Catalyst]}} 2SO_3(g)$$

Catalyst is finely divided platinum and promoter is V_2O_5.

Example-3   Formation of ethanol from CO and H_2.

$$2CO + 4H_2(g) \xrightarrow[\text{[Promoter]}]{\text{[Catalyst]}} C_2H_5OH + H_2O$$

ZnO acts as catalyst and Cr_2O_3 acts as promoter.

Retarder

Sometimes presence of a trace amount of foreign substance retards or inhibits the reaction rate significantly. The function of retarder is to slow down the reaction rate appreciably, e.g., decomposition of hydrogen peroxide (H_2O_2) is sufficiently slowed down in presence of a trace amount of orthophosphoric acid. The orthophosphoric acid is called retarder.

Inhibitors

Inhibitors are those substances, which totally stops the reaction. The reaction commences only after the inhibitor is totally consumed. The period during which the reaction does not occur at all is called inhibition period. After the inhibition period the reaction starts at the same rate as that observed without inhibitor.

Example-1 Potassium cyanide (KCN) acts as inhibitor in auto-oxidation of sodium sulphite solution, catalyzed by cupric ion, (Cu^{2+}). In presence of KCN, Cu^{2+} forms stable complex $[Cu(CN)_4]^{2-}$ and loses its catalytic activity. Under this circumstance auto-oxidation of Na_2SO_3 solution ceases to occur. However, the auto-oxidation of Na_2SO_3 solution resumes at the same rate only after the total consumption of KCN.

Example-2 Inhibitor is mostly active in chain reactions, which occur in three steps. In the first step, the catalyst decomposes to produce free radicals. In the second step, these free radicals react with the substrate molecules to execute chain reactions. Finally in the third step, free radicals are terminated. In presence of inhibitor, free radicals, generated in the first step, are consumed by inhibitor molecules and hence chain reaction does not occur. Step 2 sets in only after all inhibitor molecules are consumed. This type of inhibitor is very common in polymerization reactions. For example, in free radical polymerization of propylene, nitrobenzene or thiophenol acts as inhibitor.

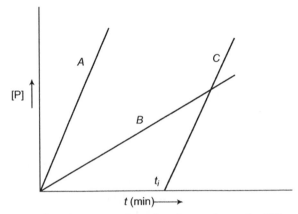

Figure 10.3 Amount of product versus time plot of a reaction under different conditions

In the Fig. 10.3, Curve-A represents rate of formation of products without any inhibitor or retarder. Curve-B shows that the same reaction proceeds with much reduced rate in presence of a retarder. In the case of Curve-C it is clear that the reaction is inhibited up to time t_i and after that the reaction proceeds with the same rate as that of Curve-A. The time t_i is known as **inhibition period** and the substance, that causes inhibition, is called **inhibitor**.

Catalyst Poison

In case of heterogeneous catalyst system, catalyst generally remains in solid state and the reaction occurs on surface of the solid catalyst through adsorption mechanism. In presence of impurity, adsorption of impurity molecules occurs first, thereby preventing the desired chemical reaction. Thus, the surface area of catalyst becomes ineffective or poisoned and the impurity is called catalyst poison. Organic functional groups and inorganic ions have a strong tendency to be adsorbed on metal surfaces, thereby prohibiting reactant molecules to be adsorbed on those sites. So catalyst is sufficiently deactivated, reducing the reaction rate significantly. Common catalyst poisons are amines, sulphides, carbon monoxide, thiols, etc. The catalyst, Ru is very much resistant to poisoning but the other catalysts, like, Ni, Co, Pd, Pt, etc. can be poisoned easily.

10.7 THEORY OF HOMOGENEOUS CATALYST SYSTEM

From the knowledge of chemical kinetics we know that each and every reaction has a definite potential energy barrier. It is believed that in presence of a positive catalyst the energy barrier is sufficiently reduced such that reactants easily cross the energy barrier and form products. This is well illustrated in the Fig. 10.4.

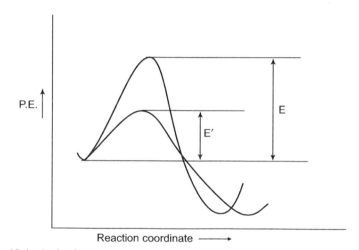

Figure 10.4   Activation energy curve of both uncatalyzed and catalyzed reactions

E = Activation energy of reaction without catalyst

E' = Activation energy of the same reaction in presence of a positive catalyst

In homogeneous catalyst system, catalyst actively participates in the chemical reaction such that reaction is accelerated with regeneration of catalyst at the end of the reaction.

Consider a 1st order reaction: $A \rightarrow$ Products

In presence of a catalyst the above reaction is believed to follow two-step mechanism as shown below:

Step-I $A + C \text{ (Catalyst)} \underset{k_2}{\overset{k_1}{\rightleftharpoons}} [AC] \text{ (Transition state complex)}$

Step-II $[AC] \xrightarrow{k_3} \text{Products} + C$

The rate of formation of product, J, is given by

$$J = k_3 C_{[AC]} \tag{10.10a}$$

As the transition state (TS) complex is short-lived, steady-state concept can be applied on it. So, we have

Rate of formation of [TS-complex] = Rate of disappearance of [TS-complex].

Or, mathematically we can write that

$$k_1 C_A C_C = k_2 C_{[AC]} + k_3 C_{[AC]}$$

Or,

$$C_{[AC]} = \frac{k_1 C_A C_C}{k_2 + k_3}$$

Putting this value of $C_{[AC]}$ in equation (10.10a) we have

$$J = \frac{k_1 k_3 C_A C_C}{k_2 + k_3} = k C_A C_C, \quad \text{where,} \quad k = \frac{k_1 k_3}{k_2 + k_3} \tag{10.10b}$$

The above equation clearly shows that reaction rate depends on catalyst concentration. However, for a given catalyst concentration C_C remains constant since catalyst remains unchanged at the end of the reaction. In that case equation (10.10b) becomes

$$J = \frac{k_1 k_3 C_A C_C}{k_2 + k_3} = k' C_A, \quad \text{where,} \quad k' = \frac{k_1 k_3 C_C}{k_2 + k_3} \tag{10.10c}$$

Example-1 Preparation of sulphur trioxide (SO_3) in presence of catalyst nitric oxide (NO).

$$2SO_2(g) + O_2(g) \xrightarrow{[NO]} 2SO_3(g)$$

Step-I

$$2NO(g) + O_2(g) \xrightarrow[k_2]{k_1} 2NO_2(g)$$

Step-II

$$\underline{2NO_2(g) + 2SO_2(g) \xrightarrow{k_3} 2SO_3(g) + 2NO(g)}$$
$$2SO_2(g) + O_2(g) \longrightarrow 2SO_3(g)$$

This is an example of gas phase homogeneous reaction.

Example-2 Oxidation of thiosulfate ion by ceric ion, catalyzed by iodide ion.

$$2Ce^{4+} + S_2O_3^{2-} \xrightarrow{[I']} 2Ce^{3+} + S_4O_6^{2-}$$

Step-I

$$2Ce^{4+} + 2I^- \xrightleftharpoons[k_2]{k_1} 2Ce^{3+} + I_2$$

Step-II

$$\underline{I_2 + S_2O_3^{2-} \xrightarrow{k_3} S_4O_6^{2-} + 2I^-}$$
$$2Ce^{4+} + S_2O_3^{2-} \longrightarrow 2Ce^{3+} + S_4O_6^{2-}$$

This is an example of liquid phase homogeneous reaction.

10.8   THEORY OF HETEROGENEOUS CATALYST SYSTEM

10.8.1   Unimolecular Reaction Where One Reactant is Involved

In any heterogeneous catalyst system, usually metal or metal oxide is chosen as catalyst, i.e., catalyst remains in solid phase. Reactant molecules are adsorbed on the catalyst surface and undergo chemical reactions. The product, Thus, formed, is desorbed from the catalyst surface. If the reactants are gases, the following processes may occur:

   Step-I   Diffusion of reactants to solid catalyst surface. The process is very fast.
   Step-II   Adsorption of reactants on the catalyst surface.
   Step-III   Reaction takes place between reactants.
   Step-IV   Desorption and diffusion of the products from the catalyst surface.

Diffusion of gases occurs very fast and adsorption $\rightleftharpoons$ desorption equilibrium sets up very quickly. So the overall reaction mostly depends on the fraction of catalyst surface covered by the reactants. Suppose p and θ be the reactant gas pressure and fraction of surface covered by reactants respectively. Thus, the rate of reaction is the rate of decrease of reactant gas pressure. As the chemical reaction occurs on the solid catalyst surface, the rate is directly proportional to θ. Thus, the rate of reaction is given by

$$-\frac{dp}{dt} \propto \theta. \quad \text{Or,} \quad -\frac{dp}{dt} = k'\theta$$

According to Langmuir concept, reactant molecules form unimolecular film thickness on catalyst surface. So, θ (fraction of surface covered by reactant molecules) is related to gas pressure (p) according to the following equation:

$$\theta = \frac{ap}{1+ap}, \quad \text{where, } a \text{ is a constant}$$

Thus, the rate equation becomes

$$-\frac{dp}{dt} = k'\theta = \frac{k'ap}{1+ap} \tag{10.11}$$

Case-I   If adsorption is small, gas pressure (p) on catalyst surface is also very low and hence, $(1 + ap) \approx 1$. Thus, equation (10.11) becomes

$$-\frac{dp}{dt} = k'ap \tag{10.12}$$

Conclusion   The catalyzed reaction follows 1st order kinetics.

Examples   (i) Decomposition of hydrogen iodide (HI) on platinum surface.

               (ii) Decomposition of phosphine (PH_3) on glass surface.

Case-II   If the adsorption is strong, gas pressure (p) on catalyst surface is very high and it may be approximated as $(1 + ap) \approx ap$. Thus, equation (10.11) becomes

$$-\frac{dp}{dt} = \frac{k'ap}{ap}k' \tag{10.13}$$

Conclusion   The catalyzed reaction follows zero-order kinetics.

Examples   (i) Dissociation of hydrogen iodide (HI) on gold surface.

(ii) Dissociation of ammonia on tungsten surface.

10.8.2   Bimolecular Reaction Where Both Reactants are Adsorbed

Bimolecular reaction   In the previous section unimolecular heterogeneous reaction is discussed. Heterogeneously catalyzed bimolecular reaction is more complicated. Let us consider a general reaction, where two different reactants take part in a chemical reaction in presence of a heterogeneous catalyst.

$$A + B \xrightarrow{\text{Catalyst}} \text{Products} + \text{Catalyst}$$

Mechanism   Both the reactants are believed to be adsorbed on catalyst surface and readily form a transition state complex, which is ultimately desorbed and breaks into products. The schematic diagram of the mechanism is shown in the Fig. 10.5.

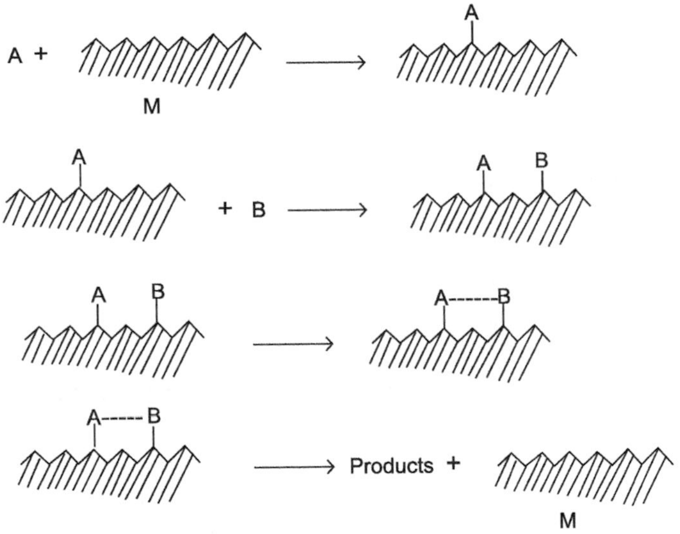

Figure 10.5   Schematic diagram showing bimolecular heterogeneously catalyzed reaction. M indicates the metal catalyst surface. Both the reactants A and B are adsorbed

The rate equation is given by

$$-\frac{dp}{dt} \propto \theta_1 \theta_2 = k\theta_1 \theta_2 \tag{10.14}$$

where, θ_1 = Fraction of the surface covered by the reactant, A; θ_2 = Fraction of the surface covered by the reactant, B; p = Total pressure = partial pressure of the reactant A + partial pressure of the reactant B; k = Reaction rate constant.

r_a^A = Rate of adsorption of reactant A. $r_a^A \propto (1 - \theta_1 - \theta_2)p_1$. Or, $r_a^A = k_1(1 - \theta_1 - \theta_2)p_1$

r_d^A = Rate of desorption of reactant A. $r_d^A \propto \theta_1$. Or, $r_d^A = k_1' \theta_1$

r_a^B = Rate of adsorption of reactant B. $r_a^B \propto (1 - \theta_1 - \theta_2)p_2$. Or, $r_a^B = k_2(1 - \theta_1 - \theta_2)p_2$

r_d^B = Rate of desorption of reactant B. $r_d^B \propto \theta_2$. Or, $r_d^B = k_2'\theta_2$

According to the principle of adsorption, we have

Rate of adsorption of reactant $A(r_a^A)$ = Rate of desorption of reactant $A(r_d^A)$ and

Rate of adsorption of reactant $B(r_a^B)$ = Rate of desorption of reactant $B(r_d^B)$.

Hence, $\qquad k_1(1 - \theta_1 - \theta_2)p_1 = k_1'\theta_1$ $\qquad\qquad$ (10.15a)

And $\qquad k_2(1 - \theta_1 - \theta_2)p_2 = k_2'\theta_2$ $\qquad\qquad$ (10.15b)

where, $(1 - \theta_1 - \theta_2)$ = Fraction of the surface still vacant for further adsorption.

Simplification of equations (10.15a) and (10.15b) gives rise to

$$\frac{k_1 p_1}{k_2 p_2} = \frac{k_1'\theta_1}{k_2'\theta_2}. \text{ Or, } \theta_1 = \left(\frac{k_2' k_1 p_1}{k_2 k_1' p_2}\right)\theta_2 \text{ and } \theta_2 = \left(\frac{k_2 k_1' p_2}{k_2' k_1 p_1}\right)\theta_1$$

Putting these values of θ_2 in the equation (10.15a) and θ_1 in the equation (10.15b), we have

$$k_1\left(1 - \theta_1 - \frac{k_2 k_1' p_2}{k_2' k_1 p_1}\theta_1\right)p_1 = k_1'\theta_1. \text{ Or, } \theta_1 = \frac{k_1 p_1}{k_1' + k_1 p_1 + \dfrac{k_2 k_1'}{k_2'}p_2} \qquad (10.16)$$

$$(k_2 - k_2\theta_1 - k_2\theta_2)p_2 = k_2'\theta_2. \text{ Or, } k_2 p_2 - k_2 p_2\frac{k_2' k_1 p_1}{k_2 k_1' p_2}\theta_2 - k_2 p_2\theta_2 = k_2'\theta_2$$

Or, $$\theta_2 = \frac{k_2 p_2}{k_2' + k_2 p_2 + \dfrac{k_2' k_1}{k_1'}p_1} \qquad\qquad (10.17)$$

Using equations (10.16) and (10.17) the rate equation [the equation (10.14)] becomes

$$-\frac{dp}{dt} \propto \theta_1\theta_2 = k\theta_1\theta_2 = \frac{k(k_1 p_1)(k_2 p_2)}{\left[k_1' + k_1 p_1 + \dfrac{k_2 k_1'}{k_2'}p_2\right]\left[k_2' + k_2 p_2 + \dfrac{k_2' k_1}{k_1'}p_1\right]}$$

Case-I If adsorption of both the gases are very weak, i.e., p_1 and p_2 are very low, the following approximations can be made.

$$k_1' + k_1 p_1 + \frac{k_2 k_1'}{k_2'}p_2 \approx k_1' \text{ and } k_2' + k_2 p_2 + \frac{k_2'}{k_1'}k_1 p_1 \approx k_2'$$

Thus, the rate equation becomes

$$-\frac{dp}{dt} = \frac{k k_1 k_2}{k_1' k_2'}p_1 p_2 \qquad\qquad (10.18)$$

Thus, the reaction follows **1st order** kinetics with respect to each reactant. However, the total order of the reaction is 2.

Examples (a) Hydrogenation of ethylene on copper surface.

(b) Formation of nitrogen pentoxide from nitric oxide and oxygen on glass.

Case-II Reactant A is more strongly adsorbed than reactant B, i.e., p_2 can be neglected with respect to p_1. Thus, the rate equation becomes

$$-\frac{dp}{dt} = \frac{kk_1k_2p_1p_2}{(k_1'+k_1p_1)\left[k_2'+\dfrac{k_2'k_1}{k_1'}p_1\right]} = \left[\frac{kk_1k_2}{k_1'k_2'}\right]\left[\frac{p_1p_2}{\left(1+\dfrac{k_1}{k_1'}p_1\right)^2}\right] = \frac{k_0p_1p_2}{(1+ap_1)^2} \quad (10.19)$$

where, $k_0 = \dfrac{k\,k_1k_2}{k_1'k_2'}$ and $a = \dfrac{k_1}{k_1'}$

Example Reaction between carbon dioxide and hydrogen on platinum surface.

Case-III Reactant A is very strongly adsorbed and $\theta_1 \approx 1$ and $p_1 \gg p_2$. Thus, the rate equation becomes

$$-\frac{dp}{dt} = k\theta_2 = \frac{kk_2p_2}{\left[k_2'+\dfrac{k_2'k_1}{k_1'}p_1\right]} \approx \left(\frac{kk_2k_1'}{k_2'k_1}\right)\left(\frac{p_2}{p_1}\right), \text{ since, } \left(\frac{k_2'k_1}{k_1'}p_1\right) \gg k_2'$$

or, $\qquad -\dfrac{dp}{dt} = k_0'\left(\dfrac{p_2}{p_1}\right), \quad \text{where, } k_0' = \left(\dfrac{k\,k_2\,k_1'}{k_2'\,k_1}\right) \qquad\qquad (10.20)$

Example Reaction between carbon monoxide and oxygen on platinum surface.

10.8.3 Bimolecular Reaction Where One Reactant is Adsorbed

One of the reactant is adsorbed on the catalyst surface and gets activated. This activated reactant molecule then reacts with another reactant molecule as shown in the Fig. 10.6.

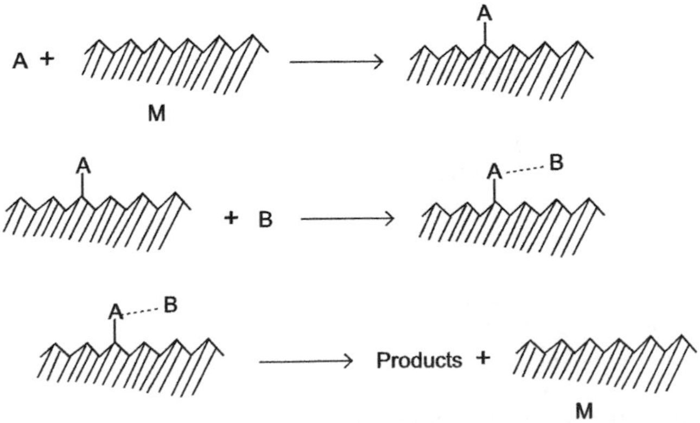

Figure 10.6 Schematic diagram showing bimolecular heterogeneously catalyzed reaction. M indicates the metal catalyst surface. Only reactant A is adsorbed

The reaction kinetics is same as that of unimolecular reaction since the rate of reaction is independent of the unadsorbed reactant B.

10.9 PRIMARY KINETIC SALT EFFECT

In the case of ionic reactions the rate constant depends on the ionic strength (I) of the solution. This effect of ionic strength (I) on rate constant is known as primary kinetic salt effect, established by Brönsted and Bjerrum.

Consider the following bimolecular ionic reaction:

$$A^{z_1} + B^{z_2} \underset{k_2}{\overset{k_1}{\rightleftharpoons}} [AB]^{z_1+z_2} \xrightarrow{k_3} \text{Products}$$

There is an equilibrium between reacting ions and the complex. k_1 and k_2 are the forward and backward reaction rates respectively of the equilibrium.

If the rate equation follows 1st order kinetics with respect to each reactant, the rate of the reaction is given by

$$\text{Rate} = \frac{d[\text{Products}]}{dt} = k_3 C_{[AB]^{z_1+z_2}} \tag{10.21}$$

Again, from the steady-state concept, we have
Rate of formation of TS complex, $[AB]$ = Rate of dissociation of TS complex.
Thus, in terms of activity we have

$$k_1 a_{A^{z_1}} \cdot a_{B^{z_2}} = (k_2 + k_3) a_{[AB]^{z_1+z_2}}. \quad \text{Or,} \quad a_{[AB]^{z_1+z_2}} = \left(\frac{k_1}{k_2+k_3}\right)(a_{A^{z_1}})(a_{B^{z_2}})$$

$a_{A^{z_1}}$ and $a_{B^{z_2}}$ and $a_{[AB]^{z_1+z_2}}$ are the activity terms of reactants A, B and TS complex respectively.

Again we know that $a = c \cdot \gamma$, where, c and γ is called concentration and activity coefficient respectively of a particular species.

Hence,
$$C_{[AB]^{z_1+z_2}} = \left(\frac{k_1}{k_2+k_3}\right)\frac{(C_{A^{z_1}} \cdot \gamma_{A^{z_1}})(C_{B^{z_2}} \cdot \gamma_{B^{z_2}})}{\gamma_{[AB]^{z_1+z_2}}} \tag{10.22}$$

Combining equations (10.21) and (10.22), we have

$$\text{Rate} = \frac{d[\text{Products}]}{dt} = \left(\frac{k_1 k_3}{k_2+k_3}\right)(C_{A^{z_1}} \cdot C_{B^{z_2}})\left(\frac{\gamma_{A^{z_1}} \cdot \gamma_{B^{z_2}}}{\gamma_{[AB]^{z_1+z_2}}}\right) = k(C_{A^{z_1}} \cdot C_{B^{z_2}}) \tag{10.23}$$

where,
$$k = k'\left(\frac{\gamma_{A^{z_1}} \cdot \gamma_{B^{z_2}}}{\gamma_{[AB]^{z_1+z_2}}}\right), \quad \text{and} \quad k' = \left(\frac{k_1 k_3}{k_2+k_3}\right) \tag{10.24}$$

Taking $\log_{10}$ on both sides of the equations (10.24), we get

$$\log k = \log k' + \left[\log \gamma_{A^{z_1}} + \log \gamma_{B^{z_2}} - \log \gamma_{[AB]^{z_1+z_2}}\right] \tag{10.25}$$

10.9.1 Brönsted-Bjerrum Equation

According to Debye-Hückel relation, we have

$$\log \gamma = -0.509 z^2 \sqrt{I} \quad (\text{at } 25°C)$$

where, I is the ionic strength of the solution and z is valency of the ion. $\gamma_{A^{z_1}}$, $\gamma_{B^{z_2}}$, and $\gamma_{[AB]^{z_1+z_2}}$ are the activity coefficients of reactants A, B and TS complex respectively. So, we can write

$$\log \gamma_{A^{z_1}} = -0.509 z_1^2 \sqrt{I}, \quad \log \gamma_{B^{z_2}} = -0.509 z_2^2 \sqrt{I}, \quad \text{and}$$

$$\log \gamma_{[AB]^{z_1+z_2}} = -0.509(z_1 + z_2)^2 \sqrt{I}$$

where, z_1 is the valency of the cation and z_2 is the valency of the anion. Hence,

$$\left[\log \gamma_{A^{z_1}} + \log \gamma_{B^{z_2}} - \log \gamma_{[AB]^{z_1+z_2}} \right] = -0.509 \left[z_1^2 + z_2^2 - (z_1 + z_2)^2 \right] \sqrt{I} = 1.018 z_1 z_2 \sqrt{I}$$

Valency of the TS complex is $(z_1 + z_2)$. So, $[AB] = [AB]^{z_1+z_2}$, $\gamma_{[AB]} = \gamma_{[AB]^{z_1+z_2}}$ and $a_{[AB]} = a_{[AB]^{z_1+z_2}}$

Thus, the equation (10.25) becomes

$$\log k = \log k' + 1.018 z_1 z_2 \sqrt{I} \tag{10.26}$$

The equation (10.26) is known as Brönsted-Bjerrum equation. The plot of log k against $\sqrt{I}$ yields a straight line of slope $1.018 \, z_1 z_2$.

Case-I Both the reactants, A and B are positively charged, or negatively charged. Slope $(1.018 z_1 z_2)$ is positive since, z_1 and z_2 are of same sign.

Example $[NiCl_4]^{2-} + 4CN^- \rightarrow [Ni(CN)_4]^{2-} + 4Cl^-$

Case-II The reactants, A and B are oppositely charged, i.e. $(z_1 z_2)$ is negative and hence the slope, $1.018 z_1 z_2$ is also negative.

Example $Pb^{4+} + S_2O_3^{2-} \rightarrow Pb^{2+} + S_4O_6^{2-}$

If one of the reactants, A or B is neutral, i.e., z_1 or z_2 is zero and the magnitude of the slope $1.018 z_1 z_2$ is zero, which means that the rate constant of the reaction is independent of ionic strength, I.

Example $C_{12}H_{22}O_{11} + H^+ \longrightarrow C_6H_{12}O_6 + C_6H_{12}O_6$
(Sucrose)   (Glucose)   (Fructose)

EXERCISES

1. Explain what is meant by homogeneous and heterogeneous as applied to catalysts.

2. The reaction between persulfate ions and iodide ions is very slow in the absence of a catalyst.

$$S_2O_8^{2-} + 2I^- \longrightarrow 2SO_4^{2-} + I_2$$

It can be speeded up considerably by the presence of either iron(II) ions or iron(III) ions in the solution.

   (a) What type of catalysis is this?

   (b) Why is the reaction so slow in the absence of a catalyst?

   (c) Use equations to help you to explain what happens in the presence of iron(II) ions.

3. What is called autocatalysis? Illustrate it with a suitable example.

4. "In homogeneous catalyst system, catalyst takes part in chemical reaction". – Explain it with a suitable example.

5. Deduce Michaelis-Menten equation. State the assumption made to deduce the equation.

6. What is called isoelectric point? State its significance in enzyme catalysis system.

7. What is called ionic strength? How is it related to rate constant (k) for a bimolecular reaction.

8. Explain induced catalyst with a suitable example.

11 Electrochemistry

11.1 THEORY OF ELECTROLYTES

For an electrolytic solution the osmotic pressure π is given by

$$\pi = icRT \tag{11.1}$$

where c = molarity of the solution and i = Van't Hoff factor.

S. Arrhenius proposed that total number ions and neutral molecules present per mole of a solute constitute the factor i. Consider an electrolyte AB which dissociates as follows

$$AB \rightleftharpoons v_+ A^{z+} + v_- B^{z-}$$

where α is the degree of dissociation. Then the factor i is given by

$$i = (1 - \alpha) + v_+\alpha + v_-\alpha = 1 + (v - 1)\alpha, \quad [\text{where, } v = v_+ + v_-] \tag{11.2}$$

where α is the degree of dissociation of electrolyte AB.

11.1.1 Shortcomings of Arrhenius Theory

(a) According to Arrhenius theory, ions in a solution behave as ideal gas molecules. Thus, ions execute erratic motion in solution. Interionic attraction is not considered. In actual practice strong electrostatic force of attraction exists among the ions, as a result of which, each ion is surrounded by a definite number of oppositely charged ions, forming an ionic atmosphere. This ionic atmosphere, surrounding each ion, restricts the erratic motion of that ion.

(b) According to Arrhenius theory, the dissociation constant of any strong electrolyte is not constant but changes with concentration. In actual practice every strong electrolyte remains in completely dissociated form at any concentration.

(c) Actual degree of dissociation of any electrolyte is much higher than that calculated from equation (11.2).

11.1.2 Activity (a) and Activity Coefficient (γ) of Electrolytes

Electrolytes are dissociated into ions in solution. The chemical potentials of positive and negative ions are given by

$$\mu_+ = \mu_+^0 + RT \ln a_+ = \mu_+^0 + RT \ln m_+ + RT \ln \gamma_+ \tag{11.3a}$$

$$\mu_- = \mu_-^0 + RT \ln a_- = \mu_-^0 + RT \ln m_- + RT \ln \gamma_- \tag{11.3b}$$

where, $a_+ = m_+ \gamma_+$ and $a_- = m_- \gamma_-$. m_+ and m_- are the molalities of positive ions and negative ions respectively. γ_+ and γ_- are the activity coefficients of positive ions and negative ions respectively.

Consider an electrolyte AB, which is dissociated in solution as follows

$$AB(m) \longrightarrow v_+ A^{z+} + v_- B^{z-}$$

$$\text{Molality} \longrightarrow (m) \qquad (v_+ m) \qquad (v_- m)$$

So, $m_+ = v_+ m$ and $m_- = v_- m$. If n_1 and n_2 are the number of moles of solvent and electrolyte respectively, we can write that

$$n_1 d\mu_1 + n_2 d\mu_2 = 0$$

Again we have $\qquad d\mu_1 = RTd \ln a_1$ and $d\mu_2 = RTd \ln a_2$

Hence, $\qquad n_1 d \ln a_1 + n_2 d \ln a_2 = 0$

For 1000 g of solvent, $n_1 = (1000/M_1)$ and $n_2 = m$, where, M_1 = Molecular weight of the solvent.

Putting the values of n_1 and n_2 in the above equation, we get

$$\frac{1000}{M_1} d \ln a_1 + md \ln a_2 = 0 \tag{11.4}$$

As the electrolyte remains in completely dissociated form in solution, we can write that

$$\frac{1000}{M_1} d \ln a_1 + m_+ d \ln a_+ + m_- d \ln a_- = 0 \tag{11.5}$$

From equations (11.4) and (11.5) we get

$$m_+ d \ln a_+ + m_- d \ln a_- - md \ln a_2 = 0.$$

Or, $\qquad (v_+ m) d \ln a_+ + (v_- m) d \ln a_- - md \ln a_2 = 0$

Or, $\qquad v_+ d \ln a_+ + v_- d \ln a_- - d \ln a_2 = 0.$ Or, $d \ln \left(\dfrac{a_+^{v_+} \cdot a_-^{v_-}}{a_2} \right) = 0$

On integration, we get

$$\left(\frac{a_+^{v_+} \cdot a_-^{v_-}}{a_2} \right) = \text{Constant} = K$$

The unit is so chosen that the value of the constant is unity.

So, $$a_2 = a_+^{v_+} \cdot a_-^{v_-} \qquad (11.6)$$

If $a_\pm$ is the mean activity, we have

$$a_\pm^v = a_+^{v_+} \cdot a_-^{v_-}. \quad \text{Or,} \quad a_2 = a_\pm^v$$

Similarly, we can write that $m_2 = m = m_\pm^v$ and $\gamma_2 = \gamma_\pm^v$,

$$a_2 = a_\pm^v, \quad m_2 = m = m_\pm^v \quad \text{and} \quad \gamma_2 = \gamma_\pm^v \qquad (11.7)$$

$$m = m_+^{v_+} \cdot m_-^{v_-} = (v_+ m)^{v_+} (v_- m)^{v_-} = (v_+^{v_+})(v_-^{v_-}) m^v$$

Again, $m = m_\pm^v$. Hence, $m_\pm^v = (v_+^{v_+})(v_-^{v_-}) m^v$. Or, $m_\pm = (v_+^{v_+} \cdot v_-^{v_-})^{\frac{1}{v}} \cdot m = L \cdot m$.

where, $$L = (v_+^{v_+} \cdot v_-^{v_-})^{\frac{1}{v}} = \text{Constant}$$

Or, $$m_\pm = L \cdot m \qquad (11.8)$$

Thus we can write that

$$a_\pm = m_\pm \gamma_\pm = Lm\gamma_\pm \qquad (11.9)$$

For uni-univalent salt like *NaCl*: $v_+ = 1, v_- = 1$ and $v = 2$. Hence, $L = 1$.

For di-univalent salt like, $CaCl_2$: $v_+ = 2, v_- = 1$ and $v = 3$. Hence, $L = (1^1 2^2)^{\frac{1}{3}} = 1.588$

Thus, for a solution containing a mixture of salts, it is difficult to express the concentration of solution by activity term. Usually the concentration of an electrolytic solution containing a mixture of salts, is expressed by *ionic strength (I)*, which is expressed as

$$I = \frac{1}{2} \Sigma m_i z_i^2 \qquad (11.10)$$

where, m_i and z_i are the molality and valency of ith ionic species respectively.

11.2 ELECTROLYSIS

11.2.1 Conductor

Any medium, which conducts electricity, is called conductor.

Types of conductors There are two types of conductors: (a) Electronic conductor, (b) Electrolytic conductor.

(a) **Electronic conductor** The material which conducts electricity through flow of electrons, is called electronic conductor, e.g., copper, silver, etc.

(b) **Electrolytic conductor** The material, which conducts electricity in solution, where flow of ions is the origin of current, is called electrolytic conductor, e.g., sodium chloride.

Electrochemical reaction When a chemical reaction takes place at the interface of an electrode (electronic conductor) and electrolytic solution, the reaction is called an electrochemical reaction.

For example, following reaction takes place between Cu salt solution and Cu electrode.
$$Cu^{2+} + 2e^- \longrightarrow Cu$$
The process is known as electrolysis.

11.2.2 Faraday's Laws

Laws of Electrolysis If current is passed through an electrolytic solution metals are deposited on the respective electrodes.

1st Law Amount of substance deposited at an electrode is directly proportional to the quantity of electricity passed through it. Mathematically, it can be written as $Q \propto m$. Or, $Q = F.m$

where, Q = Quantity of electricity passed; F = Proportionality constant, also known as Faraday constant.

m = Mass of substance deposited in terms of gm–equivalent.

If $m = 1$ gm. equiv., $Q = F$. Thus, F is the amount of electricity required to deposit 1 gm. equiv. of any substance. It is constant for any electrolyte.

2nd Law Same quantity of electricity is required to deposit 1 gm. equiv. of any substance at any electrode. It is called 1 Faraday (F) of electricity. The value of F is given by: $1F = 96,500$ Coulomb/gm.eqiv.

Electrochemical Equivalent (ECE) We know, $Q = F.m$

$$\text{If, } Q = 1C, \quad m = \frac{1}{F} \text{gm.equiv.} = \frac{1000E}{F} \text{mg}$$

where, E = Equivalent weight of electrolyte.

Amount of substance (in mg) deposited at an electrode by passing 1 Coulomb of electricity into its solution, is called the electrochemical equivalent (ECE) of that substance.

ECE values of silver, copper, and hydrogen have been found 1.118 mg, 0.3293 mg, and 0.010446 mg respectively. In the case of silver, 1.118 mg silver is deposited after passing 1 Coulomb of electricity into silver salt solution. Thus,

$$1.118 = \frac{1000 \times E_{Ag}}{F} = \frac{1000 \times 107.87}{F}. \quad \text{Or,} \quad F = 96,484 \text{ Coulomb/gm.equiv.}$$

In the case of copper, 0.3293 mg copper is deposited after passing 1 Coulomb of electricity into copper salt solution. Thus,

$$0.3293 = \frac{1000 \times E_{Cu}}{F} = \frac{1000 \times 31.77}{F}. \quad \text{Or,} \quad F = 96,481 \text{ Coulomb/gm.equiv.}$$

In the case of hydrogen, 0.010446 mg hydrogen is deposited after passing 1 Coulomb of electricity into an acid solution. Thus,

$$0.010446 = \frac{1000 \times E_{H_2}}{F} = \frac{1000 \times 1.00797}{F}. \quad \text{Or,} \quad F = 96,494 \text{ Coulomb/gm.equiv.}$$

The average value of F is taken as 96,500 Coulomb/gm. equiv.

For univalent ion like H^+, equivalent weight is equal to molecular weight. Thus, 1 mole of a univalent ion carries 96,500 Coulomb. As 1 gm.mole univalent ion contains 6.023×10^{23} number of ions, which is equivalent 6.023×10^{23} number of electrons, since charge carried by a univalent ion is equal to that carried by an electron.

Thus, 6.023×10^{23} number of electrons carry 96,500 Coulomb of charge. Or, 1 electron carries $(96500/6.023 \times 10^{23})$ Coulomb *(C)* of charge = $1.602 \times 10^{-19} C = 4.8063 \times 10^{-10}$ esu of charge.

11.3 ELECTROLYTIC CONDUCTANCE OF ELECTROLYTES

For any conductor relation between resistance and resistivity is given by

$$R = \rho \times \frac{l}{A}$$

R = Resistance of the conductor (Ω); ρ = Resistivity of the conductor (Ω–cm),

A = Area of each electrode (cm^2); immersed in the electrolytic solution. In case of an electronic conductor; A is the cross-sectional area of the conductor.

l = Distance of separation of two electrodes (cm). In case of an electronic conductor, l is the length of the conductor considered.

$$G = \frac{1}{R} = \text{Conductance, } \Omega^{-1}, \quad \kappa = \frac{1}{\rho} = \text{Conductivity or specific conductance, } \Omega^{-1}\text{cm}^{-1}$$

Hence,
$$G = \kappa \times \left(\frac{A}{l}\right)$$

Table 11.1 Conductivity data of some selected substances

Substance	κ, $\Omega^{-1}cm^{-1}$
Silver	615,000
Aluminium	360,000
$NaCl$ (fused)	3.40 at $750°C$
KCl (1N)	9.789×10^{-2}
Water	4.3×10^{-8}

It is evident from the above table that, conductivity of an electronic conductor is much higher than that of an electrolytic conductor.

11.3.1 Electrolytic Conductivity

As we know

$$G = \kappa \times \frac{A}{l}. \quad \text{Or,} \quad \kappa = G \times \frac{l}{A}$$

When $l = 1$ cm and $A = 1$ am^2, Volume of electrolytic solution is 1 cm^3 or 1 ml and $\kappa = G$. Thus, specific conductance or conductivity of an electrolytic solution is nothing but the conductance of that electrolytic solution where electrodes of area 1 cm^2 are separated by a distance of 1 cm. Or in other words, conductivity of an electrolytic solution is the total conductance of 1 ml of that solution.

11.3.1.1 Factors Influencing Conductivity

There are several factors influencing the conductivity of an ion. The following are the major factors influencing the conductivity of an ion in aqueous solution:

(a) **Ionic velocity**   Higher the ionic velocity, higher will be the ionic conductivity.

(b) **Ionic concentration**   With the increase in ionic concentration number of ions increases and hence ionic conductivity also increases.

(c) **Temperature**   With the increase in temperature, ionic velocity increases and hence ionic conductivity also increases.

11.3.1.2 Retarding Forces

In aqueous solution movement of ions is sufficiently restricted due to retarding forces, offered by the ionic atmosphere, surrounding the central ions.

Electrophoretic effect   In aqueous solution strong electrolyte undergoes total dissociation to form corresponding cations and anions. Each positive ion is surrounded by a negative ion atmosphere symmetrically. Similarly, each negative ion is surrounded by a positive ion atmosphere. During electrolysis central ion is trying to move forward but its ionic atmosphere offers a drag force to reduce the velocity of central ion. This retarding force is known as electrophoretic effect.

Asymmetry effect   Ionic atmosphere is symmetrical surrounding the central ion. When the central ion moves, the symmetry is lost and the ionic atmosphere vanishes and a new ionic atmosphere is created. There is a time lag between destruction and creation of ionic atmosphere during which the central ion cannot move. Hence, the velocity of central ion decreases. This retarding force is known as asymmetry effect.

11.3.1.3   Effect of Concentration on Conductivity

Conductivity of an electrolyte depends on concentration of that electrolyte. Fig. 11.1 shows the dependence of conductivity on concentration of different electrolytes.

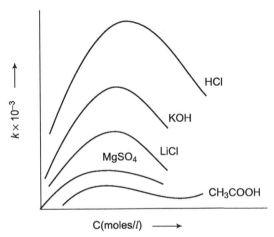

Figure 11.1   Conductivity versus concentration plot of different electrolytes

From Fig. 11.1, it is evident that conductivity of a strong electrolyte increases with the increase in concentration and reaches a maximum. As we know in aqueous solution strong electrolyte remains completely dissociated into corresponding cation and anion. Thus with the increase in electrolyte concentration, number of ions **increases** and hence conductivity **increases**. However, conductivity is sufficiently reduced due to the presence of active retarding forces, namely electrophoretic effect and asymmetry effect. These retarding forces appreciably **reduce** the velocity of the central ion.

At low concentration, effect of ionic atmosphere is low and overshadowed by the increase in number of ions. Hence, ionic conductivity increases sharply with increase in concentration of electrolytes. However, at higher concentration the effect of ionic atmosphere dominates over the effect of increase in ionic concentration and hence conductivity shows a decrease at higher concentration.

In the case of a weak electrolyte, with the increase in electrolyte concentration increase in ionic concentration is only marginal, since weak electrolyte is partially dissociated in aqueous solution. Furthermore the effect of ionic atmosphere is negligible due to low concentration of ions. Thus the concentration of electrolyte has negligible effect on its conductivity as is evident from the Fig. 11.1.

11.3.2 Equivalent Conductivity

Equivalent conductivity of an electrolyte is the total conductance of such a volume of solution that contains exactly 1 gm.equiv. of solute and the electrodes are 1 cm apart. It is represented as Λ_N or simply Λ. If v ml of an electrolytic solution of strength $c(N)$ contains 1 gm. equiv. of that electrolyte, Λ is given by

$$\Lambda = \kappa.\, v \tag{11.11}$$

Since, total conductance of 1 ml of an electrolytic solution is k. The unit of v is ml/gm.equiv. Again from the definition of Normality, we can write that

1000 ml of the solution contains c gm.equiv. of electrolyte.

Or, c gm.equiv. of electrolyte is present in 1000 ml of solution.

1 gm.equiv. of electrolyte is present in (1000/c) ml of solution.

Hence,
$$v = \frac{1000}{c} \text{ ml}$$

Thus, equation (11.11) becomes

$$\Lambda = \kappa \cdot v = \frac{1000\kappa}{c}\; \frac{\Omega^{-1}\,cm^2}{gm.equiv.} \tag{11.12}$$

Molar conductivity   Molar conductivity of an electrolyte is the conductivity of a solution containing 1 mole of solute and the electrodes are 1 cm apart. It is represented as Λ_M, which is expressed as

$$\Lambda_M = \frac{1000\kappa}{M}\; \Omega^{-1}\,cm^2/\text{mole} \tag{11.13}$$

where M is the molarity of the electrolytic solution.

Λ_M and Λ_N are related according to the following equation

$$\Lambda_M = z\Lambda_N, \quad \text{where,}\quad z = \frac{MW}{EW} \text{ (for a given electrolyte)}$$

As we have discussed earlier conductivity of a strong electrolyte depends on concentration. So, equivalent conductivity, Λ, also depends on concentration. Fig. 11.2 illustrates how Λ changes with concentration, c.

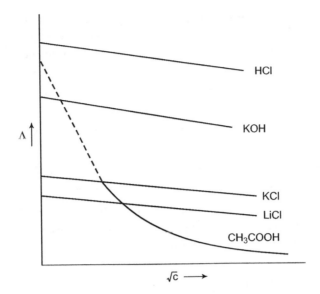

Figure 11.2 Λ versus $\sqrt{c}$ plot of some selected electrolytes

From the Fig. 11.2, it is evident that the equivalent conductivity of a *strong electrolyte* decreases linearly but only slightly on increasing the concentration. The mathematical expression of this linear relationship was first proposed by Kohlrausch as follows

$$\Lambda_c = \Lambda_0 - A\sqrt{c} \qquad (11.14)$$

where Λ_0 is the equivalent conductivity at infinite dilution, i.e., at $c \to 0$. As we know from equation (11.12) that

$$v = \frac{1000}{c}. \quad \text{So, as } c \to 0, \ v \to \infty$$

So, Λ_0 may also be represented as Λ_∞, i.e., equivalent conductivity at infinite dilution. A is a constant and Λ_c is the equivalent conductivity at any concentration c. The equation (11.14) is valid only for strong electrolytes.

The above equation does not hold good for weak electrolytes due to partial dissociation of the latter. However, it is evident from the Fig. 11.2 that, equivalent conductivity of weak electrolyte sharply increases with the increase in dilution and reaches a very high value, equal to that of a strong electrolyte, at infinite dilution.

He also observed that difference between conductivities of sodium and potassium salts is fairly constant irrespective of the associated ions. His observations are tabulated in Table 11.2.

Table 11.2 Equivalent conductivity values of some salts

Salt	Λ_N $(\Omega^{-1}\ cm^2/gm.eqv.)$	Difference
NaCl	108.90	
KCl	130.10	21.20
NaBr	111.10	
KBr	132.30	21.17
NaNO$_3$	105.33	
KNO$_3$	126.50	21.20

Thus at infinite dilution both strong and weak electrolytes behave in a similar manner. Bearing this observation in mind, the scientist Kohlrausch proposed the following law, which holds good for all electrolytes at infinite dilution only.

11.4 KOHLRAUSCH'S LAW

At infinite dilution any electrolyte is completely dissociated and all interionic attractions and repulsions vanish so that ions are moving independently. Under this condition, each ion makes definite contribution to the equivalent conductivity of the electrolyte irrespective of the nature of other ions associated with it. So, equivalent conductivity of any electrolyte is the sum of equivalent conductivities of cations and anions.

In case of uni-univalent electrolyte like NaCl, equivalent conductivities of the electrolyte at infinite dilution can be expressed as

$$\Lambda_\infty = \Lambda_0 = \lambda_+^0 + \lambda_-^0 \tag{11.15}$$

where, λ_+^0 and λ_-^0 are the equivalent conductivities of cations and anions respectively at infinite dilution. As all interionic attractions and repulsions vanish at infinite dilution, the total current, I, carried by an electrolyte is the sum of currents carried by cations and anions, i.e., I_+ and I_- respectively. Thus we can write that

$$I = I_+ + I_- \text{ (at infinite dilution)} \tag{11.16a}$$

Consider a uni univalent strong electrolyte AB which dissociates in solution as follows

$$AB \longrightarrow A^+ + B^-$$

Concentration $c(N)$ $c(N)$ $c(N)$

Thus, concentration of cations = concentration of anions = concentration of electrolyte

Concentration is expressed in gm.equiv./ml. Thus,

$$c_+ \text{ gm.equiv./ml} = c_- \text{ gm. equiv./ml} = c \text{ gm. equiv./ml}$$

Let us consider a cylinder through which cations and anions are flowing independently without any interionic forces. Suppose, q_+ and q_- are the velocities of cations and anions respectively in cm/s and A is the area of cross-section through which the ions are flowing [Fig. 11.3].

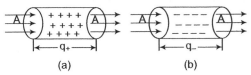

(a)   (b)

Figure 11.3   (a) Volumetric flow rate of cations, (b) Volumetric flow rate of anion

Let us consider two cylinders of length q_+ and q_- respectively. Thus (Aq_+) is the volume of cations (in c.c. or ml) flowing per sec or volumetric flow rate of cations through the area of cross-section A [Fig. 11.3a]. Similarly, volumetric flow rate of anions through the area of cross-section A is (Aq_-) with unit ml/s.

So, mass flow rate of cation $= Aq_+$ ml/s $\times c_+$ gm.equiv./ml $= (Ac_+q_+)$ gm.equiv./s.

We know that 1 gm–equiv. of ions carry F Coulomb of charge (C).

Thus, current carried by cation $= (Ac_+q_+)$ gm. equiv./s $\times$ (F) C/gm. equiv.

$$= (Ac_+q_+F)\ \text{C/s or Amp.}$$

Or, $I_+ = (Acq_+\ F)\ \text{Amp.}$ [since, $c_+ = c$]

Current carried by anion $= (Ac_-q_-)$ gm. equiv./s $\times$ (F) C/gm. equiv.

$$= (Ac_-q_-F)\ \text{C/s or Amp.} = I_-$$

Or, $I_- = (Acq_-\ F)\ \text{Amp.}$ [since, $c_- = c$]

$$I_+ = (Acq_+\ F)\ \text{Amp. and}\ I_- = (Acq_-\ F)\ \text{Amp.} \tag{11.16b}$$

If E is the emf of electrolytic cell and l is the distance of separation of two electrodes, (E/l) is the potential gradient of the cell. The velocity of any ion largely depends on potential gradient. Higher the value of (E/l) velocity of an ion increases proportionately. Thus velocity of any ion is directly proportional to the potential gradient. Hence, we can write that

$$q_+ \propto \frac{E}{l} = u_+ \frac{E}{l}\ \text{ and }\ q_- \propto \frac{E}{l} = u_- \frac{E}{l} \tag{11.16c}$$

The proportionality constants u_+ and u_- are called the absolute mobilities of cations and anions respectively.

Definition of mobility   During electrolysis, ions are moving towards their corresponding electrodes and the velocity of cations or anions under unit potential gradient (i.e., $E/l = 1$) is called mobility of that cation or anion respectively. The unit of mobility is

$$u_+ = \frac{q_+}{(E/l)} = \frac{\text{cm/ s}}{\text{volt/ cm}} = \frac{\text{cm}^2}{\text{volt}-\text{s}}$$

Combining equations (11.16b) and (11.16c), we get

$$I_+ = AcF\left(u_+ \frac{E}{l}\right)\ \text{ and }\ I_- = AcF\left(u_- \frac{E}{l}\right) \tag{11.16d}$$

Or, $I_+ \propto u_+$ and $I_- \propto u_-$ [For a given cell and cell emf] (11.16e)

Combining equations (11.16a) and (11.16d), we get

$$I = I_+ + I_- = AcF\left(\frac{E}{l}\right)(u_+ + u_-) = \left(\frac{EAc}{l}\right)(Fu_+ + Fu_-)$$ (11.16f)

The unit of Fu_+ is

$$\left(\frac{C}{\text{gm. equiv.}}\right)\left(\frac{\text{cm}^2}{\text{volt} - \text{s}}\right) = \text{cm}^2\left(\frac{\text{amp}}{\text{volt}}\right)\left(\frac{1}{\text{gm. equiv.}}\right) = \frac{\Omega^{-1}\,\text{cm}^2}{\text{gm. equiv.}}$$

Thus the unit of Fu_+ is same as that of equivalent conductivity of cation, i.e., λ^0_+, which is also known as limiting mobility of cation, i.e., the mobility of cation at infinite dilution. Similarly, the unit of Fu_- is same as that of equivalent conductivity of anion, i.e., λ^0_-, which is also known as limiting mobility of anion, i.e., the mobility of anion at infinite dilution.

Thus, $\lambda^0_+ = Fu_+$ and $\lambda^0_- = Fu_-$ (11.16g)

Thus the equation (11.16f) becomes

$$I = \left(\frac{EAc}{l}\right)(\lambda^0_+ + \lambda^0_-)$$ (11.16h)

Again, from Ohm's law, we have

$$I = \frac{E}{R} = E \cdot G = E \cdot \left(\kappa \frac{A}{l}\right) = \left(\frac{EA}{l}\right)\left(\frac{\Lambda}{v}\right) \quad [\text{since, } \Lambda = \kappa \cdot v]$$

Unit of v is ml/gm. equiv. and that of c is gm. equiv./ml. Thus, v.c. = 1. Putting this value of I in the equation (11.16h), we have

$$\left(\frac{EA}{l}\right)\left(\frac{\Lambda}{v}\right) = \left(\frac{EAc}{l}\right)(\lambda^0_+ + \lambda^0_-). \quad \text{Or, } \Lambda = \lambda^0_+ + \lambda^0_- \quad [\text{since, } v \cdot c = 1]$$

As the above equation is valid at infinite dilution, we can write that

$$\Lambda_\infty = \Lambda_0 = \lambda^0_+ + \lambda^0_-$$ (11.16i)

The equation (11.16i) is same as equation (11.15). The equation (11.16i) is the mathematical expression of **Kohlrausch's law** for uni-univalent electrolyte, like, NaCl, CH_3COOH, etc.

(a) As strong electrolyte always remains in dissociated form in solution, the equation (11.16i) is valid at any concentration c. Thus, for strong electrolyte, we have

$$\Lambda_c = \lambda^0_+ + \lambda^0_-$$ (11.16j)

(b) For a weak electrolyte the dissociation is incomplete at any concentration as shown below

$$AB \rightleftharpoons A^+ \quad B^-$$

Concentration $\quad c(N) \qquad \alpha c \quad \alpha c$ [α is the degree of dissociation of the weak electrolyte]

Hence, $c_+ = \alpha c$ and $c_- = \alpha c$. From the equation (11.16b), we have

$$I_+ = AFc_+ q_+ \text{ and } I_- = AFc_- q_- \tag{11.16b}$$

Putting these values of c_+ and c_- in the above equation, we get

$$I_+ = AF(\alpha c)q_+ \text{ and } I_- = AF(\alpha c)q_-$$

Using these values of I_+ and I_- the equation (11.16j) becomes

$$\Lambda_c = \alpha(\lambda_+^0 + \lambda_-^0) \tag{11.16k}$$

(c) If AB is a multi-multivalent strong electrolyte, the dissociation of AB in aqueous solution is given by

$$AB \rightleftharpoons v_+ A^{z+} + v_- B^{z-}$$

Concentration $\qquad c \qquad (v_+ c) \qquad (v_- c) \qquad$ Unit of c is gm. equiv./ml

where v_+ and v_- are the number of moles of cations and anions respectively. Thus, we have

$c_+ = v_+ c$ and $c_- = v_- c$. Putting these values of c_+ and c_- in the equation (11.16b), we get

$$I_+ = AF(v_+ c)q_+ = (AFc)\, v_+ q_+ \text{ and } I_- = AF(v_- c)q_- = (AFc)\, v_- q_-$$

Using these values of I_+ and I_- the equation (11.16j) becomes

$$I_c^M = v_+ I_+^M + v_- I_+^M \tag{11.16l}$$

Putting the values of I, I_+ and I_-, we have

$$\Lambda_c^M = v_+ \lambda_+^M + v_- \lambda_-^M \tag{11.16m}$$

where, Λ_c^M represents molar conductivity of electrolytic solution.

11.5   MOBILITY

11.5.1   Mathematical Expression

Mobility of an ion is the velocity of that ion at unit field gradient or potential gradient. As an ion is moving, its motion is restricted by the surrounding oppositely charged ionic atmosphere. Thus, the motion of an ion in an electrolytic solution is very much similar to the motion of a free falling macroscopic sphere in a viscous medium. So, Stokes formula may be applied to correlate the ionic mobility and its radius as given below

$$u_+ = \frac{z_+ e}{6\pi\eta r_+} \text{ and } u_- = \frac{z_- e}{6\pi\eta r_-} \tag{11.17a}$$

z_+ = valency of the cation; z_- = valency of the anion.
e = electronic charge = 1.602×10^{-19} C.
η = viscosity factor or resistance offered by the ionic atmosphere.
r_+ = radius of the cation; r_- = radius of the anion.
u_+ = Mobility of the cation = $q_+/(E/l)$.
Putting this value of u_+ in the equation (11.17a), we get

$$q_+ = \frac{z_+e}{6\pi\eta r_+}\left(\frac{E}{l}\right). \text{ Similarly, } q_- = \frac{z_-e}{6\pi\eta r_-}\left(\frac{E}{l}\right) \quad (11.17b)$$

The quantity (E/l) is the driving force for an ion to move. To calculate absolute mobilities of several ions the value of driving force, (E/l), is taken to be unity, i.e., 1 volt/cm.

Hence,
$$u_+ = q_+ = \frac{z_+e}{6\pi\eta r_+} \quad \text{and} \quad u_- = q_- = \frac{z_-e}{6\pi\eta r_-} \quad (11.17c)$$

From equation (11.17c) it is evident that

$$u_+ \propto \frac{1}{r_+} \quad \text{and} \quad u_- \propto \frac{1}{r_-}. \quad \text{Or, } \lambda_+ \propto \frac{1}{r_+} \quad \text{and} \quad \lambda_- \propto \frac{1}{r_-} \quad [\text{using equation (11.16g)}]$$

11.5.2 Mobility and Hydration Number of Ions

In a given crystal lattice, the ideal radius of a cation is represented as r_c, which can be easily computed from crystal unit cell dimension, e.g., unit cell of NaCl crystal is Face-centred cubic or FCC structure. Thus we have, $r_c = a/2\sqrt{2}$, where, a is the edge length of the unit cell cube. Thus, r_c can be evaluated from the crystallographic analysis.

The ions remain hydrated in aqueous solution and hydrated radius is represented as r_H, which can be calculated using Stokes equation as follows.

$$u = \frac{ze}{6\pi\eta r_H} \quad [\text{Ionic radius is expressed in Å.}]$$

Volume of hydrated ionic cell $= \frac{4}{3}\pi r_H^3$. Volume of actual ionic cell $= \frac{4}{3}\pi r^3$.

[Where, r is the true radius of the ion.]

So, volume of hydration $= V_H = \frac{4}{3}\pi r_H^3 - \frac{4}{3}\pi r^3 = $ Total volume of n molecules of water

Volume of one molecule of water is 30×10^{-30} m^3.

So, hydration number $= n = \dfrac{V_H}{30}$

r is the true radius of the ion and it is related to r_H as follows

$$r = \frac{ze(E/l)F}{6\pi\eta\lambda}\left(\frac{r_H}{r_c}\right) \quad (11.17d)$$

$$F = 96,500 \text{ C/gm.equiv.}, \quad e = 1.602 \times 10^{-19} \text{ C}, \quad \text{and} \quad \frac{E}{l} = 1 \text{ V/cm}$$

So, $$e \times \left(\frac{E}{l}\right) = 1.602 \times 10^{-19} \text{ J/cm} = 1.602 \times 10^{-12} \text{ ergs/cm}$$

$$r = \frac{1.602 \times 10^{-12} \times zF}{6 \times 3.14 \times \eta\lambda}\left(\frac{r_H}{r_c}\right) \text{cm} = \frac{0.820z}{\eta\lambda}\left(\frac{r_H}{r_c}\right) \text{Å}$$

Using these values of r and r_H, V_H can easily be calculated and hence also the hydration number, n.

η of water = $1.005 \, cp = 1.005 \times 10^{-2} \, p$.

Thus, $$r = \frac{0.820Z}{\eta\lambda \times 10^{-2}}\left(\frac{r_H}{r_c}\right) = \frac{82Z}{\eta\lambda}\left(\frac{r_H}{r_c}\right) \text{Å}$$

For Na^+ ion, we have $r = 3.3$Å, $r_c = 0.97$Å, $r_H = 1.83$Å.

$$V_H = \frac{4\pi}{3}(r^3 - r_c^3) \approx 150 \text{ Å}^3. \quad \text{So, } n = \frac{V_H}{30} = \frac{150}{30} = 5$$

Hydration number of other ions may be evaluated in the same way (see Table 11.9).

Due to hydration, ionic radius changes and accordingly mobility also changes. This is well illustrated in the following example.

Example 11.5.1

According to equation (11.17c), down the group of the periodic table ionic radius increases and hence mobility should decrease. So, it is expected that

$$\lambda_{Rb^+} < \lambda_{K^+} < \lambda_{Na^+} < \lambda_{Li^+}.$$

In actual practice the reverse is true, i.e., $\lambda_{Rb^+} > \lambda_{K^+} > \lambda_{Na^+} > \lambda_{Li^+}$.

The explanation of this fact lies within the radius of cation. In aqueous solution all the ions get hydrated. Thus the radius, given in the equation (11.7c) is the hydrated radius, represented as $(r_H)_+$ and in general mobility is given by

$$\lambda_+ = Fu_+ = \frac{zeF}{6\pi\eta(r_H)_+} \quad \text{and} \quad \lambda_- = Fu_- = \frac{zeF}{6\pi\eta(r_H)_-} \qquad (11.17e)$$

Smaller the ionic radius, higher the electrostatic force of attraction and higher the degree of hydration is. Thus hydrated radius is bigger in size for smaller cation or anion. So, for Group I cations, it is expected that

$$(r_H)_{Li^+} > (r_H)_{Na^+} > (r_H)_{K^+} > (r_H)_{Rb^+}.$$

Hence, $\lambda_{Rb^+} > \lambda_{K^+} > \lambda_{Na^+} > \lambda_{Li^+}$ using equation (11.17e).

11.5.3   Effect of Solvent on Mobility

Mobility or equivalent conductivity of an ion largely depends on the polarity of solvent. It is expected that higher the polarity of the solvent higher will be the degree of dissociation and hence higher will be the concentration of ions. Hence higher will be the mobility.

Furthermore, higher the polarity of the solvent higher will be the effective ionic radius and hence lower will be the mobility. In any polar solvent these two opposing factors come into effect. Among the usual solvents water is the most polar solvent and it has been found that mobility of an ion is less in non-aqueous solution than in aqueous solution. However, it has been established that mobility of an ion (except H^+ and OH^-) is much higher in acetone and hydrogen cyanide than any other solvent (see Table 11.8). In a dilute solution the equation (11.17d) becomes

$$\eta\lambda = \frac{zeF}{6\pi r} = \text{Constant and } \frac{E}{l} = 1 \text{ V/cm} \qquad (11.18)$$

Provided the effective radius r is constant for different solvents.

Thus the product of mobility and viscosity of an ion or salt is constant. This is known as **Walden's rule**. Thus λ and hence mobility of a given ion decreases with increase in viscosity of solvent.

11.5.4 Mobility of Hydroxonium (H_3O^+) and Hydroxyl (OH^-) Ions

In aqueous solution, equivalent conductivities of cations and anions usually fall within 30–90 Ω^- cm^2 with two exceptions, hydrogen ion (H^+) and hydroxyl ion (OH^-). $\lambda^0_{H^+} = 350$ Ω^{-1} cm^2 and $\lambda^0_{OH^-} = 198$ Ω^{-1} cm^2.

In aqueous solution, H^+ readily gets hydrated to form H_3O^+. Water acts as an excellent carrier to transfer H^+ ion from one end to the other as illustrated in the following:

$$H^+ + H_2O \longrightarrow H_3O^+$$

$$H_3O^+ + H_2O \longrightarrow H_2O + H_3O^+$$

$$\cdots \cdots \cdots \cdots \cdots \cdots \cdots \cdots \cdots$$

$$\overset{+}{\underset{H}{H_2 O}} \longrightarrow \underset{H}{OH} \longrightarrow \underset{H}{OH} \longrightarrow \underset{H}{OH} \longrightarrow \underset{H}{OH} \longrightarrow \underset{H}{OH} \longrightarrow \underset{H}{OH}$$

Thus in aqueous medium H^+ ion has very high mobility. This type of conductance is called relay or chain conductance. In nitrobenzene the relay conductance is hardly possible and hence equivalent conductivity of H^+ ion is very less (only 23 Ω^{-1} cm^2) in nitrobenzene.

The high mobility of hydroxyl ion can be explained in the same way as shown below

$$OH^- + H_2O \longrightarrow [OH \cdots H - O - H]^- \text{ (Ion complex } A)$$

$$[OH \cdots H - O - H]^- + H_2O \longrightarrow H_2O + [OH \cdots H - O - H]^- \text{ } (A)$$

In this case hydroxyl ion first forms an ion complex, A, which then transfers OH^- to the next water molecule and hence the relay process continues. So here also, the chain conductance is the primary reason for high mobility of hydroxyl ion. However, the size of H_3O^+ is much smaller than that of the ion complex, A and hence the transfer of H_3O^+ from one water molecule to another is much easier and faster than the ion complex, A. That is why, the mobility of H^+ ion is much higher than that of hydroxyl ion and hence $\lambda^0_{H^+} \gg \lambda^0_{OH^-}$.

11.5.5 Ionic Mobility and Concentration

The equivalent conductivity, Λ_c is related to concentration, c, as follows

$$\Lambda_c = \Lambda_0 - A\sqrt{c} \qquad \text{[Equation (11.14)]}$$

In case of uni-univalent strong electrolyte using Kohlrausch's law we can write $\Lambda_c = \Lambda_0 = \lambda_+ + \lambda_-$ at low concentration. $\lambda_+ \approx \lambda_+^0$ and $\lambda_- \approx \lambda_-^0$ [Neglecting the retarding forces] At high concentration, equation (11.14) becomes

$$\lambda_+ = \lambda_+^0 - B_1\sqrt{c} \text{ and } \lambda_- = \lambda_-^0 - B_2\sqrt{c} \text{ and } B_1 + B_2 = A \qquad (11.19)$$

Thus it is evident that with the increase in ionic concentration mobility decreases. P. Debye and E. Hückel explained it with the help of ionic atmosphere theory, according to which, each ion is surrounded by oppositely charged ionic atmosphere. This oppositely charged ionic atmosphere substantially reduces the velocity of the central ion and hence the absolute mobility of the latter decreases. The phenomenon is known as **electrophoretic effect**. Higher the concentration, higher the density of ionic atmosphere and hence lower the mobility of the ion.

While moving, an ion leaves the ionic atmosphere but a new one is immediately developed surrounding the central ion. This destruction and creation of ionic atmosphere occurs due to simultaneous movements of ionic atmosphere and central ion in opposite direction. The time-lag between destruction and creation of ionic atmosphere is very short. This time-lag is known as relaxation time, represented by τ. The phenomenon is called relaxation or asymmetry effect. Thus the retardation of an ion occurs mainly due to **electrophoretic effect** and asymmetry effect.

11.5.5.1 Wien Effect

According to **Debye-Hückel** theory, the mobility decreases with the increase in concentration due to increase in density of ionic atmosphere, but at the same time the stability of ionic atmosphere depends on field intensity, i.e., on (E/l). With the increase in field intensity the ionic atmosphere becomes more and more unstable and at a certain value of (E/l), called **critical field intensity**, the ionic atmosphere no longer exists. The ions are moving freely as if there is no interionic attraction and under this circumstance, they have maximum mobility. The phenomenon was first established by **M. Wien** in 1927 and hence the name. It has been found that at a field gradient of 2,00,000 V/cm, the equivalent conductivity of a salt at any concentration, c, is equal to Λ_0 of that salt. Thus, Wien effect merely substantiates the Debye-Hückel theory.

11.5.5.2 Debye-Falkenhagen Effect/Frequency Effect/Dispersion of Electrical Conductivity

As we have discussed earlier in asymmetry effect, there is a definite time-lag, τ, between destruction and creation of ionic atmosphere surrounding the central ion. If a periodic electric field is applied in an electrolytic solution, the existence of ionic atmosphere depends on frequency (v) of A/C supply. If $\tau < 1/v$, the asymmetry effect exists but there is a frequency, called **critical frequency, v_{cr}**, at which point $\tau > 1/v_{cr}$ and hence after destruction of one ionic atmosphere no new ionic atmosphere develops. Thus asymmetry effect vanishes at v_{cr}. The phenomenon is known as **Debye-Falkenhagen effect**. Thus at $v > v_{cr}$, the equivalent conductivity of a salt increases sharply. The frequency v_{cr} and τ are related according to the following equation

$$v_{cr}\tau = 1 \quad \text{and} \quad v_{cr} = \frac{1}{\tau} = \frac{cZ\Lambda_c}{71.3} \times 10^3 \tag{11.20}$$

where, c = concentration of ion in (N), z = valency of ion and Λ_c = equivalent conductivity of electrolyte at concentration, c.

For a uni-univalent salt at $c = 0.001$ N asymmetry effect vanishes at 10^9 Hz.

Considering both electrophoretic effect and asymmetry effect L.Onsager in 1926 deduced an equation correlating equivalent conductivity, Λ_c, with concentration, c, as given below

$$\Lambda_c = \Lambda_0 - \left[\frac{8.2 \times 10^5}{(\varepsilon_r T)^{3/2}}\Lambda_0 + \frac{82.4}{(\varepsilon_r T)^{1/2} \times \eta}\right]\sqrt{c} \tag{11.21}$$

where, ε_r = Relative permittivity, η = Coefficient of viscosity, and T = Absolute temperature.

The equation (11.21) is very much similar with that proposed by Kohlrausch [equation (11.14)], i.e.,

$$\Lambda_c = \Lambda_0 - A\sqrt{c}$$

The first term in the bracket of equation (11.21) appears due to asymmetry effect and the second term indicates the electrophoretic effect. Experimental studies reveal that the factor due to electrophoretic effect is almost ***double*** to that due to asymmetry effect.

11.5.6 Ionic Mobility and Temperature

In the case of electronic conductor, conductivity decreases with the increase in temperature. In the case of electrolytic conductor, with the increase in temperature, kinetic energy of ions increase and hence velocity (or mobility) of ions increase. The temperature coefficient of mobility is given by

$$\mu_M = \frac{1}{u_{298}}\left(\frac{\Delta u}{\Delta T}\right). \quad \text{Or,} \quad u_T = u_{298}[1 + \mu_M(T - 298)]$$

where, u_{298} and u_T are the mobility values of ion at **298°K and** T°K respectively.

For most ions the value of μ_M in aqueous solution is about 0.02, i.e., mobility and hence equivalent conductivity of an ion increases by 2% with increase in temperature by just 1°K.

Again we know that, with the increase in temperature, viscosity of a solution decreases. If the relative increase in mobility is comparable to relative decrease in viscosity with increase in temperature, the quantity $\eta\lambda$ should be independent of temperature, which means that equation (11.18) is true for any electrolytic solution. Or, in other words Walden's rule is applicable for all electrolytic solutions. Experimental studies reveal that Walden's rule is true for most of the ions with a little deviation. However, a large deviation has been observed in the cases of H^+ and OH^- ions.

11.6 APPLICATIONS OF KOHLRAUSCH'S LAW

11.6.1 Determination of Equivalent Conductivity of Weak Electrolytes at Infinite Dilution

Strong electrolytes always remain in dissociated form in aqueous solution. Thus, Kohlrausch's law [Equation (11.15)] is applicable at low concentration for strong electrolytes. However, the law is not applicable at high concentration since at high concentration retarding forces are dominating. So, at low concentration, Λ_c of any strong electrolyte is the sum of equivalent conductivities of cations and anions.

Weak electrolytes dissociate only partially in aqueous solution. Only at infinite dilution complete dissociation is assumed to take place but infinite dilution cannot be achieved in actual practice. Thus, Λ_0 of a weak electrolyte can be computed by using Λ_0 values of two or more suitable strong electrolytes. The following example illustrates how Λ_0 of acetic acid (CH_3COOH) is calculated.

$$\Lambda^0_{CH_3COOH} = \lambda^0_{H^+} + \lambda^0_{CH_3COO^-} = (\lambda^0_{H^+} + \lambda^0_{Cl^-}) - (\lambda^0_{Na^+} + \lambda^0_{Cl^-}) + (\lambda^0_{Na^+} + \lambda^0_{CH_3COO^-})$$

Or, $$\Lambda^0_{CH_3COOH} = \Lambda^0_{HCl} - \Lambda^0_{NaCl} + \Lambda^0_{CH_3COONa}$$

Thus, by measuring equivalent conductivities of strong electrolytes HCl, NaCl, and CH_3COONa at a low concentration, c, one can easily calculate the $\Lambda^0_{CH_3COOH}$. Its value has been calculated to be around 391 Ω^{-1} cm^2. As we know $\lambda^0_{H^+} = 350 \ \Omega^{-1}$ cm^2, $\Lambda^0_{CH_3COO^-} = \Lambda^0_{CH_3COOH} - \lambda^0_{H^+} = (391 - 350) = 41 \ \Omega^{-1}$ cm^2, at 25°C.

11.6.2 Determination of Degree of Dissociation (α) of a Weak Electrolyte

Equivalent conductivity of a weak electrolyte at any concentration, c, is given by $\Lambda_c = \alpha(\lambda^0_+ + \lambda^0_-)$ [Equation (11.16k)], where α is the degree of dissociation. Again, at infinite dilution equivalent conductivity is given by

$$\Lambda_0 = \lambda^0_+ + \lambda^0_- \quad \text{[Equation (11.15)]}$$

Dividing Λ_c by Λ_∞ we get

$$\frac{\Lambda_c}{\Lambda_0} = \alpha \tag{11.22a}$$

Λ_0 can be computed easily by using Kohlrausch's law as discussed above. Λ_c can be determined experimentally. Thus the degree of dissociation can easily be calculated from equation (11.22a).

Let us consider a weak acid HA, which dissociates in aqueous solution as follows:

$$HA \rightleftharpoons H^+ + A^-$$
$$(1-\alpha)c \qquad \alpha c \qquad \alpha c$$

$$K_a = \frac{a_{H^+} \cdot a_{A^-}}{a_{HA}} = \frac{\alpha^2 c}{(1-\alpha)}$$

Solving α, we get

$$\alpha = -\frac{K_a}{2c} \pm \frac{K_a}{2c}\left(1 + \frac{4c}{K_a}\right)^{\frac{1}{2}}. \quad \text{Or,} \quad \alpha = \frac{K_a}{2c}\left[\left(1 + \frac{4c}{K_a}\right)^{\frac{1}{2}} - 1\right]$$

$$\frac{1}{\alpha} = 1 + \frac{\alpha c}{K_a} \tag{11.22b}$$

According to equation (11.22a), we have

$$\alpha = \frac{\Lambda_M}{\Lambda_M^0}. \quad \text{Or,} \quad \frac{1}{\alpha} = \frac{\Lambda_M^0}{\Lambda_M}$$

Hence, the equation (11.22b) becomes

$$\frac{\Lambda_M^0}{\Lambda_M} = 1 + \left(\frac{c}{K_a}\right)\left(\frac{\Lambda_M}{\Lambda_M^0}\right). \quad \text{Or,} \quad \frac{1}{\Lambda_M} = \frac{1}{\Lambda_M^0} + \frac{c\Lambda_M}{K_a(\Lambda_M^0)^2} \tag{11.22c}$$

Thus if we plot $(1/\Lambda_M)$ versus $(c \times \Lambda_M)$, a straight line is obtained with intercept $(1/\Lambda_M^0)$. It is illustrated in the Fig. 11.4.

Experimental results verifies the equation (11.22c) at low and medium concentration. However positive deviation is observed at high concentration of electrolyte.

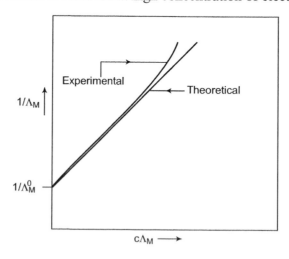

Figure 11.4   Plot of $(1/\Lambda_M)$ versus $(c\Lambda_M)$ for a weak electrolyte

11.6.3 Determination of Solubility of a Sparingly Soluble Salt

Any sparingly soluble salt has a limited solubility in water and the solubility does not change even at infinite dilution. Thus equivalent conductivity of a sparingly soluble salt remains constant at any concentration even at infinite dilution. So, we can write that

$$\Lambda_c = \Lambda_0 = \lambda_+^0 + \lambda_-^0 \quad \text{[for uni-univalent salt]}$$

Again, we have, $\Lambda_c = \Lambda_0 = \kappa v$, where κ is the conductivity and v is the volume in cm^3, containing 1 gm–equivalent of the salt.

So, v ml of salt solution contains 1 gm–equivalent of the salt.

Hence, 1000 ml salt solution contains $(1000/v)$gm–equivalent of the salt. So, concentration of the salt solution is $(1000/v)$ gm.eqiv./lit $= (1000\ EW/v)$ gm/lit, where, EW is the equivalent weight of the salt.

Again, concentration of salt in aqueous solution is nothing but the solubility, S, of the salt. So, we can write that

$$S = \frac{1000\ EW}{v}. \quad \text{Or,} \quad v = \frac{1000\ EW}{S}$$

Putting this value of v in $\Lambda_0 = \kappa v$, we get

$$\Lambda_0 = \frac{1000\kappa(EW)}{S}\Omega^{-1}\ cm^2/\text{gm.eqv.} \quad \text{Or,} \quad S = \frac{1000\kappa(EW)}{\Lambda_0} = \frac{1000\kappa(EW)}{(\lambda_+^0 + \lambda_-^0)}\ \frac{\text{gm}}{\text{lit}}$$

For multi-multivalent electrolyte

$$S = \frac{1000\kappa(EW)}{(v_+\lambda_+^0 + v_+\lambda_-^0)}\ \frac{\text{gm}}{\text{lit}} \tag{11.23}$$

where v_+ and v_- are the number of moles of cations and anions respectively produced after dissociation of 1 gm mole of electrolyte.

All the quantities of RHS of equation (11.23) are known except κ, which can easily be determined experimentally. Thus by measuring the conductivity, κ, of a sparingly soluble salt one can easily evaluate the solubility, S, of that salt.

11.6.4 Determination of Transport Number

It will be discussed in detail in the sub-section 11.7.4.

11.7 TRANSPORT NUMBERS OF IONS

In 1854, **J. Hittorf** noticed that during electrolysis concentration of electrolyte changed at the electrodes. Quantity of electricity carried by any ion depends on the concentration and mobility of that ion. As electrolytic solution is electrically neutral, the concentrations of cations and anions are same but mobilities of cations and anions differ with each other. If the cationic mobility is higher than anionic mobility, concentration of electrolyte should be higher at cathode than that at anode. Similarly, if the anionic mobility is higher than cationic mobility, concentration of electrolyte is higher at anode than that at cathode. On the basis of this observation the scientist Hittorf proposed the following law:

"Loss of electrolyte around cathode is proportional to anionic mobility and loss around anode is proportional to cationic mobility".

Mathematically, it can be represented as

Loss around cathode $\propto u_-$ and Loss around anode $\propto u_+$

Hence, $$\frac{\text{Loss around cathode}}{\text{Loss around anode}} = \frac{u_-}{u_+}.$$

Or, $$\frac{\text{Loss around anode}}{\text{Loss around cathode} + \text{Loss around anode}} = \frac{u_+}{u_- + u_+}$$

According to J. Hittorf, left hand side of the above equation is known as transport number of cation, represented by t_+.

Hence, $$t_+ = \frac{u_+}{u_+ + u_-}. \quad \text{Similarly,} \quad t_- = \frac{u_-}{u_+ + u_-} \qquad (11.24a)$$

Thus the transport number of a given ion is the ratio of mobility of that ion to the sum of mobilities of the cations and anions during electrolysis.

Again, combining equations (11.16e) and (11.24a), we have

$$t_+ = \frac{I_+}{I} \quad \text{and} \quad t_- = \frac{I_-}{I} \qquad (11.24b)$$

Thus transport number of a given ion in an electrolyte is the fraction of electricity carried by that ion during electrolysis. Where, I is the total current passed through the solution such that $I = I_+ + I_-$.

Definition of transport number

(a) The transport number of a given ion is the ratio of mobility of that ion to the sum of mobilities of the cations and anions during electrolysis.

(b) The transport number of a given ion in an electrolyte is the fraction of electricity carried by that ion during electrolysis.

Again, combining equations (11.16g) and (11.24a), we have

$$t_+ = \frac{u_+}{u_+ + u_-} = \frac{\lambda_+/F}{\lambda_+/F + \lambda_-/F} = \frac{\lambda_+}{\lambda_+ + \lambda_-} \qquad (11.24c)$$

According to Kohlrausch's law, we have $\Lambda_0 = \lambda_+^0 + \lambda_-^0$. Hence the above equation becomes

$$t_+ = \frac{\lambda_+}{\Lambda_0}. \quad \text{Or,} \quad \lambda_+ = t_+ \times \Lambda_0. \quad \text{Similarly,} \quad \lambda_- = t_- \times \Lambda_0 \qquad (11.24d)$$

11.7.1   Uni-univalent Weak Electrolyte

Let us consider a uni-univalent weak electrolyte, dissociated in aqueous solution as shown below:
$AB \rightleftharpoons A^+ + B^-$. The expression of λ_+ and λ_- are given by $\lambda_+ = \alpha F u_+$ and $\lambda_- = \alpha F u_-$, where, α is the degree of dissociation.

$$t_+ = \frac{\alpha u_+}{\alpha(u_+ + u_-)} = \frac{u_+}{u_+ + u_-} \quad \text{and} \quad t_- = \frac{\alpha u_-}{\alpha(u_+ + u_-)} = \frac{u_-}{u_+ + u_-} \tag{11.24e}$$

Thus transport number of a given ion varies from salt to salt. For example, t_+ of Na^+ in NaCl is given by

$$t_{Na^+} = \frac{u_{Na^+}}{u_{Na^+} + u_{Cl^-}}$$

t_+ of Na^+ in $NaNO_3$ is given by

$$t_{Na^+} = \frac{u_{Na^+}}{u_{Na^+} + u_{NO_3^-}}$$

As u_{Cl^-} and $u_{NO_3^-}$ are different, t_{Na^+} in case of NaCl is different from that in case of $NaNO_3$.

11.7.2   Multi-multivalent Strong Electrolyte

Let us consider a multi-multivalent strong electrolyte, dissociated in aqueous solution as shown below:

$$AB \longrightarrow v_+ A^{z+} + v_- A^{z-}, \text{ where, } |v_+| = |z_-| \text{ and } |v_-| = |z_+|.$$

where, z_+ and z_- are the valencies of cations and anions respectively.
   The expression of λ_+ and λ_- are given by $\lambda_+ = v_+ F u_+$ and $\lambda_- = v_- F u_-$
   The expressions of transport numbers are

$$t_+ = \frac{v_+ u_+}{v_+ u_+ + v_- u_-} \quad \text{and} \quad t_- = \frac{v_- u_-}{v_+ u_+ + v_- u_-} \tag{11.24f}$$

From the above equation it is evident that transport number of an ion depends on v_+, v_-, u_+, and u_-. Thus, transport number of a given ion varies from salt to salt, e.g., t_+ of Ca^{2+} in $Ca_3(PO_4)_2$ solution is given by

$$t_{Ca^{2+}} = \frac{3u_{Ca^{2+}}}{3u_{Ca^{2+}} + 2u_{PO_4^{3-}}}$$

But t_+ of Ca^{2+} in $CaCl_2$ solution is given by

$$t_{Ca^{2+}} = \frac{u_{Ca^{2+}}}{u_{Ca^{2+}} + 2u_{Cl^-}}$$

11.7.3 True Transport Number

During electrolysis cations are leaving anode and moving towards cathode while anions are leaving cathode and moving towards anode (Fig. 11.5). Thus change in concentration of electrolyte takes place in the vicinity of both the electrodes. So it may be concluded that, loss of electrolyte at cathode is directly proportional to the mobility of anion and loss of electrolyte at anode is directly proportional to the mobility of cation.

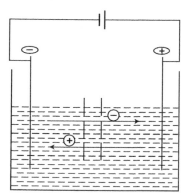

Figure 11.5 Movement of ions during electrolysis

Hence, we can write that

$$\frac{\text{Loss around cathode}}{\text{Loss around anode}} = \frac{u_-}{u_+}. \quad \text{Or,} \quad 1 + \frac{\text{Loss around cathode}}{\text{Loss around anode}} = 1 + \frac{u_-}{u_+}$$

Or,

$$t_+ = \frac{u_+}{u_+ + u_-} = \frac{\text{Loss around anode}}{\text{Total loss around both the electrodes}}$$

$$t_- = \frac{u_-}{u_+ + u_-} = \frac{\text{Loss around cathode}}{\text{Total loss around both the electrodes}}$$

Thus by measuring individual loss of electrolyte at each electrode one can easily calculate the transport numbers of ions. This is the basic principle adopted by J. Hittorf to determine transport numbers but in these losses he did not consider the loss of water molecules, which are also transferred with the ions, since the ions always remain in hydrated form in aqueous solution. Thus t_+ and t_- are called apparent transport numbers. Suppose τ_+ and τ_- are taken to be true transport numbers and n_+ and n_- are the number of moles of water associated with one mole of cation and one mole of anion respectively.

At cathode Gain of water in moles = $\tau_+ n_+$, since cation moves towards the cathode.

Loss of water in moles = $\tau_- n_-$, since anion leaves the cathode.

Net gain of water molecules = $(\tau_+ n_+ - \tau_- n_-) = y$ moles of water and $y > 0$.

At anode Gain of water in moles $= \tau_- n_-$, since anion moves towards the anode.

Loss of water in moles $= \tau_+ n_+$, since cation leaves the anode.
Net loss of water molecules $= (\tau_+ n_+ - \tau_- n_-) = y$ moles of water and $y > 0$.

So, net gain of water at cathode $=$ Net loss of water at anode.

If the electrolytic solution contains n_1 moles of solvent and n_2 moles of solute, we can write that

n_1 moles of solvent (water) is associated with n_2 moles of solute.

So, y moles of solvent (water) is associated with $(n_2/n_1)\, y$ moles of solute.

$$\left(\frac{n_2}{n_1}\right) y = \left(\frac{\dfrac{n_2}{n_1 + n_2}}{\dfrac{n_1}{n_1 + n_2}}\right) y = \left(\frac{x_2}{x_1}\right) y \text{ moles of solute.}$$

where, x_1 and x_2 are the mole fractions of solvent and solute respectively.

As, net gain of water at cathode $=$ net loss of water at anode, we can write that net gain of solute with y moles of water at cathode $=$ net loss of solute with y moles at anode

$$= \left(\frac{x_2}{x_1}\right) y \text{ moles of solute}$$

Therefore, the true transport numbers of the cation (τ_+) and anion (τ_-) are given by

$$\tau_+ = t_+ + \left(\frac{x_2}{x_1}\right) y \quad \text{and} \quad \tau_- = t_- - \left(\frac{x_2}{x_1}\right) y \tag{11.24g}$$

If, $y > 0$, $(x_2/x_1)y > 0$, $\tau_+ > \tau_-$. If $y < 0$, $(x_2/x_1)y < 0$, $\tau_+ < \tau_-$.

11.7.4   Method of Determination of Transport Number

The most widely used method to determine transport number is the moving boundary method. In this method two solutions of different densities are usually taken in a graduated tube and the two solutions are separated by a boundary *ab* as shown in the Fig. 11.6.

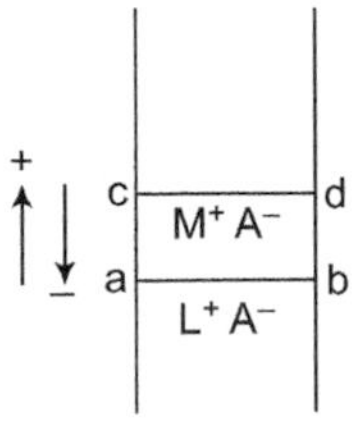

Figure 11.6   Determination of t_+ by moving boundary method. Both the electrolytes are uni-univalent

The electrolytic solutions are so chosen that they meet the following criteria:
 (a) The two strong electrolytic solutions are taken in a graduated cylinder such that the solutions are separated by an immiscible coloured boundary layer *ab*.
 (b) The electrolytes have a common anion A^-.
 (c) The mobility of one cation L^+ is higher than that of other cation M^+.
 (d) Concentration of electrolytic solution *MA* is c gm.equiv./ml. So, $[M^+]$ = c gm.equiv./ml.
 (e) Electrolyte having higher mobility cation, i.e., *LA* must have higher density than the other electrolyte, i.e, *MA*. The lighter electrolyte, i.e., *MA* solution is placed above the boundary layer *ab* and the heavier electrolyte, i.e., *LA* solution is placed below the boundary *ab*. *LA* solution is called ***indicator solution***.

Now the current I amp is passed through the system for *t* sec. So, the amount of electricity passed through the system is $q = I.t$ Coulomb. Current is passed in such a direction that cations move upward. As the velocity of L^+ is higher than that of M^+, the boundary *ab* moves to *cd* (Fig. 11.6) in time *t*. It is obvious that as the boundary moves from *ab* to *cd*, volume of M^+ ions within the region [*abcd*] must cross *cd*, since M^+ ions always remain above the boundary. The volume of the region [*abcd*] is measured from the graduation of the tube.

If *V* ml is the volume of M^+ ions within the boundary [*abcd*], the amount of M^+ ions within the boundary [*abcd*] is c*V* gm.equiv., where, c is the concentration of the electrolyte *MA* in gm.equiv./ml.

Thus, the total electricity carried by M^+ ions in time, *t* is q_+ Coulomb.

$q_+ = cV$ gm.equiv. $\times F$ Coulomb/gm. equiv. $= cVF$ Coulomb.

From the definition of transport number, we have

$$t_+ = \frac{\text{Electricity carried by } M^+ \text{ ions}}{\text{Total electricity passed}} = \frac{q_+}{q} = \frac{cVF}{q} \qquad (11.25)$$

c, F and q are known. Thus by measuring the volume *V*, one can easily calculate the transport number of M^+. As we know, $t_+ + t_- = 1$, transport number of A^- can easily be determined.

True (τ_+) and apparent (t_+) transport numbers of some selected electrolytes are listed in the Table 11.3.

Table 11.3 Values of τ_+ and t_+ of some selected electrolytes (concentration c = 1.3 N) at 25°C

Salt	y^a	t_+	τ_+
HCl	0.24	0.820	0.844
LiCl	1.50	0.278	0.304
NaCl	0.76	0.366	0.383
KCl	0.60	0.482	0.495
NaCl	0.33	0.485	0.491

y^a has the same meaning as given in equation (11.24g), i.e., it indicates the net gain in water (in moles) at cathode.

It is evident from the Table 11.3 that the difference between t_+ and τ_+ is small even at high concentration and this difference is negligible on further dilution. Thus universally accepted transport number is t_+ and hence no correction factor is introduced into the values of t_+ determined experimentally by moving boundary method.

11.7.5 Dependence of Transport Number on Concentration of Electrolyte and Temperature

Effect of concentration Usually transport number responds only marginally with the change in concentration as shown in the Table 11.5. It is evident that the transport number, having value less than 0.5 decreases slightly with the increase in concentration but the transport number, having value greater than 0.5 increases slightly with the increase in concentration. In some exceptional cases transport number decreases drastically with increase in concentration and may be equal to or less than zero, e.g., t_+ of CdI_2 has a negative value at high concentration of CdI_2. This is because of the formation of complex ion $[CdI_4]^{2-}$ at high concentration.

Sample question 11.7.1 State the differences between transport number and equivalent conductivity.

Answer The differences are summarized in the Table 11.4.

Table 11.4 Differences between transport number and equivalent conductivity

Transport number	*Equivalent conductivity*
It has no unit.	It has a definite unit, $\Omega^{-1}.cm^2/gm.eq.$
It is a fraction whose value lies within 0 and 1.	It has a large value, much greater than 1.
Transport number of a cation or an anion varies from one salt to the other.	Equivalent conductivity of a cation or anion is independent of type of salt.
Effect of concentration and temperature on transport number is marginal.	Effect of concentration and temperature on equivalent conductivity is remarkable.

Table 11.5 Dependence of transport numbers of cations (t_+) on concentrations of selected electrolytes in aqueous solutions at 25°C

Concentration gm.equiv./lit	t_+			
	HCl	*KNO₃*	*LiCl*	*BaCl₂*
0.01	0.8251	0.5084	0.3829	0.4400
0.02	0.8263	0.5087	0.3261	0.4375
0.05	0.8292	0.5093	0.3211	0.4317
0.10	0.8314	0.5103	0.3168	0.4253
0.20	0.8337	0.5120	0.3112	0.4162
0.50	–	–	0.3000	0.3986
1.00	–	–	0.2870	0.3792

Effect of temperature   According to the definition of transport number, it is expressed as

$$t_+ = \frac{u_+}{u_+ + u_-} \quad \text{and} \quad t_- = \frac{u_-}{u_+ + u_-}$$

So with increase in temperature, mobilities of both cations and anions increase proportionately and hence proportionately factor is cancelled out. Thus temperature has marginal effect on transport number. Usually transport number approaches to 0.5 with the increase in temperature (Table 11.6).

Table 11.6   Dependence of transport number on temperature at infinite dilution, i.e., $c \to 0$

Temperature (°C)	t_+	
	KCl	NaCl
15	0.4928	0.3929
25	0.4905	0.3962
35	0.4889	0.4002
45	0.4872	0.4039

Table 11.7   Limiting mobilities of ions in water at 25°C. Unit of mobility is Ω^{-1} cm^2/gm. equiv.

Ion	λ^0_+	Ion	λ^0_+	Ion	λ^0_+	Ion	λ^0_+
H_3O^+	349.8	Sr^{2+}	59.5	OH^-	197.6	$[Fe(CN)_6]^{3-}$	99.1
Li^+	38.6	Ba^{2+}	63.6	F^-	55.4	$[Fe(CN)_6]^{4-}$	111
Na^+	50.1	Cu^{2+}	56.6	Cl^-	76.4	HCO_3^-	44.5
K^+	73.5	Zn^{2+}	52.8	Br^-	78.1	CNS^-	66.5
Rb^+	77.8	Cd^{2+}	54.0	I^-	76.8		
Cs^+	77.2	Co^{2+}	49.0	N_3^-	69.0		
Ag^+	61.9	Fe^{2+}	53.5	NO_3^-	71.4		
Tl^+	74.7	Fe^{3+}	68.0	ClO_3^-	64.6		
NH_4^+	73.5	Al^{3+}	63.0	BrO_3^-	55.8		
$[NMe_4]^+$	44.9			ClO_4^-	67.3		
$[NEt_4]^+$	32.6			IO_3^-	54.5		
$[NPr_4]^+$	23.5			SO_4^{2-}	80.0		
$[NBu_4]^+$	19.4			$C_2O_4^{2-}$	74.1		
Be^{2+}	45.0			CO_3^{2-}	69.3		
Mg^{2+}	53.0			$HCOO^-$	54.6		
Ca^{2+}	59.5			CH_3COO^-	40.9		

Table 11.8   Limiting mobilities of ions in selected solvents at 25°C

Ion	λ^0 $(\Omega^{-1}cm^2/gm.equiv.)$				
	Water	*Methyl alcohol*	*Ethyl alcohol*	*Acetone*	*Nitrobenzene*
H^+	349.8	143.0	59.5	88.0	23.0
Li^+	38.6	39.8	*a*	*a*	*a*
Na^+	50.1	45.2	18.7	80.0	17.2
K^+	73.5	52.4	22.0	82.0	19.2
NH_4^+	73.5	57.9	19.3	98.0	*a*
Ag^+	61.9	50.3	17.5	88.0	18.5
Cl^-	76.4	52.9	24.3	111.0	17.3
Br^-	76.1	55.5	25.8	113.0	19.6
I^-	76.8	62.7	*a*	*a*	*a*
ClO_4^-	67.3	70.9	33.8	117.0	19.9
NO_3^-	71.4	60.8	27.9	120.0	*a*
OH^-	197.6	53.0	22.5	*a*	*a*
CNS^-	66.5	61.0	29.2	123.0	*a*

a: Data not available.

Table 11.9   Hydration number of ions (*n*). Formula $n = \dfrac{V_H}{30}$, $V_H = \dfrac{4}{3}\pi(r^3 - r_c^3)\,Å^3$

Ion	U_+^∞	$r_H(Å)$	$r_c(Å)$	$r(Å)$	$V_H(Å^3)$	*n*
Na^+	50.10	1.83	0.97	3.3	147	5
Li^+	36.68	2.37	0.60	3.7	211	7
Be^{2+}	45.00	4.08	0.55	4.6	407	13–14
Mg^{2+}	53.05	3.46	0.65	4.4	355	12
Ca^{2+}	59.50	3.09	0.99	4.2	306	10
Sr^{2+}	59.45	3.09	1.13	4.2	304	10
Ba^{2+}	63.63	2.88	1.35	4.1	278	9–10
Zn^{2+}	53.00	3.46	0.74	4.4	355	12
La^{2+}	69.75	3.95	1.15	4.6	401	13–14

11.8   CONDUCTOMETRIC TITRATIONS

Conductivity of an electrolytic solution changes noticeably in presence of another electrolytic solution. By measuring the change in conductivity one can easily evaluate the strength or concentration of an electrolytic solution. The method is called conductometric titration. Conductometric titration can be divided into two broad categories:

   (a) Acid-base titration,   (b)   Precipitation titration.

11.8.1 Acid-Base Titrations

11.8.1.1 Titration of a Strong Acid by a Strong Base

The conductivity of an acidic solution is very high due to the presence of H^+ ions, which has very high mobility (see Table 11.7). Let us take a hydrochloric acid solution. If we add sodium hydroxide solution dropwise, the following reaction takes place:

$$(H^+ + Cl^-) + (Na^+ + OH^-) \longrightarrow (Na^+ + Cl^-) + H_2O$$

As NaOH is consumed, Na^+ ions replaces H^+ ions. Since the mobility of Na^+ ions is much lower than that of H^+ ions (see Table 11.7), conductivity of HCl solution decreases sharply with the addition of NaOH solution, till the end point is reached.

At the end point there is no more H^+ ions present in the solution. Thus further addition of NaOH solution increases the conductivity sharply due to the presence of high mobile free OH^- ions. The nature of the curve is shown in the Fig. 11.7.

Two straight lines are obtained before and after the end point. The straight line before the end point, i.e., *ab*, is steeper than that obtained after the end point, i.e., *cd*.

Along *ab* conductivity decreases. $(\lambda^\infty_{H^+} - \lambda^\infty_{Na^+}) \approx 300~\Omega^{-1}~cm^2/gm.equiv.$

Along *cd* conductivity increases. $(\lambda^\infty_{Na^+} + \lambda^\infty_{OH^-}) \approx 250~\Omega^{-1}~cm^2/gm.equiv.$

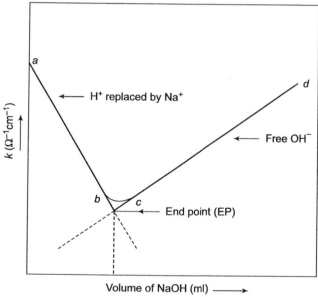

Figure 11.7   Conductometric titration of HCl solution by a strong base NaOH solution

Thus rate of decrease in conductivity along *ab* is greater than the rate of increase in conductivity along *cd*. That is why, curve *ab* is steeper than the curve *cd*. Intersecting region between the two curves *ab* and *cd*, i.e., *bc* is usually flat due to dilution effect. To minimize the dilution effect the titre (base) is taken as approximately five times stronger than the titrant (acid). The equivalence point can easily be found out by extrapolating the curves *ab* and *cd* (Fig. 11.7).

11.8.1.2 Titration of a Weak Acid by a Strong Base

If a weak acid (say CH_3COOH) is titrated by a strong base (say NaOH), the nature of the curve obtained is shown in the Fig. 11.8.

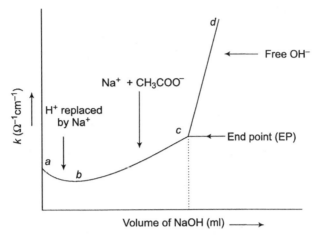

Figure 11.8 Conductometric titration of CH_3COOH solution by a strong base NaOH solution

The overall reaction, occurring during titration is given below:

$$CH_3COOH \rightleftharpoons H^+ + CH_3COO^- \text{ [Partially dissociated]}$$

$$(CH_3COOH) + (Na^+ + OH^-) \longrightarrow (Na^+ + CH_3COO^-) + H_2O$$

Acetic acid is weakly dissociated. So initially H^+ is replaced by Na^+ and hence an initial drop in conductivity is observed. The region *ab* of the curve shown in Fig. 11.8 represents the initial drop in conductivity. The neutralization reaction shows that acetate ions and sodium ions are produced continuously. Thus after sometime, concentrations of CH_3COO^- ions and Na^+ ions are significantly high and overshadow the effect of replacement of H^+ by Na^+. Hence, conductivity increases on titration (*bc* of the curve shown in Fig. 11.8). After complete neutralization further addition of NaOH solution increases the conductivity sharply due to the presence of free OH^- ion (*cd* of the curve shown in the Fig. 11.8). The point *c*, intersection of *bc* and *cd*, indicates the end point of titration.

11.8.1.3 Titration of a Strong Acid by a Weak Base

If a strong acid (say HCl) solution is titrated with a weak base (say NH_4OH), initial sharp drop in conductivity is observed due to the replacement of H^+ ions by NH_4^+ ions according to the following reaction:

$$(H^+ + Cl^-) + (NH_4OH) \longrightarrow (NH_4^+ + Cl^-) + H_2O$$

The nature of the curve is shown in the Fig. 11.9.

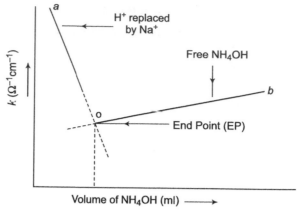

Figure 11.9 Conductometric titration of a strong acid (HCl) by a weak base (NH$_4$OH)

After the end point the conductivity increases slowly since [OH$^-$] increases only slightly as NH$_4$OH is dissociated in aqueous solution only partially curve *ob* in the Fig.11.9.

$$NH_4OH \rightleftharpoons NH_4^+ + OH^-$$

The intersecting point, *o*, indicates the end point of titration.

11.8.1.4 Titration of a Weak Acid by a Weak Base

If a weak acid (say CH$_3$COOH) solution is titrated by a weak base (say NH$_4$OH) solution, an initial drop in conductivity is observed due to the replacement of H$^+$ ions by NH$_4^+$ ions (Curve *ab* in the Fig. 11.10). After that conductivity increases due to salt formation according to the following reaction:

$$(CH_3COOH) + (NH_4OH) \longrightarrow (NH_4^+ + CH_3COO^-) + H_2O \text{ (curve } cd \text{ in the Fig. 11.10)}$$

After neutralization any further addition of NH$_4$OH solution increases the conductivity only marginally since NH$_4$OH is a weak base and almost remains in undissociated form. This is shown in the Fig. 11.10.

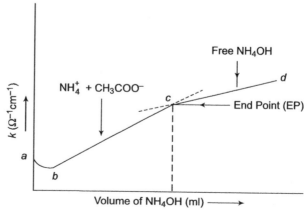

Figure 11.10 Conductometric titration of a weak acid (CH$_3$COOH) by a weak base (NH$_4$OH)

Intersecting point, *c*, of the two curves, *bc* and *cd* indicates the end point of titration. Weak acid-weak base titration cannot be performed by volumetric method.

11.8.2 Precipitation Titration

In many cases one of the products is continuously precipitated during titration. This type of titration is called precipitation titration. Two such examples are discussed here.

11.8.2.1 Titration of Potassium Chloride Solution by Silver Nitrate Solution

The reaction, taking place during titration, is given below:

$$(K^+ + Cl^-) + (Ag^+ + NO_3^-) \longrightarrow AgCl \downarrow + (K^+ + NO_3^-)$$

The reaction shows that nitrate ion replaces chloride ion but they have almost same mobility (see Table 11.7). Thus conductivity almost remains constant till the end point is reached (curve *ab* in the Fig. 11.11). After the end point there is no KCl in the system. So conductivity increases sharply with the addition of $AgNO_3$ solution due to increase in free Ag^+ and NO_3^- concentration (curve *cd* in the Fig. 11.11). Intersecting point of the two curves indicates the end point of titration.

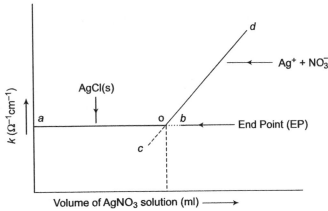

Figure 11.11   Precipitation titration of KCl solution by $AgNO_3$ solution

11.8.2.2 Titration of BaCl$_2$ Solution by Na$_2$SO$_4$ Solution

The following reaction is taking place during titration:

$$(Ba^{2+} + 2Cl^-) + (2Na^+ + SO_4^{2-}) \longrightarrow BaSO_4 \downarrow + (2Na^+ + 2Cl^-)$$

The reaction shows that 1 mole of Ba^{2+} is replaced by 2 moles of Na^+.

Equivalent conductivity of Ba^{2+} is 63.6 Ω^{-1} cm^2 (see Table 11.7). Molar conductivity of Ba^{2+} is $2 \times 63.6 = 127.2\ \Omega^{-1}$ cm^2, since Ba^{2+} is bivalent.

Equivalent conductivity of Na^+ = Molar conductivity of Na^+ = 50.1 Ω^{-1} cm^2 (see Table 11.7).

Thus the decrease in conductivity during titration = $(127.2 - 100.2) = 27\ \Omega^{-1}$ cm^2

Thus the titration shows a decrease in conductivity, (Fig. 11.12) till the end point is reached. After the end point, conductivity shows an increase due to increase in Na^+ and SO_4^{2-} ion concentrations. The point of intersection, o, of two curves, ab and cd, indicates the end point.

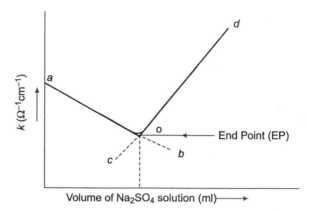

Figure 11.12 Precipitation titration of $BaCl_2$ solution by Na_2SO_4 solution

11.8.3 Advantages of Conductometric Titration Over Volumetric Titration

1. In volumetric titration an indicator is required to detect the end point but in case of conductometric titration no indicator is required to detect the end point.
2. Detection of end point in conductometric titration is far more accurate than that in volumetric titration since end point in conductometric titration is determined graphically without any personal error.
3. Weak acid-weak base type of titration can be performed by conductometric method only.
4. Precipitation type of titration can be performed by conductometric method only.

SOLVED PROBLEMS

Problem 11.1 Explain why transport number of a given ion varies from salt to salt.

Solution

Let us consider two salts having common cation: $NaCl$ and Na_2SO_4. The transport number of Na^+ ion for these two salts are given below

$$t_{Na^+} = \frac{u_{Na^+}}{u_{Na^+} + u_{Cl^-}} \quad \text{and} \quad t_{Na^+} = \frac{u_{Na^+}}{u_{Na^+} + u_{SO_4^{2-}}}$$

As u_{Cl^-} and $u_{SO_4^{2-}}$ are different in magnitudes, t_{Na+} is different for NaCl and Na$_2$SO$_4$.

Problem 11.2   Would the ionic strength increase, decrease or go unchanged with the addition of NaOH to a dilute solution of (i) MgCl$_2$, (ii) HCl, (iii) acetic acid?

Solution

(i) MgCl$_2$ The following reaction occurs on addition of NaOH.

$$MgCl_2 + NaOH \longrightarrow Mg(OH)_2 \downarrow + 2NaCl$$

Mg(OH)$_2$ is precipitated and **divalent** Mg is replaced by equivalent amount of **univalent** Na. Hence, ionic strength decreases.

(ii) HCl   The following reaction occurs on addition of NaOH.

$$H^+Cl^- + Na^+OH^- \longrightarrow Na^+Cl^- + H_2O$$

Equivalent amounts of HCl and NaCl are produced and all are **univalent** ions. So, ionic strength remains unchanged.

(iii) CH$_3$COOH   The following reaction occurs on addition of NaOH.

$$CH_3COOH + Na^+OH^- \longrightarrow Na^+CH_3COO^- + H_2O$$

Acetic acid almost remains undissociated. So ionic strength increases due to increase in number of Na$^+$ ions and CH$_3$COO$^-$ ions.

Problem 11.3 Λ_c of weak acid (HA) solution is 10 S-cm^2/g.eq. at 0.01 M and $\Lambda_0 = 410$ S-cm^2/ g.eq. Calculate pH of the solution.

Solution

$$\Lambda_c = 10\,S\text{-}cm^2/g.eq., \quad \Lambda_0 = 410\,S\text{-}cm^2/g.eq., \quad \alpha = \frac{\Lambda_c}{\Lambda_0} = 0.024$$

$$C_{H^+} = \alpha C_{HA} = 0.024 \times 0.01\,M = 2.4 \times 10^{-4}\,N, \quad pH = -\log C_{H^+} = 3.62$$

Problem 11.4 Λ_c of weak acid (HA) solution is 70 S-cm^2/g.eq. at 0.007 M and $\Lambda_0 = 405$ S-cm^2/g.eq. Calculate dissociation constant of the weak acid.

Solution

$$\Lambda_c = 70\,S\text{-}cm^2/g.eq., \quad \Lambda_0 = 405\,S\text{-}cm^2/g.eq., \quad \alpha = \frac{\Lambda_c}{\Lambda_0} = 0.17$$

$$C_{H^+} = C_{A^-} = \alpha C_{HA} = 0.17 \times 0.007\,M = 1.19 \times 10^{-3}\,M$$

$$K = \frac{C_{H^+} \times C_{A^-}}{C_{HA}} = 2.023 \times 10^{-4}$$

Problem 11.5   At 18°C, the Λ_∞ values of NH_4Cl, NaOH, and NaCl are 129.8, 217.4, and 108.9 Ω^{-1} cm²/g.eq., respectively. If the equivalent conductivity of (N/100) NH_4OH solution is 9.33 Ω^{-1} cm²/g.eq., calculate pH of NH_4OH at this dilution. Also calculate K_b.

Solution

$$\Lambda_\infty (NH_4OH) = \Lambda_\infty (NH_4Cl) + \Lambda_\infty (NaOH) - \Lambda_\infty (NaCl) = 238.3 \text{ S-cm}^2/\text{g.eq.}$$

$$\alpha = \frac{\Lambda_c}{\Lambda_\infty} = \frac{9.33}{238.3} = 0.0392, \quad C_{OH^-} = \alpha \cdot C_{NH_4OH} = 0.039 \times 10^{-2} \text{ N}$$

$$pOH = -\log C_{OH^-} = 3.41, \quad pH = 14 - pOH = 10.59. \quad K_b = \frac{\alpha^2 c}{(1-\alpha)} = 1.6 \times 10^{-5}.$$

Problem 11.6   Calculate Λ_0^N value for water.

Solution

Using Kohlrausch's law, we have
$$\Lambda_{H_2O}^0 = \Lambda_{HCl}^0 + \Lambda_{NaOH}^0 - \Lambda_{NaCl}^0 = 426 + 250 - 126 = 550 \text{ S.cm}^2/\text{gm}\cdot\text{eq.}$$

EXERCISES

1. An aqueous solution of $CuSO_4$ is electrolysed using a current of 0.15A for 5 hours. Calculate the mass of copper deposited at the cathode.

2. Electrolysis can be used to determine the gold content of a sample. Find the mass of gold that will be deposited from a solution containing Au^{3+} ion using 1.50 A for 1 hour.

3. How long will it take to produce 5.4 kg of aluminium metal by the reduction of Al^{3+} ions in an electrolytic cell using a current of 500 A?

4. A current of 8A is passed through molten aluminium oxide for 100 minutes using inert electrodes. Calculate the approximate volume of gas liberated, measured at STP. (The molar volume of a gas at STP = 22.4 dm³).

5. During the passage of a current of 1.5 A for 30 minutes through a solution of metal salt which ionises to form M^{3+} cations, 1.07 g of the metal was produced at the cathode. Determine the relative atomic mass of the metal.

6. A given HCl solution shows conductivity of 3.5×10^{-3} S/cm. $\lambda_{H^+} = 350$ S.cm²/gm.eq. $\lambda_{Cl^-} = 70$ S.cm²/g.eq. What is the pH of the solution?   [*Ans.:* 2.08]

   Hint : $\lambda_c = \dfrac{1000\kappa}{c_{HCl}}, \quad c_{HCl} = c_{H^+}, \quad pH = -\log c_{H^+}$

7. At 25°C, λ_M values of Li^+, Na^+, and K^+ are 3.87, 5.01, and 7.35 mil $S.m^2/gm.mole$ respectively. Calculate their mobilities. $F = 96,500$ C/gm.eq.

$[u_{Li^+} = 4.01 \times 10^{-8}\ m^2/V.s,\ u_{Na^+} = 5.19 \times 10^{-8}\ m^2/V.s,\ u_{K^+} = 7.62 \times 10^{-8}\ m^2/V.s]$

8. At 25°C, the Λ_M^0 of a weak acid HA is 386.6 $S.cm^2/mole$. Ionization constant is 1.4×10^{-5}. Calculate κ and Λ_M at 0.05 M HA solution at 25°C.

$[Ans.:\ \kappa = 3.235 \times 10^{-5}\ S.cm^{-1},\ \Lambda_C^M = 6.47\ S.cm^2/gm.mole]$

$$\left[\text{Hint}: \quad \alpha = \sqrt{\frac{K}{c}},\ \Lambda_M = \alpha \Lambda_M^0 \right]$$

9. A solution of $AgNO_3$ was electrolyzed between silver electrodes. The speed ratio of Ag^+ and NO_3^- ions is 0.916. Calculate t_{Ag^+}, $t_{NO_3^-}$, λ_{Ag^+} and $\lambda_{NO_3^-}$.
Given: $\Lambda_{AgNO_3}^0 = 120$ S. $cm^2/g.$ mole.

$$\left[\text{Hint}: \quad \frac{u_+}{u_-} = 0.916,\ \lambda_+ = t_+ \Lambda_0 = Fu_+ \right]$$

10. A certain conductivity cell, filled with 0.02 M KCl solution, shows a conductivity of $0.2768\ \Omega^{-1}\ m^{-1}$ and a resistance of $82.5\ \Omega$ at 25°C. The same cell, filled with 0.0025 M K_2SO_4, shows a resistance of $3260\ \Omega$. Calculate conductivity and molar conductivity of 0.0025 M K_2SO_4 solution.

11. The conductivity of $\left(\dfrac{N}{50} \right)$ KCl solution at 25°C is $0.2768\ \Omega^{-1}m^{-1}$. The resistance of this solution at 25°C is $250.2\ \Omega$. The resistance of the same cell, filled with 0.01 M $CuSO_4$ solution is $8331\ \Omega$ at 25°C. Calculate Λ_M of $CuSO_4$.
$[Ans.:\ 8.3135 \times 10^{-4}\ \Omega^{-1}m^2/g.mole]$

12. The conductivity of a standard solution of Ag_3PO_4 is $9 \times 10^{-6}\ \Omega^{-1}m^{-1}$ and $\Lambda_c = 1.5 \times 10^{-4}\ \Omega^{-1}\ m^2/gm.$ eq. Calculate K_{sp} of Ag_3PO_4. $[6.912 \times 10^{-9}]$

13. Λ_c of a weak acid solution (HA) is $10\ \Omega^{-1}\ cm^2/gm$ eq. at 0.1 M and $\Lambda_0 = 200\ \Omega^{-1}\ cm^2/gm$ eq. Calculate pH of the solution. $[Ans.:\ 2.3]$

14. A saturated solution of $AgA(K_{sp} = 3 \times 10^{-14})$ and $AgB(K_{sp} = 1 \times 10^{-14})$ has conductivity of $375 \times 10^{-10}\ \Omega^{-1}\ cm^{-1}$. $\lambda_{Ag^+}^0 = 60\ \Omega^{-1}\ cm^2/gm$ mole, $\lambda_{A^-}^0 = 80\ \Omega^{-1}\ cm^2/gm$ mole. Calculate $\lambda_{B^-}^0$. $[Ans.:\ 270\ \Omega^{-1}\ cm^2/gm\ mole]$

15. How many gm of Cu are plated out from a Cu^{2+} solution if a current of 1 amp is passed for 2 hours? Given F = 96500 C/gm.eq. $[Ans.:\ 0.0373\ moles]$

16. The equivalent conductivity of a $CuSO_4$ solution with concentration 2.54 gm/lit. is $91\ \Omega^{-1}cm^2/gm.$ eq. Calculate resistance of this solution. Given: $A = 1\ cm^2$ and $l = 1$ cm. $[Ans.:\ 346\ \Omega]$

17. Λ_c of a weak acid solution (HA) is $150\ \Omega^{-1}cm^2/gm.eq.$ at 0.007 M and $\Lambda_0 = 200\ \Omega^{-1}cm^2/gm.eq.$ Calculate dissociation constant of the weak acid. $[Ans.:\ 5.25 \times 10^{-3}]$

18. The equivalent conductance of 0.05 (N) KCl, NaCl, and K_2SO_4 are 149.9, 126.5, and 153.3 Ω^{-1} cm^2/gm. eq. respectively. Calculate an approximate value for the equivalent conductance of Na_2SO_4 of the same concentration. [*Ans.:* 130 ohm^{-1} cm^2/gm. eq.]

19. At 18°C, the Λ_∞ values of NH_4Cl, NaOH, and NaCl are 129.8, 217.4, and 108.9 Ω^{-1} cm^{-2}/g.eq. respectively. If the equivalent conductivity of (N/120) NH_4OH solution is 9.33 Ω^{-1} cm^2/gm.eq. Calculate the degree of dissociation of NH_4OH at this dilution.
[*Ans.:* 3.92%]

20. A conductivity cell, filled with (N/10) acetic acid at 25°C, shows a resistance of 9950 Ω. The same cell, filled with (N/100) KCl shows a resistance of 3735 Ω and resistivity 721 Ω cm at 25°C. Calculate the specific conductance and equivalent conductance of acetic acid. [*Ans.:* $k = 5.21 \times 10^{-4}$ Ω^{-1} cm^{-1}, $\Lambda_N = 5.21$ $\Omega^{-1}cm^2$/gm. eq.]

21. The conductivity of saturated solution of AgCl is 1.24×10^{-6} Ω^{-1} m^{-1}. The mobilities of Ag^+ and Cl^- ions are 53.8 and 65.3 Ω^{-1} cm^2 g. $eq.^{-1}$. Calculate the solubility of AgCl in g/l. Given: AW of Ag is 108 and MW of Cl_2 is 71. [*Ans.:* 1.494×10^{-5} g/l]

22. Calculate the specific conductance and equivalent conductance of 0.0469(N)NaOH solution. Given: resistance of the cell is 60.2 Ω and cell constant is 89.6 m^{-1}.
[*Ans.:* $k = 0.0149$ Ω^{-1} cm^{-1}, $\Lambda_N = 317.33$ Ω^{-1} cm^2 g $\cdot$ $eq.^{-1}$]

23. Calculate the transport number of Na^+ in 0.1 (M) NaCl at 25°C.
Given: $\lambda_{Na+} = 50.1$ Ω^{-1} cm^2 g $\cdot$ $eq.^{-1}$ and $\lambda_{Cl^-} = 76.34$ Ω^{-1} cm^2 g $\cdot$ $eq.^{-1}$ [*Ans.:* 0.3962]

24. The molar conductivity of 0.025 M acetic acid solution is 46.1 S.cm^2/gm $\cdot$ eq. Calculate the degree of dissociation and dissociation constant using the following information
$$\lambda_{H^+} = 350 \ \Omega^{-1} \ cm^2 \ g \cdot eq.^{-1}, \ \lambda_{OAc^-} = 50.1 \ \Omega^{-1} \ cm^2 \ g \cdot eq.^{-1}$$
[*Ans.:* 0.114, 3.67×10^{-4}]

12 Electrochemical Cell

12.1.1 Electrochemical Cell

Definition The cell in which electrical energy is produced at the expense of chemical energy or vice versa, i.e., chemical reaction takes place at the expense of electrical energy, is called electrochemical cell.

There are two types of electrochemical cells:

(a) **Galvanic cell** The cell, which generates current at the expense of spontaneous chemical reaction occurring in the cell, is called a galvanic cell.

(b) **Electrolytic cell** Any cell, where current is passed into the cell to bring about chemical reaction within the cell, is called an electrolytic cell.

Electrochemical cells can also be classified in the following ways.

(a) **Reversible cell** If the cell reaction exactly reverses when the direction of current is reversed, the cell is called reversible cell, e.g., **Daniel-Jacobi cell**, represented as

$$Zn|^+ ZnSO_4 \text{ (aq)} \parallel CuSO_4\text{(aq)}|^- Cu$$

The overall cell reaction is given by

$$Zn + Cu^{2+} \rightleftharpoons Zn^{2+} + Cu$$

When current is drawn from the cell the forward reaction occurs, i.e.,

$$Zn \longrightarrow Zn^{2+} + 2e^- \quad \text{(at the LHE)}$$
$$Cu^{2+} + 2e^- \longrightarrow Cu \quad \text{(at the RHE)}$$

When current is passed into the cell the reverse reaction occurs, i.e.,

$$Zn^{2+} + 2e^- \longrightarrow Zn \quad \text{(at the LHE)}$$
$$Cu \longrightarrow Cu^{2+} + 2e^- \quad \text{(at the RHE)}$$

(b) Irreversible cell   If the cell reaction does not reverse when the direction of current is reversed, the cell is called irreversible cell, e.g., the **Voltaic cell**, represented as

$$Zn|^+ H_2SO_4(aq)|^- Cu$$

When current is drawn from the cell, the following reactions takes place

$$Zn \longrightarrow Zn^{2+} + 2e^- \quad \text{(at the Zn electrode)}$$

$$2H^+ + 2e^- \longrightarrow H_2 \quad \text{(at the Cu electrode)}$$

When current is passed into the cell, the following reactions takes place

$$2H^+ + 2e^- \longrightarrow H_2 \quad \text{(at the Zn electrode)}$$

$$Cu \longrightarrow Cu^{2+} + 2e^- \quad \text{(at the Cu electrode)}$$

Two important characteristics of Galvanic cell

Let us consider a Daniel-Jacobi cell.

(i) Current is drawn from the cell as shown in the Fig. 12.1. Chemical energy is converted into electrical energy.

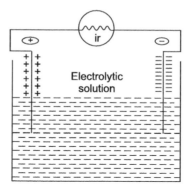

Figure 12.1   An electrochemical cell where electric current is drawn from the cell in the voltaic cell

(ii) Left Hand Electrode (LHE) is anode and positively charged. So, oxidation takes place.

$$Zn \longrightarrow Zn^{2+} + 2e^-$$

Right Hand Electrode (RHE) is cathode and negatively charged. So, reduction takes place.

$$Cu^{2+} + 2e^- \longrightarrow Cu$$

Two important characteristics of electrolytic cell

Consider the same Daniel-Jacobi cell but the configuration is changed. Cathode plate is made negative by external battery. Similarly anode plate is made positive by the same external battery.

(i) Current is sent to the cell as shown in the Fig. 12.2. Electrical energy is converted into chemical energy.

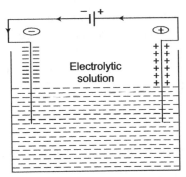

Figure 12.2   An electrochemical cell where current is sent to the cell

(ii) Left Hand Electrode (LHE) is cathode and negatively charged. So, reduction takes place.

$$Zn^{2+} + 2e^- \longrightarrow Zn$$

Right Hand Electrode (RHE) is anode and positively charged. So, oxidation takes place.

$$Cu \longrightarrow Cu^{2+} + 2e^-$$

emf and pd   In an electrochemical cell, cathode plate becomes negative and anode plate becomes positive. Cathode plate is called negative electrode and anode plate is called positive electrode. When the cell is externally disconnected, the potential difference between the two electrodes is called *emf*, represented as E.

When the electrodes are externally connected, as shown in the Fig. 12.1, the net potential difference between the two electrodes is $V = E - ir$, where, i is the current drawn from the cell and r is the external resistance. This V is called potential difference or **pd** of the cell.

Anode and Cathode   According to the latest convention anode is that electrode where oxidation takes place and cathode is that electrode where reduction takes place.

Sign of emf   When current is drawn from the cell, i.e., chemical energy is converted into electrical energy the cell emf is taken as **positive**. On the other hand when current is passed to the cell to bring about chemical reaction in the cell electrolyte, the cell emf is taken as **negative**.

12.2   THERMODYNAMICS OF ELECTROCHEMICAL CELLS

For any electrochemical cell two parts are clearly distinguished:
  (a) External source or battery or output voltage, as shown in Figs. 12.1 and 12.2.
      The free energy change = ΔG_{out}.
  (b) Internal source of current within the cell, i.e., chemical reaction continuously occurring within the cell.

The free energy change $= \Delta G_{cell}$

The net free energy change $= \Delta G = \Delta G_{cell} + \Delta G_{out}$

At equilibrium the net current produced within the cell is equal to the current drawn out externally. Thus applying thermodynamic condition of equilibrium, we have

$\Delta G = 0 = \Delta G_{cell} + \Delta G_{out}$. Or, $\Delta G_{cell} = -\Delta G_{out}$. Again, $\Delta G_{out} = nFE$ Joules, where, $E =$ emf of the cell in volt, $F = 96,500$ C/gm.equiv. and $n =$ Number of gm.equiv. of electrons exchanged within the electrodes. Thus, $\Delta G_{cell} = -nFE$. Usually, ΔG_{cell} is written as ΔG. So, the above equation becomes

$$\Delta G = -nFE \qquad (12.1a)$$

Differentiating the above equation with respect to T at constant P, we get

$$\left[\frac{\partial(\Delta G)}{\partial T}\right]_P = -nF\left(\frac{\partial E}{\partial T}\right)_P \qquad (12.2)$$

$\left(\frac{\partial E}{\partial T}\right)_P$ is known as temperature coefficient of the cell. From Gibbs Helmholtz relation,

we have

$$\Delta S = -\left[\frac{\partial(\Delta G)}{\partial T}\right]_P = nF\left(\frac{\partial E}{\partial T}\right)_P \qquad (12.1b)$$

Again, from Helmholtz equation, we have $\Delta G = \Delta H - T\Delta S$.
Hence we have,

$$\Delta H = \Delta G + T\Delta S = -nFE + nFT\left(\frac{\partial E}{\partial T}\right)_P = -nF\left[E - T\left(\frac{\partial E}{\partial T}\right)_P\right] \qquad (12.1c)$$

The equations (12.1a), (12.1b), and (12.1c) show the relation between thermodynamic properties and emf of an electrochemical cell.

12.2.1 Significance of Temperature Coefficient of the Cell

For any cell $\left(\frac{\partial E}{\partial T}\right)_P$ signifies the change in emf with temperature. The cell for which $\left(\frac{\partial E}{\partial T}\right)_P$ is close to zero, the corresponding cell emf remains constant within a wide range of temperature. That type of cell is called **standard cell**. One of such standard cell is **Weston-Cadmium cell** (Fig. 12.3).

$$\text{Cd}(12.5\%\,\text{Hg})|^+ \text{CdSO}_4 \cdot \frac{8}{3}\text{H}_2\text{O}(s)\big|\text{CdSO}_4(sat.soln.)\big|\text{Hg}_2\text{SO}_4(s)|^- \text{Hg}(l)$$

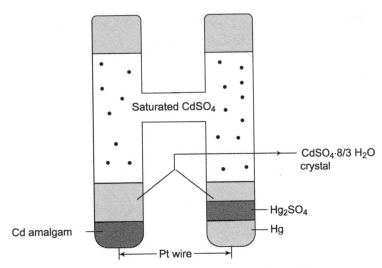

Figure 12.3 Standard Weston-Cadmium cell

Chemical reaction occurring at anode (LHE)

The LHE consists of 12.5% cadmium amalgam followed by crystals of $CdSO_4 \cdot \frac{8}{3}H_2O(s)$. Electrolyte is saturated solution of $CdSO_4$.

$$Cd \longrightarrow Cd^{2+} + 2e^-$$

$$Cd^{2+} + SO_4^{2-} + \frac{8}{3}H_2O \longrightarrow CdSO_4 \cdot \frac{8}{3}H_2O(s)$$

$$\overline{Cd + SO_4^{2-} + \frac{8}{3}H_2O \longrightarrow CdSO_4 \cdot \frac{8}{3}H_2O(s) + 2e^-}$$

Chemical reaction occurring at cathode (RHE)

The RHE consists of mercury followed by crystals of mercurous sulphate and $CdSO_4 \cdot \frac{8}{3}H_2O(s)$. The electrolyte is saturated solution of $CdSO_4$.

$$2Hg(l) \longrightarrow Hg_2^{2+} + 2e^-$$

$$\frac{Hg_2^{2+} + SO_4^{2-} \longrightarrow Hg_2SO_4(s)}{2Hg(l) + SO_4^{2-} \longrightarrow Hg_2SO_4(s) + 2e^-}$$

The overall cell reaction = LHE – RHE

$$Cd + Hg_2SO_4(s) + \frac{8}{3}H_2O \longrightarrow CdSO_4 \cdot \frac{8}{3}H_2O(s) + 2Hg(l)$$

The cell emf at 20°C is 1.018 V and $(\partial E/\partial T)_P = -5 \times 10^{-5}$ V/°K

12.3 NERNST EQUATION FOR SINGLE ELECTRODE POTENTIAL AND ELECTROCHEMICAL CELL

Consider a metal/metal ion electrode of type M/M^{n+}. The electrode reaction is

$$M \longrightarrow M^{n+} + ne$$

The oxidation potential is ϕ_{ox}.

According to Gibbs-Duhem equation, we have

$G = \sum \mu_i n_i$, where n_i is the number of moles of ith species and μ_i is the chemical potential of ith species.

Thus for 1 mole of metal the free energy change is given by

$$\Delta G = (\mu_{M^{n+}} - \mu_M) \tag{12.2}$$

Again, we know that,

$$\mu_i = \mu_i^0 + RT \ln a_i, \text{ where, } a_i \text{ is the activity of } i^{th} \text{ species.}$$

Thus the above equation becomes

$$\Delta G = (\mu_{M^{n+}}^0 + RT \ln a_{M^{n+}}) - (\mu_M^0 + RT \ln a_M)$$

Activity of all solids and metals are taken as unity. Thus, $a_M = 1$ and $\ln a_M = 0$.

Thus, $\qquad \Delta G = (\mu_{M^{n+}}^0 + RT \ln a_{M^{n+}}) - \mu_M^0 = (\mu_{M^{n+}}^0 - \mu_M^0) + RT \ln a_{M^{n+}}$

Or, $\qquad \Delta G = \Delta G^0 + RT \ln a_{M^{n+}} \tag{12.3a}$

$$[\text{where,} \quad \Delta G^0 = (\mu_{M^{n+}}^0 - \mu_M^0)]$$

Again, we know, $\Delta G = -nF\phi_{ox}$ [Equation (12.1a)]. Combining the above two equations, we get

$$-nF\phi_{ox} = (\mu_{M^{n+}}^0 - \mu_M^0) + RT \ln a_{M^{n+}}$$

Or, $\qquad \phi_{ox} = -\left(\dfrac{\mu_{M^{n+}}^0 - \mu_M^0}{nF} \right) - \dfrac{RT}{nF} \ln a_{M^{n+}}$

$$\phi_{ox}^0 = -\left(\dfrac{\mu_{M^{n+}}^0 - \mu_M^0}{nF} \right). \text{ So, } \phi_{ox} = \phi_{ox}^0 - \dfrac{RT}{nF} \ln a_{M^{n+}} \tag{12.3b}$$

where, ϕ_{ox} and ϕ_{ox}^0 are the oxidation potential and standard oxidation potential of the single electrode of type M/M^{n+}. Number of electrons exchanged is n. Equation (12.3b) is the **Nernst equation** for the single electrode potential in oxidation form.

ϕ_{ox}^0 is defined as the **oxidation potential** at unit activity of metal ion. The electrode reaction is oxidation reaction since electrons are released. If the electrode reaction is reduction type, i.e., $M^{n+} + ne^- \longrightarrow M$ (electrons are accepted), the corresponding potential is written as ϕ_{red}, which is expressed as $\phi_{ox} = -\phi_{red}$ and $\phi_{ox}^0 = -\phi_{red}^0$. Putting these values of ϕ_{ox} and ϕ_{ox}^0 in the equation (12.3b) we have

$$\phi_{red} = \phi_{red}^0 + \frac{RT}{nF} \ln a_{M^{n+}} \tag{12.3c}$$

Equation (12.3b) is the **Nernst equation** for the single electrode potential in reduction form.

12.3.1   Nernst Equation for a Cell

An electrochemical cell consists of two electrodes, namely, cathode and anode. Let us consider **Daniell-Jacobi cell.**

$$Zn \,|^+ ZnSO_4(aq) \,\|CuSO_4(aq) \,|^- Cu$$

LHE : $Zn \longrightarrow Zn^{2+} + 2e^-$. Applying equation (12.3b) we get

$$\phi_{Zn} = \phi_{Zn}^0 - \frac{RT}{2F} \ln a_{Zn^{2+}} \qquad [\text{Since, } a_{Zn} = 1]$$

RHE : $Cu \longrightarrow Cu^{2+} + 2e^-$. Applying equation (12.3b) we get

$$\phi_{Cu} = \phi_{Cu}^0 - \frac{RT}{2F} \ln a_{Cu^{2+}} \qquad [\text{Since, } a_{Cu} = 1]$$

E_{ox} = The cell emf = LHE emf – RHE emf = $\phi_{Zn} - \phi_{Cu}$

Or, $\qquad E_{ox} = \phi_{Zn} - \phi_{Cu} = (\phi_{Zn}^0 - \phi_{Cu}^0) - \frac{RT}{2F} \ln \frac{a_{Zn^{2+}}}{a_{Cu^{2+}}}$

Or, $\qquad E_{ox} = E_{ox}^0 - \frac{RT}{2F} \ln \frac{a_{Zn^{2+}}}{a_{Cu^{2+}}}$ $\qquad\qquad$ (12.4a)

where, $E_{ox}^0 = (\phi_{Zn}^0 - \phi_{Cu}^0)$ is the standard cell emf, i.e., emf of the cell at unit concentration of ions.

The overall cell reaction = LHE – RHE.   Or, $Zn + Cu^{2+} \longrightarrow Zn^{2+} + Cu$

Number of electrons exchanged is 2.

Thus, in the equation (12.4a) the **numerator** of ln contains the activity terms of right hand side of the cell reaction and the **denominator** of ln consists of activity terms of left hand side of the cell reaction.

Let us consider a general cell reaction of the following type:

$$lL + mM + \cdots \longrightarrow pP + qQ + \cdots$$

The oxidation emf of the cell is given by

$$E_{ox} = E_{ox}^0 - \frac{RT}{nF} \ln \frac{a_P^p . a_Q^q \cdots \cdots}{a_L^l . a_M^m \cdots \cdots} \tag{12.4b}$$

where, n is number of electrons exchanged in the above chemical reaction.
The reduction emf of the cell is given by

$$E_{red} = E_{red}^0 + \frac{RT}{nF} \ln \frac{a_P^p . a_Q^q \cdots \cdots}{a_L^l . a_M^m \cdots \cdots} \tag{12.4c}$$

$$E_{red} = -E_{ox}, \ E_{red}^0 = -E_{ox}^0$$

Equations (12.4b) and (12.4c) are known as **Nernst equation** for an electrochemical cell. E_{ox}^0 is called the **standard cell emf,** which is the cell emf at unit concentration of the participants.

At 25°C the equation (12.4b) can be reduced to the following form

$$E_{ox} = E_{ox}^0 - \frac{RT}{nF} \ln \frac{a_P^p . a_Q^q \cdots \cdots}{a_L^l . a_M^m \cdots \cdots}. \ \text{Or,} \ E_{ox} = E_{ox}^0 - \frac{2.303 RT}{nF} \log \frac{a_P^p . a_Q^q \cdots \cdots}{a_L^l . a_M^m \cdots \cdots}$$

$$\frac{2.303 RT}{F} = \frac{2.303 \times 8.314 \times 298}{96,500} \frac{J}{C} \ [\text{at } 25°C] = 0.0592 \ V \ [\text{at } 25°C], \ 1J = 1V - C$$

Thus, $\quad E_{ox} = E_{ox}^0 - \frac{0.0592}{n} \log \frac{a_P^p . a_Q^q \cdots \cdots}{a_L^l . a_M^m \cdots \cdots} \tag{12.4d}$

Equation (12.4d) is the **Nernst equation** of an electrochemical cell at 25°C.

12.4   CLASSIFICATION OF ELECTRODES

Electrodes can be broadly classified into three categories:

12.4.1   1st Kind of Electrode

It represents metal/metal ion type of electrode. If a metal plate is immersed into a solution of same metal salt, a potential is developed surrounding the metal plate. The electrode of this type is called 1st kind electrode, e.g., $Zn|Zn^{2+}$ (aq). The electrode reaction is $Zn \longrightarrow Zn^{2+} + 2e^-$. The electrode potential is given by

$$\phi_{Zn} = \phi_{Zn}^0 - \frac{RT}{2F} \ln a_{Zn^{2+}} \ [\text{Since,} \ a_{Zn} = 1] \tag{12.5a}$$

12.4.2 2nd Kind of Electrode

It represents metal/sparingly soluble salt/saturated salt solution type of electrode. A metal plate, coated with its sparingly soluble salt, is immersed in a solution containing a trace amount of same sparingly soluble salt and an easily soluble salt containing the same anion. One example is $Ag|AgCl(s)$, KCl. The electrode reaction is

$$Ag(s) \longrightarrow Ag^+ + e^-$$
$$\underline{Ag^+ + Cl^- \relbar\joinrel\dashrightarrow AgCl(s)}$$
$$Ag(s) + Cl^- \longrightarrow AgCl(s) + e^-$$

The electrode potential is given by

$$\phi_{Ag|AgCl(s)} = \phi^0_{Ag|AgCl(s)} - \frac{RT}{F}\ln\frac{a_{AgCl(s)}}{a_{Ag(s)} \cdot a_{Cl^-}} = \phi^0_{Ag|AgCl(s)} + \frac{RT}{F}\ln a_{Cl^-} \qquad (12.5b)$$

$$[\text{Since, } a_{Ag(s)} = a_{AgCl(s)} = 1]$$

12.4.3 3rd Kind of Electrode

It represents **redox electrode.** If a platinum plate is immersed into a solution containing a multivalent metal ion, redox electrode is formed. Pt-plate is chosen since it does not participate in chemical reaction but serves only as a carrier of electrons. One example of redox electrode is Pt – plate dipped into a solution containing both Fe^{2+} and Fe^{3+} ions.

The electrode reaction is $Fe^{2+} \longrightarrow Fe^{3+} + e^-$.

The electrode potential is given by

$$\phi_{Fe^{2+}|Fe^{3+}} = \phi^0_{Fe^{2+}|Fe^{3+}} - \frac{RT}{F}\ln\frac{a_{Fe^{3+}}}{a_{Fe^{2+}}} \qquad (12.5c)$$

12.5 REFERENCE ELECTRODES

An electrode of known potential is called **reference electrode**. The potential of any unknown electrode can be determined with the help of a reference electrode.

12.5.1 The Hydrogen Electrode

The basic reference electrode is the hydrogen electrode, also known as standard hydrogen electrode (SHE) since the value of standard potential (ϕ^0) of hydrogen electrode is universally accepted as **Zero.** So, $\phi^0_{H_2|H^+} = 0$.

12.5.1.1 Preparation of Hydrogen Electrode

A platinized Pt-plate is immersed in a weak acid solution containing sulphuric acid or hydrochloric acid. Pure hydrogen gas is passed into the solution under a constant pressure.

Hydrogen gas is first adsorbed on the Pt-surface and then diffuses into the solution. Pt-surface is further platinised to increase the efficiency of hydrogen-adsorption. The overall electrode reaction is given by

$$\frac{1}{2}H_2(g) \rightarrow H^+ + e^-$$

The electrode potential is given by

$$\phi_{H_2|H^+} = \phi^0_{H_2|H^+} - \frac{RT}{F} \ln \frac{a_{H^+}}{a_{H_2}^{1/2}} = \phi^0_{H_2|H^+} - \frac{RT}{F} \ln \frac{a_{H^+}}{p_{H_2}^{1/2}}, \text{ since, } a_{H_2} = p_{H_2} \text{ for gaseous state.}$$

Again, $\phi^0_{H_2|H^+} = 0$. Thus, the above equation becomes

$$\phi_{H_2|H^+} = -\frac{RT}{F} \ln \frac{a_{H^+}}{p_{H_2}^{1/2}} \tag{12.6a}$$

a_{H^+} is the activity of H^+ ion in aqueous solution. Usually it is maintained at 1 N. p_{H_2} is gas pressure of hydrogen on the electrode. Thus the above equation becomes

$$\phi_{H_2|H^+} = \frac{RT}{2F} \ln p_{H_2} = \frac{0.0592}{2} \log p_{H_2} \quad [\text{at } 25°C]$$

12.5.1.2 Disadvantages of Hydrogen Electrode

(a) Calculation shows that a deviation of gas pressure by 0.1 atm results in a change in potential by 0.0012 V. As the gas pressure cannot be maintained accurately, the electrode potential of standard hydrogen electrode, $\phi_{H_2|H^+}$, is often fluctuating. Obviously the potential of an unknown electrode, evaluated with the help of hydrogen electrode, is erroneous.

(b) Presence of impurities changes the potential significantly. That is why, hydrogen electrode is seldom used as a reference electrode.

(c) Hydrogen electrode is unstable under corrosive environment, particularly in strong acidic or strong alkaline medium.

12.5.2 Calomel Electrode

Mercurous chloride, Hg_2Cl_2, is known as **calomel.** When a mercury electrode is dipped in a solution of potassium chloride (KCl) of a definite concentration saturated with calomel, i.e., Hg_2Cl_2, the resultant electrode is called **calomel electrode.** It is represented as $Hg(l)|Hg_2Cl_2(s), KCl(a)$

The electrode reaction is given by

$$2Hg(l) \longrightarrow Hg_2^{2+} + 2e^-$$
$$\underline{Hg_2^{2+} + 2Cl^- \longrightarrow Hg_2Cl_2(s)}$$
$$2Hg(l) + 2Cl^- \longrightarrow Hg_2Cl_2(s) + 2e^-$$

The electrode potential is given by

$$\phi_{cal} = \phi_{cal}^0 - \frac{RT}{2F} \ln \frac{a_{Hg_2Cl_2(s)}}{a_{Hg(l)} \cdot a_{Cl^-}^2} = \phi_{cal}^0 + \frac{RT}{F} \ln a_{Cl^-} \tag{12.6b}$$

Since, $a_{Hg_2Cl_2(s)} = a_{Hg(l)} = 1$

The value of ϕ_{cal} depends on KCl concentration. Thus, it has a definite potential at a definite concentration of KCl, e.g.,

(a) $a_{Cl^-} = m_{Cl^-} = (N/10)$ $\phi_{cal} = -0.3338$ V at 25°C

(b) $a_{Cl^-} = m_{Cl^-} = (N)$ $\phi_{cal} = -0.2800$ V at 25°C

(c) $a_{Cl^-} = m_{Cl^-} = $ Saturated KCl solution $\phi_{cal} = -0.2415$ V at 25°C

12.5.3   Silver-Silver Chloride Electrode

This electrode is similar to the calomel electrode. The electrode is represented as

      Ag|AgCl(s), KCl

A silver electrode is immersed into a solution of potassium chloride (KCl), saturated with silver chloride (AgCl). The electrode reaction is given by

$$Ag(s) \longrightarrow Ag^+ + e^-$$
$$Ag^+ + Cl^- \longrightarrow AgCl(s)$$
$$\overline{Ag(s) + Cl^- \longrightarrow AgCl(s) + e^-}$$

The electrode potential is given by

$$\phi_{Ag|AgCl(s)} = \phi_{Ag|AgCl(s)}^0 - \frac{RT}{F} \ln \frac{a_{AgCl(s)}}{a_{Ag(s)} \cdot a_{Cl^-}} = \phi_{Ag|AgCl(s)}^0 + \frac{RT}{F} \ln a_{Cl^-} \tag{12.6c}$$

$[$Since, $a_{Ag(s)} = a_{AgCl(s)} = 1]$

Table 12.1   Standard electrode potential of different electrodes in aqueous solution

Electrode	$\phi^0(V)$	*Electrode*	$\phi^0(V)$	*Electrode*	$\phi^0(V)$
Li|Li$^+$	+3.045	Ni|Ni^{2+}	+0.250	Au|Au^{3+}	−1.290
K|K$^+$	+2.925	Pb|Pb^{2+}	+0.126	H$_2$|H$_3$O$^+$	0.000
Na|Na$^+$	+2.714	Cu|Cu^{2+}	−0.337	OH$^-$|O$_2$	−0.401
Mg|Mg^{2+}	+2.37	Cu|Cu$^+$	−0.521	I$^-$|I$_2$	−0.536
Al|Al^{3+}	+1.66	Pb|Pb^{4+}	−0.700	Br$^-$|Br$_2$	−1.065
Zn|Zn^{2+}	+0.763	Hg|Hg$_2^{2+}$	−0.789	Cl$^-$|Cl$_2$	−1.360
Fe|Fe^{2+}	+0.440	Ag|Ag$^+$	−0.799	F$^-$|F$_2$	−2.650
Cd|Cd^{2+}	+0.403	Hg|Hg^{2+}	−0.854		

Table 12.2   Some standard redox potentials in aqueous solution

Electrode	Electrode Process	$\phi^0(V)$
$Cr^{2+}, Cr^{3+} \mid Pt$	$Cr^{2+} \rightarrow Cr^{3+} + e^-$	+0.410
$Sn^{2+}, Sn^{4+} \mid Pt$	$Sn^{2+} \rightarrow Sn^{4+} + 2e^-$	−0.150
$(ClO_3 + 2OH^-), ClO_4^- \mid Pt$	$(ClO_3^- + 2OH^-) \rightarrow ClO_4^- + H_2O + e^-$	−0.360
$MnO_4^{2-}, MnO_4^- \mid Pt$	$MnO_4^{2-} \rightarrow MnO_4^- + e^-$	−0.564
$Fe^{2+}, Fe^{3+} \mid Pt$	$Fe^{2+} \rightarrow Fe^{3+} + e^-$	−0.771
$Hg_2^{2+}, Hg^{2+} \mid Pt$	$Hg_2^{2+} \rightarrow 2Hg^{2+} + 2e^-$	−0.920
$Mn^{2+}, Mn^{4+} (4OH^-) \mid Pt$	$Mn^{2+} + 2H_2O \rightarrow MnO_2 + 4H^+ + 2e^-$ $4H^+ + 4OH^- \rightarrow 4H_2O$	−1.230
$Pb^{2+}, Pb^{4+} (4OH^-) \mid Pt$	$Pb^{2+} + 2H_2O \rightarrow PbO_2 + 4H^+ + 2e^-$	−1.455
$Mn^{2+}, Mn^{3+} \mid Pt$	$Mn^{2+} \rightarrow Mn^{3+} + e^-$	−1.510
$PbSO_4, PbO_2 \mid Pt$	$PbSO_4 + 2H_2O \rightarrow PbO_2 + H_2SO_4 + 2H^+ + 2e^-$	−1.685
$Pb^{2+}, Pb^{4+} \mid Pt$	$Pb^{2+} \rightarrow Pb^{4+} + 2e^-$	−1.700
$Co^{2+}, Co^{3+} \mid Pt$	$Co^{2+} \rightarrow Co^{3+} + e^-$	−1.820
$Cr^{3+}, Cr^{6+} \mid Pt$	$2Cr^{3+} + 7H_2O \rightarrow Cr_2O_7^{2-} + 14H^+ + 6e^-$	−1.330
$Mn^{2+}, Mn^{7+} \mid Pt$	$Mn^{2+} + 7H_2O \rightarrow MnO_4^- + 8H^+ + 5e^-$	−1.520
$Ce^{3+}, Ce^{4+} \mid Pt$	$Ce^{3+} \rightarrow Ce^{4+} + e^-$	−1.610

12.6 APPLICATIONS OF SINGLE ELECTRODE POTENTIAL

12.6.1 Determination of pH of an Unknown Solution

12.6.1.1 Determination of pH by SHE

(i) Let us consider the following cell using silver electrode as reference electrode.

$$Pt \mid^+ H_2(g) \mid \text{Unknown buffer solution } (H^+) \parallel KCl(a), AgCl \mid^- Ag(s)$$

LHE $1/2\, H_2 \longrightarrow H^+ + e^-$, Electrode potential is ϕ_L, which is expressed as

$$\phi_L = \phi^0_{H_2 \mid H^+} - \frac{RT}{F} \ln \frac{a_{H^+}}{p_{H_2}^{1/2}} = -0.0592 \log a_{H^+} \quad [\text{at } 25°C], \text{ since, } p_{H_2} \text{ is maintained 1 atm.}$$

RHE Electrode potential is ϕ_R

$$Ag(s) \longrightarrow Ag^+ + e^-$$

$$Ag^+ + Cl^- \longrightarrow AgCl(s)$$

$$\overline{Ag(s) + Cl^- \longrightarrow AgCl(s) + e^-}$$

The electrode potential is given by

$$\phi_R = \phi_{Ag|AgCl(s)} = \phi^0_{Ag|AgCl(s)} - \frac{RT}{F} \ln \frac{a_{AgCl(s)}}{a_{Ag(s)} \cdot a_{Cl^-}}$$

$$= \phi^0_R + \frac{RT}{F} \ln a_{Cl^-} = \phi^0_R + 0.0592 \log m_{Cl^-} \quad [\text{at } 25°C]$$

[Assuming that $\gamma_{Cl^-} = 1$, $a_{Cl^-} = a_{KCl} = m_{Cl^-} \times \gamma_{Cl^-} = m_{Cl^-}$]
Cell emf (E) is given by

$$E = \phi_L - \phi_R = -0.0592 \log a_{H^+} - \phi^0_R - 0.0592 \log m_{Cl^-}$$

$$= E_0 - 0.592 \log m_{Cl^-} + 0.0592 \text{ pH}_a$$

[Where, $E^0 = -\phi^0_R$ and pH$_a = -0.0592 \log a_{H^+}$]

Or, $$\frac{E - E^0}{0.0592} = pH_a - \log m_{Cl^-}$$

E^0 is known and constant and E is a measurable quantity. Thus, different $E-$ values are recorded at different chloride ion concentrations, i.e., m_{Cl^-}. Thus, by plotting $\left(\dfrac{E - E^0}{0.0592}\right)$ against $\log m_{Cl^-}$ a straight line of negative slope is obtained. The intercept, obtained at $m_{Cl^-} = 1$ N, gives the value of pH$_a$ of the buffer solution.

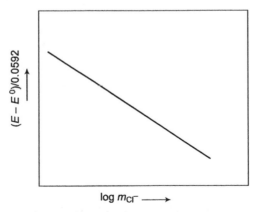

Figure 12.4 Plot of $\left(\dfrac{E - E^0}{0.0592}\right)$ versus $\log m_{Cl^-}$ of a buffer solution. Intercept $= p$Ha

(ii) Let us consider the following cell using calomel electrode as reference electrode.

$$Pt|^+ H_2(g)| \text{Unknown buffer solution (H}^+) \parallel KCl(a), Hg_2Cl_2 |^- Hg(l)$$

LHE $1/2\ H_2 \longrightarrow H^+ + e^-$, Electrode potential is ϕ_L, which is expressed as

$$\phi_L = \phi^0_{H_2|H^+} - \frac{RT}{F}\ln\frac{a_{H^+}}{p^{1/2}_{H_2}} = -0.0592\log a_{H^+}\ \ [\text{at } 25°C],\ \text{since } p_{H_2}\ \text{is maintained 1 atm.}$$

Or, $\phi_L = 0.0592\ \text{pH}_a$, since, $\text{pH}_a = -0.0592\ a_{H^+}$

RHE Electrode potential is ϕ_{cal}

$$2Hg(l) \longrightarrow Hg_2^{2+} + 2e^-$$

$$\underline{Hg_2^{2+} + 2Cl^- \longrightarrow Hg_2Cl_2(s)}$$

$$2Hg(l) + 2Cl^- \longrightarrow Hg_2Cl_2(s) + 2e^-$$

The electrode potential is given by

$$\phi_{cal} = \phi^0_{cal} - \frac{RT}{2F}\ln\frac{a_{Hg_2Cl_2(s)}}{a^2_{Hg(l)}\cdot a^2_{Cl^-}} = \phi^0_{cal} + \frac{RT}{F}\ln a_{Cl^-} = \phi^0_{cal} + 0.0592\log a_{Cl^-}\ \ [\text{at } 25°C]$$

As it is a reference electrode, ϕ_{cal} is known at a given concentration of KCl.

Cell emf is given by $E = \phi_L - \phi_{cal}$

Putting the value of ϕ_L in the above equation, we get $E = 0.0592\ \text{pH}_a - \phi_{cal}$

Or, $$pH_a = \frac{E + \phi_{cal}}{0.0592}$$

E is measurable quantity and ϕ_{cal} is known. Hence, pH_a of the unknown solution can easily be determined.

12.6.1.2 Determination of pH Using Quinhydrone Electrode

Quinol or hydroquinone (often represented as H_2Q) is oxidized reversibly in presence of $FeCl_3$ to produce *p*-benzoquinone or simply quinone (often represented as Q) as shown below

Quinol/Hydroquinone   *p*-benzoquinone

Symbolically, the above equation may be represented as

$$H_2Q \rightleftharpoons Q + 2H^+ + 2e^-$$

Thus, using Nernst equation, we can write that

$$\phi_Q = \phi^0_Q - \frac{RT}{2F}\ln\frac{a_Q}{a_{H_2Q}}\cdot a^2_{H^+} = \phi^0_Q - \frac{RT}{2F}\ln\frac{a_Q}{a_{H_2Q}} - \frac{RT}{F}\ln a_{H^+}$$

The concentrations of quinone and hydroquinone are so maintained that $a_Q / a_{H_2Q} = 1$. Hence, the above equation becomes

$$\phi_Q = \phi_Q^0 - \frac{RT}{F} \ln a_{H^+} = \phi_Q^0 - \frac{2.303 RT}{F} \log a_{H^+} = \phi_Q^0 + \frac{2.303 RT}{F} pH$$

At 25°C, the value of 2.303 RT/F is 0.0592. Putting this value in the above equation, we get

$$\phi_Q = \phi_Q^0 + 0.0592 \, pH$$

To measure the pH of an unknown solution, $(Q|H_2Q)$ system is added to the unknown solution. A gold foil is immersed into the solution. Platinum foil cannot be used since platinum catalyzes the oxidation of quinol and hence the equilibrium is disturbed. The quinhydrone electrode is connected with a standard calomel electrode as shown below:

$$Hg\,|\,Hg_2Cl_2(s), \text{ saturated KCl} \,\|\, \text{Unknown solution} + (Q|H_2Q)\,|\,Au$$

Calomel electrode is used as anode and quinhydrone electrode is used as cathode. The cell emf can be written as

$$E = \phi_{cal} - \phi_Q = \phi_{cal} - \phi_Q^0 - 0.0592 \, pH \qquad \text{(at 25°C)}$$

Hence, $$pH = \frac{\phi_{cal} - \phi_Q^0 - E}{0.0592} \qquad \text{(at 25°C)}$$

At 25°C, the value of $\phi_Q^0 = -0.699$ V and $\phi_{cal}|_{sat} = -0.241$ V. Cell emf (E) is a measurable quantity and hence pH can be calculated from the above equation.

Advantages of quinhydrone electrode
(a) The electrode is simple and easy to handle.
(b) The equilibrium between quinone and hydroquinone is quickly attained.
(c) A little amount of $(Q|H_2Q)$ system is enough to determine pH of an unknown solution.
(d) The electrode is stable under corrosive environment.
(e) The electrode can be safely used in presence of other ions or organic substances.

Disadvantages of quinhydrone electrode
(a) Platinum foil or wire cannot be used in this electrode since Pt catalyzes oxidation of hydroquinone, thereby disturbing the equilibrium.
(b) The electrode can be safely used in alkaline solution of pH up to 8 but cannot be used for solutions having pH greater than 8.5.

12.6.1.3 Determination of pH Using Glass Electrode

Principle

Glass electrode is widely used for determining pH of an unknown solution. If two aqueous solutions of different pH are separated by a thin glass membrane, a potential is developed across the membrane due to concentration difference. If pH of one solution is kept constant while pH of other solution varies, the electrode potential can be represented as

$$\phi_G = \phi_G^0 - \frac{RT}{F} \ln a_{H^+} = \phi_G^0 - \frac{2.303RT}{F} \log a_{H^+} = \phi_G^0 + 0.0592 \text{ pH} \qquad \text{(at 25°C)}$$

In glass electrode, a glass membrane is used to separate $Ag|AgCl(s)$, 0.1(N)HCl solution and the unknown solution, whose pH is to be determined. To measure the pH of an unknown solution, glass electrode is connected with a standard calomel electrode as shown below

$$Ag|AgCl(s), 0.1 \text{ (N) } HCl|Glass|Unknown \text{ solution } (a_{H^+})\|Standard \text{ Calomel electrode}$$

The emf of the cell is given by

$$E = \phi_G - \phi_{cal} = \phi_G^0 + 0.0592 \text{ pH} - \phi_{cal} \qquad \text{(at 25°C)}$$

E_G^0 is not known. Thus pH of unknown solution is determined against known pH of a standard solution. Suppose, pH(S) is the pH of the standard solution and E_S is the emf of the corresponding cell. Then we can write

$$E_S = \phi_G^0 + 0.0592 \text{ pH(S)} - \phi_{cal} \qquad \text{(at 25°C)}$$

From the above two equations, we get

$$\Delta E = E - E_S = 0.0592[\text{pH} - \text{pH(S)}] \qquad \text{(at 25°C)}$$

pH(S) is known, E and E_S are measurable quantities. So, pH of unknown solution can be calculated from the above equation.

Advantages of glass electrode
 (a) It is simple and easy to handle.
 (b) Equilibrium across the glass membrane is quickly achieved.
 (c) Accurate pH value can be obtained.
 (d) It can be used to measure very high pH quite accurately.
 (e) The glass electrode is not affected in presence of impurities, ions, acid, alkali, oxidizing agents, and reducing agents.
 (f) The electrode is stable under corrosive environment.

Disadvantages of glass electrode
 (a) Special glass membranes are required to measure high pH especially for pH >10.
 (b) Ordinary potentiometer cannot be used due to high resistance of the glass membrane.
 (c) The glass electrode is safe to measure pH up to 12 but for solutions, having pH >12, cations affect the glass membrane and the results are erroneous.

12.6.2   Determination of Solubility Product of a Sparingly Soluble Salt

Solubility and solubility product of any sparingly soluble salt can be determined by choosing suitable cell. Let us find out the solubility product of silver chloride [$AgCl(s)$] which dissociates in aqueous solution as given:

$AgCl(s) \rightleftharpoons Ag^+ + Cl^-$. Solubility product $= K_{aP} = a_{Ag^+} \cdot a_{Cl^-} = S^2$, where, S is the solubility of the salt in aqueous solution. Let us consider the following cell to determine solubility of the salt.

$$Ag(s)|^+ AgCl(s), 0.1 \text{ N KCl}, Hg_2Cl_2(s)|^- Hg(l)$$

The RHE is the calomel electrode. The emf of calomel electrode at 0.1 N KCl, is –0.334 V. The cell emf is given by

$$E = \phi_{Ag} - \phi_{cal} = \phi_{Ag} + 0.334.$$

The cell emf E is measured and its value is 0.0455 V.

Hence, $\quad \phi_{Ag} = E - 0.334 = 0.0455 - 0.334 = -0.289 \text{ V}$

AgCl is sparingly soluble in aqueous solution giving rise to Ag^+ and Cl^-. This Ag^+ forms the LHE with metallic Ag.

$$a_{Ag^+} = a_{Cl^-} = S \text{ and } K_{aP} = a_{Ag^+} \cdot a_{Cl^-} = S^2$$

Thus at the LHE, we have $Ag \longrightarrow Ag^+ + e^-$

$$\phi_{Ag} = \phi_{Ag}^0 - \frac{RT}{F} \ln a_{Ag^+} = \phi_{Ag}^0 - 0.0592 \log \frac{a_{Ag^+} \cdot a_{Cl^-}}{a_{Cl^-}} \qquad \text{[at } 25°C\text{]}$$

Or, $\quad \phi_{Ag} = -0.799 - 0.0592 \log \dfrac{K_{aP}}{S}$, since, $\phi_{Ag}^0 = -0.799\text{V}$

In presence of KCl, $a_{Cl^-} = (S + a_{KCl}) \approx a_{KCl}$, since $a_{KCl}, \gg S$

Thus the above equation becomes

$$\phi_{Ag} = -0.799 - 0.0592 \log \frac{K_{aP}}{a_{KCl}}$$

$$a_{Cl^-} = a_{KCl} = m_{KCl} \cdot \gamma_{Cl^-} = 0.1 \times 0.769 \text{ (N)} = 0.0769\text{(N)}$$

since, $\quad \gamma_{Cl^-} = 0.769 \text{ at } m_{Cl^-} = [\text{KCl}] = 0.1 \text{ N}$

Putting all these values in the above equation and solving K_{aP}, we get $K_{aP} = 1.75 \times 10^{-10}$.

Again, $\quad K_{aP} = S^2$. Hence, $S^2 = 1.75 \times 10^{-10}$ and $S = 1.323 \times 10^{-5}$ unit.

12.6.3 Determination of Ionic Product of Water

To determine ionic product of water, the following cell is considered

$$H_2(g) \ (1 \text{ atm})|^+ NaOH(m_1) \| NaCl(m_2), AgCl(s)|^- Ag(s)$$

Water is dissociated as follows $H_2O \rightleftharpoons H^+ + OH^-$. Thus, LHE reaction can be written as

LHE $\dfrac{1}{2}H_2(g) \longrightarrow H^+ + e^-$. The electrode potential is given by

$$\phi_L = \phi^0_{H_2|H^+} - \frac{RT}{F}\ln\frac{a_{H^+}}{a_{H_2}^{1/2}} = -\frac{RT}{F}\ln\frac{a_{H^+}}{p_{H_2}^{1/2}} = -0.0592\log\frac{a_{H^+}\cdot a_{OH^-}}{a_{OH^-}} = -0.0592\log\frac{K_w}{a_{OH^-}}$$

[at 25°C]

$$[a_{H_2} = p_{H_2} = 1 \text{ atm}, \phi^0_{H_2|H^+} = 0 \text{ V and } K_w = a_{H^+}\cdot a_{OH^-}]$$

In presence of NaOH $a_{OH^-} = a_{NaOH} = m_1$. Thus, the above equation becomes

$$\phi_L = -0.0592\log\frac{K_w}{m_1} \text{ [at 25°C]}$$

RHE

$$Ag(s) \longrightarrow Ag^+ + e^-$$
$$\underline{Ag^+ + Cl^- \longrightarrow AgCl(s)}$$
$$Ag(s) + Cl^- \longrightarrow AgCl(s) + e^-$$

The electrode potential is given by

$$\phi_R = \phi_{Ag|AgCl(s)} = \phi^0_{Ag|AgCl(s)} - \frac{RT}{F}\ln\frac{a_{AgCl(s)}}{a_{Ag(s)}\cdot a_{Cl^-}} = \phi^0_R + 0.0592\log a_{Cl^-} \text{ [at 25°C]}$$

$$[a_{Ag(s)} = a_{AgCl(s)} = 1, \phi^0_{Ag|AgCl(s)} = \phi^0_R]$$

The cell emf $= E = \phi_L - \phi_R = -0.0592\log\frac{K_w}{m_1} - \phi^0_R - 0.0592\log a_{Cl^-}$ [at 25°C]

$$E = E^0 - 0.0592\log\frac{a_{Cl^-}\cdot K_w}{a_{OH^-}} = E^0 - 0.0592\log K_w - 0.0592\log\frac{m_2}{m_1}, [E^0 = -\phi^0_R]$$

$[a_{Cl^-} = a_{NaCl} = m_2]$. Hence, $\log K_w$ is given by

$$\log K_w = -\left(\frac{E - E^0}{0.0592}\right) - \log\frac{m_2}{m_1} \text{ [at 25°C]} \tag{12.7}$$

m_1 and m_2 are the known concentrations of caustic soda (NaOH) and sodium chloride (NaCl). E^0 is also a known quantity. Thus simply by measuring the cell emf, E, the ionic product of water, K_W, can easily be determined by using equation (12.7).

12.6.4   Determination of Equilibrium Constant

Using the concept of thermodynamics, for a given reversible chemical reaction, we know that
$$\Delta G^0 = -RT\ln K_a \tag{12.8a}$$
where, ΔG^0 is the standard free energy change and K_a is the equilibrium constant of that reversible reaction.

Again, if that given reversible chemical reaction is occurring in an electrochemical cell with cell emf E, we know that, $\Delta G = -nFE$ [Equation (12.1a)]

If E^0 is the standard cell potential, ΔG^0 is given by

$$\Delta G^0 = -nFE^0 \tag{12.8b}$$

where, n is the number of electrons exchanged in that chemical reaction.

Comparing equations (12.8a) and (12.8b), we get

$$E^0 = \frac{RT}{nF}\ln K_a = \frac{0.0592}{n}\log K_a \quad [\text{at } 25°C] \tag{12.8c}$$

For any reversible cell, n and E^0 are known. Hence the equilibrium constant, K_a, can easily be determined from the equation (12.8c).

Example 12.6.1

Consider the **Daniell-Jacobi cell.** The cell reaction is given by

$$Zn(s) + Cu^{2+} \longrightarrow Zn^{2+} + Cu(s) \ (n = 2)$$

$$E^0 = \phi^0_{Zn} - \phi^0_{Cu} = 0.763 - (-0.337) = 1.1 \text{ V} \quad\quad [\text{From Table 12.1}]$$

The equilibrium constant, K_a, is given by

$$K_a = \frac{a^{eq.}_{Zn^{2+}} \cdot a^{eq.}_{Cu(s)}}{a^{eq.}_{Cu^{2+}} \cdot a^{eq.}_{Zn(s)}} = \frac{a^{eq.}_{Zn^{2+}}}{a^{eq.}_{Cu^{2+}}}$$

where, $a^{eq.}_i$ is the activity of the ith species at equilibrium condition. From equation (12.8c), we have

$$\log K_a = \frac{2 \times 1.1}{0.0592} = 37.162. \text{ Or, } K_a = 1.45 \times 10^{37} \quad [\text{at } 25°C]$$

Example 12.6.2

Consider the following redox reaction

$$Sn^{2+} + 2Fe^{3+} \rightarrow Sn^{4+} + 2Fe^{2+}$$

The equilibrium constant, K_a, is given by $K_a = \dfrac{a^{eq.}_{Sn^{4+}} \cdot (a^{eq.}_{Fe^{2+}})^2}{a^{eq.}_{Sn^{2+}} \cdot (a^{eq.}_{Fe^{3+}})^2}$

$$E^0 = \phi^0_{Sn} - \phi^0_{Fe} = -0.150 - (-0.771) = 0.621 \text{ V}$$

Putting these values of n and E^0 in the equation (12.8c), we get

$$\log K_a = \frac{2 \times 0.621}{0.0592} = 20.98. \text{ Or, } K_a = 9.54 \times 10^{20} \quad [\text{at } 25°C]$$

12.6.5 Potentiometric Titration

At first a suitable indicator electrode is placed in the titrant (the solution to be titrated). So a potential is developed around the electrode, denoted by ϕ_a. Then this electrode is connected with a reference electrode, ϕ_R (say, calomel electrode) to form a complete cell. The cell emf $= E = \phi_a - \phi_R$.

As ϕ_R is fixed, any change in ϕ_a results in a change in E. The titre (the solution used for titration) is added dropwise from a burette and the cell emf, E, is measured continuously. A slight change in E is observed up to the end point. After the end point titrant does not exist because it is totally consumed but addition of titre is continued. Hence, the nature of the electrode also changes to ϕ_b, i.e., potential developed around the electrode is ϕ_b not ϕ_a and the cell emf is given by $E = \phi_b - \phi_R$. Thus at the end point a sharp change in potential occurs as shown in the Fig. 12.5.

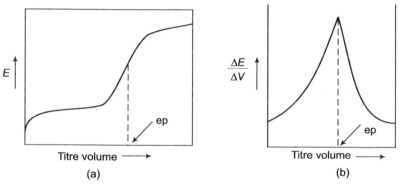

Figure 12.5 (a) Cell emf (E) versus volume of titre plot of a typical redox reaction

(b) $\left(\dfrac{\Delta E}{\Delta V}\right)$ versus volume plot of the same redox reaction. ΔV = Change in titre volume

The titre volume at which the last trace of titrant is consumed by the titre, is known as the **equivalence point** and the corresponding potential is called **equivalence potential**, denoted by E_p. The equivalence point is difficult to identify in Fig. 12.5(a) but it is very sharp in Fig. 12.5(b).

12.6.5.1 Selection of Indicator Electrode

Selection of indicator electrode depends on type of titration as given below:
   (a) In the case of *redox titration* the indicator electrode is usually Pt-electrode.
   (b) In the case of *precipitation titration* like titration of sodium chloride (NaCl) by silver nitrate ($AgNO_3$), silver electrode is chosen as an indicator electrode.
   (c) In the case of *acid-base titration* a suitable hydrogen electrode is used as an indicator electrode.

12.6.5.2 Calculation of Equivalence Potential, E_p, of a Redox Reaction

A redox reaction can be written as

$$aOx_1 + bRed_2 \rightleftharpoons aRed_1 + bOx_2$$

Ox_1 = Oxidizing agent, converted to corresponding reduced form, Red_1.

Red_2 = Reducing agent, converted to corresponding oxidizing form, Ox_2.

Considering only oxidation form at both the electrodes, we have

At anode $\quad Red_2 \longrightarrow Ox_2 + be$

$$\phi_a = \phi_a^0 - \frac{RT}{bF} \ln \frac{[Ox_2]}{[Red_2]}$$

At cathode $\quad Red_1 \longrightarrow Ox_1 + ae$

$$\phi_c = \phi_c^0 - \frac{RT}{aF} \ln \frac{[Ox_1]}{[Red_1]}$$

At the equivalence point $\phi_c = \phi_a = E_p$

$$E_p = \phi_a^0 - \frac{RT}{bF} \ln \frac{[Ox_2]}{[Red_2]} = \phi_c^0 - \frac{RT}{aF} \ln \frac{[Ox_1]}{[Red_1]}$$

Or, $\quad bE_p = b\phi_a^0 - \frac{RT}{F} \ln \frac{[Ox_2]}{[Red_2]}$ and $aE_p = a\phi_c^0 - \frac{RT}{F} \ln \frac{[Ox_1]}{[Red_1]}$

$$(a+b)E_p = (a\phi_c^0 + b\phi_a^0) - \frac{RT}{F} \ln \left[\frac{[Ox_1]}{[Red_1]} \times \frac{[Ox_2]}{[Red_2]} \right]$$

Again, at the equivalence point $\quad a[Ox_1] = b[Red_2]$ and $a[Red_1] = b[Ox_2]$

Hence, $\quad \dfrac{[Ox_1]}{[Red_1]} = \dfrac{[Red_2]}{[Ox_2]}$. Or, $\dfrac{[Ox_1]}{[Red_1]} \times \dfrac{[Ox_2]}{[Red_2]} = 1$

Using this condition in the above equation, we get

$$(a+b)E_p = (a\phi_c^0 + b\phi_a^0). \text{ Or, } E_p = \frac{a\phi_c^0 + b\phi_a^0}{a+b} \tag{12.9}$$

12.7 CONCENTRATION CELLS

12.7.1 Expression of E.M.F. and Transport Number

An important type of electrochemical cell is **concentration cell**. In this special type of cell, electrolyte is the same in both the electrodes. There is no net chemical reaction in the cell. So the current produced in this cell is not due to chemical reaction but due to concentration difference of electrolyte in two electrodes. Let us consider a concentration cell, where two similar electrodes are separated by a reference electrode. Exchange of ions between two electrodes is prevented due to the presence of reference electrode. This type of electrochemical cell is known as concentration cell without transference. The cell emf is represented as E.

$$\text{Pt, } H_2(g)(1 \text{ atm})|^+ HCl(a_1), AgCl(s)|Ag(s)|AgCl(s), HCl(a_2)|^- H_2(g)(1 \text{ atm), Pt}$$

Each electrode is an elementary cell

LHE

$$\frac{1}{2}H_2(g) \longrightarrow H^+ + e^-$$

$$\underline{AgCl(s) + e^- \longrightarrow Ag(s) + Cl^-}$$

$$\frac{1}{2}H_2(g) + AgCl(s) \longrightarrow Ag(s) + H^+ + Cl^-$$

Or, $\qquad \dfrac{1}{2}H_2(g) + AgCl(s) \longrightarrow Ag(s) + HCl(a_1)$ $\qquad\qquad$ (12.10a)

Thus 1 mole of HCl of activity a_1 is produced at the LHE

RHE

$$H^+ + e^- \longrightarrow \frac{1}{2}H_2(g)$$

$$\underline{Ag(s) + Cl^- \longrightarrow AgCl(s) + e^-}$$

$$Ag(s) + H^+ + Cl^- \longrightarrow \frac{1}{2}H_2(g) + AgCl(s)$$

Or, $\qquad Ag(s) + HCl(a_2) \longrightarrow \dfrac{1}{2}H_2(g) + AgCl(s)$ $\qquad\qquad$ (12.10b)

Thus, 1 mole of HCl of activity a_2 is consumed at the RHE

The net cell reaction is obtained by adding the equations (12.10a) and (12.10b).

$$HCl(a_2) \longrightarrow HCl(a_1)$$

Free energy change, ΔG, is given by

$$\Delta G = \mu_{HCl(a_1)} - \mu_{HCl(a_2)} = (\mu_{HCl}^0 + RT \ln a_1) - (\mu_{HCl}^0 + RT \ln a_2). \text{ Or, } \Delta G = RT \ln \frac{a_1}{a_2}$$

Again, we have $\Delta G = -FE$, where, E is the cell emf. Thus,

$$-FE = RT \ln \frac{a_1}{a_2}. \text{ Or, } E = \frac{RT}{F} \ln \frac{a_2}{a_1}$$

Again, in aqueous solution HCl remains in dissociated form, i.e., $HCl \longrightarrow H^+ + Cl^-$. So, activity of HCl can be written as $a_{HCl} = a_{H^+} \cdot a_{Cl^-}$ and $a_1 = (a_+)_1 \cdot (a_-)_1$. If $a_\pm$ is the mean activity, we can write that

$(a_\pm)_1^2 = (a_+)_1 \cdot (a_-)_1$, where, $(a_+)_1 = (a_\pm)_1$ and $(a_-)_1 = (a_\pm)_1$. Thus, we can write that

$$a_1 = (a_\pm)_1^2 \text{ and } a_2 = (a_\pm)_2^2 \tag{12.11}$$

Putting these values of a_1 and a_2 in the above equation we get

$$E = \frac{RT}{F}\ln\frac{(a_\pm)_2^2}{(a_\pm)_1^2} = \frac{2RT}{F}\ln\frac{(a_\pm)_2}{(a_\pm)_1} \tag{12.12}$$

Now we consider another cell where the reference electrode is absent and the electrodes are separated by a membrane, which allows free movement of ions between the electrodes. This type of electrochemical cell is known as concentration cell with transference. The cell emf is represented as E_t.

The two hydrochloric acid solutions are separated by a membrane, which allows ions to diffuse from one compartment to another. So, t_+ gm.equiv. of H^+ ion moves towards the cathode and t_- gm.equiv. of Cl^- ion moves towards the anode. The transfer of ions at two electrodes are given below

LHE	**RHE**
$\frac{1}{2}H_2(g) \longrightarrow H^+ + e^-$	$H^+ + e^- \longrightarrow \frac{1}{2}H_2(g)$
+1 mole of H^+ (appears)	-1 mole of H^+ (disappears)
$-t_+$ mole of H^+ (disappears)	$+t_+$ mole of H^+ (appears)
$+t_+$ mole of Cl^- (appears)	$-t_+$ mole of Cl^- (disappears)
$+(1-t_+)$ mole of H^+ (appears)	$-(1-t_+)$ mole of H^+ (disappears)
$+t_-$ mole of Cl^- (appears)	$-t_-$ mole of Cl^- (disappears)
$+t_-$ mole of H^+ (appears)	$-t_-$ mole of H^+ (disappears)
$+t_-$ mole of Cl^- (appears)	$-t_-$ mole of Cl^- (disappears)
$+t_-$ mole of HCl of activity a_1 (appears)	$-t_-$ mole of HCl of activity a_2 (disappears)

Thus free energy change (ΔG) is given by

$$\Delta G = t_-\mu_{HCl(a_1)} - t_-\mu_{HCl(a_2)} = t_-[\mu_{HCl(a_1)} - \mu_{HCl(a_2)}]$$

$$\Delta G = t_-[(\mu_{HCl}^0 + RT\ln a_1) - (\mu_{HCl}^0 + RT\ln a_2)] = t_- RT\ln\frac{a_1}{a_2}$$

If E_t is the cell emf, we have $\qquad \Delta G = -FE_t$.

Or, $\qquad -FE_t = t\_RT \ln \dfrac{a_1}{a_2}$. Or, $E_t = \dfrac{t\_RT}{F} \ln \dfrac{a_2}{a_1}$

Putting the values of a_1 and a_2 from equation (12.11) in the above equation we get

$$E_t = \frac{t\_RT}{F} \ln \frac{(a_\pm)_2^2}{(a_\pm)_1^2} = \frac{2t\_RT}{F} \ln \frac{(a_\pm)_2}{(a_\pm)_1} \tag{12.13}$$

From equations (12.12) and (12.13), we have

$$\frac{E_t}{E} = t\_ \tag{12.14}$$

Thus, transport number can easily be determined by measuring E_t and E.

12.7.2   Liquid Junction Potential or Diffusion Potential

If the same electrolytic solutions with different concentrations are separated by a membrane such that ions are not allowed to pass through, the above cell becomes

$$\text{Pt, } H_2(g)(1 \text{ atm}) \overset{+}{|} HCl(a_1) \overset{M}{|||} HCl(a_2) \overset{-}{|} H_2(g)(1 \text{ atm}), \text{Pt}$$

The above cell is called **concentration cell without transference**.

LHE $\dfrac{1}{2} H_2(g) \longrightarrow H^+ + e^- \qquad\qquad$ **RHE** $H^+ + e^- \longrightarrow \dfrac{1}{2} H_2(g)$

So, 1 mole of H^+ is consumed at RHE and 1 mole of H^+ is generated at LHE. So, free energy change, ΔG, is given by

$$\Delta G = (\mu_{H^+})_1 - (\mu_{H^+})_2 = [\mu_{H^+}^0 + RT \ln(a_{H^+})_1] - [\mu_{H^+}^0 + RT \ln(a_{H^+})_2]$$

$$= RT \ln \frac{(a_{H^+})_1}{(a_{H^+})_2} = RT \ln \frac{(a_\pm)_1}{(a_\pm)_2}$$

$$[(a_{H^+})_1 = (a_\pm)_1 \quad \text{and} \quad (a_{H^+})_2 = (a_\pm)_2]$$

$(a_\pm)_1$ and $(a_\pm)_2$ are the mean activities of HCl solutions of anode and cathode respectively. Again, we know that $\Delta G = -FE_w$, where E_w is the cell emf. Thus, we have

$$-FE_w = RT \ln \frac{(a_\pm)_1}{(a_\pm)_2}. \quad \text{Or,} \quad E_w = \frac{RT}{F} \ln \frac{(a_\pm)_2}{(a_\pm)_1} \tag{12.15}$$

The same cell with transference has the cell emf E_t given by equation (12.13).

$$E_t = \frac{2t_- RT}{F} \ln \frac{(a_{\pm})_2}{(a_{\pm})_1} \quad \text{[Equation (12.13)]}$$

The difference $(E_t - E_w)$ is known as **diffusion potential or liquid junction potential.** This potential difference arises due to transference of ions or diffusion of ions through the membrane.

$$E_l = (E_t - E_w) = (2t_- - 1)\ln\frac{(a_{\pm})_2}{(a_{\pm})_1} \tag{12.16}$$

Liquid junction potential

An electrolyte is dissolved in two immiscible solvents with different concentrations. When these two electrolytic solutions come in contact, a liquid junction potential sets in at the boundary of the solutions. Let us consider the following cell:

$$\overset{-}{\text{Pt, H}_2(g)(1\text{atm}) \mid} \text{HCl}(a_1)(aq) \mid \text{HCl}(a_1)(\text{phenol}) \overset{+}{\mid} \text{H}_2(g)(1\text{ atm}), \text{ Pt}$$

Hydrochloric acid is dissolved in both water and phenol. These two solutions are separated by a semipermeable membrane, through which ions can diffuse. A potential difference is automatically developed at the junction of two solutions. This potential is called **liquid junction potential, E_l.** The net emf of the above concentration cell is given by

$$E = (\phi_L + E_l) - \phi_R.$$

When the diffusion is near to complete there is no transfer of solute from one phase to another phase. So equilibrium is reached and hence $\Delta G = 0$.
As we know, $\Delta G = -FE$ for 1 mole of solute transfer, we have $E = 0$.
Thus we can write that

$$(\phi_L + E_l) - \phi_R = 0 \text{ (at equilibrium). Or, } E_l = \phi_R - \phi_L$$

$$\phi_R = \phi_R^0 - \frac{RT}{F} \ln a_{H^+(\text{phenol})} \quad \text{and} \quad \phi_L = \phi_L^0 - \frac{RT}{F} \ln a_{H^+(aq)}$$

$$E_l = \phi_R - \phi_L = \phi_R^0 - \phi_L^0 - \frac{RT}{F} \ln \frac{a_{H^+(\text{phenol})}}{a_{H^+(aq)}} = E_l^0 - \frac{RT}{F} \ln \frac{a_{H^+(\text{phenol})}}{a_{H^+(aq)}} \tag{12.17}$$

As the solvents are different, ϕ_L^0 and ϕ_R^0 are different and unknown. Thus one cannot evaluate E_l by using equation (12.17). To solve the problem, different kind of cells are chosen as given:

$$\overset{+}{\text{Hg} \mid} \overset{I}{\text{Hg}_2\text{Cl}_2(s),\mid} \overset{II}{\text{HCl}(a_1)(aq) \mid} \overset{III}{\text{HCl}(a_2)(\text{phenol}) \parallel} \overset{IV}{\text{HCl}(a_3)(\text{phenol}) \mid} \overset{V}{\text{HCl}(a_4)(aq) \parallel} \overset{VI}{\text{HCl}(a_1)(aq),\mid} \overset{-}{\text{Hg}_2\text{Cl}_2(s) \mid \text{Hg}}$$

The net emf of the above cell is given by $E = (\phi_L + E_d^1 + E_l^1 + E_l^2) - (E_d^2 + \phi_R)$.

The phases I and VI are equal. So, $\phi_L = \phi_R$. E_d^1 and E_d^2 are the diffusion potentials, i.e., potential at the junction of two similar solutions, phenol-phenol solutions and aqueous-

aqueous solutions respectively. E_l^1 and E_l^2 are the liquid junction potentials, i.e., potential at the boundary of two dissimilar solutions, i.e., phenol-water solutions. The values of E_d^1 and E_d^2 are very small as compared to E_l^1 and E_l^2. Thus, cell emf is given by

$$E = E_l^1 + E_l^2. \qquad E_l^1 = E_{1(aq/phenol)}^0 - \frac{RT}{F} \ln \frac{a_{H^+(aq)}}{a_{H^+(phenol)}} = E_{1(aq/phenol)}^0 - \frac{RT}{F} \ln \frac{a_1}{a_2}$$

$$E_l^2 = E_{1(phenol/aq)}^0 - \frac{RT}{F} \ln \frac{a_{H^+(phenol)}}{a_{H^+(aq)}} = E_{1(phenol/aq)}^0 - \frac{RT}{F} \ln \frac{a_3}{a_4}$$

$$= -E_{1(aq/phenol)}^0 - \frac{RT}{F} \ln \frac{a_3}{a_4}$$

$$[E_{1(phenol/aq)}^0 = -E_{1(aq/phenol)}^0]$$

$$E = E_l^1 + E_l^2 = -\frac{RT}{F} \ln \frac{a_1}{a_2} - \frac{RT}{F} \ln \frac{a_3}{a_4} = \frac{RT}{F} \ln \frac{a_2 \cdot a_4}{a_1 \cdot a_3} = 0.0592 \log \frac{a_2 \cdot a_4}{a_1 . a_3} \qquad (12.18)$$

Putting the values of a_1, a_2, a_3, and a_4 in the above equation E, the liquid junction potential, can easily be calculated.

Definition of liquid junction potential The potential developed at the boundary of any two electrolytic solutions, prepared with two immiscible solvents, is called liquid junction potential. If the solvent of both the solutions is same, the liquid junction potential is specially called diffusion potential.

12.7.3 Determination of Activity and Activity Coefficient

Let us consider the cell

$$Pt, H_2(g)(1\,atm) \overset{+}{\underset{\frown}{|}} HCl(a_1), AgCl(s) | Ag(s) | AgCl(s), HCl(a_2) \overset{-}{\underset{\frown}{|}} H_2(g)(1\,atm), Pt$$

The emf of the cell is given by equation (12.12), which can be modified at 25°C as follows

$$E = \frac{2RT}{F} \ln \frac{(a_\pm)_2}{(a_\pm)_1} = 2 \times 0.0592[\log(a_\pm)_2 - \log(m_\pm)_1 - \log(\gamma_\pm)_1], \quad (a_\pm)_1 = (m_\pm)_1.(\gamma_\pm)_1$$

where, $(a_\pm)_1$ and $(a_\pm)_2$ are the mean activities of HCl solutions of anode and cathode respectively.

$$[E + 0.1184 \log (m_\pm)_1] = 0.1184 \log (a_\pm)_2 - 0.1184 \log (\gamma_\pm)_1.$$

Or, $\qquad Y = I - 0.1184 \log (\gamma_\pm)_1 \qquad (12.19a)$

where, $\qquad Y = E + 0.1184 \log (m_\pm)_1$ and $I = 0.1184 \log (a_\pm)_2$.

Plot of Y versus $\sqrt{(m_\pm)_1}$ is shown in the Fig. 12.6.

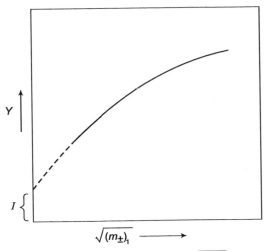

Figure 12.6 Plot of Y versus $\sqrt{(m_\pm)_1}$

As $\qquad (m_\pm)_1 \to 0, (\gamma_\pm)_1 \to 1.$

Hence, $\qquad I = \lim_{(m_\pm)_1 \to 0} Y = \lim_{(m_\pm)_1 \to 0} [E + 0.1184 \log(m_\pm)_1]$ $\qquad\qquad$ (12.19b)

Thus extrapolation of the graph on y–axis gives the value of I.

Again, $\qquad I = 0.1184 \log (a_\pm)_2$

Thus $(a_\pm)_2$ can easily be determined, and hence $(\gamma_\pm)_2$ can be obtained from the following relation

$$(a_\pm)_2 = (m_\pm)_2 \cdot (\gamma_\pm)_2 \qquad [(m_\pm)_2 \text{ is known}]$$

Note that $(\gamma_\pm)_2$ varies with $(m_\pm)_2$. Again, from equation (12.19a), we have

$$\log(\gamma_\pm)_1 = \frac{I - Y}{0.1184}$$

I is known. At a particular value of $(m_\pm)_1$, the value of Y can be determined from the Fig.12.6. Thus, $(\gamma_\pm)_1$ can be easily computed from the above equation. Hence, $(a_\pm)_1$ can be determined from the relation:

$$(a_\pm)_1 = (m_\pm)_1 \cdot (\gamma_\pm)_1$$

So here also, $(\gamma_\pm)_1$ varies with $(m_\pm)_1$. Thus activity coefficient is not constant but varies with concentration.

12.8 DECOMPOSITION VOLTAGE

Exchange current   If an electrode is immersed in an electrolytic solution, exchange of charged particles takes place between electrodes and solution. Due to this exchange some quantity of electricity is passed from electrode to the solution and vice versa. This quantity of exchanged electricity per unit time is called exchange current.

12.8.1   Concentration Polarization

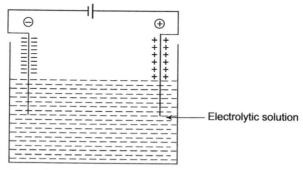

Electrolytic solution

Figure 12.7   Electrolysis of an electrolytic solution C_0 = Concentration of electrolyte in the bulk of the solution. C_s = Concentration of electrolyte near cathode and on the surface of cathode

During electrolysis cations get deposited on cathode and hence concentration of electrolyte decreases in the vicinity of cathode, though the concentration of cation remains unchanged in the bulk of the solution. Let C_0 and C_s be the concentrations of cation in the bulk and near the cathode respectively. Thus a concentration gradient sets up, resulting in a flow of current in the opposite direction. Due to the existence of concentration gradient cations diffuse from bulk of the solution to cathode. If n_i is the number of gm.equiv. of cations diffused per unit time, we can write in accordance with the **Fick's law** that

$$n_i = DA\left(\frac{C_0 - C_s}{l}\right)$$

where, D = Diffusion coefficient of discharging ion; A = Surface area of electrode;

l = Thickness of the diffused layer.

As we know, F is the electricity carried by 1 gm.equiv. of ions, the total current carried due to mass flow rate of cation n_i is

$$I = n_i \cdot F = FDA\left(\frac{C_0 - C_s}{l}\right) \tag{12.20a}$$

The value of diffusion current (I) becomes maximum when $C_s = 0$, i.e., all cations get deposited at cathode.

$$I = I_0 = FDA\left(\frac{C_0}{l}\right). \text{ Or, } C_0 = \frac{I_0 \cdot l}{FDA}$$

Again, from equation (12.20a), we have

$$(C_0 - C_s) = \frac{I \cdot l}{FDA}. \text{ Or, } C_0\left(1 - \frac{C_s}{C_0}\right) = \frac{I \cdot l}{FDA}$$

Putting the value of C_0 in the above equation we get

$$\frac{I_0 \cdot l}{FDA}\left(1 - \frac{C_s}{C_0}\right) = \frac{I \cdot l}{FDA}. \text{ Or,}\left(1 - \frac{C_s}{C_0}\right) = \frac{I}{I_0}. \text{ Or, } C_s = C_0\left(1 - \frac{I}{I_0}\right) \qquad (12.20b)$$

According to **Nernst equation** we can write that

$$\phi_{bulk} = \phi^0 - \frac{RT}{Z_+ F}\ln C_0 \quad \text{and} \quad \phi_{surface} = \phi^0 - \frac{RT}{Z_+ F}\ln C_s$$

where, Z_+ is the valency of the cation. Thus,

$$\Delta\phi_c = \phi_{bulk} - \phi_{surface} = -\frac{RT}{Z_+ F}\ln\frac{C_0}{C_s} = \frac{RT}{Z_+ F}\ln\frac{C_s}{C_0} = \frac{RT}{Z_+ F}\ln\left(1 - \frac{I}{I_0}\right)$$

$$\text{[From equation (12.20}b\text{)]}$$

Or, $$\Delta\phi_c = \frac{RT}{Z_+ F}\ln\left(1 - \frac{I}{I_0}\right) \qquad (12.20c)$$

$\Delta\phi_c$ is called potential due to concentration polarization at cathode. Similarly, we have $\Delta\phi_a$, which is the potential due to concentration polarization at anode. Thus, emf of concentration polarization is given by

$$E_{conc} = \Delta\phi_c + \Delta\phi_a$$

As the direction of E_{conc} is opposite to the applied voltage, E_{appl}, the net emf required to execute the electrolysis is given by

$$E = E_{conc} + E_{appl}$$

In industrial application the value of E_{conc} is sufficiently high and hence considerable amount of electrical energy is required to overcome the concentration emf. To avoid this loss of electrical energy, continuous stirring is employed to counter in setting up concentration gradient.

12.8.2   Electrochemical Polarization

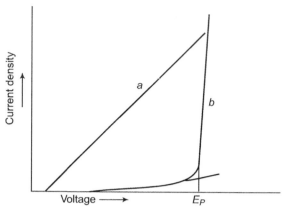

Figure 12.8   Current density versus voltage curves for electronic conductors
(Curve *a*) and for electrolytic conductors (Curve *b*)

In the case of electronic conductor, current density is directly proportional to the applied voltage, i.e. $I = \dfrac{E}{R}$ (Ohm's law). In the case of electrolytic conductor, the curve characteristics shows that the current realized in the circuit remains insignificant until a certain voltage is reached. After that growth of current with voltage is very fast and the curve rises very steeply. The cross-over voltage is known as **polarization voltage,** denoted by E_p (shown in the Fig. 12.8). Thus the current (I) is related to the applied voltage (E) in the following way

$$I = \frac{E - E_p}{R} \tag{12.21}$$

where, E_p is the polarization voltage.

In absence of any external emf, ionic equilibrium is established at both the electrodes. On application of an external emf cathode receives extra negative charge and anode receives extra positive charge. As a result of this, equilibrium is disturbed and a potential difference sets in against the direction of applied emf. This potential difference is our well-known polarization voltage (E_p), which depends on electrolytic solution. The extra negative charge received by cathode is shifted to the surrounding cations, e.g. in the case of HCl solution the following reaction takes place at the cathode

$$2H_3O^+ + 2e^- \longrightarrow 2H_2O + H_2(\text{absorbed on Pt}) \longrightarrow 2H_2O + H_2(g) \uparrow$$

The process is called discharging of ions, which is the origin of polarization. Hence discharging of ions is also known as **electrochemical polarization,** and the potential due to discharging of ions, is called **discharge potential**. Thus discharge potential is the potential required to discharge an ion to the corresponding element. The sum of discharge potentials of ions at two electrodes is known as **decomposition voltage** of the electrolyte. Thus the decomposition voltage of an electrolyte is the minimum potential difference required to begin electrolysis. Thus it is clear that the decomposition voltage of an electrolyte is equal to the emf of polarization.

12.8.3 Rule of Discharging of Ion

If more than one cation is present at cathode, the cation having the lowest oxidation potential (or the highest discharge potential) will get deposited **first**. When cathode potential increases and reaches next higher oxidation potential (or lower discharge potential) the next cation, having second lowest oxidation potential will start depositing and so on. For example, in aqueous solution of KCl two species of cations namely, K^+ and H_3O^+ are present, but discharge potential of K^+ ion is -2.925 V and that of H_3O^+ ion is around **zero** volt.

$$K^+ + e^- \longrightarrow K, \ \phi = -2.925 \text{ V (discharge of } K^+ \text{ ion) [see Table 12.1]}$$

Hence liberation of hydrogen always takes place at cathode during electrolysis of KCl solution according to the following reaction

$$2H_3O^+ + 2e^- \longrightarrow 2H_2O + H_2(g) \uparrow$$

If the concentration of hydroxonium ion, H_3O^+, is sufficiently low the following reaction takes place at the cathode.

$$2H_2O + 2e^- \longrightarrow 2OH^- + H_2(g) \uparrow$$

If more than one anion is present at anode, that anion having highest oxidation potential will get deposited first. For example, if OH^-, H_2O and SO_4^{2-} are present, following anodic reactions may occur

$4OH^- \longrightarrow 2H_2O + O_2(g) + 4e^-$	(a) $\phi^0 = -1.23$ V
$6H_2O \longrightarrow 4H_3O^+ + O_2(g) + 4e^-$	(b) $\phi^0 = -1.00$ V
$2SO_4^{2-} \longrightarrow 2SO_4 + 4e^-, 2SO_4 + 6H_2O \longrightarrow 2SO_4^{2-} + 4H_3O^+ + O_2(g)$	(c) $\phi^0 = -2.00$ V

Thus, in alkaline solution hydroxyl ion will get deposited first and in acidic solution hydroxonium ion (H_3O^+) will get deposited first but in no case sulfate ion (SO_4^{2-}) will be deposited at the anode.

Thus, during electrolysis of an electrolyte only selected ions are deposited at cathode and anode while the remaining ions maintain sufficient electrical conductance to continue the electrolysis.

12.8.4 Electrolysis of Hydrochloric Acid Solution

At anode, both OH^- and Cl^- ions are present. Following two reactions may proceed at anode.

$$4OH^- \longrightarrow 2H_2O + O_2(g) + 4e^- \ \cdots \ \phi^0 = -1.23 \text{ V}$$

$$2Cl^- \longrightarrow Cl_2(g) + 2e^- \ \cdots \ \phi^0 = -1.36 \text{ V}$$

$$\phi_{Cl^-/Cl_2} = \phi^0_{Cl^-/Cl_2} - \frac{RT}{F} \ln \frac{p_{Cl_2}^{1/2}}{a_{Cl^-}}$$

The electrode is so maintained that $p_{Cl_2} = 1$ atm. Hence the equation becomes

$$\phi_{Cl^-/Cl_2} = \phi^0_{Cl^-/Cl_2} + \frac{RT}{F} \ln a_{Cl^-} = -1.36 + 0.0592 \log a_{Cl^-} \text{ [at } 25°C]$$

At high concentration of Cl^- ions, a_{Cl^-} is high and hence net electrode potential, ϕ_{Cl^-/Cl_2}, is higher than -1.23 V. Hence chlorine is liberated at the anode, whereas in dilute solutions, a_{Cl^-} is low enough so that the value of ϕ_{Cl^-/Cl_2} falls below -1.23 V and hence oxygen is liberated at the anode.

Decomposition voltage depends on several factors
  (a) Shape and dimension of the electrodes.
  (b) Nature of the electrode surface.
  (c) Removal of products formed during electrode processes.
  (d) Concentration of ions.

The nature of electrode process depends on the net electrode potential of ions. At cathode that electrode process, having **lowest** oxidation potential, takes place but at anode that electrode process, having **highest** oxidation potential, takes place.

Consider a cell where the following electrode processes take place

At cathode $M^{p+} + pe^- \longrightarrow M(s)$ (discharge of cation)

At anode $N^{q-} \longrightarrow N(g) + qe^-$ (discharge of anion)

Where, p, q are the valencies of cation and anion respectively.
Discharge potential of cation $= \phi_c$. It is reduction potential.

$$\phi_c = \phi^0_c + \frac{RT}{pF} \ln a_+, \quad a_+ = \text{activity of } M^{p+}, a_{M(s)} = 1$$

Discharge potential of anion $= \phi_a$. It is oxidation potential.

$$\phi_a = \phi^0_a + \frac{RT}{qF} \ln a_-, \quad a_- = \text{activity of } N^{q-} \text{ and } a_{N(g)} = p_{N(g)} = 1 \text{ atm}$$

Decomposition voltage $= E_{dc} = \phi_c + \phi_a$, $E^0_{dc} = (\phi^0_c + \phi^0_a)$

$$E_{dc} = \phi_c + \phi_a = E^0_{dc} + \frac{RT}{pF} \ln a_+ + \frac{RT}{qF} \ln a_- \tag{12.22}$$

It is to be noted here that, ϕ_c and ϕ^0_c are the **reduction potentials** and ϕ^0_a and ϕ^0_a are the **oxidation potentials.**

12.9 OVERVOLTAGE

According to the concept of decomposition voltage, it is the minimum potential difference (pd) applied from an external source to begin electrolysis. In actual practice the minimum pd required to begin electrolysis is always **greater** than the decomposition voltage. This **excess voltage**, required to begin electrolysis, is called **overpotential** or **overvoltage**.

If η_c and η_a are the overpotentials at cathode and anode respectively, we can write that

$$\eta_c = \phi_c - \phi_{dc} \text{ and } \eta_a = \phi_a - \phi_{da} \tag{12.23a}$$

ϕ_{dc} and ϕ_{da} are the decomposition potentials of cathode and anode respectively.

The total overvoltage of the cell $= \eta = \eta_c + \eta_a = (\phi_c + \phi_a) - (\phi_{dc} + \phi_{da})$.

It is to be noted that the excess voltage, i.e., $[(\phi_c + \phi_a) - (\phi_{dc} + \phi_{da})]$ is actually the sum of overvoltage (η) and Ohmic voltage drop in solution. The Ohmic voltage drop is small enough to be neglected as compared to overvoltage. Hence we can write that

$$\eta = (\phi_c + \phi_a) - (\phi_{dc} + \phi_{da}) \tag{12.23b}$$

The magnitude of the overvoltage on an electrode depends on several factors:

(a) Nature of the electrode

(b) Density of the current (J)

(c) Composition of the solution.

J. Tafel (1905) was the first scientist who proposed the following relation between overvoltage (η) and current density (J).

$$\eta = a + b \log J \tag{12.24a}$$

Where, a and b are the constants. The value of a for any electrochemical process, largely depends on the materials of electrode but the value of b slightly depends on materials and is approximately equal to $2 \times 0.0592 = 0.1184$ at 25°C. Again, at very low current density, $J \to 0$ or, $J \to e^{-\infty}$. So, $\log J \to -\infty$. Hence, $\eta \to -\infty$. In actual practice, $\eta \to 0$, while $J \to 0$. Thus the equation (12.24a) is not applicable at very low current density.

12.9.1 Theories of Hydrogen Overvoltage

Discharge of an ion usually proceeds through multiple steps. If all the steps proceed at comparable rate there will be no overvoltage but if one of the steps proceeds at relatively low rate, overvoltage is resulted. The discharge of hydroxonium ion consists of five steps:

Step-I   **Diffusion**   In this step hydroxonium ion, H_3O^+, diffuses from aqueous solution to electrode (cathode).

Step-II   **Dehydration**   In this step dehydration of hydroxonium ion takes place according to the reaction: $H_3O^+ \longrightarrow H^+ + H_2O$

Step-III   **Discharge and adsorption**   In this step H^+ ion gets converted to atomic hydrogen which is then adsorbed on the electrode surface.

$$H^+ + e^- \longrightarrow H(\text{nascent})$$

$$H(\text{nascent}) + Me \longrightarrow H(Me), \text{ where, } Me = \text{a metal or glass electrode.}$$

Step-IV **Recombination** In this step adsorbed atomic hydrogen recombines to molecular hydrogen which still remain adsorbed on the surface. $2H(Me) \longrightarrow H_2(Me)$

Step-V **Gas liberation step** In this step adsorbed hydrogen is liberated from the electrode as gas.

$$H_2(Me) \longrightarrow H_2(g)\uparrow + Me$$

Among these five steps it is very difficult to identify the **slowest** one. According to Tafel, **step-IV** is the slowest one but according to Volmer et. al discharge of ion, i.e., **step-III** is the slowest step.

12.9.1.1 Recombination Theory

According to this theory, **step-IV** is the slowest step.

$$2H(Me) \longrightarrow H_2(Me)$$

$$\text{Rate} = \frac{dn}{dt} = k[H]^2$$

where, $(dn/dt) = $ number of moles of hydrogen produced per second. If current density, j, is applied for time dt, we can write that

$jdt = F(zdn)$, where, z is the number of electrons transferred per mole of hydrogen. In this case, $z = 2$.

$zdn = $ Number of gm.equiv. or moles of electrons.

Hence, $\dfrac{j}{2F} = \dfrac{dn}{dt}$, since, $2H^+ \rightarrow 2H + 2e^-$ and $j = \dfrac{I}{A}$

Combining the above two equations we get

$$k[H]^2 = \frac{j}{2F}. \text{ Or, } [H]^2 = \frac{j}{2kF}. \text{ Or, } [H] = k_1\sqrt{j} \tag{12.24b}$$

where, $\qquad k_1 = \left(\dfrac{1}{2kF}\right)^{1/2}$

Thus, the amount of hydrogen adsorbed is always greater than the amount of hydrogen recombined, since the latter is the **slowest** process. The overvoltage appears due to increase in adsorbed hydrogen on the electrode.

If $[H_c]$ is the amount of adsorbed hydrogen converted to molecular hydrogen, the overvoltage is given by

$$\eta = \phi_H - \phi_C = \left\{\phi^0 + \frac{RT}{F}\ln[H]\right\} - \left\{\phi^0 + \frac{RT}{F}\ln[H_c]\right\} = \frac{RT}{F}\ln\frac{[H]}{[H_c]}$$

Putting the value of [H] from equation (12.24b) in the above equation we get

$$\eta = \frac{RT}{F} \ln \frac{k_1 \sqrt{j}}{[H_c]} = a + \frac{RT}{2F} \ln j = a + 0.0296 \log j \qquad (12.24c)$$

where, $a = \frac{RT}{F}(\ln k_1 - \ln[H_c])$

The equation (12.24c) is similar to the equation (12.24a) but the value of b is 0.0296 in equation (12.24b). In actual practice the value of b is around 0.1184. This is the serious **drawback** of Tafel's recombination theory.

12.9.1.2 Theory of Slow Discharge of Ions

According to this theory discharge of ions is the **slowest** step. The discharge of hydroxonium ion consists of two steps:

$$H_3O^+ \xrightarrow{slow} H^+ + H_2O \quad \text{(dehydration)}$$

$$H^+ + e^- \xrightarrow{fast} H \qquad \text{(discharge)}$$

Dehydration step is the **slowest** step and hence rate $= J$

$$J = k[H_3O^+] = Ae^{-E/RT}[H_3O^+], \text{ since we know that } k = Ae^{-E/RT}$$

$E = (E^0 - \alpha F\eta)$, where, $E^0 =$ Activation energy of an unpolarized electrode and $(aF\eta)$ is the activation energy due to overvoltage and a is a constant. Again, we have

$$\text{Rate} = \frac{dn}{dt} = \frac{j}{2F}$$

where, j is the current density, (I/A) in Amp/cm^2.

Hence we can write that

$$\text{Rate} = \frac{j}{2F} = A[H_3O^+]e^{-(E^0 - \alpha F\eta)/RT} = A[H_3O^+]e^{-E^0/RT} \cdot e^{\alpha F\eta/RT}. \text{ Or, } \frac{j}{2F} = Ce^{\alpha F\eta/RT}.$$

Or, $\quad j = 2FCe^{\alpha F\eta/RT}$, where, $C = A[H_3O^+]e^{-E^0/RT}$

Taking ln on both sides we get

$$\ln j = \ln(2FC) + \frac{\alpha F\eta}{RT}.$$

Or, $\quad \eta = a + \frac{RT}{\alpha F}\ln j = a + \frac{0.0592}{\alpha}\log j \quad [\text{at } 25°C] \qquad (12.25)$

where, $\quad a = -\frac{RT}{\alpha F}\ln(2FC)$

The coefficient of $\log j$ in the above equation is the constant b in equation (12.24a).

Or, $\quad b = \frac{0.0592}{\alpha}$. At $\alpha = 0.5$, $b = 0.1154$

Thus the value of b calculated using equation (12.25) is same as that found in actual practice.

12.10 ACCUMULATORS

In an electrochemical cell chemical energy is the source of current or electricity. In many cells electricity is obtained only once due to irreversible nature of chemical reactions. These types of cells are called galvanic cells. In other type of cells energy is stored and current is drawn from the cell with repeated action. This type of cell is called storage cell or accumulator.

Principle of accumulator   At first electric current is passed from an external source to execute electrochemical reactions. The free energy of reaction is stored within the cell. The process is known as charging.

During charging, electrons are accepted from LHE and electrons are rejected to RHE. In the process of discharge current is withdrawn from the cell. So in the discharging process electrons are rejected to LHE and electrons are accepted from RHE. Some widely used accumulators are:
  (i) Lead-acid accumulator.
  (ii) Cadmium-Nickel alkaline accumulator.
  (iii) Silver-Zinc alkaline accumulator.

12.10.1   Lead-acid Accumulator

The cell is under discharging condition given by

$$\text{Pb, PbSO}_4(s) \mid^+ \text{H}_2\text{SO}_4(32 - 34\%) \mid^- \text{PbO}_2(s), \text{Pb}$$

Current is drawn under discharging condition and current is passed under charging condition. LHE is anode under discharging condition and cathode under charging condition.

Charging at LHE

$$\text{PbSO}_4(s) \longrightarrow \text{Pb}^{2+} + \text{SO}_4^{2-}$$

$$\text{Pb}^{2+} + 2e^- \longrightarrow \text{Pb}(s)$$

$$\overline{\text{PbSO}_4(s) + 2e^- \longrightarrow \text{Pb}(s) + \text{SO}_4^{2-}}$$

Discharging at LHE

$$\text{Pb}(s) \longrightarrow \text{Pb}^{2+} + 2e^-$$

$$\text{Pb}^{2+} + \text{SO}_4^{2-} \longrightarrow \text{PbSO}_4(s)$$

$$\overline{\text{Pb}(s) + \text{SO}_4^{2-} \longrightarrow \text{PbSO}_4(s) + 2e^-}$$

Charging at RHE

$$\text{PbSO}_4(s) \longrightarrow \text{Pb}^{2+} + \text{SO}_4^{2-}$$

$$\text{Pb}^{2+} + 2\text{H}_2\text{O} \longrightarrow \text{PbO}_2(s) + 4\text{H}^+ + 2e^-$$

$$\overline{\text{PbSO}_4(s) + 2\text{H}_2\text{O} \longrightarrow \text{PbO}_2(s) + 4\text{H}^+ + \text{SO}_4^{2-} + 2e^-}$$

Discharging at RHE

$$\text{PbO}_2(s) + 4\text{H}^+ + 2e^- \longrightarrow \text{Pb}^{2+} + 2\text{H}_2\text{O}$$

$$\text{Pb}^{2+} + \text{SO}_4^{2-} \longrightarrow \text{PbSO}_4(s)$$

$$\overline{\text{PbO}_2(s) + 4\text{H}^+ + \text{SO}_4^{2-} + 2e^- \longrightarrow \text{PbSO}_4(s) + 2\text{H}_2\text{O}}$$

The cell reaction during charging process is given by

$$2\text{PbSO}_4(s) + 2\text{H}_2\text{O} \longrightarrow \text{Pb}(s) + \text{PbO}_2(s) + 4\text{H}^+ + 2\text{SO}_4^{2-}$$

During discharging process the overall cell reaction is given by

$$\text{Pb}(s) + \text{PbO}_2(s) + 4\text{H}^+ + 2\text{SO}_4^{2-} \longrightarrow 2\text{PbSO}_4(s) + 2\text{H}_2\text{O}$$

The combined cell reaction is

$$Pb(s) + PbO_2(s) + 4H^+ + 2SO_4^{2-} \xrightleftharpoons[\text{Charging}]{\text{Discharging}} 2PbSO_4(s) + 2H_2O$$

Calculation of cell emf during discharging process

$$\phi_L^0 = \phi_{Pb|Pb^{2+}}^0 = +0.126\,V \quad [\text{Cf. Table 12.1}].$$

EMF at the LHE $\quad \phi_L = \phi_L^0 - \dfrac{RT}{2F} \ln a_{Pb^{2+}}$

Again, $PbSO_4(s) \rightleftharpoons Pb^{2+} + SO_4^{2-}.$ $\quad K_{aP} = a_{Pb^{2+}} \cdot a_{SO_4^{2-}} = 2.2 \times 10^{-8}$

$$\phi_L = +0.126 - \frac{0.0592}{2} \log \frac{a_{Pb^{2+}} \cdot a_{SO_4^{2-}}}{a_{SO_4^{2-}}} = +0.126 - \frac{0.0592}{2} \log \frac{K_{aP}}{a_{SO_4^{2-}}} \quad [\text{at } 25°C]$$

Putting the value of K_{aP} in the above equation, we get

$$\phi_L = +0.126 + 0.228 + \frac{0.0592}{2} \log a_{SO_4^{2-}} = +0.354 + \frac{0.0592}{2} \log a_{SO_4^{2-}} \quad [\text{at } 25°C]$$

EMF at the RHE $\quad \phi_R = \phi_R^0 - \dfrac{RT}{2F} \ln \dfrac{a_{H^+}^4 \cdot a_{SO_4^{2-}}}{a_{H_2O}^2}$

$$= -1.685 - \frac{0.0592}{2} \log \frac{a_{H^+}^4 \cdot a_{SO_4^{2-}}}{a_{H_2O}^2} \quad [\text{at } 25°C]$$

$[\phi_R^0 = \phi_{Pb^{2+}|Pb^{4+}}^0 = -1.685\,V]$. Cell emf $= E = \phi_L - \phi_R$

Or, $\quad E = (0.354 + 1.685) + \dfrac{0.0592}{2} \log \dfrac{a_{H^+}^4 \cdot a_{SO_4^{2-}}^2}{a_{H_2O}^2} = 2.037 + 0.0592 \log \dfrac{a_{H^+}^2 \cdot a_{SO_4^{2-}}}{a_{H_2O}} \quad [\text{at } 25°C]$

Again, $H_2SO_4(m) \longrightarrow 2H^+ (2m) + SO_4^{2-} (m)$. Hence, $a_{H_2SO_4} = a_{H^+}^2 \cdot a_{SO_4^{2-}}$. If, $a_{\pm}$ and m are the mean activity and molality respectively, we can write that

$a_{H_2SO_4} = a_{\pm}^3 = \gamma_{\pm}^3 \cdot 4m^3$, where, $\gamma_{\pm}$ is the mean activity coefficient. Thus the above equation becomes

$$E = 2.037 + 0.0592 \log \frac{a_{H_2SO_4}}{a_{H_2O}} = 2.037 + 0.0592 \log \frac{\gamma_{\pm}^3 \cdot 4m^3}{a_{H_2O}} \tag{12.26}$$

12.10.2 Nickel-Cadmium Alkaline Accumulator

The cell is under discharging condition given by

$$Cd \mid^+ Cd(OH)_2, KOH\ (20\%) \parallel KOH\ (20\%), Ni(OH)_2, Ni(OH)_3 \mid^- Ni$$

Current is drawn under discharging condition and current is passed under charging condition. LHE is anode under discharging condition and cathode under charging condition.

Discharging at LHE

$$Cd \longrightarrow Cd^{2+} + 2e^-$$

$$\underline{Cd^{2+} + 2\ OH^- \longrightarrow Cd(OH)_2}$$

$$Cd + 2OH^- \longrightarrow Cd(OH)_2 + 2e^-$$

Discharging at RHE

$$Ni^{3+} + e^- \longrightarrow Ni^{2+}$$

$$\underline{2Ni(OH)_3 + 2e^- \longrightarrow 2Ni(OH)_2 + 2OH^-}$$

$$2Ni(OH)_3 + 2e^- \longrightarrow 2Ni(OH)_2 + 2OH^-$$

Charging at LHE

$$Cd^{2+} + 2e^- \longrightarrow Cd$$

$$\underline{Cd(OH)_2 \longrightarrow Cd^{2+} + 2OH^-}$$

$$Cd(OH)_2 + 2e^- \longrightarrow Cd + 2OH^-$$

Charging at RHE

$$Ni^{2+} \longrightarrow Ni^{3+} + e^-$$

$$\underline{2Ni(OH)_2 + 2OH^- \longrightarrow 2Ni(OH)_3 + 2e^-}$$

$$2Ni(OH)_2 + 2OH^- \longrightarrow 2Ni(OH)_3 + 2e^-$$

The overall cell reaction is given by

$$Cd + 2Ni(OH)_3 \longrightarrow Cd(OH)_2 + 2Ni(OH)_2 \quad \text{(Discharge)}$$

$$Cd(OH)_2 + 2Ni(OH)_2 \longrightarrow Cd + 2Ni(OH)_3 \quad \text{(Charge)}$$

Combining above two equations we can write

$$Cd + 2Ni(OH)_3 \underset{\text{Charging}}{\overset{\text{Discharging}}{\rightleftharpoons}} Cd(OH)_2 + 2Ni(OH)_2$$

The emf of Nickel-Cadmium accumulator is about 1.36 V.

12.10.3 Zinc-Silver Alkaline Accumulator

Porous Zn–plate forms the anode and (silver and silver oxide) forms the cathode under discharging condition. The cell is under discharging condition given by

$$Zn \mid^+ K_2ZnO_2 + KOH\ (40\%) \mid AgO(\text{or } Ag_2O) \mid^- Ag$$

Discharging at LHE

$$Zn \longrightarrow Zn^{2+} + 2e^-$$

$$Zn^{2+} + 2OH^- \longrightarrow Zn(OH)_2$$

$$\overline{Zn + 2\ OH^- \longrightarrow Zn(OH)_2 + 2e^-}$$

Discharging at RHE

$$AgO + \frac{1}{2}H_2O + e^- \longrightarrow \frac{1}{2}Ag_2O + OH^-$$

$$\frac{1}{2}Ag_2O + \frac{1}{2}H_2O + e^- \longrightarrow Ag + OH^-$$

$$\overline{AgO + H_2O + 2e^- \longrightarrow Ag + 2OH^-}$$

The cell reaction, at discharging state, is

$$Zn + AgO \xrightarrow{\text{Discharging}} ZnO + Ag$$

Charging at LHE

$$Zn^{2+} + 2e^- \longrightarrow Zn$$

$$Zn(OH)_2 \longrightarrow Zn^{2+} + 2OH^-.$$

$$\overline{Zn(OH)_2 + 2e^- \longrightarrow Zn + 2OH^-}$$

Charging at RHE

$$\frac{1}{2}Ag_2O + OH^- \longrightarrow AgO + \frac{1}{2}H_2O + e^-$$

$$Ag + OH^- \longrightarrow \frac{1}{2}Ag_2O + \frac{1}{2}H_2O + e^-$$

$$\overline{Ag + 2OH^- \longrightarrow AgO + H_2O + 2e^-}$$

The cell reaction, at charging state, is

$$ZnO + Ag \xrightarrow{\text{Charging}} Zn + AgO$$

Combining the above two equations, we can write

$$Zn + 2AgO \underset{\text{Charging}}{\overset{\text{Discharging}}{\rightleftarrows}} ZnO + 2Ag$$

If only Ag_2O/Ag constitutes the anode, the overall cell reaction can be written as

$$Zn + Ag_2O \underset{\text{Charging}}{\overset{\text{Discharging}}{\rightleftarrows}} ZnO + 2Ag$$

The cell emf with RHE AgO/Ag is 1.86 V but the same is around 1.60 V if the RHE is Ag_2O/Ag. The specialty of Zn–Ag accumulator is that unlike lead and nickel accumulators the electrolyte of Zn–Ag accumulator does not precipitate either in the charge or in the discharge state. This criterion makes it possible to place the electrodes very close to each other and are separated by a thin layer of cellophane.

The electrolyte is freely flowing through the porous electrodes. So, the Zn–Ag accumulator is available in a very small size also. Furthermore, it has high capacity, high energy and high power per unit mass and volume. That is why Zn–Ag accumulator has wide application area particularly where small size is desired.

12.11 FUEL CELL

Fuel cell is an electrochemical device which converts chemical energy into electrical energy and heat without combustion. Continuous supply of fuel and oxidant is the primary requirement of a fuel cell to produce dc electric power.

In the case of battery, chemical energy is converted to electrical energy without need of any external fuel. Furthermore, the electrical energy is stored in the battery but after complete discharge the battery needs recharging. But in the case of a fuel cell, recharging is not at all required.

12.11.1 Working Principle

The cell consists of two permeable Ni-electrodes between which a conducting electrolytic solution is taken. Hydrogen gas is supplied at the positive electrode, i.e., anode, whereas oxygen gas is supplied at the negative electrode, i.e., cathode. The electrodes are connected through an external circuit. At the anode electrons are generated from hydrogen and these electrons move to the cathode through the external circuit and finally consumed by oxygen. The schematic diagram of the fuel cell is shown in the Fig. 12.9.

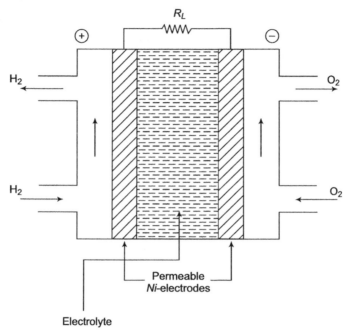

Figure 12.9 Schematic diagram of a fuel cell

The following reactions take place at the two electrodes:

$$2H_2 \longrightarrow 4H^+ + 4e^- \qquad \text{(at anode)}$$

$$O_2 + 4H^+ + 4e^- \longrightarrow 2H_2O \qquad \text{(at cathode)}$$

Hydrogen molecule is dissociated to form ions at the anode due to catalytic action of Ni-electrode. Oxygen at cathode reacts with H^+ ions of electrolyte by absorbing electrons to form water.

In presence of alkaline electrolyte water is produced at the anode with the release of electrons, which move to the cathode through the external circuit. At cathode OH^- ions are regenerated in presence of oxygen. The following reactions are believed to take place:

$$2H_2 + 4OH^- \longrightarrow 4H_2O + 4e^- \quad \text{(at anode)}$$

$$O_2 + 2H_2O + 4e^- \longrightarrow 4OH^- \quad \text{(at cathode)}$$

In a fuel cell water and heat are always generated along with electrical energy.

12.11.2 Types of Fuel Cells

Depending on the electrolyte used in the cell, fuel cells can be classified into several categories:

(a) Phosphoric acid fuel cell

(b) Alkaline fuel cell

(c) Direct methanol fuel cell

(d) Proton/polymer exchange membrane fuel cell

(e) Molten carbonate fuel cell

(f) Solid oxide fuel cell

(g) Zinc-air fuel cell

(h) Regenerative fuel cell.

12.11.2.1 Acid and Alkaline Fuel Cell

In alkaline fuel cell generally caustic potash (KOH) is used as an electrolyte and in acid fuel cell 100% concentrated phosphoric acid (H_3PO_4) is used as an electrolyte. As the ionic conductivity of phosphoric acid is low at low temperature, the cell is usually operated at high temperature (around 150°–200°C). The waste heat of acid fuel cell is used to generate steam at atmospheric pressure. It can be used for plants of capacity 50–200 kW. Overall efficiency may be achieved as high as 85%.

12.11.2.2 Molten Carbonate Fuel Cell

In this type of cell a mixture of molten Li_2CO_3 and K_2CO_3 or Na_2CO_3 is used. High operating temperature (around 650°C) is required to have the molten salt mixture. The carbonate ions (CO_3^{2-}) are consumed at the anode and regenerated at the cathode. During this process electrons are generated at the anode and flowing to the cathode through the external circuit. The following reactions are believed to take place:

At anode: $\quad CO_3^{2-} + H_2 \longrightarrow H_2O + CO_2 + 2e^-$

At cathode: $\quad CO_2 + \dfrac{1}{2}O_2 + 2e^- \longrightarrow CO_3^{2-}$

The efficiency of the cell may be as high as 60%.

12.11.2.3 Direct Methanol Fuel Cell

In this fuel cell liquid methanol is oxidized at the anode with the release of electrons, which then move to the cathode through the external circuit. At the cathode oxygen consumes electrons to form water. As methanol is a volatile liquid, the operating temperature of the cell is kept low (50°–120°C). The following reactions are believed to take place:

At anode: $\quad CH_3OH + H_2O \longrightarrow CO_2 + 6H^+ + 6e^-$

At cathode: $\quad \dfrac{3}{2}O_2 + 6H^+ + 6e^- \longrightarrow 3H_2O$

Active catalyst is required to execute oxidation of methanol at the anode, making the cell costlier. The efficiency of the cell is low (around 40%).

12.11.2.4 Solid Oxide Fuel Cell (SOFC)

In this cell non-porous metal oxide is used as an electrolyte through which oxygen ions move. At the anode O^{2-} reacts with hydrogen with the release of electrons, which are consumed at the cathode. The following reactions are believed to take place:

At anode: $\quad 2H_2 + 2O^{2-} \longrightarrow 2H_2O + 4e^-$

At cathode: $\quad O_2 + 4e^- \longrightarrow 2O^{2-}$

Passage of oxygen ions through metal oxide is realized only at high temperature at around $650°–1000°C$.

12.11.2.5 Proton Exchange Membrane Fuel Cell (PEMFC)

In this type of fuel cell a solid polymer membrane (usually a thin plastic film) is placed between cathode and anode so that only protons, not electrons, are allowed to pass through. Protons are generated at the cathode and move towards the anode through the membrane. The following reactions are believed to take place:

At anode: $\quad 2H_2 \longrightarrow 4H^+ + 4e^-$

At cathode: $\quad O_2 + 4H^+ + 4e^- \longrightarrow 2H_2O$

Operating temperature is below $100°C$. The method is less expensive. The cell has long life and the electrolyte is less corrosive than others. Low power generation is the main disadvantage of this type of cell.

12.11.2.6 Zinc-Air Fuel Cell

In this type of fuel cell cathode acts as a permeable membrane through which atmospheric oxygen diffuses into the solution. The oxygen reacts with H^+ ions to produce OH^- ions and water. At anode methane gas is introduced as fuel and Zn consumes OH^- ions to produce $Zn(OH)_2$. The reactions at cathode and anode are given below:

At anode: $\quad CH_4 + 2H_2O \longrightarrow CO_2 + 6H^+ + 6e^-$

$\quad\quad\quad\quad Zn + 2OH^- \longrightarrow Zn(OH)_2 + 2e^-$

At cathode: $2O_2 + 4H^+ + 4e^- \longrightarrow 4OH^-$

$\quad\quad\quad\quad O_2 + 4H^+ + 4e^- \longrightarrow 2H_2O$

However, high operating temperature (around $750°C$) is the main disadvantage of this type of cell.

12.11.2.7 Regenerative Fuel Cell

In this type of cell water acts as electrolyte. Hydrogen and oxygen are used as fuel. Water is split into H^+ and OH^- ions by using solar or wind energy. Hydrogen and oxygen are regenerated and recycled. Hence it is called regenerative fuel cell.

12.11.3 EMF and Efficiency of a Fuel Cell

Emf generated in the fuel cell is related with free energy change ΔG of the overall reaction. According to the following equation:

$$\Delta G = -nFE, \text{ where,}$$

n = Number of electrons exchanged

and F = Faraday's constant = 96,500 C/g.equiv.

Efficiency (η) of a fuel cell is given by:

$$\eta = \frac{\text{Power output}}{\text{Power input}}, \text{ where, } \text{Power input} = \frac{\Delta G}{M_{H_2}}, \quad M_{H_2} = \text{Molar mass of hydrogen}$$

The value of η depends on the following factors:

(a) Activation losses (b) Transportation losses
(c) Resistance losses (d) Fuel losses.

12.11.4 Choice of Fuel

Hydrogen is the best choice of fuel for any fuel cell because of its low cost, low density, and sufficient abundance. Besides hydrogen, hydrocarbons, ammonia, hydrazine or carbon monoxide can be used as fuel. E^0 values of different fuels are listed in the following Table 12.3. Operating temperature of different electrolytes are listed in Table 12.4. However, choice of fuel mainly depends on availability of the type of fuel in the area where the fuel cell is to be implemented.

Table 12.3 E^0 values of different fuels

Fuel	E^0 at 25°C (in V)
H_2	1.23
CH_3OH	1.21
C_3H_8	1.09
CO	1.29
NH_3	1.17
N_2H_4	1.61

Table 12.4 Operating temperature of different electrolytes

Electrolyte	Operating temperature (°C)
Phosphoric acid	220
Alkali	50–250
Molten carbonate mixture	600
Solid oxide	500–1000
Solid polymer	50–100
Methanol	50–120

12.11.5 Advantages, Limitations and Applications

Advantages

(a) Efficiency is high. 90% efficiency can be realized.
(b) Fuel cells are simple, compact, portable, noiseless, and smokeless.
(c) Fuel cells are environmentally friendly.
(d) A wide variety of fuels, including hydrocarbons, LPG, biogas, coal gas can be used as fuel.
(e) Fuel cells can easily be installed even in remote areas.

Limitations

(a) Development cost is high.
(b) Operating temperature is high.
(c) Low output voltage. Output power is usually less than 1 kW.
(d) Short service life.

Applications

It is mainly used in vehicles, submarines, locomotives, defence and different portable power plants.

SOLVED PROBLEMS

Problem 12.1   The standard emf for the following cell reaction is 0.337 V at 25°C. What is the value of ΔG^0? If the cell reaction is doubled, what would be the value of ΔG^0 and E^0?

$$Cu^{2+}(aq) + H_2(g) \rightleftharpoons 2H^+(aq) + Cu$$

Solution

E^0 remains constant because it is an intensive property.
$\Delta G^0 = -2FE^0 = -65041$ J. If the cell reaction is doubled, number of electrons exchanged is 4. So, $\Delta G^0 = -4FE^0 = -130082$ J.

Problem 12.2

$$Cu^{2+} + 2e^- \longrightarrow Cu \qquad E^0 = +0.337 \text{ V}$$
$$Cu^{2+} + e^- \longrightarrow Cu^+ \qquad E^0 = +0.153 \text{ V}$$

Calculate E^0 for the reaction: $Cu^+ + e^- \longrightarrow Cu$

Solution

$$Cu \longrightarrow Cu^{2+} + 2e^- \quad E_1^0 = -0.337 \text{ V}, \Delta G^0 = +2 \times 0.337 \text{ FJ} \qquad \text{(i)}$$
$$Cu^+ \longrightarrow Cu^{2+} + e^- \quad E_2^0 = -0.153 \text{ V}, \Delta G^0 = +1 \times 0.153 \text{ FJ} \qquad \text{(ii)}$$

(i) – (ii) Cu $\longrightarrow$ Cu$^+$ + e^-, $\Delta G^0 = +0.521$ FJ $= -1 \times FE_3^0$. Or, $E_3^0 = -0.521$ V.

For Cu$^+$ + e^- $\longrightarrow$ Cu, $E^0 = +0.521$ V.

Problem 12.3 The emf of the following cell is 0.85 V.

$$Pt[H_2(g)]\left|H^+(aq)\right|\left|Hg_2^{2+}(aq)\right|Hg(l).$$ Evaluate the maximum work of

this cell when 1 gm of $H_2(g)$ is consumed.

Solution

$Hg_2^{2+}(aq) + H_2 \longrightarrow 2Hg + 2H^+(aq)$. So, $n = 2$ and 2 gm of $H_2(g)$ is consumed.

$W_{max} = -\Delta G = nFE = 164050$ J/mole. Hence, W_{max} for 1 gm of $H_2(g) = 82025$ J.

Problem 12.4 Can we use an iron vessel to store 1M AgNO$_3$ solution?

Given: $\varphi_{Fe^{2+}|Fe}^0 = -0.44$ V and $\varphi_{Ag^+|Ag}^0 = 0.799$ V.

Solution

If AgNO$_3$ reacts with iron metal, it is not possible to store AgNO$_3$ solution in the vessel. The electrochemical cell will be $Ag\left|Ag^+\right|\left|Fe^{2+}\right|Fe$. The cell reaction is

$$2AgNO_3 + Fe \longrightarrow 2Ag + Fe(NO_3)_2$$

$$E^0 = \varphi_{Fe|Fe^{2+}}^0 - \varphi_{Ag|Ag^+}^0 = 0.44 - (-0.799) = 1.038\,V$$

$\Delta G^0 = -2FE^0 = -2F \times 1.038$ J. As ΔG° is negative, the reaction is spontaneous and hence we cannot store 1 M AgNO$_3$ solution in an iron vessel.

EXERCISES

1. Write down the cell reaction and expression of cell emf of the following cells:
   (a) $Zn\,|\,ZnSO_4\,\|\,KCl, AgCl(s)\,|\,Ag$
   (b) $Al\,|\,AlCl_3(aq)\,\|\,Cu(NO_3)_2\,|\,Cu$
   (c) $Pt\,|\,H_2(g), HCl\,\|\,KCl, Hg_2Cl_2(s)\,|\,Hg(l)$

2. $Ag|AgCl(s),\ 0.1$ N KCl, $Hg_2Cl_2(s)|\ Hg(l),\ E = 0.0455$ V, $\phi_{cal} = -0.334$ V, $\gamma_{Cl^-} = 0.769$. Calculate K_{sp} and S of AgCl. $\phi_{Ag}^0 = -0.799$ V.

   [*Ans.*: $K_{sp} = S^2 = 1.75 \times 10^{-10}$, $S = 1.323 \times 10^{-5}$]

3. $Zn\,|\,ZnSO_4\,\|\,CuSO_4\,|\,Cu$, $\phi_{Zn}^0 = 0.763$ V, $\phi_{Cu}^0 = -0.337$ V. Calculate equilibrium constant.

   [*Ans.*: $K_a = 1.45 \times 10^{37}$]

4. Determine the pH of the solution at 25°C using the following data:
   $Pt|H_2(1 \text{ atm})|H^+(a) \parallel$ Calomel electrode. The emf of the cell is 0.6346 V and $\phi_{cal} = -0.28$ V. *[Ans.: pH = 6]*

5. Given, $E^0_{Pb^{2+}/Pb} = -0.126$ V, $E^0_{Zn^{2+}/Zn} = -0.763$ V. Calculate the emf of the cell
   $Zn|Zn^{2+}$ (0.1 M)$\|Pb^{2+}$ (1 M)|Pb *[Ans.: 0.667 V]*

6. $Cu^1(aq)$ is unstable in solution and undergoes simultaneous oxidation and reduction, according to the reaction $2Cu^+(aq) \rightleftharpoons Cu^{2+}(aq) + Cu(s)$. Calculate E^0 of the reaction.

   Given: $\phi^0_{Cu^{2+}/Cu} = +0.34$ V and $\phi^0_{Cu^{2+}/Cu^+} = +0.15$ V. *[Ans.: $E^0 = 0.38$ V]*

7. A galvanic cell made of $Hg|Hg^{2+}$ and $Fe^{2+}|Fe^{3+}$ electrodes. Calculate $[Hg^{2+}]$ at 25°C at which $E = 0$ and $[Fe^{2+}] = [Fe^{3+}]$. Given: $\phi^0_{Hg|Hg^{2+}} = -0.85$ V and $\phi^0_{Fe^{2+}|Fe^{3+}} = -0.77$ V. *[Ans.: 1.983×10^{-3} M]*

8. Is it possible to store $Pb(NO_3)_2$ solution in Ag–coated container?

   Given: $\phi^0_{Ag|Ag^+} = -0.799$ V and $\phi^0_{Pb|Pb^{2+}} = +0.13$ V

9. Calculate emf of the following cell: *[Ans.: 0.3296 V]*

   $Fe|Fe^{2+}(0.2M)\| Sn^{2+}(2M)|Sn.$ Given: $\phi^0_{Fe^{2+}|Fe} = -0.44$ V, $\phi^0_{Sn^{2+}|Sn} = -0.14$ V

10. Calculate equilibrium constant of the following cell reaction: *[Ans.: 4.7388]*

    $2Au^+ + Pb^{2+} \rightleftharpoons 2Au(s) + Pb^{4+}$. Given: $\phi^0_{Pb^{4+}|Pb^{2+}} = 1.67$ V, $\phi^0_{Au^+|Au} = 1.69$ V.

11. Calculate E^0 of the given electrode reaction: $Fe \longrightarrow Fe^{3+} + 3e^-$ *[Ans.: 0.0433 V]*
    Given: $Fe \longrightarrow Fe^{2+} + 2e^-$, $\phi^0 = 0.45$ V, $Fe^{2+} \longrightarrow Fe^{3+} + e^-$, $\phi^0 = -0.77$ V

    [Hint: $\Delta G^0_1 = -2F \times 0.45$, $\Delta G^0_2 = -1F \times (-0.77)$, $\Delta G^0_3 = \Delta G^0_1 + \Delta G^0_2 = -3FE^0$]

12. For a spontaneous electrochemical process which one of the following is true?
    (a) $E > 0$, (b) $E < 0$, (c) $E = 0$. *[Ans.: $\Delta G = -nFE$. So, $E > 0$]*

13. State the standard condition for E^0 value. *[Ans.: 25°C, 1 atm, 1 M]*

14. Calculate $\Delta G°$ in *kJ* for the following cell $Ag|^+ AgNO_3 \| AuNO_3 |^- Au$

    Given: $\phi^0_{Au^+|Au} = 1.69$ V, $\phi^0_{Ag^+|Ag} = 0.799$ V. *[Ans.: 85.9515 kJ]*

13 Ionic Equilibria

13.1 DEBYE-HÜCKEL LIMITING LAW

An ion remains in solvated form in aqueous solution. Thus the effective radius of an ion is always greater than the actual ionic radius. Furthermore, each ion is surrounded by oppositely charged ions. Thus the potential around an ion consists of the potential of the central ion and the potential of the surrounding ionic atmosphere. So, we can write that

$\phi = \phi_i + \phi_a$, where, ϕ_i and ϕ_a are the potentials of central ion and ionic atmosphere respectively. Considering this concept, Debye-Hückel deduced an expression of activity coefficient, γ_i, of any ionic species, i, as given below

$$\log \gamma_i = - AZ_i^2 \sqrt{I} \tag{13.1a}$$

where A is constant and has a value of 0.5117 at 25°C for pure water. Z_i is the valency of the ion, i, and I is the **ionic strength** of the solution. Let us consider an electrolyte MN which dissociates in aqueous solution as follows

$MN(m) = v_+ M^{2+} (v_+ m) + v_- M^{2-} (v_- m)$, where, m is the molality of the electrolytic solution.

The mean activity, $a_\pm$, is given by

$$a_\pm^v = (\gamma_+^{v_+} \cdot \gamma_-^{v_-})(m_+^{v_+} \cdot m_-^{v_-}) = \gamma_\pm^v \cdot m_\pm^v$$

where, $\qquad m_+ = (v_+ m)$ and $m_- = (v_- m)$, $v = v_+ + v_-$

Hence, $\gamma_\pm^v = \gamma_+^{v_+} \cdot \gamma_-^{v_-}$. Or, $v \log \gamma_\pm = v_+ \log \gamma_+ + v_- \log \gamma_-$

$$\log \gamma_\pm = \frac{v_+ \log \gamma_+ + v_- \log \gamma_-}{(v_+ + v_-)} \tag{13.1b}$$

where, $a_\pm$ = Mean activity of the electrolytic solution, $m_\pm$ = Mean molality of the electrolytic solution.

$\gamma_{\pm}$ = Mean activity coefficient of the electrolytic solution.

Replacing $\log \gamma_{+}$ and $\log \gamma_{-}$ [using equation (13.1a)] in equation (13.1b), we have

$$\log \gamma_{\pm} = - A\left(\frac{v_{+}z_{+}^{2} + v_{-}z_{-}^{2}}{v_{+} + v_{-}}\right)\sqrt{I} \qquad (13.1c)$$

As the solution is electrically neutral, we have $\qquad v_{+}z_{+} = v_{-}z_{-}$

Or, $\qquad \dfrac{v_{+}z_{+}^{2} + v_{-}z_{-}^{2}}{v_{+} + v_{-}} = \dfrac{(v_{-}z_{-})z_{+} + (v_{+}z_{+})z_{-}}{(v_{+} + v_{-})} = z_{+}z_{-}$

Thus the equation (13.1c) becomes

$$\log \gamma_{\pm} = - Az_{+}z_{-}\sqrt{I} \qquad (13.1d)$$

Considering the radius of ionic atmosphere the above equation becomes

$$\log \gamma_{\pm} = \frac{-Az_{+}z_{-}\sqrt{I}}{1 + \sqrt{I}} \text{ (in aqueous solution)} \qquad (13.1e)$$

As we know, each ion remains in hydrated form, the polarization of solvent molecules must be taken into account. Considering this fact into account, the ultimate equation of mean activity coefficient becomes

$$\log \gamma_{\pm} = \frac{-Az_{+}z_{-}\sqrt{I}}{1 + \sqrt{I}} + CI \qquad (13.1f)$$

where C is a constant.

The equations (13.1d), (13.1e), and (13.1f) are the Debye-Hückel equations for mean activity coefficient.

13.2 ACIDS AND BASES

13.2.1 Properties of Acids and Bases

Properties of acids

(a) It tastes sour.
(b) It turns blue litmus to red.
(c) It reacts with metals to liberate hydrogen gas (H_2).
(d) It reacts with base to form a salt and hence loses acidic character.
(e) It reacts with salt to give a new acid and a new salt.
(f) It conducts electricity.

Properties of bases

(a) It tastes bitter.

(b) It turns red litmus to blue.

(c) It forms lather in aqueous solution.

(d) It reacts with acid to form a salt and hence loses basic character.

(e) It conducts electricity.

13.2.2 Arrhenius Concept

(i) According to Svante August Arrhenius, any species, which releases H^+ ion in aqueous solution, is called an acid and any species, which releases OH^- in aqueous solution, is called a base. This is illustrated in the following example:

$$HCl \longrightarrow H^+ + Cl^- \quad \text{and} \quad NaOH \longrightarrow Na^+ + OH^-$$

(ii) An acid neutralizes a base to form water: $\quad H^+ + OH^- \longrightarrow H_2O$

(iii)Strong acids and bases are completely dissociated in aqueous solution but weak acids and bases are only partly dissociated in aqueous solution as illustrated below:

$$HA + H_2O \rightleftharpoons H_3O^+ + A^- \quad \text{and} \quad K_a = \frac{[H_3O^+][A^-]}{[HA]}$$

$$A^- + H_2O \rightleftharpoons HA + OH^- \quad \text{and} \quad K_b = \frac{[HA][OH^-]}{[A^-]}$$

Water also remains in partially dissociated form as follows:

$$H_2O \rightleftharpoons H^+ + OH^- \quad \text{and} \quad K_a = \frac{[H^+][OH^-]}{[H_2O]}$$

As water is present in profuse amount, $[H_2O]$ is taken as constant.

$$K_w = [H^+][OH^-] = [H_3O^+][OH^-] = K_a K_b \qquad (13.2a)$$

$$pK_a = -\log K_a, pK_b = -\log K_b, pK_w = -\log K_w = 14$$

Hence, $\qquad pK_w = pK_a + pK_b \qquad (13.2b)$

13.2.2.1 Drawbacks of Arrhenius Concept

(a) It is applicable only for aqueous solutions. It does not consider other solvents. Water is not involved in acid-base reactions.

(b) It is applicable only when the substance is dissociated in aqueous solution.

(c) It cannot explain acid character of CO_2 in aqueous solution.

(d) It cannot explain basic character of ammoniacal solution.

(e) It cannot explain why the mixture of HCl and NH_3 in aqueous solution is acidic.

(f) It cannot explain the acid characteristics of the substances having no H atom like BF_3, $AlCl_3$, etc. It also cannot explain basic character of the substances having no OH group like NH_3, Na_2CO_3, etc.

13.2.3 Bronsted-Lowry Concept

In 1923, the scientist duo J. Bronsted and T. Lowry proposed that acid-base definitions should be based on proton transfer. According to this concept acid is a substance which can donate a proton whereas base is a substance which can accept a proton. Furthermore, an acid or a base may be either a molecule or an ion. Thus Cl^-, NH_3 are all bases as shown below:

$$Cl^- + H^+ \longrightarrow HCl, \quad NH_3 + H^+ \longrightarrow NH_4^+.$$

Conversely, we can say, both HCl and NH_4^+ are acids. Hence, each and every acid has its conjugate base. Conversely each and every base has its conjugate acid. In the above reactions, HCl is the conjugate acid of the base Cl^- or Cl^- is the conjugate base of the acid HCl. Similarly, NH_4^+ and NH_3 is the conjugate acid-base pair. If the conjugate base is stable, it does not accept a proton easily to form its conjugate acid, which means that conjugate base is weak and hence corresponding acid is strong.

13.2.3.1 Conclusion

(a) Weaker a base is, stronger the conjugate acid is. Conversely, weaker an acid is, stronger the conjugate base is.

(b) Thus a weak acid acts as a base in presence of a strong acid as described below:

$HCl + HCO_3^- \longrightarrow Cl^- + H_2CO_3$ [Here, HCl acts as an acid and HCO_3^- acts as a base.]

$HCO_3^- + OH^- \longrightarrow CO_3^{2-} + H_2O$ [Here, HCO_3^- acts as an acid and OH^- acts as a base.]

Thus HCO_3^- can act as both acid and base. This type of species is called **amphiprotic species**.

(c) The acid or basic character of a species does not depend on the solvent and hence shortcoming of Arrhenius concept is removed.

(d) It cannot explain acid or basic character of those species, having no H–atom.

13.2.4 Lewis Concept

In 1923, G.N. Lewis introduced the concept of transfer of a pair of electrons to define acids and bases. According to this concept any species, which can donate a pair of electrons, is called base and conversely, any species, which can accept a pair of electrons, is called acid.

Thus BF_3, $AlCl_3$ are acids, often called Lewis acids and Cl^-, CO_3^{2-} are considered as bases, often called Lewis bases.

$$AlCl_3 + Cl^- \longrightarrow AlCl_4^-$$ [Here, $AlCl_3$ and Cl^- are the Lewis acid and Lewis base respectively.]

$$
\begin{array}{ccc}
\text{F} & \text{H} & \text{F} \quad \text{H} \\
| & | & | \quad | \\
\text{F---B} \longleftarrow \text{:N---H} & \longrightarrow & \text{F---B} \longleftarrow \text{N---H} \\
| & | & | \quad | \\
\text{F} & \text{H} & \text{F} \quad \text{H} \\
\text{(acid)} & \text{(base)} & \text{(salt)}
\end{array}
$$

13.3 LEVELLING EFFECT OF SOLVENT

Solvent plays an important role in acid-base characteristics. The acidic and basic characters are strongly influenced by the protic solvent like water, which dissociates as follows:

$$H_2O + H_2O \rightleftharpoons H_3O^+ + OH^-$$

Any strong acid in aqueous solution reacts with water to form H_3O^+ and hence there is no acid stronger than H_3O^+. Thus the strength of all strong acids, stronger than H_3O^+, are apparently equal in strength in aqueous solution. This property of solvent to level off all the strong acids is called levelling effect of water.

Similarly, OH^- is a strong base. Any strong base, stronger than OH^-, reacts with water to produce OH^- and hence all the strong bases are apparently equivalent.

In case of ammonia as solvent, NH_4^+ appears as the strongest acid and NH_2^- as the strongest base.

$$NH_3 + NH_3 \rightleftharpoons NH_4^+ + NH_2^-$$

As ammonia is a basic solvent, it has a strong tendency to help dissociation of weak acid completely. Thus an acid, partly dissociated in aqueous solution, is completely dissociated in ammonia.

In aqueous solution HCl is much stronger acid than acetic acid, as the latter is partly dissociated in aqueous solution. However, both HCl and acetic acid are strong acids in ammonia.

Thus in neutral or weakly basic solution strong and weak acids are distinguishable but they are indistinguishable in strongly basic solution.

This property of solvent to level off the strength of acids or bases is called levelling property of solvent.

On this basis solvents can be divided into four categories:

(a) **Protophilic solvent**   Any solvent, having the property of accepting a proton, is called protophilic solvent, e.g., water, alcohol, ammonia, amine, ether, acetone, etc.

(b) **Protogenic (Protophobic) solvent**   Any solvent, having the property of giving up a proton, is called protogenic (or protophobic) solvent, e.g., sulphuric acid, formic acid, acetic acid, etc.

(c) **Amphiprotic (amphoteric) solvent**   Any solvent, capable of either accepting or giving up a proton, is called amphiprotic or amphoteric solvent, e.g., water, alcohol, etc.

(d) **Aprotic solvent** Any solvent, incapable of accepting or giving up a proton, is called aprotic solvent. One aprotic solvent is benzene. Other examples are carbon tetrachloride, DMSO, DMF, etc.

13.4 IONIC PRODUCT OF WATER

Water is a unique **amphoteric** solvent, which dissociates into hydroxonium ion and hydroxyl ion, according to the following equation:

$$H_2O + H_2O \rightleftharpoons H_3O^+ + OH^-$$

$$H_2O \rightleftharpoons H^+ + OH^-$$

Thus water contains equimolar quantities of H^+ and OH^- ions and hence water is a perfectly neutral solvent. The dissociation constant, K_{H_2O}, is given by

$$K_{H_2O} = \frac{a_{H^+} \cdot a_{OH^-}}{a_{H_2O}}. \text{ Or, } K_w = K_{H_2O} \cdot a_{H_2O} = a_{H^+} \cdot a_{OH^-} \quad (13.3a)$$

As water molecules mostly remain in undissociated form, a_{H_2O} may be taken as constant. K_w is known as ionic product of water. Neglecting the activity coefficient terms, we can write that

$$a_{H^+} = C_{H^+} = [H^+] \text{ and } a_{OH^-} = C_{OH^-} = [H^-]. \text{ Thus, } K_w = [H^+].[OH^-] \quad (13.3b)$$

K_w can be measured by emf method. The value of K_w is found to be 1.0×10^{-14} at 25°C. For distilled water, $a_{H^+} = a_{OH^-}$. So, $K_w = a_{H^+}^2 = 1.0 \times 10^{-14}$. Or, $a_{H^+} = a_{OH^-} = 1.0 \times 10^{-7}$.

Or, $\qquad\qquad [H^+] = [OH^-] = 1.0 \times 10^{-7}$ unit.

13.5 pH

Strong acids and bases always remain in dissociated form in aqueous solution. Thus strength of acids or bases can be represented by the corresponding concentrations or acids or bases. For example, HCl in aqueous solution dissociates as follows:

$$HCl(m) \longrightarrow H^+(m) + Cl^-(m)$$

So, HCl solution of molality m has the concentration of $H^+ = [H^+] = m$ molal. In case of weak acids the concentration of acid solution does not represent the concentration of H^+ since weak acids are partly dissociated in aqueous solution, e.g., acetic acid.

$$CH_3COOH(m) \rightleftharpoons H^+(\alpha m) + CH_3COO^-(\alpha m)$$

If α be the degree of dissociation of acetic acid in aqueous solution, $[H^+] = m$, where, m is the molality of acid solution. The value of α is different for different acids. Furthermore it varies with dilution. To avoid this difficulty S. Sorensen in 1909 introduced a term pH to express hydrogen ion concentration in case of weak acids.

Definition of pH It is a measure of acid strength of any weak acid solution. Mathematically, it is expressed as $pH = -\log[H^+]$. In terms of activity, we can write that $pH_a = -\log a_{H^+}$.

pH is also known as **power of hydrogen**.

In case of pure water we know that $[H^+] = 10^{-7}$.

Hence, $pH = -\log 10^{-7} = 7$. So, pH of pure demineralized water is 7. As pure water is neutral in nature, pH of any neutral solution is 7. In acidic solution $[H^+]$ is greater than 10^{-7} and hence pH value falls below 7 whereas in case of basic solution $[OH^-]$ is greater than 10^{-7}.

As we know $K_w = [H^+].[OH^-] = 10^{-14}$, $[H^+] < 10^{-7}$, since, $[OH^-] > 10^{-7}$.

Or, $\log[H^+] < -7$. Or, $pH > 7$.

Thus, pH value is **greater than** 7 in **basic** solution. When, $[OH^-] = 10^0$, $[H^+] = 10^{-14}$. Thus **maximum** value of pH is 14 and **minimum** value of pH is 0. Negative pH has no meaning.

A solution is said to be **acidic** when pH of that solution lies within ≤ 6 and a solution behaves **basic** in nature when pH lies within 8 to 14. Like pH, pOH may be expressed as: $pOH = -\log[OH^-]$. We know that, $K_w = [H^+].[OH^-] = 10^{-14}$. Taking *log* on both sides, we get $\log K_w = \log[H^+] + \log[OH^-]$.

Or, $-\log[H^+] - \log[OH^-] = -\log K_w$. Or, $pH + pOH = pK_w$, where, $pK_w = -\log K_w = 14$

Or, $pH + pOH = 14$. Hence, $pOH = 14 - pH$ (13.4)

13.6 DISSOCIATION CONSTANT

13.6.1 Dissociation of Weak Acid

As discussed earlier any weak acid reversibly dissociates in aqueous solution at a low degree. Let us consider the dissociation of acetic acid (HOAC), which is dissociated as follows:

$$HOAC + H_2O \rightleftharpoons H_3O^+ + OAC^-$$

The equilibrium constant, K, is given by

$$K = \frac{[H_3O^+][OAC^-]}{[HOAC][H_2O]}$$

As the concentration of water does not change, we can write that

$$K_a = K[H_2O] = \frac{[H_3O^+][OAC^-]}{[HOAC]} = \frac{[H^+][OAC^-]}{[HOAC]}$$

If c is the molarity of acetic acid solution and α is the degree of dissociation,

$$[HOAC] = (1-\alpha)c,\ [H^+] = \alpha c \text{ and } [OAC^-] = \alpha c$$

Thus,

$$K_a = \frac{[H^+][OAC^-]}{[HOAC]} = \frac{(\alpha c)(\alpha c)}{(1-\alpha)c} = \frac{\alpha^2 c}{(1-\alpha)} \qquad (13.5a)$$

The equation (13.3a) is known as **Ostwald's dilution law**, first formulated by W. Ostwald in 1888. It is evident from equation (13.5a) that, greater the value of α, higher the value of K_a and hence stronger the acid is. Thus dissociation constant, K_a, is a **measure** of acid strength of a weak acid.

Considering that α is very low, we can assume that $1 - \alpha \approx 1$. Thus equation (13.5a) becomes

$$K_a = \alpha^2 c. \text{ Or, } (\alpha c)^2 = K_a \cdot c. \text{ Or, } [H^+] = \alpha c = \sqrt{K_a \cdot c}.$$

Taking $\log_{10}$ on both sides we get

$$\log[H^+] = \frac{1}{2}\log K_a + \frac{1}{2}\log c.$$

Or,

$$pH = \frac{1}{2}pK_a - \frac{1}{2}\log c \qquad (13.5b)$$

$$[pK_a = -\log K_a]$$

Conclusion   Higher the value of K_a, lower the value of pK_a and stronger the acid is.

The above equation is the general expression of pH of any monobasic weak acid solution.

Again,

$$K_a = \alpha^2 c. \text{ Or, } \alpha = \sqrt{\frac{K_a}{c}}$$

α can easily be determined by measuring conductivity and using Kohlrausch's law $\alpha = \Lambda_c/\Lambda_\infty$, where, Λ_c and Λ_∞ are the equivalent conductivities of a given weak acid solution at concentration, c, and at infinite dilution respectively. Using this value of α one can easily calculate K_a of that weak acid solution.

13.6.2   Dissociation of Weak Base

Let us consider a weak base hydroxide ammonium hydroxide, NH_4OH, which dissociates reversibly as follows:

$$NH_4OH(1 - \alpha)c \rightleftharpoons NH_4^+(\alpha c) + OH^-(\alpha c)$$

$$\text{Equilibrium constant} = K_b = \frac{[NH_4^+][OH^-]}{[NH_4OH]} = \frac{\alpha^2 c}{(1 - \alpha)}$$

Considering α to be very low, we can write that $K_b = \alpha^2 c. \text{ Or, } (\alpha c)^2 = K_b \cdot c$

Conclusion   Greater the value of α, higher the value of K_b, stronger the base is.

Thus, $[OH^-] = \alpha c = \sqrt{K_b \cdot c}$. Taking $\log_{10}$ on both sides we get

$$\log[OH^-] = \frac{1}{2}\log K_b + \frac{1}{2}\log c.$$

Or, $$pOH = \frac{1}{2}pK_b - \frac{1}{2}\log c \qquad (13.6a)$$

$$[pK_b = -\log K_b]$$

Conclusion   Higher the value of K_b, lower the value of pK_b and stronger the base is.

Again, $pOH = pK_w - pH$. Putting this value of pOH in the above equation, we get

$$pH = pK_w - \frac{1}{2}pK_b + \frac{1}{2}\log c \qquad (13.6b)$$

The equation (13.6b) is the general expression of pH of any monoacidic weak base solution.

13.6.3   Dissociation of Polybasic Acids

A polybasic acid dissociates in steps, e.g., phosphoric acid. It dissociates in three steps:

Step-I $H_3PO_4 + H_2O \rightleftharpoons H_3O^+ + H_2PO_4^-$.

$$K_{d1} = \frac{[H_3O^+][H_2PO_4^-]}{[H_3PO_4]} = 7.6 \times 10^{-3} \text{ at } 18°C.$$

Step-II $H_2PO_4^- + H_2O \rightleftharpoons H_3O^+ + HPO_4^{2-}$.

$$K_{d2} = \frac{[H_3O^+][HPO_4^{2-}]}{[H_2PO_4^-]} = 5.9 \times 10^{-8} \text{ at } 18°C.$$

Step-III $HPO_4^{2-} + H_2O \rightleftharpoons H_3O^+ + PO_4^{3-}$.

$$K_{d3} = \frac{[H_3O^+][PO_4^{3-}]}{[HPO_4^{2-}]} = 3.5 \times 10^{-13} \text{ at } 18°C.$$

The value of first dissociation constant is **maximum,** since presence of H_3O^+ shifts the equilibrium towards the left in step-II. In step-III, the concentration of H_3O^+ further increases and hence the value of K_{d3} decreases sharply.

13.6.3.1   Diprotic Acid

There are some acids, which give rise to protons in two separate reactions as shown below:

$$H_2A \rightleftharpoons H^+ + HA^-. \quad K_1 = \frac{[H^+][HA^-]}{[H_2A]}$$

$$HA^- \rightleftharpoons H^+ + A^{2-}. \quad K_2 = \frac{[H^+][A^{2-}]}{[HA^-]}, \quad K_2 \ll K_1$$

It may be assumed that first reaction is the major source of H^+ ion. So, $[H^+] \approx [HA^-]$

$$K_1 \times K_2 = \frac{[H^+]^2[A^{2-}]}{[H_2A]} \tag{13.7}$$

Example 13.6.1 Estimate the concentrations of conjugate species present in a 0.01M solution of sulfurous acid in pure water. $K_1 = 0.017$ and $K_2 = 10^{-7.19}$.

$$H_2SO_3 \rightleftharpoons H^+ + HSO_3^-$$

$[H^+] \approx [HSO_3^-] = \alpha c$ and $\alpha = $ Degree of dissociation, $c = 0.01M$

$$K_1 = \frac{[H^+][HSO_3^-]}{[H_2SO_3]} = 0.017. \quad \text{Or,} \quad \frac{\alpha^2 c}{1-\alpha} = 0.017. \quad \text{Or,} \quad \alpha = 0.7064$$

$$K_1 = \frac{[H^+][HSO_3^-]}{[H_2SO_3]} = \frac{[H^+]^2}{[H_2SO_3]}$$

$$K_1 \times K_2 = \frac{[H^+]^2[SO_3^{2-}]}{[H_2SO_3]}. \quad \text{Or,} \quad [SO_3^{2-}] = K_2 = 10^{-7.19} = 6.456 \times 10^{-8}.$$

13.6.3.2 Ampholyte Salt

An ampholyte salt may be represented as NaHA, a salt made of H_2A and NaOH. The salt undergoes following equilibrium reactions in aqueous solution:

$$H_2A \rightleftharpoons H^+ + HA^-. \quad K_1 = \frac{[H^+][HA^-]}{[H_2A]}$$

$$HA^- \rightleftharpoons H^+ + A^{2-}. \quad K_2 = \frac{[H^+][A^{2-}]}{[HA^-]}$$

$$HA^- + H_2O \rightleftharpoons H_2A + OH^-. \quad K_3 = \frac{[H_2A][OH^-]}{[HA^-][H_2O]}, \quad K_3 = \frac{K_w}{K_1}$$

Amount of acid generated during the reaction = Amount of bases generated during the reaction.

Or, $\quad [H^+] + [H_2A] = [OH^-] + [A^{2-}]$

Or,
$$[H^+] = [OH^-] + [A^{2-}] - [H_2A] = \frac{K_w}{[H^+]} + \frac{K_2[HA^-]}{[H^+]} - \frac{[H^+][HA^-]}{K_1}$$

Or,
$$[H^+]^2 \left\{ 1 + \frac{[HA^-]}{K_1} \right\} = K_w + K_2[HA^-]. \quad \text{Or,} \quad [H^+] = \sqrt{\frac{K_w + K_2[HA^-]}{1 + \frac{[HA^-]}{K_1}}}$$

$$K_w \ll K_2[HA^-] \quad \text{and} \quad K_w + K_2[HA^-] \approx K_2[HA^-].$$

Also,
$$1 + \frac{[HA^-]}{K_1} \approx \frac{[HA^-]}{K_1} \quad \text{as} \quad K_1 \ll 1$$

Hence,
$$[H^+] = \sqrt{K_1 K_2}.$$

Or,
$$pH = \frac{1}{2} pK_1 + \frac{1}{2} pK_2 \tag{13.8}$$

13.6.3.3 Zwitterion

Amino acids are the building blocks of proteins. Each amino acid has one proton donor group (–COOH) and one proton acceptor group (–NH$_2$). Thus both positive and negative charges are developed on amino acid molecule through transfer of H$^+$ ion from acid group to amino group. The simplest amino acid is glycine (Gly).

$$\underset{(H_2Gly^+)}{H_3N^+CH_2COOH} \rightleftharpoons \underset{(HGly)}{H_3N^+CH_2COO^-} \rightleftharpoons \underset{(Gly^-)}{H_2NCH_2COO^-}$$

So the molecule exists as "double ion", which is known as Zwitterion. Two equilibrium reactions are developed:

$$H_2Gly^+ \rightleftharpoons H^+ + HGly. \quad K_1 = \frac{[H^+][HGly]}{[H_2Gly^+]} = 10^{-2.35}$$

$$HGly \rightleftharpoons H^+ + Gly^-. \quad K_2 = \frac{[H^+][Gly^-]}{[HGly]} = 10^{-9.78}$$

If glycine is dissolved in water, all positive and negative charges are balanced.

$$[H_2Gly^+] + [H^+] = [Gly^-] + [OH^-]$$

Substituting [H$_2$Gly$^+$] and [Gly$^-$] from above equations, we get

$$\frac{[H^+][HGly]}{K_1} + [H^+] = \frac{K_2[HGly]}{[H^+]} + \frac{K_w}{[H^+]}$$

Solving $[H^+]$ we get

$$[H^+] = \sqrt{\frac{K_2[\text{HGly}] + K_w}{\frac{[\text{HGly}]}{K_1} + 1}} \qquad (13.9)$$

Example 13.6.2 Calculate the concentrations of various species of aqueous solution of glycine with concentration 0.1 M. Given: $K_1 = 10^{-2.35}$, $K_2 = 10^{-9.78}$ and $K_w = 10^{-14}$.

$$[\text{HGly}] = 0.1\,\text{M}, \quad [H^+] = \sqrt{\frac{K_2[\text{HGly}] + K_w}{\frac{[\text{HGly}]}{K_1} + 1}} = 10^{-6.08}$$

$$K_1 = \frac{[H^+][\text{HGly}]}{[\text{H}_2\text{Gly}^+]}. \quad \text{Or,} \quad [\text{H}_2\text{Gly}^+] = 10^{-4.73}$$

$$K_2 = \frac{[H^+][\text{Gly}^-]}{[\text{HGly}]}. \quad \text{Or,} \quad [\text{Gly}^-] = 10^{-4.70}$$

13.7 SALT HYDROLYSIS

13.7.1 Salt of Strong Acid and Strong Base

This type of salt does not hydrolyze at all. It remains in completely dissociated form and the corresponding ions get hydrated in aqueous solution, e.g., sodium chloride (NaCl). NaCl is a salt produced from HCl (strong acid) and NaOH (strong base).

$$(\text{Na}^+ + \text{OH}^-) + (\text{H}^+ + \text{Cl}^-) \longrightarrow (\text{Na}^+ + \text{Cl}^-) + \text{H}_2\text{O}$$

$$\text{Na}^+ + n\text{H}_2\text{O} \longrightarrow [\text{Na}(\text{H}_2\text{O})_n]^+ \text{ and } \text{Cl}^- + m\text{H}_2\text{O} \longrightarrow [\text{Cl}(\text{H}_2\text{O})_m]^-$$

13.7.2 Salt of Weak Acid and Strong Base

This type of salt readily hydrolyzes in aqueous solution and the corresponding salt solution is alkaline in nature, e.g., sodium acetate (NaOAC). It is produced from acetic acid (weak acid), and sodium hydroxide (strong base) according to the following equation:

$$(\text{Na}^+ + \text{OH}^-) + (\text{HOAC}) \longrightarrow (\text{Na}^+ + \text{OAC}^-) + \text{H}_2\text{O}$$

$$(\text{Na}^+ + \text{OAC}^-) + \text{H}_2\text{O} \rightleftharpoons (\text{Na}^+ + \text{OH}^-) + (\text{HOAC})$$

As HOAC is a weak acid, it remains in undissociated form and the solution is alkaline due to free OH⁻ ions. It is to be noted that salt hydrolyses reversibly. The salt NaOAC is continuously produced during titration of HOAC by NaOH. At the beginning only HOAC is present and the following equilibrium exists:

$$HOAC \rightleftharpoons H^+ + OAC^- \qquad (13.10a)$$

The concentration of acid is a, where,

a = Concentration of undissociated acid (HOAC), i.e., C_{HOAC} + Concentration of dissociated acid which is equivalent to C_{OAC^-}.

Or, $$a = C_{HOAC} + C_{OAC^-} \qquad (13.10b)$$

Now, NaOH solution is added in little quantity. NaOH readily reacts with undissociated acid to form salt according to the following reaction:

Step-I $\qquad (Na^+ + OH^-) + (HOAC) \longrightarrow (Na^+ + OAC^-) + H_2O$

Step-II $\qquad OAC^- + H_2O \rightleftharpoons OH^- + HOAC,$

OAC^-, produced in step-I, undergoes hydrolysis in step-II to form undissociated HOAC. Otherwise equilibrium in equation (13.10a) will be disturbed. Thus the solution contains two types of cations Na^+, H^+ and two types of anions OH^-, OAC^-. As the solution is electrically neutral, we can write that

Concentration of cation = Concentration of anion

Or, $\qquad C_{Na^+} + C_{H^+} = C_{OH^-} + C_{OAC^-}$

NaOH remains fully dissociated form in aqueous solution. So, $C_{NaOH} = C_{Na^+}$. If b be the concentration of sodium hydroxide solution, $C_{Na^+} = b$.

Thus, $\qquad C_{OAC^-} = b + C_{H^+} - C_{OH^-} \qquad (13.10c)$

Combining equations (13.10b) and (13.10c), we get

$$C_{HOAC} = a - C_{OAC^-} = a - b - C_{H^+} + C_{OH^-} \qquad (13.10d)$$

The equilibrium of equation (13.10a) is still maintained in aqueous solution. The equilibrium constant, K_a is given by

$$K_a = \frac{C_{H^+} \cdot C_{OAC^-}}{C_{HOAC}}. \quad \text{Or,} \quad C_{H^+} = \frac{K_a \cdot C_{HOAC}}{C_{OAC^-}}$$

Putting the values of C_{OAC^-} and C_{HOAC} from equations (13.10c) and (13.10d) in the above equation, we get

$$C_{H^+} = \frac{K_a(a - b - C_{H^+} + C_{OH^-})}{b + C_{H^+} - C_{OH^-}} \qquad (13.11)$$

(i) At the beginning of titration or before adding any NaOH solution,

b = concentration of NaOH solution added = 0 and C_{OH^-} = Concentration of OH^- in the mixture = 0.

Thus equation (13.11) becomes

$$C_{H^+} = K_a\left(\frac{a - C_{H^+}}{C_{H^+}}\right). \quad \text{Or,} \quad C_{H^+}^2 + K_a C_{H^+} - K_a \cdot a = 0$$

Or,
$$C_{H^+} = \frac{-K_a \pm \sqrt{K_a^2 + 4aK_a}}{2} = -\frac{1}{2}K_a \pm \sqrt{\frac{1}{4}K_a + aK_a}.$$

Or,
$$C_{H^+} = -\frac{1}{2}K_a + \sqrt{\frac{1}{4}K_a + aK_a}$$

The other root of C_{H^+} is negative and hence is not considered. Again, concentration of acid, a, is far greater than K_a. Or, $a \gg K_a$. Thus, $\frac{1}{4}K_a + aK_a \approx aK_a$. Or, $C_{H^+} = -\frac{1}{2}K_a + \sqrt{aK_a}$.

Again, $a \gg K_a$. Or, $aK_a \gg K_a^2$.

Or, $\sqrt{aK_a} \gg K_a$. Hence, $C_{H^+} = -\frac{1}{2}K_a + \sqrt{aK_a} \approx \sqrt{aK_a}$

Taking $\log_{10}$ on both sides, we get $\log C_{H^+} = \frac{1}{2}\log K_a + \frac{1}{2}\log a$

$$-\log C_{H^+} = -\frac{1}{2}\log K_a - \frac{1}{2}\log a. \text{ Or, } pH = \frac{1}{2}pK_a - \frac{1}{2}\log a \quad (13.12)$$

The equation (13.12) is identical with the equation (13.5b).

(ii) Near the end point of titration or near the **equivalence point** $C_{H^+} \approx C_{OH^-}$ and $b < a$. So, all base is consumed. Thus the equation (13.11) becomes

$$C_{H^+} = K_a\left(\frac{a-b}{b}\right)$$

Taking $\log_{10}$ on both sides we get

$$\log C_{H^+} = \log K_a + \log\left(\frac{a-b}{b}\right)$$

$(a-b)$ = Concentration of acid remaining after consumption of the base.
b = Concentration of base added = Concentration of salt since all base is converted to salt.

So, when the salt, $(Na^+ + OAC^-)$, hydrolyses to produce free OH^-, the acid, HOAC, readily consumes that free OH^-.

$$(Na^+ + OAC^-) + H_2O \rightleftharpoons (Na^+ + OH^-) + HOAC$$

$$HOAC + OH^- \longrightarrow H_2O + OAC^-$$

So,
$$-\log C_{H^+} = -\frac{1}{2}\log K_a - \frac{1}{2}\log\left(\frac{a-b}{b}\right).$$

Or,
$$pH = pK_a + \log\left(\frac{b}{a-b}\right)$$

Or,
$$pH = pK_a + \log\left(\frac{b}{a-b}\right) = pK_a + \log\frac{[\text{Salt}]}{[\text{Acid}]} \qquad (13.13)$$

This equation is known as **Henderson-Hasselbalch equation**, representing the pH of a buffer solution, based on weak acid and its salt with strong base. pH of this buffer solution is less than 7 and hence this type of buffer is known as acid buffer solution.

13.7.3 Buffer Solution

13.7.3.1 Definition of a Buffer Solution

Buffer solution is a solution which possesses a definite pH and offers strong resistance to change the pH whenever a little amount of a strong acid or a strong base is added to it. Buffer solution is usually prepared by mixing a weak acid and its salt with strong base or by mixing a weak base and its salt with strong acid. The equation (13.13) represents the pH of a buffer solution prepared by mixing weak acid and its salt with strong base.

13.7.3.2 Mechanism of Buffer Action

pH of a buffer solution, prepared from weak acid and its salt, is given by

$$pH = pK_a + \log\frac{[\text{Salt}]}{[\text{Acid}]} \qquad (13.13)$$

where, K_a is the dissociation constant of weak acid, taken. Let us consider a buffer solution prepared from acetic acid (HOAC) and its salt sodium acetate (NaOAC).

For acetic acid, $K_a = 1.76 \times 10^{-5}$ and $p_{Ka} = 4.76$.
The buffer solution is so prepared that [Salt] = [Acid] = 1 mole/lit.
So according to equation (13.13), we have pH = pK_a = 4.76.

Case-I HCl is added into a buffer solution of pH = 4.76.

Now HCl solution containing 0.9 mole of HCl is added into the buffer solution. 1 mole HCl reacts with 1 mole salt to form 1 mole undissociated acid according to the following equation

$$(H^+ + Cl^-) + (Na^+ + OAC^-) \longrightarrow (Na^+ + Cl^-) + HOAC$$

So, H^+ ions are consumed to form weak acid and hence pH changes very little. Thus, 0.9 mole HCl produces 0.9 mole HOAC.

Thus, [Acid] = (1 + 0.9) moles/lit = 1.9 moles/lit and [Salt] = (1 − 0.9) moles/lit = 0.1 mole/lit.
Thus, pH is given by

$$pH = pK_a + \log\frac{0.1}{0.9} = 3.48$$

Thus, 0.9 mole HCl is required to change the pH from 4.76 to 3.48.
Let us consider an acidic solution of pH = 4.76.

Or, $\qquad\qquad$ pH $= -\log_{C_{H^+}} = 4.76$. Hence, $C_{H^+} = 10^{-4.76}$ moles/lit.

pH of the acid solution is changed to 3.48 after addition of HCl solution.

$$\text{pH} = -\log C_{H^+} = 3.48. \text{ Hence, } C_{H^+} = 10^{-3.48} \text{ moles/lit.}$$

Thus, moles of HCl added per litre of solution $= (10^{-3.48} - 10^{-4.76})$ moles $= 3.1 \times 10^{-4}$ mole.

Thus, only 3.1×10^{-4} mole of HCl is sufficient to change pH from 4.76 to 3.48 but in case of buffer solution 0.9 mole HCl is required to change pH from 4.76 to 3.48.

Conclusion   A buffer solution offers a resistance to change its pH on addition of strong acid.

Case-II   NaOH is added into a buffer solution of pH = 4.76.

Instead of HCl, NaOH solution containing 0.9 mole of NaOH, is added to the buffer solution. Then free OH⁻ reacts with undissociated acid to form equivalent amount of salt according to the following equation:

$$(Na^+ + OH^-) + HOAC \longrightarrow (Na^+ + OAC^-) + H_2O$$

Thus OH⁻ ions are consumed to form salt and hence pH of the buffer solution changes very little.

So, 0.9 mole of NaOH reacts with 0.9 mole of HOAC to form 0.9 mole of salt.

So, [Salt] $= (1 + 0.9)$ mole/lit and [Acid] $= (1 - 0.9)$ mole/lit $= 0.1$ mole/lit.

$$pH = pK_a + \log\frac{[Salt]}{[Acid]} = 4.76 + \log\frac{1.9}{0.1} = 6.039, \text{ since } pK_a = 4.76$$

Thus 0.9 mole NaOH is required to change pH of buffer solution from 4.76 to 6.039. If we take a weak acid solution of pH $= 4.76$, concentration of acid is given by $C_{H^+} = 10^{-4.76}$ moles/lit.

Now pH of that solution is changed to 6.039 after addition of NaOH solution. Thus the amount of NaOH required to change the pH from 4.76 to 6.039 is $= (10^{-4.76} - 10^{-6.039})$ mole $= 16.48 \times 10^{-6}$ mole.

So only 16.48×10^{-6} mole of NaOH is enough to change the pH from 4.76 to 6.039. However, in case of buffer solution 0.9 mole of NaOH is required to realize the change of pH form 4.76 to 6.039.

Conclusion   A buffer solution offers a resistance to change its pH on addition of strong base.

13.7.3.3   Buffer Capacity or Buffer Index

Amount of strong acid or base, in moles, required to change the pH of a buffer solution by unity, is called buffer capacity or buffer index of that buffer solution. Let us take a buffer solution, containing acetic acid (0.8 mole) and its salt sodium acetate (0.2 mole). Thus pH of the buffer solution is given by

$$pH = pK_a + \log\frac{[Salt]}{[Acid]} = 4.76 + \log\frac{0.2}{0.8} = 4.158, \text{ since, } pK_a = 4.76$$

Now pH of buffer solution decreases by unity upon addition of x moles of HCl. So, pH $= 3.158$.

$$pH = 3.158 = 4.76 + \log\left(\frac{0.2 - x}{0.8 + x}\right).$$

Or, $\qquad \log\left(\frac{0.2 - x}{0.8 + x}\right) = 3.158 - 4.76 = -1.602$

Or, $\qquad \log\left(\frac{0.8 + x}{0.2 - x}\right) = 1.602.$ Or, $\left(\frac{0.8 + x}{0.2 - x}\right) = 10^{1.602} = 39.994 \approx 40$

Or, $\qquad (0.8 + x) = 40(0.2 - x).$ Hence, $x = 0.176$ mole.

Thus, 0.176 mole of HCl is required to decrease the pH of that buffer solution by unity. So, the **acidic buffer capacity** $= 0.176$.

Now pH of buffer solution increases by unity upon addition of y moles of NaOH. So, pH $= 5.158$.

$$pH = 5.158 = 4.76 + \log\left(\frac{0.2 + y}{0.8 - y}\right)$$

Or, $\qquad \log\left(\frac{0.2 + y}{0.8 - y}\right) = 5.158 - 4.76 = 0.398$

Or, $\qquad \left(\frac{0.2 + y}{0.8 - y}\right) = 10^{0.398} = 2.5.$ Or, $(0.2 + y) = 2.5(0.8 - y).$

Hence, $\qquad\qquad\qquad y = 0.514$ mole.

So, the **alkaline buffer capacity** $= 0.514$.

13.7.3.4 Expression of Buffer Index (β)

The pH of a weak acid/salt buffer solution is given by

$$pH = pK_a + \log\frac{[\text{Salt}]}{[\text{Acid}]} \qquad \text{(equation 13.13)}$$

$a = $ Number of moles of weak acidic present in the solution at the beginning.

$b = $ Number of moles of base added in the acidic solution $=$ Number of moles of salt present in the buffer solution.

Assume that db moles of base is added to a buffer solution to increase the pH by dpH. So, dpH is positive. Addition of db moles of base means buffer solution contains $(b + db)$ moles of salt and $[a - (b + db)]$ moles of acid. So, base content or salt content of buffer solution increases by db moles.

Definition of Buffer Index   The amount of alkali required to change the pH of a buffer solution by unity is called buffer index, represented by β. If, db mole of alkali is required to change the pH by dpH, buffer index is given by

$$\alpha = \frac{dpH}{db}. \quad \text{Buffer index} = \beta = \frac{1}{\alpha} = \frac{db}{dpH}$$

For weak acid/salt buffer solution, we know that

$$pH = pK_a + \log\left(\frac{b}{a-b}\right) \quad \text{(equation 13.13)}$$

Differentiating both sides of the above equation with respect to b we get

$$\frac{dpH}{db} = \left(\frac{a-b}{b}\right)\left[\frac{(a-b+b)}{(a-b)^2}\right] = \frac{a}{b(a-b)}$$

$$\beta = \frac{db}{dpH} = \frac{b(a-b)}{a} \quad (13.14)$$

a, the initial acid content, is constant, So, β of a buffer solution depends on the base content, b. The β vs. b curve shows a maximum, where the value of β is β_{max}. To determine the value of β_{max}, 1st derivative of β, i.e., $\frac{d\beta}{db}$ is set to zero, i.e., $\frac{d\beta}{db} = 0$. Differentiating the equation (13.14) with respect to b, we get

$$\frac{d\beta}{db} = \frac{[(a-b)-b]}{a} = \frac{a-2b}{a} \quad \text{and} \quad \frac{d^2\beta}{db^2} = -\frac{2}{a}$$

Or, $$\frac{d^2\beta}{db^2} < 0, \quad \text{since} \quad a > 0$$

At, $$\beta = \beta_{max}, \frac{d\beta}{db} = 0.$$

Or, $$\frac{a-2b}{a} = 0. \quad \text{Or,} \quad (a-2b) = 0. \quad \text{Or,} \quad b = \frac{a}{2}$$

Putting the value of b in equation (13.14), we get

$$\beta_{max} = \frac{\frac{a}{2}\left(a - \frac{a}{2}\right)}{a}. \quad \text{Or,} \quad \beta_{max} = \frac{a}{4}$$

Thus buffer index reaches to **maximum** value of $\frac{a}{4}$ at $b = \frac{a}{2}$.

Table 13.1 Some buffer solutions with their compositions and pH range

S. No.	Composition	pH range	Types of buffer
1.	Acetic acid and Sodium acetate	3.7 – 5.6	Acidic buffer
2.	Potassium hydrogen phthalate and Sodium hydroxide	4.0 – 6.2	Basic buffer
3.	Dibasic sodium citrate and Sodium hydroxide	5.0 – 6.3	Basic buffer
4.	Monobasic potassium phosphate and Sodium hydroxide	5.8 – 8.0	Basic buffer
5.	Dibasic sodium phosphate and Sodium hydroxide	11.0 – 12.0	Basic buffer
6.	Borax and Hydrochloric acid	7.6 – 9.2	Acidic buffer
7.	Borax and Sodium hydroxide	9.2 – 11.0	Basic buffer
8.	Boric acid and Sodium hydroxide	7.8 – 10.0	Basic buffer
9.	Ammonium hydroxide and Ammonium chloride	10.0 – 11.0	Basic buffer

13.7.4 Expression of pH for Salt of Weak Acid and Strong Base

We have seen in the earlier section that buffer solution is formed near the end point of titration of weak acid (HOAC) by a strong base (NaOH) and $b < a$.

At the end point of titration or just beyond the end point, all acid is consumed and there is no free H^+ ion. So, $b = a$ and $C_{H^+} = 0$. Hydrolysis of salt results in the formation of free OH^- and the solution becomes alkaline.

$$(Na^+ + OAC^-) + H_2O \rightleftharpoons (Na^+ + OH^-) + HOAC$$

Putting the values of b and C_{H^+} in the equation (13.11), we get

$$C_{H^+} = K_a \frac{[(a-a) - 0 + C_{OH^-}]}{a + 0 - C_{OH^-}} = K_a \frac{C_{OH^-}}{a - C_{OH^-}}$$

Again, we know that

$$C_{H^+} \cdot C_{OH^-} = K_w. \quad \text{Or,} \quad C_{OH^-} = \frac{K_w}{C_{H^+}}$$

Putting this value of C_{OH^-} in the above equation we get

$$C_{H^+} = K_a \frac{\dfrac{K_w}{C_{H^+}}}{a - \left(\dfrac{K_w}{C_{H^+}}\right)} = \frac{K_a K_w}{(aC_{H^+} - K_w)}$$

Or,
$$aC_{H^+}^2 - K_w C_{H^+} - K_a K_w = 0$$

Or,

$$C_{H^+} = \frac{K_w \pm \sqrt{K_w^2 + 4aK_aK_w}}{2a}.$$

As,

$$K_a \gg K_w, \quad K_aK_w \gg K_w^2. \quad \text{Or,} \quad K_w \ll \sqrt{K_aK_w}$$

So,

$$K_w^2 + 4aK_aK_w = 4aK_aK_w.$$

Thus,

$$C_{H^+} = \frac{K_w \pm \sqrt{4aK_aK_w}}{2a} = \frac{K_w}{2a} \pm \sqrt{\frac{K_aK_w}{a}}$$

Again,

$$K_w \ll \sqrt{K_aK_w}.$$

So,

$$\frac{K_w}{2a} \ll \sqrt{\frac{K_aK_w}{a}}. \quad \text{So,} \quad C_{H^+} = \pm\sqrt{\frac{K_aK_w}{a}} = \sqrt{\frac{K_aK_w}{a}}$$

[Since, C_{H^+} cannot be negative]
Taking $\log_{10}$ on both sides we get

$$\log C_{H^+} = \frac{1}{2}\log K_a + \frac{1}{2}\log K_w - \frac{1}{2}\log a.$$

Or,

$$-\log C_{H^+} = -\frac{1}{2}\log K_a - \frac{1}{2}\log K_w + \frac{1}{2}\log a$$

Or,

$$pH = \frac{1}{2}pK_a + \frac{1}{2}pK_w + \frac{1}{2}\log a \tag{13.15}$$

The above equation holds good when the end point of titration of a weak acid by a strong base just reaches. So the solution is assumed to contain only the salt. So the equation(13.15) is the expression of pH of solution of a salt of weak acid and strong base, where, a is the initial concentration of weak acid.

As $\frac{1}{2}pK_w = 7$, pH > 7 according to equation (13.15) and hence the solution is **alkaline**.

13.7.5 Salt of Weak Base and Strong Acid

One example of this type of salt is ammonium chloride (NH_4Cl), which hydrolyses in aqueous solution according to the following equation:

$$NH_4OH + (H^+ + Cl^-) \longrightarrow (NH_4^+ + Cl^-) + H_2O$$
$$(NH_4^+ + Cl^-) + H_2O \rightleftharpoons NH_4OH + (H^+ + Cl^-)$$

Thus the aqueous solution of ammonium chloride is acidic in nature. The salt NH_4Cl is continuously produced during titration of NH_4OH by HCl. At the beginning of titration only NH_4OH is present and the following equilibrium exists:

$$NH_4OH \rightleftharpoons NH_4^+ + OH^-$$

$b = C_{NH_4OH} +$ Concentration of dissociated base, equivalent to $C_{NH_4^+}$ (13.16a)

Now, a little quantity of HCl is added to this solution. HCl reacts with NH_4OH to form NH_4Cl, which on hydrolysis gives

$$(NH_4^+ + Cl^-) + H_2O \rightleftharpoons NH_4OH + (H^+ + Cl^-)$$

As the solution is electrically neutral, we can write that
Concentration of cations = Concentration of anions. Thus, we have

$$C_{H^+} + C_{NH_4^+} = C_{OH^-} + C_{Cl^-}.$$

$C_{Cl^-} = C_{HCl} =$ Amount of acid added $= a$ [since HCl completely dissociated in aqueous solution].

$$C_{NH_4^+} = (a + C_{OH^-}) - C_{H^+} \qquad (13.16b)$$

Using equation (13.16b) in equation (13.16a), we have

$$C_{NH_4OH} = b - C_{NH_4^+} = b - (a + C_{OH^-}) + C_{H^+} = (b - a) - C_{OH^-} + C_{H^+} \qquad (13.17)$$

The equilibrium constant, K_b, for NH_4OH is given by

$$K_b = \frac{C_{NH_4^+} \cdot C_{OH^-}}{C_{NH_4OH}}$$

Putting these values of $C_{NH_4^+}$ and C_{NH_4OH} in the above equation, we get

$$K_b = \frac{[(a + C_{OH^-}) - C_{H^+}]C_{OH^-}}{[(b - a) - C_{OH^-} + C_{H^+}]}. \text{ Or, } C_{OH^-} = K_b \left[\frac{(b - a) - C_{OH^-} + C_{H^+}}{(a + C_{OH^-}) - C_{H^+}} \right] \qquad (13.18a)$$

(i) At the beginning of titration there is no HCl in the mixture. So, $a = 0$ and $C_{H^+} = 0$. Thus equation (13.18a) becomes

$$C_{OH^-} = K_b \left(\frac{b - C_{OH^-}}{C_{OH^-}} \right). \text{ Or, } C_{OH^-}^2 + K_b C_{OH^-} - bK_b = 0$$

Solving C_{OH^-} we get,

$$C_{OH^-} = \frac{-K_b \pm \sqrt{K_b^2 + 4bK_b}}{2}.$$

As, $b \gg K_b, \; bK_b \gg K_b^2.$ Hence, $K_b^2 + 4bK_b \approx 4bK_b$

So, $C_{OH^-} = \dfrac{-K_b \pm \sqrt{4bK_b}}{2}.$

Again, $-K_b \pm \sqrt{4bK_b} \approx \pm\sqrt{4bK_b} = \pm 2\sqrt{bK_b}$

Or, $C_{OH^-} = \dfrac{\pm 2\sqrt{bK_b}}{2} = \pm\sqrt{bK_b} = \sqrt{bK_b}$, since, C_{OH^-} cannot be negative.

Taking $\log_{10}$ on both sides, we get

Or,
$$\log C_{OH^-} = \frac{1}{2} \log K_b + \frac{1}{2} \log b.$$

Or,
$$-\log C_{OH^-} = -\frac{1}{2} \log K_b - \frac{1}{2} \log b$$

Or, $pOH = \frac{1}{2} pK_b - \frac{1}{2} \log b$. Or, $pH = pK_w - pOH = pK_w - \frac{1}{2} pK_b + \frac{1}{2} \log b$ (13.18b)

This is the expression of pH of solution of a salt of weak base and strong acid, b is the concentration of weak base. As $\frac{1}{2} pK_w = 7$, pH > 7 and hence the solution is **basic**.

(ii) Near the end point of titration, $C_{OH^-} \approx C_{H^+}$ and $a < b$. So all acid is consumed to form salt. So moles of acid added = moles of salt formed = a moles. Moles of remaining weak base in the mixture is $(b - a)$ moles. Thus, the equation (13.18a) becomes

$$C_{OH^-} = K_b \left(\frac{b-a}{a} \right) = K_b \frac{[base]}{[salt]}$$

Taking $\log_{10}$ on both sides we get

$$\log C_{OH^-} = \log K_b + \log \frac{[base]}{[salt]}. \quad \text{Or,} \quad pOH = pK_b + \log \frac{[salt]}{[base]} \quad (13.18c)$$

Again, we know that $pH + pOH = pK_w$. Or, $pOH = pK_w - pH$. Putting the value of pOH in the above equation, we get

$$pH = pK_w - \left(pK_b + \log \frac{[salt]}{[base]} \right) = pK_w - pK_b - \log \frac{[salt]}{[base]} \quad (13.18d)$$

Equation (13.18d) is the expression of pH of a buffer solution prepared from weak base and strong acid.

According to equation (13.18c), pOH < 7 or pH > 7. Hence, corresponding buffer solution is also known as **basic buffer solution**.

(iii) At the end point of titration or just beyond the end point, all base is consumed and there is no free OH^- ion. So, $b = a$ and $C_{OH^-} = 0$. Hydrolysis of salt results in the formation of free H^+ ion and the solution becomes acidic, e.g., NH_4Cl on hydrolysis forms free H^+ ion according to the following equation

$$(NH_4^+ + Cl^-) + H_2O \rightleftharpoons NH_4OH + (H^+ + Cl^-)$$

Putting these values of b and C_{OH^-} in the equation (13.18a), we get

$$C_{OH^-} = K_b \left(\frac{b - b - 0 + C_{H^+}}{b - C_{H^+}} \right) = K_b \left(\frac{C_{H^+}}{b - C_{H^+}} \right).$$

Again, $\qquad C_{H^+} \cdot C_{OH^-} = K_w.$ Or, $C_{H^+} = \dfrac{K_w}{C_{OH^-}}$

$$C_{OH^-} = K_b \left(\frac{\dfrac{K_w}{C_{OH^-}}}{b - \dfrac{K_w}{C_{OH^-}}} \right) = \frac{K_w K_b}{bC_{OH^-} - K_w}.$$

Or, $\qquad bC_{OH^-}^2 - K_w C_{OH^-} - K_w K_b = 0$

Solving C_{OH^-} we get

$$C_{OH^-} = \frac{K_w \pm \sqrt{K_w^2 + 4bK_b K_w}}{2b}.$$

As, $\qquad K_b \gg K_w, \;\; K_b K_w \gg K_w^2. \;$ Hence, $\; K_w^2 + 4bK_b K_w \approx 4bK_b K_w$

Or, $\qquad C_{OH^-} = \dfrac{K_w \pm \sqrt{4bK_b K_w}}{2b}.$

As, $\qquad K_b K_w \gg K_w^2, \;\; \sqrt{K_b K_w} \gg K_w, \;\; K_w \pm \sqrt{4bK_b K_w} \approx \pm \sqrt{4bK_b K_w}$

Or, $\; C_{OH^-} = \dfrac{\pm\sqrt{4bK_b K_w}}{2b} = \dfrac{\sqrt{4bK_b K_w}}{2b} = \sqrt{\dfrac{K_b K_w}{b}}, \;$ (since, C_{OH^-} cannot be negative)

Taking $\log_{10}$ on both sides, we get

$$\log C_{OH^-} = \frac{1}{2}\log K_w + \frac{1}{2}\log K_b - \frac{1}{2}\log b. \; \text{Or,} \; pOH = \frac{1}{2}pK_w + \frac{1}{2}pK_b + \frac{1}{2}\log b \quad (13.18e)$$

Again, we know that $\quad pH + pOH = pK_w$ and $pH = pK_w - pOH$

Hence, $\qquad pH = pK_w - \left(\dfrac{1}{2}pK_w + \dfrac{1}{2}pK_b + \dfrac{1}{2}\log b \right) = \dfrac{1}{2}pK_w - \dfrac{1}{2}pK_b - \dfrac{1}{2}\log b$

Or, $\qquad pH = \dfrac{1}{2}pK_w - \dfrac{1}{2}pK_b - \dfrac{1}{2}\log b \; \text{[Usually } |b| > 1 \text{]} \qquad (13.18f)$

The above equation holds good when the end point of titration of a weak base by a strong acid just reaches. So the solution is assumed to contain only the salt. So the equation(13.18f) is the expression of pH of solution of a salt of weak base and strong acid, where, b is the initial concentration of weak base.

As $\dfrac{1}{2}pK_w = 7$, pH < 7 according to equation (13.18f) and hence the solution is **acidic**.

13.7.6   Salt of Weak Acid and Weak Base

NH_4OH is a weak base and HOAC is a weak acid. So ammonium acetate (NH_4OAC) is formed during the reaction of NH_4OH and HOAC according to the following equation:

$$NH_4OH + HOAC \longrightarrow (NH_4^+ + OAC^-) + H_2O$$

$$(NH_4^+ + OAC^-) + H_2O \rightleftharpoons NH_4OH + HOAC$$

Thus the hydrolysis of salt gives rise to the corresponding weak acid and weak base. In aqueous solution NH_4OH and HOAC dissociate as follows

$$NH_4OH \rightleftharpoons NH_4^+ + OH^- \qquad \text{[Equilibrium constant is } K_b\text{]}$$

$$HOAC \rightleftharpoons H^+ + OAC^- \qquad \text{[Equilibrium constant is } K_a\text{]}$$

The expressions of K_b and K_a are given as

$$K_b = \frac{C_{NH_4^+} \cdot C_{OH^-}}{C_{NH_4OH}} \quad \text{and} \quad K_a = \frac{C_{H^+} \cdot C_{OAC^-}}{C_{HOAC}} \tag{13.19a}$$

The salt, NH_4OAC, is continuously formed during titration of weak acid (HOAC) by weak base (NH_4OH). At the beginning of titration only acetic acid is present.

a = concentration of acid = concentration of undissociated acid (HOAC) + concentration of dissociated acid, which is equivalent to concentration of OAC^-.

$$\text{Or, } a = C_{HOAC} + C_{OAC^-} \tag{13.19b}$$

Now a little amount of NH_4OH solution is added to this. The mixture contains H^+, NH_4^+, and OAC^- ions. As the solution is electrically neutral, concentration of cations is equal to concentration of anions. So, we have

$$C_{OAC^-} = C_{H^+} + C_{NH_4^+} \tag{13.19c}$$

C_{NH_4OH} = Amount of base added = b moles. Again, from equation (13.19b), we have

$$C_{HOAC} = a - C_{OAC^-} = a - C_{H^+} - C_{NH_4^+} \tag{13.19d}$$

Again, the equilibrium constant, K_a is given by $K_a = \dfrac{C_{H^+} \cdot C_{OAC^-}}{C_{HOAC}}$. Putting the values of C_{OAC^-} and C_{HOAC} in the above equation we get

$$K_a = \frac{C_{H^+}(C_{H^+} + C_{NH_4^+})}{(a - C_{H^+} - C_{NH_4^+})}$$

Again, from equation (13.19a) we have

$$C_{NH_4^+} = K_b\left(\frac{C_{NH_4OH}}{C_{OH^-}}\right) = K_b\left(\frac{b}{C_{OH^-}}\right)$$

Putting the value of $C_{NH_4^+}$ in the above equation, we get

$$K_a = \frac{C_{H^+}\left[C_{H^+} + \dfrac{bK_b}{C_{OH^-}}\right]}{\left[a - C_{H^+} - \dfrac{bK_b}{C_{OH^-}}\right]} = \frac{C_{H^+}(K_w + bK_b)}{(aC_{OH^-} - K_w - bK_b)}, \text{ since, } K_w = C_{H^+} \cdot C_{OH^-}$$

Or,

$$K_a = \frac{\left(\dfrac{K_w}{C_{OH^-}}\right)(K_w + bK_b)}{\left[aC_{OH^-} - (K_w + bK_b)\right]} = \frac{K_w(K_w + bK_b)}{\left[aC_{OH^-}^2 - C_{OH^-}(K_w + bK_b)\right]}$$

Or,

$$K_a \cdot aC_{OH^-}^2 - K_a C_{OH^-}(K_w + bK_b) - (K_w^2 + bK_b K_w) = 0$$

Solving C_{OH^-}, we get

$$C_{OH^-} = \frac{K_a(K_w + bK_b) \pm \sqrt{K_a^2(K_w + bK_b)^2 + 4aK_a(K_w^2 + bK_b K_w)}}{2aK_a}$$

$K_b \gg K_w$. So, $K_b K_w \gg K_w^2$ and $K_a K_b K_w \gg K_a K_w^2$. Again, $K_b^2 \ll K_b$

So, $K_a^2(K_w + bK_b)^2 \ll 4aK_a(K_w^2 + bK_b K_w)$ and $4aK_a(K_w^2 + bK_b K_w) \approx 4abK_a K_b K_w$.

Thus the above equation reduces to

$$C_{OH^-} = \frac{K_a K_w + bK_a K_b \pm \sqrt{4abK_a K_b K_w}}{2aK_a} \approx \frac{\sqrt{4abK_a K_b K_w}}{2aK_a},$$

since, $\sqrt{4abK_a K_b K_w} \gg K_a K_w + bK_a K_b$

$$C_{OH^-} = \sqrt{\frac{4abK_a K_b K_w}{4a^2 K_a^2}} = \sqrt{\frac{bK_b K_w}{aK_a}}, \text{ since, } C_{OH^-} \text{ cannot be negative.}$$

Taking $\log_{10}$ on both sides, we get

$$\log C_{OH^-} = \frac{1}{2}\log\frac{b}{a} + \frac{1}{2}\log K_w + \frac{1}{2}\log K_b - \frac{1}{2}\log K_a \qquad (13.19e)$$

At the end of titration or just beyond the end point all acid is consumed by all base. Hence, $b = a$. Thus, the above equation becomes

$$\log C_{OH^-} = \frac{1}{2}\log K_w + \frac{1}{2}\log K_b - \frac{1}{2}\log K_a.$$

Or,
$$pOH = \frac{1}{2} pK_w + \frac{1}{2} pK_b - \frac{1}{2} pK_a$$

Again, we know that $pH + pOH = pK_w$.

Or,
$$pH = pK_w - pOH = pK_w - \left(\frac{1}{2} pK_w + \frac{1}{2} pK_b - \frac{1}{2} pK_a \right)$$

$$= \frac{1}{2} pK_w + \frac{1}{2} pK_a - \frac{1}{2} pK_b \quad \left[\frac{1}{2} pK_w = 7 \right]$$

Or,
$$pH = \frac{1}{2} pK_w + \frac{1}{2} pK_a - \frac{1}{2} pK_b \tag{13.19f}$$

This is the expression of pH of a salt solution produced from weak acid and weak base. It is interesting to note that the above expression does not contain any concentration term. So, pH of weak acid/weak base salt solution is **independent of concentration of either acid or base**.

(i) If acid is **stronger** than base, $K_a > K_b$ and hence $pK_a < pK_b$. Thus, pH < 7 hence the salt solution is **acidic**.

(ii) If acid is **weaker** than base, $K_a < K_b$ and hence $pK_a > pK_b$. Thus, pH > 7 and hence the salt solution is **basic**.

13.8 INDICATORS

In any acid-base titration it is difficult to identify the end point. So some specific substances are used to identify the end point of titration. These substances have special characteristics to change their colour within a certain pH range. These substances are called indicators. Thus an indicator changes its colour depending on the concentration of hydrogen ion in the solution. Some indicators change their colour in the acidic pH-range and the others change their colour in the alkaline pH-range (see Table 13.2). Several researchers tried hard to establish the theory according to which an indicator changes its colour. In 1891, W. Ostwald first established a theory according to which in alkaline medium each indicator forms an anion, which co-exists with its **tautomeric** form with different colours. Concentration of one tautomeric form dominates over the other in acidic medium and the solution appears in the colour of that tautomeric form. Conversely, in alkaline medium the other tautomeric form dominates and hence the colour of the solution changes near the end point of titration. For example, phenolphthalein dissociates in aqueous solution according to the equation given:

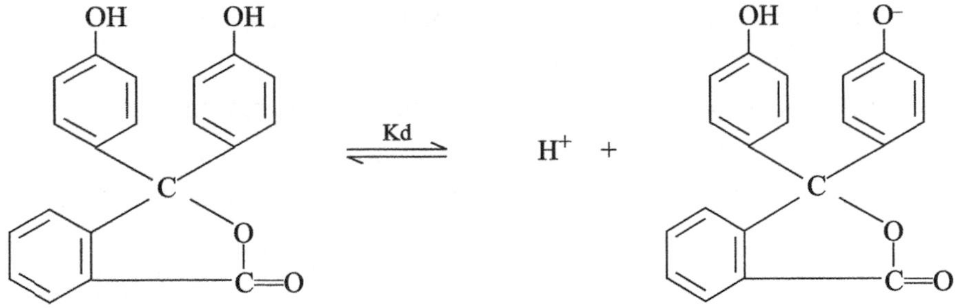

Phenolphthalein
(HIn)

(In_I^-)

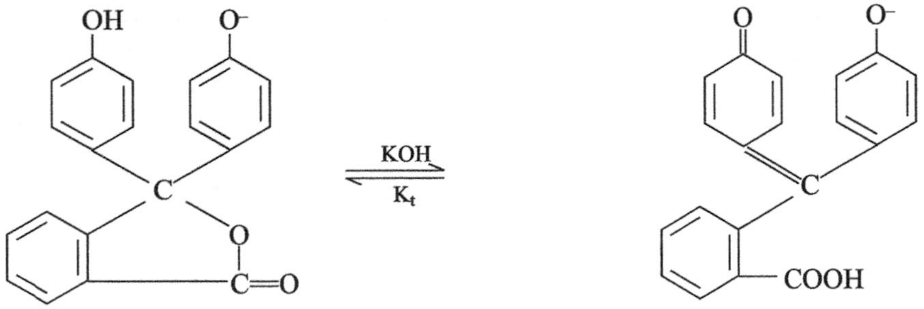

(In_I^-)

(In_{II}^-)

K_d and K_t are given by

$$K_d = \frac{C_{H^+} \cdot C_{In_I^-}}{C_{HIn}} \quad \text{and} \quad K_t = \frac{C_{In_{II}^-}}{C_{In_I^-}}$$

Multiplying K_d and K_t gives the indicator constant K_{In} as shown below

$$K_{In} = K_d \cdot K_t = \frac{C_{H^+} \cdot C_{In_{II}^-}}{C_{HIn}}. \quad \text{Or,} \quad C_{H^+} = K_{In} \frac{C_{HIn}}{C_{In_{II}^-}}$$

Taking $\log_{10}$ on both sides we get

$$\log C_{H^+} = \log K_{In} + \log \frac{C_{HIn}}{C_{In_{II}^-}} \tag{13.20a}$$

Or,
$$pH = pK_{In} + \log \frac{[In_{II}^-]}{[HIn]} \tag{13.20b}$$

where, $pK_{In} = -\log K_{In}$.

It has been established experimentally that when $[In_{II}^-] = 10[HIn]$, the colour of $[In_{II}^-]$ predominates over that of [HIn] and vice versa.

Thus at $pH = pK_{In} + \log_{10} = pK_{In} + 1$, the colour of $[In_{II}^-]$ predominates.

At $pH = pK_{In} - \log 10 = pK_{In} - 1$, the colour of HIn predominates.

Thus the colour change takes place within the pH range $(pK_{In} - 1)$ to $(pK_{In} + 1)$.

Table 13.2 Selected indicators and their characteristic properties

Indicator	pK_{In}	pH range for colour change	Colour	
			In acid	In alkali
Thymol blue	1.51	1.2 – 2.8	Red	Yellow
Methyl orange	3.70	3.1 – 4.4	Light Pink	Yellow
Methyl red	5.10	4.4 – 6.0	Light Pink	Yellow
Bromocresol green	4.67	3.8 – 5.4	Yellow	Blue
Bromocresol red	5.10	4.2 – 6.3	Red	Yellow
Bromocresol purple	6.30	5.2 – 6.8	Yellow	Purple
Bromocresol blue	3.98	3.0 – 4.6	Yellow	Blue
Bromophenol	6.16	5.2 – 6.8	Yellow	Red
Chlorophenol red	5.98	4.8 – 6.4	Yellow	Red
Phenol red	7.9	6.8 – 8.4	Yellow	Red
Cresol red	8.3	7.2 – 8.8	Yellow	Red
Cresolphthalein	9.4	8.2 – 9.8	Colourless	Red
Phenolphthalein	9.4	8.3 – 10.0	Colourless	Dark Pink
Thymolphthalein	9.4	9.2 – 11.6	Colourless	Blue

13.9 pH-METRIC TITRATION

Selection of indicator in acid-base titration is an important factor but the selection will be easier if one considers the pH curve. So we first consider pH curve of different acid-base titrations.

(a) Titration of a Strong and Weak Acid by a Strong Base (KOH)

pH of strong acid lies around 1. However, pH of weak acid is high as shown in the Fig. 13.1a. pH slowly increases upon addition of strong base. Near equivalence point pH sharply increases to around 11. The change in pH is instantaneous and the curve runs almost vertically as shown in the Fig. 13.1a.

In case of strong acids, sharp change in pH is observed within the pH range 3–11. Thus, any indicator can be used.

In case of weak acids, sharp change in pH is observed within the pH range 7–11. Thus basic indicator, changing colour above pH = 7, is suitable for this type of titration. Phenolphthalein is the right choice for weak acid-strong base type of titration.

The titration results a salt, which is basic in nature. Thus before reaching the end point pH value rises above 7. So if methyl red is used instead of phenolphthalein, the pink colour of weak acid solution (before addition of strong base) changes to light yellow colour well before the end point. So, use of methyl red as an indicator in the weak acid-strong base titration only **misguides** us in detecting the end point.

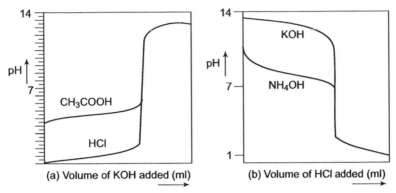

Figure 13.1   Titration curves: (a) Titration of strong and weak acid by a strong base (KOH), (b) Titration of strong and weak base by a strong acid (HCl)

(b) Titration of Strong Base and Weak Base by a Strong Acid (HCl)

pH of strong base lies around 11. However, pH of weak base is low as shown in the Fig. 13.1b. pH slowly decreases upon addition of a strong acid. Near equivalence point pH sharply decreases to around 3. The change in pH is instantaneous and the curve runs almost vertically as shown in the Fig. 13.1b.

In case of strong base, sharp change in pH is observed within the pH range 11–3. Thus, any indicator can be used.

In case of weak base, sharp change in pH is observed within the pH range 5–2. Thus, acidic indicator, changing colour below pH = 7, is suitable for this type of titration. Methyl orange or methyl red indicator is the right choice for weak base-strong acid titration.

The titration results a salt, which is acidic in nature. Thus, before reaching the end point pH value falls below 7. So, if phenolphthalein is used instead of methyl red, the weak base solution gives light pink colour, which changes to colourless well before the end point of titration. So use of phenolphthalein as an indicator in the weak acid-strong base titration only **misguides** us in detecting the end point.

(c) Titration of Weak Acid by Weak Base

During titration of a weak acid by a weak base the pH changes very slowly and usually lies near about 7. The pH curve does not have any vertical section, that means, there is **no sharp change in pH** near the end point of titration. Thus practically no indicator is suitable to detect the end point. So, weak acid-weak base titration **cannot be performed** accurately by volumetric method. In this case usually **conductometric method** is used to detect the end point of titration.

13.10 SOLUBILITY PRODUCT

There are many salts, which are sparingly soluble in water. These type of salts are known as **sparingly soluble salts.** The part of the salt, which gets dissolved in water, dissociates completely into corresponding ions. When salt achieves its **maximum** solubility in water the solution is said **saturated solution** of that salt. In any saturated solution the solid salt remains in equilibrium with its dissolved ions. Let us consider a sparingly soluble salt of formula A_xB_y which remains in equilibrium with its ions in aqueous solution as shown below:

$$A_xB_y(s) \rightleftharpoons xA^{y+} + yB^{x-} \tag{13.21a}$$

The free energy change (ΔG) is given by

$$\Delta G = (x\mu_{A^{y+}} + y\mu_{B^{x-}}) - \mu_{A_xB_y(s)}$$

At equilibrium, i.e., at saturated condition $\Delta G = 0$. Hence we have

$$\mu_{A_xB_y(s)} = (x\mu_{A^{y+}} + y\mu_{B^{x-}}) \tag{13.21b}$$

Again we know that $\mu_i = \mu_i^0 + RT \ln a_i$

Hence, $$\mu_{A_xB_y(s)} = \mu_{A_xB_y(s)}^0 + RT \ln a_{A_xB_y(s)} = \mu_{A_xB_y(s)}^0$$

$$= \text{Constant [at a particular temperature]} = C$$

[Since $a_{A_xB_y(s)} = 1$]

Similarly we can write

$$\mu_{A^{y+}} = \mu_{A^{y+}}^0 + RT \ln a_{A^{y+}} \quad \text{and} \quad \mu_{B^{x-}} = \mu_{B^{x-}}^0 + RT \ln a_{B^{x-}}$$

Thus the equation (13.21b) becomes

$$C = (x\mu_{A^{y+}}^0 + RTx \ln a_{A^{y+}}) + (y\mu_{B^{x-}}^0 + RTy \ln a_{B^{x-}})$$

Or, $$(x\mu_{A^{y+}}^0 + y\mu_{B^{x-}}^0) + RT \ln a_{A^{y+}}^x \cdot a_{B^{x-}}^y = C \tag{13.21c}$$

As $(x\mu_{A^{y+}}^0 + y\mu_{B^{x-}}^0)$ is constant, $(a_{A^{y+}}^x \cdot a_{B^{x-}}^y) = \text{Constant} = K_{aP}$

Or, $$K_{aP} = a_{A^{y+}}^x \cdot a_{B^{x-}}^y = (C_{A^{y+}}^x \cdot C_{B^{x-}}^y)(\gamma_{A^{y+}}^x \cdot \gamma_{B^{x-}}^y) = K_{SP}(\gamma_{A^{y+}}^x \cdot \gamma_{B^{x-}}^y) \tag{13.21d}$$

K_{aP} is known as solubility product. γ is the activity coefficient and K_{SP} is the solubility product of a sparingly soluble salt in terms of concentration.

The solution of a weakly soluble salt may be considered as a dilute solution. In that case we can write that

$\gamma_{A^{y+}} = \gamma_{B^{x-}} = 1$ and hence the equation (13.21d) becomes

$$K_{aP} = K_{SP} = C_{A^{y+}}^x \cdot C_{B^{x-}}^y = [A^{y+}]^x [B^{x-}]^y \tag{13.21e}$$

If S is the solubility of the salt $A_xB_y(s)$, we can write $[A^{y+}] = xS$ and $[B^{x-}] = yS$ [Equation (13.21a)]. Hence, the equation (13.21e) can be written as

$$K_{SP} = (xS)^x (yS)^y = S^{x+y} x^x y^y \qquad (13.21f)$$

13.10.1 Solubility Product of Sparingly Soluble Salts

(a) Consider a uni-univalent sparingly soluble salt, AgCl, which dissociates in aqueous solution according to the following equation:

AgCl(s) $\rightleftharpoons$ Ag$^+$ + Cl$^-$. According to equation (13.21f), $x = y = 1$.

Thus, $K_{SP} = S^2$

(b) Consider another sparingly soluble salt, Ag_2S, which dissociates in aqueous solution according to the following equation:

$Ag_2S(s) \rightleftharpoons 2Ag^+ + S^{2-}$. According to equation (13.16f), $x = 2$, $y = 1$. $[S^{2-}] = S$, $[Ag^+] = 2S$.

Thus, $K_{SP} = (2S)^2(S) = 4S^3$

13.10.2 Common Ion Effect

Presence of common ion in aqueous solution markedly influence the solubility of a sparingly soluble salt. Consider the equation (13.21a)

$$A_xB_y(s) \rightleftharpoons xA^{y+} + yB^{x-} \qquad (13.21a)$$

If S is the solubility of the salt in aqueous solution, solubility product is

$$[A^{y+}] = xS \text{ and } [B^{x-}] = yS. \text{ Hence, } K_{SP} = (xS)^x (yS)^y \qquad (13.21f)$$

Now a metal salt M_xB, having a common ion B^{x-}, with concentration C gm mole/l, is added to the solution. Thus according to the Le Chatelier's principle the backward reaction is favoured due to excess anion concentration and hence the solubility, S gm mole/l, decreases to S' gm mole/l. Thus

$$[A^{y+}] = xS' \text{ and } [B^{x-}] = (yS' + C). \text{ Again, } (yS' + C) \approx C, \text{ as } C \gg S'.$$

$$K_{SP} = (xS')^x (C)^y \qquad (13.22)$$

Comparing equations (13.21f) and (13.22), we get

$$(xS')^x (C)^y = (xS)^x (yS)^y. \quad \text{Or,} \quad (S')^x \cdot C^y = S^{x+y} \cdot y^y \qquad (13.23)$$

The equation (13.23) shows how the solubility of a sparingly soluble salt is influenced in presence of a common ion.

Case-I Solubility of Ag_2S in presence of common anion S^{2-}.

In case of Ag_2S, $x = 2$ and $y = 1$. $[Ag^+] = 2S$ and $[S^{2-}] = S$.

After addition of Na_2S, $[Ag^+] = 2S'$ and $[S^{2-}] = S' + C \approx C$ as $C \gg S'$.

Thus equation (13.23) becomes

$$(S')^2 \cdot C = S^{2+1}. \quad \text{Or}, \quad S' = \frac{S^{3/2}}{\sqrt{C}} \tag{13.24a}$$

where, S = solubility of the salt Ag_2S before adding Na_2S.

S' = solubility of the salt Ag_2S after adding Na_2S; C = Concentration of Na_2S.

Case-II Solubility of Ag_2S in presence of common cation Ag^+.

In case of Ag_2S, $x = 2$ and $y = 1$. $[Ag^+] = 2S$ and $[S^{2-}] = S$

After addition of $AgNO_3$, $[Ag^+] = (2S' + C) \approx C$ as $C \gg S'$ and $[S^{2-}] = S'$.

Thus equation (13.23) becomes

$$(C)^2 \cdot S' = (2S)^2 \cdot S. \quad \text{Or}, \quad S' = \frac{4S^3}{C^2} \tag{13.24b}$$

where, S = solubility of the salt Ag_2S before adding Na_2S.

S' = solubility of the salt Ag_2S after adding Na_2S; C = Concentration of $AgNO_3$.

Case-III Solubility of Ag_2S in presence of complexing agent $Na_2S_2O_3$.

In presence of $Na_2S_2O_3$ the cation Ag^+ readily forms complex according to the following equation:

$$Ag^+ + 2S_2O_3^{2-} \longrightarrow \left[Ag(S_2O_3^{2-})_2\right]^{3-}. \text{ The complex dissociates into ions reversibly}$$

as follows:

$$\{Ag(S_2O_3^{2-})_2\}^{3-} \rightleftharpoons Ag^+ + 2S_2O_3^{2-}. \text{ Instability constant}, K_{IC} \text{ is given by}$$

$$K_{IC} = \frac{[Ag^+][S_2O_3^{2-}]^2}{\left[\{Ag(S_2O_3^{2-})_2\}^{3-}\right]} = 1.02 \times 10^{-13},$$

$$[Ag^+] = \frac{\left[\{Ag(S_2O_3^{2-})_2\}^{3-}\right]}{[S_2O_3^{2-}]^2} \times 1.02 \times 10^{-13}$$

As the value of K_{IC} is very low, it may be assumed that all Ag^+ ions remain as $\{Ag(S_2O_3^{2-})_2\}^{3-}$. The solubility of Ag_2S changes from S gm/lit to S' gm/lit.

$$[Ag^+] = [\{Ag(S_2O_3^{2-})_2\}^{3-}] = 2S', \quad [S^{2-}] = S' \text{ and } K_{SP} = [Ag^+]^2[S^{2-}]$$

Putting the value of $[Ag^+]$ in the above equation, we get

$$K_{SP} = \left\{\frac{\left[\{Ag(S_2O_3^{2-})_2\}^{3-}\right]^2}{[S_2O_3^{2-}]^4} \times 1.02^2 \times 10^{-26}\right\}[S^{2-}]$$

Or, $$K_{SP} = (2S')^2(S')\frac{1.02^2 \times 10^{-26}}{[S_2O_3^{2-}]^4} \tag{13.25a}$$

$[S_2O_3^{2-}]$ = Concentration of free thiosulfate ion = $(C - 4S') \approx C$, since, $C \gg 4S'$.

where, C is the concentration of $Na_2S_2O_3$ added in the system.

$4S'$ = Concentration of bound thiosulfate ion as $\{Ag(S_2O_3^{2-})_2\}^{3-}$ since, 1 mole salt contains 2 moles of $S_2O_3^{2-}$. So, $2S'$ moles of $\{Ag(S_2O_3^{2-})_2\}^{3-}$ contains $4S'$ moles of $S_2O_3^{2-}$.

Thus, equation (13.25a) becomes

$$K_{SP} = (2S')^2 (S') \times \frac{1.02^2 \times 10^{-26}}{C^4}. \text{ Or, } 4S'^3 = \frac{K_{SP} \times C^4}{1.02^2 \times 10^{-26}} \qquad (13.25b)$$

K_{SP} and C are known. Hence, S' can easily be calculated using equation (13.25b).

Case-IV Solubility of Ag_2S in presence of electrolyte, having no common ion.

S_0 = Solubility of the salt Ag_2S in absence of any foreign ions.

$Ag_2S(s) \rightleftharpoons 2Ag^+ + S^{2-} \cdot K_{SP} = (2S_0)^2 \, S_0 = 4S_0^3$. In terms of activity solubility product is given by

$$K_{aP} = (2S_0\gamma_0)^2 (S_0\gamma_0) = 4S_0^3\gamma_0^3 \qquad (13.26a)$$

where γ_0 is the mean ionic activity.

Now an electrolyte, having no common ion, is added. Solubility S_0 changes to $S_\pm$ and the mean ionic activity γ_0 changes to $\gamma_\pm$ but K_{aP} remains constant. Thus we can write that

$$K_{aP} = (2S_\pm\gamma_\pm)^2 (S_\pm\gamma_\pm) = 4S_\pm^3\gamma_\pm^3 \qquad (13.26b)$$

Combining the above two equations (13.26a) and (13.26b) we get

$$4S_\pm^3\gamma_\pm^3 = 4S_0^3\gamma_0^3. \text{ Or, } \frac{S_\pm}{S_0} = \frac{\gamma_0}{\gamma_\pm}$$

Taking $\log_{10}$ on both sides, we get

$$\log\frac{S_\pm}{S_0} = \log\gamma_0 - \log\gamma_\pm \qquad (13.26c)$$

Again, according to Debye-Hückel equation, we have

$$\log\gamma_0 = -AZ_+Z_-\sqrt{I_0} \text{ and } \log\gamma_\pm = -AZ_+Z_-\sqrt{I_\pm}$$

Hence, $\log\gamma_0 - \log\gamma_\pm = -AZ_+Z_-(\sqrt{I_0} - \sqrt{I_\pm})$ \qquad (13.26d)

With the addition of an electrolyte, **ionic strength** of the solution increases and hence $I_\pm > I_0$.

Thus, $(\log\gamma_0 - \log\gamma_\pm) > 0$ and hence, $S_\pm > S_0$ according to equation (13.26c).

Inference Solubility of a sparingly soluble salt **increases** with the addition of a foreign electrolyte having **no common ion**.

The variation of solubility of a sparingly soluble salt with the addition of an electrolyte is well illustrated in the Fig. 13.2.

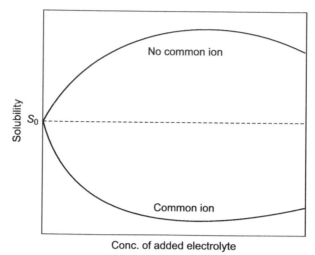

Figure 13.2   Effect of addition of an electrolyte to the solution of a sparingly soluble salt

SOLVED PROBLEMS

Problem 13.1   Calculate the concentrations of OH⁻ and H_3O^+ ions in 0.01 M solution of HCl.

Solution

Dissociation of H_2O and HCl are as follows:

$$H_2O + H_2O \rightleftharpoons H_3O^+ + OH^- \text{ and } HCl + H_2O \rightarrow H_3O^+ + Cl^-$$

α = Degree of dissociation of water. $[H_3O^+] = (0.01 + \alpha)$ and $[OH^-] = \alpha$

$K_w = [H_3O^+][OH^-] = (0.01 + \alpha)(\alpha) = 0.01\alpha$, since, $0.01 + \alpha \approx 0.01$

Or, $0.01\alpha = K_w = 10^{-14}$. Hence, $[OH^-] = \alpha = 10^{-12}$ M and $[H_3O^+] = 0.01$ M.

Problem 13.2   Will a precipitate form when 0.15 lit. of 3.0×10^{-2} M $Pb(NO_3)_2$ solution is added to 300 ml of a 8.0×10^{-2} M NaCl solution? Given K_{sp} of $PbCl_2$ is 2.4×10^{-4}.

Solution

$[Pb(NO_3)_2] = 3.0 \times 10^{-2}$ M. Hence, number of moles of $Pb^{2+} = 3.0 \times 10^{-2} \times 0.15 = 4.5 \times 10^{-3}$

$[NaCl] = 8.0 \times 10^{-2}$. Hence, number of moles of $Cl^- = 8.0 \times 10^{-2} \times 0.30 = 2.4 \times 10^{-3}$.

Total volume is 450 ml = 0.45 lit.

$$[Pb^{2+}] = \frac{4.5 \times 10^{-3}}{0.45} M = 0.010 \text{ M}, [Cl^-] = \frac{2.4 \times 10^{-3}}{0.45} M = 0.053 \text{ M}$$

$[Pb^{2+}] \times [Cl^-]^2 = 0.01 \times 0.053^2 = 2.8 \times 10^{-5}$. So, $[Pb^{2+}] \times [Cl^-]^2 < K_{sp}$. Hence, no precipitate will be formed.

Problem 13.3 What is the solubility of silver phosphate in a 0.20 M silver nitrate solution? Given $K_{sp} = 1.1 \times 10^{-16}$.

Solution

$$Ag_3PO_4(s) \rightleftharpoons 3Ag^+ + PO_4^{3-}. \; K_{sp} = [Ag^+]^3 \, [PO_4^{3-}] = 1.1 \times 10^{-16}$$

S = Solubility in presence of common cation Ag^+. Thus, $[PO_4^{3-}] = S$ and $[Ag^+] = (0.2 + 3S) \approx 0.2$ since, $3S \ll 0.2$

Thus, $(0.2)^3 \times S = 1.1 \times 10^{-16}$. Or, $S = 1.4 \times 10^{-14} \, M$.

Problem 13.4 What is the solubility of Fe^{2+} in a solution with a pH of 9.00?
Given: K_{sp} of $Fe(OH)_2$ is 7.9×10^{-15}.

Solution

pH = 9. Or, $pOH = 5$. Hence, $[OH^-] = 10^{-5} \, N = 10^{-5} \, M$

$Fe(OH)_2(s) \rightleftharpoons Fe^{2+} + 2OH^-. \; K_{sp} = [Fe^{2+}] \, [OH^-]^2 = 7.9 \times 10^{-15}$

S = Solubility of $Fe(OH)_2(s)$. $[Fe^{2+}] = S$, $[OH^-] = (2S + 10^{-5}) \approx 10^{-5}$

$S \times (10^{-5})^2 = 7.9 \times 10^{-15}$. Or, $S = 7.9 \times 10^{-5}$.

Problem 13.5 Calculate the pH of a solution prepared by adding 0.10 mole of NaOH to a solution of 0.20 M iodic acid, HIO_3.
Given: K_a of HIO_3 is 0.2.

Solution

pH of an aqueous solution, composed of a weak acid and strong base, is given by

$$pH = pK_a + \log \frac{[salt]}{[acid]}$$

According to the question [salt] = [acid] = 0.1 M. Thus, pH = 0.7. But this is wrong. The problem is that HIO_3 is rather a "strong" weak acid, so the value of $[H^+]$ should be calculated using the following polynomial equation:

$$[H^+]^2 + ([acid] + K_a) \, [H^+] - [acid] \, K_a = 0$$

Putting the values of [acid] and K_a in the above equation, we get $[H^+] = 0.056 \, M$. Hence, pH = 1.2.

Problem 13.6

(i) Calculate the pH of a solution containing 0.01 mole of ammonium chloride and 0.02 mole of ammonia in 100 ml of solution. Given: $pK_b = 4.7$.

(ii) What will be the change in the pH if 10 ml of 0.100 M HCl is added to 100 ml of the above solution?

Solution

(i)
$$\frac{[NH_4^+][OH^-]}{[NH_3]} = 10^{-4.7}$$

The nominal concentrations of the acid and conjugate base are respectively [acid] = 0.1 M and [base] = 0.2 M.

Considering the mass balance equation, we have

$$[NH_4^+] + [NH_3] = [acid] + [base] = 0.3 \text{ and } [Cl^-] = [acid] = 0.1 \ M$$

The charge balance equation gives

$$[Cl^-] + [OH^-] = [NH_4^+]. \text{ Or, } [NH_4^+] = 0.1 + [OH^-] \approx 0.1$$

$$[NH_3] = 0.3 - [NH_4^+] = 0.2.$$

Hence, $\dfrac{0.1[OH^-]}{0.2} = 10^{-4.7}$. Or, $[OH^-] = 4.0 \times 10^{-5}$.

Or, $pOH = 4.4.$ Or, $pH = 9.6.$

(ii) $C_b = \dfrac{20 \text{ mmole} - 1 \text{ mmole}}{110} = 0.173 \text{ M} \approx [NH_3],$

$C_a = \dfrac{10 \text{ mmole} + 1 \text{ mmole}}{110} = 0.1 \text{ M} \approx [NH_4^+]$

$\dfrac{0.1[OH^-]}{0.173} = 10^{-4.7}.$ Or, $[OH^-] = 1.73 \times 2 \times 10^{-5}.$

Or, $pOH = 4.5.$ Or, $pH = 9.5.$

Change in pH = 0.1 unit.

EXERCISES

1. Compute the degree of dissociation and percent dissociation of acetic acid in its 0.1 M solution.

    Given: $K_a = 1.8 \times 10^{-5}$. [*Ans.:* 1.34%]

2. Calculate the pH of 0.1 M aqueous solution of acetic acid.

    Given K_a of acetic acid is 1.85×10^{-5} and $\alpha = 0.0134$. [*Ans.:* 2.87]

3. Calculate the pH and degree of dissociation of a solution of 0.1 M acetic acid containing 0.1 M sodium acetate. Given K_a of acetic acid is 1.85×10^{-5} mole/dm^3.

    [*Ans.:* pH = 4.73, $\alpha = 0.000185$]

4. Calculate the pH of ammonium hydroxide–ammonium chloride buffer solution, prepared by 0.1 M ammonium hydroxide solution and 0.01 M ammonium chloride solution. Given pK$_b$ of NH$_4$OH is 9.25. [*Ans.:* pOH = 8.25]

5. The solubility of calcium sulphate in water is 4.9×10^{-3} mole/dm^3 at 298°K. Calculate K_{sp} for CaSO$_4$ at this temperature. [*Ans.:* 2.4×10^{-5} mole^2 dm^{-6}]

6. Calculate the molar solubility of AgI in a solution containing 0.1M AgNO$_3$. The solubility product of silver iodide, (AgI) is 8.5×10^{-17} mole^{2}dm^{-6} at 298°K. Also calculate the molar solubility of silver iodide in absence of any common ion.

    [*Ans.:* 8.5×10^{-16} mole/dm^3, 9.2×10^{-9} mole/dm^3]

7. Calculate the solubility of Bi$_2$S$_3$ in water at 298°K. Given $K_{sp} = 1.0 \times 10^{-97}$ mole^5 dm^{-5}.

8. Calculate the solubility of AgI in 0.10 M NaI at 298°K. Given $K_{sp} = 8.5 \times 10^{-7}$ at this temperature.

9. Calculate the pH of a 0.01 M solution of ammonium chloride in pure water.

    Hint: $[NH_4^+] + [H^+] = [OH^-] + [Cl^-] \approx [Cl^-] = 0.01\ M$

    $$NH_4^+ \rightleftharpoons H^+ + NH_3 \text{ and } [NH_3] = [H^+]$$

    $$\frac{[H^+][NH_3]}{[NH_4^+]} = K_a = 5.0 \times 10^{-10}.$$

    Or, $\qquad [H^+] = K_a \dfrac{[NH_4^+]}{[NH_3]}.$ Or, $[H^+]^2 = K_a\{0.01 - [H^+]\} \approx K_a \times 0.01$

    $$[H^+] = 2.2 \times 10^{-6}.$$

10. Estimate the pH of a 0.01 M solution of ammonium format in water.

    Given $K_{NH_4^+} = K_1 = 10^{-9.3}$, $K_{HCOOH} = K_a = 10^{-3.7}$

    Hint: $NH_4^+ + H_2O \rightleftharpoons NH_3 + H_3O^+ \qquad K_1 = 10^{-9.3}$

    $HCOO^- + H_2O \rightleftharpoons HCOOH + OH^- \qquad K_a = 10^{-3.7}$

$$NH_4^+ + HCOO^- \rightleftharpoons NH_3 + HCOOH \quad K_3 = \frac{K_1}{K_a} = 10^{-5.6}$$

$$[NH_4^+] = [HCOO^-] \text{ and } [NH_3] = [HCOOH]$$

$$K_3 = \frac{[NH_3][HCOOH]}{[NH_4^+][HCOO^-]} = \frac{[HCOOH]^2}{[HCOO^-]^2} = 10^{-5.6}$$

Or, $$\frac{[HCOOH]}{[HCOO^-]} = \frac{[H^+]}{K_a} = 10^{-2.8}$$

Or, $$[H^+] = K_a \times 10^{-2.8} = 10^{-6.5}.$$

14 Chemical Equilibrium

14.1 LAW OF MASS ACTION

According to the principle of chemical kinetics, rate of a reaction is directly proportional to the active mass of the reactants. Let us consider a general reversible reaction of type shown below

$$lL + mM + \cdots \rightleftharpoons yY + zZ + \cdots$$

The reaction is carried out in a closed container in order to reach the equilibrium. The rate expressions for the forward and backward reactions, r_f and r_b, are given by

$$r_f = k_f a_L^l \cdot a_M^m \cdots \text{ and } r_b = k_b a_Y^y \cdot a_Z^z \cdots$$

When equilibrium is established, the rate of forward reaction is equal to the rate of backward reaction. Thus we can write that

$$r_f = r_b. \quad \text{Or,} \quad k_f a_L^l \cdot a_M^m \cdots = k_b a_Y^y \cdot a_Z^z \cdots \text{ Hence, we have}$$

$$K_a = \frac{k_f}{k_b} = \frac{a_Y^y \cdot a_Z^z \cdots}{a_L^l \cdot a_M^m \cdots} \tag{14.1}$$

The term, K_a, is known as equilibrium constant.

For a homogeneous gas phase reaction, activity term may be replaced by partial pressure term. Thus the above equation can be written as

$$K_p = \frac{k_f}{k_b} = \frac{p_Y^y \cdot p_Z^z \cdots}{p_L^l \cdot p_M^m \cdots} \tag{14.2}$$

Assuming that all the reactants and products are in gas phase and behave ideally, we can write

$p_i = c_i RT$, where, p_i and c_i are the partial pressure and concentration of ith component respectively. Thus at constant temperature,

$$p_L^l \cdot p_M^m \cdots = (c_L^l \cdot c_M^m \cdots)(RT)^{n_1}, \text{ where, } n_1 = l + m + \cdots$$

$$p_Y^y \cdot p_Z^z \cdots = (c_Y^y \cdot c_Z^z \cdots)(RT)^{n_2}, \text{ where, } n_2 = y + z + \cdots \quad \Delta n = (n_2 - n_1)$$

Putting the values of partial pressure terms in the equation (14.2), we get

$$K_p = \frac{(c_Y^y \cdot c_Z^z \cdots)(RT)^{n_2}}{(c_L^l \cdot c_M^m \cdots)(RT)^{n_1}} = \frac{(c_Y^y \cdot c_Z^z \cdots)}{(c_L^l \cdot c_M^m \cdots)} \times (RT)^{\Delta n} = K_c \times (RT)^{\Delta n} \quad (14.3)$$

where, $$K_c = \frac{(c_Y^y \cdot c_Z^z \cdots)}{(c_L^l \cdot c_M^m \cdots)} \quad (14.4)$$

Conclusion For any chemical reaction performed at constant temperature,
if $\Delta n > 0$, $K_p > K_c$, if $\Delta n < 0$, $K_p < K_c$ and if $\Delta n = 0$, $K_p = K_c$.
Again from Avogadro's law, we have
$p_i = x_i P$ where, p_i and x_i are the partial pressure and mole fraction of ith component respectively at equilibrium and P is the total pressure.

$$p_L^l \cdot p_M^m \cdots = (x_L^l \cdot x_M^m \cdots)(P)^{n_1}, \text{ where, } n_1 = l + m + \cdots$$

$$p_Y^y \cdot p_Z^z \cdots = (x_Y^y \cdot x_Z^z \cdots)(P)^{n_2}, \text{ where, } n_2 = y + z + \cdots \quad \Delta n = (n_2 - n_1)$$

Thus replacing partial pressure terms by mole fraction terms in equation (14.2), we get

$$K_p = \frac{(X_Y^y \cdot X_Z^z \cdots)(P)^{n_2}}{(X_L^l \cdot X_M^m \cdots)(P)^{n_1}} = \frac{(X_Y^y \cdot X_Z^z \cdots)}{(X_L^l \cdot X_M^m \cdots)} \times P^{\Delta n} = K_x \times (P)^{\Delta n} \quad (14.5)$$

where, $$K_x = \frac{(X_Y^y \cdot X_Z^z \cdots)}{(X_L^l \cdot X_M^m \cdots)} \quad (14.6)$$

Conclusion For any chemical reaction performed at constant pressure,
if $\Delta n > 0$, $K_p > K_X$, if $\Delta n < 0$, $K_p < K_X$ and if, if $\Delta n = 0$, $K_p = K_X$.
Combining equations (14.3) and (14.5), we get

$$K_c \times (RT)^{\Delta n} = K_x \times (P)^{\Delta n}. \text{ Or, } K_c = \frac{K_x}{(RT/P)^{\Delta n}}. \text{ Or, } K_x = K_c(\bar{V})^{\Delta n} \quad (14.7)$$

where, $\bar{V}$ is the molar volume of the system.

14.2 HOMOGENEOUS SYSTEM

14.2.1 Calculation of Equilibrium Constant (ICE Method)

Example 14.2.1

Let us consider the following reaction:

$$2SO_2(g) + O_2(g) \rightleftharpoons 2SO_3(g).$$

$$K_c = \frac{[SO_3]^2}{[SO_2]^2[O_2]} \quad \text{and} \quad K_p = \frac{p_{SO_3}^2}{p_{SO_2}^2 \cdot p_{O_2}} = \frac{X_{SO_3}^2}{X_{SO_2}^2 \cdot X_{O_2}} \times P^{-1}$$

Or, $\quad K_p = \dfrac{n_{SO_3}^2}{n_{SO_2}^2 \cdot n_{O_2}} \times \left(\dfrac{n_0}{P}\right)$. Or, $\quad \dfrac{n_{SO_3}}{n_{SO_2}} = \sqrt{K_p \cdot P \left(\dfrac{n_{O_2}}{n_0}\right)}$

Assuming that the ratio $\left(\dfrac{n_{O_2}}{n_0}\right)$ is constant, we have

$$\frac{n_{SO_3}}{n_{SO_2}} \propto \sqrt{P}$$

Thus, the yield of sulphur trioxide increases with increase in total pressure.

Table 14.1 ICE Table

Concentration	[SO₂] in M	[O₂] in M	[SO₃] in M
Initially	a	b	0
Change	$-2x$	$-x$	$+2x$
At Equilibrium	$(a-2x)$	$(b-x)$	x

Total number of moles is $= (a-2x) + (b-x) + 2x = a + b - x$

$$[SO_3] = 2x, \quad [SO_2] = (a-2x), \quad [O_2] = (b-x)$$

$$X_{SO_3} = \frac{2x}{(a+b-x)}, \quad X_{SO_2} = \frac{(a-2x)}{(a+b-x)}, \quad X_{O_2} = \frac{(b-x)}{(a+b-x)}$$

$$K_c = \frac{[SO_3]^2}{[SO_2]^2[O_2]} = \frac{4x^2}{(a-2x)^2(b-x)}, \quad K_x = \frac{4x^2(a+b-x)}{(a-2x)^2(b-x)}$$

Example 14.2.2

A closed system initially containing 0.001 (M) of $H_2(g)$ and 0.002 (M) of $I_2(g)$ to execute the reaction: $H_2(g) + I_2(g) \rightleftharpoons 2HI(g)$. The equilibrium is reached at 448°C.

Equilibrium concentration of HI(g) is observed 0.00187 (M). Calculate K_c at 448°C for the reaction.

Table 14.2 ICE Table

Concentration	[H₂] in M	[I₂] in M	[HI] in M
Initially	a	b	0
Change	$-x$	$-x$	$+2x$
At Equilibrium	$(a-x)$	$(b-x)$	$+2x$

Total number of moles $= (a-x)+(b-x)+2x = (a+b)$

$$X_{H_2} = \frac{(a-x)}{(a+b)}, \quad X_{I_2} = \frac{(b-x)}{(a+b)}, \quad X_{HI} = \frac{2x}{(a+b)}$$

$$K_P = K_X = \frac{4x^2}{(a-x)(b-x)}$$

$a = 0.001, \quad b = 0.002, \quad [HI] = 2x = 1.87 \times 10^{-3}, \quad \text{or,} \quad x = 9.35 \times 10^{-4}$

$[H_2] = (a-x) = 6.5 \times 10^{-5}, \quad [I_2] = (b-x) = 1.065 \times 10^{-3}$

$$K_c = \frac{[HI]^2}{[H_2][I_2]} = \frac{[1.87 \times 10^{-3}]^2}{[6.5 \times 10^{-5}][1.065 \times 10^{-3}]} = 51$$

Example 14.2.3

$$N_2O_4(g) \rightleftharpoons 2NO_2(g).$$

Table 14.3 ICE Table

Concentration	[N₂O₄] in M	[NO₂] in M
Initially	a	0
Change	$-x$	$+2x$
At Equilibrium	$(a-x)$	$+2x$

Total number of moles $= (a-x)+2x = (a+x)$

$$[N_2O_4] = (a-x), \quad [NO_2] = 2x$$

$$X_{N_2O_4} = \frac{(a-x)}{(a+x)}, \quad X_{NO_2} = \frac{2x}{(a+x)}, \quad K_X = \frac{X_{NO_2}^2}{X_{N_2O_4}} = \frac{4x^2}{(a^2-x^2)}$$

$$K_P = \frac{p_{NO_2}^2}{p_{N_2O_4}}, \quad K_P = K_X \cdot P = \frac{4x^2}{(a^2-x^2)} \times P$$

Assume that $x \ll a$, we can write that $(a^2 - x^2) \approx a^2$.

Hence

$$\frac{x}{a} = \sqrt{\frac{K_P}{4P}}$$

$$K_c = \frac{k_f}{k_b} = \frac{[NO_2]^2}{[N_2O_4]} = \frac{(2x)^2}{(a-x)}$$

14.3 HETEROGENEOUS SYSTEM

14.3.1 Solid-Gas System

Example 14.3.1

If more than one phase is present in the equilibrium system, the latter is called heterogeneous equilibrium system. If the system contains solid phase, activity term of that solid phase is taken as unity. One example is dissociation of $CaCO_3$ in a closed system.

$$CaCO_3(s) \rightleftharpoons CaO(s) + CO_2(g)$$

$$a_{CaCO_3} = a_{CaO} = 1 \text{ and } a_{CO_2} = p_{CO_2}$$

$$K_a = \frac{a_{CaO} \cdot a_{CO_2}}{a_{CaCO_3}} = a_{CO_2} = p_{CO_2} = K_P$$

Example 14.3.2 Conversion of steam to hydrogen.

$$3\,Fe(s) + 4\,H_2O(g) \rightleftharpoons Fe_3O_4(s) + 4\,H_2(g)$$

$$K_a = \frac{a_{Fe_3O_4} \cdot a_{H_2}^4}{a_{Fe}^3 \cdot a_{H_2O}^4} = \frac{a_{H_2}^4}{a_{H_2O}^4} = \frac{p_{H_2}^4}{p_{H_2O}^4} = K_P$$

Example 14.3.3 Production of water gas.

$$C(s) + H_2O(g) \rightleftharpoons CO(g) + H_2(g)$$

$$K_a = \frac{a_{CO} \cdot a_{H_2}}{a_C \cdot a_{H_2O}} = \frac{p_{CO} \cdot p_{H_2}}{p_{H_2O}} = \frac{p_{H_2}^2}{p_{H_2O}} \quad [\text{since, } p_{CO} = p_{H_2}]$$

14.3.2 Liquid-Liquid System

Example 14.3.4

$$Cd^{2+}(aq) + 4Br^-(aq) \rightleftharpoons CdBr_4^{2-}(aq)$$

Table 14.4 ICE Table

Concentration	$[Cd^{2+}]$ in M	$[Br^-]$ in M	$[CdBr_4^{2-}]$ in M
Initially	a	b	0
Change	$-x$	$-4x$	$+x$
At Equilibrium	$(a-x)$	$(b-4x)$	$+x$

$$[CdBr_4^{2-}] = x, \quad [Cd^{2+}] = (a-x), \quad [Br^-] = (b-4x)$$

$$K_c = \frac{k_f}{k_b} = \frac{[CdBr_4^{2-}]}{[Cd^{2+}][Br^-]^4} = \frac{x}{(a-x)(b-4x)^4}$$

14.4 LE CHATELIER'S PRINCIPLE

Any chemical reaction, occurring in a closed container, reaches an equilibrium state, at which forward reaction rate is same as backward reaction rate. However, reaction does not stop at all and it occurs continuously on both sides at equal rate. Hence the equilibrium is often called **dynamic equilibrium**.

Statement 1 For a given chemical reaction, dynamic equilibrium is established under certain magnitude of parameters, i.e., pressure, temperature, and concentration of products and reactants. If the equilibrium is disturbed by changing any of the parameters slightly, the system automatically adjusts itself to restore the dynamic equilibrium within a short time period. This is known as Le Chatelier's Principle.

Statement 2 If a system, at equilibrium, receives an external stress, the system will react in that direction, relieving the external stress to maintain the value of K_a constant.

Let us consider a general equation: $aA + bB \rightleftharpoons lL + mM$.

The forward reaction rate: $R_f = k_f C_A^a \cdot C_B^b$

The backward reaction rate: $R_b = k_b C_L^l \cdot C_M^m$

At equilibrium, $R_f = R_b$. Or, $k_f C_A^a \cdot C_B^b = k_b C_L^l \cdot C_M^m$

Equilibrium constant $= K_c = \dfrac{k_f}{k_b} = \dfrac{C_L^l \cdot C_M^m}{C_A^a \cdot C_B^b}$

The equilibrium is dynamic, i.e., both forward and backward reactions are occurring spontaneously with equal rates.

14.4.1 Significance of K_c

(a) For a given equilibrium reaction, if $K \gg 1$, the forward reaction is much more favourable than the backward reaction, which means, concentration of products are dominating over that of reactants.

(b) For a given equilibrium reaction, if $K \ll 1$, the backward reaction is much more favourable than the forward reaction, with means, concentrations of reactants are dominating over that of products.

(c) The expression of K_c depends on how the reaction is expressed, e.g., formation of HI(g) from its elements.

$$H_2(g) + I_2(g) \rightleftharpoons 2HI(g), \quad K_c = \frac{[HI]^2}{[H_2][I_2]}$$

Dissociation of HI(g) into its elements

$$2HI(g) \rightleftharpoons H_2(g) + I_2(g), \quad K_c' = \frac{[H_2][I_2]}{[HI]^2} = \frac{1}{K_c}$$

Formation of 1 mole of HI(g) from its elements

$$\frac{1}{2}H_2(g) + \frac{1}{2}I_2(g) \rightleftharpoons HI(g), \quad K_c'' = \frac{[HI]}{[H_2]^{1/2}[I_2]^{1/2}} = \sqrt{K_c}$$

(d) If a chemical reaction occurs in two or more steps, equilibrium constant of the reaction may be calculated from the knowledge of equilibrium constants of individual steps. It is described in the following reactions:

$$2CO_2(g) \rightleftharpoons 2CO(g) + O_2(g), \quad K_1 = \frac{[CO]^2[O_2]}{[CO_2]^2} = 1.6 \times 10^{-11}$$

$$2H_2O(g) \rightleftharpoons 2H_2(g) + O_2(g), \quad K_2 = \frac{[H_2]^2[O_2]}{[H_2O]^2} = 1.3 \times 10^{-10}$$

$$CO(g) + H_2O(g) \rightleftharpoons CO_2(g) + H_2(g)$$

$$K_3 = \frac{[CO_2][H_2]}{[CO][H_2O]} = \sqrt{\frac{K_2}{K_1}} = \sqrt{8.125} = 2.85$$

(e) In case of heterogeneous chemical reaction, equilibrium constant does not contain concentrations of solids and pure liquids, since activities of all solids including metals as well as pure liquids are considered as unity, i.e.,
$a_{solid} = a_{pure\ liquid} = 1$. Hence, corresponding concentration terms are also taken as unity. One example is solubility product of AgCl(s).

$$AgCl(s) \rightleftharpoons Ag^+(aq) + Cl^-(aq), \quad K_a = a_{Ag^+} \cdot a_{Cl^-} = K_{sp} = \text{Solubility product} = S^2$$

K_{sp} is known as solubility product and S is called solubility. $a_{AgCl(s)} = 1 = C_{AgCl(s)}$. $\gamma_{AgCl(s)}$, where, γ is called activity coefficient. In actual practice equilibrium constant for solutions is written in terms of activity.

(f) **Effect of temperature** Chemical reactions are broadly classified into two categories: (i) Exothermic reaction, where, heat is produced in the forward reaction and $\Delta H < 0$. Thus if heat is added to the system, temperature of the system will increase and hence the system will try to consume this additional heat by favouring backward reaction. (ii) Endothermic reaction, where, heat is absorbed during execution of forward reaction and $\Delta H > 0$. Thus if heat is added to the system, temperature of the system will increase and hence the system will try to consume this additional heat by favouring forward reaction.

(g) **Effect of concentration** Let us consider the following reaction in a closed system: $A + B \rightleftharpoons C + D$. At equilibrium, concentrations of all the reactants and products are fixed. Any change in concentration of either reactant or product will drive the system to move in the forward direction or backward direction accordingly. For example, if some amount of reactant A is added into the system, position of equilibrium is shifted towards the right, i.e., forward reaction is favoured in order to consume the excess reactant. Similarly, if some amount of product C is added into the system, position of equilibrium is shifted towards the left, i.e., backward reaction is favoured in order to consume the excess product.

(h) **Effect of pressure** From equation (14.5), we have $K_p = K_x \times (P)^{\Delta n}$.
If the volume of the system is compressed, pressure of the system will increase. Conversely if the volume of the system is expanded, corresponding pressure will decrease.

Case-I $\Delta n > 0$. If the pressure of the system is increased, backward reaction rate will be favoured so that K_x decreases to make K_p constant. Conversely, if the pressure of the system is decreased, forward reaction rate will be favoured so that K_x increases to make K_p constant. One example is

$$PCl_5(g) \rightleftharpoons PCl_3(g) + Cl_2(g).$$

If a reaction occurs with increase in number of moles, forward reaction is favoured at low pressure conditions.

Case-II $\Delta n < 0$. If the pressure of the system is increased, forward reaction rate will be favoured so that K_x increases to make K_p constant. Conversely, if the pressure of the system is decreased, backward reaction rate will be favoured so that K_x decreases to make K_p constant. One example is

$$2SO_2(g) + O_2(g) \rightleftharpoons 2SO_3(g).$$

If a reaction occurs with decrease in number of moles, forward reaction is favoured at high pressure conditions.

Case-III $\Delta n = 0$. The equilibrium remains unchanged with the change in total pressure.

(i) **Effect of catalyst** Presence of catalyst does not change the position of chemical equilibrium since a catalyst speeds up both the forward and backward reaction rates at equal extent so that the ratio of reaction rates remains constant. In some cases reaction occurs at a very slow rate so that the system requires several years to reach the dynamic equilibrium. In that case presence of catalyst helps the system reach the dynamic equilibrium within a short period of time. The effect of catalyst on energy diagram is shown in the Fig. 14.1.

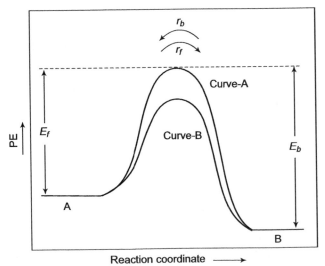

Figure 14.1 Energy diagram of a reaction. Curve-A: In absence of catalyst; Curve-B: In presence of catalyst

14.4.2 Influence of Inert Gas

Let us consider the following reaction:

$$lL + mM \rightleftharpoons dD + eE$$

X_L = Mole fraction of the component, $L = \dfrac{n_L}{n_0}$, X_M = Mole fraction of the component,

$\quad M = \dfrac{n_M}{n_0}$

X_D = Mole fraction of the component, $D = \dfrac{n_D}{n_0}$, X_E = Mole fraction of the component,

$\quad E = \dfrac{n_E}{n_0}$

n_L = Equilibrium concentration of the component, L in moles

n_M = Equilibrium concentration of the component, M in moles

n_D = Equilibrium concentration of the component, D in moles

n_E = Equilibrium concentration of the component, E in moles

n_0 = Total number of moles at equilibrium

Thus at equilibrium, we can write

$$n_0 = n_L + n_M + n_D + n_E$$

$$K_p = K_X P^{\Delta v}, \quad \Delta v = (d + e) - (l + m)$$

Or,
$$K_P = \left(\frac{X_D^d \cdot X_E^e}{X_L^l \cdot X_M^m} \right) P^{\Delta v} = \left(\frac{n_D^d \cdot n_E^e}{n_L^l \cdot n_M^m} \right) \left(\frac{P}{n_0} \right)^{\Delta v}$$

Introduction of any inert gas in the system increases the value of n_0. Inert gas may be added either at constant pressure or at constant volume.

Case-I   Inert gas is added at constant pressure.

(a) For a reaction $\Delta v = 0$. So,

$$K_P = \left(\frac{n_D^d \cdot n_E^e}{n_L^l \cdot n_M^m} \right)$$

Thus addition of inert gas does not disturb the equilibrium system.

(b) For a reaction $\Delta v > 0$. So, on addition of inert gas into the system, n_0 increases and hence $n_0^{\Delta v}$ also increases. Thus to maintain K_p constant, forward reaction is favoured so that n_D, n_E increase and n_L, n_M decrease. So forward reaction rate is increased.

(c) For a reaction $\Delta v < 0$. So, on addition of inert gas into the system, n_0 increases and hence $n_0^{\Delta v}$ decreases. Thus to maintain K_p constant, backward reaction is favoured so that n_D, n_E decrease and n_L, n_M increase. So backward reaction rate is increased.

One example is formation of ammonia in Haber's process.

$$N_2(g) + 3H_2(g) \rightleftharpoons 2NH_3(g). \quad \Delta v = -2$$

Addition of inert gas Ar into the system sufficiently suppresses the formation of ammonia.

Case-II   Inert gas is added at constant volume.

As volume of reaction vessel is fixed, total pressure will increase on addition of inert gas since $P \propto n_0$. Or, $\dfrac{P}{n_0}$ is constant. Thus the value of K_p remains constant. Equilibrium is not at all disturbed.

14.4.3 Effect of Addition of One of the Products

Case-I Addition of one of the products at constant pressure.

Let us consider the following reaction:

$$lL + mM \rightleftharpoons dD + eE$$

$$K_P = K_X P^{\Delta v}, \quad \Delta v = (d + e) - (l + m)$$

Or,

$$K_P = \left(\frac{X_D^d \cdot X_E^e}{X_L^l \cdot X_M^m} \right) P^{\Delta v} = \left(\frac{n_D^d \cdot n_E^e}{n_L^l \cdot n_M^m} \right) \left(\frac{P}{n_0} \right)^{\Delta v}$$

On addition of one of the product, say E, n_E increases as well as n_0.

Thus if $\Delta v = 0$, number of moles of reactants should increase to maintain K_P constant. So, backward reaction rate increases.

If, $\Delta v > 0$, $\left(\dfrac{P}{n_0} \right)$ decreases on addition of the product, E. So, equilibrium system remains undisturbed.

If, $\Delta v < 0$, $\left(\dfrac{P}{n_0} \right)$ increases on addition of the product, E. So, number of moles of reactants should increase to maintain K_P constant. So, backward reaction rate increases.

Case-II Addition of one of the products at constant volume.

$$K_P = \left(\frac{n_D^d \cdot n_E^e}{n_L^l \cdot n_M^m} \right) \left(\frac{P}{n_0} \right)^{\Delta v} = \left(\frac{n_D^d \cdot n_E^e}{n_L^l \cdot n_M^m} \right) \left(\frac{RT}{V} \right)^{\Delta v} \quad \text{since,} \quad PV = n_0 RT$$

Addition of any one of the products in the system results in increasing the value of numerator. So to maintain K_p constant, denominator should also increase and hence backward reaction rate will be favoured.

14.5 COMBINATION OF EQUILIBRIA

Example 14.5.1

Consider the following two separate equilibrium systems:

$$2 H_2O(g) \rightleftharpoons 2 H_2(g) + O_2(g), \quad K_{P_1} = \frac{p_{H_2}^2 \cdot p_{O_2}}{p_{H_2O}^2}$$

$$2CO_2(g) \rightleftharpoons 2CO(g) + O_2(g), \; K_{P_2} = \frac{p_{CO}^2 \cdot p_{O_2}}{p_{CO_2}^2}$$

If steam and carbon dioxide are mixed in a closed chamber at high temperature, water gas reaction occurs as shown below

$$CO_2(g) + H_2(g) \rightleftharpoons CO(g) + H_2O(g)$$

$$K_{P3} = \frac{p_{CO} \cdot p_{H_2O}}{p_{CO_2} \cdot p_{H_2}} = \sqrt{\frac{K_{P2}}{K_{P1}}}. \;\; \text{Or,} \;\; K = \sqrt{\frac{K_2}{K_1}}. \;\; \text{Or,} \;\; \ln K = \frac{1}{2}\ln K_2 - \frac{1}{2}\ln K_1$$

Again, we know that $\Delta G^0 = -RT \ln K$. Hence, we can write that

$$-RT \ln K = \frac{1}{2}(-RT \ln K_2 + RT \ln K_1). \;\; \text{Or,} \;\; \Delta G^0 = -\frac{1}{2}(\Delta G_2^0 - \Delta G_1^0)$$

Example 14.5.2

Consider the following two separate equilibrium systems:

$$2H_2O(g) \rightleftharpoons 2H_2(g) + O_2(g), \; K_{P1} = K_1 = \frac{p_{H_2}^2 \cdot p_{O_2}}{p_{H_2O}^2}$$

$$2HCl(g) \rightleftharpoons H_2(g) + Cl_2(g), \; K_{P2} = K_2 = \frac{p_{H_2} \cdot p_{Cl_2}}{p_{HCl}^2}$$

Now consider oxidation of HCl(g)

$$4HCl(g) + O_2(g) \rightleftharpoons 2H_2O(g) + 2Cl_2(g), \; K_P = K = \frac{p_{H_2O}^2 \cdot p_{Cl_2}^2}{p_{HCl}^4 \cdot p_{O_2}} = \frac{K_2^2}{K_1}$$

Or, $\ln K = 2 \ln K_2 - \ln K_1$. Or, $-RT \ln K = (-2RT \ln K_2 + RT \ln K_1)$

Or, $\Delta G^0 = (2\Delta G_2^0 - \Delta G_1^0)$

14.6 VAN'T HOFF ISOCHORE

For a given thermodynamic system ΔG^0 and K_p are correlated as

$$\Delta G^0 = -RT \ln K_p$$

Again, from Gibbs-Helmholtz relation, we have

$$\left[\frac{\partial(\Delta G^0/T)}{\partial T} \right]_P = -\frac{\Delta H^0}{T^2}$$

From the above two equations, we can write that

$$\left(\frac{\partial \ln K_P}{\partial T}\right)_P = \frac{\Delta H^0}{RT^2}$$

At constant pressure

$$\frac{d \ln K_P}{dT} = \frac{\Delta H^0}{RT^2} \tag{14.8a}$$

Or,

$$\frac{d \log K_P}{dT} = \frac{\Delta H^0}{2.303 RT^2} \tag{14.8b}$$

The equation (14.8a) or equation (14.8b) is known as Van't Hoff equation.
Assume that ΔH^0 is independent of temperature, integrating equation (14.8b), we get

$$\log K_P = -\frac{\Delta H^0}{2.303 RT} + I \tag{14.8c}$$

Again, we know that $K_P = K_c (RT)^{\Delta v}$

Or,

$$\ln K_P = \ln K_c + \Delta v \ln R + \Delta v \ln T$$

Or,

$$\frac{d \ln K_P}{dT} = \frac{d \ln K_c}{dT} + \frac{\Delta v}{T}. \quad \text{Or,} \quad \frac{d \ln K_c}{dT} = \frac{d \ln K_P}{dT} - \frac{\Delta v}{T} = \frac{\Delta H^0}{RT^2} - \frac{\Delta v}{T}$$

[Using equation (14.8a)]

Or,

$$\frac{d \ln K_c}{dT} = \frac{\Delta H^0 - \Delta v RT}{RT^2} = \frac{\Delta U}{RT^2} \tag{14.9}$$

[Since from the knowledge of thermodynamics, we have $\Delta H = \Delta U + \Delta v RT$ for homogeneous gas phase reaction at constant pressure].

The equation (14.9) is known as Van't Hoff isochore.

Assume that ΔH^0 is independent of temperature. So, integration of the equation (14.8b) within the limits, we get

$$\log\frac{K_{P2}}{K_{P1}} = -\frac{\Delta H}{2.303R}\left[\frac{1}{T_2} - \frac{1}{T_1}\right] = \frac{\Delta H}{2.303R}\left(\frac{T_2 - T_1}{T_1 T_2}\right) \tag{14.10}$$

For exothermic reaction, $\Delta H < 0$ and hence, $K_{P2} < K_{P1}$. So with increase in temperature equilibrium is shifted on the left direction (see Fig. 14.2).

For endothermic reaction, $\Delta H > 0$ and hence, $K_{P2} > K_{P1}$. So with increase in temperature equilibrium is shifted on the right direction (see Fig. 14.2).

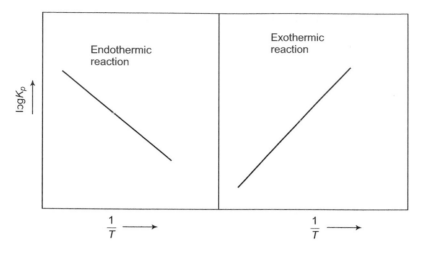

Figure 14.2   Plot of log K_P versus $1/T$ for endothermic and exothermic reactions

SOLVED PROBLEMS

Problem 14.1   Calculate the solubility of $Pb(IO_3)_2$ in aqueous solution of $Pb(NO_3)_2$ with concentration 0.1 M. Given $K_{sp} = 2.5 \times 10^{-13}$.

Solution

$Pb(IO_3)_2$ dissociates in aqueous solution reversibly as follows:

$$Pb(IO_3)_2 \rightleftharpoons Pb^{2+} + 2IO_3^-$$

Let S be the solubility of $Pb(IO_3)_2$ in presence of common ion Pb^{2+}.

ICE Table

Concentration	[Pb(IO₃)₂]	[Pb²⁺] in M	[IO₃⁻] in M
Initially	solid	0.1	0
Change	solid	+x	+2x
At Equilibrium	solid	(0.1 + x)	−2x

$K_{sp} = [Pb^{2+}][IO_3^-]^2 = (0.1 + S)(2S)^2$. Or, $4S^2(0.1 + S) = 2.5 \times 10^{-13}$. Assume that $(0.1 + S) \approx 0.1$. So, the above equation becomes $0.4S^2 = 2.5 \times 10^{-13}$. Or, $S = 7.91 \times 10^{-7}$.

Problem 14.2   Methanol is produced according to the following equation:

$$CO(g) + 2H_2(g) \rightleftharpoons CH_3OH(g)$$

In an experiment, 1 mole each of CO and H_2 were allowed to react in a sealed 10 litre reaction vessel at 500°K. When the equilibrium was established, the mixture was found to contain 0.0892 mole of CH_3OH. What are the equilibrium concentrations of CO, H_2, and CH_3OH?

Calculate the equilibrium constants K_c and K_p for this reaction at 500°K?

Given: $R = 0.0821$ lit–atm/mole°K.

Solution $[CO] = 0.0911$ M, $[H_2] = 0.0822$ M, $[CH_3OH] = 0.0892$ M

$$K_c = 14.5 \text{ and } K_p = 8.6 \times 10^{-3}.$$

Problem 14.3 Calculate the molar solubility of $Mg(OH)_2$ in 1M NH_4Cl.

Given: K_{sp} of $Mg(OH)_2(s)$ is 1.8×10^{-11}, K_b of NH_4OH is 1.8×10^{-5}.

Solution

$$Mg(OH)_2(s) \rightleftharpoons Mg^{2+} + 2OH^- \quad K_1 = K_{sp}$$

$$2NH_4^+ + 2OH^- \rightleftharpoons 2NH_4OH \quad K_2 = 1/K_b^2$$

$$K = \frac{K_{sp}}{K_b^2}$$

$$Mg(OH)_2(s) + 2NH_4^+ \rightleftharpoons Mg^{2+} + 2NH_4OH$$
$$\quad\quad\quad\quad\quad x \quad\quad\quad\quad 2x$$

x = Molar solubility of $Mg(OH)_2(s)$ in 1M NH_4Cl solution.

$$K = \frac{K_{sp}}{K_b^2} = \frac{1.8 \times 10^{-11}}{(1.8 \times 10^{-5})^2} = \frac{[Mg^{2+}][NH_4OH]^2}{[NH_4^+]^2} = \frac{x \times (2x)^2}{1^2}. \text{ Or, } 4x^3 = \frac{1}{18}$$

Or, $x = 0.24$ M.

Problem 14.4 An aqueous solution of a metal bromide $MBr_2(0.05M)$ is saturated with H_2S. What is the minimum pH at which metal sulphide, MS, will precipitate?

Given: $K_{sp}(MS)=6\times10^{-21}, [H_2S(aq)]=0.1\,M, K_{a1}(H_2S)=1\times10^{-7}, K_{a2}(H_2S)=1.3\times10^{-13}$.

Solution

$$MS(s) \rightleftharpoons M^{2+} + S^{2-}, \quad [S^{2-}] = \frac{K_{sp}(MS)}{[M^{2+}]} = \frac{6 \times 10^{-21}}{0.05} = 1.2 \times 10^{-19}$$

Dissociation of H_2S is given by

$$H_2S \rightleftharpoons H^+ + HS^- \text{ and } HS^- \rightleftharpoons H^+ + S^{2-}. \text{ Or, } H_2S \rightleftharpoons 2H^+ + S^{2-}$$

Hence, $K_a(H_2S) = \frac{[H^+]^2[S^{2-}]}{[H_2S]} = K_{a1}(H_2S) \times K_{a2}(H_2S) = 1.3 \times 10^{-20}$

Or, $1.3 \times 10^{-20} = \dfrac{[H^+]^2 \times 1.2 \times 10^{-19}}{0.1}$.   Hence, $[H^+] = 0.109$ and $pH = 0.96$.

Problem 14.5   How much AgBr could be dissolved in 1 lit. of 0.4 M NH_3 solution? Assume that Ag^+ remains as $[Ag(NH_3)_2]^+$. Given: $K(\text{complex}) = 1.0 \times 10^8$ and $K_{sp}(AgBr) = 5 \times 10^{-13}$.

Solution   Assume that solubility of AgBr is S. $AgBr(s) \rightleftharpoons Ag^+ + Br^-$.

$$[\{Ag(NH_3)_2\}^+] = [Br^-] = S, \ K_{sp}(AgBr) = [Ag^+][Br^-]$$

Again, $Ag^+ + 2NH_3 \rightleftharpoons \{Ag(NH_3)_2\}^+$.   Or, $K(\text{complex}) = \dfrac{\left[\{Ag(NH_3)_2\}^+\right]}{[Ag^+][NH_3]^2}$

$$[Ag^+] = \frac{[\{Ag(NH_3)_2\}^+]}{K(\text{complex}) \times [NH_3]^2} = \frac{S}{1.0 \times 10^8 \times (0.4)^2}$$

$$K_{sp}(AgBr) = [Ag^+][Br^-] = \left\{ \frac{S}{1.0 \times 10^8 \times (0.4)^2} \right\}(S). \ \ \text{Or,} \ S = 2.8 \times 10^{-3} \text{ M}.$$

Problem 14.6   HCl reacts with O_2 in a closed vessel according to the following equilibrium reaction:

$$4HCl + O_2 \rightleftharpoons Cl_2 + 2H_2O$$

Initial concentration of HCl and O_2 are 0.5 and 0.05 M respectively. Equilibrium concentration of Cl_2 is 0.048 M. Given: Final temperature of the system is 25°C.
(a) Calculate the equilibrium concentrations of other components.
(b) Calculate K_c and K_p of the reaction.

Solution

ICE Table

Concentration	[HCl]	[O₂] in M	[Cl₂] in M	[H₂O] in M
Initially	0.5	0.05	0	0
Change	-4×0.048	-0.048	$+0.048$	$+0.048$
At Equilibrium	0.308	0.002	0.048	0.096

$$K_c = \frac{(0.048 \text{ M})(0.096 \text{ M})^2}{(0.308)^4(0.002 \text{ M})} = 25 \text{ M}^{-2} \ \text{ and } \ K_p = K_c(RT)^{-2} = 0.041 \text{ atm}^{-2}.$$

EXERCISES

1. The dissociation constant of $CaCO_3$ at 900°C and 1000°C are 770 mm and 2940 mm respectively. Calculate the heat of dissociation within this temperature range.

   [*Ans.:* 39646 cal]

2. Consider the reaction:

   $$C(s) + H_2O(g) \rightleftharpoons CO(g) + H_2(g), \Delta H = 31.36 \text{ kcal.}$$

   Predict the effect of the following on the equilibrium:

   (a) The amount of $C(s)$ is increased without altering the total pressure.

   (b) The temperature of the system is increased.

   (c) 1 gm.mole neon gas is introduced into the system at constant pressure.

3. At 2000°K, ΔG^0 for the following reaction is given by: $\Delta G^0 = 22000 - 2.5$ T

   $$N_2(g) + O_2(g) \rightleftharpoons 2NO(g)$$

   Estimate K_p at 2000°K.   [*Ans.:* 1.4×10^{-2}]

4. Consider the following gas phase reaction:

   $$2A(g) + B(g) \rightleftharpoons A_2B(g)$$

   $$\Delta G^0 = -1200 \text{ cal at } 227°C.$$

   What would total pressure be necessary to produce 60% conversion of B to A_2B?
   Mole ratio of A:B = 2:1.   [*Ans.:* 1.513 atm]

5. A mixture of SO_2 and O_2 in molar ratio (2:1) is allowed to react in a reaction vessel at 650°C in presence of *Pt* catalyst. The total pressure was 10 atm. If 60% SO_2 be converted to SO_3 in this process. Calculate K_p for the reaction:

   $$2SO_2(g) + O_2(g) \rightleftharpoons 2SO_3(g)$$

   State with reason the effect of following on the above equilibrium process:

   (a) Volume of the reaction vessel is increased at constant temperature.

   (b) N_2 is introduced into the vessel at constant T and V.

   (c) The catalyst is removed at constant T and V without disturbing the gases.

   (d) The reaction vessel is suddenly cooled to 0°C.

6. Consider a reaction: $F_2 \rightleftharpoons 2F$. Calculate

   (a) the degree of dissociation and density of fluorine at 0.4 atm and 1000°K.
   Given: $K_p = 1.4 \times 10^{-2}$ atm.

   (b) ΔH^0 for the dissociation of fluorine, if $K_{P(160)} = 2 \times 10^{-5}$ atm and $K_{P(960)} = 4 \times 10^{-3}$ atm.   [*Ans.:* 38405 cal]

7. A flask is charged with 2 atm of nitrogen dioxide and 1 atm of dinitrogen tetroxide at 25°C and allowed to reach equilibrium. When equilibrium is established, the partial pressure of NO_2 has decreased by 1.24 atm.

   (a) What are the partial pressures of NO_2 and N_2O_4 at equilibrium?

(b) Calculate K_P and K_c for the following reaction at 25°C:

$$2NO_2(g) \rightleftharpoons N_2O_4(g)$$

[*Ans.: $K_P = 2.80$ and $K_c = 68.6$*]

8. The reaction: $N_2(g) + 3H_2(g) \rightleftharpoons 2NH_3(g)$, has equilibrium constant, $K_c = 0.0602$ at 500°K. Calculate the equilibrium constant for the following reaction:

$$NH_3(g) \rightleftharpoons \frac{1}{2}N_2(g) + \frac{3}{2}H_2(g)$$

9. During rusting of iron the following equilibrium is established:

$$3\,Fe(s) + 4\,H_2O \rightleftharpoons Fe_3O_4(s) + 4\,H_2(g) + Q\,cal$$

Predict the probable shift in equilibrium according to the following changes:
(a) Increase in amount of water, (b) Decrease the total volume of the system to half,
(c) Continuous removal of $Fe_3O_4(s)$, (d) Adding hydrogen in the system,
(e) Increase the temperature.

[*Ans.:*   (a) Shifting forward direction, (b) No shift, (c) No shift,
(d) Shifting backward direction, (e) Shifting backward direction]

10. Let us consider a chemical equation with equilibrium constant, K. It has been observed that $K > 1$. Then which one of the following is true and why?
(a) There are more reactants than products at equilibrium.
(b) There are more products than reactants at equilibrium.
(c) There are the same amount of products and reactants at equilibrium.
(d) The reaction is not at equilibrium.

11. Let us consider a closed system, containing equal amount of two reactants. After certain time the reactants may be converted almost entirely to products. Equilibrium constant is K. Then which one of the following is true and why?
(a) $K < 1$ (b) $K > 1$ (c) $K = 1$ (d) $K = 0$.

12. In the Haber process of manufacturing of ammonia in a closed container, the following reaction occurs:

$$N_2(g) + 3H_2(g) \rightleftharpoons 2NH_3(g)$$

If hydrogen gas is added into the equilibrium system, which one of the following is true and why?
(a) The reaction shifts to the right to produce more products.
(b) The reaction shifts to the left to produce more reactants.
(c) The reaction stops. All the nitrogen gas has already been used up.
(d) Need more information.

15 Phase Equilibria

15.1 PHASE RULE

(i) Phase It is a physically distinct, chemically homogeneous, and mechanically separable region, separated by other parts of the system by a definite boundary. The number of phases is represented by P. A system, containing one phase, is called homogeneous system and a system, containing more than one phase, is called heterogeneous system.

Examples (a) Gases are miscible in all proportions. So, gas phase is always a homogeneous system since mixture of two or more gases always constitutes a single gas. (b) In some cases mixture of two liquids form a homogeneous single phase, e.g., water-ethanol system. However, mixture of two immiscible liquids (e.g., benzene and water) will form two distinct phases. The mixture is called homogeneous system. (c) When a solid is dissolved in its solvent, the solution is called heterogeneous system, e.g., NaCl solution. Here the salt, NaCl, cannot maintain its separate identity and crystal structure in solution. If AgCl is added in water, the salt maintains its separate identity and crystal structure, forming heterogeneous system. (d) There are several solid phases depending on the crystal structure. For example, decomposition of calcium carbonate as shown below:

$CaCO_3(s) \rightleftharpoons CaO(s) + CO_2(g)$. Here there are two solid phases and one gas phase. So it is a three phase system.

(ii) Component In an equilibrium system, the minimum number of independent chemical constituents with the help of which composition of all phases in equilibrium can be described successfully, are called components. The number of components is represented by C. Some examples are given below:

Example 15.1.1 $Ice(s) \rightleftharpoons Water(l) \rightleftharpoons Water\ vapour(g)$

The chemical constituent H_2O successfully describes the composition of all the three phases. So the above system is a single component system.

Example 15.1.2 $CaCO_3(s) \rightleftharpoons CaO(s) + CO_2(g)$ in a closed vessel.

The number of components is apparently 3 and the number of phases are 3 but 3 components are linked by a chemical reaction. Thus at least one component is a dependent variable. The composition of the phase $CO_2(g)$ can be expressed by means of $CaCO_3$ and CaO. So the system is a two-component system.

Phase	Component
$CaCO_3(s)$	$CaCO_3$
$CaO(s)$	CaO
$CO_2(g)$	$CaCO_3 - CaO$

We know that $\Sigma x_i = 1$. In this case $x_1 + x_2 + x_3 = 1$. If x_1 and x_2 are known, x_3 can easily be evaluated $[x_3 = 1 - (x_1 + x_2)]$. Thus the third component is dependent variable. So the total number of components is $(3 - 1) = 2$.

Restriction Any chemical reaction among the components restricts the number of independent chemical constituents.

Example 15.1.3 $NH_4Cl(s) \rightleftharpoons NH_3(g) + HCl(g)$ in a closed vessel.

The above system has two phases (one is solid NH_4Cl and the other is the gas phase). The composition of gas phase can be expressed in terms of $NH_4Cl(s)$. So only NH_4Cl is enough to express the composition of both the phases. Thus it is a one-component system.

Example 15.1.4 $2KClO_3(s) \rightleftharpoons 2KCl(s) + 3O_2(g)$

Phase	Composition
$KClO_3(s)$	$KCl(s) + \dfrac{3}{2}O_2(g)$
$KCl(s)$	$KCl(s) + 0 \times O_2(g)$
$O_2(g)$	$0 \times KCl(s) + O_2(g)$

Example 15.1.5 $Na_2SO_4 - H_2O$ system.

The following phases are in equilibrium $NaSO_4(s)$, $Na_2SO_4 \cdot 7H_2O(s)$, ice, liquid solution and vapour.

Number of components are 2, since all the phases can be expressed by only two constituents Na_2SO_4 and H_2O.

Example 15.1.6 $PCl_5(g) \rightleftharpoons PCl_3(g) + Cl_2(g)$

$$K_X = \frac{X_{PCl_3} \cdot X_{Cl_2}}{X_{PCl_5}}. \text{ Or, } X_{PCl_5} = \frac{(X_{PCl_3} \cdot X_{Cl_2})}{K_X}$$

Two of the constituents can be altered independently while the third one is fixed. So, the number of components is 2.

Number of constituents are 3 but composition can be expressed using only two constituents. Thus, number of components are 2.

(iii) Degrees of Freedom It is the minimum number of independent variables, including concentrations of components, pressure and temperature, which are required to describe an equilibrium system perfectly. The number of degrees of freedom is represented by F.

Phase Rule For an equilibrium system, uninfluenced by external forces like gravitational force, electrical or magnetic forces, etc., the total number of degrees of freedom is related to the number of components (C) and the number of phases (P) according to the following equation

$$F = C - P + 2 \qquad (15.1)$$

where the factor 2 in the right hand side indicates the two independent variables, pressure and temperature. It was first proposed by Willard Gibbs in 1874.

Proof As we know for any equilibrium system the chemical potential of ith component is equal in all phases at equilibrium. If there are P number of phases, represented by α_1, α_2, $\alpha_3 \cdots \alpha_p$ etc., in equilibrium, for the 1st component, we can write that

$$\mu_1^{\alpha_1} = \mu_1^{\alpha_2} = \mu_1^{\alpha_3} = \mu_1^{\alpha_4} = \cdots \mu_1^{\alpha_p}$$

where, $\mu_1^{\alpha_1}$ is the chemical potential of the 1st component in α_1 phase. Again $\sum \alpha_i = P$. Similarly, we can write that

$$\mu_2^{\alpha_1} = \mu_2^{\alpha_2} = \mu_2^{\alpha_3} = \mu_2^{\alpha_4} = \cdots \mu_2^{\alpha_p} \qquad \text{for the } 2^{nd} \text{ component}$$

$$\mu_3^{\alpha_1} = \mu_3^{\alpha_2} = \mu_3^{\alpha_3} = \mu_3^{\alpha_4} = \cdots \mu_3^{\alpha_p} \qquad \text{for the } 3^{rd} \text{ component}$$

$$\cdots\cdots\cdots\cdots\cdots\cdots\cdots\cdots\cdots\cdots\cdots\cdots\cdots\cdots$$

$$\cdots\cdots\cdots\cdots\cdots\cdots\cdots\cdots\cdots\cdots\cdots\cdots\cdots\cdots$$

$$\mu_C^{\alpha_1} = \mu_C^{\alpha_2} = \mu_C^{\alpha_3} = \mu_C^{\alpha_4} = \cdots \mu_C^{\alpha_p} \qquad \text{for the component } 'C'$$

where C is the total number of components. Thus for phase α_1, the total number of independent components is $(C - 1)$, since $\sum x_i = 1$, where x_i is the mole fraction of the ith component. As there are number of phases, the total number of independent variables is $[P(C - 1)]$. Considering pressure and temperature are the two independent variables, the total number of independent variables become $[P(C-1)+2]$. Each component is distributed in P number of phases in equilibrium. So the number of restrictions for each component is $(P - 1)$. As there are C number of components, the total number of restrictions or chemical equations is $C(P - 1)$. Thus the total number of independent variables = degrees of freedom (DOF) = F.

$$F = [P(C - 1) + 2] - C(P - 1) = C - P + 2 \qquad (15.2)$$

For $F = 0$, the system is said to be ***invariant***, which means that all the independent variables are fixed at that point. For $F = 1$, the system is said to be ***univariant***, which means that one independent variable may be varied. For $F = 2$, the system is said to be **bivariant**, which means that two independent variables may be varied.

Advantages of Phase Rule

(a) It connects all the phases in heterogeneous system in equilibrium.
(b) The phase rule can predict the behaviour of the equilibrium system on changing the variables like pressure, temperature or compositions.
(c) It is applicable to physical as well as chemical phase reactions.
(d) It is a very important tool to investigate complex heterogeneous equilibria.
(e) It predicts the conditions under which some given substances may remain in equilibrium. It also predicts whether interconversion among the given substances occurs or not.

Limitations of Phase Rule

(a) Composition of different phases in equilibrium cannot be obtained using phase rule. Only degrees of freedom can be calculated.
(b) The phase rule is applicable to heterogeneous system in equilibrium. Hence it cannot be applicable to non-equilibrium heterogeneous systems.
(c) The phase rule considers only pressure, temperature, and compositions but does not consider magnetic, electrical and gravitational influences. Considering all these influences the factor 2 should be adjusted accordingly.
(d) In case of solid phase, normal appearance is considered. If it is in finely divided state, corresponding vapour pressure will differ from the normal value.

15.2 ONE-COMPONENT SYSTEM

15.2.1 Water System

It is the simplest example of a single component system. The following equilibrium holds good:

Ice $(s) \rightleftharpoons$ Water$(l) \rightleftharpoons$ Water vapour (g), as shown in the Fig. 15.1(a).

$C = 1$ for water system and hence, $F = C - P + 2 = 1 - P + 2 = 3 - P$.

The region bound by the curves *AOB* or *BOC* or *COA* contains vapour phase or solid phase or liquid phase respectively. So, $P = 1$ and hence, $F = 3 - 1 = 2$. The system is called bivariant. Both pressure and temperature vary independently.

Along the curve *OA*, both liquid and vapour phases coexists. So, $P = 2$ and hence, $F = 3 - P = 3 - 2 = 1$. So if pressure varies independently, temperature variation depends on pressure variation. So temperature is a dependent variable. Similarly if temperature varies independently, pressure will be dependent variable. The system is called univariant system and the curve is called boiling point curve.

Along the curve *OC* both liquid and solid phases coexist and the curve is called melting point curve. Here also F = 1 and only one variable, either pressure or temperature varies independently. The system is called univariant system. Interestingly, it is observed that the slope of the curve is negative because volume decreases on melting of ice. Thus, $\dfrac{dP}{dT}$ is negative according to Clausius-Clapeyron equation

$$\frac{dP}{dT} = \frac{L_f}{T(V_l - V_s)}, \quad V_l < V_s$$

Along the curve OB both solid and vapour coexist and the curve is called sublimation curve. Here also $F = 1$ and only one variable, either pressure or temperature varies independently. The system is called univariant system.

At the point O, three phases solid, liquid, and vapour coexist and hence, $F = 3 - P = 3 - 3 = 0$. The system is called invariant system. All the variables are fixed at O and no variable can vary independently. The point O is known as triple point. Pressure and temperature at O are 0.00603 atm (4.58 mm of Hg) and 0.0075°C respectively.

The point D in the Fig. 15.1a indicates the boiling point of water at normal pressure, i.e., 1 atm. The point A indicates the highest point where two phases (liquid and vapour) coexist and is known as ***critical point*** above which the liquid and vapour phases are indistinguishable from each other.

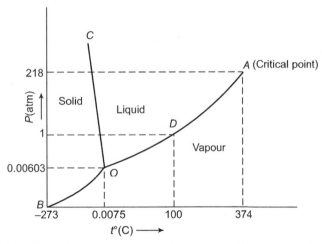

Figure 15.1a Phase diagram (pressure vs. temperature) of water system (not to scale)

15.2.2 Carbon Dioxide System

The phase diagram of carbon dioxide system is same as water system as shown in the Fig. 15.1b. In this case slope of the melting point curve is positive. Triple point appears at –56°C and 4 atm pressure. If cooling of carbon dioxide is carried out at 1 atm pressure, i.e., below its triple point, the gas is converted to solid carbon dioxide, which appears as ice-like structure. This solid carbon dioxide is popularly known as dry ice. The point A corresponds to critical point of carbon dioxide with coordinate (31.1°C, 73 atm).

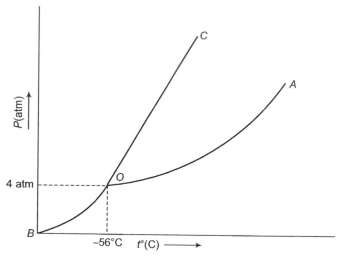

Figure 15.1b   Phase diagram (pressure vs. temperature) of carbon dioxide system (not to scale)

15.2.3 Sulphur System

Polymorphism   There are some substances, which exist in more than one crystalline form. The phenomenon is called polymorphism. There are two types of polymorphism: (a) enantiotropy, and (b) monotropy. In enantiotropy polymorphism system there is a critical temperature above which one crystalline form is stable and below which the other crystalline form is stable.

One example of enantiomer is sulphur. It exists in two crystalline forms: Orthorhombic and monoclinic. The phase diagram is shown in the Fig. 15.2a. There are four phases in sulphur system: orthorhombic sulphur, monoclinic sulphur, liquid sulphur, and gaseous sulphur. There are four triple points:

(i) The point B: Coordinate (95.5°C, 0.01 mm Hg). Three phases: rhombic, monoclinic, and vapour are in equilibrium.

(ii) The point C: Coordinate (119.2°C, 0.025 mm Hg). Three phases: monoclinic, liquid, and vapour are in equilibrium.

(iii) The point E: Coordinate (151°C, 1290 atm). Three phases: rhombic, monoclinic, and liquid are in equilibrium.

(iv) The point O: Coordinate (114.5°C, 0.03 mm Hg). Three phases: rhombic, liquid, and vapour are in equilibrium. This is metastable triple point.

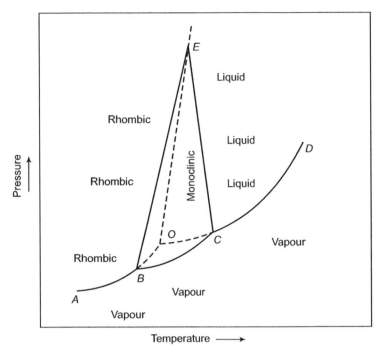

Figure 15.2a Phase diagram of sulphur system

15.2.4 Phosphorus System

Phosphorus system is another example of polymorphism. There are four phases: white phosphorus (I), white phosphorus (II), red phosphorus, liquid and vapour. The phase diagram of phosphorus system is shown in the Fig. 15.2b.

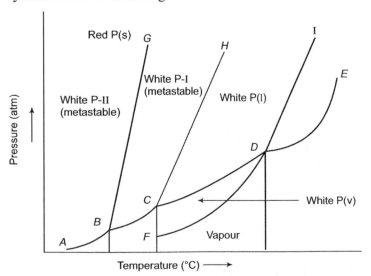

Figure 15.2b Phase diagram of phosphorus system

The point B is the transition temperature, above which white $P(I)$ is stable and below which white $P(II)$ is stable. Thus the transition of white $P(I) \rightleftharpoons$ white $P(II)$ is an enantiotropic change. The transition temperature is $-76.9°C$ (point B in the Fig. 15.2b).

The point D is the melting point of red P. The coordinate of the point D is ($588.5°C$, 43.1 atm). The curve DI represents melting point curve of red P. The point D is the triple point, where, red phosphorus, liquid and vapour are in equilibrium.

DC is the vapour pressure curve of supercooled liquid phosphorus. At the point C the solid phase, white $P(I)$ first appears. The coordinate of the point C is ($44°C$, 0.181 atm).

BG and CH are the melting point curves of white $P(II)$ and white $P(I)$ respectively.

AB and BC represent sublimation curves of white $P(II)$ and white $P(I)$ respectively.

15.3 IDEAL LIQUID-VAPOUR SYSTEM

Figure 15.3 shows the phase diagram of a pure liquid in equilibrium with its vapour. It is a two component system, where homogeneous liquid phase is separated from the corresponding vapour phase by a heterogeneous loop. The component B has higher boiling point and hence is richer in the liquid whereas the lower boiling component A is richer in corresponding equilibrium vapour. This is well illustrated by isothermal line lm, also known as tie line, connecting the equilibrium liquid and vapour phase at t°C. The equilibrium holds good only when the system is observed in a closed volume, e.g., in a fractional distillation column. Now the temperature is raised to t'°C so that all liquid is converted to vapour. Now the system is allowed to settle at this temperature so that equilibrium is attained and corresponding tie line is $l'm'$. The liquid at the point m' contains higher concentration of B than that at the point m. The liquid at m' may be heated up again like before and ultimately pure liquid B will be obtained.

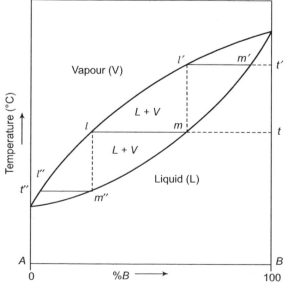

Figure 15.3 Phase diagram of an ideal liquid-vapour system

Similarly if the equilibrium vapour at *l* is condensed as the temperature *t* decreases to *t''*, the tie line *l'' m''* is obtained. The vapour at *l''* is richer in the component *A* than that at *l*. If this condensation process is continued under equilibrium condition ultimately vapour of pure liquid *A* will be obtained. Further condensation gives rise to pure liquid *A*.

Examples of such systems are: (i) oxygen and nitrogen, (ii) carbon disulphide and benzene, (iii) benzene and toluene, (iv) acetone and water, (v) acetone and ether, (vi) chloroform and carbon tetrachloride, (vii) chloroform and bromobenzene.

15.4 GAS-LIQUID SYSTEM

15.4.1 Solubility of Gas in a Liquid

Solubility of gases in liquids is expressed by two parameters: (i) adsorption coefficient, and (ii) solubility coefficient.

(i) **Absorption coefficient:** Suppose *v ml* of a gas with partial pressure *p* atm is dissolved in *V ml* of a solvent at room temperature. *v ml* of the gas is equivalent to v_0 *ml* of that gas at STP (0°C, 1 atm). Then absorption coefficient, α, is given by

$$\alpha = \frac{v_0}{V \cdot p}$$

(ii) **Solubility coefficient:** Suppose *v ml* of a gas with partial pressure *p* atm is dissolved in *V ml* of a solvent at temperature T°K, solubility coefficient, β, is given by $\beta = v/V$. Assuming ideal gas behaviour, we have

$$\frac{p.v}{T} = \frac{1.v_0}{T_0}. \text{ Or, } v_0 = \frac{p.v.T_0}{T}, \text{ where, } T_0 = 273°K$$

Putting this value of v_0 in the above expression, we have

$$\alpha = \frac{\beta.T_0}{T} = \frac{\beta \times 273}{T}$$

Solubility of any gas in a solvent or liquid is largely influenced by temperature and pressure. Generally with increase in temperature solubility of a gas decreases (Le Chatelier's principle) and hence concentration of the gas in the liquid phase decreases with increase in temperature. This decrease in concentration can be correlated with temperature using Clausius-Clapeyron equation as given below

$$\frac{d \ln c}{dT} = -\frac{\Delta H}{RT^2} \qquad (15.3)$$

ΔH is the differential heat of solution when 1 mole of the gas is dissolved at room temperature and pressure. Further, ΔH is assumed to be independent of temperature. Thus integrating equation (15.3) gives

$$\ln \frac{c_2}{c_1} = \frac{\Delta H}{R} \left(\frac{1}{T_2} - \frac{1}{T_1} \right) \tag{15.4}$$

where, c_1 and c_2 are the concentrations of the solution in moles/litre at T_1 and T_2 respectively. As concentration is directly proportional to the absorption coefficient, equation (15.4) becomes

$$\ln \frac{\alpha_2}{\alpha_1} = \frac{\Delta H}{R} \left(\frac{1}{T_2} - \frac{1}{T_1} \right) \tag{15.5}$$

Influence of pressure   The influence of pressure on solubility of gas in a solvent is well defined by Henry's law, which states that:

"**The mass of a gas dissolved in a given volume of solvent, at constant temperature, is proportional to the partial pressure of the gas with which it is in equilibrium**".

Mathematically, it can be expressed as $m \propto p$. Or, $m = kp$, where, m is the mass of the gas dissolved in a solvent, p is the partial pressure of the gas and k is the Henry constant, which is also known as distribution ratio because p represents concentration of the gas in surrounding atmosphere. Thus k can be written as

$$k = \frac{m}{p} = \frac{\text{Concentration of the gas in liquid phase}}{\text{Concentration of the gas in surrounding atmosphere}}$$

The values of k of carbon dioxide gas in different solvents is shown in Table 15.1.

Table 15.1 k values of carbon dioxide gas in different solvents at $-59°C$

Equilibrium pressure	k value in different solvents		
	Methyl alcohol	*Acetone*	*Methyl acetate*
100 mm	42.5	67.2	75.8
200 mm	42.7	68.0	77.1
400 mm	43.1	69.2	77.6
700 mm	43.3	72.8	79.0

15.4.2   Duhem-Margules Equation

According to Gibbs-Duhem relation, we have

$$\Sigma n_i d\mu_i = 0 \tag{15.6}$$

Let us consider a liquid mixture, having two components such that equilibrium vapour pressure of each component follows ideal gas behaviour. n_1 and n_2 are the number of moles of the components, p_1 and p_2 are the corresponding vapour pressures of the components. Then chemical potentials of the components, μ_1 and μ_2, are given by:

$$\mu_1 = \mu_0^1 + RT \ln p_1 \quad \text{and} \quad \mu^2 = \mu_0^2 + RT \ln p_2.$$

On differentiating we have

$$d\mu_1 = RTd \ln p_1 \quad \text{and} \quad d\mu_2 = RTd \ln p_2$$

Using equation (15.6), we can write that

$$n_1 \, d\mu_1 + n_2 d\mu_2 = 0. \text{ Or, } n_1 d\mu_1 = -n_2 d\mu_2 \tag{15.7}$$

Dividing both sides of equation (15.7) by $(n_1 + n_2)$ we get

$$x_1 d\mu_1 = -x_2 d\mu_2 \tag{15.8}$$

where, x_1 and x_2 are the mole fractions of the components 1 and 2 respectively.

Again, $x_1 + x_2 = 1$ and $dx_1 = - dx_2$. Dividing both sides of equation (15.8) by dx_1, we get

$$\frac{x_1 d\mu_1}{dx_1} = -\frac{x_2 d\mu_2}{dx_1} = \frac{x_2 d\mu_2}{dx_2}. \text{ Or, } \frac{RTd \ln p_1}{d \ln x_1} = \frac{RTd \ln p_2}{d \ln x_2}. \text{ Or, } \frac{d \ln p_1}{d \ln x_1} = \frac{d \ln p_2}{d \ln x_2} \tag{15.9}$$

The equation (15.9) is called Duhem-Margules equation. The only assumption during derivation of Duhem-Margules equation is that vapour behaves ideally.

15.4.3 Composition of Liquid and Vapour: Konowaloff Rule

From equation (15.9), we have

$$\left(\frac{x_1}{p_1}\right)\frac{dp_1}{dx_1} = \left(\frac{x_2}{p_2}\right)\frac{dp_2}{dx_2}. \text{ Or, } \left(\frac{x_1}{p_1}\right)\frac{dp_1}{dx_1} - \left(\frac{x_2}{p_2}\right)\frac{dp_2}{dx_2} = 0. \text{ Or, } \left(\frac{x_1}{p_1}\right)\frac{dp_1}{dx_1} + \left(\frac{x_2}{p_2}\right)\frac{dp_2}{dx_1} = 0$$

Or,
$$\frac{dp_1}{dx_1} = -\left(\frac{x_2}{p_2}\right)\left(\frac{p_1}{x_1}\right)\frac{dp_2}{dx_1} = -\left(\frac{p_1 x_2}{p_2 x_1}\right)\frac{dp_2}{dx_1}$$

Again, $x_1 + x_2 = 1$ and $dx_1 = -dx_2$

The total pressure, P, is given by $P = p_1 + p_2$. Differentiating with respect to x_1 and putting the value of $\frac{dp_1}{dx_1}$, we get

$$\frac{dP}{dx_1} = \frac{dp_1}{dx_1} + \frac{dp_2}{dx_1} = \frac{dp_2}{dx_1}\left[1 - \left(\frac{p_1 x_2}{p_2 x_1}\right)\right] = -\frac{dp_2}{dx_2}\left[1 - \left(\frac{p_1 x_2}{p_2 x_1}\right)\right]$$

As we know with increase in mole fraction of one component in the solution, corresponding vapour pressure of the same component also increases. So, $\frac{dp_2}{dx_2} > 0$.

Case-I

Thus if $\dfrac{dP}{dx_1} > 0, \left(\dfrac{p_1 x_2}{p_2 x_1}\right) > 1.$ Or, $\dfrac{p_1}{p_2} > \dfrac{x_1}{x_2}$ (15.10a)

Conclusion-1 If the solution is richer with the 1st component, the corresponding vapour is also richer with the 1st component. Furthermore, % concentration of the 1st component is more in vapour than that in liquid.

Case-II

If $\dfrac{dP}{dx_1} < 0, \left(\dfrac{p_1 x_2}{p_2 x_1}\right) < 1.$ Or, $\dfrac{p_1}{p_2} < \dfrac{x_1}{x_2}$ (15.10b)

Conclusion-2 If the solution is richer with the 1st component, the corresponding vapour is richer with the 2nd component.

Statement of Konowaloff Rule

For any given solution, the corresponding vapour is richer with that component whose addition in the liquid solution results in increase in total vapour pressure.

15.4.4 Raoult's Law and Henry's Law

Assumptions

(a) Liquid phase and vapour phase are in equilibrium.
(b) Vapour phase behaves as an ideal gas.
(c) Liquid phase behaves as an ideal solution.

Let us consider an ideal homogeneous solution, consisting of one non-volatile solute and its solvent. The solution remains in equilibrium with its vapour, containing the solvent vapour. The solute decreases the number of solvent molecules on the liquid surface and hence vapour pressure of the pure solvent decreases.

x_1 = Mole fraction of the solvent in solution. y_1 =Mole fraction of the solvent in vapour.

As the system is in equilibrium we can write for its component that $\mu_1(l) = \mu_1(v)$. Thus, we can write that

$$\mu_{1,l}^0(T,P) + RT \ln x_1 = \mu_{1,v}^0(T,P) + RT \ln y_1 = \mu_{1,v}^0(T,P) + RT \ln \dfrac{p_1}{P} \qquad (15.11a)$$

where, p_1 and P are the partial vapour pressure of the solvent and total vapour pressure respectively. If the above equilibrium exists between pure solvent and its vapour, we have $x_1 = 1, p_1 = P^0$ and $\mu_{1,l}^0 (T, P) = \mu_{1,l}^0 (T, P^0)$, where, P^0 is the vapour pressure of the pure solvent at temperature T. Thus the equation (15.11a) becomes

$$\mu_{1,l}^0(T, P^0) = \mu_{1,v}^0(T,P) + RT \ln \dfrac{P^0}{P} \qquad (15.11b)$$

Subtracting the equation (15.11b) from the equation (15.11a) we get

$$\mu_{1,l}^0(T,P) - \mu_{1,l}^0(T,P^0) + RT \ln x_1 = RT \ln \frac{p_1}{p^0}$$

As the pressure difference $(P - P^0)$ is very low, we can write that

$$\mu_{1,l}^0(T, P) = \mu_{1,l}^0(T, P^0)$$

Thus the above equation becomes

$$RT \ln x_1 = RT \ln \frac{p_1}{p^0}. \text{ Or, } p_1 = x_1 P^0 \qquad (15.11c)$$

Equation (15.11c) is the mathematical form of Raoult's law, which states that

"For a given ideal solution, vapour pressure of that solution is directly proportional to the mole fraction of the solvent present in that solution".

15.4.5 Ideal Gas-Liquid System

In case of an ideal gas-liquid mixture, x_2 represents the mole fraction of gas remained dissolved in the solution. The equation (15.11a) can be written as

$$\mu_{2,l}^0(T, P) - \mu_{2,v}^0(T, P) = RT \ln p_2 - RT \ln x_2 - RT \ln P$$

Or,

$$RT \ln \frac{p_2}{x_2} = \mu_{2,l}^0(T,P) - \mu_{2,v}^0(T,P) + RT \ln P$$

Or,

$$\ln \frac{p_2}{x_2} = \frac{1}{RT}\left[\mu_{2,l}^0(T,P) - \mu_{2,v}^0(T,P) + RT \ln P\right]$$

Let

$$\ln K' = [\mu_{2,l}^0(T, P) - \mu_{2,v}^0(T, P) + RT \ln P]/RT$$

Or,

$$\ln \frac{p_2}{x_2} = \ln K'. \text{ Or, } \ln \frac{x_2}{p_2} = \ln K \text{ where, } K = \frac{1}{K'} = \text{Constant.}$$

Or,

$$x_2 = K p_2 \qquad (15.12)$$

At constant pressure and temperature only, the value of K is constant.

Equation (15.12) is the mathematical form of Henry's law which states that

"For any given ideal gas-liquid system, mole fraction of gas, present in that solution, is directly proportional to the partial vapour pressure of gas provided temperature and total pressure are constant".

Henry's law holds good for any ideal gas-liquid mixture.

Inference

(i) Any ideal liquid-liquid mixture, obeying Raoult's law, is called ideal solution.

(ii) Vapour pressure of the solvent depends on the amount of non-volatile solute present in the solution but independent of the quality of the solute.

(iii) In an ideal dilute solution the solvent obeys Raoult's law and each solute obeys Henry's law.

15.5 HOMOGENEOUS TWO COMPONENT LIQUID-LIQUID SYSTEM

Homogeneous two component liquid-liquid system may be of two types: (i) Ideal solution, obeying Raoult's law, (ii) Non-ideal solution, showing deviation from Raoult's law.

15.5.1 Ideal Homogeneous Mixture

Consider an ideal liquid mixture, e.g., water and ammonia. Boiling point of water is 100°C while that of ammonia is –33.3°C. Temperature-composition diagram of this mixture is shown in the Fig. 15.4a. It shows that with the addition of ammonia boiling point of water decreases till it reaches the boiling point of ammonia. Within the envelope *ocado*, heterogeneous mixture of two phases (liquid and vapour) exists. Only vapour phase exists above the curve *oca* and liquid phase exists below the curve *oda*. At any particular composition, x_2, the point *d* indicates the temperature at which the liquid starts boiling. The point *d* is called bubble point and the locus of the point *d*, i.e., the curve *oda* is known as bubble point curve. At the same composition x_2, the point *c* indicates the temperature at which the vapour starts condensing. The point *c* is called dew point and the locus of the point *c*, i.e., the curve *oca* is known as dew point curve. The temperature difference $(t_c - t_d)$ is known as boiling range or condensing range.

The isothermal line *LV* is known as tie line. Along the tie line the liquid and vapour phases remain in equilibrium. At *L*, mole fraction of *B* in the liquid phase is x_2 and at *V*, mole fraction of *B* in the gas phase is y_2.

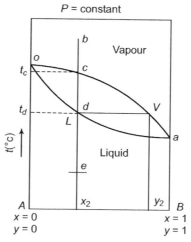

Figure 15.4a   Two component liquid-vapour phase diagram of completely miscible water/ ammonia system. Component *A*: higher boiling liquid (water) and Component *B*: lower boiling liquid (ammonia)

Total vapour pressure (P) is given by $P = p_A + p_B$, where, p_A and p_B are the partial vapour pressures of the components *A* and *B* respectively. As the solution obeys Raoult's law, we can write that $p_A = x_A p_0^A = (1 - x_B) p_B$ and $p_B = x_B p_0^B$, where, p_0^A and p_0^B are the vapour pressures of the pure liquids, *A* and *B* respectively. The corresponding pressure-composition diagram is shown in the Fig. 15.4b.

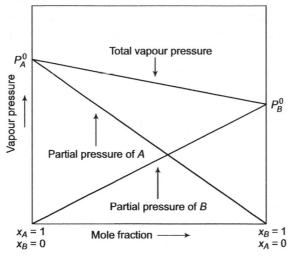

Figure 15.4b   Pressure-Composition diagram of a two-component liquid-vapour system

15.5.2   Non-ideal Homogeneous Solution (Azeotropic Mixture)

In the case of an ideal liquid mixture, e.g., in liquid water/ammonia system Raoult's law holds good but in the case of a real mixture, deviation of Raoult's law is a common phenomenon. The phase diagrams of positive and negative deviation of Raoult's law are shown in the Figs. 15.5 and 15.6 respectively.

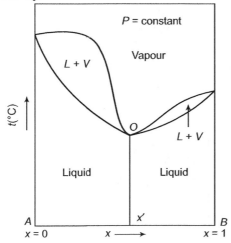

Figure 15.5a   Two component liquid–vapour phase diagram, showing positive deviation of Raoult's law. System: $CHClF_2$ and CCl_2F_2 mixture

In the case of positive deviation of Raoult's law, vapour pressure is higher than the corresponding ideal vapour pressure and hence boiling point usually decreases with mole fraction giving rise a minimum (Fig. 15.5a). At the minimum (the point O in the Fig. 15.5a) bubble point curve and the dew point curve merge and the mixture, like a pure substance, boils or condenses at a constant pressure with a constant composition. The mixture is called azeotrope. As the boiling point of azeotrope is lower than that of either component, it is called **_minimum boiling azeotrope_**. The temperature-composition phase diagram is shown in the Fig. 15.5a.

The nature of vapour pressure-composition diagram of such system is shown in the Fig. 15.5b.

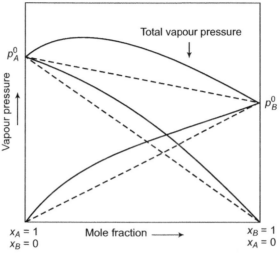

Figure 15.5b   Pressure-Composition diagram of a two-component liquid-vapour phase system, showing positive deviation of Raoult's law. System: $CHClF_2$ and CCl_2F_2 mixture

In the case of negative deviation of Raoult's law vapour pressure is lower than the corresponding ideal mixture and hence boiling point usually increases with the mole fraction, giving rise a maximum (Fig. 15.6a). At the maximum (the point O in the Fig. 15.6a) bubble point curve and dew point curve merge and the corresponding mixture is called **_maximum boiling azeotrope_**. The temperature-composition phase diagram is shown in the Fig. 15.6a.

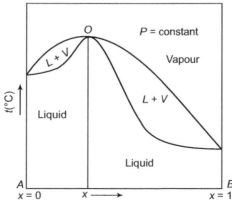

Figure 15.6a   Two component liquid-vapour phase diagram, showing negative deviation of Raoult's law. System: acetone/chloroform mixture

The nature of vapour pressure-composition diagram of such system is shown in the Fig. 15.6b.

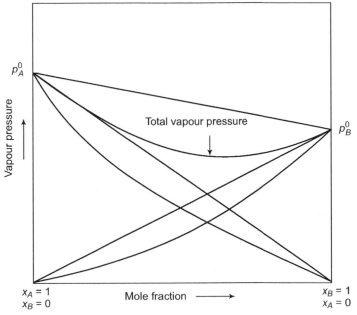

Figure 15.6b   Pressure-Composition diagram of a two-component liquid-vapour phase system, showing negative deviation of Raoult's law. System: acetone/chloroform mixture

15.6 PARTIALLY MISCIBLE LIQUID SOLUTIONS

15.6.1   Upper Consolute Temperature and Lower Consolute Temperature

In case of binary liquid system, it may be possible that both the components are partially miscible in each other, forming two phases such that each phase contains both the components in equilibrium. Examples are phenol-water system, ether-water system, aniline-water system, etc. There are two distinct liquid layers in each case. Aqueous layer is the saturated solution of organic compound and organic layer is the saturated solution of water. These two liquid layers in equilibrium are called conjugate solutions. In some examples both organic layers are formed, e.g., aniline-hexane system, carbon disulphide-methanol system, etc.

If the properties of the two liquid layers, e.g., internal pressure, polarity, association etc., widely differ, corresponding vapour pressure is expected to be greater than that of ideal system as described by Raoult's law. So, the deviation from ideal behaviour is positive as shown in the Fig. 15.7.

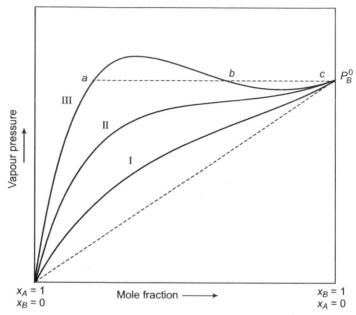

Figure 15.7 Pressure-Composition diagram of a two-component partially miscible liquid-liquid system

The curves I, II and III represent three different types of systems. However, curve III is particularly interesting as it is clearly observed that there are three points with different compositions, *a*, *b* and *c* with same vapour pressure. This signifies that four phases (3 liquid phases and vapour phase) remain in equilibrium, the phenomenon which is impossible to happen according to Gibbs phase rule $F = C - P + 2$. As temperature is fixed, $F = 1$. It is a two component system. So, $C = 2$. Hence, $P = 3$. So, maximum three phases may remain in equilibrium. In actual practice, there is no existence of the liquid phase *b*. Vapour pressure remains constant from the point *a* to the point *c*.

This type of partially miscible two phase liquid-liquid system can be classified into two categories:

(i) There are some systems where miscibility of both the components into each other increases with increase in temperature and finally a temperature is reached when two phases disappear to form a single phase. That temperature is called upper critical solution temperature (UCST) or upper consolute temperature, at which the following reaction takes place:

$$\text{Liquid } (L) \rightleftharpoons \text{Liquid } (L_1) + \text{Liquid } (L_2)$$

As the effect of pressure on liquid is considered negligible, the phase rule may be revised as:

$F = C - P + 2 - 1 = C - P + 1$. At UCST, $C = 2$ and $P = 3$ and hence, $F = 0$. Thus, UCST is an invariant point.

Examples of such systems are: (a) phenol-water system, (b) methanol-cyclohexane, (c) aniline-hexane system, etc. The phase diagram of such system is shown in the Fig. 15.8.

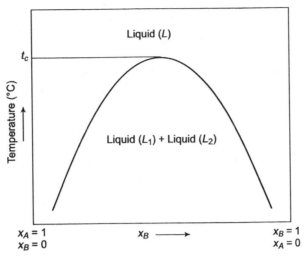

Figure 15.8 Temperature versus composition diagram of a UCST system. x_A and x_B are the mole fractions of the components A and B respectively

The phase diagram of aniline-hexane system is shown in the Fig. 15.9. The point C is the invariant point. $L_1 L_2$ is the tie line connecting the two conjugate solutions. At the point L_1, conjugate solution is enriched with the component A, i.e., aniline and at the point L_2, conjugate solution is enriched with the component B, i.e., hexane. Compositions of these two conjugate solutions at the point X can be calculated by using lever rule.

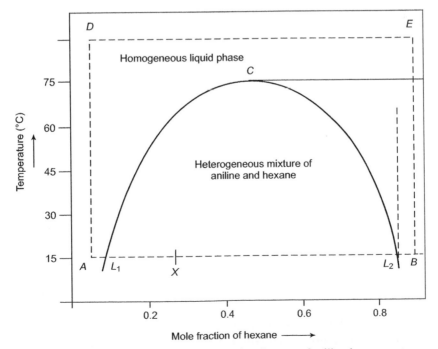

Figure 15.9 Temperature versus composition diagram of aniline-hexane system

$$\frac{\text{Conjugate solution enriched with } A}{\text{Conjugate solution enriched with } B} = \frac{XL_2}{XL_1}$$

The length of the tie line L_1L_2 decreases as the temperature increases and eventually merges at a point C, which is known as UCST, at which the two liquid layers are just miscible.

(ii) There are some other systems where both the components are completely miscible at low temperature but partially miscible at higher temperature, forming two distinct phases. That temperature is called lower critical solution temperature (LCST) or lower consolute temperature, at which the following reaction takes place

$$\text{Liquid } (L) \rightleftharpoons \text{Liquid } (L_1) + \text{Liquid } (L_2)$$

As the effect of pressure on liquid is considered negligible, the phase rule may be revised as

$F = C - P + 2 - 1 = C - P + 1$. At LCST, $C = 2$ and $P = 3$ and hence, $F = 0$. Thus, LCST is an invariant point.

One example of such system is: triethylamine-water system. The phase diagram of such system is shown in the Fig. 15.10.

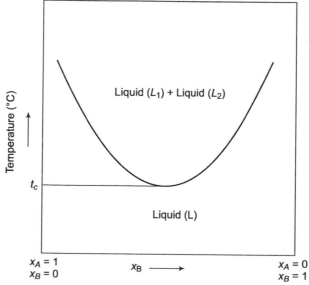

Figure 15.10 Temperature versus composition diagram of a LCST system. x_A and x_B are the mole fractions of the components A and B respectively

(iii) There are other systems, which possess both lower and upper consolute temperatures. The corresponding phase diagram shows a closed loop as shown in the Fig. 15.11. The figure shows that the components are immiscible only in the closed loop and the system remains as homogeneous single phase outside the loop.

One such example is nicotine-water system. t_{c1} and t_{c2} are the LCST and UCST respectively. Both t_{c1} and t_{c2} are considered invariant points.

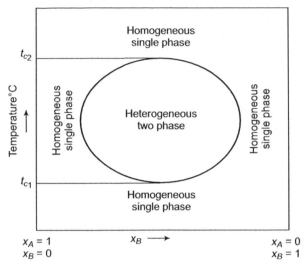

Figure 15.11 Temperature versus composition diagram of a UCST-LCST system. x_A and x_B are the mole fractions of the components A and B respectively

If a LASER beam is passed through any of such systems, described earlier and the temperature of the system is increased regularly, it will be observed that the emergent LASER beam is continuous till the system remains homogeneous single phase but the same emergent LASER beam appears as scattered when solution becomes heterogeneous two phase system.

15.7 PARTITION COEFFICIENT OR DISTRIBUTION COEFFICIENT

If a solute is added in a mixture of two immiscible solvents, the solute has a natural tendency to distribute itself in the two solvents. The distribution process proceeds through diffusion and as diffusion is a slow process, it takes long time to reach the equilibrium. However, the mixture is often subjected to shaking, which accelerates the diffusion process so that the equilibrium is reached quickly. At equilibrium, chemical potential of the solute is same in both the solvents. A and B represent the two immiscible solvents and C_A, C_B represent concentrations of the solute in the solvents, A and B respectively. The chemical potentials of the solute in both the solvents are μ_A and μ_B which can be written as:

$\mu_A = \mu_A^0 + RT \ln C_A$ and $\mu_B = \mu_B^0 + RT \ln C_B$. At equilibrium $\mu_A = \mu_B$.

Or, $\mu_A^0 + RT \ln C_A = \mu_B^0 + RT \ln C_B$. μ_A^0 and μ_B^0 are constants and the distribution process is carried out at constant temperature. So, we can write that

$$\ln\left(\frac{C_A}{C_B}\right) = \frac{\mu_B^0 - \mu_A^0}{RT} = \text{Constant. Or,} \ \frac{C_A}{C_B} = \text{Constant} = K_D \tag{15.13}$$

This K_D is known as partition coefficient or distribution coefficient of the solute.

15.7.1 Characteristics of K_D

(a) For a given solute the value of K_D depends on type of solvents chosen. Changing the type of solvent, K_D value changes.

(b) The value of K_D depends on temperature as evident from the equation (15.13). With increase in temperature, diffusion process is faster and less time is required to reach the equilibrium.

(c) The K_D has no unit.

(d) The expression of K_D in equation (15.13) is valid only if the solute maintains its chemical formula in both the solvents without any association or dissociation.

(e) If the solute (S) remains partially dissociated in solvent B, the following equilibrium occurs:

$S \rightleftharpoons S_1 + S_2$. If α is the degree of dissociation, concentration of undissociated solute is $(1 - \alpha)C_B$. Thus the equation (15.13) becomes

$$\frac{C_A}{(1-\alpha)C_B} = K_D \tag{15.14a}$$

If the solute remains partially dissociated in both the solvents and α_1 and α_2 are the degree of dissociation of the solute in solvents A and B respectively, the equation (15.13) becomes

$$\frac{(1-\alpha_1)C_A}{(1-\alpha_2)C_B} = K_D \tag{15.14b}$$

The K_D has no unit.

(f) If the solute remains associated in one of the solvents, say solvent B, the following equilibrium exists $nS \rightleftharpoons S_n$, where, n is the degree of association. If α is the fraction of solute, remained associated, concentration of unassociated solute is $C_S = (1 - \alpha)C_B$ and concentration of associated solute is $C_{S_n} = \alpha C_B/n$. Hence, the equilibrium constant, K, is given by

$$K = \frac{C_{S_n}}{C_S^n} = \frac{\alpha C_B}{n(1-\alpha)^n . C_B^n}. \text{ Or, } (1-\alpha)C_B = \left(\frac{\alpha C_B}{n.K}\right)^{\frac{1}{n}}$$

Thus the concentration of unassociated solute is

$$C_S = (1-\alpha)C_B = \left(\frac{\alpha C_B}{n.K}\right)^{\frac{1}{n}} = K'C_B^{\frac{1}{n}}, \text{ where, } K' = \left(\frac{\alpha}{n.K}\right)^{\frac{1}{n}}$$

Thus the equation (15.13) becomes

$$\frac{C_A}{C_S} = \frac{C_A}{K'C_B^{\frac{1}{n}}} = K_D'. \text{ Or, } \frac{C_A}{C_B^{\frac{1}{n}}} = K'.K_D' = K_D \tag{15.14c}$$

The unit of K_D is $\left(\dfrac{\text{moles}}{\text{litre}}\right)^{1-\frac{1}{n}}$.

15.7.2 Applications of K_D

(a) To determine equilibrium constant

Let us consider the following equilibrium constant $A + B \rightleftharpoons AB$ and one of the reactants, say reactant A has two immiscible solvents, I and II. The another reactant, B, is soluble in solvent II only. The reaction is carried out in that solvent, where the reactants and products, A, B and AB are soluble. The equilibrium constant is given by $K = \dfrac{C_{AB}}{C_A.C_B}$, where, C_A and C_B are the concentrations of free reactants, A and B respectively, C_{AB} is the concentration

of the product. C_B^0 is the initial concentration of B. Initially the reactant A is allowed to distribute in a system of two solvents, I and II. The distribution coefficient is K_D. Under this condition, C_A^0 is the concentration of A in solvent II. Now the reactant B with concentration C_B^0 is added into the system. The reactant B is dissolved in solvent II and reacts with the other reactant A to form the product AB and the free reactant A redistributes itself into both the solvents. Finally the equilibrium is established in both the solvents.

C_A^I is the concentration of the reactant A in the solvent I and can be estimated. $K_D = C_A^I/C_A$, where, C_A is the concentration of free reactant A in the solvent II. K_D value is known and hence, $C_A = C_A^I/K_D$.

C_{AB} = Concentration of the product in solvent II = (Concentration of A in solvent II) – (Concentration of free reactant A in the solvent II) = $(C_A^0 - C_A) = (C_A^0 - C_A^I/K_D)$

[Since 1 mole of reactant A consumed in solvent II $\equiv$ 1 mole of the product AB]

C_B = Concentration of free reactant B in the solvent II = (Initial concentration of the reactant B) – (Concentration of the product) = $(C_B^0 - C_A^0 + C_A^I/K_D)$

[Since 1 mole of reactant B consumed in solvent II $\equiv$ 1 mole of the product AB]

Hence, equilibrium constant of the above reaction, K, can be expressed as

$$K = \frac{C_{AB}}{C_A.C_B} = \frac{(C_A^0 - C_A^I/K_D)}{(C_A^I/K_D)(C_B^0 - C_A^0 + C_A^I/K_D)} \tag{15.15}$$

Let us consider the equilibrium reaction: $I^- + I_2 \rightleftharpoons I_3^-$.

Now two immiscible solvents of I_2 are chosen, i.e., water and CCl_4. A given amount of iodine is added in KI solution of known concentration C_{I^-}. The other solvent, CCl_4 is added into it so that iodine is distributed in both aqueous layer and organic layer. C_A and C_O are the concentrations of I_2 in aqueous layer and organic layer respectively. One fraction of I_2 reacts with I^- to form I_3^- according to the above reaction.

$$K_c = \frac{[I_3^-]}{[I^-][I_2]} = \frac{C_{I_3^-}}{C_{I^-}^{free} \cdot C_{I_2}^{free}} \quad \text{and} \quad C_{I_2} = C_{I_2}^{free} + C_{I_3^-}, \quad C_{I^-} = C_{I^-}^{free} + C_{I_3^-}$$

C_{I^-} = Initial concentration of KI solution and is known.

C_{I_2} = The total concentration of iodine in aqueous solution. In a separate experiment the same amount of iodine is dissolved in same volume of distilled water instead of KI solution. The iodine can be estimated by iodometric titration using standard $Na_2S_2O_3$ solution.

C_O can be determined by titration of organic layer. $C_{I_2}^{free}$ can be estimated using partition coefficient, K_D, which is given by

$$K_D = \frac{C_O}{C_{I_2}^{free}} \quad [K_D \text{ value is known}]$$

$C_{I_3^-} = C_{I_2} - C_{I_2}^{free}$. C_{I_2} can be estimated by titration of aqueous layer. $C_{I_2}^{free}$ is already

determined. Hence, $C_{I_3^-}$ can be estimated.

$C_{I^-}^{free} = C_{I^-} - C_{I_3^-}$. Here, $C_{I_3^-}$ is already estimated and C_{I^-} is known. Hence, $C_{I^-}^{free}$ can

be calculated. Thus K_c can be determined easily.

(b) Extraction of solute by using second solvent, also called leaching

Let us consider v *ml* of a solution (Phase-I), containing w *gm* of solute. In order to extract the solute from the solution, l *ml* of second solvent (Phase-II), immiscible to the first one, is added into it and shaken the mixture well till equilibrium is established. At equilibrium, suppose, Phase-I contains w_1 *gm* of the solute. So, $(w - w_1)$ gm of solute diffuses into the second solvent or Phase-II. Or, in other words, $(w - w_1)$ gm of the solute is extracted or leached out from the 1st solvent. So, concentration of the solute in the 1st solvent is w_1/v (gm/ml), while that in the 2nd solvent is $(w - w_1)/l$ (gm/ml).

Thus, the distribution coefficient, K_D can be written as

$$K_D = \frac{w_1/v}{(w - w_1)/l}. \quad \text{Hence after 1}^{st} \text{ extraction,} \quad w_1 = w\left(\frac{K_D \times v}{K_D v + l}\right)$$

After the 2nd extraction, the Phase-II contains $(w_1 - w_2)$ gm of the solute and 1st solvent contains w_2 gm. Thus w_2 can be written as

$$w_2 = w_1\left(\frac{K_D \times v}{K_D v + l}\right) = w\left(\frac{K_D \times v}{K_D v + l}\right)^2$$

So after n^{th} extraction, amount of solute, still remaining in Phase-I is

$$w_n = w\left(\frac{K_D \times v}{K_D v + l}\right)^n \tag{15.16}$$

To achieve high efficiency in extraction process, the value of n should be kept large while the value of v is to be maintained small.

(c) To determine the degree of association of solute

Let us consider a solute, which remains unassociated in aqueous solution with concentration C_A and remains associated in a given organic solvent with concentration C_O. So the K_D is given by

$$K_D = \frac{C_A}{C_O^{\frac{1}{n}}}. \quad C_A = K_D . C_O^{\frac{1}{n}}.$$

Taking ln on both sides of the above equation, we get

$$\ln C_A = \ln K_D + \frac{1}{n} \ln C_O \qquad (15.17)$$

By changing the solute concentration, both C_A and C_O change and hence plotting of $\ln C_A$ versus $\ln C_O$ will give rise to a straight line with slope $(1/n)$ and intercept $\ln K_D$. Hence, both degree of association (n) and K_D can be evaluated from the graph.

15.7.3 Scheil-Gulliver Equation

Assumptions

(a) There is no diffusion of solute in the solid.

(b) The solute is completely soluble in the liquid.

(c) Equilibrium exists at the interface so that equilibrium phase diagram is applicable at the interface.

(d) At the solid-liquid interface it is assumed that f_L fraction of liquid is in equilibrium with f_S fraction of solid such that:

$$f_L + f_S = 1 \text{ and hence, } f_L = (1 - f_S) \qquad (15.18)$$

Now the fraction of solid increased at the interface is df_S in time dt as shown in the Fig. 15.12.

C_L and C_S are the equilibrium concentrations of solute in liquid phase and solid phase respectively at time t.

dC_L = Change in solute concentration in liquid phase.

$f_L dC_L$ = Amount of solute, initially present in the liquid, converted to solid phase in time dt.

$(C_L - C_S)df_S$ = Amount of solute, present in the solid front, developed in time dt.

Using the conservation of mass, we can write

$$(C_L - C_S) df_S = f_L dC_L = (1 - f_S) dC_L \qquad (15.19)$$

[Using equation (15.18)]

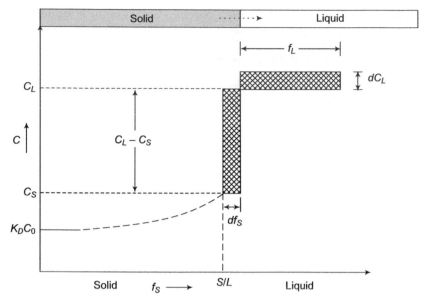

Figure 15.12   Solidification of an alloy following Scheil equation

Solute is distributed in both liquid and solid phases and partition coefficient is K_D, which is expressed as

$$K_D = \frac{C_S}{C_L}. \text{ Or, } C_S = K_D C_L \tag{15.20}$$

Combining equations (15.19) and (15.20), we get

$$\frac{df_S}{(1-f_S)} = \frac{1}{(1-K_D)} \frac{dC_L}{C_L} \tag{15.21}$$

When $f_S = 0$, $C_L = C_0$ and $f_S = f_S$, $C_L = C_L$. Integrating the equation (15.21) using these boundary conditions we get

$$\int_0^{f_S} \frac{df_S}{(1-f_S)} = \frac{1}{(1-K_D)} \int_{C_0}^{C_L} \frac{dC_L}{C_L}$$

Or, $C_L = C_0(1-f_S)^{K_D-1}$. Or, $C_L = C_0 f_L^{K_D-1}$ (15.22)

Using equation (15.20), we get

$$C_S = K_D C_0 (1-f_S)^{K_D-1} \tag{15.23}$$

Both the equations (15.22) and (15.23) are popularly known as Scheil-Gulliver Equation or simply Scheil equation. This equation is very much useful in metallurgy in describing the solute redistribution during solidification of an alloy.

15.8 LIQUID-SOLID BINARY SYSTEM

15.8.1 Binary System where Components are Miscible in Both Liquid and Solid Phases

There are several solid-liquid binary systems, where the components are miscible in both liquid and solid phases to form homogeneous liquid and solid solutions. However, three different types are quite discernible.

(i) Freezing and melting curves are continuous without touching each other.

The phase diagram of such system is shown in the Fig. 15.13.

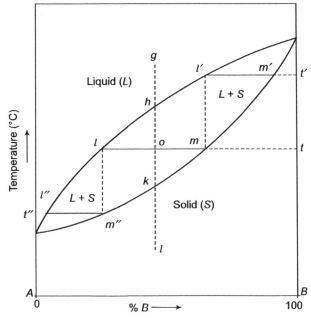

Figure 15.13 Phase diagram of a binary system containing two phases. Freezing point and melting point curves do not touch each other

The lower curve is called freezing curve and the upper curve is called melting curve. Component *A* has lower melting point than that of component *B*. Addition of lower melting component, *A*, lowers the freezing point of the liquid solution and conversely addition of higher melting component, *B*, raises the freezing point of the liquid solution. Fractional crystallization at any point on the melting curve gives rise to the pure component with higher freezing point, i.e., component, *B*. Within the loop, liquid and solid phases coexist. *lm* is the tie line, representing isothermal transformation of liquid phase to solid phase. *ghokl* is the cooling curve, where the point *g* represents homogeneous liquid phase. On cooling the point *h* is reached where first solid phase appears. The cooling curve meets the tie line, *lm*, at the point *o*, where the composition of equilibrium liquid and solid phases can be calculated using *lever rule* as given:

$$\% \text{ solid phase} = \frac{ol}{lm} \times 100, \quad \% \text{ liquid phase} = \frac{om}{lm} \times 100$$

Examples of such system are: Cu/Ni, Co/Ni, Au/Ag, Au/Pt, AgCl/NaCl, etc.

(ii) The freezing and melting curves touch each other at the maximum freezing point. The phase diagram of such system is shown in the Fig. 15.14.

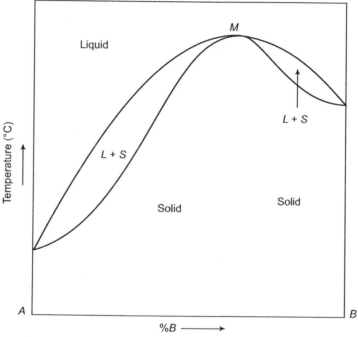

Figure 15.14 Phase diagram of a binary system containing two phases. Azeotropic mixture is formed with maximum freezing point

The phase diagram shows that addition of any component in the mixture raises the freezing point of the mixture. As a result of this, melting and freezing curves meet at the point *M*, at which fractional crystallization will lead only a solid solution with constant composition. This solid solution is known as ***azeotropic mixture with maximum freezing point***, which has same melting and freezing point like pure component. In order to continue the fractional crystallization: composition of the solid solution needs to be changed.

Examples are: *d* – and *l* – carvoxime, *d* – and *l* – *a* – bromocamphor systems, etc.

(iii) Freezing and melting curves touch each other at the minimum freezing point. The phase diagram of such system is shown in the Fig. 15.15.

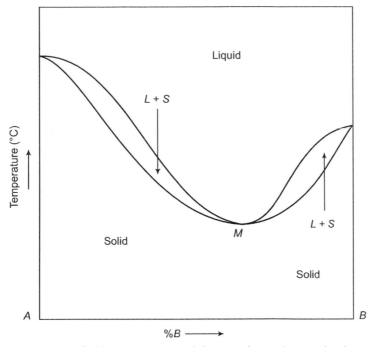

Figure 15.15 Phase diagram of a binary system containing two phases. Azeotropic mixture is formed with minimum freezing point

The phase diagram shows that addition of any component in the mixture reduces the freezing point of the mixture. As a result of this melting and freezing curves meet at the point *M*, at which fractional crystallization will lead only a solid solution with constant composition. This solid solution is known as ***azeotropic mixture with minimum freezing point***, which has same melting and freezing point like pure component. In order to continue the fractional crystallization, composition of the solid solution needs to be changed.

Examples are: Cu/Mn, Cu/Au, Mn/Ni, Co/Mn systems, etc.

15.8.2 Binary System where Components are Miscible in Both Liquid and Solid Phases and form a Solid Compound, which is Miscible in Either Component

The phase diagram of such system is shown in the Fig. 15.16.

In this system the compound *AB* is miscible in both the components so that the solid solution contains the components *A* and *B* as well as the compound *AB*. The compound *AB* is formed at the point *C*, at a temperature, higher than the melting point of *A* but lower than the melting point of *B*. Hence the point *C* is called transition point. Examples of such system are: (i) iodine and bromine, (ii) magnesium and cadmium.

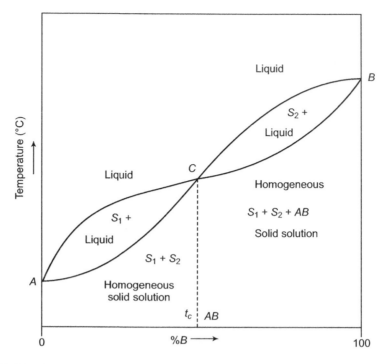

Figure 15.16   Phase diagram of a binary system where a solid compound is formed. The compound is miscible in either component

15.8.3   Binary System where Components form a Compound with a Congruent Melting Point and the Compound is Immiscible in Both the Components

Type-I   Components form one stable compound with a congruent melting point. The phase diagram of such system is shown in the Fig. 15.17.

Formula of the compound depends on the composition. If the mole ratio of the components is 50:50, the formula of the compound is AB. The components A and B form an intermediate compound AB. There are two eutectic points, E_1 and E_2. At E_1, component A and compound AB are in equilibrium with the liquid. At E_2, component B and compound AB are in equilibrium with the liquid. The tie line ab indicates the equilibrium mixture of the pure component A and the homogeneous liquid. The tie lines $a'b'$ and lm indicate the equilibrium mixture of the compound and the homogeneous liquid. However, both tie lines ($a'b'$ and lm) merge at the point C, where solid and liquid compositions are same. Hence the temperature corresponding to the point C is called congruent melting point of the compound, AB. The tie line $l'm'$ indicates the equilibrium mixture of the pure component B and the homogeneous liquid.

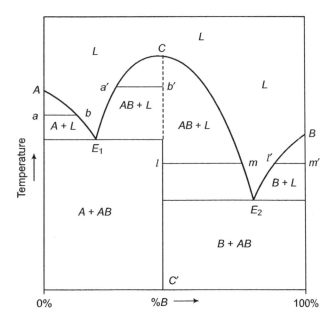

Figure 15.17 Phase diagram of a binary system where a stable solid compound with congruent melting point is formed

At E_1, $L \rightleftharpoons A(s) + AB(s)$. At E_2, $L \rightleftharpoons B(s) + AB(s)$

At the point C, the compound AB partially dissociates into the pure components A and B as follows:

$AB(1 - \alpha) \rightleftharpoons A(\alpha) + B(\alpha)$, where, α is the degree of dissociation and $(1 + \alpha)$ is the total number of moles. Hence, we have

$$x_A = x_B = \frac{\alpha}{1 + \alpha}, \quad x_{AB} = \frac{1 - \alpha}{1 + \alpha}, \quad K = \frac{x_A \times x_B}{x_{AB}} = \frac{\alpha^2}{(1 - \alpha)(1 + \alpha)} = \frac{\alpha^2}{1 - \alpha^2}$$

The mole fraction x_{AB} is related to freezing point depression, ΔT as follows

$$\ln x_{AB} = -\frac{L_f}{R}\left[\frac{1}{T_0} - \frac{1}{T}\right] = -\frac{L_f}{R}\left(\frac{\Delta T}{T.T_0}\right),$$

where, T_0 = Theoretical melting temperature of AB and $\Delta T = T - T_0$

Thus by measuring ΔT and x_{AB}, Latent heat of fusion (L_f) of the compound, AB, can be determined.

One example of this type of system is Pb/Mg system. The phase diagram is shown in the Fig. 15.18.

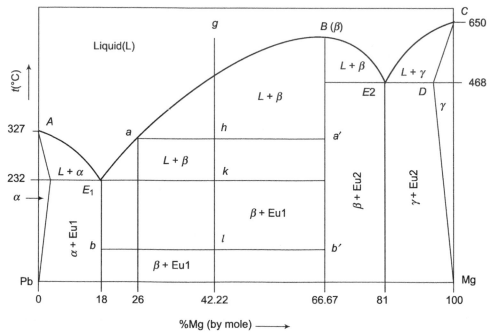

Figure 15.18 Phase diagram of Pb/Mg system

Eutectic point at E1 Temperature is 232°C, composition is 26.67% (by mole) $Mg_2Pb(s)$ and 73.33% (by mole) α-phase. At the point A, Pb-content is 100% and at the point B, Mg_2Pb-content is 100%. Reaction is given by

$$\text{Melt} \underset{\text{Heating}}{\overset{\text{Cooling}}{\rightleftarrows}} \underset{(\alpha\text{-phase})}{Pb(s)} + \underset{(\beta\text{-phase})}{Mg_2Pb(s)}$$

Eutectic point at E2 Temperature is 468°C, composition is 52.78% γ- phase and 47.22% $Mg_2Pb(s)$-phase. At the point B, $Mg_2Pb(s)$-content is 100%, at the point C, Mg(s)-content is 100% and at the point D γ-phase is 100%. Reaction is given by

$$\text{Melt} \underset{\text{Heating}}{\overset{\text{Cooling}}{\rightleftarrows}} \underset{(\gamma\text{-phase})}{Mg(s)} + \underset{(\beta\text{-phase})}{Mg_2Pb(s)}$$

γ-phase at the point D is a homogeneous solid solution, containing 93% Mg with 7% Pb by mole basis. Consider the Pb/Mg alloy homogeneous melt at the point g and it is allowed to cool along *ghkl*. Draw two tie lines *aha'* and *blb'* through the points h and l respectively.

At a' and b' $Mg_2Pb(s)$ (β-phase) is 100% (by mole basis).

At h and l $Mg_2Pb(s)$ (β-phase) is 57.78% (by mole basis).

At a $Mg_2Pb(s)$ (β-phase) is 39% (by mole basis).

At b $Mg_2Pb(s)$ (β-phase) is 26.67% (by mole basis).

Consider the lever *aha'*

$$\% \ \beta\text{-phase} = \frac{ah}{aa'} \times 100 = \frac{(57.78 - 39)}{(100 - 39)} \times 100 = 30.79\%$$

$$\% \ \text{melt phase} = \frac{a'h}{aa'} \times 100 = \frac{(100 - 57.78)}{(100 - 39)} \times 100 = 69.21\%$$

Consider the lever *blb'*

$$\% \ \beta\text{-phase} = \frac{bl}{bb'} \times 100 = \frac{(57.78 - 26.67)}{(100 - 26.67)} \times 100 = 42.42\%$$

$$\% \ \text{eutectic} = \frac{b'l}{aa'} \times 100 = \frac{(100 - 57.78)}{(100 - 26.67)} \times 100 = 57.78\%$$

Some examples of this type of system are listed in the Table 15.2.

Table 15.2 Systems having congruent melting point

Component 1		Component 2		Compound	
A	*Melting point (°C)*	*B*	*Melting point (°C)*	*Type*	*Melting point (°C)*
Gold	1064	Tin	232	AB	425
Urea	132	Phenol	43	AB_2	61
Aluminium	657	Magnesium	650	A_3B_4	463

Type-II Components form more than one stable compound with congruent melting points. The phase diagram of such system is shown in the Fig. 15.19.

In some cases two components of a binary system form more than one compound with congruent melting point. Corresponding phase diagram has three distinct eutectic points as shown in the Fig. 15.19.

At E_1, $L \rightleftharpoons A(s) + A_2B(s)$

At E_2, $L \rightleftharpoons A_2B(s) + AB_2(s)$

At E_3, $L \rightleftharpoons B(s) + A_2B(s)$

The compounds A_2B and AB_2 have congruent melting points at the points D and F respectively.

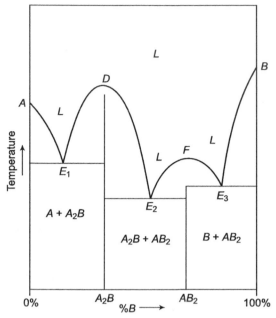

Figure 15.19   Phase diagram of a binary system where more than one stable solid compound with congruent melting point is formed

15.8.4   Binary System where Components are Immiscible in Solid State but Partially Miscible in Liquid State

The phase diagram of this type of system is shown in the Fig. 15.20.

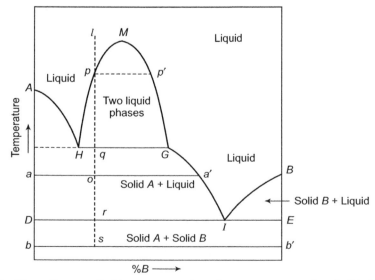

Figure 15.20   Phase diagram of a binary system where the components are immiscible in solid phase but partially miscible in liquid phase

Along AH liquid phase is homogeneous and below the curve AH both liquid and solid components A coexist. At the point H, two distinct liquid phases first appear and remain in equilibrium with solid and hence degrees of freedom, $F = 0$. The system is invariant. With increase in %B, the heterogeneous liquid phase forms a phase diagram HMG, where M is the UCST. At the point G, heterogeneity almost disappears and further increase in B gives rise to only two phases, liquid and solid A and the system becomes univariant. At the point I, the liquid phase remains in equilibrium with two solid phases, A and B. Again the system becomes invariant and the point I is the eutectic point. Temperature remains constant until all liquid condenses.

lpqors is the cooling curve at a particular composition of B. At the point l, only homogeneous liquid phase exists. On cooling the point p is reached where another liquid phase appears. On further cooling %heterogeneity in liquid phase increases and the point q is reached where the system becomes invariant. Temperature is constant till all heterogeneity of liquid phase disappears. The tie line aoa' represents the composition of solid and liquid phases.

$$\% \text{ Solid } A = \left(\frac{oa'}{aa'}\right) \times 100, \quad \% \text{ Liquid} = \left(\frac{oa}{aa'}\right) \times 100$$

On further cooling, the point r is reached where the following equilibrium exists and the temperature is called eutectic temperature. $L \rightleftharpoons A(s) + B(s)$. Temperature remains constant till all the liquid condenses.

$$\% \text{ Solid } A = \left(\frac{rI}{ID}\right) \times 100, \quad \% \text{ Eutectic} = \left(\frac{rD}{ID}\right) \times 100$$

Again temperature decreases to the point s, where another tie line bsb' is obtained. Composition of two solid phases can be calculated as follows:

$$\% \text{ Solid } A = \left(\frac{sb'}{bb'}\right) \times 100, \quad \% \text{ Solid } B = \left(\frac{sb}{bb'}\right) \times 100.$$

15.8.5   Salt Hydrate with Congruent Melting Point

There are some salts, which remain as hydrated form. Some of them may remain as different type of hydrates. One example is ferric chloride-water system, which forms four stable hydrates with congruent melting points. The corresponding phase diagram is shown in the Fig. 15.21.

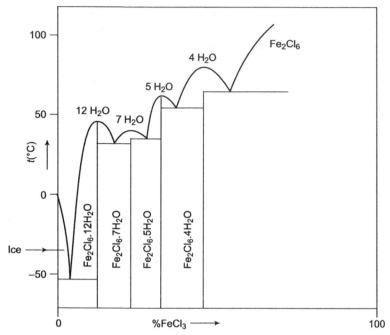

Figure 15.21   Phase diagram of ferric chloride-water system

Figure 15.22 shows the partial phase diagram of $FeCl_3$– hydrates where an isothermal phase transformation line *al* is observed.

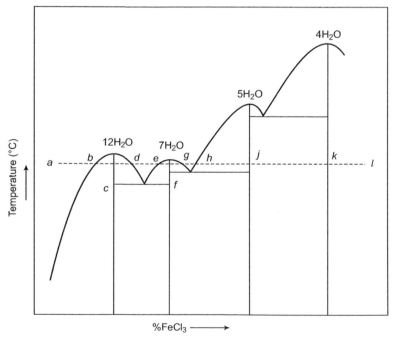

Figure 15.22   Partial phase diagram of ferric chloride-water system. *al* is the isothermal phase transformation line

At the point a complete homogeneous $FeCl_3$ solution is obtained. On increasing $FeCl_3$ concentration, the point b is reached, where first solid phase $FeCl_3 \cdot 12H_2O$ develops. Further increase in $FeCl_3$ concentration %solid phase increases and %liquid phase decreases and the point c is reached, where only solid hydrate, $FeCl_3 \cdot 12H_2O$ exists. Further increase in $FeCl_3$ concentration %solid phase decreases along with increase in %liquid phase. The point d is reached, where only homogeneous $FeCl_3$ solution exists. The homogeneous solution remains unchanged on increase in $FeCl_3$ concentration till the point e is reached, where the 2nd solid hydrate, $FeCl_3 \cdot 7H_2O$ appears. Further increase in $FeCl_3$ concentration %solid phase increases along with decrease in %liquid phase. The point f is reached, where only solid hydrate, $FeCl_3 \cdot 7H_2O$ exists. Further increase in $FeCl_3$ concentration %solid phase decreases along with increase in %liquid phase. The point g is reached, where only homogeneous $FeCl_3$ solution exists. The homogeneous solution remains unchanged on increase in $FeCl_3$ concentration till the point h is reached, where the 3rd solid hydrate, $FeCl_3 \cdot 5H_2O$ appears. At the point j only solid hydrate, $FeCl_3 \cdot 5H_2O$ exists. Similarly at the point k, the 4th solid hydrate, $FeCl_3 \cdot 4H_2O$ exists only. Finally at the point l homogeneous solution of $FeCl_3$ is obtained. The peaks of the curve at points c, f, j and k represent the melting points of the hydrates, $FeCl_3 \cdot 12H_2O$, $FeCl_3 \cdot 7H_2O$, $FeCl_3 \cdot 5H_2O$ and $FeCl_3 \cdot 4H_2O$ respectively.

15.8.6 Binary System Containing Salt and Water

There are several salt-water systems. One such example is *KI*-water system. The freezing point of pure water is 0°C but as *KI* is continuously added into water, freezing point decreases along AC and reaches a minimum at –20°C. At the point C, $F = 0$ and hence it is called eutectic point. The scientist, F. Rüdorff, first studied this type of salt-water system and designated the eutectic point as ***cryohydric point***. Further increase in *KI* concentration increases the freezing point of the mixture along CB. The hydrate, separated along the curve CB is called ***cryohydrate***. As the solubility of *KI* in water is very high, the curve CB never touches the *KI* axis. At the point C there should be three distinct phases, salt solution, solid *KI* and Ice but in actual practice only two phases, cryohydrate and salt solution, are obtained. Thus, cryohydric point cannot be actually called eutectic point. The phase diagram of *KI*-water system is shown in the Fig. 15.23.

Two important points are to be noted:
(a) Ice can be melted below 0°C in presence of *KI*. The minimum temperature at which ice can be melted using *KI* is –20°C.
(b) Below –20°C only solid phases exist and hence ice cannot be melted below –20°C.

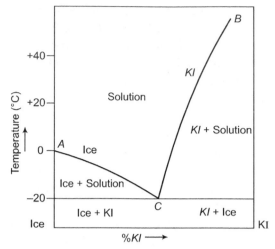

Figure 15.23   Phase diagram of potassium iodide-water system

15.8.7   Binary System where Solubility of Two Components Changes Abruptly with Temperature

In general, in a binary system components may be: (i) completely miscible in solid phase, (ii) partially miscible in solid phase, (iii) immiscible in solid phase. However, there are some binary systems, where components are completely miscible in solid phase at high temperature but immiscible at low temperature to form heterogeneous solid phase. The phase diagram of such a system is shown in the Fig. 15.24.

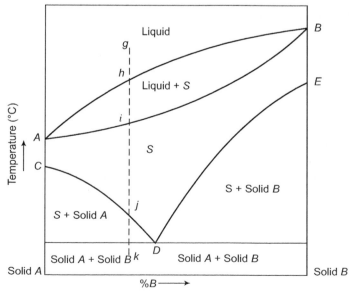

Figure 15.24   Phase diagram of a binary system where components are miscible at high temperature but immiscible at low temperature. *S* represents solid solution

Components A and B form homogeneous liquid and solid phases at high temperature. As the temperature decreases component A is slowly separating out from the solid solution resulting a heterogeneous mixture of solid solution and solid A. Freezing point decreases as the concentration of B increases till the eutectic point D is reached where the solid B starts separating out from the solid solution. *ghijk* is the cooling curve at a given concentration of B. At the point g homogeneous liquid phase exists. On cooling, the point h is reached where the solid phase appears first. On further cooling the point i is reached where solidification is near to complete. Further decrease in temperature gives rise to homogeneous solid solution only. On continuous cooling the point j is reached, where the solid A starts separating out till the point eutectic temperature is reached. At eutectic temperature, solid B starts separating out and temperature remains constant till all the liquid phase gets solidified. Then temperature decreases further to reach the point k where heterogeneous solid phase, containing both the components, exists.

15.9 BINARY PHASE DIAGRAM FOR SOLID SOLUTIONS

15.9.1 Solid Solution

If an alloy contains more than one component but exists as homogeneous single phase, the alloy is called solid solution. Solid solution can be divided into two categories:

(a) Substitutional solid solution There are some solid solutions where some lattice sites of the parent crystal are substituted by the atoms of another metal, giving rise to a homogeneous alloy. This homogeneous alloy is called substitutional solid solution. One such example is Cu/Ni alloy, where the type of crystal lattice remains unchanged but some Cu atoms are replaced by Ni atoms. However, the formation of solid solution depends on the following conditions:

  (i) The size difference between atoms of two elements must be less than 15%.

  (ii) The valency of two elements must be same.

 (iii) The electronegativity difference between two elements should be close to zero.

 (iv) Crystal structures of the two elements must be same.

The above four conditions are popularly known as ***Hume-Rothery conditions*** for the formation of substitutional solid solution. The solid solution, Cu/Ni alloy, is shown in the Fig. 15.25a.

(b) Interstitial solid solution There are some solid solutions, where foreign atoms slowly diffuse into the voids of another metallic crystal, giving rise to a homogeneous alloy. This type of solid solution is called interstitial solid solution. One such example is austenitic steel, where C atoms diffuse into the voids of iron crystal. Figure 15.25b illustrates the interstitial solid solution.

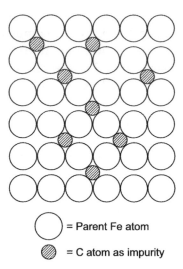

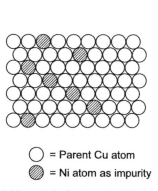

○ = Parent Cu atom

◎ = Ni atom as impurity

○ = Parent Fe atom

◎ = C atom as impurity

Figure 15.25a   Substitutional solid solution: Cu/Ni alloy

Figure 15.25b   Interstitial solid solution: Austenitic steel

However, strength of the solid solution depends on per cent variation of atomic size. Higher the size difference, higher the strength of the solid homogeneous alloy. Furthermore, strength of the homogeneous alloy is further increased if the size of the foreign atom is smaller than that of the parent atom. It is well reflected in the Fig. 15.25c. Size difference between different alloying elements and copper is shown in Table 15.3.

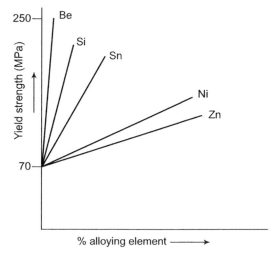

Figure 15.25c   Effect of size difference on yield strength of Cu alloy

Table 15.3   Size difference between copper and other alloying elements

Alloying elements	*Size difference*
Be	−10.6%
Si	−8.0%
Sn	+9.9%
Ni	−2.7%
Zn	+4.2%

It is evident that yield strength of Cu alloy increases as the size difference increases. Furthermore, if the size difference is negative, i.e., size of foreign atom is smaller than that of parent atom, strengthening mechanism is more pronounced. However, only Ni is miscible with Cu at any proportion. This type of system where both the metals are completely miscible to form homogeneous solid solution at any proportion is called "Isomorphous system". The best example of this kind is Cu/Ni system.

15.9.2   Completely Miscible Solid Solutions or Isomorphous System

If the system consists of only liquid and solid solution or solid phases, effect of pressure may be neglected as there is no gaseous phase. Hence, the degrees of freedom is given by:

$$F = C - P + 2 - 1 = C - P + 1.$$

Figure 15.26a shows the binary phase diagram of a completely miscible solid solution (Cu/Ni) system. From the Fig. 15.26a it is evident that freezing point of Cu increases with the addition of Ni or conversely freezing point of Ni decreases with the addition of Cu. The upper curve, *lheak*, is known as *liquidus or freezing curve* and the lower curve, *lcfgk*, is known as *solidus or melting curve*. In between the solidus and the liquidus curve two phases coexist and $F = 1$ but above the liquidus curve and below the solidus curve one phase exists with $F = 2$. *oabcd* is the cooling curve of an alloy solution of a particular composition. Along *oa* the liquid is homogeneous. At the point *a* the solid phase starts separating out and the composition of solid and liquid phases can easily be evaluated by using the tie line *ag*. If the solution is cooled further, the point *b* is reached. The corresponding compositions of liquid and solid phases can be evaluated by using the tie line *ef*. At the point *e* the phase is liquid, enriched with copper and at the point *f* the phase is solid, enriched with nickel. C_l and C_s represent %Ni in the liquid and solid phases respectively. The point *c* on the solidus curve is the last point of heterogeneity. Further cooling reaches the point *d* where homogeneous solid phase results. Other examples of this class of solid solutions are Ag/Cu, Ge/Si, etc.

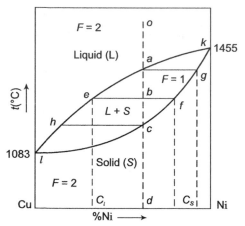

Figure 15.26a   Binary phase diagram of Cu/Ni system

15.9.2.1   Structural Composition/Lever Rule

The isothermal transformation line or tie line is called lever. At any point on the tie line two phases coexist and composition of the phases at that point can be calculated by using lever rule. That particular point on the lever is called fulcrum. In the Fig. 15.26a ef is the tie line, also called lever and the point b is called fulcrum. The liquid phase appears at e and the solid phase appears at f. According to the lever rule, the amount of solid phase is calculated using the portion be of the lever ef and the amount of liquid phase is calculated using the portion bf of the lever ef.

Assume that w gm be the net amount of molten alloy Cu/Ni, at a. The melt is cooled to the point b.

$$\text{Ni – content at } b \text{ is} = 60\% \text{ of the alloy} = 0.60w \qquad (15.24a)$$

At the point b, both liquid phase and solid phase coexist. The liquid phase appears at e and the amount of liquid phase is x gm. The solid phase appears at f and its amount is y gm. So, $w = x + y$.

At the point e, Ni – content is 30% and at the point f, Ni – content is 80%. Considering both the phases

$$\text{Total Ni – content} = (0.30x + 0.80y) \qquad (15.24b)$$

Total Ni – content at the point e + Total Ni – content at the point f = Total Ni–content at the point b or at the point O.

Thus, $0.30x + 0.80y = 0.60w = 0.60\,(x + y)$ [since, $w = x + y$]

Or, $(0.60 - 0.30)x = (0.80 - 0.60)y \qquad (15.24c)$

Or, $x = \dfrac{(0.80 - 0.60)y}{(0.60 - 0.30)} \text{ and } y = \left(\dfrac{0.60 - 0.30}{0.80 - 0.60}\right)x$

Or, $x + y = \dfrac{(0.80 - 0.30)y}{(0.60 - 0.30)} = \dfrac{(0.80 - 0.30)x}{(0.80 - 0.60)} \qquad (15.24d)$

From equations (15.24c) and (15.24d) we have

$$\frac{x}{x+y} = \frac{(0.80-0.60)}{(0.80-0.30)}. \quad \text{Or,} \quad \% \text{ liquid phase} = \frac{(0.80-0.60)}{(0.80-0.30)} \times 100 = \frac{bf}{ef} \times 100 = 40\%$$

Again using equations (15.24c) and (15.24d), we get

$$\frac{y}{x+y} = \frac{(0.60-0.30)}{(0.80-0.30)}. \quad \text{Or,} \quad \% \text{ solid phase} = \frac{(0.60-0.30)}{(0.80-0.30)} \times 100 = \frac{be}{ef} \times 100 = 60\%$$

The cooling curve of Cu/Ni system is shown below.

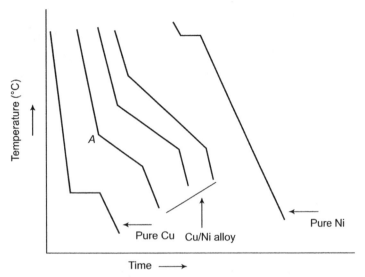

Figure 15.26b   Cooling curve of Cu/Ni system. At the point '*A*' heterogeneous phase appears first

15.9.3   Components Miscible in Liquid Phase but Immiscible in Solid Phase

In the earlier examples metals are completely miscible in both liquid and solid phases. But there are some other systems where metals are completely miscible in liquid phase but totally immiscible in solid phase.

Example-1   One of such systems is Cd/Bi system. The phase diagram of that system is shown in the Fig. 15.27. The line *AE* represents freezing point curve of *Cd*. According to colligative properties freezing point of *Cd* decreases with addition of *Bi*. Both liquid phase and solid *Cd* phase coexist within the triangle AEX. Similarly, the line *BE* represents the freezing point of *Bi*, which shows that with the increase in *Cd* content the freezing point of *Bi* decreases. Within the triangle BEY liquid phase and solid *Bi* phase coexist. At the point *E* two freezing point curves *AE* and *BE* meet and the following reaction takes place at *E*:

$$\text{Liquid (L)} \quad \underset{\text{Heating}}{\overset{\text{Cooling}}{\rightleftharpoons}} \quad \text{Cd(s)} + \text{Bi(s)}$$

Thus, three phases coexist at E and hence $F = C - P + 1 = 2 - 3 + 1 = 0$. The point E is known as *eutectic point* and the line XEY is known as *eutectic line*. At the eutectic point eutectic temperature is 140°C and composition 60%Bi/40% Cd. Consider the point G, where the molten alloy contains 26% Bi. If the alloy is cooled, the point o is reached.

At o both melt and solid Cd(s) coexist. Further cooling touches the eutectic line XEY, where both Cd(s) and eutectic mixture, Eu, starts separating out. The cooling ends at the point H, which is a mixture of Cd(s) and eutectic mixture, Eu only.

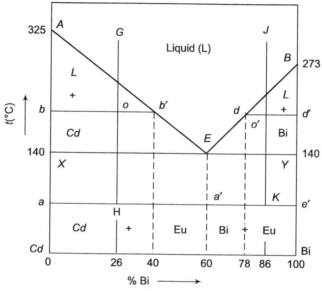

Figure 15.27   Binary phase diagram of Cd/Bi system

To find out the composition at o draw a line parallel to composition axis through the point o. The line bob' is called *lever*.

$$\% \text{ melt phase} = \frac{ob}{bb'} \times 100 = \frac{(26-0)}{(40-0)} \times 100 = 65\%$$

$$\% \text{ Cd-phase} = \frac{ob'}{bb'} \times 100 = \frac{(40-26)}{(40-0)} \times 100 = 35\%$$

To find out the composition at H draw a lever aHa'. At a' 100% Eu and at the point a 100% Cd(s) phase exist. Thus at H composition can be calculated using lever rule.

$$\% Eu = \frac{aH}{aa'} \times 100 = \frac{(26-0)}{(60-0)} \times 100 = 43.33\%$$

$$\% \text{ Cd}(s) \text{ phase} = \frac{Ha'}{aa'} \times 100 = \frac{(60-26)}{(60-0)} \times 100 = 56.67\%$$

Similarly to find out the composition at o' draw a lever $do'd'$. At the point d' 100% $Bi(s)$ phase and at the point d 100% melt exists. Thus at o', composition of the heterogeneous phase is given by

$$\% \text{ melt phase} = \frac{o'd'}{dd'} \times 100 = \frac{(100 - 86)}{(100 - 78)} \times 100 = 63.64\%$$

$$\% \text{ Bi-phase} = \frac{o'd}{dd'} \times 100 = \frac{(86 - 78)}{(100 - 78)} \times 100 = 36.36\%$$

Now consider the point J, where homogeneous molten alloy exists. Bi – content is around 86%. If the molten alloy is cooled from the point J, the point o' is reached. At o' both melt and $Bi(s)$ phases coexist. On further cooling to 140°C liquid phase disappears and eutectic mixture (Eu) and $Bi(s)$ phases are separated out at constant composition. At the point K both Eu –phase and $Bi(s)$ –phase coexist.

To find out the composition at K draw a lever $a'Ke'$. At the point e' 100% $Bi(s)$ phase and at the point a' 100% Eu phase exist. Thus at K, composition of the heterogeneous phase is given by

$$\% Eu = \frac{Ke'}{a'e'} \times 100 = \frac{(100 - 86)}{(100 - 60)} \times 100 = 35\%$$

$$\% Bi(s) \text{phase} = \frac{a'K}{a'e'} \times 100 = \frac{(86 - 60)}{(100 - 60)} \times 100 = 65\%$$

Example-2 Another example of this type of system is Ag/Pb system, binary phase diagram of which is shown in the Fig. 15.28.

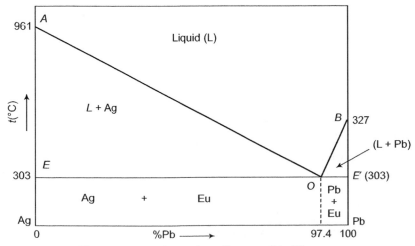

Figure 15.28 Binary phase diagram of Ag/Pb system

The curve AO represents the decrease in melting point of Ag on addition of Pb. The curve BO represents the decrease in melting point of Pb on addition of Ag. EOE′ represents eutectic line. The point O is the eutectic point where three phases are in equilibrium and $F = 0$. The equilibrium reaction is given below

$$\text{Liquid (L)} \underset{\text{Heating}}{\overset{\text{Cooling}}{\rightleftharpoons}} \text{Ag(s)} + \text{Pb(s)}$$

Eutectic temperature is $303°C$ and eutectic composition is 97.4% Pb.

15.9.4 Components are Miscible in Liquid Phase but Only Partially Miscible in Solid Phase

Example-1 Pb/Sn system: In this system Sn is partially miscible in Pb phase to form homogeneous α-phase, containing a maximum of 18% Sn. Similarly Pb is partially miscible in Sn phase to form β-phase, containing a maximum of 3% Pb. The binary phase diagram of Pb/Sn system is shown in the Fig. 15.29a.

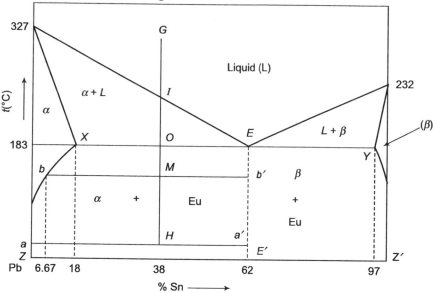

Figure 15.29a   Binary phase diagram of Pb/Sn system

In the above figure XEY represents the eutectic line and E is the eutectic point, having coordinate (183°C, 62% Sn). The molten alloy having composition 38% Sn is cooled continuously. At the point G the molten alloy is homogeneous. At the point O melt phase disappears, giving rise to two solid phases α- and β-phases. At the point M draw the lever bMb'.

$$\% Eu = \frac{bM}{bb'} \times 100 = \frac{(38 - 6.67)}{(62 - 6.67)} \times 100 = 56.62\%$$

$$\% \ \alpha\text{-phase} = \frac{Mb'}{bb'} \times 100 = \frac{(62 - 38)}{(62 - 6.67)} \times 100 = 43.38\%$$

The cooling line ends at the point H where another lever aHa' is drawn.

$$\% Eu = \frac{aH}{aa'} \times 100 = \frac{(38-0)}{(62-0)} \times 100 = 61.29\,\%$$

$$\% Pb(s) = \frac{Ha'}{aa'} \times 100 = \frac{(62-38)}{(62-0)} \times 100 = 38.71\%$$

Microstructures of different points of cooling line GIOMH are shown in the Fig. 15.29b. At the point G in the Fig. 15.29a homogeneous liquid phase exists. In the same figure the line EE' represents eutectic composition. The alloy belonging to the region $EXZE'$ is known as *hypoeutectic alloys* and the alloy belonging to the region EYZ'E' is known as *hypereutectic alloys*.

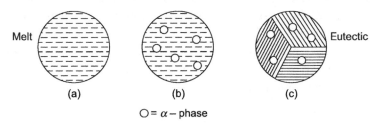

$\bigcirc = \alpha - $ phase

Figure 15.29b Microstructures of Pb/Sn system at different temperatures of cooling line **GIOMH**, (a) homogeneous liquid phase (melt) at the point G in the Fig. 15.29a, (b) heterogeneous phase (melt and α-phase) at the point I in the Fig. 15.29a, (c) heterogeneous phase, consisting of eutectic and α-phase at the point M in the Fig. 15.29a

Example-2 Ag/Cu system. The phase diagram is shown in the Fig. 15.30.

Eutectic point is at 778°C with 28% Cu. Maximum concentration of Cu in α-solid solution is 8.8%. Maximum concentration of Ag in β-solid solution is 8%. The line GOH represents cooling line of an alloy having composition of 60% Cu and 40% Ag. At the point H draw the lever aHa'.

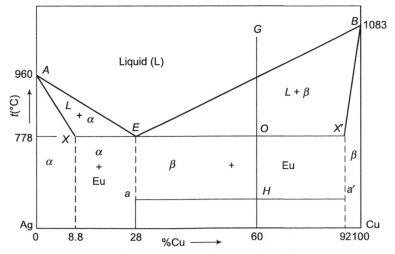

Figure 15.30 Binary phase diagram of Ag/Cu system

$$\% Eu = \frac{a'H}{aa'} \times 100 = \frac{(92-60)}{(92-28)} \times 100 = 50\%$$

$$\% \beta\text{-phase} = \frac{aH}{aa'} \times 100 = \frac{(60-28)}{(92-28)} \times 100 = 50\%$$

15.9.5 Binary Solid Solution Having Peritectic Point

There are many alloys, where the components differ widely with respect to the melting point. In that case, on addition of low melting component, melting point of the alloy decreases sharply and the solidus line, thus formed, overshadows the solidus line of low melting component. As a result of this, two solidus lines do not intersect but there exists a point where three phases are in equilibrium. The point is known as *peritectic point* and the horizontal line passing through this point is known as *peritectic line*. At eutectic point homogeneous liquid phase on cooling gives rise to two solid phases whereas at peritectic point two phases (one liquid phase and one solid phase) disappear on cooling to form one solid phase.

Example-1 Ag/Pt system. The binary phase diagram of Ag/Pt system is shown in the Fig. 15.31a.

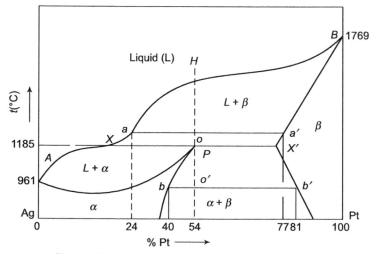

Figure 15.31a Binary phase diagram of Ag/Pt system

Draw a lever through the point o, i.e., aoa'.

$$\% \beta\text{-phase} = \frac{ao}{aa'} \times 100 = \frac{(54-24)}{(77-24)} \times 100 = 56.6\%$$

$$\% \text{ melt phase} = \frac{oa'}{aa'} \times 100 = \frac{(77-54)}{(77-24)} \times 100 = 43.4\%$$

Draw another lever through the point o', i.e., bo' b'.

$$\% \beta\text{-phase} = \frac{bo'}{bb'} \times 100 = \frac{(54-40)}{(81-40)} \times 100 = 34.15\%$$

$$\% \alpha\text{-phase} = \frac{o'b'}{bb'} \times 100 = \frac{(81-54)}{(81-40)} \times 100 = 65.85\%$$

P is the peritectic point where two phases disappear giving rise to a new phase. The overall reaction is given by

$$\text{Liquid (L)} + \beta(s) \underset{\text{Heating}}{\overset{\text{Cooling}}{\rightleftharpoons}} \alpha(s)$$

Thus three phases coexist at P and hence the point P is invariant. The line XPX' is known as peritectic line. It is to be noted that at eutectic point one phase, on cooling, gives two new phases but at peritectic point two phases disappear on cooling giving rise to a new phase. Consider the alloy containing 54% Pt at the point H, where only homogeneous liquid phase exists. On cooling to around 1482°C β-particles begin to form and the β-particles grow in size as the cooling proceeds. The heterogeneous phase is cooled further to peritectic temperature where α-phase crystals begin to form surrounding the β-phase particles. On further cooling α-phase particles grow in size through diffusion of Pt from β-phase and diffusion of Ag from α-phase. If sufficient time is allowed for diffusion the formation of homogeneous α-phase particles will be completed. The Fig. 15.31b shows the change in microstructure taking place during cooling. It is evident from the Fig. 15.31a that at room temperature only α-phase exists at 40% Pt composition but the microstructure shows that *dendritic* β-phases are entrapped within the α-phase, as shown in the Fig. 15.31c.

Reason During solidification α-phase begins to form, surrounding the β-phases. As the α-grains grow in size β-grains are entrapped and are unable to diffuse out of the α-phase. As a result of this, β-grains are entrapped within the α-phase. If the alloy is subjected to hot working, shielding of α-phase disappears, rendering to a complete diffusion of β-grains and equilibrium microstructure is reached.

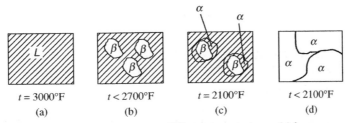

Figure 15.31b Microstructures of Ag/Pt system at different temperatures: (a) homogeneous liquid phase at 1650°C, (b) heterogeneous phase containing liquid and β-phases at below 1480°C, (c) α-phase, β-phase and liquid phase at peritectic temperature, (d) only α-phase just below peritectic temperature, e.g., at 1150°C

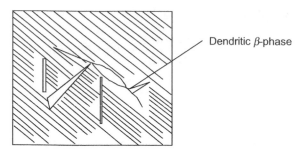

Figure 15.31c   Microstructure of Ag/Pt alloy containing 40% Pt. Dendritic β-grains are encased within the α-phase

Example-2   Fe/Ni system.   The phase diagram of this system is shown in the Fig. 15.32.

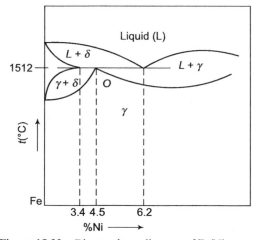

Figure 15.32   Binary phase diagram of Fe/Ni system

The point O is called peritectic point, where the following reaction takes place:

$$\text{Liquid} + \delta(s) \xrightarrow[\text{Heating}]{\text{Cooling}} \gamma(s)$$

Peritectic point appears at 1512°C and 4.5% Ni.

15.9.6   Binary System where Components form a Compound with an Incongruent Melting Point and the Compound is Immiscible in Both the Components

In some cases melting point of the intermetallic compound is far lower than that of pure metal. In that case the intermetallic compound is formed under peritectic reaction. The intermetallic compound is unstable and its melting point is not at all realized practically. The compound decomposes below its melting point giving rise to a new solid and solution. The corresponding decomposition temperature is known as *incongruent melting point* of the intermetallic compound. It is well illustrated in the Fig. 15.33. There is no formation or rupture of chemical bond during decomposition of the intermetallic compound. Thus the decomposition process is not considered as a chemical reaction. This decomposition process is called peritectic reaction.

Let us consider an alloy containing two components A and B, having melting points marked in the Fig. 15.33 as G and K respectively. The metal B has low melting point and forms a eutectic point E at 80% of B. The metal A has high melting point and GS represents melting point curve of the alloy AB. At 20% B, a compound A_mB_n is formed but the compound has far low melting point than the metal A. HS represents the peritectic line and the solid compound, A_mB_n, exists up to the point S. At the point H three phases coexist through an equilibrium reaction as follows

$$A(s) + \text{Liquid (L)} \underset{\text{Heating}}{\overset{\text{Cooling}}{\rightleftharpoons}} A_mB_n(s)$$

Thus the point H is called peritectic point. The curve SE represents melting point curve of the compound. The line CHS connects the homogeneous liquid melt (the point S), having 40% B, with pure metal A, represented by the point C. At the point H the above reaction occurs to form the compound A_mB_n of composition 20% B. The point H represents the *incongruent melting point* of the compound A_mB_n.

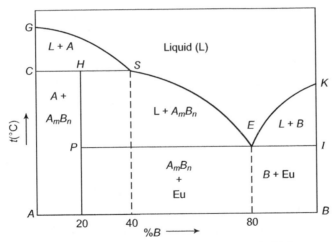

Figure 15.33 Binary system with components forming a compound with incongruent melting point

15.10 THREE PHASE REACTIONS AND INVARIANT POINT

In any heterogeneous binary system invariant point occurs in the phase diagram due to any one of the following type of reactions:

Type-1 In this type of heterogeneous binary system the given reaction occurs at the invariant point.

$$\text{Liquid (L}_1) \; \underset{\text{Heating}}{\overset{\text{Cooling}}{\rightleftharpoons}} \; \text{Liquid (L}_1) + \text{Liquid (L}_2)$$

This invariant point is called Upper Critical Solution Temperature (UCST).

Type-2 In this type of heterogeneous binary system the following reaction occurs at the invariant point.

$$\text{Liquid (L}_1) \; + \; \text{Liquid (L}_2) \; \underset{\text{Heating}}{\overset{\text{Cooling}}{\rightleftharpoons}} \; \text{Liquid (L)}$$

This invariant point is called Lower Critical Solution Temperature (LCST).

Type-3 In this type of heterogeneous binary system the following reaction occurs at the invariant point.

$$\text{Liquid (L)} \; \underset{\text{Heating}}{\overset{\text{Cooling}}{\rightleftharpoons}} \; \alpha(s) \; + \; \beta(s)$$

Here homogeneous liquid phase disappears on cooling giving rise to two solid phases and the invariant point is called eutectic point.

Type-4 In this type of heterogeneous binary system the following reaction occurs at the invariant point.

$$\text{Liquid (L)} \; + \; \alpha(s) \; \underset{\text{Heating}}{\overset{\text{Cooling}}{\rightleftharpoons}} \; \beta(s)$$

Here two phases (one liquid phase and one solid phase) disappears on cooling giving rise to one solid phase and the invariant point is called peritectic point.

Type-5 In this type of heterogeneous binary system the following reaction occurs at the invariant point.

$$\gamma(s) \; \underset{\text{Heating}}{\overset{\text{Cooling}}{\rightleftharpoons}} \; \alpha(s) + \; \beta(s)$$

Here solid solution disappears on cooling giving rise to two solid phases and the invariant point is called eutectoid point.

Type-6 In this type of heterogeneous binary system the following reaction occurs at the invariant point.

$$\gamma(s) \; + \; \delta(s) \; \underset{\text{Heating}}{\overset{\text{Cooling}}{\rightleftharpoons}} \; \alpha(s)$$

Here heterogeneous solid phase containing two solid solutions disappears on cooling giving rise to one solid phase and the invariant point is called peritectoid point. However, peritectoid reaction is seldom observed.

15.11 PRECIPITATION REACTIONS

In most cases it has been observed that one component of binary system is partially soluble in another component to form a homogeneous solid solution. However, solubility decreases with the decrease in temperature and hence soluble component is slowly precipitated out as a new solid phase from the homogeneous solid solution. One such example is Al/Ag system, phase diagram of which is shown in the Fig. 15.34.

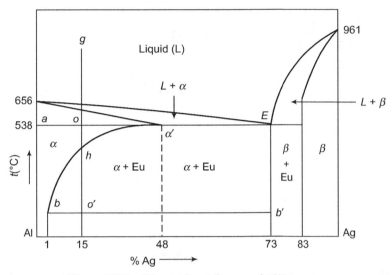

Figure 15.34 Binary phase diagram of Al/Ag system

Consider the homogeneous Al/Ag alloy melt, having composition of 15% Ag (the point *g* in the Fig. 15.34). *goho'* is the cooling line of the melt. It is slowly cooled to the eutectic temperature, i.e., at 538°C, where only homogeneous α-phase, having 15% Ag exists. At this temperature solubility of Ag in the α-phase may be increased to 48%. If the mixture is further cooled along *oo'*, % Ag in α-phase is reduced to only 1% and the remaining Ag is slowly precipitated out along the curve *hb* to form β-phase. Draw a tie line *bb'* through *o'*:

$$\% \; \alpha\text{-phase} = \frac{b'o'}{bb'} \times 100 = \frac{(73-15)}{(73-1)} \times 100 = 80.56\%$$

$$\% \; \text{eutectic} = \frac{bo'}{bb'} \times 100 = \frac{(15-1)}{(73-1)} \times 100 = 19.44\%$$

Thus the solubility of Ag decreases from 15% to only 1% on decreasing the temperature from 538°C to 82°C. As the temperature decreases, β-grains are separated out from α-grains and get precipitated as plates on α-grains.

15.12 IRON-CARBON SYSTEM

Study of iron-carbon system is valuable since, (i) it helps in explaining the properties of steel, (ii) several solid state reactions can be studied by means of Fe/C phase diagram, which is shown in the Fig. 15.35a.

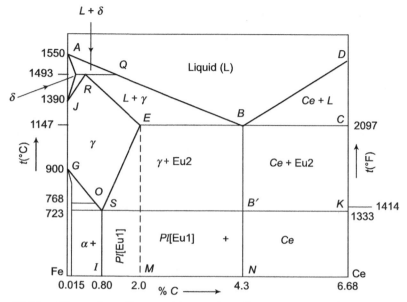

Figure 15.35a Binary phase diagram of iron-carbon alloys. Ce = Cementite, represented by Fe$_3$C, Pl = Pearlite or Eutectoid(1) or Eu(1)

The iron-carbon alloy has three invariant points, S, B and R.

(i) At S, the following reaction occurs

$$\gamma \text{ (s)} \underset{\text{Heating}}{\overset{\text{Cooling}}{\rightleftharpoons}} \alpha\text{(s)} + \text{Ce(s)}$$

All the three phases are in solid state and hence the mixture is called *eutectoid mixture*. Ce represents cementite, having formula Fe$_3$C, an intermetallic compound. Eutectoid temperature is 723°C or 1333°F. The carbon content at the eutectoid point, S is 0.8%.

(ii) At B, the following reaction occurs

$$\text{Liquid} \underset{\text{Heating}}{\overset{\text{Cooling}}{\rightleftharpoons}} \gamma\text{(s)} + \text{Ce(s)}$$

Here liquid phase breaks into two solid phases and the resultant mixture is called *eutectic mixture*. Eutectic temperature is 1147°C or 2097°F. The carbon content at the eutectic point B is 4.3%.

(iii) At *R*, the following reaction occurs

$$\text{Liquid} \ + \ \delta(s) \ \underset{\text{Heating}}{\overset{\text{Cooling}}{\rightleftharpoons}} \ \gamma(s)$$

Here two phases disappear to form a new phase and hence the mixture is called *peritectic mixture*. Peritectic temperature is 1493°C or 2719°F. The carbon content at the peritectic point *R* is 0.16%.

15.12.1 Allotropy

Iron exists in different form at different temperatures. All these forms are called allotropes of iron.

(i) *γ-iron*: Characteristic features: (a) Its crystal structure is FCC, (b) It is non-magnetic, (c) it exists above 723°C. The area under the curve JRESG in the Fig. 15.35a constitutes γ-iron, which is often called austenite, (d) The line *SE* in the Fig. 15.35a represents solubility of carbon in austenite. Maximum carbon content in austenite is 2%, (e) γ-iron is denser than α-iron, (f) Its grain size depends on temperature, time and working.

(ii) *α-iron*: Characteristic features: (a) Its crystal structure is *BCC*, (b) it is lighter than γ-iron, (c) It is popularly known as ferrite and exists from room temperature to 900°C, (d) Its maximum carbon content is 0.015%, (e) It is magnetic, (f) Forging of α-iron (or ferrite) causes distortion of grain size, (g) Its grain size does not change with time and temperature.

(iii) *δ-iron*: Characteristic features: (a) It appears above 1390°C, (b) Its crystal structure is *BCC*, (c) Its maximum carbon content is 0.10%, (d) It is non-magnetic.

(iv) *β-iron*: At temperature above 768°C (or 1414°F) crystal atoms of α-iron undergo some rearrangements, as a result of which, properties like volume, internal energy and electrical conductivity change in small scale but a remarkable change in magnetic properties is observed. However, crystal structure remains unaltered during this rearrangement. This form of iron, which exists above 768°C, possesses same *BCC* structure of α-iron but unlike α-iron does not exhibit any magnetic characteristics, is known as *β*-iron. Interestingly it has been observed that a little below 768°C, iron becomes strongly magnetic and is known as α-iron.

15.12.2 Ferrite Solubility Curve

The line *JG* in the Fig. 15.35a indicates that pure γ-iron exists at the point *J* and slowly transforms into α-iron at the point *G* on cooling. The α-iron, known as *ferrite*, contains a very little amount of carbon (0.015%) and hence is capable of holding considerable amount of various elements, such as, nickel, silicon, phosphorus, etc. This α-iron can be considered as a solvent in a solid solution. The line *GOS* in the Fig. 15.35a is the demarcation line above which all α-iron gets converted to γ-iron. The temperature of conversion from α-iron to γ-iron is maximum at *G* (900°C) and decreases with increase in carbon content, ultimately reaches a minimum at *S* (723°C). This conversion of α-iron to γ-iron is known as solubility of α-iron to γ-iron and the line *GOS* is often called *ferrite solubility curve*.

15.12.3 Cementite Solubility Curve

The eutectic temperature 1147°C (the point *E* in the Fig. 15.35a) is the temperature where α-iron holds 2% carbon. As the temperature decreases below 1147°C (or 2097°F), γ-iron is unable to retain its carbon content and hence excess carbon is converted to cementite. As cooling proceeds along *ES*, cementite phase slowly gets precipitated out from γ-phase. Thus the line *SE* represents solubility of cementite in austenite at various temperatures and is often called ***cementite solubility curve***.

15.12.4 Eutectoid Mixture or Pearlite

The point *S* in the Fig. 15.35a is the eutectoid point, where carbon content is only 0.8%. The temperature at the point *S* is 723°C. This is the minimum temperature at which 0.8% carbon remains in soluble form in the γ-iron matrix. If the temperature is further lowered below 723°C, two changes take place simultaneously. **First**, γ-phase of FCC structure is converted slowly to α-phase of BCC structure. **Second**, the excess carbon slowly diffuses out of the γ-phase to form an intermetallic compound, *cementite*, having 6.68% carbon. Thus the resultant eutectoid mixture contains both ferrite and cementite at a definite proportion (88.2% ferrite and 11.8% cementite). The eutectoid mixture is often called ***pearlite*** because its appearance is similar to that of mother-of-pearl.

15.12.5 Ledeburite

Consider an alloy melt, containing carbon content 4.3%. It is cooled continuously until the eutectic point *B* (see the Fig. 15.35a) is reached. At this stage both γ-iron and cementite start separating out from the liquid melt and the separation continues till the temperature reaches to 723°C. This mixture, thus separated at the eutectic temperature (1147°C) along *BB'* (see the Fig. 15.35a), is known as ***ledeburite***. Below 723°C γ-iron further decomposes to α-iron and cementite. Ledeburite is a mixture of austenite or γ-iron and cementite. Carbon content of ledeburite and cementite remains constant at 4.3% and 6.68% respectively but that of γ-iron varies from 723°C to 1147°C. So, composition of ledeburite also varies. Just below 1147°C, carbon content of γ-iron is around 2% and hence ledeburite is a mixture of 50.75% γ-iron and 49.25% cementite. On the other hand, just above 723°C, carbon content of γ-iron is around 0.8% and hence ledeburite is a mixture of 40.37% γ-iron and 59.63% cementite.

15.12.6 Hypo-eutectoid Steel

Consider an alloy, containing 0.25% carbon at the point *G* in the Fig. 15.35b. The alloy is allowed to cool along *GHIo*. At the point *H*, α-phase starts separating out from γ-phase and reaches the point *I* on the eutectic line. Further cooling results disappearance of γ-phase, leading to a mixture of ferrite and pearlite.

A tie line *aoa'* is drawn through the point *o*. Thus, using lever rule, we have

$$\% \text{ ferrite} = \frac{oa'}{aa'} \times 100 = \frac{(0.80 - 0.25)}{(0.80 - 0.015)} \times 100 = 70.06\%$$

$$\% \text{ pearlite} = \frac{ao}{aa'} \times 100 = \frac{(0.25 - 0.015)}{(0.80 - 0.015)} \times 100 = 29.93\%$$

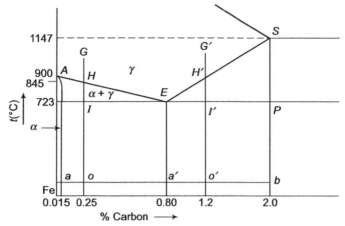

Figure 15.35b   Partial phase diagram of Fe/C alloy

When carbon content lies within 0.80%, the steel contains a mixture of ferrite and pearlite and it is known as *hypo-eutectoid steel*. The microstructure of hypo-eutectoid steel (see the 15.36a) shows that free ferrite forms a network of crystal grains around the pearlite grains.

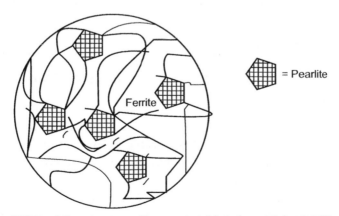

Figure 15.36a   Microstructure of hypo-eutectoid steel, containing 0.25% carbon

15.12.7   Hyper-eutectoid Steel

Steel, having composition between 0.80% and 2.0% *C*, is known as hyper-eutectoid steel, also known as *tool steel*. Consider an alloy containing 1.2% carbon at the point *G'* in the

Fig. 15.35b. The phase is γ-phase and all the carbon remains dissolved in the iron matrix. As the alloy is cooled slowly, the point H' is reached where the temperature is around 844°C (or 1550°F). γ-iron, at this stage, is unable to hold that amount of carbon and the cementite phase starts separating out. The separation of cementite phase continues till the point I' (723°C) is reached. If the alloy is cooled further, austenite (γ-iron) splits into pearlite and cementite. A tie line $a'o'b$ is drawn through the point o' (Fig. 15.35b). Thus, using lever rule, we have

$$\% \text{ pearlite} = \frac{o'b}{a'b} \times 100 = \frac{(6.68 - 1.2)}{(6.68 - 0.80)} \times 100 = 93.20\%$$

$$\% \text{ cementite} = \frac{a'o'}{a'b} \times 100 = \frac{(1.2 - 0.80)}{(6.68 - 0.80)} \times 100 = 6.80\%$$

Steel, containing carbon content within 0.80% to 20%, is a mixture of pearlite and cementite. It is known as *hyper-eutectoid steel*. The hyper-eutectoid steel consists of pearlite crystal matrix dispersed with long cementite particles as shown in the Fig. 15.36b.

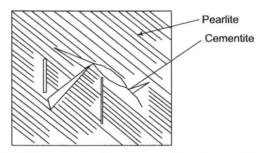

Figure 15.36b   Microstructure of hyper-eutectoid steel, containing 1.2% carbon

15.13 TERNARY SYSTEM

15.13.1 Components are Solid and Immiscible in Both Solid and Liquid States

Let us consider a three component system such that components are immiscible in both liquid and solid states. In this case $F = 4$ is possible as two of the three components may vary independently along with pressure and temperature. To illustrate such a system three pure components A, B and C are taken.

There are several characteristics of this type of system.
(a) The mixture of these three components gives rise to three different phases, which are represented by an equilateral triangle as shown in the Fig. 15.37a.

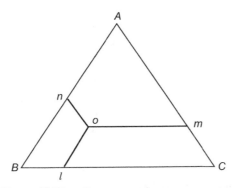

Figure 15.37a   Components in a ternary system

The vertices *A*, *B* and *C* represent pure components. Along the side *AB* only two components *A* and *B* exist. Similarly, along *AC* only two components, *A* and *C* exist and along *BC* only two components *B* and *C* exist. The point *o* lies within the triangle. Three components with three phases are in equilibrium at the point *o*. To find out composition at this point, a line *ol*, parallel to *AC* is drawn opposite to the *AC* such that the line meets the side *BC* at the point *l*. Similarly the lines *om* and *on* are drawn. The sum of the three, i.e., (*ol* + *om* + *on*) remains constant at any position within the triangle and is equal to the side of the triangle (*AB* or *BC* or *CA*). Assuming the side of the equilateral triangle is unity we can calculate the composition at any point within the triangle. For example at the point *o*.

ol = Fraction of the component *A*; *om* = Fraction of the component *B*; *on* = Fraction of the component *C*.

At the point *n* the mixture contains only *A* and *B*.

$$\frac{nB}{nA} = \frac{\text{Amount of component } A}{\text{Amount of component } B}$$

Similarly at the point *l*, we have

$$\frac{lC}{lB} = \frac{\text{Amount of component } B}{\text{Amount of component } C}$$

At the point *m*, we have

$$\frac{mA}{mC} = \frac{\text{Amount of component } C}{\text{Amount of component } A}$$

(b) The same equilateral triangle of the above three-component system is shown in the Fig. 15.37b.

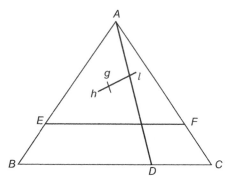

Figure 15.37b   Components in a ternary system. The line *hl* represents
series of mixture of three components

A straight line *AD* is drawn from the vertex *A* to the opposite side *BC*. At the point *A*, only pure *A* exists and at the point *D*, the component *A* does not exist in the mixture. In between *A* and *D* the ratio of components B:C remains constant but only the component *A* varies in amount.

(c) Let us consider a series of mixture of these three components, *A*, *B* and *C* within the triangle along the line *hl*, shown in the Fig. 15.37b. Consider a point *g* on the line *hl*. Thus we can write

$$\frac{lg}{hg} = \frac{\text{Amount of the mixture at the point } h}{\text{Amount of the mixture at the point } l}$$

15.13.2   Components are Miscible in Liquid State but Immiscible in Solid State

Anorthite mineral $CaAl_2Si_2O_8$ It is the calcium end-member of plagioclase feldspar. The crystal structure is triclinic. It is represented as *An*. Melting point is 1553°C.

Diopside mineral $MgCaSi_2O_6$ It is known as pyroxene mineral. The crystal structure is monoclinic. It is represented as *Di*. Melting point is 1392°C.

Forsterite mineral Mg_2SiO_4 It is known as magnesium silicate. The crystal structure is orthorhombic. It is represented as *Fo*. Melting point is 1890°C.

Ternary eutectic system with *C* = 3.

One common example is *Di–An–Fo* system. The phase diagram at 0.1 MPa is shown in the Fig. 15.38.

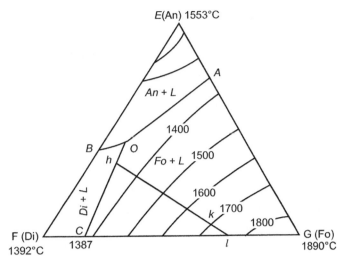

Figure 15.38 Ternary eutectic system containing three components *An*, *Di* and *Fo*

The point *O* is called ternary eutectic, where all three phases *An*, *Di* and *Fo* are in equilibrium.

EXERCISES

Multiple Choice Questions

1. The number of components in a solution of common salt is:
   (a) 0   (b) 1   (c) 2   (d) 3

2. The system with zero-degree of freedom is known as:
   (a) Monovariant (b) Bivariant   (c) Invariant   (d) None of these

3. For a three phase system with one component, the degree of freedom according to phase rule is:
   (a) 0   (b) 1   (c) 2   (d) 3

4. The transition temperature of a substance is that temperature at which:
   (a) One enantiomer changes into another enantiomer
   (b) One allotropic form changes to another
   (c) All the three phases (solid, liquid, and gas) can co-exist in equilibrium
   (d) None of the above

5. The sulphur system has four phases: rhombic, monoclinic, liquid, and vapour sulphur. It is:
   (a) One-component system   (b) Two-component system
   (c) Three-component system   (d) Four-component system

6. For a two-component system in a single phase, the degree of freedom is:
   (a) Zero   (b) One   (c) Two   (d) Three

7. A one component system has four phases. Can the four phases co-exist in equilibrium?

    (a) No   (b) Yes   (c) Sometimes   (d) None of these

8. Which one of the following is not true about a phase diagram?

    (a) It gives information on transformation rates

    (b) Relative amount of different phases can be found under given equilibrium conditions

    (c) It indicates the temperature at which different phases start to melt

    (d) Solid solubility limits are depicted by it

9. An invariant reaction that produces a solid on cooling two liquids:

    (a) Eutectic   (b) Peritectic   (c) Monotectic   (d) Syntectic

10. Gibbs phase rule for general system:

    (a) $P + F = C - 1$ (b) $P + F = C + 1$

    (c) $P + F = C - 2$ (d) $P + F = C + 2$

11. Liquid phase always exists for all compositions in a phase diagram:

    (a) above tie line   (b) above solvus line

    (c) above solidus line   (d) above liquidus line

12. The Gibbs phase rule:

    (a) Holds only for systems with more than one component

    (b) Predicts that a maximum of three phases can coexist in a one-component system

    (c) Predicts that a two-component three-phase system will have three degrees of freedom

    (d) Does not count phase compositions as intensive variables

ANSWERS

1(c),   2(c),   3(a),   4(b),   5(a),   6(d),   7(a),   8(a),   9(d),   10(d),   11(c), 12(b).

Short and Long Answer Type Questions

1. Define phase, component, and DOF. Hence establish Gibbs phase rule.
2. Explain triple point, eutectic point and eutectoid point with examples.
3. What is called solid solution? State the conditions necessary to form substitutional solid solution.
4. (a) State lever rule. (b) On cooling at 1 atm. carbondioxide gas is converted to solid ice instead of liquid—explain.
5. Apply phase rule to the two phase field of a binary isomorphous diagram. What conclusion can be drawn?

Answer

Phase rule is given by $F = C - P + 2$. There are two phases and two components. So, $P = 2$ and $C = 2$. Pressure is kept constant. So, $F = 1$. So, it has only one degree of freedom. A tie line may be drawn at a given temperature and the corresponding compositions of equilibrium solid and liquid phases are fixed.

6. Two metals A (melting point 800°C) and B (melting point 600°C) form a binary isomorphous system. An alloy having 35% B has 75% solid and rest liquid whereas an alloy having 55% B has 25% solid at 700°C. Estimate the composition of solidus and liquidus at the above temperature.

7. What is the difference between the states of phase equilibrium and metastability?

Answer

In case of phase equilibrium $\Delta G = 0$. The phases are stable and the phase parameters are constant. In case of metastability, the system is not at equilibrium and unstable. The system is disturbed significantly even on slight change in phase parameters.

8. For a given ternary system there are three components. Pressure is constant but temperature is variable. Calculate the maximum number of phases in equilibrium for that ternary system.

Answer

$C = 3$. Thus using Gibbs phase rule, we have $F = C - P + 1$ (as pressure is constant)

Hence, $P = C - F + 1$. Thus P will be maximum when $F = 0$. Hence, $P = 4$.

9. At 26°C vapour pressures of chloroform and carbon tetrachloride are 200 mm of Hg and 115 mm of Hg respectively. Calculate the weight% of chloroform in the vapour phase in equilibrium with a liquid mixture of 1 mole of each of the pure liquids.

[*Ans.*: 57%]

10. The system KI/H_2O is a simple eutectic with eutectic temperature –23°C, at which concentration of *KI* is 52%. Given at 20°C and 40°C saturated solution of *KI* contains 65% and 75% *KI* respectively. Draw and explain an approximate phase diagram of the system.

11. Illustrate the following:
    (a) enantiotropism   (b) retroflex solubility

12. How many triple points are there in sulphur system? State the temperature and composition of each triple point.

13. "A eutectic mixture has a definite composition and sharp melting point but still it is not a compound"—explain.

Answer

In a eutectic mixture components are present without any chemical bonding, i.e., without any stoichiometric proportion. That is why a eutectic mixture cannot be considered as a compound.

14. Explain the terms.
    (a) Eutectic point
    (b) Eutectic mixture
    (c) Peritectic point
15. Distinguish between congruent melting point and incongruent melting point.

Answer

There are some systems where components form a definite solid compound, which melts at a particular temperature to form liquid of same composition as that of the solid compound. Such melting point is known as congruent melting point.

There are some other systems, where components form a definite solid compound, which melts at a particular temperature to form liquid of different composition as that of the solid compound. Such melting point is known as incongruent melting point.

16. Discuss the desilverisation of lead with the help of a phase diagram.
17. Define the terms efflorescence and deliquescence with examples.

Answer

There are some salts, which remains as hydrated crystal form. For example, $CuSO_4 \cdot 5H_2O$, $Na_2SO_4 \cdot 10H_2O$, $Na_2CO_3 \cdot 10\ H_2O$, etc. However, these hydrated salts exist only when pressure of water vapour reaches a certain value. If the ambient vapour pressure of water is less than the dissociation pressure, the hydrated salt undergoes dehydration. The phenomenon is known as efflorescence. In case of $CuSO_4 \cdot 5H_2O$, dissociation pressure is less than the ambient vapour pressure and hence it is stable and does not undergo efflorescence. In the cases of $Na_2SO_4 \cdot 10H_2O$ and $NaCO_3 \cdot 10H_2O$ dissociation pressure is greater than the ambient vapour pressure and hence undergo dehydration or effloresce.

There are some saturated salt solutions for which vapour pressure of salt solution is less than that of ambient vapour pressure of water. The salt on dry condition absorbs moisture from surroundings and gets hydrated. The phenomenon is known as deliquescence. One example is $CaCl_2$. Vapour pressure of saturated $CaCl_2$ solution is less than that of ambient vapour pressure of water. So it always absorbs moisture from surroundings or undergo deliquesce to form $CaCl_2 \cdot 6H_2O$.

18. (a) State the limitations of phase rule.
    (b) Distinguish between triple point and eutectic point.
    (c) In the phase diagram of water system explain why melting curve of ice has negative slope.
19. Draw the phase diagram of $Na_2SO_4 - H_2O$ system. State the number of phases and components present in the system. What is called cryohydric point?
20. Draw the phase diagram of Pb/Ag system and discuss its salient features.

21. Determine the number of components and phases for the following systems:
    (a) $NH_4Cl(s) \rightleftharpoons NH_4Cl(g) \rightleftharpoons NH_3(g) + HCl(g)$
    (b) Ice $\rightleftharpoons$ Water $\rightleftharpoons$ Water vapour
22. (a) In case of phase diagram of water system explain the following:
    (i) bivariant system, (ii) univariant system, and (iii) invariant system.
    (b) "At 1 atm ice melts at 0°C but camphor does not melt at all at any temperature but only sublimes to vapour"—explain.
23. A binary alloy having 28 wt % Cu and balance Ag solidifies at 779°C. The solid consists of two phases α and β. α-phase has 9% Cu whereas β-phase has 8% Ag at 779°C. At room temperature these are pure Ag and Cu respectively. Sketch the phase diagram. Label all fields and lines. Melting points of Cu and Ag are 1083°C and 960°C respectively. Estimate the amount of α- and β-phases in the above alloy at 779°C.
24. Two alloys belonging to a binary system have the following microstructures. One having 25% *B* consists of 50% α-phase and 50% eutectic and the other having 75% *B* has 50% β-phase and 50% eutectic. Microstructural examination shows that eutectic is made of 50% α-phase and 50% β-phase. Estimate the composition of α, β and eutectic.

16   Dilute Solutions

16.1   INTRODUCTION

Any pure liquid remains in equilibrium with its vapour. The corresponding vapour pressure determines the boiling point and freezing point of the liquid. The relation between vapour pressure and boiling point or freezing point is best described by Clausius-Clapeyron equation given below:

$$\ln \frac{p_2}{p_1} = -\frac{L}{R} \left[\frac{1}{T_2} - \frac{1}{T_1} \right]$$

where, p_1 and p_2 are the vapour pressures of the liquid at temperatures T_1 and T_2 respectively. L represents latent heat per gm-mole of the liquid and R is the molar gas constant. Thus any change in vapour pressure changes the boiling point and freezing point of the liquid. The above equation suggests that if $p_2 < p_1$, $T_2 < T_1$. So elevation of boiling point and depression of freezing point are observed on lowering of vapour pressure.

Vapour pressure of a liquid depends on the number of vapour molecules produced per ml of that liquid. In presence of a solute, liquid acts as a solvent. Each ml of the solution contains both solute molecules and solvent molecules. Hence number of solvent vapour molecules produced per ml of solution is less than that produced per ml of the pure liquid. Thus vapour pressure of any pure liquid decreases in presence of solute. This is called 'lowering of vapour pressure'. The extent of lowering of vapour pressure depends on the **number of solute molecules** present in the system. Hence it is called colligative property. As boiling point and freezing point also change due to lowering of vapour pressure, elevation of boiling point and depression of freezing point are also called **colligative properties**. The following properties are considered as colligative properties:

(a) Lowering of vapour pressure.

(b) Elevation of boiling point.

(c) Depression of freezing point.

(d) Osmotic pressure.

Following conclusions can be made regarding colligative properties:

(a) Colligative properties of any solution depend on the number of solute particles but independent of the characteristics of solute particles.

(b) Colligative properties of any solution depend on escaping tendency of solvent particles in presence of non-volatile solute particles.

16.2 LOWERING OF VAPOUR PRESSURE

Each and every liquid remains in equilibrium with its vapour and the vapour pressure depends on the mole fraction of the solvent present in the vapour. Furthermore, mole fraction of solvent in gas phase depends on the number of solvent particles present on the surface of the liquid. Lesser the number of solvent particles on the surface of the liquid, less is the mole fraction of solvent in gas phase and hence less is the vapour pressure of the solvent.

In presence of a non-volatile solute, mole fraction of solvent decreases on the surface of the liquid, as a result of which, vapour pressure of the solvent decreases. So, addition of any non-volatile solute in its solvent always decreases the vapour pressure of solvent.

In case of an ideal solution the relation between lowering of vapour pressure and concentration of non-volatile solute is governed by Raoult's law.

Ideal solution   Ideal solution is that solution where enthalpy of mixing (ΔH^M) of all the components is zero. Mathematically, we can write that $\Delta H^M = 0$.

Raoult's law   The vapour pressure of an ideal solution depends on vapour pressure of each component as well as mole fraction of each component present in the solution.

Thus at equilibrium the total vapour pressure (p) of an ideal solution is given by

$$p = x_A p_A^* + x_B p_B^* + x_C p_C^* + \cdots = \Sigma x_i p_i^* = p_1 + p_2 + p_3 + \cdots = \Sigma p_i, \text{ where, } p_i = x_i p_i^*$$

p_i^* = Vapour pressure of the pure component, i; x_i = Mole fraction of that component in the solution.

In the case of two component ideal solution, we have $p = x_A p_A^* + x_B p_B^*$.

The corresponding vapour pressure versus concentration diagram is shown in the Fig. 16.1.

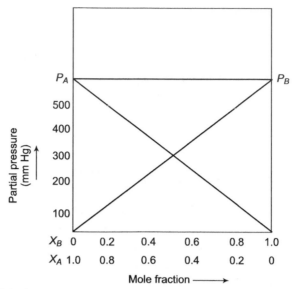

Figure 16.1 Vapour pressure versus concentration diagram of an ideal solution

16.2.1 Derivation of Raoult's Law

From the knowledge of thermodynamics, the chemical potential of ith component in a given mixture is given by

$$\mu_i = \mu_i^0 + RT \ln x_i \qquad (16.1)$$

where, μ_i and μ_i^0 are the chemical potential and standard chemical potential of ith component, x_i is the mole fraction of ith component present in the mixture. In case of liquid solution the above equation becomes

$$\mu_l = \mu_{i,l}^0 + RT \ln x_i \qquad (16.2)$$

where, x_i = mole fraction of ith component present in the solution.

In case of vapour mixture the above equation becomes

$$\mu_v = \mu_{i,v}^0 + RT \ln y_i \qquad (16.3)$$

In case of pure solvent $\qquad \mu_{i,l}^0 = \mu_{i,v}^0 \qquad (16.4)$

Any solvent or solution always remains in equilibrium with its vapour. Thus chemical potentials of both liquid (μ_l) and vapour (μ_v) are equal. Thus in case of a pure solvent, we can write

$$\mu_l = \mu_v = \mu_{i,v}^0 + RT \ln y_i = \mu_{i,v}^0 + RT \ln \frac{p_i^0}{P}, \qquad y_i = \frac{p_i^0}{P} \qquad (16.5)$$

where, $\mu_{i,v}^0$ = Standard chemical potential of ith component in vapour

y_i = Mole fraction of pure i^{th} component in vapour

$p_i^{\,0}$ = Vapour pressure of pure i^{th} component and P is the total vapour pressure.
Thus

$$\mu_l = \mu_{i,v}^0 + RT \ln \frac{p_i^{\,0}}{P} \tag{16.6}$$

Comparing equations (16.2) and (16.6), we have

$$\mu_{i,l}^0 + RT \ln x_i = \mu_{i,v}^0 + RT \ln \frac{p_i^{\,0}}{P} \tag{16.7}$$

From equations (16.4) and (16.7), we have

$$RT \ln x_i = RT \ln \frac{p_i}{p_i^{\,0}}. \quad \text{Or,} \quad p_i = x_i p_i^{\,0} \tag{16.8}$$

The above equation is the mathematical form of Raoult's law. The equation shows that vapour pressure of the solvent increases linearly with its mole fraction.

In colligative properties, concentration of solute is expressed in terms of *molality*. For a given solution, molality is the number of moles of solute dissolved in 1000 gm of solvent. Hence, we can write that

$$\text{Molality} = \frac{\text{Number of moles of solute}}{1000 \text{ gm of solvent}} = m$$

16.2.1.1   Limitations of Raoult's Law

Adhesive forces are not considered in deriving the above relation. However, the above relation depends upon adhesive forces between solute and solvent molecules. At high concentration of solute, adhesive forces have significant influence and the curve deviates from linear relationship as given in equation (16.8). However, in dilute solutions, the influence of adhesive forces may be assumed to be negligible and hence equation (16.8) is successfully applicable.

16.3   ELEVATION OF BOILING POINT

Any pure liquid starts boiling at that temperature at which vapour pressure of the liquid equals to the atmospheric pressure. In presence of non-volatile solute vapour pressure of the solvent decreases and hence the vapour pressure reaches the atmospheric pressure at higher temperature. The Fig. 16.2a shows phase diagram of pure water. Melting point and boiling point of water are 0°C and 100°C respectively. The Fig. 16.2b shows that vapour pressure of solvent decreases in presence of non-volatile solute and hence during freezing, 1 atm. pressure is achieved below 0°C and during boiling 1 atm. pressure is achieved above 100°C.

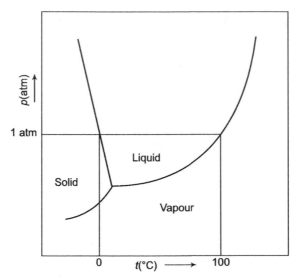

Figure 16.2a Normal boiling point and melting point of pure water

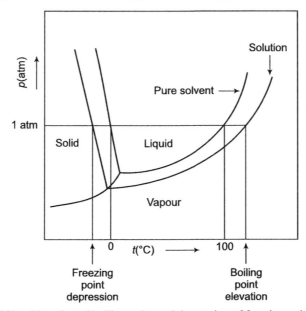

Figure 16.2b Elevation of boiling point and depression of freezing point of water

From the definition of Gibbs free energy of thermodynamics, we have

$$G = H - TS \quad \text{and} \quad \Delta G = \Delta H - T\Delta S \quad (16.9a)$$

From Gibbs-Helmholtz relation, we have

$$S = -\left(\frac{\partial G}{\partial T}\right)_P \text{ and } \Delta S = -\left(\frac{\partial \Delta G}{\Delta T}\right)_P \quad (16.9b)$$

For one mole of pure solvent A, we have

$$\mu_A^{vap} = \mu_A^{liq} = \mu_A^0 + RT \ln X_A \quad \text{and} \quad \Delta\mu_A = \Delta G_A \qquad (16.9c)$$

where, $\quad \Delta\mu_A = \mu_A^{liq.} - \mu_A^0$

So, $\quad \left[\dfrac{\partial(\Delta\mu_A/T)}{\partial T} \right]_P = \left[\dfrac{\partial(\Delta G_A/T)}{\partial T} \right]_P = \dfrac{1}{T}\left(\dfrac{\partial\Delta G_A}{\partial T} \right)_P - \dfrac{\Delta G_A}{T^2}$

Using equation (16.9b), we have

$$\left[\frac{\partial(\Delta\mu_A/T)}{\partial T} \right]_P = -\frac{\Delta S_A}{T} - \frac{\Delta G_A}{T^2} = -\frac{\Delta S_A}{T} - \frac{(\Delta H_A - T\Delta S_A)}{T^2} = -\frac{\Delta H_A}{T^2}$$

Or, $\quad \left[\dfrac{\partial(\Delta\mu_A/T)}{\partial T} \right]_P = -\dfrac{\Delta H_A}{T^2} \qquad (16.10)$

Again from equation (16.9c), we have

$$\Delta\mu_A = RT \ln X_A \quad \text{Or,} \quad \frac{\Delta\mu_A}{T} = R \ln X_A$$

Differentiating both sides with respect to T, we get

$$\left[\frac{\partial(\Delta\mu_A/T)}{\partial T} \right]_P = \left[\frac{\partial(R \ln X_A)}{\partial T} \right]_P \qquad (16.11)$$

From equations (16.10) and (16.11), we get

$$\left[\frac{\partial(R \ln X_A)}{\partial T} \right]_P = -\frac{\Delta H_A}{T^2}$$

At constant pressure, we can write that

$$Rd \ln X_A = -\frac{\Delta H_A}{T^2} dT$$

On integration, we get

$$\int_1^{X_A} d \ln X_A = -\frac{\Delta H_A}{R} \int_{T_b}^T \frac{dT}{T^2}. \quad \text{Or,} \quad \ln X_A = \frac{\Delta H_A}{R}\left[\frac{1}{T} - \frac{1}{T_b} \right] \qquad (16.12)$$

when $X_A = 1$, the solution is actually pure solvent and $T = T_b$ = boiling temperature of the solvent. In presence of non-volatile solute, $X_A < 1$ and $T > T_b$.

X_A and X_B are the mole fractions of solvent and solute respectively. Thus

$$X_A + X_B = 1. \quad X_A = 1 - X_B. \quad \text{Hence,} \quad \ln X_A = \ln(1 - X_B) \approx -X_B \qquad (16.13)$$

From equations (16.12) and (16.13), we get

$$-X_B = \frac{\Delta H_A}{R}\left[\frac{1}{T} - \frac{1}{T_b}\right]. \quad \text{Or,} \quad -\frac{\Delta H_A}{R}\left[\frac{1}{T} - \frac{1}{T_b}\right] = X_B \tag{16.14}$$

$\Delta H_A = L_e$ = Heat of vapourisation or latent of evaporation of pure solvent/mole.

M_A = Molecular weight of pure solvent. n_A and n_B are the number of moles of solvent and solute in the solution respectively. In case of dilute solution, $n_A \gg n_B$ and $n_A + n_B \approx n_A$.

l_e = latent of vapourisation or evaporation of pure solvent/gm and $l_e > 0$.

So,
$$L_e = le \times M_A \tag{16.15}$$

$$X_B = \frac{n_B}{n_A + n_B} \approx \frac{n_B}{n_A} = \frac{n_B}{w_A/M_A} = \frac{n_B.M_A}{w_A} = \left(\frac{1000 n_B.M_A}{w_A}\right)\left(\frac{1}{1000}\right)$$

$$\left(\frac{1000 n_B}{w_A}\right) = \frac{\text{Number of moles of solute}}{1000 \text{ gm of solvent}} = \text{molality} = m$$

So,
$$X_B = \frac{m \times M_A}{1000} \tag{16.16}$$

From equations (16.14) and (16.16) we get

$$\frac{m \times M_A}{1000} = -\frac{L_e}{R}\left[\frac{1}{T} - \frac{1}{T_b}\right] = \frac{L_e}{R}\left[\frac{1}{T_b} - \frac{1}{T}\right] = \frac{L_e}{R}\left[\frac{T - T_b}{T \cdot T_b}\right] = \left(\frac{L_e}{R}\right)\left(\frac{\Delta T_b}{T_b^2}\right)$$

$$\left[\text{since, } (T - T_b) = \Delta T_b \text{ and } (T \cdot T_b) \cong T_b^2\right]$$

So,
$$X_B = \frac{m \times M_A}{1000} = \left(\frac{L_e}{R}\right)\left(\frac{\Delta T_b}{T_b^2}\right) \tag{16.17a}$$

Or,
$$\Delta T_b = \left(\frac{RT_b^2 M_A}{1000 L_e}\right)m = \left(\frac{RT_b^2}{1000 l_e}\right)m = K_b \cdot m \quad [\text{Since, } L_e = l_e \times M_A]$$

So,
$$\Delta T_b = K_b \cdot m \quad \text{and} \quad K_b = \left(\frac{RT_b^2}{1000 l_e}\right) \tag{16.17b}$$

For a given solvent, T_b and l_e are known. Hence, K_b, popularly known as **Ebullioscopic constant**, can be easily calculated by using equation (16.17b). Values of K_b of some solvents are given in the Table 16.1.

Table 16.1 K_b values of some selected solvents

Solvent	Formula	Boiling point (°C)	Ebullioscopic constant (K_b) (°K.kg/mole)
Water	H_2O	100	0.512
Acetic acid	CH_3COOH	118	3.07
Phenol	C_6H_5OH	182	3.04
Benzene	C_6H_6	80	2.53
Carbon disulfide	CS_2	46	2.37
Naphthalene	$C_{10}H_8$	218	5.8
Carbon tetrachloride	CCl_4	77	4.95
Acetone	CH_3COCH_3	56.2	2.67
Aniline	$C_6H_5NH_2$	184.3	3.69
Chloroform	$CHCl_3$	61.2	3.88
Ethanol	C_2H_5OH	78.4	1.19
Formic acid	$HCOOH$	101	2.4
Nitrobenzene	$C_6H_5NO_2$	210.9	5.24
Cyclohexane	C_6H_{12}	80.74	2.79
Diethyl ether	$C_2H_5OC_2H_5$	34.5	2.16
Bromobenzene	C_6H_5Br	156	6.26
Camphor		204	5.95

Characteristics

(a) At the boiling point vapour pressure of the liquid is equal to 1 atm.

(b) Presence of any non-volatile solute decreases the vapour pressure of the solvent and hence boiling point of solution increases since availability of solvent molecules on the surface decreases.

(c) Elevation of boiling point is directly proportional to the molality of the solution.

16.4 DEPRESSION OF FREEZING POINT

From Gibbs-Helmholtz relation, we have

$$\Delta S = -\left(\frac{\partial \Delta G}{\partial T}\right)_P \tag{16.18a}$$

For one mole of pure solvent A, we have

$$\mu_A^{solid} = \mu_A^{liq} = \mu_A^0 + RT \ln X_A \quad \text{and} \quad \Delta\mu_A = \Delta G_A \tag{16.18b}$$

where, $\Delta\mu_A = \mu_A^{liq} - \mu_A^0$

So, $\left[\dfrac{\partial(\Delta\mu_A/T)}{\partial T}\right]_P = \left[\dfrac{\partial(\Delta G_A/T)}{\partial T}\right]_P = \dfrac{1}{T}\left(\dfrac{\partial \Delta G_A}{\partial T}\right)_P - \dfrac{\Delta G_A}{T^2}$

Using equation (16.18a), we have

$$\left[\frac{\partial(\Delta\mu_A/T)}{\partial T}\right]_P = -\frac{\Delta S_s}{T} - \frac{\Delta G_s}{T^2} = -\frac{\Delta S_s}{T} - \frac{(\Delta H_s - T\Delta S_s)}{T^2} = -\frac{\Delta H_s}{T^2}$$

Or,
$$\left[\frac{\partial(\Delta\mu_A/T)}{\partial T}\right]_P = -\frac{\Delta H_s}{T^2} \tag{16.19}$$

$$\Delta H_s = (H_A^{\text{solid}} - H_A^0) \quad \text{and} \quad \Delta H_s < 0$$

[Since latent heat is released during solidification process]

Again from equation (16.18b), we have

$$\Delta\mu_A = (\mu_A^{\text{solid}} - \mu_A^0) = RT \ln X_A. \quad \text{Or,} \quad \frac{\Delta\mu_A}{T} = R \ln X_A$$

Differentiating both sides with respect to T, we get

$$\left[\frac{\partial(\Delta\mu_A/T)}{\partial T}\right]_P = \left[\frac{\partial(R\ln X_A)}{\partial T}\right]_P \tag{16.20}$$

From equations (16.19) and (16.20), we get

$$\left[\frac{\partial(R\ln X_A)}{\partial T}\right]_P = -\frac{\Delta H_s}{T^2}$$

At constant pressure, we can write that

$$R\, d\ln X_A = -\frac{\Delta H_s}{T^2} dT$$

On integration, we get

$$\int_1^{X_A} d\ln X_A = -\frac{\Delta H_s}{R} \int_{T_f}^T \frac{dT}{T^2} \tag{16.21}$$

when $X_A = 1$, the solution is actually pure solvent and $T = T_f =$ melting temperature of the solvent. In presence of non-volatile solute, $X_A < 1$ and $T < T_f$.

On integration within limits, we get

$$\ln X_A = \frac{\Delta H_s}{R}\left[\frac{1}{T} - \frac{1}{T_f}\right] \tag{16.22}$$

X_A and X_B are the mole fractions of solvent and solute respectively. Thus

$$X_A + X_B = 1. \quad X_A = (1 - X_B)$$

Hence,
$$\ln X_A = \ln(1 - X_B) \approx -X_B \tag{16.23}$$

Putting the value of $\ln X_A$ in the equation (16.23), we get

$$-X_B = \frac{\Delta H_s}{R}\left[\frac{1}{T} - \frac{1}{T_f}\right]. \quad \text{Or, } X_B = -\frac{\Delta H_s}{R}\left[\frac{1}{T} - \frac{1}{T_f}\right] \tag{16.24}$$

$\Delta H_s = L_f = $ Latent of fusion of pure solvent/mole and $L_f < 0$

$M_A = $ Molecular weight of pure solvent. n_A and n_B are the number of moles of solvent and solute in the solution. In case of dilute solution, $n_A \gg n_B$ and $n_A + n_B \approx n_A$.

$L_f = $ Latent heat of fusion of pure solvent/gm and $l_f < 0$

So, $$L_f = l_f \times M_A \tag{16.25}$$

$$X_B = \frac{n_B}{n_A + n_B} \approx \frac{n_B}{n_A} = \frac{n_B}{w_A / M_A} = \frac{n_B.M_A}{w_A} = \left(\frac{1000 n_B.M_A}{w_A}\right)\left(\frac{1}{1000}\right)$$

$$\left(\frac{1000 n_B}{w_A}\right) = \frac{\text{Number of moles of solute}}{1000 \text{ gm of solvent}} = \text{molality} = m$$

So, $$X_B = \frac{m \times M_A}{1000} \tag{16.26}$$

Putting the value of X_B in the equation (16.24), we get

$$\frac{m \times M_A}{1000} = -\frac{L_f}{R}\left[\frac{1}{T} - \frac{1}{T_f}\right] = \frac{L_f}{R}\left[\frac{1}{T_f} - \frac{1}{T}\right] = \frac{L_f}{R}\left[\frac{T - T_f}{T \cdot T_f}\right] = \left(\frac{L_f}{R}\right)\left(\frac{\Delta T_f}{T_f^2}\right)$$

$$[\text{Since, } (T - T_f) = \Delta T_f \text{ and } (T \cdot T_f) \cong T_f^2]$$

So, $$X_B = \frac{m \times M_A}{1000} = \left(\frac{L_f}{R}\right)\left(\frac{\Delta T_f}{T_f^2}\right) \tag{16.27a}$$

Or, $$\Delta T_f = \left(\frac{RT_f^2 \times M_A}{1000 L_f}\right) m = \left(\frac{RT_f^2}{1000 l_f}\right) m = K_f \cdot m \quad [\text{Since, } L_f = l_f \times M_A] $$

So, $$\Delta T_f = K_f \cdot m \quad \text{and} \quad K_f = \left(\frac{RT_f^2}{1000 l_f}\right) \tag{16.27b}$$

For a given solvent, T_f and l_f are known. Hence, K_f, popularly known as **Cryoscopic constant** can be easily calculated by using equation (16.27b). Values of K_f of some solvents are given in the Table 16.2.

Table 16.2 K_f values of some selected solvents

Solvent	Formula	Freezing point (°C)	Cryoscopic constant (K_f) (°K.kg/mole)
Water	H_2O	0	1.86
Acetic acid	CH_3COOH	16.6	3.90
Phenol	C_6H_5OH	40.5	7.27
Benzene	C_6H_6	5.5	5.12
Carbon disulfide	CS_2	−112	3.83
Naphthalene	$C_{10}H_8$	80.26	6.9
Carbon tetrachloride	CCl_4	−22.8	29.8
Chloroform	$CHCl_3$	−63.5	4.68
Aniline	$C_6H_5NH_2$	−5.96	5.87
Ethanol	C_2H_5OH	−114.6	1.99
Formic acid	$HCOOH$	8	2.77
Nitrobenzene	$C_6H_5NO_2$	5.85	7.00
Cyclohexane	C_6H_{12}	6.4	20.2
Diethyl ether	$C_2H_5OC_2H_5$	−114.3	1.79
Camphor		179.8	39.7

16.5 OSMOTIC PRESSURE

When a solvent is separated from its solution by a semipermeable membrane, only solvent, not the solute, is allowed to flow through the membrane. The phenomenon is called osmosis. The process of osmosis through semipermeable membranes was first observed in 1748 by Jean-Antoine Nollet.

However, when the solvent is separated from its solution by a completely permeable membrane, solute molecules move through the membrane due to concentration difference. The phenomenon is called diffusion. Thus a semipermeable membrane is an essential component to execute osmosis. Concentration difference is the driving force for osmosis and the pressure developed due to this driving forced is called osmotic pressure, due to which the solvent is flowing from dilute solution (also known as hypotonic solution) to the concentrated solution (also known as hypertonic solution) through semipermeable membrane. The mechanism of osmosis is illustrated in the Fig. 16.3.

When a pure solvent is separated from its solution by a semipermeable membrane, solvent molecules have a natural tendency to move into the solution chamber through the semipermeable membrane. The excess pressure, which has to be applied to the solution chamber to resists the inflow of solvent till equilibrium is established, is called osmotic pressure of the solution. In other words, osmotic pressure is the excess pressure, which must be applied to a concentrated solution in order to resist the inflow of solvent through the semipermeable membrane from a dilute solution.

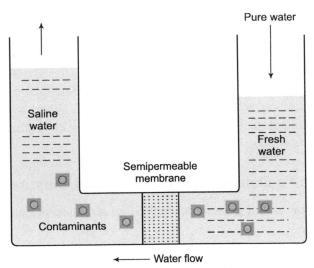

Figure 16.3   Osmosis: Natural flow of solvent from low concentration to high concentration through semipermeable membrane

16.5.1   Osmotic Potential

Let us take a dilute solution, also known as **hypotonic solution** in one compartment while a concentrated solution, also known as **hypertonic solution** is taken in another compartment. Now these two compartments are separated by a semipermeable membrane. A potential is developed across the membrane due to concentration difference. This potential is known as osmotic potential, due to which, water molecules are moving from a dilute solution compartment to the concentrated solution compartment across the semipermeable membrane.

Examples of semipermeable membrane

(a) Plant cell walls.

(b) Plasma membrane.

(c) Tonoplast membrane.

(d) Parchment paper.

(e) Phospholipid bilayer, where, a group of phospholipids are arranged to form a double layer such that phosphate forms the head and fatty acid forms the tail of each molecule.

Plant cells are always surrounded by cell walls, which act as semipermeable membranes, since only water is allowed to pass through the cell walls. When water flows into the cell through the wall, excess pressure is developed and the cell is inflated and pressed against the walls. Cell osmotic pressure is high and the phenomenon is called **turgidity**. When water flows from the cells through the walls, cell osmotic pressure drops and hence cells shrink. Cells get detached from the cell walls. The phenomenon is called **plasmolysis**.

16.5.2 Isotonic Solution

Osmotic pressure is a colligative property. It depends only on number or concentration of solute and independent of type of solute. Thus equimolecular solutions of different solutes exert equal osmotic pressure at same temperature. Such solutions which have the same osmotic pressure are termed isotonic or iso-osmotic. When two isotonic solutions are separated by a semipermeable membrane, no flow of solvent molecules is observed on either side.

16.5.3 Van't Hoff's Conclusions on Osmotic Pressure

(a) The magnitude of osmotic pressure depends on solute concentration. So it is a colligative property.

(b) If the solution is sufficiently dilute, volume of solute is negligible compared to that occupied by the solvent. Then osmotic pressure exerted by the solute is similar to the ideal gas pressure of an ideal solution, containing same volume of gas.

(c) Osmotic pressure, exerted by any non-electrolyte solute in a solution, is assumed to follow ideal gas laws.

16.5.4 Laws of Osmotic Pressure

Law-1 Osmotic pressure of a given solution is directly proportional to the molar concentration of the solution provided temperature is constant. Mathematically, it can be written as $\pi \propto c$ (T is constant), where, π is called osmotic pressure and c is the molar concentration of the solution. T is the absolute temperature.

Law-2 Osmotic pressure of a given solution increases with absolute temperature proportionately provided molar concentration of the solution is constant. Mathematically, it can be written as

$$\pi \propto T \, (c \text{ is constant})$$

Combining law-1 and law-2, we get

$$\pi \propto cT. \quad \text{Or,} \quad \pi = cRT \tag{16.28}$$

where, R is the molar gas constant. Again, $c = (n/V)$, where, n and V are the number of moles of solute and volume of the solution respectively. Putting the value of c in the above equation, we get

$$\pi = \left(\frac{n}{V}\right)RT. \quad \text{Or,} \quad \pi V = nRT \tag{16.29}$$

The above equation is popularly known as Van't Hoff's equation.

Law-3 At constant temperature and concentration, the osmotic pressure, exerted by a solution, is independent of the type of solute, dissolved in the solution, provided that solute remains in undissociated monomeric from in the solution.

16.5.5 Proof of Van't Hoff's Equation

Gibbs free energy (G) is related with pressure and temperature as follows:

$$dG = VdP - SdT$$

At constant temperature, the above equation becomes $dG = VdP$

For 1 mole $d\overline{G} = d\mu = \overline{V}dP$, where, $\overline{V}$ and μ are the molar volume and chemical potential of the solution.

Let us assume that the solution is in equilibrium with its vapour and the vapour behaves ideally. So, we can write that $\mu_l = \mu_v$, where, μ_l and μ_v are the chemical potentials of the solution and its vapour at equilibrium respectively. Now the vapour pressure p is slightly changed to $(p + dP)$. Consequently, μ_l and μ_v are also changed to $(\mu_l + d\mu_l)$ and $(\mu_v + d\mu_v)$. Hence, solution pressure P is also changes to $(P + dP)$.

$$d\mu_l = d\overline{G}_l = \overline{V}_l dP \text{ and } d\mu_v = d\overline{G}_v = \overline{V}_v dp = \frac{RT}{p}dp = RTd\ln p$$

$\overline{V}_l$ and $\overline{V}_v$ are the molar volumes of the solvent and its vapour respectively.
At equilibrium, we have

$$d\mu_l = d\mu_v. \text{ Or, } \overline{V}_l dP = RTd\ln p$$

On integration at constant temperature, we have

$$\overline{V}_l \int_{p^0}^{P} dP = RT \int_{p^0}^{p} d\ln p. \text{ Or, } \overline{V}_l(P - P^0) = RT\ln\frac{p}{p^0} \tag{16.30}$$

P^0 and P are the pressures of the pure solvent and the solution respectively. Similarly, p^0 and p are the corresponding vapour pressures. $P^0 > P$ and $(P^0 - P) = \pi$(osmotic pressure). Also $p^0 > p$. Again, mole fraction of solute, i.e., x_2 is related with vapour pressure according to Raoult's law as follows

$$x_2 = \frac{p^0 - p}{p^0}. \text{ Or, } \frac{p}{p^0} = 1 - x_2$$

Thus the equation (16.30) becomes

$$\overline{V}_l(P^0 - P) = -RT\ln\frac{p}{p^0} = -RT\ln(1 - x_2)$$

As the solution is dilute, $x_2 \ll 1$ and $\ln(1 - x_2) \approx - x_2$. Thus we can write

$$\pi \overline{V_l} = RTx_2 \tag{16.31}$$

Again,
$$x_2 = \frac{n_2}{n_1 + n_2} \approx \frac{n_2}{n_1} \text{ (for dilute solution)}$$

Thus,
$$\pi \overline{V_l} = \left(\frac{n_2}{n_1} \right) RT \tag{16.32}$$

n_1 and n_2 are the number of moles of solvent and solute respectively.

$n_1 \overline{V_l}$ = Total volume of the solvent in litre = V. Thus the equation (16.32) becomes

$$\pi (n_1 \overline{V_l}) = n_2 RT. \quad \text{Or,} \quad \pi V = n_2 RT$$

$$\frac{n_2}{V} = \frac{\text{Number of moles of solute}}{\text{Volume of solvent (litre)}} = \text{Molarity of the solution} = c$$

So,
$$\pi = \left(\frac{n_2}{V} \right) RT = cRT \tag{16.33}$$

The equation (16.33) is known as Van't Hoff equation.

Again,
$$n_2 = \frac{w_2}{M_2}$$

where, w_2 and M_2 are the weight and molecular weight of the solute respectively. So, putting the value of n_2 in the equation (16.33), we get

$$\pi = \left(\frac{n_2}{V} \right) RT = \left(\frac{w_2}{M_2 V} \right) RT. \quad \text{Or,} \quad \pi V = \left(\frac{w_2}{M_2} \right) RT \tag{16.34}$$

The value of R is known. Temperature, T, is fixed and known. w_2 and V are also known. Thus by measuring the osmotic pressure, π, we can calculate the molecular weight, M_2, of the solute.

16.5.6 Relation between π and ΔT_b or ΔT_f

From equation (16.31), we have $\pi \overline{V_l} = RTx_2$

Or,
$$x_2 = \frac{\pi \overline{V_l}}{RT} \tag{16.35}$$

Again, from equations (16.17a) and (16.27a), we have

$$x_2 = \frac{L_e}{R} \frac{\Delta T_b}{T_b^2} \quad \text{and} \quad x_2 = -\frac{L_f}{R} \frac{\Delta T_f}{T_f^2} \tag{16.36}$$

From equations (16.35) and (16.36), we have

$$\frac{\pi \overline{V_l}}{RT} = \frac{L_e}{R} \frac{\Delta T_b}{T_b^2} \quad \text{and} \quad \frac{\pi \overline{V_l}}{RT} = -\frac{L_f}{R} \frac{\Delta T_f}{T_f^2}$$

Or,

$$\pi = \left(\frac{L_e T}{\overline{V_l}}\right)\left(\frac{\Delta T_b}{T_b^2}\right) \quad \text{and} \quad \pi = -\left(\frac{L_f T}{\overline{V_l}}\right)\left(\frac{\Delta T_f}{T_f^2}\right) \tag{16.37}$$

Thus at a given temperature, ΔT_b and ΔT_f can be computed by measuring osmotic pressure, π, since other parameters are known.

16.5.7 Van't Hoff Factor

Colligative property depends on number of moles of solute present in the solution. In case of non-electrolytic solute, number of moles of solute remains unchanged in solution. However, in case of electrolytic solute, number of moles of solute increases due to dissociation of solute in solution. For example, NaCl dissociates in solution giving rise to two ions, Na^+ and Cl^-. Thus for non-electrolytic solute, expression of osmotic pressure is $\pi_0 = cRT$ but for electrolytic solute, the expression should be rewritten as

$\pi = icRT$, where, i represents Van't Hoff factor and it can be expressed as

$$i = \frac{\pi}{\pi_0} \tag{16.38a}$$

where, π_0 and π are the theoretical and experimental values of osmotic pressure respectively. Again, $c = w/M$, where, w and M are the weight and molecular weight of solute respectively. So, we can write

$$\pi_0 = \left(\frac{w}{M_0}\right)RT \quad \text{and} \quad \pi = \left(\frac{w}{M}\right)RT$$

where, M_0 and M are the molecular weight of undissociated/unassociated solute and apparent molecular weight of the solute in solution.

Hence,

$$i = \frac{\pi}{\pi_0} = \frac{M_0}{M} \tag{16.38b}$$

Van't Hoff factor can be used for any colligative property. Thus in general, we can write that

$$i = \frac{\pi}{\pi_0} = \frac{M_0}{M} \tag{16.38c}$$

where, Δ_0 and Δ are the theoretical and experimental values of a colligative property.

In aqueous solution NaCl and Na_2CO_3 are almost completely dissociated, producing 2 moles of ions and 3 moles of ions respectively as given below

$$NaCl \longrightarrow Na^+ + Cl^- \quad \text{and} \quad Na_2CO_3 \longrightarrow 2Na^+ + CO_3^{2-}$$

So, Van't Hoff factor i is close to 2 in case of NaCl and 3 in case of Na_2CO_3. However, this limiting value can be achieved only at high dilution and the value of i decreases from the limiting value as the concentration increases. At high concentration, number of free ions decreases due to strong interionic attractions. Consequently the value of i also decreases.

16.5.7.1 Partial Dissociation of Solute

In case of weak electrolyte, e.g., acetic acid, dissociation is incomplete in solution.

$$CH_3COOH \rightleftharpoons CH_3COO^- + H^+$$
$$(1-\alpha)c \qquad \alpha c \qquad \alpha c$$

α is the degree of dissociation and c is the concentration of acetic acid. Acetic acid produces 2 ions in solution. Thus total number of moles in 1 litre of solution is $(1-\alpha)c + \alpha c + \alpha c = [1 + (2-1)\alpha]c$. If a weak electrolyte produces n number of ions in solution, total number of moles in 1 litre of solution will be $(1-\alpha)c + n\alpha c = [1 + (n-1)\alpha]c$. Thus the expression of osmotic pressure is

$\pi = [1 + (2-1)\alpha]cRT$ for acetic acid solution
$\pi = [1 + (n-1)\alpha]cRT$ for general weak electrolytic solution.

So, $i = 1 + (2-1)\alpha$ for acetic acid and $i = 1 + (n-1)\alpha$ in general. Hence, degree of dissociation α can be computed as

$$\alpha = \frac{i-i}{n-1} \tag{16.39}$$

16.5.7.2 Partial Association of Solute

In some cases, association of solute molecules takes place in solution, e.g., acetic acid dimerises in *n*-butanol as shown below

$$CH_3COOH \rightleftharpoons \frac{1}{2}(CH_3COOH)_2$$

$$(1-\beta)c \qquad \frac{1}{2}\beta c$$

where, β is the degree of association.

So, total number of moles/litre $= (1-\beta)c + \frac{1}{2}\beta c = \left[1 - \left(1 - \frac{1}{2}\right)\beta\right]c$

In general we can write

Total number of moles/litre $= (1-\beta)c + \frac{1}{n}\beta c = \left[1 - \left(1 - \frac{1}{n}\right)\beta\right]c$

Thus the expression of osmotic pressure is

$$\pi = \left[1 - \left(1 - \frac{1}{2}\right)\beta\right]cRT \text{ for acetic acid in } n\text{-butanol}$$

$$\pi = \left[1 - \left(1 - \frac{1}{n}\right)\beta\right]cRT \quad \text{for general weak electrolytic solution}.$$

So, $i = \left[1 - \left(1 - \frac{1}{2}\right)\beta\right]$ for acetic acid and $i = \left[1 - \left(1 - \frac{1}{n}\right)\beta\right]$ in general.

Hence, degree of association β can be computed as

$$\beta = \frac{1 - i}{1 - \frac{1}{n}} \tag{16.40a}$$

16.5.7.3 Degree of Association and Molecular Weight

Let us consider a solute, S, which remains in dimeric form in a solvent. Then the following equilibrium is assumed to hold good. $S \rightleftharpoons \frac{1}{2}S_2$

If β is the degree of association, $(1 - \beta)$ is the number of moles of the monomer S and $\frac{1}{2}\beta$ is the number of moles of the dimer, S_2. So, the total number of moles

$$= \left(1 - \beta + \frac{1}{2}\beta\right) = \left(1 - \frac{1}{2}\beta\right).$$

Suppose average MW is M, MW of monomer is M_0. So, MW of dimer is $2M_0$. Considering the mass balance, we have

$$M\left(1 - \frac{1}{2}\beta\right) = (2M_0)\frac{1}{2}\beta + M_0(1 - \beta)$$

Hence, $$\beta = \frac{2(M - M_0)}{(2 - 1)M} = \frac{2(M - M_0)}{M} \tag{16.41a}$$

If the solute remains as S_n, the above expression becomes

$$\beta = \frac{n(M - M_0)}{(n - 1)M} \tag{16.41b}$$

16.5.7.4 Physical Significance of i

(a) If there is no association or dissociation of solute in the solution, value of i is 1. One example is glucose solution.

(b) If solute particles remain associated in solution, value of i is less than 1. Examples are: (i) acetic acid in *n*-butanol, (ii) benzoic acid in benzene, etc.

(c) If solute particles remain dissociated in solution, value of i is greater than 1. Examples are: (i) aqueous solution of NaCl, (ii) aqueous solution of KCl, (iii) any strong acid, alkali or salt solution, etc.

16.6 DETERMINATION OF ACTIVITY COEFFICIENT

The term concentration indicates the number of moles or molecules of solute present in a given volume of solution. However, a fraction of solute molecules is active to establish colligative properties of solution or to bring about a chemical reaction while the other fraction remains inactive. To separate the active molecules from the inactive ones, the term *activity* is used instead of concentration and designated by the symbol a. Obviously, activity is always less than concentration and directly proportional to the concentration (c). Mathematically, we can write

$a \propto c$. Or, $a = \gamma c$, where, γ is known as activity coefficient and its value is less than 1. However, at infinite dilution, $\gamma \to 1$ and $a = c$.

From the knowledge of thermodynamics, the chemical potential of ith component in a solution is given by

$$\mu_i = \mu_0^i + RT \ln a_i$$

where, μ_0^i is the standard chemical potential of ith component, i.e., the chemical potential at $a_i = 1$.

16.6.1 Definition of Standard Chemical Potential

If 1 kg of a solvent contains 1 mole of solvent such that the solution possesses the properties of dilute solution. So, $\gamma \to 1$ and $a = m$. For that hypothetical solution chemical potential of the solute is called standard chemical potential of that solute.

Let us consider a dilute solution containing a non-electrolytic solute. a_1 and a_2 represent chemical potentials of the solvent and solute respectively. For the pure solvent, $a_1 = 1$ and the corresponding vapour pressure is p_1^0. As the concentration of the solute increases activity of the solvent, a_1 decreases and hence also the vapour pressure. From Raoult's law, we have

$$p_1 = x_1 p_1^0 \text{ and } x_1 = \frac{n_1}{n} = \frac{n_1 / V}{n / V} = \frac{m_1}{n / V} = \frac{a_1}{n / V}, \text{ since, } m_1 \approx a_1 \text{ for dilute solution.}$$

where, n and n_1 are the total number of moles (considering solvent and solute) and number of moles of solvent respectively. If n_2 represents number of moles of solute, $n_1 + n_2 = n$. Again, V is the volume of the solvent = 1000 ml. So equation (16.12) can be rewritten as

$$\ln x_1 = \frac{L_e}{R} \left[\frac{1}{T} - \frac{1}{T_b} \right] = -\frac{L_e \cdot \Delta T_b}{RT_b^2}$$

Or,
$$\ln \frac{a_1}{n / V} = -\frac{L_e \cdot \Delta T_b}{RT_b^2} \tag{16.42}$$

Thus by measuring ΔT_b, a_1 can be calculated.

We know that

$$\mu_i = \mu_0^i + RT \ln a_i \text{ and } d\mu_i = RTd \ln a_i$$

From Gibbs-Duhem equation, we have

$$n_1 d\mu_1 + n_2 d\mu_2 = 0. \quad \text{Or} \quad n_1 RTd \ln a_1 + n_2 RTd \ln a_2 = 0$$

Or,
$$n_1 d \ln a_1 + n_2 d \ln a_2 = 0 \tag{16.43}$$

a_1 is already calculated using equation (16.42). n_1 and n_2 are known. Hence, a_2 can be calculated by using the above equation (16.43).

Again, $a_2 = m_2 \gamma_2$. Molality (m_2) is known. a_2 is already calculated. Hence, γ_2 can be calculated.

16.7 REVERSE OSMOSIS (RO)

It is also known as pressurized filtration method. In this method the system consists of two compartments, separated by a semipermeable membrane. In one compartment, pure solvent, preferably water, is taken while the other one contains highly concentrated solution. The semipermeable membrane does not allow large ions or solute molecules to pass through but allows the solvent molecules to pass through the membrane. Under normal osmosis condition, water molecules are moving from pure solvent chamber to the concentrated solution chamber through the semipermeable membrane. But in reverse osmosis, high pressure is applied on the concentrated solution compartment so that solvent molecules, preferably water molecules, are forced to move from concentrated solution compartment to pure solvent compartment. As a result of this, all solutes, ions, etc. are left on the concentrated solution chamber and the pure solvent or water is collected in the other chamber. The mechanism of reverse osmosis illustrated in the Fig. 16.4.

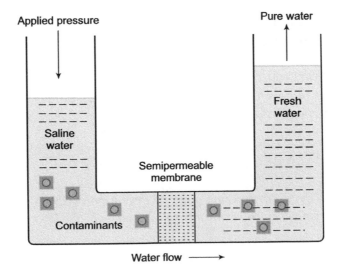

Figure 16.4 Reverse osmosis: pressurized flow of solvent from high concentration to low concentration through semipermeable membrane

Definition Reverse osmosis is the artificial process by which solvent is forced to flow from a region of concentrated solution through a semipermeable membrane to dilute solution by applying a pressure in excess of the osmotic pressure.

There are different types of filters: (a) Particle filter, which removes particles, having diameter ≥ 1000 nm. (b) Microfilter, which removes particles, having diameter ≥ 50 nm. (c) Ultrafilter, which removes particles, having diameter ≥ 3 nm. (d) Hyperfilter, which removes particles, having diameter ≥ 0.1 nm.

The semipermeable membrane is made of multiple layer of polymer matrix. Pressure applied in reverse osmosis depends on the concentration of the solution. For example, in drinking water purification plant 2–20 atm pressure is required but in desalination plant 40–70 atm pressure is required. The required membrane pore size is around 0.1 nm. Thus RO is often called hyperfiltration technique. Membranes are made of either cellulose triacetate or polyamide. Usually a two layer membrane is used. The first layer is polyamide membrane and the second one is the carbon filter, which absorbs other chemicals passing through the first membrane.

6.8 KEY CONCEPTS ON OSMOSIS

1. Osmotic pressure arises when two solutions of different concentrations, or a pure solvent and a solution, are separated by a semipermeable membrane. Molecules such as solvent molecules that can pass through the membrane will migrate from the side of lower concentration to the side of higher concentration in a process known as osmosis.
2. The pressure required to stop osmosis is called the **osmotic pressure**.
3. In dilute solutions, osmotic pressure (π) is directly proportional to the **molarity** of the solution and its **temperature** in °K.
4. **Van't Hoff Equation:** $\pi = cRT$, where, π is the osmotic pressure, c is the molarity of the solution, i.e., number of moles of solute dissolved in 1000 ml of solvent, R is the molar gas constant and T is the temperature in °K.
5. Solvent can be removed from a solution using a pressure greater than the osmotic pressure. This is known as **reverse osmosis**.

SOLVED PROBLEMS

Problem 16.1 What is the freezing point of a solution of 250 g of $CaCl_2$ (which is a strong electrolyte) in 1.0 kg of water? (K_f for H_2O = 1.86 °k · kg/mole)

Solution

Freezing point depression (ΔT) is given by

$$\Delta T = iK_f m, \quad i \text{ of } CaCl_2 = 3, \quad \text{MW of } CaCl_2 = 111$$

$$m = \frac{250}{111} = 2.25 \text{ moles/(1000 gm) of } H_2O.$$

$$\Delta T = 3 \times 1.86 \times 2.25 = 12.555° \approx 12.5°$$

Problem 16.2   Calculate the osmotic pressure exhibited by a 0.10 M sucrose solution at 20°C.

Solution

$$\pi = cRT = 0.10 \times 0.082 \times 293 = 2.4 \text{ atm.}$$

Problem 16.3   0.500 g Haemoglobin was dissolved in enough water to make 100 ml of solution. At 25°C the osmotic pressure was found to be 1.78×10^{-3} atm. Calculate the molecular mass (formula weight) of the haemoglobin.

Solution

$\pi = cRT$, $\pi = 1.78 \times 10^{-3}$ atm. Putting the values of π, R and T, we get

$$c = 7.28 \times 10^{-5} \text{ M} = 7.28 \times 10^{-5} \text{ mole/l}$$
$$= 7.28 \times 10^{-6} \text{ mole/(100 ml)}$$

So, number of moles of Haemoglobin present in 100 ml of solution is $n = 7.28 \times 10^{-6}$.

$$n = \frac{\text{Weight}}{MW} = \frac{0.500}{MW}. \text{ Hence, } MW = \frac{0.500}{7.28 \times 10^{-6}} = 68,681 \text{ gm/gm.mole.}$$

Problem 16.4   The osmotic pressure of a 1.0×10^{-2} M solution of cyanic acid (HOCN) is 217.2 torr at 25°C. Calculate the K_a HOCN.

Solution

$\pi = icRT$, Given $\pi = 217.2$ torr $= 0.2858$ atm, $c = 10^{-2}$ M
$R = 0.08206$ lit.atm/mole.°K, $T = 298$°K. Hence, $i = 1.17$. So the actual concentration

$$HOCN \rightleftharpoons H^+ + OCN^-. K_a = \frac{[H^+][OCN^-]}{[HOCN]}$$

10^{-2} M is the concentration of HOCN if we consider that HOCN remains as undissociated form. But HOCN is partially dissociated and hence,

$$[H^+] + [OCN^-] + [HOCN] = ic = 0.0117 \text{ M}$$

$$[H^+] = [OCN^-] = \alpha c, \ [HOCN] = (1 - \alpha)c$$

where, α is the degree of dissociation.
Thus, $\alpha c + \alpha c + (1 - \alpha)c = 0.0117$. Hence, $\alpha c = 0.0017$ M, $(1 - \alpha)c = 0.0083$M

$$K_a = \frac{[H^+][OCN^-]}{[HOCN]} = \frac{(\alpha c)(\alpha c)}{(1-\alpha)c} = 3.5 \times 10^{-4}.$$

Problem 16.5   A 0.035 M aqueous nitrous acid (HNO_2) solution has an osmotic pressure of 0.93 atm at 22°C. Calculate the percent ionization of the acid.

Solution

$\pi = icRT$, Given $\pi = 0.93$ atm, $c = 0.035$M $R = 0.08206$ (lit.atm)/(mole.°K), $T = 295$°K. Hence, $i = 1.1$.

$$HNO_2 \rightleftharpoons H^+ + NO_2^-. \quad [H^+] = [NO_2^-] = \alpha c, \quad [HNO_2] = (1-\alpha)c$$

Thus, $\alpha c + \alpha c + (1-\alpha)c = ic = 1.1 \times 0.035$ M $= 0.0384$M, $\alpha c = 0.0034$M

$$\% \text{ Ionization} = \frac{0.0034M}{0.035M} \times 100 = 9.8\%.$$

Problem 16.6   A solid mixture contains $MgCl_2$ and NaCl. When 0.5 g of this mixture is dissolved in enough water to form 1 lit. of solution, the osmotic pressure at 25°C is observed to be 0.399 atm. What is the mass percent of $MgCl_2$ in the solid? (Assume ideal behaviour for the solution.)

Solution

We know $\pi = icRT$, $\pi = 0.399$ atm, $T = 298$°K, $R = 0.08206$ (lit.atm)/(mole.°K), $i_c = 0.0163$ M.

For 100% $MgCl_2$, $\quad [MgCl_2] = 3 \times \left(\dfrac{0.5}{95}\right) M = 0.01578 \ M$

since, $MgCl_2$ dissociates into 3 ions in aqueous solution.

For 100% NaCl, $\quad [NaCl] = 2 \times \left(\dfrac{0.5}{58.5}\right) M = 0.0171 \ M$

since, NaCl dissociates into 2 ions in aqueous solution.

Suppose the mixture contains x mole fraction of $MgCl_2$ and $(1-x)$ mole fraction of NaCl. Hence we can write

$$0.01578x + 0.0171 \ (1-x) = 0.0163. \text{ Or, } x = \frac{0.0008}{0.00132} = 0.61.$$

So the mixture contains 61% $MgCl_2$ and 39% NaCl.

Problem 16.7   A solution is prepared by dissolving 35 gm of haemoglobin in enough water to make up 1 lit. in volume. The osmotic pressure of the solution is found to be 10.0 mm Hg at 25°C. Calculate the molar mass of haemoglobin.

Solution

We know $\pi = icRT$, $\pi = \left(\dfrac{10}{760}\right)$ atm $= 0.01316$ atm, $T = 298°K$.

$$R = 0.082 \text{ (lit.atm)/(mole.°K)}, \ i = 1.$$

Hence, $$c = 0.0005385 \text{ M}$$

Thus, 0.0005385 moles of haemoglobin remains dissolved in 1 lit. of solution. So, we have

$$n = \frac{w}{MW} = 0.0005385, \ w = 35 \text{ gm. Or, } MW = 64995 \approx 65000$$

Problem 16.8 For rehydration therapy of cholera patients, the World Health Organization uses an aqueous solution with the following concentration.

[NaCl] = 3.5 gm/l, [NaHCO$_3$] = 2.5 gm/l, [KCl] = 1.5 gm/l, [glucose] = 20 gm/l

Calculate the osmolarity of the resulting solution.

Solution

Osmolarity of any solution is ic.

For NaCl, $\qquad$ $[\text{NaCl}] = 2 \times \left(\dfrac{3.5}{58.5}\right) M = 0.1196\,M$

For NaHCO$_3$, $\quad$ $[\text{NaHCO}_3] = 2 \times \left(\dfrac{2.5}{84}\right) M = 0.05952\,M$

For KCl, $\qquad$ $[\text{KCl}] = 2 \times \left(\dfrac{1.5}{74.5}\right) M = 0.04026\,M$

For glucose, $\quad$ $[\text{glucose}] = \left(\dfrac{20}{180}\right) M = 0.1111\,M$

So, osmolarity of the solution $= (0.1196 + 0.05952 + 0.04026 + 0.1111)M$
$$= 0.33048 \ M$$

Problem 16.9 A biochemical engineer isolates a bacterial gene fragment and dissolves a 17.6 mg sample of the material in enough water to make 31.5 ml of solution. The osmotic pressure of the solution is 0.34 torr at 25°C. (i) What is the molar mass of the gene fragment? (ii) Calculate the freezing point depression if the density of the solution is 0.997 gm/ml. Given $K_f = 1.86°K$. kg/mole.

Solution

We know $\pi = icRT$. $\pi = 0.34$ torr $= (0.34/760)$ atm $= 0.0004474$ atm

$T = 298°K$, R $= 0.082$ (lit.atm)/(mole.°K). Hence, $ic = 1.831 \times 10^{-5}$ M, $i = 1$. Number of gm-moles present in 31.5 ml solution is $n = 1.831 \times 10^{-5} \times 0.0315 = 5.768 \times 10^{-7}$.

$$n = \frac{w}{MW} = 5.768 \times 10^{-7}, \quad w = 0.0176 \text{ gm. Or, } MW = 3.051 \times 10^4$$

$\Delta T_f = K_f.m$. Mass of the solution is 0.997×31.5 gm $= 31.4$ gm. Mass of sample is 0.0176 gm. So, mass of water is $(31.4 - 0.0176)$ gm $= 31.3824$ gm.

$$m = \left(\frac{5.768 \times 10^{-7}}{31.3824}\right) \times 1000 = 1.838 \times 10^{-5}. \text{ Hence, } \Delta T_f = 3.42 \times 10^{-5} \text{ °K.}$$

Problem 16.10   Consider a 4% starch solution and a 10% starch solution separated by a semipermeable membrane. Which starch solution will decrease in volume as osmosis occurs?

Solution
Diffusion is an entropically favorable spontaneous movement of molecules to create equal distributions. Without a membrane present, starch would diffuse from the 10% solution toward the 4% solution, and water would diffuse from the 4% solution (where water content is high) to the 10% solution (where water content is low). However, since starch molecules are too large to pass through a semipermeable membrane, only the water will move across the membrane, causing the 4% solution to decrease in volume.

Problem 16.11   A 0.102 gm sample of an unknown compound dissolved in 100 ml of water has an osmotic pressure of 28.1 mm of Hg at 20°C. Calculate the molar mass of the compound.

Solution
We know $\pi = icRT$. $\pi = \left(\frac{28.1}{760}\right)$ atm $= 0.037$ atm.

$T = 293$°K, $R = 0.082$ (lit.atm)/(mole.°K). Hence, $ic = 0.00154$ M, $i = 1$. Number of gm-moles present in 100 ml water is $n = 0.000154 = (w/MW)$. $w = 0.102$ gm. Hence, $MW = 662.34$.

Problem 16.12   To determine the molar mass of an unknown protein, 1 mg was dissolved in 1 ml of water. At 25°C the osmotic pressure of this solution was found to be 1.47×10^{-3} atm. What is the molar mass of this protein?

Solution
We know $\pi = icRT$. $i = 1$, $\pi = 1.47 \times 10^{-3}$ atm.
$$T = 298\text{°K}, R = 0.082 \text{ (lit.atm)/(mole.°K). Hence, } c = 6.01 \times 10^{-5} \text{ M.}$$
Number of moles present $= n = 6.01 \times 10^{-8} = w/MW$. $w = 0.001$ gm. Hence, $MW = 1.664 \times 10^4$.

Problem 16.13 2 solutions, a 0.1% (*m*/*v*) albumin solution (compartment *A*) and a 2% (*m*/*v*) albumin solution (compartment *B*), are separated by a semi-permeable membrane. (Albumins are colloidal proteins).

(a) Which compartment will have the higher osmotic pressure?

(b) Which compartment will lose water?

(c) If some NaCl is added to compartment A, will the sodium and chloride ions stay in compartment A, or will some cross the membrane to compartment *B*?

Solution

(a) Compartment *B* due to the higher albumin concentration in *B*.

(b) Compartment *A*. Since it has the lower albumin concentration, it has the higher water "concentration."

(c) As long as the membrane isn't selective against Na^+ and Cl^- ions, then they will flow to B so they attain equilibrium. Why can the Na^+ and Cl^- flow through the membrane but not albumin? Albumin is much larger than the Na^+ or Cl^- ions. That's why in part *B*, compartment *A* loses water to compartment *B* instead of compartment *B* losing albumin to compartment *A*.

EXERCISES

1. Calculate the osmotic pressure exhibited by a 0.42M KOH solution at 30°C.

   [*Ans.:* $KOH(aq) \longrightarrow K^+(aq) + OH^-(aq)$. $\pi = icRT$, $i = 2$, $c = 0.42$ M, $T = 303°K$
   $R = 0.082$ (lit.atm)/(mole,°K). Hence, $\pi = 20.9$ atm.]

2. Add 10 gm of NaCl to 1 lit. of water and 20 gm of glucose to another 1 lit. of water. Which one has a higher osmolarity? Explain your answer.

   [*Ans.:* $[NaCl] = 2 \times \left(\dfrac{10}{58.5}\right) M = 0.3419 M$, $[glucose] = \left(\dfrac{20}{180}\right) M = 0.1111 M$

   So, NaCl solution has higher osmolarity.]

3. Osmosis is the process responsible for carrying nutrients and water from groundwater supplies to the upper parts of trees. The osmotic pressures required for this process can be as high as 18.6 atm. What would the molar concentration of the tree sap have to achieve this pressure on a day when the temperature is 27°C?

   [*Ans.:* We know $\pi = icRT$, $\pi = 18.6$ atm, $T = 300°K$, $R = 0.082$ (lit.atm)/(mole.°K)
   Hence, $ic = 0.7561$ M]

4. Isotonic saline solution, which has the same osmotic pressure as blood, can be prepared by dissolving 0.923 gm of NaCl in enough water to produce 100 ml of solution. What is the osmotic pressure, in atmospheres of this solution at 25°C?

   [*Ans.:* $[NaCl] = (0.923/58.5)$ gm/(100 ml) = 0.1578 M. $i = 2$. Thus we have
   $\pi = icRT = 2 \times 0.1578 \times 0.082 \times 298 = 7.712$ atm]

5. Arrange the following solutions in order of decreasing osmotic pressure:
   (a) 0.10 M urea (b) 0.06 M NaCl (c) 0.05 M $Ba(NO_3)_2$ (d) 0.06 M sucrose

6. The colligative properties of a 1.26 M solution of calcium chloride shows the same colligative properties of a sucrose solution. Calculate the concentration of sucrose solution.

7. What is the value of the van't Hoff factor for a 0.64 molar aqueous solution of acetone?

8. The mixture of benzene and toluene is assumed to obey the Raoult's law. A given mixture, containing 0.34 mole fraction of toluene boils at 88°C. The vapour pressures of benzene and toluene at this temperature are 960 mm and 380 mm respectively. Calculate the composition of the vapour formed on its boiling.

   [*Ans.:* Mole fraction of benzene is 0.83]

9. At 26°C vapour pressures of $CHCl_3$ and CCl_4 are 200 mm and 115 mm respectively. The liquid mixture contains 1 mole each of the pure liquids. What is the weight percentage of $CHCl_3$ in the vapour phase, remaining equilibrium with that liquid mixture.

   [*Ans.:* 57%]

10. At 25°C the vapour pressure of water is 23.55 mm. What would be the vapour pressure of the solution containing 6 gm of urea (MW is 60) in 100 gm of water at the same temperature? [*Ans.:* 23.13 mm]

    Hint: $\left[\dfrac{p^0 - p}{p^0} = x_2, \ p^0 = 23.55 \text{ mm} \right]$

11. The bp of acetic acid is 118.1°C and its $l_v = 121$ cal/gm. A solution containing 0.4344 gm anthracene in 44.16 gm acetic acid boils at 118.24°C. Calculate the MW of anthracene.

    [*Ans.:* MW = 178]

    Hint: $\left[\Delta T_b = K_b.m, \ m = \dfrac{w_2 \times 1000}{w_1 \times M_2}, \ K_b = \dfrac{RT_b^2}{1000 l_v} \right]$

12. The mp of phenol is 40°C. A solution containing 0.172 gm acetanilide in 12.54 gm phenol freezes at 39.25°C. Calculate K_f and l_f of phenol. Acetanilide, $C_6H_5NHCOCH_3$, MW is 135. [*Ans.:* $K_f = 7.38$ °K·kg/mol, $l_f = 26.5$ cal/gm]

    Hint: $\left[\Delta T_f = K_f.m, \ m = \dfrac{w_2 \times 1000}{w_1 \times M_2}, K_f = \dfrac{RT_f^2}{1000 l_f} \right]$

13. Calculate the fp of a cane sugar solution which has an osmotic pressure of 5 atm at 50°C. [*Ans.:* $T_f = 272.648$ °K]

    Hint: $\left[\pi = \dfrac{\rho RT \Delta T_f}{1000 K_f}, \ \rho = 1 \text{ gm/ml} \right]$

14. Acetic acid forms dimers in benzene. 1.65 gm of acetic acid, dissolved in 100 gm of benzene, raised the bp by 0.36°C. $K_b = 2.57$ unit. Calculate the degree of association.

    [*Ans.:* 0.983]

    Hint: $\left[\alpha = \dfrac{2(M'_2 - M_2)}{M_2}, \quad M'_2 = K_b \dfrac{w_2 \times 1000}{w_1 \times T_b} = 118, \quad M_2 = 60 \right]$

15. Human blood is isotonic with 0.9% NaCl solution at 27°C. Calculate the osmotic pressure of the human blood. [*Ans.:* $\pi = 7.6$ atm]

16. The osmotic pressure of 3% aqueous solution of A at 25°C is 3.632 atm. What is the vapour pressure of the solution? p_{H_2O} of pure water is 23.7 mm Hg at 25°C.

    [*Ans.:* $p = 23.64$ mm Hg]

17. 150 ml of an aqueous solution, containing 3.6 gm of solute, shows an osmotic pressure of 4.48 atm at 17°C. What is the bp of the solution? Given $K_b = 0.52$ unit.

    [Answer: $t_b = 100.098$°C]

18. A solution, containing 0.684 gm NaCl in 100 gm of water freezes at –0.4°C. Given K_f of water is 1.85 unit. Estimate apparent MW of NaCl and Van't Hoff factor.

    [MW = 31.65, $i = 1.848$]

19. The bp of ether is 36°C. A solution of 1.6 gm naphthalene (MW = 128) in 20 gm ether boils at 36.31°C. Calculate K_b and l_v of ether.

    [*Ans.:* $K_b = 2.1$ unit, $l_v = 90.4$ cal/gm]

20. A sample of camphor ($K_f = 40$) melts at 176°C. A solution of 0.0205 gm of a hydrocarbon in a 0.261 gm camphor melts at 156°C. The hydrocarbon contains 92.3% carbon. Determine the molecular formula of the hydrocarbon. [*Ans.:* $C_{12}H_{12}$]

21. 0.142 gm naphthalene in 20.25 gm benzene lowered the fp by 0.284°C. Calculate the MW of naphthalene. Given $K_f = 5.12$. [*Ans.:* MW = 126]

22. 100 gm of benzene contained 10 gm of non-volatile solute (A). The loss in weight of the solution in bubbling in a slow stream of air when passed directly through pure benzene was 1.273 gm. What is the MW of A? [*Ans.:* MW = 138.2]

17 Surface Chemistry—Adsorption

17.1 INTRODUCTION

Adsorption isotherm Adsorption of a gas on a solid adsorbent is generally carried out at constant temperature. There is a relation between gas pressure and the amount of gas adsorbed. This relation is known as adsorption isotherm.

Example of adsorption If a gas like SO_2 is passed over finely divided charcoal, gas pressure is found to be reduced indicating that gas is adsorbed on charcoal surface.

Whenever an adsorbate comes in contact with a surface, i.e., an adsorbent, change in concentration of the adsorbent is observed along the surface. As concentration changes surface tension of the adsorbent also changes. For a binary system, the Gibbs adsorption equation is given by:

$$-d\gamma = \Gamma_1 d\mu_1 + \Gamma_2 d\mu_2 \tag{17.1}$$

where, γ is the surface tension, Γ_1 and Γ_2 are surface excess of component-1 and component-2 respectively. μ_1 and μ_2 are the chemical potentials of the component-1 and component-2 respectively. The above equation is known as Gibbs adsorption equation, correlating surface tension and concentration of the solutes.

Surface excess If a component, preferably solute, is mixed with another component, preferably solvent, the solute is distributed in the solvent in an uneven manner. In some cases, the concentration of the solute is more in the bulk phase than on the surface. For example, sugar solution. Top layer of water is less sweet than that of the bulk. The phenomenon is called *negative surface excess*. However, in other cases, the concentration of the solute is more on the surface than in the bulk phase. For example, soap solution. Soap concentration on the surface is much higher than that of the bulk. The phenomenon is called *positive surface excess*.

If a system has two distinct phases, α and β, there is an interfacial surface, separating the two phases as shown in the Fig. 17.1.

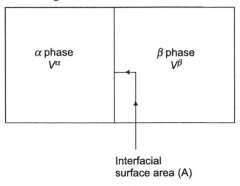

Figure 17.1 Gibbs ideal interface

According to the Gibbs model, volume of the interface, $V^S = 0$ and the infinitesimal thin boundary layer is called Gibbs dividing plane. Thus the total volume of the system is given by

$$V = V^\alpha + V^\beta$$

However, the Gibbs dividing plane possesses a certain number of molecules with internal energy, U^S.

Thus the total internal energy, U, is given by

$$U = U^\alpha + U + U^S$$

$c_i^\alpha = $ Concentration of ith component in phase α in (Number of moles or molecules/volume)

$c_i^\beta = $ Concentration of ith component in phase β in (Number of moles or molecules/volume)

$N_i^\alpha = $ Number of molecules of ith component in phase $\alpha = c_i^\alpha \times V^\alpha$

$N_i^\beta = $ Number of molecules of ith component in phase $\beta = c_i^\beta \times V^\beta$

$N_i^S = $ Number of molecules of ith component in the Gibbs dividing plane.

Thus the total number of molecules, N_i is given by

$$N_i = N_i^S + c_i^\alpha \times V^\alpha + c_i^\beta \times V^\beta$$

Or, $$N_i^S = N_i - (c_i^\alpha \times V^\alpha + c_i^\beta \times V^\beta)$$

If A represents area of the Gibbs dividing plane, the interfacial excess or surface excess is given by

$$\Gamma_i = \frac{N_i^S}{A} \quad \text{Number of moles or molecules/area}$$

Surface Tension In the bulk of a solution or material each atom is attracted equally from all sides. Attractive forces are balanced and hence there is no residual attraction. However, on the surface the atoms are attracted inward direction only. So attractive forces are not balanced, thereby generating a pseudo-attractive force throughout the surface. This pseudo-attractive force is the origin of surface tension, having unit N/m^2 or J/m^3. However, the value of surface tension depends on the presence of the second component or solute.

Case-1 Surface excess of the second component is positive.

In this case, composition of the second component or solute is higher on the surface than that in the bulk. So with the increase in concentration of the solute, the latter starts diffusing from bulk to the surface, thereby increasing the chemical potential of the solute. So surface tension is lowered. When the surface excess of a component is positive, chemical potential of that component increases, thereby, reducing the surface tension.

Case-2 Surface excess of the second component is negative.

In this case, composition of the second component or solute is lower on the surface than that in the bulk. So with increase in concentration of the solute, the solute molecules try to move into the bulk but the movement is not possible due to high concentration of solute in the bulk. So surface tension increases. When the surface excess of a component is negative, increasing the chemical potential of that component increases the surface tension.

17.2 GIBBS EQUATION FOR ADSORPTION ISOTHERM

For a two-component system, two phases (α and β) are separated by a surface. Thus the Gibbs free energy (G) is given by:

$$G = G^\alpha + G^\beta + G^S$$

where, G^α, G^β and G^S are the Gibbs free energy for α-phase, β-phase and surface of separation of the two phases. The change of Gibbs free energy is given by

$$dG = VdP - SdT + \sum \mu_i dn_i + \sum n_i d\mu_i + Ad\gamma + \gamma dA \qquad (17.2)$$

Taking surface as constant, $dA = 0$ the above equation becomes

$$dG = VdP - SdT + \sum \mu_i dn_i + \sum n_i d\mu_i + Ad\gamma \qquad (17.3)$$

From the knowledge of thermodynamics, the expression of dG for multicomponent system is given by

$$dG = VdP - SdT + \sum \mu_i dn_i \qquad (17.4)$$

Combining equations (17.3) and (17.4) we have

$$\sum n_i d\mu_i + Ad\gamma = 0 \qquad (17.5)$$

Thus considering α-phase, β-phase and interfacial surface the above equation becomes

$$\sum n_i^\alpha d\mu_i + \sum n_i^\beta d\mu_i + \sum n_i^S d\mu_i + Ad\gamma = 0 \tag{17.6}$$

where, n_i^α and n_i^β represent number of moles of ith component in α and β phases respectively and n_i^S represents number of moles of ith component at the interface region between α and β phases.

At equilibrium distribution of each component in both α-phase and β-phase reaches a constant value. Hence, we can write that

$$\sum n_i^\alpha d\mu_i = \sum n_i^\beta d\mu_i = 0 \tag{17.7}$$

Thus the above equation becomes

$$\sum n_i^S d\mu_i + Ad\gamma = 0 \tag{17.8}$$

$\Gamma_i = \dfrac{n_i^S}{A}$ = Surface excess of the ith component in (moles/m²). Thus for binary system the equation (17.8) becomes

$$-d\gamma = \Gamma_1 d\mu_1 + \Gamma_2 d\mu_2 \tag{17.9}$$

The above equation is well known **Gibbs adsorption equation** in terms of surface excess.

The chemical potential of ith component is given by

$$\mu_i = \mu_i^0 + RT \ln a_i = \mu_i^0 + RT \ln fc_i \tag{17.10}$$

where, a_i = Activity of ith component, c_i = Concentration of ith component,

f = Activity coefficient of ith component. μ_i^0 = Standard chemical potential of ith component.

$$\mu_i = \mu_i^0 + RT \ln fc_i = \mu_i^0 + RT \ln f + RT \ln c_i \tag{17.11}$$

Differentiating both sides, we get

$$d\mu_i = RT\,d\ln c_i \tag{17.12}$$

Putting the values of $d\mu_1$ and $d\mu_2$ in the equation (17.9), we get

$$-d\gamma = \Gamma_1 RT\,d\ln c_1 + \Gamma_2 RT\ln c_2 \tag{17.13}$$

So, $\quad \Gamma_1 = -\dfrac{1}{RT}\left(\dfrac{\partial\gamma}{\partial\ln c_1}\right)_{T,P,c_2}$ and $\quad \Gamma_2 = -\dfrac{1}{RT}\left(\dfrac{\partial\gamma}{\partial\ln c_2}\right)_{T,P,c_1}$ $\qquad$ (17.14)

Area per molecule is given by

$$a = \frac{1\times10^{20}}{N_A \times \Gamma}\ (\text{Å})^2, \quad [N_A = \text{Avogadro Number}] \tag{17.15}$$

17.2.1   Characteristic Features of Gibbs Adsorption Equation

(a) $d\gamma/dc$ is negative for that solute, which reduces surface tension of the solvent. In that case, solute concentration is higher at the surface than in the bulk. The value of Γ is *positive*. The adsorption is called positive adsorption.

(b) $d\gamma/dc$ is positive for that solute, which increases surface tension of the solvent. In that case, solute concentration is lower at the surface than in the bulk. The value of Γ is *negative*. The adsorption is called negative adsorption.

(c) When a finely porous substance, such as charcoal/silica gel, is added, Gibbs adsorption equation is applied to the surface of contact between the solid and the solution.

(d) The solute, which reduces the surface tension of the solvent, will be adsorbed from solution by a solid adsorbent. Water has a very high surface tension and most of the solute reduces the surface tension of water. Thus majority of substances are adsorbed from aqueous solution by a solid adsorbent with positive value of Γ. One such substance is charcoal. The surface tension of the ethanol is much lower than that of water.

17.2.2   Applications of Gibbs Adsorption Isotherm

(a) Adsorption of a gas (or vapour) on the surface of a liquid

Case-1   Gas is not soluble in the liquid.

One example is saturated hydrocarbon in water. It is a two component system, where component-2, i.e., saturated hydrocarbon, is a vapour and component-1, i.e., water, is a liquid. For two component system the Gibbs equation is

$$-d\gamma = \Gamma_1 d\mu_1 + \Gamma_2 d\mu_2 \tag{17.16}$$

μ_1 and μ_2 are the chemical potentials of water and saturated hydrocarbon respectively. As the saturated hydrocarbon is completely insoluble in water, chemical potential of water, i.e., μ_1 remains unchanged in the bulk phase as well as surface phase. So, $d\mu_1 = 0$. Thus the Gibbs equation becomes

$$-d\gamma = \Gamma_2 d\mu_2$$

γ is the surface tension on water-vapour interface. The expression of chemical potential of a gas is

$$\mu_2 = \mu_2^0 + RT \ln p. \text{ Hence, } d\mu_2 = RT\, d \ln p$$

Putting the value of $d\mu_2$ in the above equation, we get

$$-d\gamma = \Gamma_2 (RT\, d \ln p). \text{ Hence, } \Gamma_2 = -\frac{1}{RT}\frac{d\gamma}{d \ln p} = -\frac{p}{RT}\left(\frac{\partial \gamma}{\partial p}\right) \tag{17.17}$$

Case-2   The 2nd component is soluble in the 1st component.

In this case, concentration of the 2nd component is different on the surface with respect to the bulk phase. So, μ_1 changes as we move from the surface to the bulk. Assuming $\Gamma_1 = 0$ on the surface, the Gibbs equation becomes

$$-d\gamma = \Gamma_2 d\mu_2$$

The chemical potential (μ_2) changes with concentration as follows:

$$\mu_2 = \mu_2^0 + RT \ln c_2. \quad \text{Hence,} \quad d\mu_2 = RTd \ln c_2 = RT \cdot \frac{dc_2}{c_2} \qquad (17.18)$$

Putting the value of $d\mu_2$ in the above equation, we get

$$-d\gamma = \Gamma_2 RT \frac{dc_2}{c_2}. \quad \text{Or,} \quad \Gamma_2 = -\left(\frac{c_2}{RT}\right)\left(\frac{\partial \gamma}{\partial c_2}\right) \qquad (17.19)$$

(b) Adsorption of surfactant on water surface

Each molecule of a surfactant has two parts: (i) hydrocarbon part, which remains suspended on the water surface since it is non-polar and does not react with water molecules, (ii) active polar part, i.e., functional groups, like, –OH, –COOH, –COOR, –NH$_2$, etc. This active part interacts with polar water molecules. So presence of surfactants decreases the intermolecular interactions among the water molecules on the surface, thereby reducing the surface tension. So, the surface tension (γ) decreases with concentration c_2. This decrease in surface tension with c_2 at infinite dilution is called **surface activity**. Mathematically it can be written as

$$\text{Surface activity} = -\left(\frac{\partial \gamma}{\partial c_2}\right)_{c_2 \to 0}$$

Substances, that reduce the surface tension of the solvent, are called surface-active substances or surfactants. In that case $\left(\dfrac{\partial \gamma}{\partial c_2}\right) < 0$ and $\Gamma_2 > 0$. Examples are: fatty acids, alcohols, amines, etc. Initially after adsorption, these molecules are irregularly oriented on the water surface but as the concentration increases orientation of molecules takes place at the cost of surface energy. Polar groups are facing to water molecules while the non-polar hydrocarbon parts are arranged vertically, forming a close-packed chain. In this case the homogeneous interaction among water molecules is much greater than the heterogeneous interaction between surfactant molecules and water molecules.

(c) Inorganic salt solution

In the case of aqueous solution of inorganic salts heterogeneous interaction between ions and water molecules is much greater than the homogeneous interaction among water molecules. So the ions are mainly distributed in the bulk of the solution. So, $\left(\dfrac{\partial \gamma}{\partial c_2}\right) > 0$ and $\Gamma_2 < 0$.

So surface tension of solvent increases in presence of these substances. These are generally called surface-inactive substances.

The effect of surfactants (e.g., sodium lauryl sulphate) and surface-inactive substances (e.g., potassium nitrate) on the surface tension of water is illustrated in the Fig. 17.2.

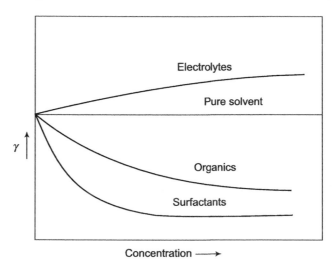

Figure 17.2   Effect of organic and inorganic salts on surface tension of water

17.3 CHANGE IN WORK FUNCTION IN ADSORPTION PROCESS

Let us consider a system, containing a pure adsorbent and a pure adsorbate kept separately. The work function A^0 is given by

$$A^0 = \gamma_1 + \Gamma_1^0 \mu_1^0 + n_2 \mu_2^0 \tag{17.20}$$

where, $\Gamma_1^0 = $ Excess concentration of the adsorbent at the interface. μ_1^0 and μ_2^0 are the chemical potentials of the adsorbent and the adsorbate respectively. In case of multicomponent system, we have

$$A = \gamma + \sum \mu_i \Gamma_i, \quad G = \sum \mu_i n_i \tag{17.21a}$$

From the knowledge of thermodynamics, we have

$$A = U - TS, \ G = H - TS = (U + PV) - TS = (U - TS) + PV = A + PV$$

Or, $A = G - PV$ \hfill (17.21b)

Considering only adsorbate before adsorption and assuming the gas (adsorbate) behaves ideally, we have

$$A = G - PV = G - n_2' RT = n_2' (\mu_2 - RT) \tag{17.22}$$

$\Gamma_2 = $ Number of moles of adsorbate per unit surface area of the adsorbent remains adsorbed on the surface of the adsorbent.

$n'_2 = (n_2 - \Gamma_2)$ = Number of moles of adsorbate per unit surface area of the adsorbent remaining in the gaseous phase.

After adsorption the work function is given by

$$A = \gamma + \Gamma_1\mu_1 + \Gamma_2\mu_2 + n'_2(\mu_2 - RT)$$

So, change in work function (ΔA) is given by

$$\Delta A = A - A^0 = (\gamma - \gamma^0) + \Gamma_1\mu_1 - \Gamma_1^0\mu_1^0 + \Gamma_2(\mu_2 - \mu_2^0) + n'_2(\mu_2 - \mu_2^0 - RT)$$

If the adsorbate molecules remain concentrated on the surface of the adsorbent and do not penetrate into the bulk, we can write

$$\mu_1 = \mu_1^0, \Gamma = \Gamma_1^0 \text{ and } n'_2 \approx 0$$

Thus the above equation becomes

$$A - A^0 = (\gamma - \gamma^0) + \Gamma_2(\mu_2 - \mu_2^0)$$

Or, $\quad\quad \Delta A = \Delta\gamma + \Gamma_2\Delta\mu_2 \quad\quad\quad\quad\quad\quad\quad\quad\quad\quad\quad\quad (17.23)$

If p_s and p are the saturated vapour pressure and actual vapour pressure of the adsorbate respectively above the adsorbent surface, change in chemical potential ($\Delta\mu_2$) is given by

$$\Delta\mu_2 = RT\ln\frac{p}{p_s} \quad\quad\quad\quad\quad\quad\quad\quad\quad\quad\quad\quad (17.24)$$

Putting the value of $\Delta\mu_2$ in the above equation we get

$$\Delta A = \Delta\gamma + \Gamma_2 RT\ln\frac{p}{p_s} \quad\quad\quad\quad\quad\quad\quad\quad\quad\quad (17.25)$$

Differentiating equation (17.23), we get

$$d(\Delta A) = d(\Delta\gamma) + \Gamma_2 d(\Delta\mu_2)$$

At equilibrium $d(\Delta A) = 0$. Hence, $d(\Delta\gamma) = -\Gamma_2 d(\Delta\mu_2)$

Or, $\quad\quad -\Delta\gamma = \int\Gamma_2 d(\Delta\mu_2) \quad\quad\quad\quad\quad\quad\quad\quad\quad\quad (17.26)$

Substituting this value of $\Delta\gamma$ in the equation (17.23), we get

$$\Delta A = \Gamma_2\Delta\mu_2 - \int\Gamma_2 d(\Delta\mu_2)$$

Integrating by parts and using equation (17.24), we get

$$\Delta A = \int_0^{\Gamma_2}\Delta\mu_2 d\Gamma_2 = \int_0^{\Gamma_2} RT\ln\frac{p}{p_s} d\Gamma_2 \quad\quad\quad\quad\quad (17.27)$$

Work done $= W = -\Delta A = -\int_0^{\Gamma_2}\Delta\mu_2 d\Gamma_2 = -\int_0^{\Gamma_2} RT\ln\frac{p}{p_s} d\Gamma_2 \quad\quad (17.28)$

Hence, $$\frac{\partial(\Delta A)}{\partial \Gamma_2} = \Delta \mu_2 = RT \ln \frac{p}{p_s}$$ (17.29)

W is called differential work of adsorption and $\dfrac{\partial(\Delta A)}{\partial \Gamma_2}$ is called differential change in work function due to adsorption.

17.4   FREUNDLICH ADSORPTION ISOTHERM

17.4.1   Adsorption of Gas on Solid Surface

In case of gas adsorption on solid surface the scientist Freundlich proposed the following empirical relation correlating mass of adsorbate and gas pressure.

$$\frac{w}{m} \propto P^{\frac{1}{n}}. \text{ Or, } \frac{w}{m} = kP^{\frac{1}{n}}. \text{ Hence, } \log \frac{w}{m} = \log k + \frac{1}{n} \log P$$ (17.30a)

where, w = Mass of the adsorbate, i.e., the gas; m = Mass of the adsorbent, i.e., the solid.

P = Pressure of the gas applied; k = Proportionality constant; n = An exponential constant.

Both n and k depend on the nature of adsorbate, nature of adsorbent and temperature.

The plots of $\left(\dfrac{w}{m}\right)$ versus P and $\log \dfrac{w}{m}$ versus $\log P$ are shown in the Fig. 17.3.

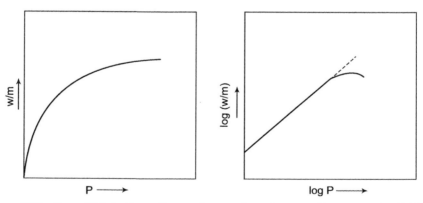

Figure 17.3   Freundlich isotherm diagram for gas adsorption according to equation (17.30a)

According to the equation (17.30a) the plot of $\log \dfrac{w}{m}$ versus $\log P$ should be a straight line. However, it is true at low pressure only. The deviation from straight line is well marked at high pressure, indicating that Freundlich proposition is approximate and not applicable at high pressure.

17.4.2 Adsorption of Solute from a Solution by Adsorbent

If a solution is passed over a porous bed of a solid surface, the dissolved solute may be adsorbed on the solid surface, thereby decreasing the concentration of the solution. This type of adsorption is similar to adsorption of gas on solid surface. The following characteristics are to be noted:

(a) For a given adsorbent amount of solute adsorbed depends on the nature of solute.

(b) As adsorption is an exothermic process increase in temperature decreases the extent of adsorption.

(c) The extent of adsorption depends on the nature of adsorbent and available surface area. So finely divided material or porous material is very much effective in adsorption process.

(d) An equilibrium is established between the amount of solute adsorbed and the concentration of the solution. With increase in amount of adsorption concentration of solution decreases.

(e) A unimolecular film is formed on the solid surface and Freundlich isotherm is applicable to determine the amount of solute adsorbed. Thus the mathematical expression of Freundlich isotherm is given below:

$$\frac{w}{m} \propto C^{\frac{1}{n}}. \text{ Or, } \frac{w}{m} = k.C^{\frac{1}{n}}$$

where, k and n are constants; C is the concentration of the solution.

w = Mass of the adsorbate; i.e., the solute; m = Mass of the adsorbent, i.e., the solid.

Taking $\log_{10}$ on both sides we get

$$\log\left(\frac{w}{m}\right) = \log k + \frac{1}{n}\log C \tag{17.30b}$$

As w increases, concentration of the solute in the solution decreases. Thus the plot of $\log\left(\frac{w}{m}\right)$ against $\log C$ results a straight line with negative slope. This is well illustrated in the Fig. 17.4.

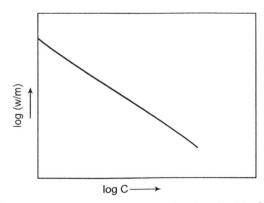

Figure 17.4   Freundlich isotherm diagram for solute adsorption from liquid according to equation (17.30b)

17.5 LANGMUIR EQUATION OF ADSORPTION ISOTHERM

17.5.1 Postulates

(a) Adsorbate molecules form a monolayer film on the adsorbent surface.

(b) The adsorption occurs uniformly throughout the surface so that the layer thickness is uniform all over the adsorbent surface.

(c) There is no interaction among adsorbate molecules on the adsorbent surface.

(d) Number of sites on a surface, occupied by adsorbate molecules, is limited. C_m = maximum concentration of an adsorbate remains adsorbed on a surface.

(e) Adsorption may be classified in two types, depending on the bond between adsorbate molecules and adsorbent. In this case adsorbate molecules can move along the surface of adsorbent. Hence it is called mobile adsorption.

  (i) **Physisorption** When adsorbate molecules are adhered on the surface of an adsorbent through weak Van der Waals type of forces, the adsorption is called physisorption. In this case the bond between adsorbate molecules and adsorbent is sufficiently strong to prevent motion of adsorbate molecules along the surface. Hence it is called localized adsorption.

  (ii) **Chemisorption** Sometimes adsorption takes place through the formation of strong chemical bond like covalent bond, hydrogen bond, etc. The adsorption is called chemisorption.

(f) Interactions among adsorbate molecules is neglected.

(g) Adsorbate molecules forms a localized adsorption complex on adsorbent surface. Influence of this localized adsorption complex on neighbouring sites is neglected.

17.5.2 Only One Type of Gaseous Adsorbate is Considered

α = Number of occupied sites; α_v = Number of vacant sites; α_m = Maximum number of sites available on the adsorbent surface.

θ = Fraction of surface covered by adsorbate molecules.

θ_v = Fraction of surface remains vacant.

$$\theta = \frac{\alpha}{\alpha_m}, \qquad \theta_v = \frac{\alpha_v}{\alpha_m}, \qquad \theta + \theta_v = 1, \tag{17.31a}$$

x = Amount of gas adsorbed per gm of the adsorbent, covering θ fraction of the surface of the adsorbent.

x_m = Maximum amount of gas adsorbed per gm of the adsorbent, covering the total surface area of the adsorbent.

On the basis of amount of gas adsorbed, θ is given by

$$\theta = \frac{x}{x_m} \tag{17.31b}$$

Rate of adsorption is given by J_a. It is directly proportional to the fraction of uncovered surface, i.e., $(1 - \theta)$ and gas pressure, p. Thus we can write

$$J_a \propto (1-\theta)p. \text{ Or, } J_a = k_a(1-\theta)p$$

where, k_a is the adsorption rate constant.

Similarly, rate of desorption is given by J_d. It is directly proportional to the fraction of covered surface, θ only. Thus we can write

$$J_d \propto \theta. \qquad \text{Or, } J_d = k_d\theta$$

where, k_d is the desorption rate constant.

At equilibrium, rate of adsorption is equal to rate of desorption. Hence, we can write

$$k_a(1-\theta)p = k_d\theta. \text{ Or, } \frac{\theta}{(1-\theta)} = Kp$$

Or, $$\theta = \frac{Kp}{1+Kp} \tag{17.32}$$

where, K = Equilibrium constant, also known as adsorption coefficient and $K = \dfrac{k_a}{k_d}$

The adsorption coefficient (K) is given by

$$K = \frac{\theta}{p(1-\theta)} = \frac{x}{p(x_m - x)} \text{ [Using equation (17.31}b)] \tag{17.33}$$

Again from equation (17.32), we have

$$p = \frac{\theta}{K(1-\theta)} \tag{17.34a}$$

Combining equations (17.31b) and (17.32), we have

$$x = \frac{x_m Kp}{1+Kp} \tag{17.34b}$$

The above equations (17.34a) and (17.34b) is collectively called the Langmuir equation of adsorption isotherm.

At very low concentration of adsorbate, $Kp \ll 1$. Hence, $\theta \approx Kp$ [Using equation (17.32)]

Hence, from equation (17.34b), we have $x = x_m Kp$

The above expression belongs to Henry's equation. Thus at low pressure region Langmuir equation changes to Henry's equation.

Again rearranging equation (17.34b), we have

$$x = \frac{x_m Kp}{1+Kp}. \text{ Or, } \frac{p}{x} = \frac{1}{Kx_m} + \frac{p}{x_m} \tag{17.35}$$

$$\frac{x}{p} = \frac{x_m K}{1 + Kp} = \frac{Kx_m(1 + Kp - Kp)}{1 + Kp}$$

$$= Kx_m - K\left(\frac{x_m Kp}{1 + Kp}\right) = Kx_m - Kx \; [\text{Using equation}(17.34b)]$$

Or, $$\frac{x}{p} = Kx_m - Kx \qquad\qquad (17.36)$$

If (p/x) is plotted against p, a straight line of slope $(1/x_m)$ is obtained as shown in the Fig. 17.5.

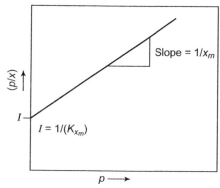

Figure 17.5 Langmuir adsorption isotherm diagram according to equation (17.35)

17.5.3 Thermodynamic Significance of K

Thermodynamically the equilibrium constant K is related with Gibbs free energy as follows:

$$\Delta G^0 = \Delta H^0 - T\Delta S^0 = -RT \ln K. \quad \text{Or,} \quad \ln K = -\frac{\Delta H^0}{RT} + \frac{\Delta S^0}{R} \qquad (17.37)$$

Or, $$K = e^{\Delta S^0/R} \cdot e^{-\Delta H^0/RT} = ge^{-\Delta H^0/RT}, \text{ where, '}g\text{' is called entropy factor.}$$

As adsorption phenomenon is an exothermic process, $\Delta H^\circ < 0$ and hence K decreases with increase in temperature. Consequently surface coverage at a given pressure decreases with increase in temperature.

17.5.4 Langmuir Equation for More Than One Type of Adsorbate

When only one type of adsorbate remains adsorbed on the adsorbent surface, the Langmuir equation is given by:

$$K = \frac{\theta}{p(1 - \theta)} \qquad\qquad (17.38)$$

However, if more than one component is present in the system, the above equation needs to be modified. If K_1 and K_2 are the equilibrium constants for the component-1 and component-2 respectively, the above equation for the two components is as follows

$$K_1 = \frac{\theta_1}{p_1(1-\theta_1-\theta_2)} \quad \text{and} \quad K_2 = \frac{\theta_2}{p_2(1-\theta_1-\theta_2)} \tag{17.39}$$

where, p_1 and p_2 are the partial pressures of the component-1 and component-2 respectively. θ_1 and θ_2 are the fraction of surface covered by the component-1 and component-2 respectively. From the above two equations we have

$$\frac{\theta_1}{\theta_2} = \frac{K_1 p_1}{K_2 p_2} \tag{17.40}$$

Using equations (17.39) and (17.40), we have

$$\theta_1 = \frac{K_1 p_1}{1 + K_1 p_1 + K_2 p_2} \quad \text{and} \quad \theta_2 = \frac{K_2 p_2}{1 + K_1 p_1 + K_2 p_2} \tag{17.41}$$

Thus in presence of multicomponent mixture of gases, the fraction of surface covered by i^{th} component, i.e., θ_i is given by

$$\theta_i = \frac{K_i p_i}{1 + K_1 p_1 + K_2 p_2 + \cdots\cdots\cdots + K_i p_i + \cdots\cdots\cdots} \tag{17.42}$$

However, in this case monolayer adsorption is considered.

17.6 BRUNAUER, EMMETT AND TELLER (BET) EQUATION

17.6.1 Postulates

(a) Adsorption of component is assumed to form multilayer film on the adsorbent surface.

(b) At first a monolayer component is formed and then other molecules are adsorbed in combination with monolayer adsorbate molecules to form single layer complex, double layer complex, etc. However, the physical and chemical bond, existing in the single layer complex, is much stronger than that existing in the double layer complex, etc.

(c) Only adsorbate-adsorbent interaction is considered.

17.6.2 Derivation

θ', θ'', θ''' etc. are the fraction of surface covered by single layer complex, double layer complex, triple layer complex, etc. Thus the corresponding Langmuir equations are

Single layer complex: $\quad K' = \dfrac{\theta'}{p\theta_0}$

Double layer complex: $K'' = \dfrac{\theta''}{p\theta'}$

Triple layer complex: $K''' = \dfrac{\theta'''}{p\theta''}$

The value of K'', K''' etc. are much smaller than that of K'. Hence it can be assumed that

$$K'' = K''' = ... = K_L$$

K_L = Equilibrium constant for saturated vapour-liquid equilibrium = $1/p_s$, where, p_s is the saturated vapour pressure.

$$\theta' = K'p\theta_0$$

$$\theta'' = K''p\theta' = K_L p\theta' = \left(\frac{p}{p_s}\right)\theta' = \left(\frac{p}{p_s}\right)K'p\theta_0$$

$$\theta''' = K'''p\theta'' = K_L p\theta'' = K_L p\left(\frac{p}{p_s}\right)\theta' = \left(\frac{p}{p_s}\right)^2\theta' = \left(\frac{p}{p_s}\right)^2 K'p\theta_0$$

$$\theta'''' = K''''p\theta''' = K_L p\theta''' = K_L p\left(\frac{p}{p_s}\right)^2\theta'' = \left(\frac{p}{p_s}\right)^3\theta' = \left(\frac{p}{p_s}\right)^3 K'p\theta_0$$

θ' = Fraction of surface covered by single layer complex.

θ'' = Fraction of surface covered by double layer complex.

θ''' = Fraction of surface covered by triple layer complex.

θ_0 = Fraction of free surface.

So we can write that

$$\theta_0 + \theta' + \theta'' + \theta''' + ... = 1 \tag{17.43}$$

$$\theta' = K'p\theta_0, \quad \theta'' = \left(\frac{p}{p_s}\right)\theta' = K'\left(\frac{p}{p_s}\right)p\theta_0, \quad \theta''' = \left(\frac{p}{p_s}\right)^2\theta' = K'\left(\frac{p}{p_s}\right)^2 p\theta_0$$

$p < p_s$,   and hence, $\dfrac{p}{p_s} < 1$

So, $$\theta_0\left\{1 + K'p\left[1 + \left(\frac{p}{p_s}\right) + \left(\frac{p}{p_s}\right)^2 + \cdots\cdots\cdots\cdots\right]\right\} = 1 \tag{17.44}$$

From the knowledge of infinite series of binomial expression we can write

$$1 + \left(\frac{p}{p_s}\right) + \left(\frac{p}{p_s}\right)^2 + \cdots\cdots\cdots\cdots = \frac{1}{1 - \dfrac{p}{p_s}} \tag{17.45}$$

Differentiating both sides with respect to (p/p_s), we get

$$1 + 2\left(\frac{p}{p_s}\right) + 3\left(\frac{p}{p_s}\right)^2 + \cdots\cdots\cdots\cdots = \frac{1}{\left(1 - \dfrac{p}{p_s}\right)^2} \tag{17.46a}$$

So,

$$\theta_0 = \left[\frac{1}{1 + \dfrac{K'p}{\left(1 - \dfrac{p}{p_s}\right)}}\right] \tag{17.46b}$$

x_m = The maximum amount of adsorbate per gm of adsorbent or in other words it is the amount of adsorption if the surface is totally covered by the adsorbate molecules to form a monolayer.

Thus the net amount of adsorbate, remaining adsorbed on the surface (a), is given by

$$x = x_m(\theta' + 2\theta'' + 3\theta''' + \cdots\cdots\cdots\cdots) \tag{17.47}$$

Putting the values of θ'', θ''' etc., we get

$$x = x_m\left[\theta' + 2\left(\frac{p}{p_s}\right)\theta' + 3\left(\frac{p}{p_s}\right)^2\theta' + \cdots\cdots\cdots\cdots\right] \tag{17.48}$$

Putting the value of θ', we get

$$x = x_m K'p\theta_0\left[1 + 2\left(\frac{p}{p_s}\right) + 3\left(\frac{p}{p_s}\right)^2 + \cdots\cdots\cdots\cdots\right] \tag{17.49}$$

Comparing equations (17.46) and (17.49), we get

$$x = \frac{x_m K'p\theta_0}{\left(1 - \dfrac{p}{p_s}\right)^2} \tag{17.50}$$

From equation (17.46b), we have

$$\theta_0\left\{1 + \frac{K'p}{\left(1 - \dfrac{p}{p_s}\right)}\right\} = 1. \quad \text{Or,} \quad \theta_0 = \frac{1}{1 + \dfrac{K'p}{\left(1 - \dfrac{p}{p_s}\right)}} = \frac{\left(1 - \dfrac{p}{p_s}\right)}{K'p + \left(1 - \dfrac{p}{p_s}\right)} \tag{17.51}$$

Putting the value of θ_0 from equation (17.50), in the equation (17.51), we get,

$$x = \frac{x_m K' p}{\left(1 - \dfrac{p}{p_s}\right)\left[K'p + \left(1 - \dfrac{p}{p_s}\right)\right]} \tag{17.52}$$

We have $K_L = 1/p_s$. Hence, we can write that

$$p = p_s\left(\frac{p}{p_s}\right) = \frac{1}{K_L}\left(\frac{p}{p_s}\right), \quad K'p = \left(\frac{K'}{K_L}\right)\left(\frac{p}{p_s}\right) = C\left(\frac{p}{p_s}\right), \quad \text{where,} \quad C = \left(\frac{K'}{K_L}\right)$$

Thus the equation (17.52) takes the form

$$x = \frac{x_m C\left(\dfrac{p}{p_s}\right)}{\left(1 - \dfrac{p}{p_s}\right)\left[1 + (C-1)\dfrac{p}{p_s}\right]} \tag{17.53a}$$

Using equation (17.31b), we get

$$\theta = \frac{x}{x_m} = \frac{C\left(\dfrac{p}{p_s}\right)}{\left(1 - \dfrac{p}{p_s}\right)\left[1 + (C-1)\dfrac{p}{p_s}\right]} \tag{17.53b}$$

The equations (17.53a) and (17.53b) is collectively called BET equation for multilayer adsorption of a vapour.

Equation (17.53a) can be rearranged as follows

$$\frac{p/p_s}{x[1 - (p/p_s)]} = \frac{1}{x_m C} + \frac{(C-1)}{x_m C}\left(\frac{p}{p_s}\right) \tag{17.54}$$

In the case of monolayer adsorption, the following approximations hold good

(i) $p/p_s \ll 1$ and hence $([1 - (p/p_s)]) \approx 1$.

(ii) $C \gg 1$ and hence $[1 + (C-1)\,p/p_s] \approx [1 + C(p/p_s)] = [1 + K'p]$, since, $C(p/p_s) = K'p$.

Using these approximations the equation (17.53a) is transformed into the Langmuir equation (17.34b)

$$x = \frac{x_m K' p}{1 + K' p}$$

17.6.3 Limitations of BET Equation

(a) BET equation assumes only adsorbate-adsorbent interaction but it does not consider adsorbate-adsorbate interaction since energy of adsorbate-adsorbent interaction is much higher than that of adsorbate-adsorbate interaction. So BET equation is not applicable to that kind of adsorption where adsorbate-adsorbate interaction is significant.

(b) BET equation is valid when the relative pressure (p/p_s) lies within 0.05 to 0.35.

7. ADSORBATE-ADSORBATE INTERACTION

The interaction between adsorbate molecules usually takes place through dispersion forces. Again, larger the surface area, higher the dispersion forces is. Hence dispersion forces is greater in case of spherical molecules rather than flat molecules. Let us consider adsorption of CCl_4 and benzene on graphitized carbon black. Dispersion forces are significant in case of spherical CCl_4 molecules but appears very less in case of flat benzene molecules. However, attractive forces between adsorbate molecules is reduced by repulsive forces.

h = Adsorbate-adsorbate interaction coefficient. When adsorbate-adsorbate interaction is significant, the adsorption equilibrium (K) depends on h according to the following relation

$$K \propto e^{-h\theta} \tag{17.55}$$

Combining equations (17.38) and (17.55), we have,

$$K = \frac{\theta e^{-h\theta}}{p(1-\theta)}. \quad \text{Or,} \quad p = \frac{\theta}{K(1-\theta)e^{h\theta}} \tag{17.56}$$

7.8 CHARACTERISTICS OF ADSORPTION

17.8.1 Factors Influencing Adsorption

The following have the significant influence on any adsorption process:

(a) **Surface area** Surface area of adsorbent plays an important role in adsorption. Higher the surface area, rate of adsorption increases. Thus finely divided metals, like *Pt*, acts as very good adsorbents.

(b) **Nature of gas** In case of gases higher the critical temperature, more easily the gas can be liquefied. It has been observed that higher the critical temperature, higher the rate of adsorption on solid adsorbent. For example, 1 gm activated charcoal adsorbs 380 ml of SO_2 (critical temperature 157°C) but the same adsorbent adsorbs only 16 ml of methane (critical temperature –83°C).

(c) **Temperature** Adsorption process is an exothermic process. Thus low temperature usually favours adsorption process. However, rate of chemisorption process increases with temperature.

(d) **Pressure** Dynamic equilibrium exists between adsorbed gas and free gas. Thus increase in gas pressure usually shifts the equilibrium towards the right, thereby increasing the adsorption rate. However, chemisorption process is independent of gas pressure.

17.8.2 Characteristics of Physical Adsorption

The nature of physical adsorption is shown in the Fig. 17.6.

(a) Van der Waals type of forces play the key role in adhering gas molecules on the solid surface. The adhesion of gas molecules on solid surface is similar to the forces of cohesion among molecules in liquid state.

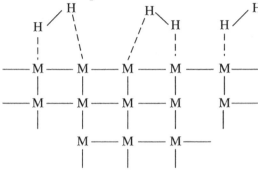

Figure 17.6 Schematic diagram of physical adsorption of hydrogen on a metal surface

(b) The physical adsorption is a reversible process and the equilibrium is established quickly. At low temperature physical adsorption takes place but desorption takes place at high temperature.

(c) Physisorption is an exothermic process, releasing heat, known as heat of adsorption. As the cohesive force is very weak, heat of adsorption is very low, around 5–10 kcal/mole.

(d) For a given adsorbent, the extent of adsorption, under given conditions is roughly related to the ease of liquefaction of the gas. As the boiling point decreases, gases, with Van der Waals forces of cohesion, are more strongly adsorbed.

(e) Physisorption depends on gas pressure. High gas pressure favours physisorption process but desorption is preferred at low gas pressure.

(f) As the physical adsorption process is exothermic, it occurs preferably at low temperature but desorption occurs at high temperature.

(g) Physical adsorption process often forms multimolecular layers on adsorbent surface.

17.8.3 Characteristics of Chemical Adsorption

The nature of chemical adsorption is shown in the Fig. 17.7.

(a) Interaction between adsorbate molecules and surface of adsorbent is significantly high, hinting at formation of strong chemical bond between adsorbate molecules and adsorbent.

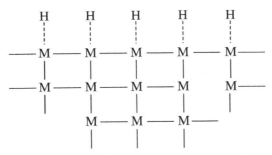

Figure 17.7 Schematic diagram of chemisorption of hydrogen on a metal surface

(b) A chemical bond is formed between adsorbate molecules and adsorbent surface resulting a compound. So, chemisorption process is irreversible.

(c) Chemisorption is also an exothermic process. As chemical bond is strong, the heat of adsorption is very high, around 20–100 kcal/ mole.

(d) Chemisorption is highly specific in nature, depending on nature of both adsorbate and adsorbent. For example, at high temperature tungsten forms tungsten oxide with oxygen and tungsten carbonyl with carbon monoxide. Thus oxygen and carbon monoxide both are chemisorbed on tungsten surface separately. Sometimes chemisorption occurs through free radical mechanism. For example, carbon black particles are chemisorbed on some rubber surface through free radical mechanism during mechanical blending of rubber and carbon black.

(e) Chemisorption process is almost independent of gas pressure.

(f) Chemisorption usually occurs at high temperature. The chemisorption rate increases with increase in temperature.

(g) Generally chemisorption process forms unimolecular layer on the adsorbent surface since chemical reaction can take place between adsorbent surface and adsorbate molecules only.

17.9 PORE SIZE DETERMINATION

17.9.1 Kelvin Equation

Usually a solid adsorbent has numerous pores. The size of pore can be calculated using Kelvin equation. In this method nitrogen gas is used as adsorbate. Temperature is decreased below the boiling point of nitrogen ($T = 77°K$) so that nitrogen gas is converted to liquid. It is assumed that liquid-vapour surface is a hemi-spherical surface of radius r_m. Thus the excess pressure, ΔP is given by the equation of Young and Laplace

$$(P^0 - P) = \frac{2\gamma}{r_m}, \text{ where, } \gamma = \text{Surface tension,} \quad r_m = \text{Mean radius of curvature}$$

$(P^0 - P) = $ Excess pressure and $(P^0 - P) > 0$.

The relation between pore radius r and r_m is illustrated in the Fig. 17.8.

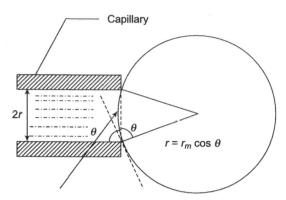

Figure 17.8 Relation between contact angle and radius of pore

r_m is related to capillary radius r as follows

$$r_m = \frac{r}{\cos\theta} \tag{17.57}$$

Again from thermodynamics, we have

$dG = VdP - SdT$. At constant temperature, we have

$$dG = VdP \tag{17.58}$$

Or, $\qquad \Delta G = V(P - P^0) = -V(P^0 - P) = -\dfrac{2\gamma V}{r_m} \tag{17.59}$

Again, for 1 mole of an ideal gas we know $V = \dfrac{RT}{P}$

Putting the value of V in the equation (17.58), we get

Hence, $\qquad dG = \left(\dfrac{RT}{P}\right)dP$. Or, $\Delta G = RT\ln\dfrac{P}{P^0} \tag{17.60}$

Where, P^0 is the saturated vapour pressure of the liquid adsorbate.
Combining equations (17.59) and (17.60), we get

$$\log\frac{P}{P^0} = -\left(\frac{2\gamma V}{r_m}\right)\left(\frac{1}{2.303RT}\right) \tag{17.61}$$

Combining equations (17.57) and (17.61) we get

$$\log\frac{P}{P^0} = -\left(\frac{2\cos\theta \times \gamma V}{r}\right)\left(\frac{1}{2.303RT}\right) = -\left(\frac{2\gamma V\cos\theta}{2.303RT}\right)\left(\frac{1}{r}\right) \tag{17.62}$$

P = Vapour pressure at which the liquid will condense in the pore of radius r. Each pore is considered as a capillary.

P^0 = Saturated vapour pressure.

V is the molar volume of the adsorbate. Thus by measuring P, we can calculate pore radius r.

θ = Contact angle.

The equation (17.62) is known as Kelvin equation.

17.9.1.1 Significances

(a) Kelvin's equation is applicable to the capillary condensation of both adsorption and desorption cycles.

(b) P^0 depends on temperature, T. As T decreases, P^0 also decreases and hence (P/P^0) term increases. Thus according to Kelvin equation, radius of the droplet, r, increases.

(c) When pore is very large we have

$$r \gg \left(\frac{2\gamma V \cos\theta}{2.303RT} \right) \text{ and hence, } \frac{P}{P^0} = 1$$

So the liquid surface is considered flat under this condition.

(d) When the pore is small in size $P < P_0$.

(e) Actual pore size (r_p) is given by $2r_p = 2r + 2t$, where, r is the apparent pore radius and t is the thickness of the adsorbed gas film as shown in the Fig. 17.9.

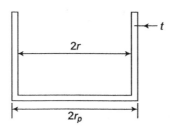

Figure 17.9 Actual pore radius

17.9.2 Cohan Modification

According to the scientist Cohan, during adsorption vapour condenses to form hemisphere but during desorption a sphere of droplet is evaporated. Thus for adsorption the Kelvin equation should be written as

$$\log\frac{P_{ads}}{P^0} = -\left(\frac{\gamma V \cos\theta}{2.303RT} \right)\left(\frac{1}{r} \right). \text{ Or, } \ln\frac{P_{ads}}{P^0} = -\left(\frac{\gamma V \cos\theta}{RT} \right)\left(\frac{1}{r} \right)$$

When θ is very small, $\cos\theta \approx 1$ and the above equation becomes

$$\ln\frac{P_{ads}}{P^0} = -\left(\frac{\gamma V}{RT} \right)\left(\frac{1}{r} \right). \text{ Or, } \frac{P_{ads}}{P^0} = e^{-\left(\frac{\gamma V}{rRT} \right)} \tag{17.63}$$

In case of desorption a sphere of droplet is considered and hence the Kelvin equation becomes

$$\log \frac{P_{des}}{P^0} = -\left(\frac{2\gamma V \cos\theta}{2.303RT}\right)\left(\frac{1}{r}\right). \text{ Or, } \ln\frac{P_{des}}{P^0} = -\left(\frac{2\gamma V \cos\theta}{RT}\right)\left(\frac{1}{r}\right)$$

When θ is very small, $\cos\theta \approx 1$ and the above equation becomes

$$\ln\frac{P_{des}}{P^0} = -\left(\frac{2\gamma V}{RT}\right)\left(\frac{1}{r}\right). \text{ Or, } \frac{P_{des}}{P^0} = e^{-\left(\frac{2\gamma V}{rRT}\right)} \tag{17.64}$$

Comparing equations (17.63) and (17.64), we get

$$\frac{P_{des}}{P^0} = \left(\frac{P_{ads}}{P^0}\right)^2$$

Inferences   Pressure required for desorption of liquid adsorbate is the square of the pressure required for adsorption of the same liquid adsorbate. For example, if reduced pressure required (P_{ads}/P^0) for adsorption is 0.8, the corresponding reduced desorption pressure is $\frac{P_{des}}{P^0} = 0.64.$

17.10 APPLICATIONS OF ADSORBENTS

(a) It is very difficult to generate high vacuum in a vessel. However, in presence of good absorbent, like cooled charcoal, most of the gas molecules can be removed, creating high vacuum in the vessel, which can be used to store liquid nitrogen, liquid oxygen, liquid hydrogen or liquid air.

(b) The adsorption principle is widely used to remove contaminants and colouring matters to produce pure substance. For example, fuller's earth is used in petroleum industries to remove colouring materials from natural gasoline. Animal charcoal is used to remove colouring materials in the manufacture of cane sugar.

(c) The finely divided metals are often used as adsorbents in heterogeneous catalyst systems.

(d) In automobiles activated charcoal is used in catalytic converter to remove harmful unburnt hydrocarbons from the exhaust.

(e) Gas masks are used in mines. High quality activated charcoal is used in manufacturing gas masks to remove pollutants ensuring pure air for breathing.

(f) Adsorption phenomenon is widely used in chromatographic analysis, where, a suitable common adsorbent, usually alumina, is taken in a tube. A mixture of solutes is dissolved in a given solvent, usually hexane. The solution is then slowly passed through the tube from the top. The strong adsorbate is adsorbed first at the top and as the solution is coming down other solutes are slowly adsorbed step by step depending on their ability to get adsorbed on alumina surface.

(g) During manufacturing of demineralized water, cation exchange resin and anion exchange resin are used. Cation exchange resin consists of macromolecular anion and H^+ ion, represented as $[RSO_3]^- H^+$, while anion exchange resin is made of a macromolecular cation and OH^-, represented as $[RN(CH_3)_3]^+ OH^-$. When the raw water is passed over cation exchange resin bed, desorption of H^+ ion and adsorption of other cations take place according to the following reaction:

$$n[RSO_3]^- H^+ + M^{n+} \rightarrow \{[RSO_3]^-\}_n M^{n+} + nH^+$$

In the next step, the water is passed over anion exchange bed, where desorption of OH^- and adsorption of other anions take place according to the following reaction:

$$n[RN(CH_3)_3]^+ OH^- + X^{n-} \rightarrow \{[RN(CH_3)_3]^+\}_n X^{n-} + nOH^-$$

Ultimately the water contains only H^+ ion and OH^- ion. This water is known as deionized water.

EXERCISES

Questions and Numerical Problems

1. Distinguish between adsorption and absorption.

   Answer

   The term *adsorption* means molecules of a particular substance, known as adsorbate, remain adhered to the surface of other substance, known as adsorbent but molecules do not penetrate within the adsorbent. For example, when a piece of chalk is immersed in an ink solution, surface of chalk only becomes coloured due to adsorption of ink but interior portion remains white.

   The *absorption* means molecules of a particular substance, penetrate through the other substance producing a homogeneous or heterogeneous matrix depending on the solubility of absorbed molecules. For example, if a sponge is immersed in an ink solution, the whole sponge material takes the colour of ink.

2. Distinguish between physisorption process and chemisorption process.

3. Define the terms: (a) Adsorption, (b) Physical adsorption, (c) Chemical Adsorption, (d) Freundlich adsorption isotherm, (e) Langmuir adsorption isotherm.

4. What is adsorption? Define the terms 'adsorbent' and 'adsorbate' giving suitable examples. Describe the phenomenon of the adsorption of solids from a solution.

5. (a) What is the effect of temperature on adsorption of gases on solids?

   (b) Write the assumptions of Langmuir adsorption isotherm and derive the equation pertaining to it.

6. Discuss the factors which affect the adsorption of a gas on a solid adsorbent.

7. Discuss in brief the type of adsorption isotherms commonly observed for the adsorption of gases on a variety of adsorbents at different temperatures.

8. (a) Write the main points of Langmuir's theory of adsorption.

   (b) What are adsorption isobars?

   (c) Draw adsorption isobars for physical adsorption and chemical adsorption.

9. Discuss Langmuir theory of adsorption and derive expression for Langmuir monolayer adsorption isotherm. Hence show that for a moderate range of pressures it reduces to Freundlich adsorption isotherm.

10. Distinguish between physical adsorption and chemical adsorption. What are adsorption isobars?

11. Discuss Freundlich adsorption isotherm of a gas on a solid surface. How are the constants in this isotherm equation determined? How will you prove that Langmuir adsorption isotherm is superior to Freundlich adsorption isotherm?

12. (a) Write down the mathematical from of Langmuir's adsorption isotherm. Also write two limitations in Langmuir's theory.

    (b) What signs of ΔH and ΔS in the case of physical adsorption are expected? Justify your answer.

    (c) How can Langmuir adsorption isotherm be used to determine the surface area of an adsorbent?

13. (a) Show diagrammatically the different types of adsorption isotherms obtained for adsorption of gases on solids.

    (b) Discuss the behaviour of Langmuir adsorption isotherm at very low and very high pressures.

14. Derive Langmuir adsorption equation in the form of

$$y = \frac{ap}{1 + bp}$$

How is this equation verified?

15. How is Langmuir adsorption isotherm related to Freundlich isotherm? How are shapes of adsorption isotherms modified when multilayer adsorption takes place.

16. (a) Give two applications of adsorption.

    (b) Explain the reason why a finally powdered substance is more effective adsorbent?

17. How can Langmuir adsorption isotherm equation be used to explain the observation— the decomposition of PH_3 gas on tungsten metal surface is first order at low pressure and zeroth-order at high pressure?

18. Derive Freundlich adsorption isotherm from the Gibbs adsorption isotherm applied to a gas.

19. Draw the typical adsorption isotherms obtained in the case of unimolecular and multimolecular adsorption.

20. 10 gm of oxygen is adsorbed on 2.5 gm of metal powder at 273°K and 1 atm pressure. Calculate the volume of the gas adsorbed per gram of adsorbent. [*Ans.:* 2798.25 ml]

21. 100 ml of 0.3 M acetic acid is shaken with 0.8 g of wood charcoal. The final concentration of the solution after adsorption is 0.125 M. Calculate the weight of acetic acid adsorbed per gram of carbon. [*Ans.:* 1.31 gm]

22. 4 gm of a gas is adsorbed on 1.5 gm of metal powder at 300°K and 0.7 atm. Calculate the volume of the gas at STP adsorbed per gram of adsorbent. [*Ans.:* 2052.5 ml]

23. For an adsorbent-adsorbate system obeying the Langmuir adsorption isotherm, $a = 0.48 \, bar^{-1}$ and $b = 0.16 \, bar^{-1}$. At what pressure will 50% of the surface be covered?

[*Ans.:* 1.25 bar]

24. Five grams of a catalyst absorb 400 cm^3 of N_2 at STP to form a monolayer. What is the surface area per gram if the area occupied by a molecule of N_2 is 16 Å?

[*Ans.:* 344 m^2 gm^{-1}]

25. The adsorption of benzene in graphite follows a Langmuir isotherm. At the pressure of 1 torr the volume of benzene adsorbed on a sample of 2 mg graphite is 4.2 mm^3 at STP. At the pressure of 3 torr corresponding volume is 8.5 mm^3. Admitting that the benzene molecule occupies 30×10^{-20} m^2/molecule, estimate the surface area of graphite.

26. Use Kelvin equation to calculate the radius of pores that correspond to the capillary condensation of nitrogen at 77°K and a relative pressure of 0.5. Consider the adsorption in multilayer's as having the thickness of 0.65 nm at this pressure.

Given for the nitrogen at 77°K, $\gamma = 8.05$ mN/m and the molar volume is 34.7 cm^3/mole.

27. Data in the table refer to the adsorption of CO (carbon monoxide) in coal at 273 K. Confirm that they obey to the Langmuir isotherm and obtain the constant K and the volume corresponding to a monolayer.

P(torr)	100	200	300	400	500	600	700
V(cm³)	10.2	18.6	25.5	31.5	36.9	41.6	46.1

28. The following data refer to the adsorption of nitrogen in a sample of 0.92 g of silica gel at 77°K, where P is the equilibrium pressure and V is the adsorbed volume. Saturated vapour pressure, $P_s = 101.3$ kPa.

P(kPa)	3.7	8.5	15.2	23.6	31.5	38.2	46.1	54.8
V(cm³)	82	106	124	142	157	173	196	227

Draw the adsorption isotherm and use the BET equation to calculate the specific area of the sample of silica gel.

Given the molecular nitrogen area is 16.2×10^{-20} m^2.

29. The following data refer to the adsorption of n-butane at 273 K for a sample of tungsten powder that has a specific area (determined by nitrogen adsorption measurements at 77 K) of 6.5 m^2/gm.

P/P_s	0.04	0.10	0.16	0.25	0.30	0.37
V(cm³/gm)	0.33	0.46	0.54	0.64	0.70	0.77

Use BET equation to calculate the area of molecular butane adsorbed in the monolayer and compare with the value of 32.1×10^{-20} m^2/molecule estimated from the density of the liquid butane.

Multiple Choice Questions

1. In physical adsorption the gas molecules are held to the solid surface by
   - (a) hydrogen bond
   - (b) sigma bond
   - (c) pi bond
   - (d) van der Waals forces

2. The adsorption of hydrogen on charcoal is
   - (a) physical adsorption
   - (b) chemical adsorption
   - (c) sorption
   - (d) none of these

3. In chromatographic analysis, the principle used is
   - (a) absorption
   - (b) adsorption
   - (c) distribution
   - (d) evaporation

4. Which of the following is not a characteristic of physical adsorption?
   - (a) adsorption is reversible
   - (b) multimolecular layer is formed
   - (c) ΔH is of the order 400 kJ
   - (d) occurs rapidly at low temperature

5. Which is incorrect statement?
   - (a) physical adsorption is irreversible in water
   - (b) physical adsorption involves multimolecular layers
   - (c) the energy evolved is small
   - (d) physical adsorption is caused by van der Waals forces

6. The efficiency of adsorbent increases with increase in
   - (a) viscosity
   - (b) surface tension
   - (c) surface area
   - (d) number of ions

7. Which of the following is not an application of adsorption?
   - (a) gas masks
   - (b) heterogeneous catalysis
   - (c) froth flotation process
   - (d) softening of water by boiling

8. Which of the following is incorrect?
   - (a) chemisorption is caused by bond formation
   - (b) chemisorption is specific in nature
   - (c) chemisorption is reversible
   - (d) chemisorption increases with increase in temperature

9. In an adsorption process unimolecular layer is formed. It is
   - (a) physical adsorption
   - (b) chemical adsorption
   - (c) ion-exchange
   - (d) chromatographic analysis

10. Adsorption takes place with
    - (a) decrease in enthalpy of the system
    - (b) increase in enthalpy of the system
    - (c) no change in enthalpy of the system
    - (d) none of these

11. Freundlich isotherm is not applicable at
    (a) high pressure
    (b) low pressure
    (c) 273°K
    (d) room temperature
12. With increase in temperature rate of chemisorption
    (a) increases
    (b) decreases
    (c) remains the same
    (d) none of these
13. At high temperature rate of physical adsorption is
    (a) low
    (b) high
    (c) absolute zero
    (d) none of these

ANSWERS

1 (d), 2 (a), 3 (b), 4 (c), 5 (a), 6 (c), 7 (d), 8 (c), 9 (b), 10 (a), 11 (a), 12 (a), 13 (a).

18 The Colloidal State

18.1 DEFINITION

The term colloid represents that category of particles, having dimension between 1 nm to 1 μm. When these colloid particles remain dispersed in a liquid medium, the solution is called colloidal solution. So colloidal solution is heterogeneous system but true solution is a homogeneous system. The differences between true solution and colloidal solution are given below.

Differences between true solution and colloidal solution

True solution	*Colloidal solution*
It is a single phase system.	It is a two phase system.
It can pass through ordinary filter papers as well as through parchment or cellophane papers.	It can pass through ordinary filter papers like true solution but the dispersed phase is not allowed to pass through parchment or cellophane papers.
As the solute is completely soluble in the solvent, solute molecules cannot be examined under electron microscope nor do they scatter light.	Colloidal particles can be examined under electron microscope. Also they respond light scattering experiment.

18.2 CLASSIFICATION OF COLLOIDS

Colloids can be classified either on the basis of solvent-loving or on the basis of spontaneous dispersion of particles.

On the basis of solvent-loving, colloids can be classified into two categories:
(a) Lyophilic colloids, and   (b) Lyophobic colloids.

Lyophilic colloid   This type of colloid is based on the dispersed phase, which is termed solvent-loving. The particles are spontaneously dispersed in the medium. For example, gelatin particles are readily dispersed in water. Similarly protein particles and detergent particles are also readily dispersed in water.

Generally lyophilic sols are unstable at isoelectric point. On the basis of stability at isoelectric point lyophilic sols may be divided into two categories: (a) Isostable and (b) Isolabile.

In some lyophilic sols, solvent interaction is strong enough to make the sol stable even at its isoelectric point. These lyophilic sols are called isostable lyophilic sols. One example is Gelatin.

The other lyophilic sols are quite unstable at their isoelectric points. These are called isolabile lyophilic sols. One example is Casein.

Lyophobic colloid This type of colloid is based on the dispersed phase, which is termed solvent-hating. The particles generally do not disperse in the medium. However, they can be dispersed mechanically or chemically. For example, dispersion of colourants in latex medium.

On the basis of spontaneous dispersion of colloidal particles, colloids can be divided into two categories:
   (a) Reversible colloids and   (b) Irreversible colloids.

Reversible colloid In this type of colloid, particulates are spontaneously dispersed in the continuous matrix by thermal energy. Gibbs free energy of dispersion is negative. So it is thermodynamically stable. Distribution of colloidal particles is uniform throughout the matrix. If the dispersed phase is separated, it can be redispersed easily. One example is protein sol.

Irreversible colloid In this type of colloid, particulates are not readily dispersed in the continuous matrix. Mechanical energy is often applied to disperse those particles in the matrix. Gibbs free energy of sol formation is positive. So it is thermodynamically unstable. Distribution of colloidal particles is not uniform throughout the matrix. Once the dispersed phase is separated, it cannot be redispersed easily. One example is gold sol.

18.3   PROPERTIES OF COLLOIDAL SOLUTION

1. **Brownian movement** The English Botanist Robert Brown first established in 1927 that tiny particles in aqueous suspension are always in constant erratic zigzag motion, which is popularly known as Brownian motion. In colloidal solutions also, particles are quite small and exhibit Brownian motion. However, Brownian movement of colloidal particles is independent of the nature of particles but depends on size of particles and solution viscosity. Smaller the particle size more pronounced the Brownian motion is. Again, the said motion is very much prominent if the viscosity of the medium is very low. Brownian motion of particles stabilizes the colloidal solution as it prevents the particles from settling due to gravity. If a colloidal solution is taken in a vertical

column, particles are distributed throughout the column on the basis of two opposing forces: (a) Brownian motion, and (b) Gravitational force. The distribution of particles is governed by the equation (18.1).

$$\left(\frac{RT}{N_0}\right)\ln\frac{n_1}{n_2}=\frac{4}{3}\pi r^3(h_2-h_1)(\rho-\rho_m)\tag{18.1}$$

N_0 = Avogadro Number; n_1 and n_2 are the number of particles at the height h_1 and h_2 respectively; ρ and ρ_m are the densities of particles and liquid medium respectively.

2. **Tyndall effect**   When a beam of light is passed through a colloidal solution, light waves are scattered by colloidal particles, due to which, path of the beam through the solution is quite visible. It was first observed by the scientist Tyndall in 1869 and hence the effect is called Tyndall effect. The illuminated path of the beam is called Tyndall cone. However, intensity of the scattered light depends on the difference between refractive indices of the dispersion medium and the dispersed phase. In case of lyophobic sol, this difference in refractive indices is large and hence the effect is more pronounced than lyophilic sol, where this difference in refractive indices is small. Tyndall effect is useful in explaining some common facts discussed below:

   (a) **Blue colour of sky**   The Earth's atmosphere contains huge amount of moisture dispersed with colloidal particles. This colloidal solution is called aerosol, which scatters blue light but absorbs other colours of solar radiation. So the sky appears in blue colour in day time. At night there is no solar radiation and hence there is no such scattering. So, the sky appears black.

   (b) The ocean water contains huge amount of salt, which scatters blue colour wavelength of solar radiation and hence the ocean appears blue.

   (c) Tyndall effect explains the visibility of the path of sunlight, passing through the small slits.

3. **Shape and colour of particles**   There are many sols, which show characteristic colour. Particle size and shape are responsible for this colour of sol. One example is gold sol, which changes colour from red to violet to blue. It is believed that colloidal particles absorb photons in the visible range and wavelength of absorbed ray is approximately proportional to the radius of the particle. In the purest form of gold sol, particle size is very small and hence particles absorb colour in the blue-violet region and the sol appears red but as precipitation proceeds, particle size increases and waves of larger wavelength are absorbed. The sol appears greenish blue. When precipitation is nearly complete, particles absorb colour in the red region and hence the sol appears blue. Shape of particles also has significant influence on absorption of colour by particles. In case of spherical particles, light is scattered in all directions equally but in case of rod-shaped particles, scattering of light along horizontal direction is different from that in vertical direction.

4. **Two phases**   A colloidal solution is made of two phases. One phase forms continuous medium, known as dispersion medium while the other phase, known as dispersed phase, remains distributed throughout the matrix discontinuously. Both the dispersion medium and dispersed phase may be solid, liquid or gas.

Case 1 Dispersion medium is liquid and dispersed phase is solid. Examples are gold sol, $Fe(OH)_3$ sol in $FeCl_3$ solution, As_2S_3 sol in H_2S solution, etc.

Case 2 Dispersion medium is gas and dispersed phase is solid. Examples are smoke, dust, etc.

Case 3 Dispersion medium is gas and dispersed phase is liquid. Examples are fog, mist, and cloud. These are also known as aerosols.

Case 4 Dispersion medium is solid and dispersed phase is gas. Examples are metal articles, rubber articles, where often gas remains as dispersed phase.

5. **Electrophoresis** A colloidal solution is taken in a beaker and two *Pt* electrodes are immersed into it. When a potential difference (electric field) is applied across two platinum electrodes immersed in a colloidal solution, the particles of dispersed phase move towards either the positive or negative electrode. This phenomenon was first discovered by **Reuss** in **1807** and was investigated later by **Linder** and **Picton**. The movement of colloidal particles under the action of electric field is known as *Electrophoresis*, also called *cataphoresis*. If the particles are positively charged, they move towards the cathode. On the other hand if the particles are negatively charged, they move towards the anode. If the colloidal particles move towards the positive electrode (Anode) they carry negative charge. On the other hand if the sol particles migrate towards the negative electrode (Cathode), they are positively charged. From the direction of movement of colloidal particles it is possible to find out the charge on colloidal particles.

During electrophoresis of a sol, without stirring bottom layer is gaining particles and becomes more and more concentrated while the top layer becomes lighter. The phenomenon is called *electrodecantation* and is often used for purification and concentrating the sol.

In absence of any electric field if particles are forced to move in a particular direction, a potential is developed due to concentration difference. The phenomenon was first observed by the scientist Dorn and the effect is called **Dorn effect**. The potential, developed due to Dorn effect, is called sedimentation potential.

Applications

(a) **Collection of Natural Rubber (NR)** When electrophoresis is carried out in NR latex, agglomeration of colloidal rubber particles sets in, resulting deposition of rubber on anode.

(b) **Purification of water** Sewage water contains colloidal particles. Using electrophoresis these colloidal impurities can be separated and deposited, resulting pure water.

(c) Smoke contains charged colloidal carbon particles, which can be separated using electrophoresis. These charged colloidal carbon particles are responsible for blue tinge of smoke.

6. **Electro-osmosis**   A colloidal solution as a whole is electrically neutral in nature, i.e., dispersion medium carries an equal and opposite charge to that of the particles of dispersed phase. When the movement of dispersed phase of colloidal solution is prevented by suitable means, the dispersion medium can be made to move under the influence of an applied electric field or potential. This phenomenon is referred to as **Electro-osmosis**. Thus electro-osmosis may be defined as *the movement of the dispersion medium under the influence of an applied electric field when the particles of dispersed phase are prevented from moving.*

   Let us consider a colloidal solution in a beaker. A semipermeable membrane is used to separate the beaker into two compartments. Under this condition an electric field is applied in the solution. Only the dispersion medium is allowed to pass through the membrane but the particles are prevented from passing through the barrier. As a result of this, concentration difference is developed in the two compartments. The phenomenon is called electro-osmosis.

   In the reverse case without application of any electric field, the dispersion medium is forced to pass through the semipermeable membrane, a potential is developed due to concentration difference. This potential is called **streaming potential**. So development of streaming potential is the opposite phenomenon of electro-osmosis.

7. **Coagulation**   Stability of a colloidal solution depends on repulsion between similarly charged colloidal particles. On addition of electrolyte, neutralization of charged particles begins, leading to drastic decrease in repulsive force. As a result of this, colloidal solution becomes unstable, leading to coagulation or flocculation. $Fe(OH)_3$ sol is stabilized by Fe^{3+} ion and can be easily coagulated on addition of negative ions like OH^-, etc. On the other hand As_2S_3 sol is stabilized by S^{2-} and can be easily coagulated on addition of positive ions like Pb^{2+}, etc.

Factors influencing coagulation

(a) A little amount of electrolyte is enough to bring about coagulation of a sol.

(b) Ions, oppositely charged to colloidal particles, are effective for coagulation.

(c) Mixing of oppositely charged sols results coagulation of both the sols. The phenomenon is called mutual coagulation.

(d) Rate of coagulation increases with increase in temperature.

(e) Coagulating power of an electrolyte depends on several factors as given below:

   (i) Coagulating power increases with increase in valency of ion (Hardy-Schulze rule).

   (ii) Lyophobic sol is more easily coagulated than lyophilic sol.

   (iii) **Peptization**   It is a process by which a precipitate is converted to colloidal solution or sol. One example is freshly prepared precipitate of $Fe(OH)_3$. On addition of a little amount of $FeCl_3$ solution precipitate is slowly transferred to reddish brown colloidal solution as Fe^{3+} ions stabilize the colloidal particles.

18.4 COAGULATION AND FLOCCULATION

Whenever a colloidal system is destabilized, dispersed phase gets agglomerated to form macroscopic particles, visible in naked eye. These macroscopic particles may be precipitated at the bottom. The phenomenon is called coagulation.

In another case, macroscopic particles form an open porous network, which is lighter than continuous medium. This open network floats either at the surface or within the continuous medium. The phenomenon is called flocculation.

Presence of ions may either stabilize the colloid or destabilize the colloid. Stabilization process is known as peptization and destabilization process is known as coagulation or flocculation.

Most of the colloidal systems are stabilized by negatively charged ions. Thus addition of any positively charged ions generally destabilizes colloidal system. Destabilization of a colloid can be done by introducing electrolyte in the system. The minimum amount of electrolyte required to coagulate a particular colloid system is called *critical colloid concentration* or *CCC* of that colloid. However, destabilization process depends on valency of ion added into the system. Higher the valency, lesser the amount of electrolyte required to coagulate or flocculate a colloid. For example, AgI-water colloid system is coagulated using KNO_3, $Ca(NO_3)_2$ or $Al(NO_3)_3$. (i) In case of KNO_3 CCC value is 1.36×10^{-1} M, (ii) In case of $Ca(NO_3)_2$, CCC value is 2.37×10^{-3} M and (iii) In case of $Al(NO_3)_3$, CCC value is 7.00×10^{-5} M.

18.4.1 Schulze-Hardy Rule

(a) Positive ions of the foreign electrolyte is usually responsible for coagulation of most of the colloidal systems.

(b) Negative ions of the foreign electrolyte has no significant role in coagulation of most of the colloidal systems.

(c) Characteristics of positive ions have little or no significance in coagulation process.

(d) CCC value of the foreign electrolyte is related to valency of positive ion as follows

$$CCC \propto \frac{1}{\text{valency}^6} \tag{18.2}$$

Equation (18.2) is known as Schulze-Hardy rule.

18.4.2 DLVO (Derjaguin, Landau, Verwey and Overbeek) Theory

The scientists Derjaguin, Landau, Verwey and Overbeek proposed a theory to explain the stability criteria of a colloidal solution.

Let us consider two colloidal particles, surrounded by ions, namely positively charged ions. The particles are separated by a distance X as shown in the Fig. 18.1, where, the plates indicate colloidal particles.

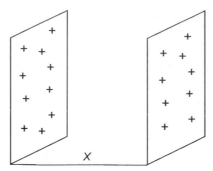

Figure 18.1   Schematic diagram showing two colloidal particles separated by a distance X

Total free energy of interaction is the sum of energy due to forces of attraction and energy due to forces of repulsion. $\Delta G = \Delta G_{\text{attractions}} + \Delta G_{\text{repulsions}}$

Considering van der Waals type of forces of attractions, $\Delta G_{\text{attractions}}$ can be written as

$$\Delta G_{\text{attractions}} = -\frac{A}{12\pi X^2},$$

where, X is the distance of separation of two successive particles.

A is known as Hamaker constant for the colloidal system.

According to Gouy-Chapman model, $\Delta G_{\text{repulsions}}$ is given by

$$\Delta G_{\text{repulsions}} = \frac{64 n_0 kT \gamma_0^2 e^{-\kappa X}}{\kappa}, \quad \text{where,} \ \frac{1}{\kappa} \ \text{is the Debye length}.$$

$$\gamma_0 = \tan h\left(\frac{Ze\psi_0}{4kT}\right) \ \text{and} \ \frac{1}{\kappa} = \left(\frac{\varepsilon kT}{2e^2 Z^2 n_0}\right)^{\frac{1}{2}}, \ \kappa = \left(\frac{2e^2 Z^2 n_0}{\varepsilon kT}\right)^{\frac{1}{2}} \tag{18.3}$$

k = Boltzmann constant; ε = Dielectric constant of the liquid medium considered.

n_0 = Number of colloidal particles per m^3 of solution; T = Temperature; Z = Valency of the ion.

γ_0 = Surface tension of the liquid; ψ_0 = Electrostatic potential at the surface of the particle.

So,

$$\Delta G = -\frac{A}{12\pi X^2} + \frac{64 n_0 kT \gamma_0^2 e^{-\kappa X}}{\kappa} = -\frac{A}{12\pi X^2} + B e^{-\kappa X}, \ B = \frac{64 n_0 kT \gamma_0^2}{\kappa} \tag{18.4}$$

The change in free energy with distance of separation, D, is shown in the Fig. 18.2.

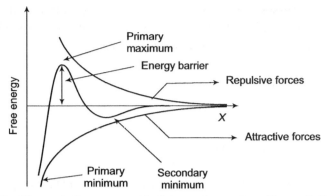

Figure 18.2   Change in free energy with distance of separation of two colloidal particles

Primary minimum is the lowest energy state of the particles and the particles are sufficiently sable at this state. Primary maximum indicates the energy barrier. Colloidal particles gain necessary activation energy through collision and agglomeration of particles sets in. Particles after crossing the energy barrier, may reach the state of secondary minimum, where, flocculation begins.

Higher the energy barrier, more stable the colloid is because coagulation or flocculation begins only when particles cross the energy barrier.

However, the energy barrier depends on concentration. As the concentration increases barrier height decreases and the colloid is more and more unstable as shown in the Fig. 18.3.

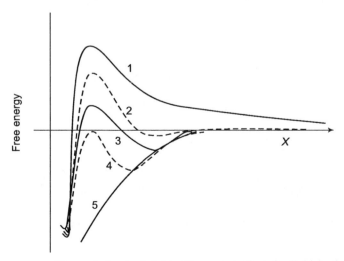

Figure 18.3   Change in barrier height with concentration of colloidal solution

Energy barrier height decreases from curve 1 to curve 4, where energy barrier is exactly zero and particles are free to cluster together, leading to coagulation spontaneously. The concentration, at which curve 4 is reached, is known as critical coagulation concentration (CCC). Curve 1 indicates that concentration is very low and the colloid is sufficiently stable.

However, stability decreases from curve 1 to curve 3 as concentration increases. In case of curve 5, concentration is sufficiently high so that very fast coagulation takes place. Even a little disturbance is enough to set in the coagulation process.

18.4.3 Calculation of CCC

We know

$$\Delta G = -\frac{A}{12\pi X^2} + Be^{-\kappa X} \tag{18.4}$$

CCC is calculated on the basis of curve 4, where, $\Delta G = 0$ at $X = X_c$.

So,

$$-\frac{A}{12\pi X_c^2} + Be^{-\kappa X_c} = 0 \tag{18.5a}$$

So, differentiating the equation (18.4) with respect to X, we get

$$\frac{d(\Delta G)}{dX} = \frac{A}{6\pi X^3} - B\kappa e^{-\kappa X} = 0 \tag{18.5b}$$

Again, at the primary maximum, we can write

$$\frac{d(\Delta G)}{dX} = 0 \text{ at } X = X_c.$$

Thus at primary maximum the equation (18.5b) becomes

$$\frac{A}{6\pi X_c^3} - B\kappa e^{-\kappa X_c} = 0 \tag{18.6}$$

From the above two equations (18.5a) and (18.6), we get

$$\kappa X_c = 2. \text{ Or, } X_c = \frac{2}{\kappa} \tag{18.7}$$

Putting the value of X_c in the equation (18.5a), we get

$$-\frac{\kappa^2 A}{48\pi} + Be^{-2} = 0. \text{ Or, } -\frac{\kappa^2 A}{48\pi} + \left(\frac{64n_0 kT\gamma_0^2}{\kappa}\right)e^{-2} = 0 \tag{18.8}$$

Substituting the value of κ from equation (18.3) in the above equation (18.8) and solving n_0, we get

$$n_0 = \left[\frac{512(48\pi)^2 \varepsilon^3 (kT)^5 \gamma_0^4}{A^2 (Ze)^6}\right]e^{-4}, \text{ where, } \gamma_0 = \tan h\left(\frac{Ze\psi_0}{4kT}\right) \tag{18.9}$$

Z is the valency of the ion.

If the surface potential, ψ_0 is sufficiently high, preferably > 100 mV, $\tan h\left(\dfrac{Ze\psi_0}{4kT}\right) \to 1$ and hence, $\gamma \to 1$.

So, $n_0 \propto \dfrac{1}{Z^6}$. The result is the mathematical form of Schulze-Hardy rule. Thus Schulze-Hardy rule is applicable at high surface potential.

If the surface potential, ψ_0 is sufficiently small, preferably > 10 mV, $\tan h\left(\dfrac{Ze\psi_0}{4kT}\right) \to \dfrac{Ze\psi_0}{4kT}$ and hence, $\gamma \propto Z\psi_0$. So, $n_0 \propto \dfrac{1}{Z^2}$.

Table 18.1   Some typical values of CCC against valency of ion

Valency of ion	CCC in (mole/lit)
1	5×10^{-2}
2	7×10^{-4}
3	9×10^{-5}

18.4.4   Kinetics of Coagulation

Coagulation of a colloidal solution depends on the energy barrier. If the energy barrier is equivalent to kT, coagulation occurs very slowly. On the other hand, coagulation is very fast when the energy barrier almost vanishes.

Assumptions

(a) Coagulation occurs through diffusion process only.
(b) Collision occurs between two particles to form doublet only. Possibility to form higher agglomerates is neglected.
(c) Particles are assumed to have same radius.

Let us consider particles are spheres with radius R and centre to centre distance between two particles is r.

N is the number of particles per unit volume, φ is the fraction of particles, crossing the energy barrier and f is the friction factor.

Consider a spherical shell surrounding a particle. The surrounding particles are diffusing into the shell to accelerate coagulation. The diffusion flux is given by

$$J \cdot A = -4\pi r^2\left(D\dfrac{dN}{dr} + \dfrac{N}{f}\dfrac{d\varphi}{dr}\right), \quad \text{where,} \quad D = \text{Diffusion coefficient}$$

Boundary condition when $r \to \infty$, $N \to N_\infty$ and when $r = 2R$, $N = 0$.

So, neglecting the second term, we get

$$J \cdot A = -4\pi r^2 D \frac{dN}{dr}. \quad \text{Or,} \quad J \cdot A \frac{dr}{4\pi r^2} = -D dN. \quad \text{Or,} \quad \frac{J \cdot A}{4\pi} \int_{\infty}^{2R} \frac{dr}{r^2} = -D \int_{N_\infty}^{0} dN$$

Or, $\quad J \cdot A = -8\pi RDN_\infty$. Or, in general $\quad J \cdot A = -8\pi RDN$

As there are two particles involved in collision, D should be replaced by $2D$. Thus the above equation becomes

$$J.A = -16\pi RDN$$

Rate of coagulation depends on the flux as well as the number of particles, N. Thus

$$\text{Rate} = -\frac{dN}{dt} = -(J \cdot A). N = 16\pi RD \cdot N^2 = k_r \cdot N^2. \quad \text{Or,} \quad k_r = 16\pi RD \quad (18.10)$$

It is assumed that two particles whenever collide, will stick together. So, there is no energy barrier for coagulation process and hence rate constant, k_r is independent of temperature.

Using Stokes-Einstein equation, diffusion coefficient, D, can be expressed as follows

$$D = \frac{kT}{f} = \frac{kT}{6\pi \eta R} \quad (18.11)$$

where, k = Boltzmann constant, η = Coefficient of viscosity.

Combining equations (18.10) and (18.11), we get

$$k_r = \frac{8kT}{3\eta} \quad (18.12)$$

$$-\frac{dN}{dt} = 16\pi RD \cdot N^2. \quad \text{Or,} \quad \int_0^t dt = -\frac{1}{16\pi RD} \int_{N_0}^{N} \frac{dN}{N^2}.$$

Or, $$t = \frac{1}{16\pi RD}\left(\frac{1}{N} - \frac{1}{N_0}\right) \quad (18.13)$$

18.4.5   Zeta Potential

Each and every particle in a colloidal system is surrounded by ions. Some ions remain adsorbed on the surface of the particle and are static. This layer of ions is called **stern layer**. Other ions are loosely associated and have a tendency to diffuse into the bulk of the system. However, the movement of ions is restricted up to a certain thickness. This layer, where ions are moving with restriction, is called **diffuse layer**. The diffuse layer has a boundary, within which ions maintain their identity. As the particle moves, all the layers with boundary also move. The magnitude of potential is maximum at the surface (ψ_0) and decreases linearly up to the shear plane and then decreases exponentially as distance from the shear plane decreases. The potential at the boundary of the diffuse layer is called ***Zeta potential***, (ξ). The potential at the stern layer is represented as ψ_s.

18.4.5.1 Factors Influencing Zeta Potential (ξ)

1. *pH* and isoelectric point Zeta potential of any colloidal system is markedly influenced by pH of the medium. Let us consider a colloidal system, stabilized by negatively charged ions. On addition of alkali pH increases and at the same time the system gains more and more negative charge, thereby increasing the magnitude of Zeta potential. However, Zeta potential is negative in this region and the colloidal system is stable. On the other hand if acid is added, pH decreases and at the same time negatively charged ions are getting neutralized, thereby decreasing the magnitude of Zeta potential. Thus if pH is plotted against Zeta potential, a curve is obtained, where $\xi = 0$. That point is called **isoelectric point**, where the colloidal system is the least stable.

Further addition of acid leads to positive Zeta potential, where the colloidal system is highly unstable. In some sols, particles are surrounded by positively charged ions and in those cases sols are stable at pH ≤ 4 while in some other sols, particles are surrounded by negatively charged ions and the sols are stable at pH $\gg 7.5$.

Isoelectric point The term isoelectric point is generally applicable to amino acids, where, zwitterions are formed. For a given amino acid, isoelectric point represents the pH, at which the net charge on that amino acid is zero.

In case of amino acids, both acid and basic characters are found due to carboxylic acid group and amino group respectively. As a result of this a zwitterion is formed such that the molecule is neutral. The pH, at which an amino acid exists as zwitterion form, is called isoelectric point of that amino acid. The zwitterion has two pK_a values due to $-CO_2^-$ and $-NH_3^+$ groups. At low pH, $-CO_2^-$ is converted to $-CO_2H$ and corresponding pK_a value is called pK_a1. At high pH, $-NH_3^+$ is converted to $-NH_2$ and corresponding pK_a value is called pK_a2.

$$
\underset{\substack{\text{Acidic media} \\ \text{low pH}}}{\overset{+}{H_3N}\!\!-\!\!\underset{R}{\overset{CO_2H}{\underset{|}{\overset{|}{C}}}}\!\!-\!\!H} \quad \underset{pKa_1}{\rightleftharpoons} \quad \underset{\substack{\text{Neutral} \\ \text{form}}}{\overset{+}{H_3N}\!\!-\!\!\underset{R}{\overset{CO_2^-}{\underset{|}{\overset{|}{C}}}}\!\!-\!\!H} \quad \underset{pKa_2}{\rightleftharpoons} \quad \underset{\substack{\text{Basic media} \\ \text{high pH}}}{H_2N\!\!-\!\!\underset{R}{\overset{CO_2^-}{\underset{|}{\overset{|}{C}}}}\!\!-\!\!H}
$$

The isoelectric point (pI) is calculated using the following formula

$$pI = \frac{1}{2}(pK_a1 + pK_a2)$$

In case of glycine, $pK_a1 = 2.34$ and $pK_a2 = 9.6$ and $pI = 5.97$.

In some amino acids more than one acid groups are present. In that case pK_a1 and pK_a3 values are observed at low pH and pK_a2 value is observed at high pH. As acid group is dominated, pI is calculated on the basis of pK_a1 and pK_a3 values as shown:

$$\underset{\substack{\text{Acidic media}\\\text{low pH}}}{\overset{\text{CO}_2\text{H}}{\underset{\text{CH}_2\text{CO}_2\text{H}}{H_3\overset{+}{N}\text{---}\text{H}}}} \underset{1.88}{\overset{pKa_1}{\rightleftharpoons}} \underset{\substack{\text{Neutral}\\\text{form}}}{\overset{\text{CO}_2^-}{\underset{\text{CH}_2\text{CO}_2\text{H}}{H_3\overset{+}{N}\text{---}\text{H}}}} \underset{3.65}{\overset{pKa_3}{\rightleftharpoons}} \overset{\text{CO}_2^-}{\underset{\text{CH}_2\text{CO}_2^-}{H_3\overset{+}{N}\text{---}\text{H}}} \underset{9.68}{\overset{pKa_2}{\rightleftharpoons}} \underset{\substack{\text{Basic media}\\\text{high pH}}}{\overset{\text{CO}_2^-}{\underset{\text{CH}_2\text{CO}_2^-}{H_2N\text{---}\text{H}}}}$$

$$pI = \frac{1}{2}(pK_a1 + pK_a3)$$

2. **Conductivity** Thickness and stability of electrical double layer depend on concentration of ions in the system or ionic strength. At higher concentration of ions the electrical double layer is sufficiently richer in ions and hence remains under compression. Degree of compression is further increased with increase in valency of ions, e.g., compression is much more in presence of Al^{3+} ions than in presence of Na^+ ions. There are some ions, which do not get adsorbed on particle surface nor do they have any effect on isoelectric point. These are called **non-specific ions**. However, some other ions are adsorbed on the particle surface, leading to a change in isoelectric point. These are called **specific ions**. If the concentration of ions is sufficient, charge reversal on the particle surface also occurs.

18.5 SEDIMENTATION OF COLLOIDAL SOLUTION

Colloids can be coagulated using either gravitational field or centrifugal field.

18.5.1 Sedimentation by Gravitational Field

Let us consider a free falling spherical particle in a liquid medium. Three different forces are acting on the particle.

(i) Gravitational force = mg, where, m is the mass of the particle.

(ii) Buoyant force = $(m/\rho) \times \rho_l \times g$, where, ρ and ρ_l are the density of the particle and the density of the liquid medium respectively.

(iii) Viscous force or drag force = $6\pi\eta ru$ in accordance with Stokes' law, where, η and u are the viscosity of the liquid medium and velocity of the particle respectively. The force balance is shown in the Fig. 18.4.

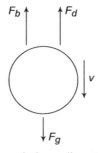

Figure 18.4 Force balance during sedimentation by gravitational field

The resultant force $= F_g - (F_b + F_d)$

$$F_g = \text{mg}, \quad F_b = \left(\frac{m}{\rho}\right) \times \rho_l \times g, \quad F_d = 6\pi\eta ru$$

Under this condition the particle is slowly moving towards the bottom with acceleration (du/dt). Thus the resultant force in the downward direction is given by

$$m\frac{du}{dt} = \text{mg} - \left[(m/\rho) \times \rho_l \times g + 6\pi\eta ru\right]$$

Assumptions

(a) Movement of particle is not influenced by any other foreign particle, present in the system. The phenomenon is called free settling. If sufficient number of foreign particles are present, collision between particles needs to be considered. The phenomenon is called hindered settling.

(b) Collision between walls of container and particle is neglected.

(c) The free falling particle is large in size compared to the mean free path of molecules so that any slipping of particle is neglected.

As the particle is moving towards the bottom, drag force is increasing and ultimately, $\frac{du}{dt} = 0$ and the particle is moving with constant velocity. Corresponding velocity is the maximum velocity, known as terminal velocity, represented as u_t, which is given as

$$\frac{du}{dt} = 0 \quad \text{at} \quad u = u_t$$

Hence, $$u_t = \frac{\text{mg}\left(1 - \dfrac{\rho_l}{\rho}\right)}{6\pi\eta r} = \frac{\left(\dfrac{4}{3}\pi r^3 \rho\right)g\left(1 - \dfrac{\rho_l}{\rho}\right)}{6\pi\eta r} = \frac{2r^2 g(\rho - \rho_l)}{9\eta} = \frac{d^2(\rho - \rho_l)g}{18\eta}$$

Or, $$u_t = \frac{d^2(\rho - \rho_l)g}{18\eta} \tag{18.14}$$

where, d is the diameter of the particle.

Case 1 If $\rho > \rho_l$, colloidal particles slowly settle at the bottom of the container in the form of precipitate, known as sediments. The phenomenon is known as sedimentation. It is often observed in case of paint.

Case 2 If $\rho < \rho_l$, colloidal particles slowly move up to the surface of the liquid in the form of cream. The phenomenon is known as creaming. It is often observed in case of milk.

18.5.2 Sedimentation by Centrifugal Field

If the size of colloidal particles is ≤ 100 nm, particles can be separated using centrifugal force of action. In centrifugal force field, g is to be replaced by $\omega^2 r$, where, ω and r are the angular velocity and radius of rotation of the particle under centrifugal force field respectively. Under balanced condition, $\dfrac{dr}{dt}$ is called terminal velocity.

$$m\omega^2 r\left(1 - \frac{\rho_l}{\rho}\right) = 3\pi\eta d\left(\frac{dr}{dt}\right).$$

Or,

$$\frac{\pi}{6}d^3(\rho - \rho_l)\omega^2 r = 3\pi\eta d\left(\frac{dr}{dt}\right) \quad \left[\text{Since, } m = \frac{\pi}{6}d^3\rho\right]$$

Or,

$$\frac{dr}{dt} = \frac{d^2(\rho - \rho_l)\omega^2 r}{18\eta} = u_t\left(\frac{\omega^2 r}{g}\right) \tag{18.15a}$$

Under centrifugal force field, particles are moving outward and hence accelerating force is increasing. So centrifugal acceleration cannot be zero or $\omega^2 r \neq 0$.

Centrifugation with very high angular velocity ($\omega = 5 \times 10^3$ to 10^5 rad/s) is called ultracentrifugation. Acceleration under this ultracentrifuge force field is 10^6 times the gravitational acceleration. Thus sedimentation process is quite rapid under ultracentrifuge force field.

Sedimentation coefficient (s) is given as

$$s = \frac{dr / dt}{\omega^2 r} \tag{18.15b}$$

18.6 ELECTRICAL DOUBLE LAYER

Colloidal stability depends on two forces, which are operating always in a colloidal system.

(a) Repulsive forces between similar ions, e.g., positive ions. This repulsive force helps in forming electrical double layer surrounding the particles. Potential due to repulsive forces is given by

$$V_{\text{repulsion}} = \psi_0^2 e^{-\kappa x}$$

where, ψ_0 = surface charge, x = distance between two particles.

(b) Intermolecular attractive forces, called van der Waals type of forces, exist among the particles. Considering colloidal particles as spheres, potential due to attractive forces is given by

$$V_{\text{Attraction}} = -\frac{A \cdot a}{12x}, \text{ where, } A = \text{Hamaker constant}.$$

Considering the above two forces, resultant free energy of the system is shown in the Fig. 18.5.

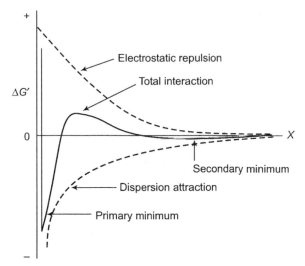

Figure 18.5   Change in free energy in a given colloidal system

These two forces are balanced to provide stability of colloidal solution. Colloidal particles are usually negatively charged and hence they attract positive ions from the surrounding medium, forming an electrostatic layer, known as ***Fixed or Stern layer***. Electrostatic potential is very high in this layer. The thickness of stern layer is small around 1 nm. Counter ions are specifically adsorbed in the inner region of the Stern layer and the thickness of this inner surface is known as inner Helmholtz plane or IHP. Potential drop in this layer is very sharp and the potential depends on the number of ions. Outer region of the Stern layer is dominated by either positive or negative ions. This layer is known as outer Helmholtz plane or OHP.

Outside this Fixed layer, electrostatic force field decreases exponentially to zero. This electrostatic region is known as ***Diffused or Gouy layer***. Thickness of the diffused layer is called Debye length, represented by $(1/\kappa)$ and κ is known as Debye-Hückel parameter.

The thickness of electrical double layer is the sum of the thickness of fixed layer and the thickness of diffused layer.

Stability of lyophilic colloids depends on solvation and charge density. There are some lyophilic sols, where solvation interaction is very strong and the colloid is stable even at its isoelectric pH. This type of lyophilic sol is called ***isostable*** colloid. One example is Gelatin, which is stable at its isoelectric pH but can be precipitated using sufficient amount of dehydrating agent like acetone or alcohol.

There are other lyophilic sols, which are unstable and get precipitated at their isoelectric pH. One example is Casein (isoelectric point is 4.6). This type of lyophilic sol is called ***isolabile*** colloid.

18.6.1   Stability Ratio

If there is no energy barrier all the collisions are successful and the rate constant of coagulation is k_d. Rate depends on diffusion-controlled inter-particle collision and k_d is independent of temperature. If there is an energy barrier, all the collisions are not successful and the rate of coagulation is slower with rate constant k_s. Rate depends on interaction-force-controlled inter-particle collision and k_s depends on temperature according to Arrhenius equation

$$k_s \propto e^{-(E_a/RT)}$$

The ratio of these two rate constant values is known as stability ratio, w

$$w = \frac{k_d}{k_s} > 1$$

The stability of dispersion can be increased by:

(i) increasing particle radius, (ii) increasing surface potential, ζ, (iii) decreasing Hamaker constant, (iv) decreasing ionic strength, and (v) decreasing temperature.

18.6.2 Electrical Double Layer Theory

The theory is based on Gouy-Chapman model (1910–13).

Assumptions

(a) Ions are distributed on the surface according to Poisson-Boltzmann distribution function.
(b) Ions are considered point charges.
(c) Ions do not interact with each other.
(d) Strong electrostatic force exists between oppositely charged ions near the surface and hence diffusion of ions begins after some distance from the surface.

However, in actual practice magnitude of potential starts decreasing from the surface and according to Stern (1924)/Grahame (1947) model, potential changes linearly up to shear plane within which charges are assumed to be static. After the shear plane potential runs exponentially till Gouy plane is reached. After the Gouy plane potential curve almost runs asymptotically to the *x*-axis. It is well illustrated in the Fig. 18.6.

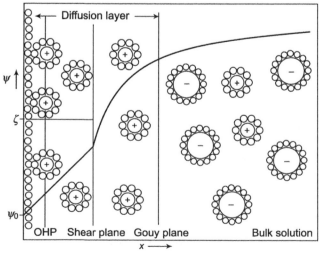

Figure 18.6   Potential curve of a colloidal system according to Gouy-Chapman model.
OHP: Outer Helmholtz Plane

18.6.2.1 Parallel Plate Capacitor Model

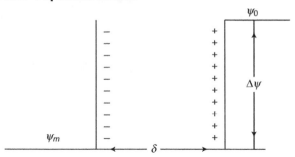

Figure 18.7 Potential curve in parallel plate capacitor model

According to parallel plate capacitor model, two oppositely charged particles, separated by a distance δ, act as parallel plate capacitor as shown in the Fig. 18.7. Electrostatic force of attraction (F) between two particles of charges q_1 and q_2 is given by equation (18.16). Assume that $q_1 = q$.

$$F = \frac{1}{4\pi\varepsilon}\frac{q_1 q_2}{r^2}, \varepsilon = \varepsilon_0\varepsilon_r, \quad E = -\frac{d\psi}{dx} = \frac{F}{q_2}. \quad \text{Or,} \quad \frac{\Delta\psi}{\delta} = \frac{1}{4\pi\varepsilon}\frac{q}{r^2} \quad (18.16)$$

According to Gaussian theorem "The total electric flux, coming out from a closed surface is equal to the charge enclosed by that surface." Mathematically, we can write that

$$\nabla \cdot E = \frac{\rho}{\varepsilon}, \quad \text{where,} \quad \varepsilon = \varepsilon_0\varepsilon_r \quad (18.17)$$

ρ = Charge density in Coulomb/m³, ε_0 and ε_r are the permittivity of the vacuum and dielectric constant of the medium. ε is the permittivity of the medium.

Again by definition of electric potential, E, we have

$$E = -\nabla \cdot \psi. \quad \text{Or,} \quad \nabla \cdot E = \nabla \cdot (-\nabla \cdot \psi) = \frac{\rho}{\varepsilon}. \quad \text{Or,} \quad \nabla^2\psi = -\frac{\rho}{\varepsilon} \quad (18.18)$$

The above equation is known as Poisson distribution of charge in three dimensions. In one dimension the above equation becomes

$$\frac{d^2\psi}{dx^2} = -\frac{\rho}{\varepsilon} \quad (18.19)$$

Assume that charge density decreases according to Boltzmann distribution function as the distance from the surface increases.

n_+ = Number of positive ions per unit volume (m³), n_- = Number of negative ions per unit volume (m³).

n_0 = Number of positive or negative ions per unit volume (m³) near the surface.

$n_+ = n_0 e^{-ze\psi/kT}$. The density of positive charge $= zen_+ = zen_0 e^{-ze\psi/kT}$. Here, $z = +z$

$n_- = n_0 e^{ze\psi/kT}$. The density of negative charge $= -zen_- = -zen_0 e^{ze\psi/kT}$. Here, $z = -z$

where, $+z$ and $-z$ are the oxidation states of positive ions and negative ions respectively.

So,
$$\rho = n_+ + n_- = -zen_0 \left(e^{\frac{ze\psi}{kT}} - e^{-\frac{ze\psi}{kT}} \right) = -2zen_0 \sin h\left(\frac{ze\psi}{kT} \right)$$

According to Debye-Hückel approximation, $\dfrac{ze\psi}{kT} \ll 1$. So, we can write that $\sin h\left(\dfrac{ze\psi}{kT} \right) \approx \dfrac{ze\psi}{kT}$.

Thus,
$$\rho = -2zen_0 \left(\frac{ze\psi}{kT} \right) = -\frac{2n_0 z^2 e^2 \psi}{kT}$$

Thus the equation (18.19) becomes

$$\frac{d^2\psi}{dx^2} = \frac{2n_0 z^2 e^2 \psi}{\varepsilon kT} = \kappa^2 \psi, \text{ where, } \kappa = \sqrt{\frac{2n_0 z^2 e^2}{\varepsilon kT}} \qquad (18.20)$$

κ is known as Debye-Hückel length, also known as decay length.

Solution of the above equation is

$$\psi = \psi_0 e^{-\kappa x} \qquad (18.21)$$

The exponential decay of ψ with x is shown in the Fig. 18.8.

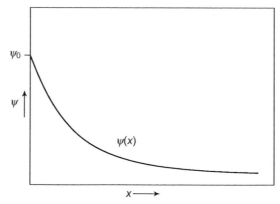

Figure 18.8   Electrical double layer potential curve

18.7   GOLD NUMBER AND PROTECTION OF COLLOIDS

In general lyophilic sols are more stable than lyophobic sols, which are coagulated even in presence of a little quantity of an electrolyte. However, when a little quantity of lyophilic sol is added into a lyophobic sol, the latter gains stability and cannot be coagulated even on addition of large quantity of electrolyte. The phenomenon is called protection of colloids. However, protecting power of different lyophilic colloids is different and this protecting power is expressed in terms of gold number. The scientist Zsigmondy first introduced the term "Gold Number" to define the protecting power of lyophilic colloids. The number is based on a reference sol, i.e., standard gold sol (0.0053% – 0.0058% Au), which turns blue from red on addition of 1 ml 10% NaCl solution. For a given lyophilic sol, gold number is the mg of that dried sol added to a standard gold sol in order to prevent the colour change from red to blue on addition of 1 ml 10% NaCl solution to the said gold sol. Lower the Gold Number, higher the protecting power of the sol. Gold Number of some lyophilic sols are listed below:

Table 18.2 Gold number values of different hydrophilic substances

Hydrophilic substance	Gold number	Hydrophilic substance	Gold number
Gelatin	0.005–0.01	Sodium oleate	0.4–1.0
Sodium caseinate	0.01	Gum tragacanth	2
Haemoglobin	0.03–0.07	Potato starch	25
Gum arabic	0.15–0.25		

18.7.1 Mechanism of Sol Protection

(a) The actual mechanism of sol protection is very complex. However it may be due to the adsorption of the protective colloid on the lyophobic sol particles, followed by its solvation. Thus it stabilises the sol via solvation effects.

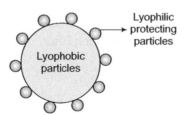

Figure 18.9 Lyophilic protection of lyophobic sol particles

(b) Solvation effects contribute much towards the stability of lyophilic systems. For example, gelatin has a sufficiently strong affinity for water. It is only because of the solvation effects that even the addition of electrolytes in small amount does not cause any flocculation of hydrophilic sols. However at higher concentration, precipitation occurs. This phenomenon is called salting out.

(c) The salting out efficiency of an electrolyte depends upon the tendency of its constituent ions to get hydrated, i.e., the tendency to squeeze out water initially tied up with the colloidal particle.

(d) The cations and the anions can be arranged in the decreasing order of the salting out power. Such an arrangement is called lyotropic series.

Cations: $Mg^{2+} > Ca^{2+} > Sr^{2+} > Ba^{2+} > Li^+ > Na^+ > K^+ > NH_4^+ > Rb^+ > Cs^+$

Anions: $Citrate^{3-} > SO_4^{2-} > Cl^- > NH_2^- < I^- < CNS^-$

Ammonium sulphate, due to its very high solubility in water, is often used for precipitating proteins from aqueous solutions.

(e) The precipitation of lyophilic colloids can also be affected by the addition of organic solvents of non-electrolytes. For example, the addition of acetone or alcohol to aqueous gelatin solution causes precipitation of gelatin. Addition of petroleum ether to a solution of rubber in benzene causes the precipitation of rubber.

18.8 EMULSIFIER

In general oil and water mixture always form completely separated discrete immiscible two-phase system. However, in presence of an emulsifier, oil phase remains dispersed in the water phase to form a continuous matrix, known as emulsion. So the function of an emulsifier is to form a bridge between two immiscible phases resulting an emulsion. An emulsifier contains a hydrophilic head, which forms bond with water molecules, and a hydrophobic tail, which forms bond with oil molecules. Examples of food emulsions are milk, butter, ice cream, etc.

18.8.1 Functions of Emulsifier

(a) It forms bond with both oil phase and water phase to form emulsion.
(b) Some emulsifiers interact with food ingredients like proteins and carbohydrates.
(c) Some emulsifiers act as aerating agent.
(d) Some emulsifiers act as starch complexing agent.
(e) Some emulsifiers act as crystallization inhibitor.

18.8.2 Stabilizers

In many cases emulsions are unstable and the oil phase starts separating out from water phase as droplets even in presence of emulsifier. In that case stabilizer is often added to the emulsion to prevent the phase separation. Examples are gelatin and carrageenan.

18.8.3 Hydrophilic-Lipophilic Balance (HLB)

Generally surfactant is used as an emulsifier. Each surfactant has hydrophilic head, which is soluble in water, and lipophilic head, soluble in lipids and oils. HLB of a surfactant refers to the relative efficiency of the hydrophilic part to that of lipophilic part of that surfactant.

18.8.3.1 Griffin's scale of HLB

According to this scale, the HLB values of different surfactants are numerically categorized between 0 and 20 depending on the size and strength of polar portion relative to the non-polar portion of the molecule. The above scale is applicable to non-ionic surfactants. For ionic surfactants the above scale ranges between 0 and 50. The following graph shows different categories of surfactants on the basis of HLB values.

HLB scale

Examples of popular emulsifiers are Spans and Tweens. Spans are the sorbitan fatty acid esters with HLB values ranging from 1.8 to 8.6. They are lipophilic in character. Tweens are the polyoxyethylene derivatives of spans with HLB values ranging from 9.6 to 16.7. They are hydrophilic in character.

18.8.3.2   Significances of HLB

(a) Higher the HLB value, more the surfactant is lyophilic in character. Or in other words, lower the HLB value, more the surfactant is lipophilic in character.

(b) Different type of surfactants can be compared with respect to their activities on the basis of HLB values.

(c) In many cases a mixture of two or more surfactants are used as emulsifiers. In that case HLB values give an idea about proper ratio of surfactants in the mixture, used as emulsifier.

18.8.3.3   Calculation of HLB

(a) According to Griffin's method, HLB of any non-ionic surfactant can be calculated as follows

$$HLB = 20\left(\frac{M_h}{M}\right)$$

where, M_h is the molecular mass of the hydrophilic portion and M is the molecular mass of the whole molecule.

Example 18.8.1   Calculate HLB value of the following using Griffin's method.

C10-8EO, having the formula $C_{10}H_{22}O - (EtO)_8$.

Solution

Formula of hydrophobic part $CH_3(CH_2)_9OH$, $MW = 158$.

Formula of hydrophilic part $(CH_2CH_2O)_8$, $MW = 352$.

$$HLB = 20 \times \left(\frac{352}{352 + 158} \right) = 13.8$$

(b) According to Davies' method HLB of any surfactant is calculated on the basis of chemical groups of the molecule using the following formula

$$HLB = \Sigma \text{ Hydrophilic group contributions} - \Sigma \text{ Hydrophobic group contributions} + 7$$

m is the number of hydrophilic groups in the molecule, n is the number of lipophilic groups in the molecule and H_i is the value of ith hydrophilic group obtained from the Table 18.3. For lipophilic group, corresponding value is -0.475.

Table 18.3   HLB contribution of different hydrophilic groups

Hydrophilic group	HLB contribution
$-SO_4^-Na^+$	38.7
$-SO_3^-Na^+$	20.7
$-OSO_3^-Na^+$	20.8
$-CONH_2$	9.6
$-COO^-K^+$	21.1
$-COO^-Na^+$	9.4
Tertiary amine	9.4
Ester (sorbitan ring)	6.8
Ester	2.4
$-COOH$	2.1
$-OH$	1.9
Hydroxyl (sorbitan)	0.5
$-O-$	1.3
$-CH_2CH_2O-$	0.33
$-CH_2CH_2OH$	0.95

18.8.3.4 HLB of a Blend

One of the important properties of HLB is that it is additive in nature. Thus in case of a blend HLB values of individual surfactants can be added to get the HLB value of the blend. Let us consider of a blend of two surfactants A and B and corresponding individual HLB values are HLB_1 and HLB_2 respectively. x and $(1-x)$ are the volume fractions of A and B respectively. Thus HLB of the mixture is

$$HLB_m = x\,HLB_1 + (1-x)HLB_2. \quad Or, \quad x = \frac{HLB_m - HLB_2}{HLB_1 - HLB_2}$$

$$Thus, \ \%A \text{ in the blend} = \left(\frac{HLB_m - HLB_2}{HLB_1 - HLB_2}\right) \times 100 = X \text{ and } \% B \text{ in the blend} = (100 - X)$$

Example 18.8.2   Calculate the total HLB for the given emulsion system.

Liquid paraffin	35%
Wool fat	1%
Cetyl alcohol	1%
Emulsifier system	7%
Water	56%

Solution

The total contribution of oily phase is 37% and hence individual contributions are

Liquid paraffin	$(35/37) = 0.946$
Wool fat	$(1/37) = 0.027$
Cetyl alcohol	$(1/37) = 0.027$

HLB contribution of liquid paraffin = $0.946 \times$ HLB of liquid paraffin = $0.946 \times 10.5 = 9.93$
HLB contribution of wool fat = $0.027 \times$ HLB of wool fat = $0.027 \times 10 = 0.3$
HLB contribution of liquid paraffin = $0.027 \times$ HLB of cetyl alcohol = $0.027 \times 15 = 0.4$
Total HLB = $9.93 + 0.3 + 0.4 = 10.63$.

18.9   DESTABILIZATION OF COLLOIDAL SOLUTION

Destabilization of a colloidal solution may be carried out in any one of the following ways:

(a) Double layer compression   On addition of charged species opposite to that of colloid results neutralization of counter ions and hence double layer thickness is significantly reduced. Van der Waals attractive forces predominate over the repulsive forces, destabilizing the colloidal solution.

(b) Adsorption and charge neutralization   In some colloids, foreign particles, oppositely charged to that of colloidal particles, are adsorbed on the colloidal particles. As a result of this, surface potential is reduced and the colloidal solution is destabilized. Coagulation may occur if sufficient amount of adsorption takes place. However, if adsorption of foreign particles occurs in significant amount, charge distribution on colloidal particles is reversed and hence again colloid is stabilized.

(c) Enmeshment in a particle, also known as sweep flocculation   There are some metal ions which form insoluble hydroxides in water, e.g., Al^{3+}, Fe^{3+}, etc. When these metal ions are added into water based colloidal solutions, corresponding hydroxides are precipitated [$Al(OH)_3$, $Fe(OH)_3$]. Then it is possible that colloidal particles are entrapped and settle within the precipitate, resulting coagulation of the colloidal solution.

(d) Adsorption and interparticle bridging   Addition of polymer also destabilizes the colloidal solution. Polymer molecules may be charged or dipolar or neutral species but they are attached to the colloidal particles through hydrogen bonding or van der Waals forces, etc. As a result of this particle size increases and these large particles are associated to form flocs. These flocs are then separated from the colloidal solution resulting destabilization of the colloidal solution.

EXERCISES

Short and Long Answer Type Questions

1.   Describe how hydrophilic and hydrophobic colloids are stabilized in water.
2.   Give a brief definition or explanation of the following concepts in colloid science:
      (a) Double layer, (b) Counter ion, (c) Isoelectric point, (d) Zeta potential,
      (e) Flocculation, (f) Electrokinetic mobility

Answer

  (a) **Double layer**   It is the layer of ions on a colloidal particle. The inner (Stern) layer consists of ions that are electrostatically attracted to the charged particle. The outer (diffuse) layer consists of counter ions.
  (b) **Counter ion**   It is an ion, oppositely charged with respect to a charged chemical species.
  (c) **Isoelectric point**   For a given amino acid sample, it is the pH value at which the net charge on the sample molecule is zero.
  (d) **Zeta potential**   For a given colloid sample, zeta potential is a measure of repulsion between adjacent, similarly charged particles. It also represents the stability of a colloid. Colloids with high zeta potentials are electrically stabilized, whilst those with low potentials tend to coagulate.
  (e) **Flocculation**   In a given colloid sample due to the presence of external reagent, colloidal particles get aggregated to form larger particles causing separation of the colloid. The phenomenon is called flocculation.
  (f) **Electrokinetic mobility**   The movement of colloidal particles in response to the application of an electric field is called electrokinetic mobility of the particles.

3. Why do phospholipids self-assemble in solution, what structures do they form, and why are they relevant to cell biology?

Answer

Each phospholipid molecule contain a hydrophilic head and 2 hydrophobic tails. These phospholipid molecules have a tendency to assemble in bilayer with the hydrophobic tails in the centre and the hydrophilic heads at the interface with the solution. Thus a bilayer structure is formed. This bilayer is highly stable.

Lipid bilayers (intercalated with proteins) make up over 50% of cell membranes in biology. This bilayer arrangement creates an effective barrier against the free passage of water and ions into and out of cells. So the lipid bilayers play an important role in cell activities.

4. Give examples of different conditions that might cause a colloidal dispersion to coagulate. In each case, explain why coagulation occurs?

Answer

Different reasons of coagulation of a colloidal dispersion are:

Heating and stirring   It increases the frequency and velocity of collisions which are responsible for coagulation of a colloidal solution.

Addition of an electrolyte   Any addition of electrolyte increases the ion concentration and these ions neutralize the surface charges, thus removing the electrostatic repulsion between colloidal particles. Zeta potential decreases fast and hence colloidal solution gets coagulated.

Changing the *pH*   Any change in pH may cause desorption of electrostatic stabilizers, accelerating coagulation process.

5. Explain how soap acts to remove oil.

Answer

Each soap molecule consists of a long hydrophobic tail and a charged hydrophilic head. The molecules are able to form micelles (see diagram below) in which the tails interact with the oil particles and the heads interact with the water molecules. In this way, the oil is dissolved in the water and can be removed.

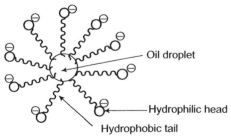

6.   Explain why surface effects are important in colloidal systems.

Answer

The small size of the colloidal particles means that they have a very large total surface area. The colloid can be stabilized by steric and/or electrostatic interactions among particles. If the particle size is large, surface area decreases and distance between two colloidal particles increases. So electrostatic interactions decrease, resulting coagulation by flocculation.

7.   Describe how the addition of an electrolyte can alter the state of a colloidal dispersion.

Answer

In case of a charged colloidal particle, a layer of oppositely charged ions will form on the surface (the Stern layer). In the region, just outside of the colloid, there is a build-up of counter ions creating a double charge layer. The charge surrounding one colloidal particle will repel the charge surrounding other particles and so coagulation is prevented. Addition of an electrolyte leads to reduction in the net charge surrounding each colloidal particle, thereby, destabilizing the electrical double layer. So zeta potential decreases and hence the colloidal particles tend to coagulate and the colloidal dispersion is lost.

8.   Describe how hydrophilic and hydrophobic colloids are stabilized in water.

Answer

Both hydrophilic and hydrophobic colloids are stabilized through electrostatic and steric stabilization.

Hydrophilic colloids may have a charge on their surface that attracts oppositely charged ions (H^+ or OH^- present in water) to form a tightly bound layer known as the Stern Layer. The Stern layer is surrounded by a diffuse layer which contains an excess of counter-ions (opposite in charge to the Stern layer) and a deficit of co-ions. The Stern layer and diffuse layer are collectively known as a double layer. Coagulation of a hydrophilic colloid is prevented by mutual repulsion of the double layers.

Hydrophobic colloids may be stabilized by the use of a surfactant, e.g., a long chain fatty acid with a polar head and a non-polar tail. When dispersed in water these molecules arrange themselves spherically so that the polar (hydrophilic) heads are interacting with the polar water molecules and the non-polar (hydrophobic) tails are interacting with each other. This arrangement is called a micelle. The hydrophobic colloid can be stabilized by dissolving in the non-polar interior of the micelle.

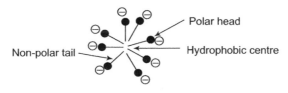

9. Define crystalloids and colloids.

Answer

*   **Crystalloids** Crystalloids are those substances which could be obtained in crystalline form and whose solutions are able to pass through an animal membrane.
*   **Colloids** Colloids are those substances which are amorphous in nature and whose solutions are unable to pass through an animal membrane.

10. What are Colloids?

Answer

A colloid can be defined as a heterogeneous system in which one substance is dispersed (dispersed phase) as very fine particles in another substance called dispersion medium. Colloids are the intermediate system between a solution and a suspension. The size of colloidal particles lies between 1 nm and 1000 nm. The colloidal particles remain suspended in the medium and do not settle down under the influence of gravity. Milk and ink are the two examples of colloids.

11. Mention the differences between colloids and suspension.

Answer

Colloids	Suspension
A colloidal system has two distinct phases and there exists strong interaction between two phases.	A suspension also has two distinct phases but apparently there is no interaction between two phases.
Colloidal particles are very small with diameter $\leq 1\,\mu m$.	In case of suspension particles are quite large, much larger than colloidal particles.
In case of a colloidal system two phases will not be separated on standing even for long time.	In case of suspension, dispersed phase particles get separated and slowly assemble at the bottom on standing.

12. Comment on the statement that "colloid is a state of matter and not considered a substance".

Answer

A colloid is not considered a substance. It is state of a matter which is dependent on the size of the dispersed phase particles. A colloid is formed when the size of the particle lies between 1 nm to 1000 nm.

For example, soap dissolves in water to form colloidal soap solution whereas it dissolves in alcohol to form a true solution. This shows that a substance can be brought into a colloidal state by different methods. Thus any substance (solid, liquid or gas) using special method can be brought into colloidal state. For example, NaCl in water forms true solution but in benzene forms colloidal solution.

13. Answer the following questions clearly.
    (a) What is salt-out? How does it differ from coagulation?
    (b) Explain the significance of Hofmeister series.
    (c) What is coacervation?
    (d) What is electroviscous effect?
    (e) Explain the terms gelation and imbibition.
    (f) Explain the significance of gold number.

Coacervation   Stability of lyophilic sol depends on several factors but when the sol is destabilized, the colloidal system gets converted into two distinct liquid phases. This phenomenon is known as *coacervation*. The liquid phase, concentrated with colloid component, is called *coacervate* and the other liquid phase is called equilibrium solution. Coacervation is usually observed in lyophilic sol and seldom observed in lyophobic sol.

If two oppositely charged colloids come in contact with each other, strong interaction takes place between the colloids resulting coacervation. The phenomenon is known as *complex coacervation*.

Electroviscous effect   Let us consider a system made of two phases. One is continuous phase and the other is the dispersed phase such that dispersed particles are uncharged and spherical with diameter close to 1 μm. Volume fraction of dispersed phase is ϕ. According to Einstein, the viscosity of that system (η) is given by

$$\eta = \eta_0 \left(1 + \frac{5}{2} \phi \right)$$

where, η_0 is the viscosity of the continuous phase.

In case of colloidal system, particles are charged and hence the above equation needs to be modified. The scientist Smoluchowski modified the above equation in the following form

$$\eta = \eta_0 \left[1 + (1 + p) \frac{5}{2} \phi \right]$$

The phenomenon is known as **primary electroviscous effect** and p is known as primary electroviscous coefficient. The magnitude of p depends on ξ–potential and relative size of the colloidal particles.

If EDLs of neighbouring colloidal particles changes, p value also changes. It is called **secondary electroviscous effect**.

If particle diameters of colloidal particles are significantly different from one another, p value also changes. The phenomenon is known as **tertiary electroviscous effect**.

Hofmeister series   In any colloidal system presence of any impurity ion changes significantly the given properties: (i) colloidal aggregation, and (ii) electrokinetic properties.

However, impact of ions on colloidal system is different for different ions. On the basis of this, impact on colloidal system ions can be arranged in descending order as follows:

$$citrate^{3-} > SO_4^{2-} > HPO_4^{2-} > F^- > CH_3COO^- > Cl^- > Br^- > NO_3^- > ClO_4^- > SCN^-$$

$$N(CH_3)_4^+ > NH_4^+ > Cs^+ > Rb^+ > K^+ > Na^+ > H^+ > Ca^{2+} > Mg^{2+}$$

The above two series are known as Hofmeister series for anions and cations respectively.

Multiple Choice Questions

1. According to **Thomas Graham** those substances which were amorphous in nature and whose solutions were unable to pass through the membrane were defined as:
   (a) crystals
   (b) crystalloids
   (c) colloids
   (d) emulsions

2. Colloidal solutions are:
   (a) homogeneous systems
   (b) heterogeneous systems
   (c) true solutions
   (d) suspensions

3. Particles in the size of colloids range:
   (a) 1–1000 nm
   (b) 100–1000 nm
   (c) 1000–10,000 nm
   (d) 1–100 nm

4. Colloidal solution:
   (a) does not diffuse
   (b) does not scatter light
   (c) is homogeneous
   (d) exhibits Tyndall effect

5. How might solid sodium carbonate be obtained from sodium carbonate solution?
   (a) Centrifugation
   (b) Filtration
   (c) Evaporation
   (d) It cannot be extracted

6. What is the best description of blood?
   (a) Sol
   (b) Foam
   (c) Solution
   (d) Aerosol

7. A suspension is formed from uniform particles of solid, of diameter 10 µm, suspended in a solvent. What is the best description of this system?
   (a) Monodisperse and coarse
   (b) Monodisperse and colloidal
   (c) Polydisperse and coarse
   (d) Polydisperse and colloidal

8. Which one of the following dispersions does not have liquid continuous phase?
   (a) Nanosuspension
   (b) Microemulsion
   (c) Gel
   (d) Foam

9. Which one of the following systems has the smallest sized domains in its dispersed phase?
   (a) Nanoemulsion
   (b) Coarse suspension
   (c) Coarse emulsion
   (d) Microemulsion

10. Which of the following sequences correctly describes the change in domain structure as more oil is added to a water-in-oil emulsion?
    - (a) Bicontinuous, spherical, cylinder-like
    - (b) Spherical, cylinder-like, bicontinuous
    - (c) Spherical, bicontinuous, cylinder-like
    - (d) Cylinder-like, spherical, bicontinuous
11. Which method for the production of dispersions involves the formation of particles from materials dissolved in true solutions?
    - (a) Bottom-up
    - (b) Top-down
    - (c) Milling
    - (d) High pressure homogenization
12. The scattering of light by coarse and colloidal dispersed systems is known as:
    - (a) Contrast matching
    - (b) DLVO theory
    - (c) Tyndall effect
    - (d) Creaming
13. Which of the following is not a mechanism for the separation of a physically unstable suspension of magnesium hydroxide in water?
    - (a) Flocculation
    - (b) Precipitation
    - (c) Aggregation
    - (d) Ostwald ripening
14. In the DLVO theory of colloids, normal thermal motion may be sufficient to overcome the energy barrier that leads to irreversible particle aggregation. The name of this energy barrier is which one of the following?
    - (a) Primary maximum
    - (b) Secondary maximum
    - (c) Primary minimum
    - (d) Secondary minimum

ANSWERS

1(c), 2(b), 3(a), 4(d), 5(c), 6(a), 7(a), 8(c), 9(d), 10(d), 11(a), 12(c), 13(d), 14(a).

19 Photochemistry

19.1 INTRODUCTION

Photochemistry is the branch of chemistry concerned with the chemical effects of light. Generally, this term is used to describe a chemical reaction caused by absorption of ultraviolet radiation (wavelength from 100–400 nm), visible light (400–750 nm) or infrared radiation (750–2500 nm).

19.2 LAWS OF PHOTOCHEMISTRY

Grotthuss-Draper law   T. Grothuss (1818) and J. Draper (1839) independently worked on different photochemical reactions and ultimately arrived at a common conclusion, which was finally formulated as Grotthuss-Draper law.

Statement   Frequencies of visible spectrum and close to them are effective to execute a chemical reaction. However, only those rays, absorbed by the reaction mixture, are chemically active.

The above law was not enough to study the mechanism of a photochemical reaction. J. Lambert in 1760 studied different photochemical reactions using monochromatic beam of light and observed that intensity of ray, emerging out of the reaction mixture, is less than that of entering into the reaction mixture. On the basis of this observation, J. Lambert proposed the following law:

Statement   Decrease in light intensity is proportional to the entering light intensity as well as thickness of the reaction cell.

Mathematically, it can be expressed as

$$-dI \propto I \text{ and } -dI \propto dt. \text{ Or, } -dI \propto I.dt$$

where, dt is the thickness of the reaction cell, expressed in cm. The negative sign is used as light intensity decreases during photochemical reaction.

Later on in 1853, A. Beer observed that concentration of the reactant also plays an important role in photochemical reaction. He proposed the following law:

Statement Decrease in light intensity is directly proportional to the number of solute particles present in the reaction cell or concentration of the reactant.

Mathematically, it can be expressed as

$-dI \propto c$, where, c is the concentration of the reactant, expressed in moles/cm^3.

Combining the above two observations, we can write

$$-\frac{dI}{I} \propto c.dt. \quad \text{Or,} \quad -\frac{dI}{I} = \varepsilon \cdot c \cdot dt \tag{19.1}$$

where, ε is known as molar extinction coefficient or molar absorptivity with unit cm^2/mole.

At $t = 0$, $I = I_0$ and at $t = t$, $I = I$. On integration of equation (19.1) using the boundary conditions, we get

$$I = I_0 e^{-\varepsilon ct} \tag{19.2a}$$

The equation (19.2a) is known as Lambert-Beer law.

In general thickness of the all 't' is replaced by the length of the cell 'l'. The equation (19.2a) becomes

$$I = I_0 e^{-\varepsilon cl} \tag{19.2b}$$

$$A = \varepsilon cl$$

The term, A is known as absorbance of a given sample under investigation.

19.3 PHOTOCHEMICAL RATE EQUATION

In 1855, R. Bunsen and H. Roscoe studied in detail the reaction between $H_2(g)$ and $Cl_2(g)$ to form HCl by passing monochromatic beam of light to the reactant mixture.

They observed that the amount of HCl formed from its elements depends on the light intensity, I, and duration of light in the reaction cell, t. For a given amount of HCl formed, the product I and t is constant.

If the light intensities, I_1 and I_2 are active for t_1 sec. and t_2 sec. respectively, we can write that

$$I_1 t_1 = I_2 t_2 \tag{19.3}$$

J. van't Hoff in 1904 generalised the above observation as follows:

"Rate of a photochemical reaction is directly proportional to quantity of light quanta absorbed by the system."

He also assumed that each reactant molecule absorbs one quanta of light, which means that

Number of quanta absorbed is equal to number of molecules reacted.

Q = Quantity of light quanta absorbed = $I_0 - I = I_0(1 - e^{-\varepsilon.c.l})$

I_0 = Intensity of light entering into the reaction cell.

I = Intensity of light emerging out of the cell.

$(I_0 - I)$ = Intensity of light absorbed by the system.

l = Length of the photochemical cell.

$$\text{Rate of photochemical reaction} = -\frac{dn}{dt}$$

$$-\frac{dn}{dt} \propto Q. \quad \text{Or,} \quad -\frac{dn}{dt} = KI_0(1 - e^{-\varepsilon.c.l}) \tag{19.4}$$

n = Number of moles of reactant present after the time interval t and K = proportionality constant.

Q = Total energy absorbed,

N_a = Number of quanta absorbed = N_r = Number of molecules reacted and transformed into product.

$$N_a = N_r = \frac{Q}{hv} \tag{19.5}$$

19.3.1 Quantum Yield

Experimental studies show that number of molecules reacted is not equal to the number of quanta absorbed. To quantify this observation, the efficiency of a photochemical process is expressed by the term quantum yield, represented by the symbol ϕ. It is defined as follows:

"For a given photochemical process γ is the ratio of number of molecules that have reacted to the number of Einstein quanta absorbed."

$$\gamma = \frac{N_r}{N_a} = \frac{N_r}{Q/hv} \tag{19.6}$$

Thus rate of photochemical reaction is given by

$$-\frac{dn}{dt} = N_r = \gamma \times N_a = \gamma \times \left(\frac{Q}{hv}\right) = \frac{\gamma I_0}{hv}(1 - e^{-\varepsilon.c.l}) \tag{19.7}$$

The above equation is the generalised rate equation combining all the laws of photochemistry.

19.4 TYPES OF PHOTOCHEMICAL PROCESSES

On the basis of quantum yield, photochemical processes can be divided into different categories as follows:

(a) There are some photochemical reactions for which $\gamma = 1$. Examples are formation of bromocyclohexane, formation of hydrogen peroxide, decomposition of hydrogen sulphide in benzene solution, etc.

(b) There are some photochemical reactions for which $\gamma < 1$. Examples are decomposition of ammonia, decomposition of acetone, decomposition of acetic acid, formation of hexabromobenzene, etc.

(c) There are some photochemical reactions for which $\gamma > 1$. Examples are formation of sulphuryl chloride, formation of ozone, decomposition of nitrogen dioxide, decomposition of hypochlorous acid, etc.

(d) There are some photochemical reactions for which $\gamma \gg 1$. Examples are reaction between chlorine and hydrogen, reaction between chlorine and carbon monoxide, etc.

Deviation of quantum yield from unity, i.e., $\gamma > 1$ signifies that reaction occurs in chain process. The light is absorbed in the initiation step, where only one quanta of light is absorbed by one reactant molecule to from radicals. These radicals in the propagation step react with other reactant molecules to form several molecules of product. Thus more than one molecule of product may be formed by absorbing one quanta of light. In those cases deviation is observed such that $\gamma > 1$. Generally type of chain reactions with $\gamma > 1$ are observed in gas phase reactions. However, in solution phase reactions, $\gamma \leq 1$ in most of the cases.

The quantum yield largely depends on secondary process, in which activated radicals may react with atoms or radicals to form products but at the same time recombination of radicals takes place, leading to formation of reactant molecules. This recombination process is often called deactivation of radicals. The quantum yield decreases due to this deactivation process. However, deactivation process is sufficiently suppressed at high temperature, thereby increasing the quantum yield. One example is dissociation of ammonia into its elements.

19.4.1 Reactions where $\gamma < 1$

Dissociation of ammonia into its elements.

The following reactions take place during photochemical dissociation of ammonia into its elements:

Primary process

$$NH_3 \xrightarrow{hv} NH_2^{\bullet} + H$$

Secondary process

$$NH_2^{\bullet} + H \longrightarrow NH^{\bullet} + H_2 \tag{a}$$

$$NH^{\bullet} + NH^{\bullet} \longrightarrow N_2 + H_2 \tag{b}$$

Recombination process

$$NH_2^{\bullet} + H \longrightarrow NH_3$$

At room temperature probability of recombination process is much greater than that of the secondary process (b). Thus most of the radicals, formed during the secondary process (a), are returned to the reactant molecules ammonia. So the quantum yield is very low, as low as 0.2 at 20°C. However, at higher temperature, the process (b) is dominating over the recombination process, as a result of which, quantum yield is significantly high. It is around 0.5 at 500°C.

19.4.2 Reactions where $\gamma > 1$

Examples are decomposition of HI, HBr and NOCl.

$$2HI \xrightarrow{hv} H_2 + I_2$$

$$2HBr \xrightarrow{hv} H_2 + Br_2$$

$$2NOCl \xrightarrow{hv} 2NO + Cl_2$$

Dissociation of HI follows the following primary and secondary processes:
Primary process

$$HI \xrightarrow{hv} H^{\bullet} + I^{\bullet}$$

Secondary processes

$H^{\bullet} + HI \longrightarrow H_2 + I^{\bullet}$	(a) Exothermic reaction
$H^{\bullet} + I_2 \longrightarrow HI + I^{\bullet}$	(b) Exothermic reaction
$I^{\bullet} + H_2 \longrightarrow HI + H^{\bullet}$	(c) Endothermic reaction
$I^{\bullet} + HI \longrightarrow I_2 + H^{\bullet}$	(d) Endothermic reaction
$H^{\bullet} + H^{\bullet} \longrightarrow H_2$	(e) Exothermic reaction
$I^{\bullet} + I^{\bullet} \longrightarrow I_2$	(f) Exothermic reaction
$I^{\bullet} + H^{\bullet} \longrightarrow HI$	(g) Exothermic reaction

The product I_2 is continuously removed from the system in order to reduce the concentration of I_2 so that reaction (b) is insignificant compared to other secondary processes. Reactions (c) and (d) are endothermic reactions. So activation energy for these two reactions is very high and hence reactions (c) and (d) are also insignificant like reaction (b). As most of the radicals $H^•$ are removed in reaction (a), concentration of $H^•$ is significantly low enough to reduce the probability of occurring reactions (e) and (g). Only the reaction (f) is significant along with the reaction (a).

Considering the above fact, the photochemical dissociation of HI can be written as:

$$HI \xrightarrow{hv} H^• + I^•$$

$$H^• + HI \longrightarrow H_2 + I^•$$

$$\underline{I^• + I^• \longrightarrow I_2}$$

$$2HI \xrightarrow{hv} H_2 + I_2$$

The above reaction clearly shows that quantum yield is 2. This is true for gas phase reaction. However, in presence of solvent, the quantum yield decreases due to termination of radicals by solvent molecules. As termination efficiency of solvent varies from one solvent to the other, quantum yield also varies, as shown in the Table 19.1.

Table 19.1 Quantum yield at different wavelength

Medium	*Wavelength (nm)*	*Quantum yield (γ)*
Gas phase	250	2
Liquid HI	300	1.84
Solution in C_8H_{14} (0.8 N)	222	1.52
Aqueous solution, (0.8 N)	222	0.078

There are some photochemical reactions for which $\gamma = 3$. One example is conversion of oxygen to ozone. The reaction is believed to follow the following mechanism:

$$O_2 \xrightarrow{hv} O_2^*$$

$$O_2^* + 2O_2 \xrightarrow{hv} 2O_3$$

19.4.3 Reactions where $\gamma \gg 1$

There are some photochemical reactions for which quantum yield is very high, i.e., $\gamma \gg 1$. One example is formation of HCl from its elements. $\gamma = 10^8$. The reaction is believed to proceed through following mechanism

Initiation step

$$Cl_2 \xrightarrow{hv} 2Cl^•$$

Propagation step

$$Cl^{\bullet} + H_2 \longrightarrow HCl + H^{\bullet}$$

$$H^{\bullet} + Cl_2 \longrightarrow HCl + Cl^{\bullet}$$

chain reaction continues

Termination step

$$H^{\bullet} + H \longrightarrow H_2$$

$$Cl^{\bullet} + Cl^{\bullet} \longrightarrow Cl_2$$

Termination step, i.e., recombination of radicals occurs with the release of huge amount of energy, which is enough to break the molecules into radicals again. Thus reaction rate of termination step is very slow, much slower than propagation step, due to which quantum yield of the above reaction is very high. However, in presence of impurities quantum yield sufficiently drops as impurity particles absorb energy released in termination step. So breaking of molecules into radicals is sufficiently restricted and hence reaction rate of termination step is high enough to reduce the quantum yield.

19.4.4 Reactions Occurring Either in Absence or in Presence of Sunlight

There are some reactions, which occur through free radical chain mechanism even in absence of any light energy or quanta. However, reaction rate has been found higher in presence of sunlight. One such example is formation of HBr from its elements.

The reaction is carried out in dark as well as in presence of sunlight separately. In both the cases mechanism is same but only difference is observed in primary process.

In the dark the following mechanism is believed to occur.

Initiation step

$$Br_2 \xrightarrow{k_1} 2Br^{\bullet} \qquad\qquad \text{[Endothermic reaction]}$$

Propagation step

$$Br^{\bullet} + H_2 \xrightarrow{k_2} HBr + H^{\bullet} \qquad\qquad \text{[Endothermic reaction]}$$

$$H^{\bullet} + Br_2 \xrightarrow{k_3} HBr + Br^{\bullet} \qquad\qquad \text{[Endothermic reaction]}$$

Termination step

$$H^{\bullet} + HBr \xrightarrow{k_4} H_2 + Br^{\bullet} \qquad\qquad \text{[Exothermic reaction]}$$

$$Br^{\bullet} + Br^{\bullet} \xrightarrow{k_5} Br_2 \qquad\qquad \text{[Exothermic reaction]}$$

Applying steady-state principle, the reaction rate is given by

$$J_1 = \frac{d[HBr]}{dt} = \frac{2\dfrac{k_2 k_3}{k_4}[H_2]\sqrt{\dfrac{2k_1}{k_5}[Br_2]}}{\dfrac{k_3}{k_4}+\dfrac{[HBr]}{[Br_2]}} \qquad (19.8a)$$

In presence of sunlight (monochromatic beam of wavelength 500 nm) only the initiation step is different.

Initiation step

$$Br_2 \xrightarrow{h\nu} 2Br^\bullet$$

Propagation step

$$Br^\bullet + H_2 \xrightarrow{k_2} HBr + H^\bullet \qquad\qquad \text{[Endothermic reaction]}$$

$$H^\bullet + Br_2 \xrightarrow{k_3} HBr + Br^\bullet \qquad\qquad \text{[Endothermic reaction]}$$

Termination step

$$H^\bullet + HBr \xrightarrow{k_4} H_2 + Br^\bullet \qquad\qquad \text{[Exothermic reaction]}$$

$$Br^\bullet + Br^\bullet \xrightarrow{k_5} Br_2 \qquad\qquad \text{[Exothermic reaction]}$$

The reaction rate is given by

$$J_2 = \frac{d[HBr]}{dt} = \frac{2\dfrac{k_2 k_3}{k_4 \sqrt{k_5}}[H_2]\sqrt{2h\nu}}{\dfrac{k_3}{k_4}+\dfrac{[HBr]}{[Br_2]}} \qquad (19.8b)$$

It has been observed that $J_2 = 500\,J_1$.

19.5 PHOTOCHEMICAL EQUIVALENT

A. Einstein proposed a fundamental concept of photochemistry. This is known as Stark-Einstein law, also known as law of photoequivalence.

Stark-Einstein law Each photon absorbed by a chemical system is used to activate only one molecule for subsequent reaction.

According to this concept, "Each and every molecule, taking part in a given photochemical reaction, absorbs one quantum of light, i.e., hv, causing the reaction."

Thus if one mole of a reactant takes part in a photochemical reaction, the amount of energy absorbed is

$$E = N_A\, hv = \frac{N_A\, hc}{\lambda} = \frac{(6.023 \times 10^{23})(6.625 \times 10^{-34})(3.0 \times 10^8)}{\lambda(\text{Å}) \times 10^{-10}} \text{ J/gm. mole}$$

Or,
$$E = \frac{11.9707 \times 10^8}{\lambda(\text{Å})} \text{ J/mole} = \frac{2.8618 \times 10^8}{\lambda(\text{Å})} \text{cal/mole} \qquad (19.9)$$

E = Energy per mole of photons = Einstein

E is often called Einstein of radiation of a given wavelength $\lambda(\text{Å})$. So the value of E depends on λ. Higher the value of λ, lesser the value of E. The following table illustrates the variation of E with λ.

$$E = \frac{2.8618 \times 10^8}{\lambda(\text{Å})} \text{cal/mole} = \frac{28618}{\lambda(\text{nm})} \text{kcal/mole}$$

Table 19.2 Variation of E with λ

Wavelength (λ) (nm)	Energy (E) (kcal/mole)
100	286.18
400	71.55
700	40.88
1000	28.62

19.6 PHOTOCHEMICAL EQUILIBRIUM

Nature of thermal equilibrium is significantly influenced in presence of sunlight or IR radiation. Presence of radiation accelerates either forward or backward process only and equilibrium is shifted accordingly. One example is dimerization of anthracene, represented by A. The forward process, i.e., dimerization process is accelerated in presence of radiations and the reverse process, i.e., decomposition process is the normal thermal process. The equilibrium, developed in presence of sunlight is called photochemical equilibrium.

$$2A \underset{\text{Thermal}}{\overset{\text{UV radiation}}{\rightleftharpoons}} A_2$$

Step 1	Excitation	$A + hv \longrightarrow A^*$	Rate $J_1 = I_{abs}$
Step 2	Dimerization	$A^* + A \longrightarrow A_2$	Rate $J_2 = k_2 C_A C_{A*}$
Step 3	Fluorescence	$A^* \longrightarrow A + hv'$	Rate $J_3 = k_3 C_{A*}$

Step 4   Decomposition $A_2 \longrightarrow 2A$ Rate $J_4 = k_4 C_{A_2}$

According to steady-state principle

$$\frac{dC_{A^*}}{dt} = J_1 - J_2 - J_3 = I_{abs} - k_2 C_{A^*} C_A - k_3 C_{A^*} = 0. \text{ Hence, } C_{A^*} = \frac{I_{abs}}{k_2 C_A + k_3}$$

Again, at photochemical equilibrium

$$\frac{dC_{A_2}}{dt} = J_2 - J_4 = k_2 C_A C_{A^*} - k_4 C_{A_2} = 0. \text{ Hence, } C_{A_2} = \frac{k_2 C_{A^*} C_A}{k_4}$$

Or, $$C_{A_2} = \frac{k_2 C_A I_{abs}}{k_4 (k_2 C_A + k_3)} = \frac{I_{abs}}{k_4 \left(1 + \dfrac{k_3}{k_2 C_A}\right)} \qquad (19.10)$$

At high concentration of anthracene number of collisions is significantly high so that step 3 is insignificant with very low value of k_3. In that case the concentration of dimer is given by

$$C_{A_2} = \frac{I_{abs}}{k_4}$$

In case of normal thermal equilibrium the concentration of dimer is given by

$$C_{A_2} = KC_A^2, \qquad \text{where,} \quad \text{K is equilibrium constant.}$$

UV absorption and emission depends on the following factors.

1. Symmetry of initial and final electronic states.
2. Multiplicity of spin [singlet (S) and triplet (T) state]. Multiplicity arises due to mixing of S and T as well as mixing of magnetic moments of electron and nucleus.

19.7   FRANCK-CONDON PRINCIPLE

(a) Excitation of electrons on absorption of radiations occurs if molecular geometry of excited state is similar to that of ground state.
(b) Only electrons are jumped from ground state to corresponding excited state but the heavier nucleus remains in the ground state without any excitation.
(c) As electronic transition is much faster than movement of nuclei, the wave function cannot change. The most likely transition is the one, which is overlapped maximum with the excited state wave function.

The excitation of electron is shown in the Fig.19.1.

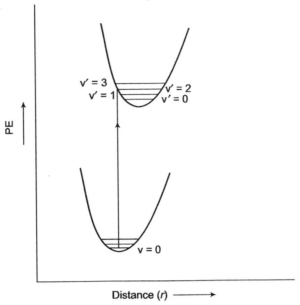

Figure 19.1 Excitation of electrons on absorption of radiations

19.7.1 Born-Oppenheimer Approximation

Considering electronic and nuclear vibrations the total wave function for a molecule can be written as $\Psi(r_1, r_2.....r_n, R_1, R_2......R_n)$, where, r_i and R_i represent electronic coordinate and nuclear coordinate respectively. According to Born-Oppenheimer, as nucleus is much heavier than electron, electronic movement is 1000 times faster than nuclear movement and both type of movement occur freely without any interference. Thus molecular wave function can be divided into two components separately as shown below:

Total wave function = Electronic wave function at each geometry × Nuclear wave function

Or, $\quad \Psi(r_1, r_2.....r_n, R_1, R_2......R_n) = \Psi^{\text{electronic}}(r_1....r_n : R_i) \times \Psi^{\text{vibration}}(R_1...R_n)$

So, $\quad E_{\text{total}} = E_{\text{electronic}} + E_{\text{vibration}}$

The energy diagram of a molecule is given.

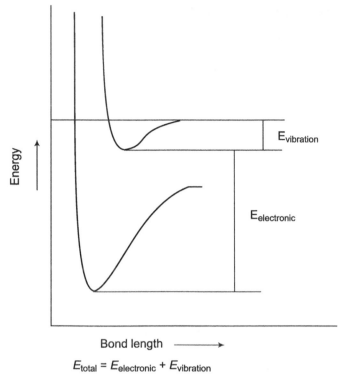

$$E_{total} = E_{electronic} + E_{vibration}$$

Figure 19.2 Energy diagram against bond length

19.8 FLUORESCENCE AND PHOSPHORESCENCE

Electrons from the excited state returns to the ground state directly with the emission of radiation. In some cases wavelength of emitted radiation is higher than that of absorbed radiation. The phenomenon is called Stokes shift. One example is dimerization of anthracene.

In other cases, wavelength of emitted radiation is lower than that of absorbed radiation. The phenomenon is called anti-Stokes shift. Example: biphenyl absorbs radiation at longer wavelength but emits at shorter wavelength.

Energy per mole of photons is known as **Einstein**.

$$E = N_A h\nu = \frac{N_A hc}{\lambda} = \frac{11.9625 \times 10^4}{\lambda(nm)} \text{kJ/mole} = \frac{28618}{\lambda(nm)} \text{kcal/mole.}$$

19.8.1 Electronic States

Molecular multiplicity (M) is related to spin multiplicity (S) as follows:

$M = 2S + 1$. Where, S represents net spin of electrons in the molecule.

In most organic molecules all the electrons are paired and hence $S = 0$. Thus $M = 1$. This is called **singlet state**.

In case of organic free radicals only one electron is present as single and hence $S = 1/2$. Accordingly,

$M = 2 \times 1/2 + 1 = 2$. This is called **doublet state**.

After absorption of light, paired electrons are unpaired without changing spin multiplicity. So the value of $M = 1$ even in excited state. This is called singlet excited state. In this state molecule shows diamagnetic behaviour. As there is no change in spin orientations, the average lifetime of excited electron is very less, of the order of 10^{-5}–10^{-8} sec.

In some cases after absorption of light paired electrons are unpaired with changing spin multiplicity. So the value of $S = \dfrac{1}{2} + \dfrac{1}{2} = 1$ and $M = 2 \times 1 + 1 = 3$. This is called **triplet excited state**. In this state molecule shows paramagnetic behaviour. In the excited state spin relaxation takes place and hence average lifetime is comparatively long, of the order of 10^{-4} sec.

All these spin orientations are illustrated in the Fig. 19.3.

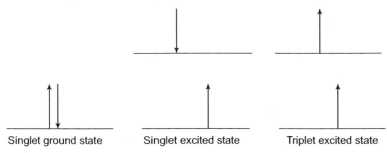

Singlet ground state   Singlet excited state   Triplet excited state

Figure 19.3   Spin orientations of electrons in ground state and excited states

In some cases electron is excited without changing spin orientation, forming singlet excited state. In that case electron returns to the ground state with emission of radiation. This is called fluorescence. Wavelength of emitted radiation is longer than that of absorbed radiation. This is called Stokes shift.

19.8.2 Jablonski Diagram

Possible electronic transitions in an organic compound is well illustrated in the Fig. 19.4, which is popularly known as Jablonski diagram.

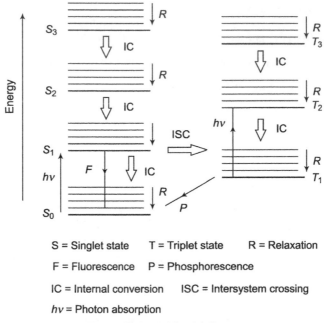

S = Singlet state   T = Triplet state   R = Relaxation

F = Fluorescence   P = Phosphorescence

IC = Internal conversion   ISC = Intersystem crossing

hv = Photon absorption

Figure 19.4   Jablonski diagram

ISC Conversion of singlet state to lower energy triplet state is called Intersystem crossing or ISC.

 IC Electrons may shift from higher energy excited states to lower energy state without changing spin multiplicity. This is known as internal conversion. Some energy is lost during this process. The process is known as relaxation.

  R For a given energy state, electrons may shift from higher level to lower one. It is known as vibrational relaxation or simply relaxation.

Approximate timescales for these different electronic transitions are shown in the Table 19.3.

Table 19.3   Timescale for different electronic transitions

Process	*Transition*	*Timescale (sec)*
Photon absorption	$S_0 \rightarrow S_n$	10^{-15}
Internal conversion	$S_n \rightarrow S_1$	$10^{-14} - 10^{-11}$
Vibrational relaxation	$S_n^* \rightarrow S_n$	$10^{-12} - 10^{-10}$
Intersystem crossing	$S_1 \rightarrow T_1$	$10^{-11} - 10^{-6}$
Fluorescence	$S_1 \rightarrow S_0$	$10^{-9} - 10^{-6}$
Phosphorescence	$T_1 \rightarrow S_0$	$10^{-3} - 10^{2}$

19.8.3 Fluorescence: Characteristic Features

1. Absorption of quanta takes place within 10^{-15} sec.
2. Singlet excited state is formed without changing spin orientation.
3. Vibrational relaxation occurs, due to which molecules are shifted to lowest vibrational state with emission of spectra in IR region. It takes only 10^{-12} sec.
4. After vibrational relaxation molecules are returned to the ground state with emission of photon in IR region. This is known as de-excitation process, which is 1000 times slower than relaxation process.
5. Emission spectrum is shifted to the higher wavelength region. It is called Stokes shift.
6. Rate of de-excitation to the ground singlet state is given by

$$\frac{dn_{exct}}{dt} = -k_F \times n_{exc}$$

k_F value is roughly proportional to the molar absorptivity as given below:

ε	k_F
10^4	10^9
10^3	10^8
10^2	10^7

Electronic transition during fluorescence is shown in the Fig. 19.5.

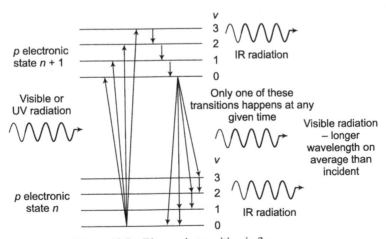

Figure 19.5 Electronic transition in fluorescence

19.8.3.1 Favourable Conditions

(a) Energy difference between excited singlet state and excited triplet state is relatively large.
(b) Energy difference between ground state and first excited singlet state is sufficiently large so that possibility of radiationless relaxation processes is negligibly small.

19.8.4 Phosphorescence: Characteristic Features

1. It occurs only when electrons from excited singlet state are shifted to excited triplet state.
2. Finally electrons return to the ground singlet state with emission of radiation.
3. Phosphorescence occurs at longer wavelength or at lower frequencies, much lower than that observed in fluorescence.
4. Rate of de-excitation to the ground singlet state is given by

$$\frac{dn_{exct}}{dt} = -k_{IS} \times n_{exc}$$

The value of k_{IS} depends on singlet-triplet gap. The smaller the gap, the larger the rate constant is. Furthermore, the value of k_{IS} can be increased with Br– and I–substitution into the double bond structure.

5. As the ground state is spin-forbidden, de-excitation from excited triplet state to singlet ground state occurs very slowly and takes longer time more than a second.
6. Phosphorescence is observed only when other deactivating processes, such as quenching, etc., are sufficiently suppressed.

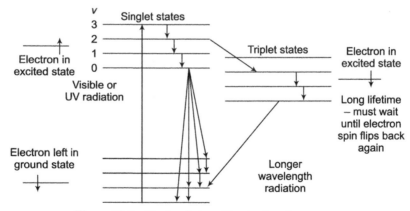

Figure 19.6 Electronic transition in phosphorescence

19.8.4.1 Favourable Conditions

(a) Energy difference between excited singlet state to excited triplet state is relatively low.
(b) The probability of radiationless transition from excited triplet state to ground state is low.
(c) Phosphorescence is generally performed in liquid nitrogen at –196°C in order to avoid deactivation process or quenching. Quenching effect is strong enough to prevent phosphorescence at room temperature.

19.8.5　Factors Influencing Photoluminescence

Structural rigidity　Rigid planar structure of organic molecule favours photoluminescence. One example are fluorene and biphenyl. Fluorene is more rigid than biphenyl and hence quantum efficiency of fluorene (1.0) is higher than that of biphenyl (0.2).

Biphenyl

Fluorene

　　Rigidity restricts rotational and vibrational motions of a molecule and hence favours π-electron transitions. Planar structure favours delocalization of π-electrons, which increases the probability of photoluminescence. However, any substitution, which increases flexibility of molecule, decreases the probability of photoluminescence.

Temperature　With increase in temperature frequency of collision among molecules increases, thereby deactivating process is dominating and hence quantum efficiency decreases. Thus probability of photoluminescence decreases with increase in temperature.

Solvent viscosity　Solvent viscosity has significant influence on fluorescence characteristics. Increase in solvent viscosity increases deactivation process and hence probability of photoluminescence decreases. Furthermore solvent, containing heavy atoms, decreases the probability of fluorescence. However, presence of heavy atoms increases spin orbital interactions and hence rate of triplet formation increases. Thus probability of phosphorescence increases.

Effect of *pH*　Molar extinction coefficient (ε) and emission intensity (I) depend on ionization and resonance characteristics of the molecule. Resonance always results a stable fixed excited state, which increases the probability of fluorescence in the UV region. One example is aniline. Aniline is sufficiently resonance stabilized and responds fluorescence but anilinium ion (formed at low pH) is not resonance stabilized and hence fluorescence characteristic is seldom observed in anilinium ion.

Aniline and its resonating structures

Anilinium ion

Effect of dissolved oxygen Oxygen molecule has paramagnetic property, due to which, dissolved oxygen accelerates the ISC process and conversion of excited molecule to triplet state. As a result of this, intensity of fluorescence decreases in aqueous solution.

19.9 CHEMILUMINESCENCE

In general electronic excitation is carried out using UV radiation. However, in many cases chemical reactions may lead to electronically excited products. Eventually excited products return to the ground state with emission of radiation in visible range. The phenomenon is known as chemiluminescence, where chemical energy is converted to radiant energy. Chemiluminescence is possible only when enthalpy change (ΔH) of the reaction is sufficiently large so that the reactant is converted to transition state, which is then shifted to electronically excited product. One example is the dissociation of cyclic peroxide into corresponding ketones. The mechanism of the reaction is shown below. ΔH of the reaction is around 90 kcal/mole, which is enough to execute electronic excitation.

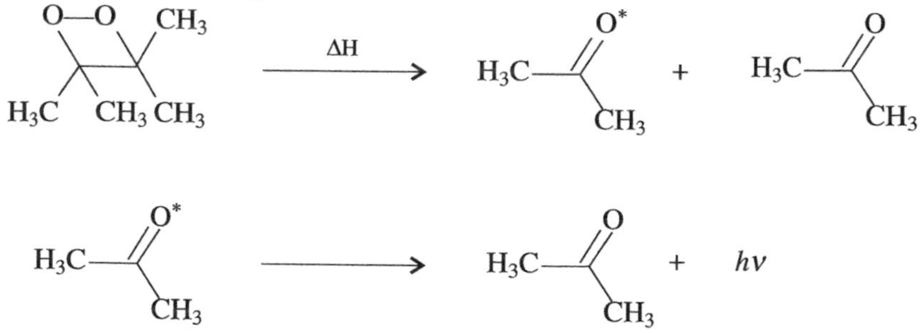

Figure 19.7 Mechanism of a chemiluminescence reaction

19.10 PHOTOCHEMICAL DISSOCIATION REACTIONS

Many organic compounds undergo dissociation reaction in presence of sunlight. These are called photochemical dissociation reactions. The mechanism of the photodissociation process was first discovered by the eminent scientist RGW Norrish.

19.10.1 Norrish Type-I and Type-II Processes

Example 1

When acetone is irradiated with a monochromatic beam of light of wavelength 313 nm, a mixture of products are obtained. Quantum yield has been found less than 0.2, indicating that possible excitation reaction is reversible and radiative (fluorescence and phosphorescence).

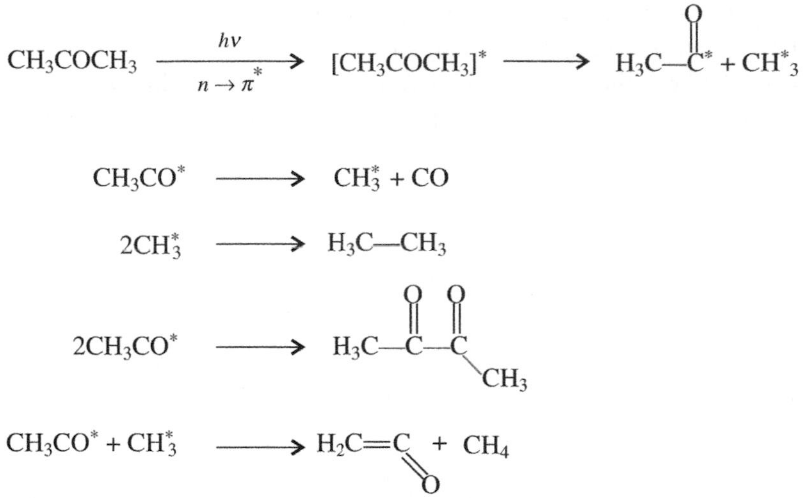

Figure 19.8   Mechanism of photodissociation reactions (Norrish type-I)

In the first step activated acetone, $[CH_3COCH_3]^*$, breaks into two radicals. Hybridisation of carbonyl carbon atom changes from sp^2 to sp. The type of cleavage is known as Norrish type-I process in the name of eminent photochemist RGW Norrish.

Example 2

In case of longer chain ketone cleavage occurs in two successive carbon atoms, next to carbonyl group, producing shorter chain ketone and alkene.

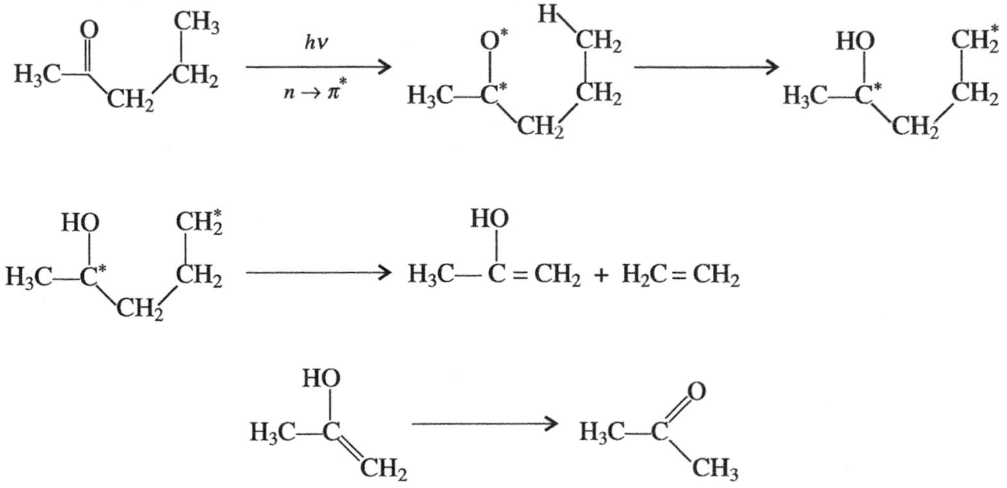

Figure 19.9   Mechanism of photodissociation reactions (Norrish type-II)

In this type of reaction hybridisation of carbonyl carbon is believed to remain unaffected. Hybridisation of carbonyl carbon atom remains unchanged to sp^2. This type of cleavage is known as Norrish type-II process.

The scientists duo RGW Norrish and G. Porter were awarded Nobel Prize in 1967 for their invaluable contribution in photochemistry.

A variety of photodissociation of different ketones are known and in all cases photodissociation process is believed to follow either Norrish type I process or Norrish type II process.

Example 3

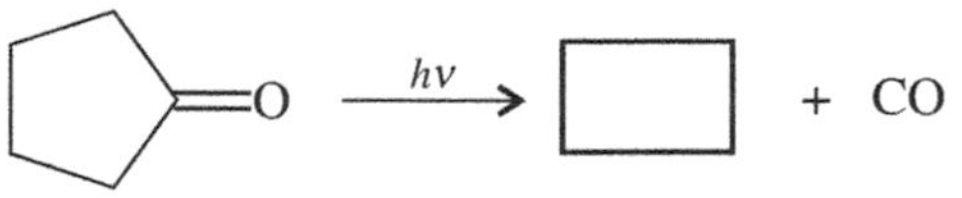

Example 4

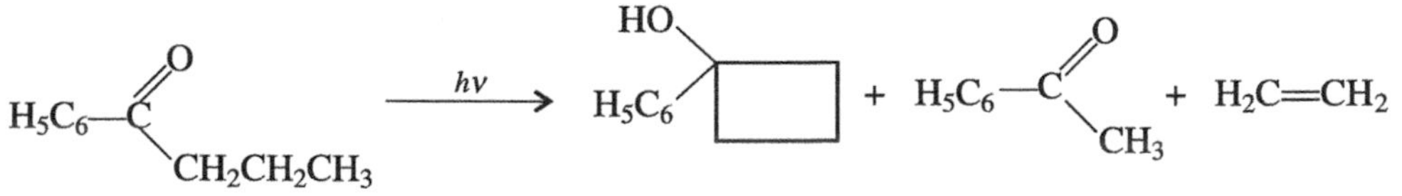

19.11 PHOTOCHEMICAL CYCLIZATION REACTIONS

Stilbene is a mixture of cis and trans isomers. On absorption of UV radiations the mixture is converted to distilbene, which is a four-member cyclic compound as shown in the Fig. 19.10. In other photochemical reaction, phenyl groups take part in chemical reaction, producing dihydroxyphenanthrene or DHP, which is then oxidised to phenanthrene as shown in the Fig. 19.10.

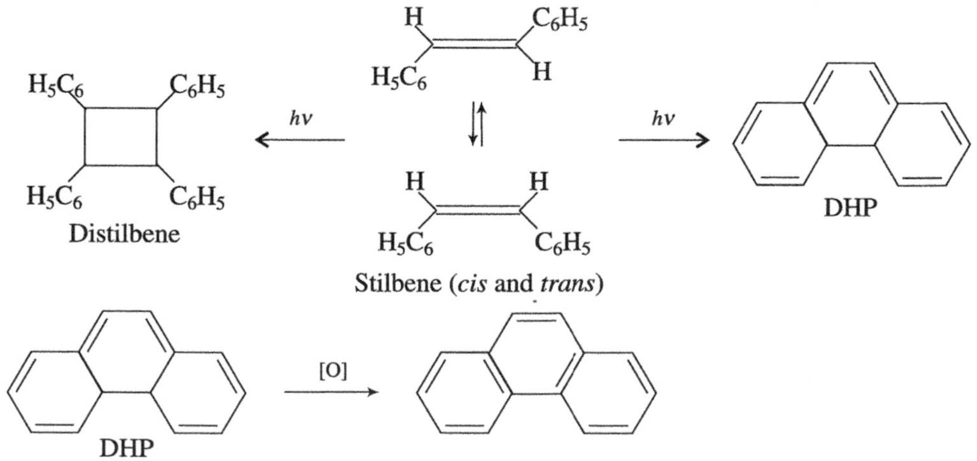

Figure 19.10   Photochemical cyclisation reactions
DHP ≡ 4a, 4b-dihydro phenanthrene

19.12 PHOTOCHEMICAL REDUCTION REACTIONS

Unlike alkyl ketones, diaryl ketones do not undergo photodissociation reactions. Instead they undergo reduction reactions in presence of UV rays. One example is the reaction between benzophenone with isopropanol. Benzophenone molecule is activated by light to form activated benzophenone, which then reacts with isopropanol to form two radicals through exchange of H-atom. Alkyl hydroxyl radical is unstable. It reduces another molecule of benzophenone to diaryl radical and itself gets oxidized to acetone. Two diaryl radicals react to from benzopinacol. The probable mechanism is shown in the Fig. 19.11.

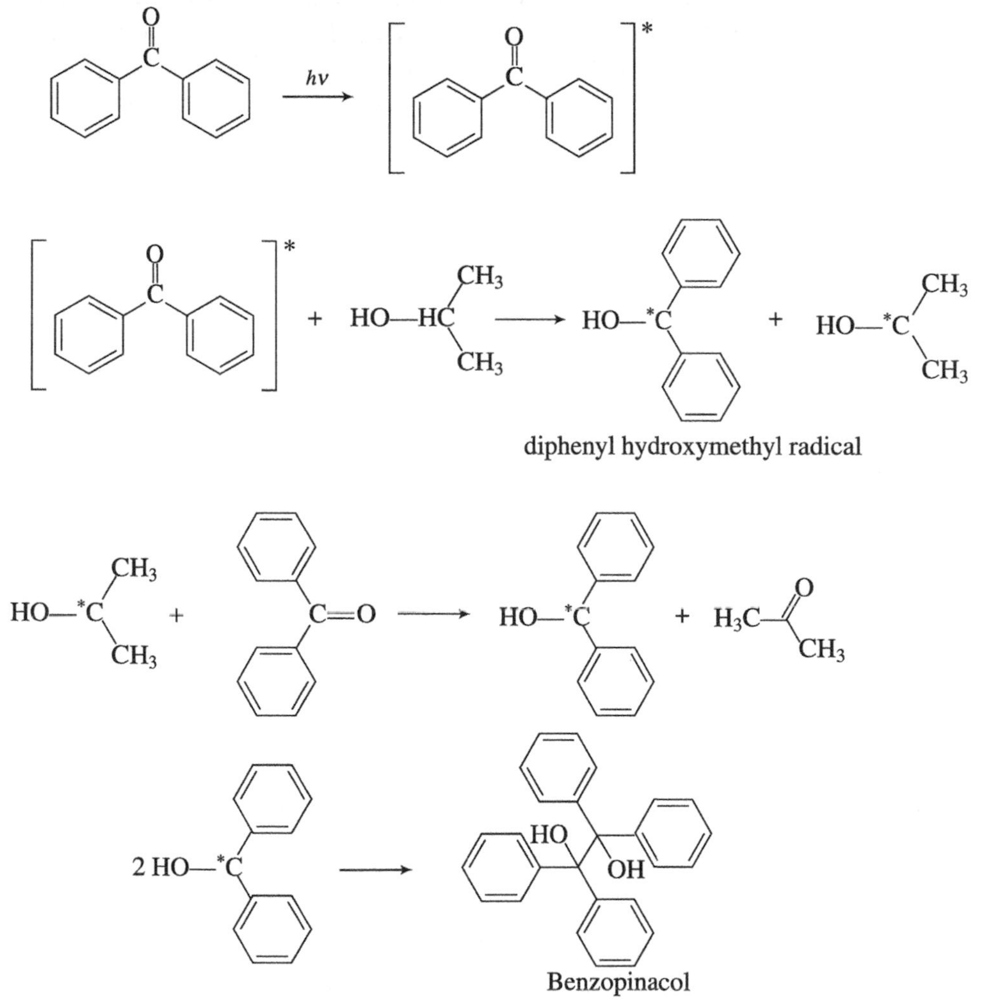

diphenyl hydroxymethyl radical

Figure 19.11 Mechanism of photochemical reduction

19.13 PHOTOSENSITIZATION REACTIONS

In some cases, reactant molecules are not sensitive enough to execute photochemical reactions but they undergo photochemical reactions in presence of particular foreign substance. This foreign substance is known as photosensitizer. At the initiation step, photosensitizer molecule, also known as **donor**, absorbs visible light and is shifted to singlet excited state. Then the molecule is further shifted from singlet excited state to triplet excited state through intersystem crossing or ISC. Life period of this excited molecule is very short. The excited molecule then transfers the energy to the reactant molecule, which is known as **acceptor**, and itself returns to the ground state. One important characteristic is that energy transfer does not change the overall spin configuration. During energy transfer process, donor molecule is transferred from excited triplet state to ground singlet state. To maintain overall spin configuration acceptor molecule, after absorbing energy from photosensitizer molecule, is transferred from ground singlet state to excited triplet state and takes part in chemical reactions. The overall mechanism is illustrated below:

$$D \xrightarrow{hv} D^{*1}$$

$$D^{*1} \xrightarrow{ISC} D^{*3}$$

$$D^{*3} + A \rightarrow A^{*3} + D$$

$$A^{*3} \rightarrow Product$$

One example is phosphorescence of naphthalene in presence of benzophenone, where the latter acts as photosensitizer or donor and naphthalene acts as acceptor. The overall photochemical reaction is shown below

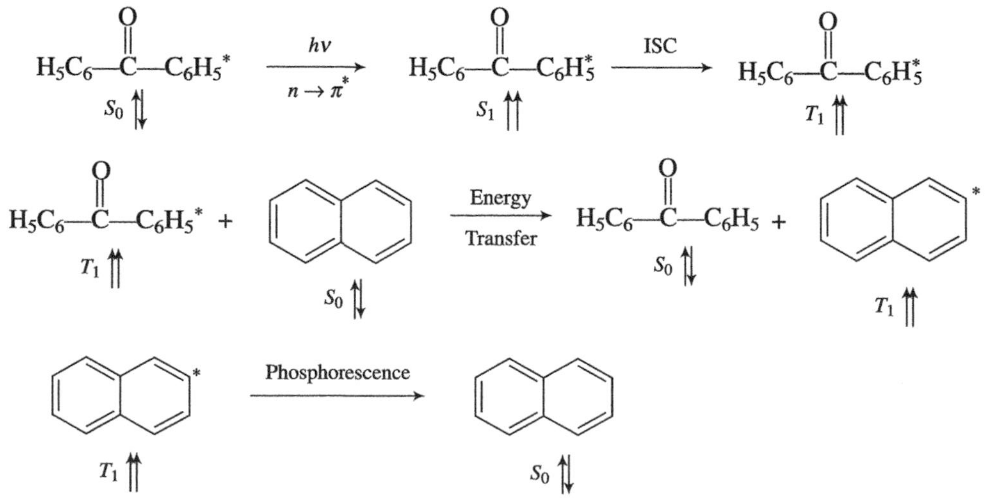

Figure 19.12 Mechanism of photosensitization reaction

19.13.1 Classification

There are two types of photosensitized reactions:

Type 1 There are some photosensitized reactions, where, either abstraction of hydrogen atom or electron transfer occurs in substrate molecule to form radicals, which then react with oxygen to form oxidation products.

Type 2 There are some other photosensitized reactions, where, energy is transferred from photosensitizer molecule to oxygen molecule, which is then transferred to excited singlet oxygen. This excited singlet oxygen molecule then reacts with reactant molecule to form oxidation products.

19.13.2 Photochemical Isomerization Reactions

During elimination of HX from alkyl or aryl halides both *cis-* and *trans-*alkenes are formed. In general *trans-*alkene is more stable and appears as major product. For example, *cis-* and *trans-*stilbene.

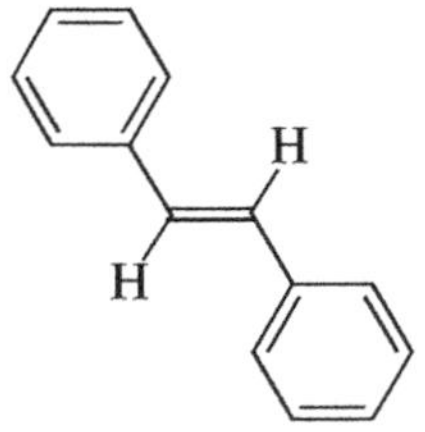

*trans-*stilbene

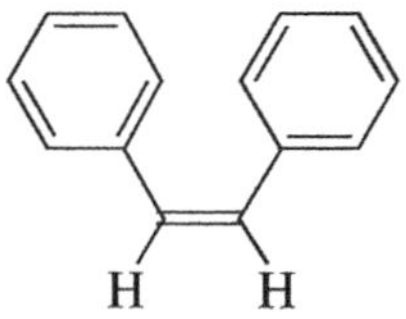

*cis-*stilbene

In another case after dehydrohalogenation Z-and E-isomers are produced as shown below.

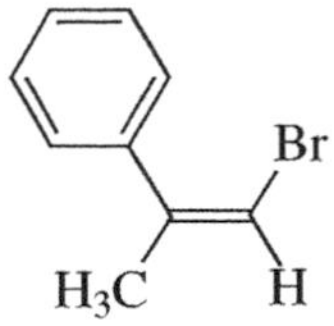

Z-isomer

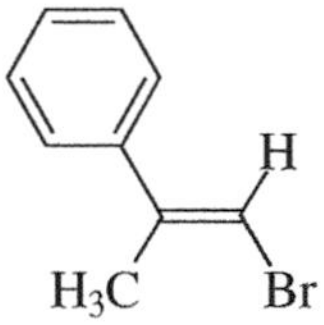

E-isomer

As sufficient steric hindrance predominates in *cis-*isomer, it is far less stable than the corresponding *trans-*isomer. Thus after dehydrohalogenation E-isomer appears as the major product (~95%). If the same reaction is carried out photochemically in presence of a sensitizer, preferably benzophenone, Z-isomer appear as the major product. The mechanism is believed to proceed as shown in the Fig. 19.13.

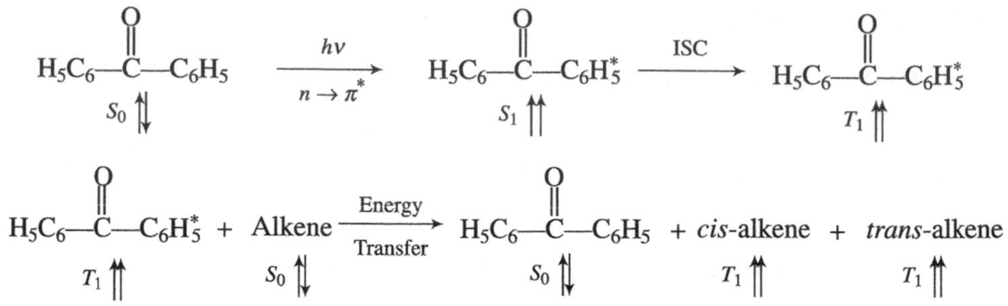

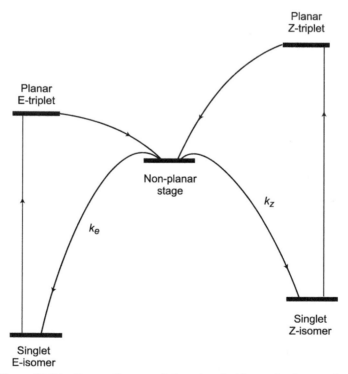

Figure 19.13   Mechanism of photochemical isomerization reaction

Both *cis-*and *trans-*triplet forms are then converted to more stable non-planar stage. The corresponding energy diagram is shown in the Fig.19.14. It clearly shows that transfer of non-planar stage to singlet Z-isomer is energetically more favourable and much faster than transfer of non-planar stage to singlet E-isomer. Thus the final product mixture contains mostly Z-isomer (85%).

Figure 19.14   Energy diagram of photochemical isomerization reaction

SOLVED PROBLEM

Problem 19.1   When propionaldehyde is irradiated with light of $\lambda = 3020$ Å, it is decomposed to form carbon monoxide according to the following reaction:

$$CH_3CH_2CHO + hv \longrightarrow C_2H_6 + CO$$

The quantum yield for the above reaction is 0.54. The light energy absorbed is 15000 erg/mole in a given time. Find the amount of carbon monoxide formed in moles in the same time.

Solution

$$1 \text{ Einstein} = E_\lambda = \frac{1.196 \times 10^{16}}{\lambda} \text{erg/mole} = \frac{1.196 \times 10^{16}}{3020} = 3.96 \times 10^{12} \text{ erg/mole}$$

$$\text{Number of Einstein absorbed} = \frac{15000}{3.96 \times 10^{12}}$$

$$\phi = 0.54 = \frac{\text{Number of moles reacted}}{\text{Number of photons absorbed}}$$

Thus, the amount of CO formed is 2.04×10^{-9} moles.

EXERCISES

1. Distinguish between photochemical and thermochemical reactions.

Photochemical reaction	*Thermochemical reaction*
This type of reaction occurs only in presence of light.	This type of reaction does not require light energy and hence occurs in presence of sunlight as well as in dark.
Molecules absorb light energy and get excited to the higher energy states.	Molecules either absorb heat energy or releases heat energy.
Molecules cross the activation energy through absorption of light energy.	Molecules cross the activation energy through collisions among themselves.
In this case active sites are formed on the molecule through electron transition and these active sites then rearrange to form products.	In this case transition state (TS) complex is formed through loose association of molecules. This TS complex eventually breaks into products.
Temperature has a very little effect on the rate of photochemical reactions.	Temperature has a significant effect on the rate of a thermochemical reaction.

ΔG value has no significance in photochemical reactions. ΔG value may be +ve or –ve.	ΔG value has great significance in thermochemical reactions. For a thermochemical reaction to occur spontaneously, ΔG value must be –ve.
Photochemical activation of molecules is highly selective. The absorbed photon excites a particular atom or group of atoms only to form active sites.	Thermochemical activation is not selective like photochemical activation. Reaction rate depends only on those excited molecules, crossing the energy barrier.

2. Calculate the energy associated with (a) one photon; (b) one Einstein of radiation of wavelength 8000 Å. Given: $h = 6.625 \times 10^{-27}$ erg-sec; $c = 3 \times 10^{10}$ cm/sec.

   [*Ans.:* (a) 2.4825×10^{-12} erg, (b) 1.4945×10^{12} erg]

3. When a substance A was exposed to light, 0.002 mole of it reacted in 1204 sec. In the same time A absorbed 2×10^6 photons of light per second. Calculate the quantum yield of the reaction.

   Given: Avogadro Number, $(N_A) = 6.023 \times 10^{23}$/g.mole.   [*Ans.:* 5×10^{11}]

4. When irradiated with light of 5000 Å wavelength, 1×10^{-4} mole of a substance is decomposed. How many photons are absorbed during the reaction if its quantum efficiency is 10. Given: Avogadro Number, $(N_A) = 6.023 \times 10^{23}$/g. mole.

   [*Ans.:* 6.023×10^{18}]

20 Nuclear Chemistry

20.1 INTRODUCTION

In 1896, French scientist Henri Becquerel first discovered radioactivity. He was experimenting fluorescence using uranium salts under sunlight and observed the fluorescence. However, he observed that X-rays were coming out from uranium salts. Then he carried out the same experiment in a dark room by covering the uranium salt using strong black paper. Even in absence of sunlight the experiment showed the emission of X-rays which clearly impressed the X-ray plates. The scientist Becquerel called it U-rays. Later on Marie Curie studied extensively on this natural emission of X-rays and coined the term "Radioactivity" to define this phenomenon. She showed that thorium is also radioactive like uranium. She along with her husband, Pierre Curie, succeeded in isolating a new element, which is million times more radioactive than uranium. She called that element "Polonium". Henri Becquerel shared the Nobel Prize with Marie Curie and Pierre Curie in 1903 in the field of physics for this discovery of radioactivity. Pierre Curie died in 1906 but Marie Curie continued her work on radioactivity and discovered a new element, known as Radium, which is 2.5 million times more radioactive than uranium. Again Marie Curie was awarded Nobel Prize in 1911 in the field of chemistry.

In 1934 Irene Curie and Frederic Joliot discovered artificial radioactivity, which was considered as a tremendous discovery since use and control of radioactivity is very much easier using artificial radioactivity. Irene Curie and Frederic Joliot were awarded Nobel prize jointly in 1935.

The famous German chemist, Otto Hahn (1879–1968), discovered nuclear fission with sufficient radiochemical proof. He is the pioneer in the field of radioactivity. He is also referred to as father of nuclear chemistry. He was awarded Nobel Prize in 1944.

Using the concept of Otto Hahn, a Manhattan project was taken up to invent atom bomb. Many world famous scientists were involved in that project.

Robert Oppenheimer, David Bohm, Leo Szilard, Eugene Wigner, Otto Frisch, Rudolf Peierls, Felix Bloch, Niels Bohr, Emilio Segre, James Franck, Enrico Fermi, Klaus Fuchs and Edward Teller.

However, Otto Hahn was deadly against to use of nuclear energy in manufacturing chemical weapons.

20.2 THE NUCLEUS

Nucleus of any atom consists mainly of protons and neutrons.

Atomic number For a given element the total number of protons present in the nucleus is called atomic number of that element.

Mass number For a given element the total number of protons and neutrons present in the nucleus is called mass number of that element.

Nuclide In nuclear science, protons and neutrons are called nucleons and mass number of an atom is often called nucleon number. Any nucleus, characterized by atomic number and nucleon number, is called nuclide. In other words, nuclide refers to a particular nucleus characterized by a definite atomic number and mass number.

Neutrino Neutrinos are similar to the electrons but unlike electrons neutrinos do not carry electric charge and hence they are not affected by the electromagnetic forces. Direction of electron path changes significantly in presence of electromagnetic forces. However, neutrinos are affected only by a "weak" sub-atomic force and hence can pass through any matter or medium for long distance without being affected by the medium. If neutrinos have mass, they also interact gravitationally with other massive particles. Neutrino is associated with electron and represented as electron neutrino, i.e., v_e.

Isotopes If an element exists in two or more different forms such that atomic number is same in all the forms but mass number differs, those different forms are called isotopes of that element. For example, protium, $_1^1H$, deuterium, $_2^1D$ and tritium, $_3^1T$ are the isotopes of hydrogen. $_{17}^{35}Cl$ and $_{17}^{37}Cl$ are the isotopes of chlorine.

Extensive properties, i.e., the properties, depending on mass or atomic mass, are different for different isotopes of a given element. Such properties are boiling point, melting point, diffusion, potential energy, kinetic energy, etc. Isotopes are placed at the same positions in the periodic table.

Isobars If atoms of two different elements have same mass number but different atomic number, atoms are called isobars with each other. One example is $_{32}^{76}Ge$ and $_{34}^{76}Se$ are isobars. Unlike isotopes, isobars are placed at different positions in the periodic table.

Isotones If atoms of two different elements have same neutron number but different atomic number, atoms are called isotones with each other. One example is $^{36}_{16}S$, $^{37}_{17}Cl$, $^{38}_{19}Ar$, $^{39}_{19}K$, $^{40}_{20}Ca$. All the nuclei possess the same number of neutron, i.e., 20. Isotones are placed at different positions in the periodic table.

20.3 NUCLEAR STABILITY

20.3.1 The Packing Fraction

Mass number of a nucleus is the sum of number of protons and neutrons. It is always a whole number. Isotopic atomic weight is the sum of masses of protons and neutrons. In most of the cases isotopic mass is less than the mass number. This difference in mass is converted to equivalent amount of energy, known as nuclear binding energy. Higher the mass difference, more stable the nucleus is. FW Aston in 1927 first proposed the term packing fraction, which is given by

$$\text{Packing Fraction} = \frac{\text{Isotopic atomic weight} - \text{Mass number}}{\text{Mass number}} \times 10^4$$

If the packing fraction is negative, some fraction of mass is converted to nuclear binding energy and the nucleus is stable. If the packing fraction is close to zero or positive, the nucleus is very much unstable. The change in packing fraction with mass number is shown in the Fig. 20.1.

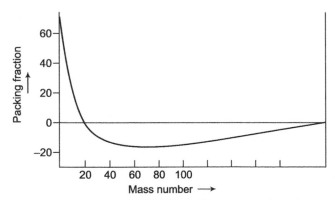

Figure 20.1 Packing fraction against mass number plot

20.3.2 Nuclear Binding Energy

A nucleus of an atom contains two types of particles, positively charged proton and neutral neutron. Two types of forces are always acting in any nucleus, one is electrostatic or electromagnetic force of repulsion, executing between two protons. Another is binding force, which strongly binds the nucleons. This binding force mainly arises due to continuous conversion of protons to neutrons and vice versa. In order to form a stable nucleus binding force must be sufficiently higher than repulsive electrostatic force so that considerable amount of energy, known as binding energy, is released. Thus a nucleus consisting of two protons only, cannot exist because only repulsive force exists in that nucleus.

The mass of nucleus should be the total mass of protons and neutrons but in actual practice the mass of a nucleus is always less than the total mass of protons and neutrons. This difference in mass is called mass defect (Δm), which is converted to energy in accordance with Einstein's relationship

$$E = \Delta mc^2 \tag{20.1}$$

This energy, E, is used to bind the nucleons tightly, forming a stable nucleus. Hence, it is called nuclear binding energy. Higher the mass defect, higher is the nuclear binding energy and more stable the nucleus is.

Nuclear binding energy is far greater than the binding energy between nucleus and electron. For example, in case of H-atom, electrostatic binding energy, calculated using Bohr's concept, is only 13.6 eV but nuclear binding energy of an α-particle is 28.2862 MeV. So, binding energy per nucleon is 7.07155 MeV.

Nuclear binding energy can be plotted against atomic number of atoms. It is shown in the Fig. 20.2. Initially the binding energy increases with increase in mass number and reaches a maximum with mass number 62 (the element Ni). Then it decreases slowly to mass number 118 but the decrease is fast after the mass number 118. Thus it may be considered that all elements up to mass number 118 are stable.

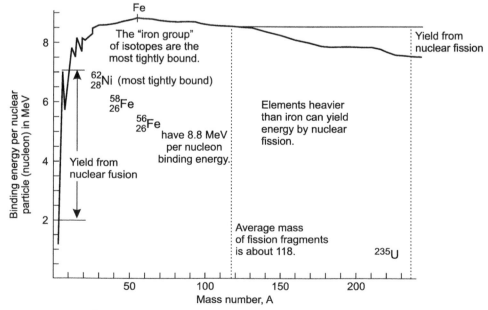

Figure 20.2 Nuclear binding energy versus atomic number plot

20.3.2.1   Calculation of Binding Energy

$$1\,eV = 1.602 \times 10^{-19}\ J.\ \ Or,\ 1\,J = 6.242 \times 10^{18}\ eV$$

Suppose, $\Delta m = 1$ kg.mole, which is equivalent to the mass of Avogadro number (6.023×10^{26}) number of atoms. Thus the mass of one atom is 1.6603×10^{-27} kg. Using the Einstein's relationship the amount of energy associated with the mass defect 1.6603×10^{-27} kg is given by

$$E = 1.66 \times 10^{-27} \times (3 \times 10^8)^2\ J = 14.94 \times 10^{-11}\ J = 93.25 \times 10^7\ eV \approx 932\ MeV$$

20.3.2.2   Atomic Nucleus

Mass of a proton (M_p)   It is the mass of a proton and its value is 1.673×10^{-27} kg.

Mass of a neutron (M_n)   It is the mass of a neutron and its value is 1.675×10^{-27} kg.

Mass of an electron (M_e)   It is the mass of an electron and its value is 9.109×10^{-31} kg.

Einstein's formula of conversion of mass and energy $E = mc^2$

$$1\,eV = 1.6029 \times 10^{-19}\ J$$

$$E_p = M_p \times c^2 = 1.673 \times 10^{-27} \times (2.998 \times 10^8)^2\ J = 1.5037 \times 10^{-10}\ J = 938.61\ MeV$$

$$E_n = M_n \times c^2 = 1.675 \times 10^{-27} \times (2.998 \times 10^8)^2\ J = 1.5055 \times 10^{-10}\ J = 939.73\ MeV$$

$$E_e = M_e \times c^2 = 9.109 \times 10^{-31} \times (2.998 \times 10^8)^2\ J = 8.1872 \times 10^{-14}\ J = 0.511\ MeV$$

Mass of an atom or Atomic Mass (AM)

$$AM = ZM_H + NM_n - BE(M, Z, N)$$

M = Mass number, Z = Atomic number, N = Number of neutrons.
BE(M, Z, N) = The mass converted to binding energy.

Hydrogen has three isotopes namely, Hydrogen ($M = 1$), Deuterium ($M = 2$) and Tritium ($M = 3$). Tritium is a radioactive isotope with half-life 12.3 years. It is not found in the nature.

Tin (*Sn*) has 10 stable isotopes $M = 112, 114, 115, 116, 117, 118, 119, 120, 122, 124.$

20.3.3   Neutron:Proton (*n:p*) Ratio

Stability of a nucleus depends on *n:p* ratio, where, *n* and *p* indicate the number of neutrons and number of protons respectively. Experimental studies show that nuclides with atomic number up to 83 are very stable. The *n-p* curve of several nuclides is shown in the Fig. 20.3. The curve shows a band of stability comprising of nuclides of atomic number up to 83. Furthermore, it is observed that stable nuclides (up to atomic number 20) are formed with *n:p* = 1:1. However, the ratio increases with increase in atomic number above 20 and finally the ratio reaches around 1.5 for atomic number 83. Nuclides with atomic number 84 and above are relatively unstable and atomic stability decreases as the atomic number increases. Two observations are important in this regard:

(i) Nuclei with even number of protons and neutrons are more stable than those with odd number of protons and neutrons.

(ii) Nuclei with proton numbers, 2, 8, 20, 28, 50 and 82 are the most stable nuclei. These numbers are called magic numbers.

The other nuclides, residing outside this band of stability, are unstable and these nuclides have a tendency to reach within this band of stability through emission of particles and γ-rays. The phenomenon is known as radioactivity and the nuclides are called radioactive nuclides. In 1896, the French physicist Henri Becquerel discovered radioactivity in Uranium. He observed that uranium containing compounds emit different types of rays and itself gets converted to other nucleus. Later on it has been established that following particles and rays are emitted from a radioactive element.

(i) α-particle, (ii) β-particle, and (iii) positron and γ-rays.

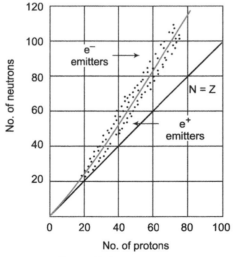

Figure 20.3 $n - p$ curve

20.4 NATURAL RADIOACTIVITY

In case of heavy nuclei n/p ratio is far greater than 1 and hence these nuclei are very much unstable. Thus to achieve stability these nuclei continuously release characteristic particles and γ-rays to form lighter and comparatively stable nuclei. The phenomenon is known as natural radioactivity. The characteristic particles, released due to this natural radioactivity are: (i) α-particle, (ii) β-particle and (iii) positron. γ-rays are always associated with this emission of characteristic particles. The products of nuclear decay are tabulated in Table 20.1.

Table 20.1 Products of nuclear decay

Particle	Symbol
Neutron	$_0^1\text{n}$
Proton	$_1^1\text{H}$
Electron	$_{-1}^0\text{e}$
α-particle	$_2^4\text{He}$
β-particle	$_{-1}^0\text{e}$
Positron	$_{+1}^0\text{e}$

20.4.1 Characteristics of Particles and Rays

α-particle It is the helium nucleus and designated as $_2^4\text{He}$. It is positively charged and total number of charge is +2. It is the heaviest particle, emitted by a radioactive element. α-particle is very large with very low penetrating ability. α-particles, when ejected from naturally occurring radioactive atoms, move with a velocity of the order of (1/20)th the velocity of light and have energies in the range of 4 to 9 MeV. It can be deflected by a thin sheet of paper also. In air, α-particle can travel up to several cm but it can penetrate Al foil with thickness $\leq 10^{-3}$ mm. Human skin is not affected by α-particle but α-particle can damage internal organs because of its size. α-particle has poor ionizing power. In presence of an electric field α-particle is deflected in a direction opposite to that of β-particle. With the emission of α-particle, position of the nucleus is shifted left two positions in the periodic table.

β-particle It is nothing but electron and designated as $_{-1}^0\text{e}$. It is negatively charged and total number of charge is -1. It has velocity (9/10)th the velocity of light. It is much smaller than α-particle but has significant penetrating power. A piece of Al foil is required to deflect β-particle. In air β-particle can travel up to 10 m but it can penetrate Al foil with thickness ≤ 0.5 mm. In presence of an electric field, β-particle is deflected in a direction opposite to that of α-particle. It has strong ionizing power, much stronger than that of α-particle. It can easily penetrate through human skin, causing severe damage to skin and internal organs.

Positron This subatomic particle was discovered in 1932 by American scientist Carl. D. Anderson, who was awarded Nobel prize in Physics in 1936. Positron is similar to β-particle but positively charged. Total number of charge is +1. It is designated as $_{+1}^0\text{e}$. The properties of positron is similar to those of β-particle. In presence of an electric field, positron particle is deviated in a direction, same as that of α-particle.

γ-rays   It is an electromagnetic radiation, generally emitted along with a charged particle. It is moving with a velocity same as that of light. It does not deviate in presence of an electric field. It has very high penetration power. γ-rays can penetrate Al foil with thickness 5–10 cm.

20.4.2   Group Displacement Law

The law was first established by Fajan, Russel and Soddy in 1913. According to this law, position of radionuclide in Periodic Table changes on emission of α- and β-particles.

On emission of α-particle only, the parent radionuclide is shifted two columns left in the Periodic Table with decrease in mass number by 4 units. One example is

$$^{238}_{92}U \xrightarrow{\alpha} {}^{234}_{90}Th$$

On emission of β-particle only, the parent radionuclide is shifted one column right in the Periodic Table without changing the mass number. One example is

$$^{214}_{82}Pb \xrightarrow{\beta} {}^{214}_{83}Bi$$

On emission of positron particle only, the parent radionuclide is shifted one column left in the Periodic Table without changing the mass number. One example is

$$^{214}_{83}Bi \xrightarrow{{}^{0}_{+1}e} {}^{214}_{82}Pb$$

20.5   RADIOACTIVE DECAY KINETICS

20.5.1   Decay of Radioactive Element Through One Reaction

The process is described as follows:

$$A \longrightarrow B$$

In radioactive decay, only radioactive atom is involved. The experimental studies show that the decay process occurs continuously and takes infinite time to complete. Thus the decay process resembles to 1st order kinetics.

Let us consider, N_0 and N are the initial number of radioactive atoms and the number of radioactive atoms present at time t respectively. According to 1st order kinetics, we can write that

$$-\frac{dN}{dt} \propto N. \ \text{ Or, } \ -\frac{dN}{dt} = \lambda N. \ \text{ Or, } \ \frac{dN}{dt} = -\lambda N$$

where, λ is known as radioactive decay constant. On integration of the above equation using the boundary condition $N = N_0$ when $t = 0$, we get

$$\lambda t = \ln\frac{N_0}{N} = 2.303\log\frac{N_0}{N}. \ \text{ Or, } \ N = N_0 e^{-\lambda t}. \ \text{ Or, } \ t = \frac{2.303}{t}\log\frac{N_0}{N} \qquad (20.2a)$$

The corresponding decay curve is shown in the Fig. 20.4.

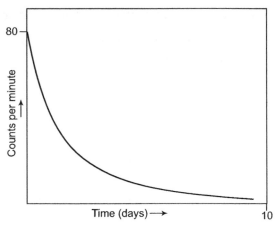

Figure 20.4 Radioactive decay curve

Half-life period of a radioactive element, $t_{1/2}$, is given by

$$t_{1/2} = \frac{2.303}{\lambda} \log 2 = \frac{0.693}{\lambda} \qquad (20.2b)$$

$t_{1/2}$ values of different radioactive elements are listed in the following Table 20.2.

Table 20.2 $t_{1/2}$ values of common radioactive isotopes

Nuclide	$t_{1/2}$	Type of change	Nuclide	$t_{1/2}$	Type of change
Rb – 87	5.7×10^{10} years	$_{-1}^{0}e$	Fe – 59	45 days	$_{-1}^{0}e$
Th – 232	1.39×10^{10} years	$_{2}^{4}He$	P – 32	14.3 days	$_{-1}^{0}e$
U – 238	4.51×10^{9} years	$_{2}^{4}He$	Ba – 131	11.6 days	$_{-1}^{0}e$ capture and $_{+1}^{0}e$
U – 235	7.13×10^{9} years	$_{2}^{4}He$	I – 131	8.06 days	$_{-1}^{0}e$
Pu – 239	2.44×10^{4} years	$_{2}^{4}He$	Rn – 222	3.82 days	$_{2}^{4}He$
C – 14	5730 years	$_{-1}^{0}e$	Au – 198	2.70 days	$_{-1}^{0}e$
Ra – 226	1622 years	$_{2}^{4}He$	Kr – 79	34.5 hrs	$_{-1}^{0}e$ capture and $_{+1}^{0}e$
Cs – 133	30 years	$_{-1}^{0}e$	C – 11	20.4 min	$_{+1}^{0}e$
Sr – 90	29 years	$_{-1}^{0}e$	F – 17	66 sec	$_{+1}^{0}e$
H – 3	12.26 years	$_{-1}^{0}e$	Po – 213	4.2×10^{-6} sec	$_{2}^{4}He$
Co – 60	5.26 years	$_{-1}^{0}e$	Be – 8	1×10^{-16} sec	$_{2}^{4}He$
Po – 210	138 days	$_{2}^{4}He$	Th – 231	24 hr	$_{2}^{4}He$
Th – 234	23 days	$_{2}^{4}He$			

20.5.2   Decay of Radioactive Element by Two or More Processes

In many cases one radionuclide undergoes two parallel decay processes, where two different radionuclides are formed. It is well described below.

λ_1 and λ_2 are the decay constants for the processes $A \rightarrow B$ and $A \rightarrow C$ respectively. Thus according to radioactive decay kinetics, we have

$$-\frac{dN}{dt} = N\lambda_1 + N\lambda_2 = N(\lambda_1 + \lambda_2) = N\lambda. \quad \text{Or,} \quad \lambda = \lambda_1 + \lambda_2$$

Using the expression of half-life period, we have

$$\frac{0.693}{t_{1/2}} = \frac{0.693}{(t_{1/2})_1} + \frac{0.693}{(t_{1/2})_2}. \quad \text{Or,} \quad \frac{1}{t_{1/2}} = \frac{1}{(t_{1/2})_1} + \frac{1}{(t_{1/2})_2}. \quad \text{Or,} \quad t_{1/2} = \frac{(t_{1/2})_1 \times (t_{1/2})_2}{(t_{1/2})_1 + (t_{1/2})_2}$$

where, $(t_{1/2})_1$ and $(t_{1/2})_2$ are called **partial half-lives** for the processes $A \rightarrow B$ and $A \rightarrow C$ respectively.

If three simultaneous decay processes occur from a radionuclide, the above equation becomes

$$\frac{1}{t_{1/2}} = \frac{1}{(t_{1/2})_1} + \frac{1}{(t_{1/2})_2} + \frac{1}{(t_{1/2})_3} \tag{20.3a}$$

$$\text{Or,} \qquad t_{1/2} = \frac{(t_{1/2})_1 \times (t_{1/2})_2 \times (t_{1/2})_3}{(t_{1/2})_1 \times (t_{1/2})_2 + (t_{1/2})_2 \times (t_{1/2})_3 + (t_{1/2})_3 \times (t_{1/2})_1} \tag{20.3b}$$

20.5.3   Radioactive Equilibrium

In a radioactive series one radionuclide decays to its daughter radionuclide, which further decays to produce another radionuclide etc. Part of the series is shown below

$$A \rightarrow B \rightarrow C.$$

Initially rate of disintegration of radionuclide A is different from that of radionuclide B but after some time period, disintegration rates of both radionuclides are same. N_A and N_B are the number of atoms of radionuclides, A and B respectively present at equilibrium. The phenomenon is known as radioactive equilibrium. At equilibrium, we can write that

$$-\frac{dN_A}{dt} = -\frac{dN_B}{dt}. \quad \text{Again,} \quad -\frac{dN_A}{dt} = \lambda_A N_A, \quad -\frac{dN_B}{dt} = \lambda_B N_B$$

$$\text{Or,} \qquad \lambda_A N_A = \lambda_B N_B. \quad \text{Or,} \quad \frac{\lambda_A}{\lambda_B} = \frac{N_B}{N_A} = \frac{t_{1/2(B)}}{t_{1/2(A)}} \tag{20.4}$$

In a radioactive series if disintegration rates of three consecutive radionuclides are same, we can write that

$$A \rightarrow B \rightarrow C \rightarrow D.$$

$\lambda_A N_A = \lambda_B N_B = \lambda_C N_C$, where, λ_A, λ_B and λ_C are the disintegration constants of radionuclides, *A*, *B* and *C* respectively and N_A, N_B and N_C are the number of atoms of radionuclides, *A*, *B* and *C* respectively present at equilibrium.

20.6 RADIOCARBON DATING

Normally in the biosphere 99% carbon exists as $^{12}_{6}C$, almost 1% carbon exists as $^{13}_{6}C$ and a trace amount of carbon, around 1 *ppt* or 1 part per trillion, exists as $^{14}_{6}C$. This ^{14}C is used to determine the age of wood and fossils. ^{14}C is sufficiently radioactive with half-life period 5730 years. Radioactive ^{14}C is produced according to the following equation

$$^{14}_{7}N + ^{1}_{0}n \longrightarrow ^{14}_{6}C + ^{1}_{1}H$$

^{14}C is readily oxidized to $^{14}CO_2$, which is then absorbed by plants. Human beings and other animals consume plants and hence ^{14}C is spread in all living beings. The ratio of ^{14}C to ^{12}C remains constant in all living beings and is given below

$$\frac{^{14}C}{^{12}C} = 10^{-12}$$

All organisms absorb carbon from their environment and maintain the same ratio of $^{14}C : ^{12}C$ as in the atmosphere. However, after death plants and animals stop absorbing fresh carbon from their environment but ^{14}C continues to decay to nitrogen according to the following equation

$$^{14}_{6}C \longrightarrow ^{14}_{7}N + ^{0}_{-1}e$$

So the ratio $^{14}C : ^{12}C$ decreases continuously with time. Thus by measuring ^{14}C concentration of a fossil and using equation (20.2a) we can easily calculate the age of the fossil using the $t_{1/2}$ value of ^{14}C, i.e., 5730 years. For example, for a given organism, if ^{14}C concentration is 0.5 ppt, the organism is assumed to have died about 5730 years ago.

20.6.1 Age of Earth

During the formation of the Earth, it is assumed that rock did not contain $^{206}_{82}Pb$ but contains radioactive $^{238}_{92}U$. That *U* is decaying since the formation of the Earth and converted to stable nucleus Pb. The overall decay is

$$^{238}_{92}U \longrightarrow ^{206}_{82}Pb + 8\,^{4}_{2}He + 6\,^{0}_{-1}e$$

$$t_{1/2} = 4.5 \times 10^{9} \text{ years}$$

For any given rock, weight ratio of $^{238}_{92}U$ and $^{206}_{82}Pb$ is measured. Using the knowledge of $t_{1/2}$ of $^{238}_{92}U$, age of the Earth can easily be determined.

20.7 RADIOACTIVE SERIES

One radioactive nuclide undergoes radioactive decay process to produce other nuclide, which is also radioactive and hence undergoes radioactive decay process to produce third radionuclide. The process continues till a stable nuclide is produced. So radioactive decay is a chain process, producing a series of elements. This series is known as radioactive decay series or nuclear disintegration series. Three such series are continuously occurring in the nature.

(i) Decay of $^{238}_{92}U$ to the stable nucleus, $^{206}_{82}Pb$.

(ii) Decay of $^{235}_{92}U$ to the stable nucleus, $^{207}_{82}Pb$.

(iii) Decay of $^{232}_{90}Th$ to the stable nucleus, $^{208}_{82}Pb$.

The decay series of $^{238}_{92}U$ is shown in the Figs. 20.5 and 20.6.

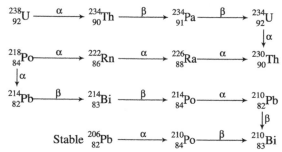

Figure 20.5 Radioactive decay series of *U-238*, showing stepwise α- and β-emissions

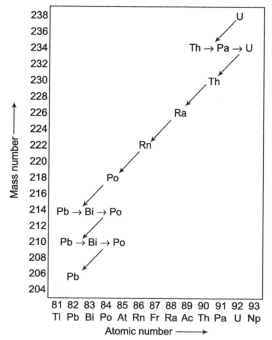

Figure 20.6 Nuclear disintegration of *U-238*, showing change in mass number and atomic number

20.8 NUCLEAR FISSION AND FUSION

20.8.1 Nuclear Fission

If a heavy radioactive nucleus is bombarded with a subatomic particle like neutron, the radioactive nucleus due its instability breaks into two stable nuclei and some more neutrons. The resultant mass defect is converted to energy according to Einstein's formula $E = \Delta mc^2$. The phenomenon is known as nuclear fission. The neutrons, thus produced due to this fission, attack other radioactive nuclei resulting fission and hence the fission process continues. The fission process stops automatically when all radioactive nuclei undergo fission. The phenomenon is known as radioactive chain reaction. One example is the fission of U_{92}^{235}. The fission process is shown in the Fig. 20.7. This fission process is executed in nuclear power plants using slow neutron to generate power.

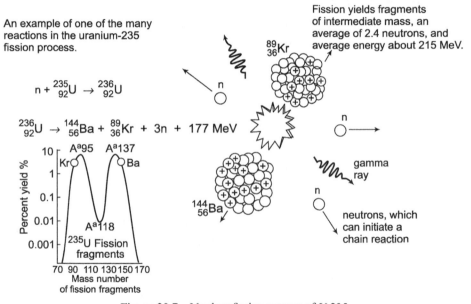

Figure 20.7 Nuclear fission process of *U-235*

In nuclear fission reaction, U_{92}^{235} is bombarded with a slow neutron in a nuclear reactor, as a result of which, heavy nucleus U_{92}^{235} splits into two smaller fragments, $_{36}^{89}Kr$ and $_{56}^{145}Ba$ along with three neutrons. Each of these three neutrons strikes another U_{92}^{235} nucleus and the chain reaction continues. The mechanism is illustrated in the Fig. 20.8. Each successful nuclear reaction between U_{92}^{235} and n_0^1 produces huge amount of energy, which can be used to generate electricity.

So the fission of U_{92}^{235} releases huge amount of energy, around 180 MeV per nuclide = 7.4×10^7 kJ/g of $_{92}^{235}U$, which is equivalent to explosion of 3×10^4 kg of TNT.

Nuclear power reactors use fission of *U* to generate electricity. Fission of *Pu* (mainly) and *U* are used to manufacture nuclear weapons.

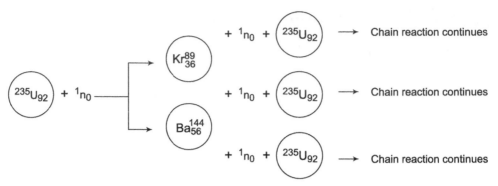

Figure 20.8   Chain mechanism of nuclear fission of U-235

20.8.2   Nuclear Fusion

Nuclear fusion was discovered by Otto Hahn. In this method unstable small nuclei can be fused together to form a stable nucleus. The most important example of this category is hydrogen fusion reaction, also known as deuterium cycle, consisting of a series of nuclear fusion reactions as shown below:

$$_1^2H + {}_1^2H \longrightarrow {}_2^3He + {}_0^1n + 3.27 \text{ MeV}$$

$$_1^2H + {}_1^2H \longrightarrow {}_1^3H + {}_1^1H + 4.03 \text{ MeV}$$

$$_1^2H + {}_1^3H \longrightarrow {}_2^4He + {}_0^1n + 17.59 \text{ MeV}$$

$$_1^2H + {}_2^3He \longrightarrow {}_2^4He + {}_1^1H + 18.30 \text{ MeV}$$

The above reactions occur at very high temperature, 40×10^6 K. This high temperature is required to overcome the Columbic repulsion barrier. However, in the Sun, tremendous gravitational force exists and this gravitational force helps nuclei to come close to each other, leading to natural nuclear fusion.

20.8.2.1   Advantages of Nuclear Fusion

(i) Amount of energy released per gm. of material during fusion reaction is much higher than that of nuclear fission reaction.

(ii) Waste of nuclear fission is very much harmful to environment. However, in *D-T* nuclear fusion the product is He nucleus, which is safe and non-toxic.

(iii) Fuel for nuclear fusion is abundant or can be made easily. Fuel for nuclear fission is very much limited.

(iv) As nuclei are fused together, chain reaction does not occur in nuclear fusion. So nuclear fusion reaction is easy to control.

(v) Electric power, generated using nuclear fusion reaction, is sufficiently cheap.

(vi) The only product of *D-T* nuclear fusion is He nucleus. No greenhouse gases are formed. So nuclear fusion reaction is environmentally friendly.

20.8.2.2 Disadvantages of Nuclear Fusion

(i) Huge amount of energy is required to overcome repulsion between nuclei. Thus very high temperature, around 40×10^6 K, is required to initiate fusion reaction.

(ii) This high temperature requirement can be achieved by using an atomic bomb to initiate the fusion process. Installation of nuclear fusion plant is very costly. Thus this fusion method is seldom used for controlled power generation.

(iii) There is no structural material, which can withstand this enormous temperature required for fusion reaction. So it is practically impossible to develop nuclear reactor for fusion reaction.

Differences between nuclear fission and nuclear fusion are discussed in the Table 20.3.

Table 20.3   Differences between nuclear fission and fusion

Nuclear fission	*Nuclear fusion*
Heavy nucleus undergoes nuclear fission reaction to produce two or more stable nuclei.	Two or more very light nuclei are fused together to form a stable nucleus.
Nuclear fission is not a natural process.	Nuclear fusion is a natural process in some stars like Sun.
Little amount of energy is enough to execute a nuclear fission reaction.	Enormous amount of energy is required to overcome electrostatic repulsions among nucleons.
Energy released in a nuclear fission reaction is million times greater than that of a chemical reaction.	Energy released during nuclear fusion reaction is three to four times greater than that released in a nuclear fission reaction.
Critical mass of heavy nucleus and high speed neutrons are the basic requirements to execute a nuclear fission reaction.	High density and high temperature are the basic requirements to execute a nuclear fusion reaction.
Nuclear power plants are based on nuclear fission reaction.	Nuclear fusion is very difficult to execute and hence generally not used in nuclear power plants.
Uranium is the primary fuel in nuclear power plants.	Deuterium and tritium are the primary fuel in nuclear fusion reaction.

20.9 ARTIFICIAL RADIOACTIVITY

Frederic Joliot and Irene Curie first discovered artificial radioactivity in 1934. They observed that a stable nucleus, when bombarded with subatomic particles, is converted to radioactive element. This phenomenon is known as artificial radioactivity. Some examples are shown below

(i) $^{10}_{5}\text{B} + ^{10}_{2}\text{He} \longrightarrow ^{13}_{7}\text{N*} + ^{1}_{0}\text{n}$. The radioactive nitrogen nucleus then decays to ^{13}C with half-life 10.1 min.

$$^{13}_{7}\text{N*} \longrightarrow ^{13}_{6}\text{C} + ^{0}_{+1}\text{e}$$

(ii) $^{27}_{13}\text{Al} + ^{4}_{2}\text{He} \longrightarrow ^{30}_{15}\text{P*} + ^{1}_{0}\text{n}$. The radioactive phosphorus nucleus then decays to ^{30}Si with half-life 3 min.

$$^{30}_{15}\text{P*} \longrightarrow ^{30}_{14}\text{Si} + ^{0}_{1}\text{e}$$

(iii) $^{14}_{7}N + ^{4}_{2}He \longrightarrow ^{17}_{8}O^* + ^{1}_{1}H$,

(iv) $^{14}_{7}N + ^{1}_{0}n \longrightarrow ^{14}_{6}C^* + ^{1}_{1}H$

Transuranium elements were discovered using artificial radioactivity. Glenn Seaborg and his team of scientists at the Lawrence Laboratory at the University of California, Berkeley, created a number of new elements, some of which—berkelium, californium, lawrencium—have been named in honour of their work.

(i) $^{238}_{92}U + ^{1}_{0}n \longrightarrow ^{239}_{92}U \longrightarrow ^{239}_{93}Np + ^{0}_{-1}e$,

(ii) $^{239}_{93}Np \longrightarrow ^{239}_{94}Pu + ^{0}_{-1}e$

(iii) $^{239}_{94}Pu + ^{4}_{2}He \longrightarrow ^{242}_{96}Cm + ^{1}_{0}n$

(iv) $^{244}_{96}Cm + ^{4}_{2}He \longrightarrow ^{244}_{96}Bk + 2^{1}_{1}H + 2^{1}_{0}n$

(v) $^{238}_{92}U + ^{12}_{6}C \longrightarrow ^{246}_{98}Cf + 4^{1}_{0}n$

(vi) $^{246}_{98}Cf + ^{10}_{5}B \longrightarrow ^{251}_{103}Lr + 5^{1}_{0}n$

(vii) $^{239}_{94}Pu + ^{1}_{0}n \longrightarrow ^{239}_{95}Am + ^{0}_{-1}e$

20.10 APPLICATIONS OF RADIOACTIVITY

20.10.1 Medical Applications

$^{60}_{28}Co$ is a well-known γ-emitter and hence is used to irradiate tumors.

^{192}Ir is implanted in tumors. γ-radiation, emitted from the radionuclide, kills the tumor cells. It is encapsulated in *Pt* to contain α- and β-rays.

^{210}Th is a well-known γ-emitter and is intentionally concentrated in the heart in order to receive the updating conditions of heart and arteries.

^{131}I is absorbed by thyroid and the activity of thyroid can easily be known through this radionuclide.

alpha emission by polonium-210, used in radiation therapy.

$$^{210}_{84}Po \longrightarrow ^{206}_{82}Pb + ^{4}_{2}He$$

beta emission by gold-198, used to assess kidney activity.

$$^{198}_{79}Au \longrightarrow ^{198}_{80}Pb + ^{0}_{-1}e$$

positron emission by nitrogen-13, is used in making brain, heart, and liver images.

$$^{13}_{7}N \longrightarrow ^{13}_{6}C + ^{0}_{+1}e$$

electron capture by gallium-67, is used to do whole body scans for tumors.

$$^{67}_{31}Ga + ^{0}_{-1}e \longrightarrow ^{67}_{30}Zn$$

20.10.2 PET (Positron Emission Tomography)

Known positron emitters ^{11}C, ^{18}F, ^{15}O, ^{13}N.

Glucose sample is labelled with a known positron emitter. So positron emission is detected and 3-D images of body organs are stored and interpreted using computer imaging.

20.10.2.1 MRI (Magnetic Resonance Imaging)

In this method, energy difference is developed when nuclear spin flips from one orientation to another in presence of a strong magnetic field. In medicine a person is used as a "sample" and inserted in a strong magnetic field with field strength 30000 Gauss. Corresponding images of body organs are stored and interpreted using computer imaging.

20.10.3 Agricultural Applications

(a) Controlling insects without using pesticides.

(b) Cobalt-60 emits gamma rays that sterilize male insects and reduce insect population.

(c) Gamma irradiation of processed food also destroys microorganisms.

(d) Cobalt-60 irradiation destroys parasites in pork (trichinosis) and chicken (salmonella).

(e) Gamma irradiation also increases shelf life without using preservatives.

20.10.4 Other Applications

Nuclide	*Nuclear change*	*Application*
Ar – 41	$^{0}_{-1}e$ emission	Measure flow of gases from smokestacks
Ba – 131	$^{0}_{-1}e$ capture	Detect bone tumors
C – 11	$^{0}_{+1}e$ emission	PET brain scan
C – 14	$^{0}_{-1}e$ emission	Archaeological dating
Cs – 133	$^{0}_{-1}e$ emission	Radiation therapy
Co – 60	$^{0}_{-1}e$ emission	Cancer therapy
Cu – 64	$^{0}_{-1}e$ emission, $^{0}_{+1}e$ emission and $^{0}_{-1}e$ capture	Lung and liver disease diagnosis
Cr – 51	$^{0}_{-1}e$ capture	Determine blood volume and red blood cell lifetime; diagnose gastrointestinal disorders
F – 18	$^{0}_{-1}e$ emission, $^{0}_{+1}e$ emission and $^{0}_{-1}e$ capture	Bone scanning; study of cerebral sugar metabolism

Ga – 67	$_{-1}^{0}e$ capture	Diagnosis of lymphoma and Hodgkin disease; whole body scan for tumours
Au – 198	$_{-1}^{0}e$ emission	Assess kidney activity
H – 3	$_{-1}^{0}e$ emission	Biochemical tracer; measurement of the water content of the body
In – 111	γ–emission	Label blood platelets
I – 125	$_{-1}^{0}e$ capture	Determination of blood hormone levels
I – 131	$_{-1}^{0}e$ emission	Measure thyroid uptake of iodine
Fe – 59	$_{-1}^{0}e$ emission	Assessment of blood iron metabolism and diagnosis of anaemia
Kr – 79	$_{+1}^{0}e$ emission and $_{-1}^{0}e$ capture	Assess cardiovascular function
N – 13	$_{+1}^{0}e$ emission	Brain, heart, and liver imaging
O – 15	$_{+1}^{0}e$ emission	Lung function test
P – 32	$_{-1}^{0}e$ emission	Leukaemia therapy, detection of eye tumours, radiation therapy and detection of breast carcinoma
Po – 210	$_{2}^{4}He$ emission	Radiation therapy
K – 40	$_{-1}^{0}e$ emission	Geological dating
Ra – 226	$_{2}^{4}He$ emission	Radiation therapy
Se – 75	$_{-1}^{0}e$ emission and $_{-1}^{0}e$ capture	Measure size and shape of pancreas
Na – 24	$_{-1}^{0}e$ emission	Blood studies and detection of blood clots
Tc – 99	γ–emission	Bone scans and detection of blood clots
Xe – 133	$_{-1}^{0}e$ emission	Lung capacity measurement

SOLVED PROBLEMS

Problem 20.1   Calculate nuclear binding energy of helium nucleus or α-particle.

Solution   There are two protons and two neutrons in the α-particle. Mass of a proton and a neutron is 1.00728 and 1.00866 respectively. So, total mass of the α-particle is $M_{apparent} = 2(1.00728 + 1.00866) = 4.03188$. Actual mass of the α-particle is $M_{actual} = 4.00153$. So, $\Delta m = 4.03188 - 4.00153 = 0.03035$. Hence, nuclear binding energy $= E = 0.03035 \times 932$ MeV $= 28.2862$ MeV.

Problem 20.2 $_{90}^{234}Th$ disintegrates to give $_{82}^{206}Pb$ as the final product. How many alpha and beta particles are emitted during this process?

Solution

Change in mass number $= 234 - 206 = 28$ amu. Decrease in mass number by emission of one α-particle is 4. Thus number of α-particles emitted during this process are 28/4 = 7.

Corresponding change in atomic number is $(90 - 2 \times 7) = 76$. But final atomic number is 82. So, Number of β-particles emitted are $(82 - 76) = 6$.

Problem 20.3   The activity of a radioisotope falls to 12.5% in 90 days. Compute the half-life and decay constant of the radioisotope.

Solution

$$\lambda = \frac{2.303}{t} \log \frac{N_0}{N}, \ t = 90 \text{ days}, \ N_0 = 100, \ N = 12.5, \ \lambda = 2.311 \times 10^{-2} \text{ day}^{-1}$$

$$t_{1/2} = \frac{0.693}{\lambda} = 30 \text{ days}.$$

Problem 20.4   Calculate in gm of 1 Curie and 1 rd of $^{214}_{82}$Pb. Given $t_{1/2} = 26.8$ min.

Solution

$$\lambda = \frac{0.693}{t_{1/2}} = \frac{0.693}{26.8 \times 60} = 4.31 \times 10^{-4} \text{ s}^{-1}$$

$$\text{Number of atoms in w gm} = \left(\frac{w}{214}\right) \times 6.023 \times 10^{23} = N$$

Thus, $$-\frac{dN}{dt} = \lambda N = 4.31 \times 10^{-4} \times \left(\frac{6.023 \times 10^{23}}{214}\right) w = (1.213 \times 10^{18} w) dps$$

$$1 \text{ Curie} = 3.7 \times 10^{10} \ dps. \ 1.213 \times 10^{18} w = 3.7 \times 10^{10}, \ w = 3.1 \times 10^{-8} \text{ gm}$$

$$1 \text{ Rutherford } (rd) = 10^6 \ dps. \ 1.213 \times 10^{18} w = 10^6. \ w = 8.244 \times 10^{-13} \text{ gm}$$

Problem 20.5   Radon-222, which is found in the air inside houses built over soil containing uranium, has a half-life of 3.82 days. How long before a sample decreases to 1/32 of the original amount?

Solution
In each half-life of a radioactive nuclide, the amount diminishes by one-half. So five half-lives are needed to reduce the sample to 1/32 of the original amount.
Thus total time required is 5×3.82 days $= 19.1$ days.

Problem 20.6   Plutonium-239 which has a half-life of 2.44×10^4 years. What fraction of plutonium-239 is left after 9.76×10^4 years?

Solution

The length of time divided by the half-life yields the number of half-lives, $n_{1/2}$

$$n_{1/2} = \frac{9.76 \times 10^4}{2.44 \times 10^4} = 4$$

In each half-life of a radioactive nuclide, the amount diminishes by one-half, so the fraction remaining would be $\frac{1}{2^4} = \frac{1}{16}$.

Problem 20.7   Discuss how $_{27}^{60}Co$ is formed from $_{26}^{58}Fe$.

Solution

$$_{26}^{58}Fe + _{0}^{1}n \longrightarrow _{26}^{59}Fe$$

$$_{26}^{59}Fe \longrightarrow _{27}^{59}Co + _{-1}^{0}e$$

$$_{27}^{59}Co + _{0}^{1}n \longrightarrow _{26}^{60}Co$$

EXERCISES

1. It has been observed that radioactive rate equation for $_{6}^{14}C$ is $N = N_0 e^{-0.000121 \times t}$. Calculate half-life of $_{6}^{14}C$.
2. Suppose an organism has 20 gm of ^{14}C at its time of death. What will be the amount of ^{14}C after 10320 years?
3. A fossil initially possessed 32 gm of ^{14}C but now ^{14}C content of the fossil decreases to 12 gm. Calculate the age of the fossil.
4. ^{14}C content of a fossil decreases to 12%. Calculate the age of the fossil.
5. A rock contains 0.313 mg of $_{82}^{206}Pb$ for 1 mg of $_{92}^{238}U$. Calculate the age of the rock or Earth. Given $t_{1/2} = 4.5 \times 10^9$ years.
6. Uranium-238 is one of the radioactive nuclides sometimes found in soil. It has a half-life of 4.51×10^9 years. What fraction of a sample is left after 9.02×10^9 years?
7. Balance the following nuclear reactions:
   (i) $_{94}^{239}Pu \longrightarrow _{2}^{4}He + ?$,
   (ii) $_{91}^{234}Pr \longrightarrow _{92}^{234}U + ?$,
   (iii) $_{77}^{192}Pr + ? \longrightarrow _{76}^{192}Os$
   (iv) $_{9}^{18}F? \longrightarrow _{8}^{18}O + ?$,
   (v) $_{8}^{20}O \longrightarrow _{9}^{18}F + ?$,

(vi) $^{26}_{12}\text{Mg} + ^{1}_{1}\text{H} \longrightarrow ^{4}_{2}\text{He} + ?$

(vii) $^{9}_{4}\text{Be} + ^{4}_{2}\text{He} \longrightarrow ^{12}_{6}\text{C} + ?$

(viii) $^{40}_{19}\text{K} \longrightarrow ? + ^{0}_{+1}\text{e}$

(ix) $^{137}_{55}\text{Cs} \longrightarrow ^{137}_{56}\text{Ba}^* + ?. \ ^{137}_{56}\text{Ba}^* \longrightarrow ^{137}_{56}\text{Ba} + \gamma$

(x) $^{18}_{9}\text{F} + ^{4}_{2}\text{He} \longrightarrow ? + ^{1}_{1}\text{H}$

(xi) $^{238}_{92}\text{U} + ? \longrightarrow ^{247}_{99}\text{Es} + 5\,^{1}_{0}\text{n}$

(xii) $? + ^{2}_{1}\text{H} \longrightarrow ^{239}_{93}\text{Np} + ^{1}_{0}\text{n}$

Appendix

Table 1   Physical properties of elements

Name of chemical element	Symbol	Atomic number	Melting point (°C)	Specific gravity	Electro-negativity (Pauling)	Ionisation energy (eV)
Actinium	Ac	89	1050	10.07	0.7	5.17
Aluminum	Al	13	660	2.7	1.61	5.9858
Americium	Am	95	994	19.84	1.36	5.9738
Antimony	Sb	51	630	6.68	1.96	8.6084
Argon	Ar	18	−189	0.86		15.7596
Arsenic	As	33	81	5.72	2.18	9.7886
Astatine	At	85	302		2.02	9.3
Barium	Ba	56	725	3.59	2.6	5.2117
Berkelium	Bk	97	986	14.78	1.3	6.1979
Beryllium	Be	4	1278	1.85	1.57	9.3227
Bismuth	Bi	83	271	9.75	1.62	7.2856
Bohrium	Bh	107				
Boron	B	5	2300	2.34	2.04	8.298
Bromine	Br	35	−7	3.12	2.96	11.8138
Cadmium	Cd	48	321	8.65	1.93	8.9938
Calcium	Ca	20	839	1.55	1.36	6.1132
Californium	Cf	98	900	15.1	1.3	6.2817
Carbon	C	6	3500	2.26	2.55	11.2603
Cerium	Ce	58	795	6.77	0.89	5.5387
Caesium	Cs	55	29	1.87	2.66	3.8939
Chlorine	Cl	17	−101	3.21	3.16	12.9676
Chromium	Cr	24	1857	7.19	1.55	6.7665

Name of chemical element	Symbol	Atomic number	Melting point (°C)	Specific gravity	Electro-negativity (Pauling)	Ionisation energy (eV)
Cobalt	Co	27	1495	8.9	1.91	7.881
Copper	Cu	29	1083	8.96	1.65	7.7264
Curium	Cm	96	1340	13.5	1.28	5.9915
Darmstadtium	Ds	110				
Dubnium	Db	105			1.3	
Dysprosium	Dy	66	1412	8.55	1.2	5.9389
Einsteinium	Es	99	860		1.3	6.42
Erbium	Er	68	1522	9.07	1.22	6.1077
Europium	Eu	63	822	5.24		5.6704
Fermium	Fm	100	1527	8.84	1.3	6.5
Fluorine	F	9	−220	1.7	3.98	17.4228
Francium	Fr	87	27		2.2	4.0727
Gadolinium	Gd	64	1311	7.9	1.17	6.1501
Gallium	Ga	31	30	5.91	1.81	5.9993
Germanium	Ge	32	937	5.32	2.01	7.8994
Gold	Au	79	1064	19.32	2.2	9.2255
Hafnium	Hf	72	2150	13.31	1.3	6.8251
Hassium	Hs	108				
Helium	He	2	−272	0.18		24.5874
Holmium	Ho	67	1470	8.8		6.0215
Hydrogen	H	1	−259	0.09	2.2	13.5984
Indium	In	49	157	7.31	1.69	5.7864
Iodine	I	53	114	6.24	2.05	10.4513
Iridium	Ir	77	2410	22.4	1.9	8.967
Iron	Fe	26	1535	7.87	1.88	7.9024
Krypton	Kr	36	−157	3.75	2.96	13.9996
Lanthanum	La	57	920	6.15	0.79	5.5769
Lawrencium	Lr	103	1627		1.3	4.9
Lead	Pb	82	327	11.35	2	7.4167
Lithium	Li	3	180	0.53	0.98	5.3917
Lutetium	Lu	71	1656	9.84	1.25	5.4259
Magnesium	Mg	12	639	1.74	1.31	7.6462
Manganese	Mn	25	1245	7.43	1.83	7.434
Meitnerium	Mt	109				
Mendelevium	Md	101			1.3	6.58
Mercury	Hg	80	−39	13.55	2.28	10.4375

Name of chemical element	Symbol	Atomic number	Melting point (°C)	Specific gravity	Electro-negativity (Pauling)	Ionisation energy (eV)
Molybdenum	Mo	42	2617	10.22	1.6	7.0924
Neodymium	Nd	60	1010	7.01	1.12	5.525
Neon	Ne	10	−249	0.9		21.5645
Neptunium	Np	93	640	18.95	1.5	6.2657
Nickel	Ni	28	1453	8.9	1.9	7.6398
Niobium	Nb	41	2468	8.57	1.33	6.7589
Nitrogen	N	7	−210	1.25	3.04	14.5341
Nobelium	No	102	827		1.3	6.65
Osmium	Os	76	3045	22.6	2.36	8.4382
Oxygen	O	8	−218	1.43	3.44	13.6181
Palladium	Pd	46	1552	12.02	2.28	8.3369
Phosphorus	P	15	44	1.82	2.19	10.4867
Platinum	Pt	78	1772	21.45	2.2	8.9587
Plutonium	Pu	94	640	13.67	1.38	6.0262
Polonium	Po	84	254	9.3	2.33	8.417
Potassium	K	19	64	1.78	0.82	4.3407
Praseodymium	Pr	59	935	6.77	1.1	5.473
Promethium	Pm	61	1100	7.3	1.13	5.582
Protactinium	Pa	91	1568	11.72	1.1	5.89
Radium	Ra	88	700	5.5	0.89	5.2784
Radon	Rn	86	−71	9.73	2	10.7485
Rhenium	Re	75	3180	21.04	1.5	7.8335
Rhodium	Rh	45	1966	12.41	2.2	7.4589
Roentgenium	Rg	111				
Rubidium	Rb	37	39	1.63	0.82	4.1771
Ruthenium	Ru	44	2250	12.37	1.9	7.3605
Rutherfordium	Rf	104			1.3	
Samarium	Sm	62	1072	7.52	1.14	5.6437
Scandium	Sc	21	1539	2.99	1.54	6.5615
Seaborgium	Sg	106				
Selenium	Se	34	217	4.79	2.55	9.7524
Silicon	Si	14	1410	2.33	1.9	8.1517
Silver	Ag	47	962	10.5	2.2	7.5762
Sodium	Na	11	98	0.97	0.93	5.1391
Strontium	Sr	38	769	2.54	0.82	5.6949
Sulfur	S	16	113	2.07	2.58	10.36
Tantalum	Ta	73	2996	16.65	1.27	7.5496

Name of chemical element	Symbol	Atomic number	Melting point (°C)	Specific gravity	Electro-negativity (Pauling)	Ionisation energy (eV)
Technetium	Tc	43	2200	11.5	2.16	7.28
Tellurium	Te	52	449	4.93		9.0096
Terbium	Tb	65	1360	8.23		5.8638
Thallium	Tl	81	303	11.85	2.54	6.1082
Thorium	Th	90	1750	15.4	0.89	6.3067
Thulium	Tm	69	1545	9.32	1.23	6.1843
Tin	Sn	50	232	7.31	1.78	7.3439
Titanium	Ti	22	1660	4.54	1.63	6.8281
Tungsten	W	74	3410	19.35	1.3	7.864
Ununbium	Uub	112				
Ununhexium	Uuh	116				
Ununoctium	Uuo	118				
Ununpentium	Uup	115				
Ununquadium	Uuq	114				
Ununseptium	Uus	117				
Ununtrium	Uut	113				
Uranium	U	92	1132	20.2	1.3	6.1941
Vanadium	V	23	1890	6.11	1.66	6.7462
Xenon	Xe	54	−112	5.9	2.1	12.1298
Ytterbium	Yb	70	824	6.9	1.24	6.2542
Yttrium	Y	39	1523	4.47	0.95	6.2173
Zinc	Zn	30	420	7.13	1.81	9.3942
Zirconium	Zr	40	1852	6.51	1.22	6.6339

Table 2 E_g-values of some compound semiconductors

Semiconductor	$E_g(eV)$
GaP	2.25
GaAs	1.47
GaSb	0.68
InP	1.27
CdS	2.42
CdSe	1.74
CdTe	1.50
ZnSe	2.67
ZnTe	2.26

Table 3   Physical constants

Avogadro's number	N	6.023×10^{23}/g–mole or 6.023×10^{26}/kg–mole
Molar gas constant	R	8.314 J/(mole –°K) = 1.98 cal/(mole –°K) = 0.082 lit –atm/(mole –°K)
Boltzmann's constant	k_B = R/N	1.38×10^{-23} J/°K
Faraday's constant	F	96,500 C/gm-equivalent of element
Electronic charge	e = F/N	1.602×10^{-19} C = 4.806×10^{-10} e.s.u of charge.
Atomic mass unit	amu = 1/N	1.66×10^{-24} gm = 1.66×10^{-27} kg
Velocity of light	c	3.0×10^8 m/s.
Electron rest mass	m_o	9.11×10^{-31} kg.
Bohr magneton (BM)	μ_B	9.273×10^{-24} A –m^2
Permittivity of free space	ε_o	8.854×10^{-12} F/m
Acceleration due to gravity	g	9.81 m/s^2
Dipole moment (1 Debye)	D	10^{-18} e.s.u. –cm
Gravitational constant	G	6.6732×10^{-11} N –m^2/kg^2
Stefan-Boltzmann constant	σ	5.67×10^{-8} W/(m^2 –°K^4)
Rydberg constant	R_B	109737 cm^{-1}

SI unit prefixes: 1 kilo = 10^3 unit, 1 mega = 10^6 unit, 1 giga = 10^9 unit, 1 tera = 10^{12} unit, 1 milli = 10^{-3} unit, 1 micro = 10^{-6} unit, 1 nano = 10^{-9} unit, 1 pico = 10^{-12} unit, 1 femto = 10^{-15} unit, 1 atto = 10^{-18} unit, 1Å = 10^{-10} m.

Table 4   Conversion factors

Length	1 m = 3.28 ft = 1.094 yd.
	1 in = 2.54 cm, 1 ft = 30.48 cm.
	1 mile = 1.6 km.
Area	1 m^2 = 10.764 ft^2
	1 in^2 = 645.16 mm^2
	1 hectare = 100 Ares = 10,000 m^2 = 2.471 acres.
Force	1 N = 0.2248 lb$_f$, 1 N = 10^5 dyne
	1 kg$_f$ = 9.81 N

Energy	1 J = 0.738 ft –lb$_f$ = 10^7 erg
	1 eV = 1.602×10^{-19} J
	1cal = 4.19 J
	1 J = 0.2392 Cal = 6.242×10^{18} eV
	1 B. T. U. = 252 Cal.
Volume	1 ft^3 = 28316.85 cm^3 or c.c. = 7.48 gallon (U.S.)
	1 gallon = 3786 c.c.
	1 m^3 = 35.288 ft^3
	1 barrel = 42 US –gallon
Pressure	1 torr = 1 mm Hg at 0°C
	1 atm = 101.325 kPa = 14.7 psi = 1.01325 bar
	1 atm = 760 mm Hg at 0°C
	1 bar = 100 kPa = 10^5 Pa
	1 kg$_f$/cm^2 = 9.810×10^4 Pa = 0.981 bar = 0.968 atm.
	1 Mpa = 145.1 psi = 1 N/mm^2 = 9.87 atm = 10.06 kg$_f$/ cm^2
	1 tsi = 15.2 MPa
Temperature	$\dfrac{t°C}{5} = \dfrac{(t'°F - 32)}{9}$. Or, $1.8t°C = (t'°F - 32)$
	T (R) = t°F + 459.67
	T (K) = t°C + 273
	T (R) = 1.8 T (K)
Power	1 kW = 239.2 cal/s = 1.341 hp.
	1 hp = 745 W = 0.745 kW
	1 lumen (lm) = 1.496×10^{-10} W
	1 lm/ft^2 = 1 foot –candle (fc) = 1.609×10^{-12} W/m^2
	1 lm/ft^2 = Luminous flux = 1 foot–candle (fc) = 1.609×10^{-12} W/m^2
Mass and Density	1 kg = 1000 gm = 2.2046 lb.
	1 g/ cm^2 = 1000 kg/m^3 = 62.43 lb/ft^3
Coefficient of Viscosity	1 poise (p) = 100 centi –poise (cp) = 0.1 Pa –sec.
	1 centi –poise (cp) = 10^{-3} Pa –sec = 6.7197×10^{-4} lb/ft –s = 2.4191 lb/ft –h.

Table 5a Standard Oxidation Electrode Potential (E^0_{298}) of selected redox systems in aqueous solution

Reduced form	*Oxidized form*	*No. of electron(s) released*	E^0_{298} *(V)*
Li	Li$^+$	1e$^-$	+ 3.05
K	K$^+$	1e$^-$	+ 2.936
Ca	Ca^{2+}	2e$^-$	+ 2.87
Na	Na$^+$	1e$^-$	+ 2.71
Mg	Mg^{2+}	2e$^-$	+ 2.36

$\longrightarrow$

Reduced form	Oxidized form	No. of electron(s) released	E^0_{298} (V)
Al	Al^{3+}	$3e^-$	+ 1.67
Mn	Mn^{2+}	$2e^-$	+ 1.18
½ $H_2(g)$ + OH^-	H_2O	$1e^-$	+ 0.828
Zn	Zn^{2+}	$2e^-$	+ 0.76
Fe	Fe^{2+}	$2e^-$	+ 0.44
Cd	Cd^{2+}	$2e^-$	+ 0.40
Pb + $2I^-$	PbI_2	$2e^-$	+ 0.365
Pb + SO_4^{2-}	$PbSO_4$	$2e^-$	+ 0.356
Sn	Sn^{2+}	$2e^-$	+ 0.14
Pb	Pb^{2+}	$2e^-$	+ 0.126
Fe	Fe^{3+}	$3e^-$	+ 0.04
½ D_2	D^+	$1e^-$	+ 0.01
½ H_2	H^+	$1e^-$	0.00
Cu	Cu^{2+}	$2e^-$	−0.34
Cu	Cu^+	$1e^-$	−0.518

Table 5b Standard Oxidation Electrode Potentials (E^0_{298}) of selected redox systems in aqueous solution

Reduced form	Oxidized form	No. of electron(s) change	E^0_{298} (V)
I^-	½ I_2	$1e^-$	−0.54
2 Hg(l) + SO_4^{2-}	Hg_2SO_4	$2e^-$	−0.615
Fe^{2+}	Fe^{3+}	$1e^-$	−0.77
Ag	Ag^+	$1e^-$	−0.80
Hg	Hg^{2+}	$2e^-$	−0.85
NO + $2H_2O$	NO_3^- + $4H^+$	$3e^-$	−0.96
Br^-	½ Br_2	$1e^-$	−1.07
Mn^{2+} + $2H_2O$	MnO_2 + $4H^+$	$2e^-$	−1.23
H_2O	½ O_2 (g) + $2H^+$	$2e^-$	−1.23
$2Cr^{3+}$ + $7H_2O$	$Cr_2O_7^{2-}$ + $14H^+$	$6e^-$	−1.33
Cl^-	½ Cl_2	$1e^-$	−1.36
Cl^- + 3 H_2O	ClO_3^- + $6H^+$	$6e^-$	−1.45
Mn^{2+} + 4 H_2O	MnO_4^- + $8H^+$	$5e^-$	−1.51
Ce^{3+}	Ce^{4+}	$1e^-$	−1.61
F^-	½ F_2	$1e^-$	−2.87
Ag + Br^-	AgBr	$1e^-$	− 0.073
SO_3^{2-} + H_2O	SO_4^{2-} + $2H^+$	$2e^-$	− 0.22
Ag + Cl^-	AgCl	$1e^-$	− 0.2222
2 Hg(l) + $2Cl^-$	Hg_2Cl_2	$2e^-$	− 0.2680

Table 6   Typical oxyacids with their acid strength

Acid	K_a	pK_a	Acid strength
HOBr	2.06×10^{-9}	8.7	Very Weak
HOCl	3.2×10^{-8}	7.5	Very Weak
HOI	2.3×10^{-11}	10.64	Very Weak
H_3AsO_3	6.0×10^{-10}	9.2	Very Weak
H_3BO_3	5.8×10^{-10}	9.24	Very Weak
H_4SiO_4	2.0×10^{-10}	9.7	Very Weak
H_6TeO_6	2.0×10^{-8}	7.7	Very Weak
$HClO_2$	1.1×10^{-2}	1.97	Weak
HNO_2	4.0×10^{-4}	3.40	Weak
H_2CO_3	1.32×10^{-4}	3.88	Weak
H_2SO_3	1.3×10^{-2}	1.87	Weak
H_3AsO_4	6.46×10^{-3}	2.19	Weak
H_5IO_6	3.1×10^{-2}	1.57	Weak
H_3PO_4	7.25×10^{-3}	2.12	Weak
$HBrO_3$	2.0×10^{-1}	0.7	Strong
$HClO_3$	10	-1	Strong
HIO_3	1.7×10^{-1}	0.77	Strong
HNO_3	43.6	-1.64	Strong
H_2MnO_4	10^{-1}	1.0	Strong
H_2SO_4	1.0×10^3	-3.0	Very Strong
$HClO_4$	10^{10}	-10.0	Very Strong
$HMnO_4$	2.0×10^2	-2.3	Very Strong

Table 7a   Coefficient of viscosities (η) of liquids in centipoise (cp) at 20°C

Liquid	η (cp)	Liquid	η (cp)
Water	1.005	Ethyl ether	0.245
Ethanol	1.194	Chloroform	0.563
Methanol	0.593	Chlorobenzene	0.637
Acetone	0.331	Benzene	0.647
Carbon tetrachloride	0.958	Acetic acid	1.222
Nitrobenzene	1.98	Glycerin	950
Mercury	1.6	Carbon disulphide	0.36
Kerosene	1.64	Blood	4.0

Table 7b   Coefficient of viscosities (η) of several liquid metals and alloys

Material	Temperature (°C)	Viscosity (η) (cp)
Aluminium	700	3.0
Copper	1200	3.2
Iron	1600	6.2
White cast iron	1300	2.4
Steel	> MP	~ 3.0
Magnesium	680	1.2
Zinc	500	3.7

Table 8   Solubility product of some barely soluble compounds in water at 25°C

Compound	Solubility product (K_{sp})	Compound	Solubility product (K_{sp})
AgBr	5.3×10^{-13}	AgCN	1.4×10^{-16}
AgCl	1.78×10^{-10}	Ag_2CrO_4	2.5×10^{-12}
AgI	8.30×10^{-17}	PbS	2.5×10^{-27}
$CaSO_4$	9.1×10^{-6}	CdS	4.0×10^{-29}
$SrSO_4$	2.8×10^{-7}	CuS	3.0×10^{-42}
$BaSO_4$	1.1×10^{-10}	HgS	3.0×10^{-54}
$PbCl_2$	2.4×10^{-4}	CoS	2.0×10^{-27}
$PbBr_2$	7.9×10^{-5}	ZnS	1.0×10^{-23}
PbI_2	8.7×10^{-9}	NiS	1.4×10^{-24}
$CaCO_3$	4.8×10^{-9}	MnS	2.0×10^{-15}
$BaCO_3$	5.1×10^{-9}	$Mg(OH)_2$	6.0×10^{-13}
$Ca_3(PO_4)_2$	2.0×10^{-29}	CaF_2	4.0×10^{-11}
$AlPO_4$	5.75×10^{-19}	CuI	5.0×10^{-12}
CuBr	1.60×10^{-11}	CuCl	1.0×10^{-6}

Table 9   Surface tension values of liquid metals and other liquids

Material	Temperature (°C)	Surface tension (γ) (Dynes/cm)
Aluminium	750	520
Copper	1200	1160
Iron	1600	1360
Lead	350	453
Magnesium	681	563
Mercury	20	465
Zinc	600	770
Fused salts	400 to 930	35 to 180
Water	20	74

Table 10   Values of C_p, C_v and γ for ideal gases at 25°C

Gas	Chemical formula	C_p (kJ/kg-°K)	C_v (kJ/kg-°K)	γ
Air	—	1.0035	0.7165	1.4
Oxygen	O_2	0.922	0.622	1.393
Nitrogen	N_2	1.042	0.745	1.4
Argon	Ar	0.520	0.312	1.667
Hydrogen	H_2	14.209	10.085	1.409
Helium	He	5.193	3.116	1.667
Carbon monoxide	CO	1.041	0.744	1.4
Carbon dioxide	CO_2	0.842	0.653	1.289
Methane	CH_4	2.254	1.736	1.299
Ethane	C_2H_6	1.7662	1.4897	1.86
Propane	C_3H_8	1.6794	1.4909	1.126
Isobutane	C_4H_{10}	1.716	1.573	1.091
Acetylene	C_2H_2	1.712	1.394	1.23
Water vapour (Steam)	H_2O	1.8723	1.4108	1.327

Table 11   Some Reference Electrodes

Electrode	Reaction	E^0_{OX} at 25°C
K_2SO_4 (sat) | Hg_2SO_4 | Hg	$Hg_2SO_4 + 2e^- \rightarrow 2Hg + SO_4^{2-}$	– 0.64 V
KCl (sat) | AgCl, Ag	$AgCl + e^- \rightarrow Ag + Cl^-$	– 0.1989 V
KCl (1.0M) | AgCl, Ag	$AgCl + e^- \rightarrow Ag + Cl^-$	– 0.237 V
KCl (0.1M) | AgCl, Ag	$AgCl + e^- \rightarrow Ag + Cl^-$	– 0.290 V

Table 12   'K' and 'α' values for some polymer-solvent systems at different temperatures

Polymer	Solvent	θ (°C)	$K \times 10^5$	α
Polyvinyl acetate	Acetone	25	18.8	0.69
Polyvinyl acetate	Acetone	30	10.2	0.72
Polyvinyl acetate	Chloroform	25	20.3	0.72
Polyvinyl alcohol	Water	25	20	0.76
Polymethylmethacrylate	Acetone	25	7.5	0.70
Polymethylmethacrylate	Acetone	30	7.0	0.70
Polymethylmethacrylate	Toluene	25	7.5	0.71
Polymethylmethacrylate	Toluene	30	7.0	0.72
Polystyrene	Benzene	25	10.2	0.74
Polystyrene	Benzene	30	11.0	0.73
Polyisobutylene	Benzene	30	61.0	0.56
Natural rubber	Benzene	30	18.5	—

Table 13   Average bond energy values in kJ/mole

Bond	Bond energy	Bond	Bond energy
C–H	415	C–Cl	328
C–C	344	C–Br	276
C–O	350	C–I	240
C–N	292	F–F	158
C–S	259	N–H	391
C–F	441	O–H	463
S–H	368	S–S	266
N–O	175	O–O	143
N–N	159	N–Cl	200
H–H	436	C=C	615
Cl–Cl	243	C=O	725
C=N	615	N=N	418
O=O	498	C≡C	812
C≡N	890	N≡N	946

Table 14   Dipole moments (μ) of different bonds

Bond	α	Bond	μ
H–O	1.5 D	C–O	0.8 D
H–N	1.3 D	C=O	2.5 D
H–C	0.4 D	C–N	0.5 D
C–Cl	1.5 D	C≡N	3.5 D
C–Br	1.4 D		

Note: 1 D = 10^{-18} esu-cm. Dipole moment of C–F bond is assumed to be 1.0 D.

The Periodic Table

H 1 1.008 $1s^1$	Gr-2							
Li 3 6.939 $2s^1$	Be 4 9.012 $2s^2$							
Na 11 22.99 $3s^1$	Mg 12 24.31 $3s^2$	Gr-3	Gr-4	Gr-5	Gr-6	Gr-7	Gr-8	Gr-9
K 19 39.10 $4s^1$	Ca 20 40.08 $4s^2$	Sc 21 44.96 $3d^1 4s^2$	Ti 22 47.90 $3d^2 4s^2$	V 23 50.94 $3d^3 4s^2$	Cr 24 52 $3d^5 4s^1$	Mn 25 54.94 $3d^5 4s^2$	Fe 26 55.85 $3d^6 4s^2$	Co 27 58.93 $3d^7 4s^2$
Rb 37 85.47 $5s^1$	Sr 38 87.62 $5s^2$	Y 39 88.91 $4d^1 5s^2$	Zr 40 91.22 $4d^2 5s^2$	Nb 41 92.91 $4d^4 5s^1$	Mo 42 95.94 $4d^5 5s^1$	Tc 43 99 $4d^6 5s^1$	Ru 44 101.1 $4d^7 5s^1$	Rh 45 102.9 $4d^8 5s^1$
Cs 55 132.9 $6s^1$	Ba 56 137.3 $6s^2$	La 57 138.9 $4f^0 5d^1 6s^2$	Hf 72 178.5 $4f^{14} 5d^2 6s^2$	Ta 73 180.9 $4f^{14} 5d^3 6s^2$	W 74 183.9 $4f^{14} 5d^4 6s^2$	Re 75 186.2 $4f^{14} 5d^5 6s^2$	Os 76 190.2 $4f^{14} 5d^6 6s^2$	Ir 77 192.2 $4f^{14} 5d^9 6s^0$
Fr 87 223 $7s^1$	Ra 88 226 $7s^2$	Ac 89 227 $6d^1 7s^2$	Rf 104 261	Db 105 262	Sg 106 266	Bh 107 264	Hs 108 277	Mt 109 268

1st transition series, known as Lanthanides ($4f^1$ to $4f^{14}$)

Ce 58 140.1 $4f^2 6s^2$	Pr 59 140.9 $4f^3 6s^2$	Nd 60 144.3 $4f^4 6s^2$	Pm 61 147.0 $4f^5 6s^2$	Sm 62 150.3 $4f^6 6s^2$	Eu 63 152.0 $4f^7 6s^2$	Gd 64 157.3 $4f^7 5d^1 6s^2$

Tb 65 158.9 $4f^8 5d^1 6s^2$	Dy 66 162.5 $4f^{10} 6s^2$	Ho 67 164.9 $4f^{11} 6s^2$	Er 68 167.3 $4f^{12} 6s^2$	Tm 69 168.9 $4f^{13} 6s^2$	Yb 70 173.0 $4f^{14} 6s^2$	Lu 71 175.0 $4f^{14} 5d^1 6s^2$

			Gr-13	Gr-14	Gr-15	Gr-16	Gr-17	He 2 4 $1s^2$
			B 5 10.81 $2s^2\,2p^1$	C 6 12.01 $2s^2\,2p^2$	N 7 14.01 $2s^2\,2p^3$	O 8 16 $2s^2\,2p^4$	F 9 19 $2s^2\,2p^5$	Ne 10 20.18 $2s^2\,2p^6$
Gr-10	Gr-11	Gr-12	Al 13 26.98 $3s^2\,3p^1$	Si 14 28.09 $3s^2\,3p^2$	P 15 30.97 $3s^2\,3p^3$	S 16 32.16 $3s^2\,3p^4$	Cl 17 35.45 $3s^2\,3p^5$	Ar 18 39.95 $3s^2\,3p^6$
Ni 28 58.71 $3d^8\,4s^2$	Cu 29 63.54 $3d^{10}\,4s^1$	Zn 30 65.37 $3d^{10}\,4s^2$	Ga 31 69.72 $4s^2\,4p^1$	Ge 32 72.59 $4s^2\,4p^2$	As 33 74.92 $4s^2\,4p^3$	Se 34 78.96 $4s^2\,4p^4$	Br 35 79.91 $4s^2\,4p^5$	Kr 36 83.80 $4s^2\,4p^6$
Pd 46 106.4 $4d^{10}\,5s^0$	Ag 47 107.9 $4d^{10}\,5s^1$	Cd 48 112.4 $4d^{10}\,5s^2$	In 49 114.8 $5s^2\,5p^1$	Sn 50 118.7 $5s^2\,5p^2$	Sb 51 121.7 $5s^2\,5p^3$	Te 52 127.6 $5s^2\,5p^4$	I 53 126.9 $5s^2\,5p^5$	Xe 54 131.3 $5s^2\,5p^6$
Pt 78 195.1 $4f^{14}\,5d^9\,6s^1$	Au 79 197 $4f^{14}\,5d^{10}\,6s^1$	Hg 80 200.6 $4f^{14}\,5d^{10}\,6s^2$	Tl 81 204.4 $6s^2\,6p^1$	Pb 82 207.2 $6s^2\,6p^2$	Bi 83 209 $6s^2\,6p^3$	Po 84 210 $6s^2\,6p^4$	At 85 211 $6s^2\,6p^5$	Rn 86 222 $6s^2\,6p^6$
110 Ds 271	111 Rg 272	112 Uub 277						

2nd transition series, known as Actinides ($5f^1$ to $5f^{14}$)

90 Th 232 $6d^2\,7s^2$	91 Pa 231 $5f^2\,6d^1\,7s^2$	92 U 238 $5f^3\,6d^1\,7s^2$	93 Np 237 $5f^5\,7s^2$	94 Pu 244 $5f^6\,7s^2$	95 Am 243 $5f^7\,7s^2$	96 Cm 247 $5f^7\,6d^1\,7s^2$

97 Bk 247 $5f^8\,6d^1\,7s^2$	98 Cf 251 $5f^{10}\,7s^2$	99 Es 254 $5f^{11}\,7s^2$	100 Fm 257 $5f^{12}\,7s^2$	101 Md 256 $5f^{13}\,7s^2$	102 No 254 $5f^{14}\,7s^2$	103 Lr 257 $5f^{14}\,6d^1\,7s^2$

Bibliography

1. Y.A. Gerasimov (Ed.), *Physical Chemistry*, English translation by Mir Publishers (1974).
2. S.M. Walas *Reaction Kinetics for Chemical Engineers*, New York (1959).
3. S. Glasstone, K.J. Laidler and H. Eyring, *The Theory of Rate Processes*, New York (1941).
4. S.W. Benson, *The Formulation of Chemical Kinetics*, New York (1960).
5. G.W. Castellan, *Physical Chemistry*, Reading Mass. (1964).
6. J.O'M Bockris (Ed.), *Modern Aspects of Electrochemistry*, New York (1966).
7. L.E. Smart and E.A. Moore, *Solid State Chemistry, An Introduction (4th Edition)*, CRC Press (2005).
8. A.R. West, *Solid State Chemistry and its Applications, Student Edition (2nd Edition)*, John Wiley & Sons Inc. (2013).
9. V.A. Kireev, *Physical Chemistry*, Moscow (1972).
10. S. Glasstone and D. Lewis, *Elements of Physical Chemistry*, The Macmillan Press Limited (1960).
11. Darrow (*Concept of entropy*), Amer. J. Phys., 12, 183 (1944).
12. A.K. Galwey, *Chemistry of solids*, Science Paperbacks and Chapman and Hall Ltd., London (1967).
13. N.N. Greenwood, *Ionic Crystals, Lattice Defects and Non-Stoichiometry*, Butterworths (1968).
14. Ira N. Levine, *Physical Chemistry*, Tata McGraw-Hill Publishing Company Limited, New Delhi (1995).
15. J.H. Espenson, *Chemical Kinetics and Reaction Mechanisms*, McGraw-Hill (1981).
16. G.S. Van Wylen and R.E. Sonntag, *Fundamentals of Classical Thermodynamics (3rd edition)*, Wiley (1985).
17. A. Mallick, *Chemistry for Engineers*, Viva Books Pvt. Ltd. (2008).

18. P.K. Nag, *Engineering Thermodynamics (2nd edition)*, Tata McGraw-Hill Publishing Company Limited (2004).
19. R.E. Sonntag, C. Borgnakke and G.J. Van Wylen, *Fundamentals of Thermodynamics (5th edition)*, John Wiley & Sons Inc. (2000).
20. P.C. Rakshit, *Physical Chemistry*, Sarat Book Distributors (2001).
21. W.C. Reynolds and H.C. Perkins, *Engineering Thermodynamics*, McGraw-Hill (1977).
22. F.A. Cotton and G. Wilkinson, *Advanced Inorganic Chemistry (3rd edition)*, Wiley (1972).
23. J.D. Lee, *Concise Inorganic Chemistry (3rd edition)*, English Language Book Society and Nostrand Reinhold Company Limited, London (1979).
24. M. Karplus and R.N. Porter, *Atoms and Molecules*, Benjamin (1964).
25. J.R. Partington, *General and Inorganic Chemistry (4th edition)*, Macmillan (1966).
26. H. Krebs, *Fundamentals of Inorganic Crystal Chemistry*, McGraw-Hill (1968).
27. H.B. Gray, *Electrons and Chemical Bonding*, Benjamin (1964).
28. J.E. Ferguson, *Stereochemistry and Bonding in Inorganic Chemistry*, Prentice Hall (1974).
29. L. Pauling, *The Nature of the Chemical Bond (3rd edition)*, Oxford University Press (1961).
30. T. Moeller, *Inorganic Chemistry*, Wiley (1952).
31. R.E. Reed-Hill and R. Abbaschian, *Physical Metallurgy Principles (3rd edition)*, Thomson Asia Pvt. Ltd., Singapore (2003).
32. A. Mallick, *Principles of Physical Metallurgy (2nd edition)*, Viva Books Pvt. Ltd. (2015).
33. Albert G. Guy, *Elements of Physical Metallurgy (2nd edition)*, Addison-Wesley Publishing Company Inc. (USA) (1962).
34. M. Hansen and K. Anderko, *Constitution of Binary Alloys (2nd edition)*, New York (1958).
35. J.E. Ricci, *The Phase Rule and Heterogeneous Equilibrium*, Dover (1966).
36. A.W. Adamson and A. P. Gast, *Physical Chemistry of Surfaces,* John Wiley, New York (1997).
37. J.C. Berg, *An Introduction to Interfaces and Colloids: The Bridge to Nanoscience*, World Scientific, Singapore (2010).
38. P. Ghosh, *Colloid and Interface Science*, PHI Learning, New Delhi (2009).
39. M. Smoluchowski, *Kolloidn. Zh.*, **18**, 194 (1916).
40. F. Hofmeister, *Arch. Exp. Pathol. Pharmakol.*, **24** 247 (1888).

Index